Vacuum Structure and QCD Sum Rules

Current Physics – Sources and Comments

Coordinating editor

H. RUBINSTEIN

1. String theory in four dimensions, *edited by* M. Dine
2. Finite-size scaling, *edited by* J.L. Cardy
3. Regge theory of low-p_T hadronic interactions, *edited by* L. Caneschi
4. Superstring construction, *edited by* A.N. Schellekens
5. CP violation, *edited by* L. Wolfenstein
6. Particle physics and cosmology: Dark matter, *edited by* M. Srednicki
7. Large-order behaviour of perturbation theory, *edited by* J.C. Le Guillou and J. Zinn-Justin
8. The standard model Higgs boson, *edited by* M.B. Einhorn
9. Quarkonia, *edited* by W. Buchmüller
10. Vacuum structure and QCD sum rules, *edited by* M.A. Shifman

NORTH-HOLLAND
AMSTERDAM · LONDON · NEW YORK · TOKYO

Vacuum Structure and QCD Sum Rules

Editor

Mikhail A. SHIFMAN

Theoretical Physics Institute
School of Physics and Astronomy
University of Minnesota
116 Church Street SE
Minneapolis, MN 55455, USA

1992
NORTH-HOLLAND
AMSTERDAM · LONDON · NEW YORK · TOKYO

North-Holland
Elsevier-Science Publishers B.V.

The Netherlands

Library of Congress Cataloging-in-Publication Data

Vacuum structure and QCD sum rules / Mikhail A. Shifman, editor.
p. cm. -- (Current physics--sources and comments ; 10)
Includes bibliographical references and index.
ISBN 0-444-89746-1 (paperback). -- ISBN
0-444-89746-1 (paperback)
1. Quantum chromodynamics. 2. Sum rules (Physics) I. Shifman,
Mikhail A. II. Series.
QC793.3.Q35V33 1992
539.7'548--dc20 92-22078
CIP

ISBN: 0 444 897461 (paperback)
0 444 89746 1 (paperback)

Printed in The Netherlands

Preface

The method of the QCD sum rules was and still remains one of the most productive tools in a wide range of problems associated with the hadronic phenomenology. In this method are accumulated a lot of heuristic ideas, computational devices, specific formulae which are useful to many theorists and – surprisingly – not only to those working in hadronic physics. Some of the results and approaches which originally have been developed in connection with the QCD sum rules can be and are successfully applied in related fields, such as supersymmetric gauge theories, nontraditional schemes of quarks and leptons, etc. Works devoted to these topics as well as more "down-to-earth" problems from hadronic physics continue to appear. Therefore, there is a need for a comprehensive source of information on this subject which would contain all relevant formulae collected in one place. (I feel that such a compilation would be practically convenient even for myself in my teaching practice.)

In this reprint volume I tried to select the papers so as to give the reader an idea of all basic elements of the sum rule approach. The number of publications devoted to this subject is so large that even limiting myself to the most important contributions I would be unable to reproduce them all in one volume. Therefore, priority was given to the works which seem most useful from the pedagogical point of view. My purpose was to make something like a textbook for those who would like to master the method of the QCD sum rules. In cases when there are several works with close results I gave preference to the papers which I knew better or which are easily understandable. I apologize to the authors who will not find their publications devoted to the QCD sum rules in the present volume. Certainly, this fact does not mean that their contribution is unimportant. For the same reason, of course, it turned out impossible to discuss all applications worked out at present. Only a few typical problems are presented here, and only those where the results are definitely settled. The reader will not find the works of the last years; as a rule they call for a further development. In this context I would like to mention the so-called instanton-liquid model and the model of the stochastic vacuum (references are given in the Introduction below), as well as a recent attempt to work out an original nontraditional approach to color confinement due to Gribov (talk at the workshop "Perturbative QCD", Lund, Sweden, 1991, [preprint LU-TP-91-7, 1991]). They definitely deserve a thorough discussion but, probably, among experts rather than in a wide audience.

In my comments I mention new elements which appeared after publication of the reproduced papers and indicate those assertions which became obsolete, turned out to be wrong or questioned in the current literature. The comments are brief, they reflect only my own point of view, not necessarily coinciding with that of other theorists. In one or two exceptional cases I failed to choose among the published papers the one which I would consider absolutely satisfactory. As a result I was forced to write corresponding mini-reviews myself especially for this volume.

Numerous illuminating discussions with my colleagues and co-authors V. Belyaev, R. Bertlmann, V. Braun, V.L. Chernyak, D.I. Diakonov, H.G. Dosch, B.L. Ioffe, A. Khodjamirian, I. Kogan, H. Leutwyler, V.A. Novikov, A.V. Radyushkin, E.V. Shuryak, Yu.A. Simonov, A.V. Smilga, A.I. Vainshtein, M.B. Voloshin, V.I. Zakharov and A. Zhitnitsky are gratefully acknowledged. I would like to thank my colleagues for reading the manuscript and making valuable remarks. Some applications of the sum rule method have been kindly summarized in chapter 9 of this volume by A. Smilga and V. Chernyak.

Acknowledgements

The following articles have been reprinted by kind permission of the publisher, The American Institute of Physics:
From Soviet Journal of Nuclear Physics:
A.V. Smilga, The calculation of power corrections in the fixed-point gauge, Sov. J. Nucl. Phys. 35 (1982) 271–277; M.B. Voloshin, On the hyperfine splitting of the S-levels of heavy quarkonium, Sov. J. Nucl. Phys. 35 (1982) 592–595; M.B. Voloshin, Pre-Coulombic asymptotic behavior of heavy-quarkonium levels, Sov. J. Nucl. Phys. 36 (1982) 143–148.
From JETP Letters:
V. Belyaev and Ya.I. Kogan, Quantum chromodynamics calculation of g_A, JETP Lett. 37 (1983) 730–733; V.A. Nesterenko and A.V. Radyushkin, Analysis of the behavior of the pion form factor at small Q^2 by the method of QCD sum rules, JETP Lett. 39 (1984) 707–710.

The following article has been reprinted from Fortschritte der Physik by kind permission of the publisher, Akademie-Verlag:
V.A. Novikov, M.A. Shifman, A.I. Vainshtein and V.I. Zakharov, Calculations in external fields in quantum chromodynamics. Technical review, Fortschr. Phys. 32 (1984) 600–622.

The following article has been reprinted from Zeitschrift für Physik by kind permission of the publisher, Springer-Verlag:
B.L. Ioffe, On the choice of quark currents in the QCD sum rules for baryon masses, Z. Phys. C 18 (1983) 67–68.

Contents

1. Introduction to QCD Sum Rules: Historical Survey and Physical Picture

1.1. Status of QCD as the theory of hadrons

The time when Quantum Chromodynamics (QCD), the generally accepted modern theory of hadrons, was not yet created already seems deep history. As usual, it is very difficult to indicate a certain date when the theory was born. Somewhat conditionally one cay say that the starting point was the talk by Gell-Mann and Fritzsch in 1972 [1] (see also ref. [2]) in which the color-octet gluons were introduced. The next step is certainly the discovery of asymptotic freedom in 1972 [3]. Thus, 19 years of the QCD era have passed.

Almost two decades is a very large period of time for high-energy physics. The front-line of theoretical investigations has shifted far ahead leaving QCD far behind. I remember well the euphoria of the first years when poor and disconnected pieces of information on the hadronic world obtained in the previous period, gave the shape of a general picture, with a lot of predictions appearing every month and with the hope shared by many theorists that very soon a complete solution of the confinement problem would be found.

Now the dust raised during the first breakthrough has settled, the situation has become more or less stable, and one can give an unbiased report on all achievements and unsolved problems.

In spite of almost two decades of intensive efforts the most global aspects of quantum chromodynamics associated with the color confinement are still not understood. Usually it is assumed that in any non-Abelian gauge theory the spectrum of the observed states consists of color-singlet composite objects (hadrons), while isolated colored objects have infinite energy. The colored quarks are connected by string-like tubes of the chromoelectric flux.

Actually, this widespread scenario is essentially a dogma since nobody has succeeded in proving it. Nobody can reliably answer even such obvious questions as

- Is the spontaneous breaking of color symmetry possible in the strong coupling regime, and if not, why?
- What would be the spectrum in the theory with non-standard matter, say color-octet quarks?
- Are all alternative scenarios ruled out, for instance, Gribov's hypothesis according to which in pure gluodynamics (with no quarks) there are no localized asymptotic states at all?

We see that the theory is far from complete and, unfortunately (or maybe fortunately) it became clear that the solution of the basic problems is a notoriously difficult task, requiring new ideas. I do not exclude that the full solution will never be found.

At the same time the progress achieved in hadronic physics within QCD, even in its modern limited version, is striking; this is especially noticeable in retrospective, in comparison with the literature of the late 1960's and early 1970's. It will be instructive to briefly discuss here a list of basic findings and approaches which will undoubtedly stay with us irrespective of any future events.

First and foremost, it is firmly established that the fundamental constituents of hadrons are three-colored quarks

$$u^\alpha, d^\alpha, s^\alpha, c^\alpha, b^\alpha, t^\alpha, \quad \alpha = 1, 2, 3 \tag{1.1}$$

and gluons belonging to the color-octet representation. Their interaction is described by the famous QCD Lagrangian,

$$\mathscr{L} = -\tfrac{1}{4} G^a_{\mu\nu} G^a_{\mu\nu} + \sum_{q=u,d,\ldots} \bar{q}(\mathrm{i}\not{\mathscr{D}} - m_q)q, \tag{1.2}$$

where $G^a_{\mu\nu}$ is the gluon field strength tensor,

$$G^a_{\mu\nu} = \partial_\mu A^a_\nu - \partial_\nu A^a_\mu + g f^{abc} A^b_\mu A^c_\nu, \tag{1.3}$$

and $\mathscr{D}_\alpha$ is the covariant derivative

$$\mathrm{i}\mathscr{D}_\alpha = \mathrm{i}\,\partial_\alpha + g A^a_\alpha T^a, \tag{1.4}$$

where T^a stands for the generators of the SU(3) color group in the fundamental representation ($T^a = \frac{1}{2}\lambda^a$ where λ^a are the Gell-Mann matrices).

The gauge constant g determines the coupling of the quark fields q with the gluon field A^a_μ and the self-interactions of the gluon fields.

The coupling constant g showing up in the Lagrangian (1.2) is the bare coupling constant. Actually, all physical processes are described by the running or effective constant

$$\alpha_s(Q^2) \equiv \frac{g^2(Q^2)}{4\pi},$$

characterizing interactions at momenta $\sim Q$ (i.e. at distances $r \sim Q^{-1}$). The famous law of asymptotic freedom [3] takes for $\alpha_s(Q^2)$ the form:

$$\alpha_s(Q^2) = \frac{2\pi}{b \ln Q/\Lambda}, \tag{1.5}$$

where b is the first coefficient in the Gell-Mann–Low function and Λ is the scale parameter of QCD. Owing to the logarithmic falloff of the effective coupling constant the structure of the theory at short distances is simple; therefore, in processes where one can isolate the short-distance contribution the dynamical analysis can be carried out in terms of quarks and gluons interacting perturbatively.

The processes in which the short-distance contribution is important are called hard. The description of hard processes is the first undoubted success of QCD. There exists an endless number of reviews devoted to this subject (e.g. ref. [4]), and we will not touch it below. As a matter of fact, the physics of jets and other hard phenomena are now a separate branch of multi-particle dynamics with its specific methods and well-developed approximation schemes.

In any reaction, whatever the characteristic distances are, experimentalists detect not quarks and gluons but, always, pions, nucleons, etc. There are no reactions in which the realm of large distances is absent at all. The fact that theorists have learned how to calculate observable quantities so that the predictions depend only on the short-distance dynamics is the greatest achievement of the theory of hard processes.

Within the limits of applicability all results here are completely controllable theoretically.

In spite of a large number of (important and useful) works, both experimental and theoretical, the development in this field conceptually does not add much to what we already knew in the pre-QCD era. More or less the same perturbation theory and the same logarithms as in electrodynamics occur. The genuine hadronic theory starts at distances ~ 0.5 fm and larger.

At such distances non-perturbative effects play the key role; the effective coupling constant α_s is large. Still, if color confinement is accepted as a starting postulate, we are able now to explain qualitatively, and in many cases quantitatively, the main regularities observed in the hadronic word. First of all, the peculiar features of the mesonic spectrum: the existence of the octet of light (massless in the chiral limit $m_q = 0$) pseudoscalar mesons; a large mass of the ninth pseudoscalar meson η'; an approximate degeneracy in masses of the vector mesons ρ and ω as well as their tensor counterparts. We understand why the baryons are typically heavier than the mesons. The old dynamical observations, such as approximate SU(6) symmetry in the family of the old hadrons, the Zweig rule, the quark–hadron duality, are no mystery any more.

Moreover, it is proven that all hadronic properties are determined by a very peculiar structure of the QCD vacuum which is densely populated by violent non-perturbative fluctuations of the quark and gluon fields. The general features of these fluctuations are known and we have an idea of global (average) vacuum characteristics. This knowledge allows us in many cases to reliably predict or at least estimate quantities not yet measured experimentally.

The theoretical methods which brought to us the lion's share of information on the vacuum and hadronic structure are discussed below.

1.1.1. The 't Hooft self-consistency condition [5]

The argument is based on the fact that the three-point function of the type $\langle VVA \rangle$ has a pole singularity at $q^2 \to 0$. (Here V and A are flavor non-singlet vector and axial-vector currents, respectively, q is the four-momentum corresponding to A.) The proof of this fact stems from the triangle anomaly relation plus kinematics. In terms of physical states the pole singularity means the presence of massless particles in the spectrum. Generally speaking, there are two possibilities:

(i) eight massless pseudoscalar mesons, the chiral invariance of the Lagrangian (1.2) with $m_q = 0$ is spontaneously broken; and,
(ii) massless nucleons, the chiral symmetry is unbroken.

One can give further arguments based on duality and counting of the color factors [6] indicating that option (ii) is actually ruled out in QCD.

1.1.2. Exact inequalities [7]

The method is based on the vector nature of the quark–gluon vertex. Owing to this fact, certain correlation functions turn out to be reducible to functional integrals with positive-definite measure which ensures, in turn, the possibility of using the Schwartz inequalities. Surprising though it is, the corresponding information of the most general nature is sufficient for proving assertions like:

- the spontaneous breaking of the vector symmetries, e.g. isotopic, is impossible in QCD;
- the lightest mesons are pseudoscalars;
- $m_\rho < M_N$;

etc.

1.1.3. The large N_c limit [8]

In our world, the number of quark colors N_c is equal to three. Many years ago 't Hooft noticed that it is extremely useful to treat N_c as a free parameter and to consider the limit $N_c \to \infty$ (in this limit $\alpha_s \sim N_c^{-1}$, see eq. (1.5), where $b = 11N_c/3 - 2n_f/3$ and n_f is the number of quark flavors). In the large N_c limit the structure of the theory simplifies. Indeed, one can easily convince oneself that only the planar graphs with the minimal number of the quark loops survive at $N_c \to \infty$. A more detailed analysis shows that, say, the three-meson constants are proportional to $N_c^{-1/2}$, the four-meson constants to N_c^{-1}, etc. Thus, the relative smallness of the meson decay widths and the suppression of the processes with production of "extra" quark flavors (the Zweig rule) find a qualitative explanation if we assume that N_c^{-1} is in a sense a small parameter.

The predictions of multicolor chromodynamics are, as a rule, of a qualitative nature. The only exception seems to be the result [9] for the η' meson. Moreover, in some special cases the powers of $1/N_c$ are compensated by large numerical factors, and the extrapolation of the asymptotic results to the physical point $N_c = 3$ leads to a picture which has nothing to do with the actual situation [10]. Nevertheless, the analysis of N_c dependence, the so-called N_c counting, is very useful in many theoretical problems, to say nothing about the fact that multicolor chromodynamics is an excellent theoretical laboratory.

1.1.4. Low-energy theorems

Exact relations between amplitudes valid at low energies have been known in the theory of hadrons even before QCD. That is, the so-called soft-pion technique which uses the Goldstone nature of these pseudoscalar mesons (for an excellent review see ref. [11]). Quantum chromodynamics provided a solid basis for this soft-pion technique, first of all explaining the very fact of the existence of the (pseudo-)Goldstone bosons. Moreover, an estimate of the accuracy of the relations emerging within this approach can be obtained only in QCD.

With the development of quantum chromodynamics a new class of low-energy theorems has been discovered [10]. The theorems belonging to this class reflect the fact that the Lagrangian (1.2) is conformally invariant at the classical level in the chiral limit ($m_q = 0$). The conformal invariance is broken by quantum corrections (anomalies). Perhaps, the most elegant example of this type is the amplitude describing the conversion of gluons into a pair of pions,

$$\langle | \alpha_s G^a_{\mu\nu} G^a_{\mu\nu} | 2\pi \rangle = \text{const.}\ q^2, \tag{1.6}$$

where q is the invariant mass of the pion pair. Equation (1.6) codes highly nontrivial information about the gluonic component of pions and leads to surprising predictions, say, for the decays of a hypothetical light Higgs particle, $\psi' \to J/\psi\pi\pi$, etc.

1.1.5. Quark–hadron duality

The concept of duality reads that the cross section of a physical process, being averaged in a certain way, approximately coincides with the cross section for production of quarks and gluons [12]. The theoretical justification for duality is the fact that at short distances all interactions switch off in QCD. Calculations based on duality are indispensible in many problems, such as estimating the hadronic widths of J/ψ, Υ, D mesons, etc. They are widely used for orientation. More accurate and reliable quantitative results, obviously require answering the following questions:

- What is the proper averaging procedure?
- What is to be understood by the quark (gluon) cross section? Should one include in this cross section perturbative corrections if they are large (contain large logarithms or other external parameters)?
- At what energy does the quark–hadron duality set in? Is this energy universal?

Some of the answers but not all are already available. In particular, the limits of applicability of the quark–hadron duality are shown to be determined by the interaction of the valence quarks with the vacuum fields.

1.1.6. QCD sum rules

At present this method [13] (my main topic) seems to have yielded the largest number of reliable quantitative results concerning dynamical properties of the low-lying hadronic states, mesonic and baryonic. In a sense, the QCD sum rules are a refined expression of duality. The starting point is the general idea of the QCD vacuum being densely populated by fluctuating fields whose amplitude is so large that they cannot be described by perturbation theory. The strength of the fluctuations is characterized on average by the vacuum condensates; the most important of them are the quark and gluon condensates

$$\left\langle 0 \left| \frac{\alpha_s}{\pi} G^a_{\mu\nu} G^a_{\mu\nu} \right| 0 \right\rangle, \quad \langle 0 | \bar{q}q | 0 \rangle. \tag{1.7}$$

A pair of quarks, say $\bar{u}u$, "injected" into the vacuum by an external photon, lives and evolves not in empty space but in the "vacuum medium". As long as the distance between the quarks is not very large, their dynamics is determined by coarse, averaged vacuum characteristics, as in eq. (1.7), and is completely controllable theoretically. The idea, or good luck, if you wish, is that this theoretically reliable domain stretches almost up to distances at which classical hadrons such as the ρ meson, etc., are formed. One needs only a minimal extrapolation which brings in a modest and quite controllable uncertainty.

Almost all hadrons present in the PDG compilation, except for the radial excitations, are already considered within the QCD sum rules. The number of different results, as well as the literature devoted to the sum rule approach, are huge. Some of the works are cited in review papers, see ref. [14] and chapter 9.

It is essential that the method works in two directions. Starting from a few vacuum expectation values we are able to approximately calculate a large set of parameters referring to low-lying hadrons. And vice versa, experimental data on mesons and baryons are immediately translatable in the language of vacuum condensates and other information about the vacuum structure. The latter can, in turn, be used in the analysis of such exotic particles as glueballs or hybrid mesons.

It is quite obvious that the method has an approximate character and has its limits of applicability, which, at the moment, are well-established. Its virtues are relative simplicity, clear and transparent physical interpretation, the possibility of revealing qualitative regularities and, finally, complete theoretical control at every step.

1.1.7. Instantons

In the quasi-classical approximation, i.e. for small coupling constants, the complicated vacuum structure of quantum chromodynamics can be demonstrated explicitly. Along with the trivial vacuum sector corresponding to vanishing vacuum fields $A^a_\mu = 0$ (small oscillations near $A^a_\mu = 0$ are accounted for by perturbation theory) there are infinitely many other sectors in which the vacuum field $(A^a_\mu)_{\rm vac}$ yields $G^a_{\mu\nu} = 0$ and, still, cannot be reduced to $A^a_\mu = 0$ by any continuous gauge transformation. These additional sectors are labeled by integer numbers, the so-called winding numbers. The corresponding classification of the non-equivalent vacuum sectors was first given in ref. [15], where the trajectory of the tunneling transition connecting the neighboring sectors has also been found. The field configuration interpolating between the classical vacua with different winding numbers is localized in space and in (imaginary) time. It has been called "pseudoparticle" in the original paper [15], but its later name, "instanton" [16], is much more popular in the current literature.

The one-instanton contribution in the action is equal to $8\pi^2/g^2$ and the corresponding amplitudes are proportional to

$$\exp(-8\pi^2/g^2). \tag{1.8}$$

This factor is, evidently, beyond perturbation theory. Unfortunately, in the quantitative aspect instanton calculus is not very useful in the theory of hadrons since neither the actual spectrum of particles nor the transition amplitudes can be described in the quasi-classical approximation. At least thus far the instanton-based vacuum models [17] have not led to a decisive breakthrough, although there are some clear successes. Consideration of instantons is useful also because at the moment they are practically the only concrete example of non-perturbative QCD fluctuations which, in a sense, serves as a probe element for different theoretical constructions. Besides that, instantons qualitatively solve [18] the famous U(1), or η' problem [19].

1.1.8. *Computer simulations in lattice QCD*

The idea of this approach is due to Wilson [20] and Polyakov. Instead of considering quarks and gluons in continuous (Euclidean) space–time one introduces a discrete set of points forming a lattice, usually cubic. The gluon fields are defined on the lattice sites while the quark fields live in the vertices. Then one rewrites the Yang–Mills action in terms of these discrete variables, a rather ambiguous procedure admitting many versions which all coincide with each other in the continuum limit. Calculation of the correlation functions in the Euclidean space–time now reduces to a standard problem from statistical mechanics in a system with a finite but large number of degrees of freedom. Theorists try to solve this problem numerically using Monte Carlo simulations. They generate in computers an ensemble of vacuum fields, each configuration having weight $\exp(-S)$, where S is the Euclidean action. For each value of the lattice site, a, and the corresponding coupling constant, $g(a)$, its own "vacuum" is generated and the correlation functions are determined in this "vacuum". Then they are extrapolated to the continuum limit $a \to 0$, $g(a) = \text{const.}/(\ln a\Lambda)$.

The sketch of the procedure given above is extremely simplified, of course. In actual computations one has to deal with an enormous amount of problems both of technical and fundamental nature. At present the lattice Monte Carlo simulations are continuing to develop. I would say that they already constitute an isolated branch of science referring to computational mathematics rather than to theoretical physics. The devices used in lattice computations are very specific. Even a brief review of the problems which theorists encounter within this approach would require a book. Such books exist, see ref. [21].

As for physical predictions, the results obtained on the lattice are obviously not settled yet, especially in the case when the dynamical fermions are included. The scale of the systematic errors is not clear, the finite-size effects are not worked out completely and not all problems associated with massless fermions and chiral symmetry are solved. Therefore, below I will not touch at all the lattice results concerning different correlation functions, masses, vacuum expectation values, etc. Anyhow, being an outsider in this field I would be unable to present a critical analysis. The interested reader may turn to the monographic reviews [21] and fresh status reports which appear frequently.

1.1.9. Models

The hadronic phenomenology in the families of old and new hadrons is vast. The fundamental theory in the existing form cannot answer all questions arising in data analysis or in planning new experiments. Therefore, in quantitative estimates for the time being one cannot avoid models which are motivated by QCD to this or that extent but are not directly derivable from first principles. Let us mention here only those which are most popular:

(i) The instanton-liquid model [17].
(ii) The model of stochastic vacuum [22].
(iii) The bag model [23], quark compound bag [24].
(iv) Potential models of heavy quarkonium [25].
(v) Skyrme model for baryons [26].

In many cases the models listed above are strikingly successful in describing the data, although they certainly do not (and cannot) reproduce all subtle details – "fine structure" of the hadronic states. It is important to realize that each model has a particular domain of applicability and should be used only within its domain. Working beyond the domain of applicability leads to paradoxes. A good example is the use of the bag model for calculating the pseudoscalar meson constants (f_π, f_K, etc.) and similar amplitudes. Some theorists do such calculations forgetting that when the quarks are annihilated by the corresponding currents the bag remains intact – in the model, but not in nature, of course!

The most unpleasant aspect of the models is the fact that typically the limits of applicability cannot be established within the model itself, and additional analysis "from outside" is needed in order to establish the accuracy of the corresponding predictions. This is Achilles' heel of, say, the potential model, bag model or the Skyrme model. On the other hand, as long as our understanding deepens and grows one may expect that such approaches as the instanton liquid or the stochastic vacuum (and, maybe, some others) will be less and less dependent on additional assumptions not stemming from fundamental QCD. Then with time they will acquire another status than now and will supplement the list of "fundamental" methods presented above. At present, however, the development of these approaches is far from completion. The interested reader is referred to the review papers [27] and original works.

1.2. Problems and perspectives

Even a brief summary of the situation and methods existing in QCD shows that this field of theoretical activity is not dead, the development here continues, although not as rapidly as we would like it to. My main task is limited; below I will dwell on QCD sum rules at the pedagogical level. Before proceeding to the sum rule approach, however, I would like to formulate the most important unsolved problems and perspectives, as they are seen now.

The global questions still awaiting answers are:

- What is the detailed structure of the confinement mechanism?
- Does this phenomenon depend qualitatively or, at least quantitatively, on the presence of massless quarks?
- What features of the theory are responsible for the chiral symmetry breaking? Can one avoid this phenomenon by introducing some new elements in the Lagrangian (1.2)?

Another direction important from the practical point of view is providing a reliable basis for already existing models. One can rely on the model-dependent results only if the limits of applicability of the models under consideration are firmly established. There is no way to determine the limits of applicability and estimate possible errors within the models themselves. This can be done only after their relation with the fundamental principles of QCD is understood.

Creation of a model or approximation scheme in which the global properties – color confinement and spontaneous breaking of the chiral symmetry – are built in from the very beginning, in the zeroth-order approximation, does not seem to be an unrealistic task. This would be a decisive breakthrough in the theory of hadrons. A possible candidate is the Polyakov string [28].

Needless to say, there can be surprises and unexpected topics. Nonstandard issues continue to appear. One of the examples of this type is the issue of quark matter, stable multi-quark objects containing $3A$ quarks where $A \gg 1$. It is not ruled out at the moment that such objects are essentially lighter than Am_N for certain values of A (m_N stands for for nucleon mass). Experimentally stable multi-quark pieces of matter have not been observed, but this might only mean that the corresponding production cross section is small. The possibility of existence of stability islands is interesting from the practical point of view.

An associated problem refers to nuclear physics. Indeed, the internucleon distances in nuclei are of the order of the nucleon size. It is absolutely unclear why in such circumstances all quarks together do not form a "commune" and why the mass of the nuclei – as follows from experiment – is very close to Am_N, where A is the atomic number. Both problems seem to have a common solution.

A related direction which also attracted much attention is strange-quark matter [29]. The starting idea is simple and elegant. If there are n quarks ($n \gg 1$) localized in spacial domain, then it is energetically expedient to convert about $\frac{1}{3}$ of the u and d quarks into strange quarks through inverse β decay. We lose, of course, some energy due to the difference between the quark masses $m_s - m_{u,d}$ (each s quark is heavier than its u or d counterpartner by ~ 150 MeV). Simultaneously, we gain in energy since the Fermi surface becomes much lower: the Pauli exclusion principle does not forbid s quarks to live on those levels which are already occupied by u and d quarks. Under certain conditions the loss due to m_s is less than the energy we gain, and then in the theory there appear stable multi-quark systems with roughly speaking equal baryonic charge and strangeness.

Among other nonstandard topics it is worth mentioning the possibility of cold catalysis of nuclear fusion by heavy (quasi-)stable quarks if they exist in Nature [30].

1.3. Physical picture underlying the sum rule approach

Although in classical field theory the vacuum configuration is usually simple – the lowest energy is achieved for vanishing fields – the situation in quantum field theory is less trivial because the fields are permanently fluctuating. Even in electrodynamics one encounters certain difficulties, namely the energy of zero-point oscillations,

$$E_{\text{vac}} = \sum_n \tfrac{1}{2}\omega_n,$$

diverges at large frequencies. The problem is solved by virtue of a subtraction. Since only the difference between the energies of the excited and the lowest states is observable, the sum can be treated as a constant; for instance, one can put it equal to zero without having any physical consequences. This does not mean, of course, that the presence of the zero-point oscillations is unimportant for measurable quantities. Introducing particles (electrons, positrons) perturbs the vacuum fields, and this perturbation – now the energy difference – results in a Lamb shift of atomic levels.

Naturally, in QED we are dealing with a very small effect, of order $\alpha^3 = (1/137)^3$. In QCD there is no small free parameter, and most global characteristics of the spectrum are determined in this way.

The standard perturbation theory in QCD determines the high-frequency component of the vacuum "wave function". The low-frequency part – the one most directly related to the low-lying hadronic states – is controlled by violent non-perturbative fluctuations of the quark and gluon fields, which can be described, on average, by a few phenomenological parameters, i.e. vacuum condensates. The condensates, thus, carry information about the infrared behavior of the quark and gluon Green functions.

More graphically one can say that the QCD vacuum is a medium. The vacuum medium consists of two ingredients. One is built from gluon fields, the other from the quark fields. It is important that the quark ingredient includes only light pairs, $\bar{u}u$, $\bar{d}d$, or $\bar{s}s$. Indeed, a virtual pair of heavy quarks, say $\bar{c}c$, from the uncertainty principle lives in the vacuum for a time of the order of $(2m_c)^{-1}$, a time much shorter than the characteristic scale inherent to the vacuum fields. As a result the effect of heavy quark pairs is negligible. Moreover, such vacuum pairs can be consistently accounted for within ordinary perturbation theory.

It is intuitively clear that the fact of existence of the "quark medium" drastically affects the spectrum of old, light hadrons built from u, d and s quarks: suffice it to mention the π meson, whose masslessness in the chiral limit is entirely due to collective phenomena. Another clear example is the nucleon mass [31]. At the same time, charmonium and bottomonium levels below the open flavor threshold are practically insensitive to the vacuum fluctuations of u, d and s fields. They depend on the structure of the gluon phase only *.

Let us consider some correlation functions of colorless currents, say

$$\Pi^{(c)}_{\mu\nu}(x) = \langle \text{vac} | T\{\bar{c}(x)\gamma_\mu c(x),\, \bar{c}(0)\gamma_\nu c(0)\} | \text{vac} \rangle, \tag{1.9}$$

where $\bar{c}\gamma_\mu c$ produces 1^{--} charmonium states from the vacuum. By studying the dependence of $\Pi^{(c)}_{\mu\nu}$ on x for large x one can extract, in principle, the charmonium spectrum, i.e. all levels with given quantum numbers. If x is very small, the injected quarks have no time to interact with the vacuum medium and so propagate as free particles. However, in accordance with the uncertainty principle, small x values mean bad energy resolution, or, in more exact terms, for small x the correlation function in eq. (1.9) is saturated by a large number of charmonium levels, each contributing roughly equally.

* That is, there is no direct dependence. The existence of the quark condensate strongly affects, beyond any doubt, the gluon phase of the vacuum medium, i.e. the numerical values of the gluon condensates.

With increasing x the (radially) excited states become less important and finally die away since their weight relative to the lowest-lying level is proportional to $\exp[-(M'-M)x]$ (here M' and M are the masses of the excited and ground levels with given quantum numbers, $J^{PC}=1^{--}$ in the case at hand). Thus we are left with the ground state and can concentrate on its investigation.

The price we must pay is obvious – now it is necessary to account for the interaction of the injected quarks with the vacuum medium. If we are not very ambitious and do not try to put $|x|\to\infty$ in the mathematical sense, the price may not be very high, however. Indeed, for moderate values of $|x|$ the expansion of the quark propagator in $|x|^2|\boldsymbol{E}^a_{\text{vac}}|$ or $|x|^2|\boldsymbol{H}^a_{\text{vac}}|$ is meaningful, where $\boldsymbol{E}^a_{\text{vac}}$, $\boldsymbol{H}^a_{\text{vac}}$ are typical vacuum color fields. Notice that the gauge potential A^a_μ will not enter the final formulas because the correlation function in eq. (1.9) is gauge invariant. Terms linear in $\boldsymbol{E}^a_{\text{vac}}$ or $\boldsymbol{H}^a_{\text{vac}}$ will not enter either since there is no preferred orientation. The leading dynamical effect in $\Pi^{(c)}_{\mu\nu}$ will reduce evidently to the gluon condensate $\langle 0|\alpha_s G^a_{\mu\nu}G^a_{\mu\nu}|0\rangle$ multiplied by some known function of x.

Thus, we can fix $\Pi^{(c)}_{\mu\nu}$ in the intermediate x domain without complete information on the vacuum structure. The matrix element $\langle 0|\alpha_s G^2|0\rangle$ measures the mean field squared and was introduced first in ref. [32]. Its value should, of course, be calculable in the final theory of color confinement. At present we consider it as a free parameter, to be extracted from phenomenology *. It is important that one and the same parameter appears in the sum rules for all low-lying mesonic and baryonic states [13].

The central question is whether the expansion in the vacuum field is applicable for large enough x to guarantee the saturation of $\Pi^{(c)}_{\mu\nu}$ and other similar correlation functions by the ground state alone. It is not evident *a priori* that the two requirements can be met simultaneously. If they are, we get a (quasi-)theory of ground states based directly on first principles.

Since there are no small parameters like $\alpha=1/137$, the answer depends crucially on numerical values. As we shall see below, the balance between both requirements can be achieved in a certain domain of x and this, in turn, ensures the success of the sum rule method.

Apart from the gluon condensate the key role in the sum rules for old mesons and especially baryons belongs to another vacuum expectation value, $\langle 0|\bar{q}q|0\rangle$. The quark condensate measures the mean vacuum density of u, d and s quark pairs. The fact that a quark condensate $\langle 0|\bar{q}q|0\rangle\neq 0$ does exist has been known for a long time from low-energy pion physics [33]. Soft-pion technique allows one to establish the following relation:

$$(m_{\text{u}}+m_{\text{d}})\langle 0|\bar{u}u+\bar{d}d|0\rangle=-f_\pi^2 m_\pi^2+\mathrm{O}(m^2_{\text{u,d}}), \tag{1.10}$$

where m_{u}, m_{d} are the quark masses, m_π is the pion mass and f_π is the $\pi\to\mu\nu$ decay constant, $f_\pi=133$ MeV. All we need now in order to fix the quark condensate is an estimate of $m_{\text{u}}+m_{\text{d}}$. It is worth emphasizing that eq. (1.10) contains the so-called current quark masses that enter the QCD Lagrangian and determine chiral symmetry-breaking effects. They are surprisingly small; one can hardly doubt this fact today. Quantitatively it is very difficult to determine $m_{\text{u}}+m_{\text{d}}$ precisely, but beyond any doubt, this quantity is small, of the order of 10 MeV. (For a detailed discussion see ref. [34].) If the current quark masses are that small we get

$$\langle 0|\bar{q}q|0\rangle\simeq(-250\ \text{MeV})^3,$$

* In principle, some rough estimates from instanton calculus or lattice Monte Carlo simulations are also available.

a typical hadronic value. This result, as any other assumption, has been checked within the sum rules themselves.

As has been already mentioned, the success of the sum rule approach crucially depends on the existence of a confidence domain where, on the one hand, the power expansion is controllable theoretically, and, on the other hand, the sum over physical intermediate states is saturated by the lowest-lying level. One should not think that the existence of such a domain is an absolute miracle. In order to better understand how this can happen it is instructive to consider a simple example from quantum mechanics, familiar to everybody. It demonstrates nicely that a power series expansion for a Green function, valid formally at short τ (τ stands for Euclidean time), being extrapolated to larger τ, fixes the position of the lowest-lying level to a good accuracy [35].

1.3.1. Quantum mechanical example (harmonic oscillator)

Of course, in quantum mechanics the Schrödinger equation allows one to find the spectrum of any system, at least in principle. Here we determine the position of the ground level for the three-dimensional harmonic oscillator in another way that may seem extremely awkward. It has a unique advantage – it can be generalized to QCD, where there is no analogue of the Schrödinger equation.

Consider a particle of mass m moving in the spherically symmetric potential

$$V(\tau) = \tfrac{1}{2}m\omega^2\boldsymbol{r}^2.$$

For S-wave states the following sum plays the role of the correlation function in eq. (1.9)

$$S(\tau) = \sum_{n=0,2,4,\dots} |R_n(0)|^2\,\mathrm{e}^{-E_n\tau}, \tag{1.11}$$

where R_n is the S-wave radial wave function and E_n is the corresponding energy eigenvalue. Odd-n levels do not contribute because their wave functions vanish at the origin. One immediately recognizes in $S(\tau)$ the so-called time-dependent Green function, namely

$$S(\tau) = 4\pi G(\boldsymbol{x}_2 = 0,\ t_2 = -\mathrm{i}\tau \,|\, \boldsymbol{x}_1 = 0,\ t_1 = 0),$$

and the exact expression for $G(2\,|\,1)$ taken from any textbook implies

$$S(\tau) = \frac{2}{\sqrt{2\pi}}\left(\frac{m\omega}{\sinh\,\omega\tau}\right)^{3/2}.$$

By analyzing the closed form for $\tau\to\infty$ and comparing it with the definition in eq. (1.11), one can readily obtain exhaustive information on the S-wave spectrum. For instance, the ground-state energy evidently reduces to the following simple expression:

$$E_0 = \lim_{\tau\to\infty}\left[-\frac{\mathrm{d}}{\mathrm{d}\tau}\ln\,S(\tau)\right] = \tfrac{3}{2}\omega\lim_{\tau\to\infty}(\coth\,\omega t) = \tfrac{3}{2}\omega.$$

We shall assume, however, that just as in QCD the large-τ asymptotics is unknown, and all we are able to calculate theoretically are the first few terms of a power expansion in the small-τ domain where formally $S(\tau)$ has contributions from a large number of levels. The zeroth-order term corresponds to free motion, the next one is the first-order correction, and so on:

$$-\frac{\mathrm{d}}{\mathrm{d}\tau}\ln\,S(\tau) = \tfrac{3}{2}\omega\left[\frac{1}{\omega\tau} + \frac{\omega\tau}{3} - \frac{(\omega\tau)^3}{45} + \frac{2(\omega\tau)^5}{945} + \cdots\right].$$

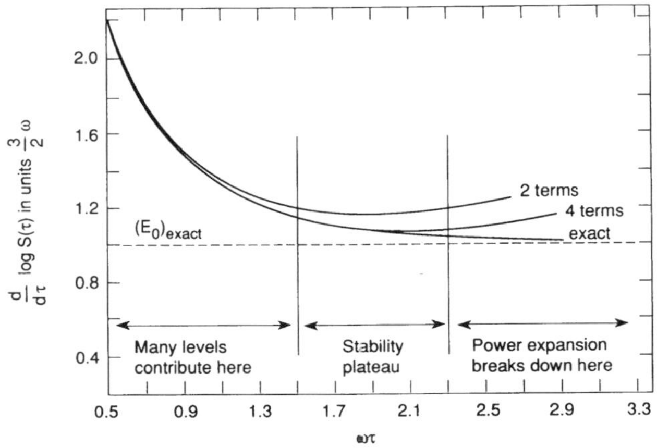

Fig. 1.1. Harmonic oscillator: power expansion of $(-d/d\tau)\ln S(\tau)$ versus $\omega\tau$. The lower curve is the exact result and the other two curves correspond to two and four terms in the expansion.

Now, it is impossible to let τ tend to infinity in the mathematical sense. Figure 1.1 shows, however, that there exists an intermediate stability plateau where, on the one hand, the perturbation expansion still works and, at the same time, the ground level almost saturates the sum. For instance, with only the first two terms the prediction for E_0 is $\frac{3}{2}\omega(2/\sqrt{3})$, 15% higher than the actual position. In the middle of the stability plateau the third (neglected) term of the expansion does not exceed 10%. If four terms are kept, $E_0 = \frac{3}{2} \times 1.06$. Moreover, even a rough idea of the contribution of higher levels (they can be accounted for in the quasi-classical approximation, say) drastically improves the estimate of E_0 and makes the accuracy better than 1%.

An analogous procedure also yields $|R_0(0)|^2$. In the case at hand, the theoretical uncertainty is evaluated by direct comparison with the exact result. In QCD one has to be more sophisticated and estimate the uncertainty by analyzing the power series themselves. It is assumed, of course, that there are no irregularities in the series and high (unknown) power corrections are not abnormally large.

1.3.2. Natural limits of applicability

My introductory remarks would be incomplete without at least a fragmentary discussion of the limits of applicability of the method.

The reader might already have noticed some similarity between the sum rule method and the old local duality concept [12], which states that a physical cross section averaged over some energy interval coincides with that for bare quarks. Likewise, the sum rules teach us that the ground-state contribution to current correlation functions is under certain conditions the same as that for the bare loop plus moderate power (non-perturbative) corrections due to gluon and quark vacuum condensates. A kinship is undeniable, although differences between the two approaches are profound. The local duality principle is rather vague since the way of averaging is ambiguous – neither the weight function nor the interval of smearing are fixed theoretically. Actually, it is an approach with an uncontrollable accuracy. In certain cases (e.g. the charmonium channel) the local duality works surprisingly well [36] but, unfortunately, one can never say beforehand whether for this or that particular meson the predictions will be accurate, what the optimal weight function is, etc.

In contrast, within the sum rule method everything is specified, and once the quark masses and basic vacuum matrix elements are determined there is no freedom left. The accuracy of the results is under theoretical control. The price we must pay for this advantage is rather high. The sum rule method, in its present form, says next to nothing about radial excitations that are strongly suppressed and represent a weak background to the dominant ground-state contribution. They are simply not seen from the Euclidean domain.

Another basic limitation prevents application to high orbital excitations, i.e. mesons with large spin. As was already mentioned, the QCD sum rules express a refined duality; they work successfully only if the groundstate (in a given channel) is more or less dual to the bare loop. Only then there exists the stability plateau where, on the one hand, non-perturbative terms are moderate and the external field expansion is applicable, and, on the other hand, the ground-state contribution dominates over all others. Also conceivable is the opposite, unlucky, situation in which resonance saturation occurs in the domain where the power terms are large and summation of the infinite non-perturbative series is necessary, a task going far beyond our present abilities. Under these circumstances we should have made a step back and confessed that the lowest-lying resonance is not resolved and all we can predict is the continuum spectral density. Unfortunately, just such a situation is realized for mesons with large spin. The sum rules practically do not constrain the properties of resonances with $J \geqslant 3$.

In order to understand why, recall the simple dynamical nature of quark–hadron duality. It merely reflects the fact that a current produces the quark pair at short times, shorter than the characteristic interaction time. If so, confinement effects on a larger time scale play the role of a large box that makes the spectrum discrete. They cannot, however, renormalize the cross section; the sum over a few close discrete levels should still reproduce the bare quark curve. In other words, the necessary condition for the quark–hadron duality to be valid is

$$\tau_{\text{prod}} \ll \tau_{\text{int}}. \tag{1.12}$$

In contrast, if $\tau_{\text{prod}} > \tau_{\text{int}}$ the confinement effects are essential in the process of quark production and there is no reason to think that the physical spectral density will coincide with the bare one after averaging.

Let us see what the implications of eq. (1.12) are. For simplicity, we consider massless quarks, although the final conclusion is independent of mass.

Thus, a $q\bar{q}$ pair with angular momentum J is injected in the vacuum with total energy E in the center-of-mass frame. It is evident that the size R of the region where the pair is produced is of order $J/p = J/E$. Then, since the quark's velocity is that of light,

$$\tau_{\text{prod}} \sim J/E. \tag{1.13}$$

On the other hand, the characteristic interaction time

$$\tau_{\text{int}} \sim 1/V_{\text{int}},$$

where V_{int} describes the dipole interaction of the $q\bar{q}$ pair with the vacuum gluon field, $V_{\text{int}} \sim \boldsymbol{d}^a \boldsymbol{E}^a_{\text{vac}}$. This estimate immediately yields

$$V_{\text{int}} \sim gRE_{\text{vac}} \sim \mu^2 J/E.$$

The dimensional parameter μ is related to the gluon condensate. Roughly speaking $\mu \sim \langle 0 | \alpha_s G^a_{\mu\nu} G^a_{\mu\nu} | 0 \rangle^{1/4}$. The interaction time is thus

$$\tau_{\text{int}} \sim E/\mu^2 J. \tag{1.14}$$

Comparing now eqs. (1.13) and (1.14) we see that the condition $\tau_{\text{prod}} \ll \tau_{\text{int}}$ actually means

$$E^2 > \mu^2 J^2.$$

In other words, quark–hadron duality in current spectral densities occurs provided that $s \gg \mu^2 J^2$.

On the other hand, if the Regge trajectories are linear, the lowest resonances are situated at $s \sim [\alpha'(0)]^{-1} J$ where $\alpha'(0)$ is the trajectory slope. The linear dependence on J means that only states with small J fall in the region of quark–hadron duality. Resonances with spin of order unity turn out to be inside the duality domain. For $J = 3$ they are close to the boundary, while for larger J the lowest-lying state becomes separated from the duality domain by a rapidly growing gap. Within this gap the physical spectral density is nonvanishing but not dual to the quark density.

This fact explains the success of the QCD sum rules in application to $\bar{q}q$ states with $J = 0$, 1, 2. The same arguments simultaneously tell us that the sum rules are useless for large-spin mesons, since the contribution due to the gluon condensate in the corresponding correlation functions increases with J too rapidly to permit saturation by the lowest state. A formal analysis of the issue, with all relevant mathematical details, is given in ref. [37].

Some other examples where the QCD sum rules fail to predict the properties of the lowest-lying state are considered in ref. [10].

References

[1] H. Fritzsch and M. Gell-Man, Proc. XVI Int. Conf. on High Energy Phys. (Batavia, 1972, v. 2) p. 135.
[2] H. Fritzsch, M. Gell-Mann and H. Leutwyler, Phys. Lett. B 47 (1973) 365.
[3] D.J. Gross and F. Wilzcek, Phys. Rev. Lett. 30 (1973) 1343.
H.D. Politzer, Phys. Rev. Lett. 30 (1973) 1346.
[4] M. Jacob, ed., Perturbative Quantum Chromodynamics (North-Holland, Amsterdam, 1982).
[5] G. 't Hooft, in: Recent Developments in Gauge Theories, eds. G. 't Hooft et al. (Plenum, New York, 1980).
[6] S. Coleman and E. Witten, Phys. Rev. Lett. 45 (1980) 100.
[7] F.T. Tomboulis, Phys. Rev. Lett. 50 (1983) 886.
D. Weingarten, Phys. Rev. Lett. 51 (1983) 1830.
E. Witten and C. Vafa, Nucl. Phys. B 234 (1984) 173.
E. Witten, Phys. Rev. Lett. 51 (1983) 2351.
S. Nussinov, Phys. Rev. Lett. 51 (1983) 1081; 52 (1983) 966.
[8] G. 't Hooft, Nucl. Phys. B 72 (1974) 461.
E. Witten, Nucl. Phys. B 160 (1979) 57.
[9] E. Witten, Nucl. Phys. B 156 (1979) 269.
G. Veneziano, Nucl. Phys. B 159 (1979) 213.
[10] V. Novikov, M. Shifman, A. Vainshtein and V. Zakharov, Nucl. Phys. B 191 (1981) 301.
[11] A.I. Vainshtein and V.I. Zakharov, Usp. Fiz. Nauk 100 (1970) 255 [Sov. Phys. Usp. 13 (1970) 737].
[12] J.H. Sakurai, Phys. Lett. B 46 (1973) 207.
[13] M. Shifman, A. Vainshtein and V. Zakharov, Nucl. Phys. B 147 (1979) 385.
M. Shifman, A. Vainshtein, M. Voloshin and V. Zakharov, Phys. Lett. B 77 (1978) 80.
M.B. Voloshin, Nucl. Phys. B 154 (1979) 365.
B.L. Ioffe, Nucl. Phys. B 188 (1981) 317.
[14] L.J. Reinders, H.R. Rubinstein and S. Yazaki, Phys. Rep. 127 (1985) 1.
B.L. Ioffe, Acta Phys. Pol. B 16 (1985) 543.
[15] A. Belavin, A. Polyakov, A. Schwarz and Y. Tyupkin, Phys. Lett. B 59 (1975) 85.
[16] G. 't Hooft, Phys. Rev. D 14 (1976) 3432.
[17] C. Callan, R. Dashen and D.J. Gross, Phys. Rev. D 19 (1979) 1826.
E. Shuryak, Nucl. Phys. B 203 (1982) 93, 117.
D.I. Dyakonov and V.Y. Petrov, Nucl. Phys. B 245 (1984) 259; Zh. Eksp. Teor. Fiz. 89 (1985) 361 [Sov. Phys. JETP 62 (1985) 204].

E. Shurak, The QCD Vacuum, Hadrons and the Superdense Matter (World Scientific, Singapore, 1988) ch. 3.
E. Shuryak, Nucl. Phys. B 328 (1989) 102.
E. Shuryak and J. Verbaarschot, Nucl. Phys. B 341 (1990) 1.
[18] G. 't Hooft, Phys. Rev. Lett. 37 (1976) 8.
[19] S. Weinberg, Phys. Rev. D 11 (1975) 3583.
[20] K. Wilson, Phys. Rev. D 10 (1974) 2445; in: New Phenomena in Subnuclear Physics, ed. A. Zichichi (Plenum, New York, 1977).
[21] M. Creutz, Quarks, Gluons and Lattices (Cambridge Univ. Press, Cambridge, 1983).
C. Rebbi, Lattice Gauge Theories and Monte Carlo Simulations (World Scientific, Singapore, 1983).
[22] H.G. Dosch, Phys. Lett. B 156 (1987) 365.
Yu.A. Simonov, Nucl. Phys. B 307 (1988) 512.
H.G. Dosch and Yu.A. Simonov, Phys. Lett. B 205 (1988) 339; Z. Phys. C 45 (1989) 147.
Yu.A. Simonov, Nucl. Phys. B 324 (1989) 67; Phys. Lett. B 226 (1989) 151; Sov. J. Nucl. Phys. 54 (1991) 115; Phys. Lett. B 249 (1990) 514; Phys. Rev. D 43 (1991) 3534; Z. Phys. C 53 (1992) 419.
[23] A. Chodos et al., Phys. Rev. D 9 (1974) 3471.
De Grand et al., Phys. Rev. D 12 (1975) 2060.
[24] R.L. Jaffe and F.E. Low, Phys. Rev. D 19 (1979) 2105.
Yu.A. Simonov, Phys. Lett. B 107 (1981) 1; Nucl. Phys. A 416 (1984) 109.
[25] T. Appelquist et al., Phys. Rev. Lett. 34 (1975) 365.
E. Eichten et al., Phys. Rev. Lett. 34 (1975) 369.
[26] T.H.R. Skyrme, Proc. R. Soc. London Ser. A 260 (1961) 237; 262; Nucl. Phys. 31 (1962) 556.
E. Witten, Nucl. Phys. B 233 (1983) 422, 433.
[27] P. Hasenfranz and J. Kuti, Phys. Rep. 40 (1978) 56.
K. Gottfried, Comments Nucl. Part. Phys. 8 (1981) 49.
T. Appelquist, N. Barnett and K. Lane, Annu. Rev. Nucl. Part. Sci. 28 (1978) 387.
J.D. Jackson, C. Quigg and J.L. Rosner, in: Proc. XIX Int. Conf. on High Energy Physics (Tokyo, 1978) p. 824.
C. Quigg and J.L. Rosner, Phys. Rep. 56 (1979) 169.
U.-G. Meißner, Phys. Rep. 161 (1988) 213.
[28] A.M. Polyakov, Phys. Lett. B 103 (1981) 207.
[29] S.A. Chin and A.K. Kerman, Phys. Rev. Lett. 43 (1979) 1292.
E. Witten, Phys. Rev. D 30 (1984) 272.
[30] G. Zweig, Science 201 (1978) 973.
B.L. Ioffe, L.B. Okun, M.A. Shifman and M.B. Voloshin, Acta Phys. Pol. B 12 (1981) 229.
[31] B.L. Ioffe, Nucl. Phys. B 188 (1981) 317.
[32] A. Vainshtein, V. Zakharov and M. Shifman, Zh. Eksp. Teor. Fiz. Pis'ma Red. 27 (1978) 60 [JETP Lett. 27 (1978) 55].
[33] M. Gell-Mann, R.J. Oakes and B. Renner, Phys. Rev. 175 (1968) 2195.
[34] J. Gasser and H. Leutwyler, Phys. Rep. 87 (1982) 77.
[35] J.S. Bell and R.A. Bertlmann, Nucl. Phys. B 177 (1981) 218.
A.I. Vainshtein, V.I. Zakharov, V.A. Novikov and M.A. Shifman, Yad. Fiz. 32 (1980) 1622 [Sov. J. Nucl. Phys. 32 (1980) 840].
[36] B.V. Geshkenbein and M.S. Marinov, Yad. Fiz. 30 (1979) 1400 [Sov. J. Nucl. Phys. 30 (1979) 726].
[37] M.A. Shifman, Yad. Fiz. 36 (1982) 1301 [Sov. J. Nucl. Phys. 36 (1982) 1290].

2. Foundations of the QCD Sum Rule Method and Sample Applications

Introduction

This chapter is the largest in this volume although it reproduces only two papers [1,2] by Shifman, Vainshtein and Zakharov giving the basis of the QCD sum rule method. After reading these papers a theorist who is a newcomer in this field will get a clear idea of all starting assumptions, approximations involved and will learn something concerning natural limitations inherent to this method. In principle, this chapter is self-contained and is quite sufficient for obtaining the initial skills allowing one to solve problems in everyday hadronic physics. Certainly, to master the subject in full it is desirable to look through other chapters as well. Chapter 3, for instance, contains a detailed discussion of the modern calculational approach, which is more efficient than that used in the original papers [1,2]. In the subsequent chapters the emphasis is shifted towards practical applications.

Twelve years after publication of refs. [1,2] the picture of the QCD vacuum outlined there remains intact. There are only one or two issues where our understanding has significantly deepened. A few assertions done in refs. [1,2] require a revision and/or comments.

First of all, the general status of the operator product expansion (OPE) discussed in subsection 3.2 of ref. [1]. It was believed that OPE is valid only for the first few terms and not for the operators of higher dimensions, due to instanton effects. Further investigations, although even now not quite completed from the practical viewpoint, showed, however, that this is not the case and the actual situation is much simpler. In dynamically nontrivial theories like QCD one can formulate, at least at a conceptual level, a systematic procedure allowing one to get a full operator product expansion, with infinitely many terms included. This procedure is based on a consistent separation of short- and large-distance contributions to all amplitudes and is usually called in the current literature the Wilson operator expansion. The large-distance contributions are ascribed to the matrix elements of composite operators while the contribution from short distances reside in the coefficient functions. Within the Wilson procedure OPE is valid for all power terms. Technical implementation of the idea requires, of course, explicit introduction of a cutoff parameter μ (normalization point) marking the boundary between these two domains. This question is treated in detail in several publications; see, e.g. paper [3] reproduced below in chapter 3 and references therein.

One should make a clear distinction between the theoretical validity of the Wilson operator expansion and the way it is practically realized in different theories. The fact that OPE is universally valid does not mean that in any theory it is as useful and successful as in QCD. Several known attempts to adapt the QCD sum rule method to two-dimensional models with a dynamically generated mass scale (e.g. O(3) σ-model) failed for this reason. The point is that, generally speaking, in each particular theory a specific definition of the separation procedure is needed. In many cases it is an extremely difficult task, and I am unaware of any fully developed construction of this type. To be absolutely consistent, one should indicate explicitly how the short- and large-distance contributions are separated, then, given a certain scheme, define the vacuum expectation values and the coefficient functions in this scheme, and only after that the sum rule mechanism can be launched. Intermediate results (the vacuum condensates, for instance) may be sensitive to the scheme chosen.

Fortunately, in QCD such a disaster does not happen, or, to be more exact, all these complications are unimportant in calculations. In QCD there is a tremendous element of luck leading to the fact that the most naive and simplified approach – it reduces to the formula: "hide all non-perturbative effects in the vacuum matrix elements, and compute all coefficient functions perturbatively" – proves to be successful in the majority of cases. Within this approximate procedure there is no need to be too accurate in what is to be called the normalization point in the exact definition of the vacuum expectation values and coefficient functions (although, of course, no purist will be happy unless such rigorous definitions are worked out). Some facts explaining why such a peculiar and favorable situation with OPE occurs in QCD are discussed in refs. [3,4]. The latter paper as well as refs. [5,6] compare different aspects of OPE in QCD and two-dimensional σ-models, an instructive exercise which I recommend to the interested reader.

Above, exceptional cases have been mentioned where the QCD sum rule method based on the simplified version of OPE turns out to be useless. We encounter such an "anomaly" in dealing with the gluonium world, especially for the scalar and pseudoscalar glueballs. Violent non-perturbative fluctuations populating the QCD vacuum seem to produce a much stronger effect on the valence gluons than on quarks. Even more important is the following distinction: if the motion of the valence quarks inside hadrons to a good approximation depends only on the averaged characteristics of the vacuum medium the valence gluons are often very sensitive to subtle features, and their dynamical properties cannot be expressed in terms of the vacuum condensates $\langle G^a_{\mu\nu}G^a_{\mu\nu}\rangle$ or $\langle\bar{\psi}\psi\rangle$. The strongest connection with an "instantaneous" vacuum field, not reducible to the gluon and quark condensates, occur in the channels with the vacuum quantum numbers: scalar and pseudoscalar gluonium and, to a lesser extent, $0^{\pm}$ light quark mesons. The corresponding mesons, effectively probing a fine structure of the vacuum medium, are singled out; they drastically differ from the majority of all other low-lying hadronic states, which feel only averaged fields and can be treated within the sum rule method. Thus, not all hadrons are alike – understanding of this fact came [7] after theorists proceeded from sample applications to a systematic analysis of the hadronic family.

The second, important topic deserving a few remarks is instantons. In refs. [1,2] the so-called dilute instanton gas approximation (DIGA) [8] has been repeatedly used for different estimates (see subsection 6.7 in ref. [1] and 6.8 in ref. [2]; references to DIGA are scattered everywhere in these papers). At present one can hardly doubt that this approximation is far from the genuine picture of the QCD vacuum and is appropriate only for qualitative arguments, if at all.

Instead of DIGA, more elaborated instanton-based vacuum models exist now. The one which is most advanced is called the instanton-liquid model and it was developed by Shuryak, Diakonov, Petrov and, later, Balitsky and Yung (see ref. [9] for a review; some other references are given in ref. [10]; for a discussion of recent progress see ref. [11]). The main idea is to use distorted instantons as building blocks for liquid-type vacuum media; this idea grew from a few marginal observations made in ref. [7]. The idea has gradually acquired flesh and blood and has materialized now as a mature semi-quantitative model with obvious virtues. In particular, all general low-energy theorems of QCD are reproduced by the instanton liquid, which is, unfortunately, not the case in DIGA. Many parameters of the low-lying hadronic states come out correctly, although not all of them. The model is still in its active stage; many theorists try to extend it and make it fully quantitative. It is worth noting that calculating the gluon condensate $\langle G^2\rangle$ with the instanton liquid one obtains roughly speaking the same numbers as the primitive estimate of ref. [1], compatible with the phenomenologic value of $\langle G^2 \rangle$.

It is quite clear that any explicit model of the QCD vacuum, when completed, would allow one to go much further than the QCD sum rules and get much more information on hadronic properties. The price we have to pay is evident: if the sum rule approach is model independent to a large extent, any vacuum model is more vulnerable, heavily relying on assumptions which thus far cannot be derived from first principles. Nevertheless, the advancement of explicit vacuum models is inevitable.

From others I would like to single out the so-called stochastic vacuum model [12] having a rapidly expanding realm of applications. The starting postulate is the dominance of stochastic type fluctuations in the QCD vacuum – with a finite correlation length. Within the framework of this hypothesis, supplemented by a few technical assumptions, a series of calculations has been carried out with promising results, including such a nontrivial quantity as α', the slope parameter of the Regge trajectories. The sum rule method in its original form certainly gives no prediction for α'.

The third point which can be considered as obsolete in ref. [2] is the discussion of the bag model in subsection 6.4. There we made an attempt to relate the bag parameter B to the value of the gluon condensate, assuming that the total vacuum energy $\epsilon \propto \langle \alpha_s G^2 \rangle$ is the same as B. Shortly it was realized that this conjecture cannot be correct. First of all, these two quantities have different dependence on N_c, the number of colors [13] (see also chapter 8). Second, there exist convincing numerical arguments indicating that B constitutes only a relatively small fraction of ϵ. We refer the reader to ref. [7] for further comments on the bag model from the point of view of the sum-rule practitioner. This paper also summarizes efforts aimed at incorporating the low-energy theorems of quantum chromodynamics in the sum rule method, a closely related issue which was essentially unknown in 1979.

A few remarks are in order now concerning technical aspects of the papers reproduced in this chapter. An indispensable element of the method ensuring its predictive power is the Borelization procedure (subsection 5.4 of ref. [1]) transforming the standard power weight function of the dispersion representation into an exponential one. Several attempts have been undertaken to introduce other weight functions which would be "better" than exponential and would possess extra desirable properties, e.g. increased sensitivity to any given resonance, not only to the lowest-lying one, etc. In particular, in ref. [14] a Gauss-type weight function is obtained. Unfortunately, this trick, being attractive conceptually, does not work as efficiently as the original Borel representation. An alternative possibility is to consider the dispersion representation for the correlation functions directly in coordinate space rather than in momentum space as it is usually done. The sum rules obtained in this way seem to be very close in structure and resolving power to the Borel version [1,2] but, probably, have a slightly more graphic interpretation [15].

I cannot help dwelling on the question of the numerical values of the condensates. In the ultimate theory they should be calculable from first principles. At present, however, just as 12 years ago, the state of the art is such that we have to rely, instead, either on models or on the sum rules themselves. A spectrum of different estimates for the quark and gluon condensates outlined in refs. [1,2] did not undergo drastic changes. The only new proposal not discussed in refs. [1,2] is the lattice calculation of the vacuum expectation value $\langle G^2 \rangle$ and other gluon condensates (see ref. [16] and the review paper [17]). The results obtained on the lattice so far do not include dynamical fermions and are plagued by some other systematic uncertainties and theoretical difficulties. Within the (relatively large) error bars they do not contradict the phenomenologic estimates [1,2]. Certainly, further progress in the lattice computations might reveal in future certain deviations.

The prediction for $\langle G^2 \rangle$,

$$\left\langle \frac{\alpha_s}{\pi} G^a_{\mu\nu} G^a_{\mu\nu} \right\rangle \approx 0.012 \text{ GeV}^4, \tag{2.1}$$

extracted in ref. [18] from the charmonium sum rules, is now usually referred to as standard. There is an ongoing controversy in the literature regarding the actual value of the gluon condensate. Some authors argue (see e.g. refs. [19–22]) that the number quoted in eq. (2.1) is considerably underestimated; the value they advocate is larger by a factor of 2 to 5. (I hasten to add that many other authors confirm eq. (2.1).)

A detailed critical report on the situation with $\langle G^2 \rangle$ goes far beyond the scope of this introductory commentary. Nor will we submerge into analysis of specific technicalities of refs. [19–22] – this is, rather, the subject for an original and very labor-consuming investigation *. On general grounds alone it is quite possible that thorough analysis could reveal some drawbacks in the procedure accepted in refs. [18,1] which were overlooked. I still believe, however, that the standard estimate, eq. (2.1), is pretty close to the exact value, and enhancement of the gluon condensate by a factor of 2 to 5 cannot take place. The point is that a whole wealth of data is accumulated now, especially in the light-quark sector, which definitely implies preference for the standard value. Basically all well established mesons and baryons from the PDG tables have been considered within the sum rule method. The precise value of $\langle G^2 \rangle$ within a factor of ~ 1.5 around the standard one is not crucial in this analysis. However, its enhancement by a factor of 2 to 5 would ruin the beautiful agreement between theory and experiment which we observe now. On the other hand, the majority of the works [19–22] are based either on toy models (quantum mechanics, two-dimensional QCD) or on such unreliable approaches as, say, finite-energy sum rules or vector dominance. The former are known to be quantitatively far from the real four-dimensional theory [4], while the latter are extremely sensitive to unknown details of the spectrum outside the low-energy domain and depend on arbitrary assumptions. Summarizing the situation and trying to avoid categoric conclusions I would say that the estimate in eq. (2.1) is still viable, and the issue of its revision remains open. The interested reader is referred to ref. [23] for further comments.

The second important condensate parameter introduced in ref. [1] is the four-quark matrix element

$$\langle \bar{\psi} \Gamma_1 \psi \bar{\psi} \Gamma_2 \psi \rangle, \tag{2.2}$$

where ψ is the light-quark field and Γ denotes any of the γ matrices. In refs. [1,2] the so-called factorization (or the vacuum saturation) hypothesis has been proposed allowing one to express $\langle \bar{\psi} \Gamma_1 \psi \bar{\psi} \Gamma_2 \psi \rangle$ in terms of a simpler matrix element, $\langle \bar{\psi} \psi \rangle$ (subsection 6.5 of ref. [1]). Some duality-type arguments substantiating the factorization hypothesis for the operator (2.2) have been given.

Since then the status of the hypothesis has been examined in many different ways. First of all, it acquired a solid theoretical justification in the limit of a large number of colors, $N_c \to \infty$, where $\langle \bar{\psi} \Gamma_1 \psi \bar{\psi} \Gamma_2 \psi \rangle$ exactly reduces to $\langle \bar{\psi} \psi \rangle^2$ (see ref. [23] and chapter 8). Second, independent estimates of different quark condensates have been obtained within the sum rule method itself. Several meson channels have been studied [24] with the special task of detecting

* As a matter of fact it might be very useful to actually carry out such an investigation.

deviations from factorization for the four-quark operators *. The conclusion common to all these works is as follows: in the vector and axial-vector cases ($\Gamma=\gamma_\mu$ or $\gamma_\mu\gamma_5$) deviations from the factorization formula

$$\langle\bar\psi\Gamma_1\Psi\bar\psi\Gamma_2\psi\rangle = -\tfrac{1}{144}\,\mathrm{Tr}(\Gamma_1\Gamma_2)\,\langle\bar\psi\psi\rangle^2 \tag{2.3}$$

cannot significantly exceed $\sim 10\%$.

Lattice people are also actively involved in the discussion of whether factorization holds [25]. This direction of research will probably give decisive results in the future; at present the systematic uncertainties of the lattice results are too high, and the best conclusion one can make is that there is no apparent contradiction between the factorization hypothesis and the lattice predictions for the four-fermion operators (see the previous footnote).

It is worth noticing that the baryon sum rules [27,28], which opened a new vast field of applications, may serve as a potentially sensitive tool for determining all quark condensates, and $\langle\bar\psi\Gamma_1\psi\bar\psi\Gamma_2\psi\rangle$ in particular. The special sensitivity of the baryon sum rules to $\langle\bar\psi\Gamma_1\psi\bar\psi\Gamma_2\psi\rangle$, $\langle\bar\psi\psi\rangle$, etc. has been emphasized in the pioneering works [27,28]. The reason is obvious: these condensates reflect the chiral symmetry breaking, the phenomenon affecting the very gross features of the baryon spectrum. A nice illustration of this sensitivity is the immediate discovery [27,28] of a suppression of $\langle\bar ss\rangle$ compared to $\langle\bar uu\rangle$ or $\langle\bar dd\rangle$. The ratio of these condensates is an important characteristic of the vacuum structure, which will be discussed at length in chapter 5.

The analysis of the baryon sum rules carried out in refs. [27,29] assumes exact factorization for $\langle\bar\psi\gamma_\mu\psi\bar\psi\gamma_\mu\psi\rangle$ and $\langle\bar\psi\gamma_\mu\gamma_5\psi\bar\psi\gamma_\mu\gamma_5\psi\rangle$. The general picture of the baryon family emerging in this way is in very good agreement with what we observe experimentally, a new and extremely impressive evidence in favor of eq. (2.3).

In spite of a significantly deeper understanding of the situation than a decade ago, I would not like to make the false impression that everything is crystally clear with the four-quark operators, and eq. (2.3) is proven. On the contrary, there are some arguments that the situation is not universal [26] and in "exotic" cases, say, $\Gamma=\gamma_5$ or $\mathbf{1}$, noticeable violations of eq. (2.3) are likely to occur. The condensates we deal with for "normal" mesons are always $\langle\bar\psi\gamma_\mu\psi\bar\psi\gamma_\mu\psi\rangle$ or $\langle\bar\psi\gamma_\mu\gamma_5\psi\bar\psi\gamma_\mu\gamma_5\psi\rangle$ while $\langle\bar\psi\gamma_5\psi\bar\psi\gamma_5\psi\rangle$ emerges in the scalar and pseudoscalar channels, which are anomalous anyhow and are not discussed in this volume.

Moreover, sometimes it is claimed that even in the vector and axial-vector cases, $\Gamma=\gamma_\mu$ or $\gamma_\mu\gamma_5$, (we will refer to them as standard, or non-exotic) factorization is off by a factor of 2 to 3 [30]. The suspicions [30] that the accuracy of eq. (2.3) in the standard, non-exotic cases can be much worse than stated above are based on the same baryon sum rules, although "cooked" in a slightly different way.

Unlike the meson channels, for the octet baryons there are two interpolating currents of the lowest dimension, and this fact results in certain ambiguities in the predictions for the mass spectrum, which might imply, in turn, noticeable deviations from factorization. The issue is discussed in more detail in chapter 5 where I explain why I still believe that the baryon sum rules do favor factorization.

Combining all arguments (see especialy ref. [24]) one can say that the evidence in favor of eq. (2.3) with possible deviations at the 10% level for $\Gamma=\gamma_\mu$ and $\gamma_\mu\gamma_5$ is stronger now than a

* Literally speaking, the works cited in ref. [24] test factorization for the four-quark operators averaged over states containing the Goldstone mesons (π, K, etc.) rather than over the vacuum state. Within the soft-pion technique, however, the former matrix elements are reducible to the latter.

decade ago, although, of course, the question cannot be considered completely closed. Accurate and reliable estimates of deviations from eq. (2.3), both in the standard case (where they are expected to be small) and in the exotic one (where it will not be surprising to get a large effect), are welcome.

New condensates of higher dimensions also start appearing on the scene, e.g. gluon operators of dimension 8 and a mixed quark–gluon operator of dimension 5. In principle, the proliferation of different vacuum matrix elements which are not fixed theoretically is in contradiction with the spirit of the sum rule method reading that the basic properties of the low-lying hadrons can be approximately expressed in terms of a few averaged vacuum characteristics, without excessive detail. A moderate number of new condensates supplementing the family listed in ref. [1] is needed to extend the limits of the method, however.

Next to nothing is known about the gluon operators of dimension 8, like $\mathrm{Tr}\,\langle G_{\mu\nu}G_{\mu\nu}G_{\alpha\beta}G_{\alpha\beta}\rangle$, etc. The exact factorization – a procedure analogous to that we apply to the four-quark operators – has been proposed in ref. [31]. Unlike the quark case, it has absolutely no theoretical justification and, moreover, there are serious reasons to think that deviations are very significant [23].

On the other hand, the mixed quark–gluon condensate is extracted [29] from the baryon sum rules with an accuracy not worse than that in $\langle G^2\rangle$ and $\langle\bar\psi\psi\rangle$. Namely [29],

$$\left\langle -g\bar\psi\sigma_{\alpha\beta}G^a_{\alpha\beta}\frac{\lambda^a}{2}\psi\right\rangle \equiv m_0^2\,\langle\bar\psi\psi\rangle, \quad m_0^2\approx 0.8\ \mathrm{GeV}^2. \tag{2.4}$$

The original estimate [29] was recently supplemented by an independent alternative calculation [32] with the result totally confirming eq. (2.4).

Concluding this brief commentary it would be nice to recommend a few reviews summarizing recent progress. Unfortunately, fresh review papers with comprehensive coverage of all theoretical aspects and applications simply do not exist. Selective topics are discussed in ref. [33] and in chapter 9 below.

References

[1] M.A. Shifman, A.I. Vainshtein and V.I. Zakharov, Nucl. Phys. B 147 (1979) 385.
[2] M.A. Shifman, A.I. Vainshtein and V.I. Zakharov, Nucl. Phys. B 147 (1979) 448.
[3] V.A. Novikov, M.A. Shifman, A.I. Vainshtein and V.I. Zakharov, Nucl. Phys. B 249 (1985) 445.
[4] V.A. Novikov, M.A. Shifman, A.I. Vainshtein and V.I. Zakharov, Phys. Rep. 116 (1984) 103.
[5] M.V. Terentyev, Yad. Fiz. 45 (1987) 589 [Sov. J. Nucl. Phys. 45 (1987) 368].
[6] M.V. Terentyev, Fortschr. Phys. 36 (1988) 117.
[7] V.A. Novikov, M.A. Shifman, A.I. Vainshtein and V.I. Zakharov, Nucl. Phys. B 191 (1981) 301.
[8] G.G. Callan, R. Dashen and D.J. Gross, Phys. Rev. D 17 (1978) 2717.
[9] E. Shuryak, Nucl. Phys. B 328 (1989) 102.
[10] E. Shuryak, Nucl. Phys. B 203 (1982) 93, 117.
D.I. Dyakonov and V.Y. Petrov, Nucl. Phys. B 245 (1984) 259; Zh. Eksp. Teor. Fiz. 89 (1985) 361 [Sov. Phys. JETP 62 (1985) 204].
E. Shuryak, The QCD Vacuum, Hadrons and the Superdense Matter (World Scientific, Singapore, 1988) ch. 3.
I.I. Balitsky and A.V. Yung, Phys. Lett. B 168 (1986) 113.
[11] J. Verbaarschot, Nucl. Phys. B 362 (1991) 33.
V.V. Khoze and A. Ringwald, Nucl. Phys. B 355 (1991) 351.
[12] H.G. Dosch, Phys. Lett. B 156 (1987) 365.
Yu.A. Simonov, Nucl. Phys. B 307 (1988) 512.
H.G. Dosch and Yu.A. Simonov, Phys. Lett. B 205 (1988) 339; Z. Phys. C 45 (1989) 147.

Yu.A. Simonov, Nucl. Phys. B 324 (1989) 67; Phys. Lett. B 226 (1989) 151; Sov. J. Nucl. Phys. 54 (1991) 115 and references therein.
Yu.A. Simonov, Phys. Lett. B 249 (1990) 514; Phys. Rev. D 43 (1991) 3534; Z. Phys. C 53 (1992) 419.
[13] W. Bardeen and V.I. Zakharov, Phys. Lett. B 91 (1980) 111.
M.A. Shifman, Nucl. Phys. B 173 (1980) 13.
[14] R.A. Bertlmann, G. Launer and E. De Rafael, Nucl. Phys. B 250 (1985) 61.
[15] E. Shuryak, Correlation functions in the QCD vacuum. Preprint SUNY-NTG-91/45, 1991 (unpublished).
[16] A. Di Giacomo and P. Rossi, Phys. Lett. B 100 (1981) 481.
A. Di Giacomo and G. Pafutti, Phys. Lett. B 108 (1982) 327.
E.M. Ilgenfritz and M. Muller-Preussker, Phys. Lett. B 119 (1982) 395.
R. Kirschner et al., Nucl. Phys. B 210 (1982) 567.
A. Di Giacomo, K. Fabricius and G. Pafutti, Phys. Lett. B 118 (1982) 129.
M. Campostrini et al., Z. Phys. C 32 (1986) 377.
N. Makhaldiani and M. Muller-Preussker, in: Proc. 1984 Georgenthal Conf. High-Energy Phys.
M. Campostrini, A. Di Giacomo and Y. Gunduc, Phys. Lett. B 225 (1989) 393.
[17] A. Di Giacomo, in: Proc. 1985 Montpellier Conf. Non-perturbative methods (World Scientific, Singapore, 1985).
[18] M.A. Shifman, A.I. Vainshtein, M.B. Voloshin and V.I. Zakharov, Phys. Lett. B 77 (1978) 80.
[19] R.A. Bertlmann, Nucl. Phys. B 204 (1982) 387.
B. Durand and J.B. Whitenton, Phys. Rev. D 28 (1983) 607.
[20] A. Bradley et al., Phys. Lett. B 102 (1981) 359.
P. Ditsas and G. Shaw, Phys. Lett. B 116 (1982) 188; Nucl. Phys. B 229 (1983) 29.
[21] V.N. Baier and Yu.F. Pinelis, Phys. Lett. B 116 (1982) 179.
J. Marrow, J. Parker and G. Shaw, Z. Phys. C 33 (1986) 237; Z. Phys. C 37 (1987) 103.
[22] G. Launer, S. Narison and R. Tarrach, Z. Phys. C 26 (1984) 433.
J. Bordes, V. Gimenez and J.A. Penarrocha, Phys. Lett. B 201 (1988) 365.
R.A. Bertlmann et al., Z. Phys. C 39 (1988) 231.
C.A. Dominguez and J. Sola, Z. Phys. C 40 (1988) 63.
V. Gimenez, J. Penarrocha and J. Bordes, Phys. Lett. B 223 (1989) 245.
B.V. Geshkenbein, Yad. Fiz. 51 (1990) 1121 [Sov. J. Nucl. Phys. 51 (1990) 719].
[23] V.A. Novikov et al., Nucl. Phys. B 237 (1984) 525.
[24] K.G. Chetyrkin et al., Phys. Lett. B 174 (1986) 104.
R. Decker, Nucl. Phys. B 277 (1986) 661.
L.J. Reinders and S. Yazaki, Nucl. Phys. B 288 (1987) 789.
K.G. Chetyrkin and A.A. Pivovarov, Nuovo Cimento A 100 (1988) 899.
[25] M.B. Gavela et al., Phys. Lett. B 211 (1988) 139; Recent results and prospects in the computation of weak amplitudes in lattice QCD, Talk at the Ringberg Workshop on Hadronic Matrix Elements and Weak Decays, Germany, 1988, in: Nucl. Phys. B (Proc. Suppl.) 7A (1989) 228.
[26] A.R. Zhitnitskii, Yad. Fiz. 41 (1985) 805, 1035, 1331 [Sov. J. Nucl. Phys. 41 (1985) 513, 664, 846].
[27] B.L. Ioffe, Nucl. Phys. B 188 (1981) 317; (E) Nucl. Phys. B 191 (1981) 591.
[28] Y. Chung, H.G. Dosch, M. Kremer and D. Shall, Phys. Lett. B 102 (1981) 175; Nucl. Phys. B 197 (1982) 57.
[29] V.M. Belyaev and B.L. Ioffe, Zh. Eksp. Teor. Fiz. 83 (1982) 876; 84 (1983) 1236 [Sov. Phys. JETP 56 (1982) 493; 57 (1983) 716].
[30] Y. Chung, H.G. Dosch, M. Kremer and D. Shall, Z. Phys. C 25 (1984) 151.
[31] S.N. Nikolaev and A.V. Radyushkin, Phys. Lett. B 124 (1983) 243.
[32] A.A. Ovchinnikov and A.A. Pivovarov, Yad. Fiz. 48 (1988) 1135 [Sov. J. Nucl. Phys. 48 (1988) 721].
[33] B.L. Ioffe, Proc. Int. Conf. High Energy Phys., Leipzig, 1984, Vol. 2, p. 176; Acta Phys. Pol. B 16 (1985) 543.
L.J. Reinders, H. Rubinstein and S. Yazaki, Phys. Rep. 127 (1985) 1.
S. Narison, in: World Sci. Lecture Series in Physics, Vol. 26 (World Scientific, Singapore, 1989).

Nuclear Physics B147 (1979) 385–447

QCD AND RESONANCE PHYSICS. THEORETICAL FOUNDATIONS

M.A. SHIFMAN, A.I. VAINSHTEIN * and V.I. ZAKHAROV
Institute of Theoretical and Experimental Physics, Moscow, 117259, USSR

Received 24 July 1978

A systematic study is made of the non-perturbative effects in quantum chromodynamics. The basic object is the two-point functions of various currents. At large Euclidean momenta q the non-perturbative contributions induce a series in (μ^2/q^2) where μ is some typical hadronic mass. The terms of this series are shown to be of two distinct types. The first few of them are connected with vacuum fluctuations of large size, and can be consistently accounted for within the Wilson operator expansion. On the other hand, in high orders small-size fluctuations show up and the high-order terms do not reduce (generally speaking) to the vacuum-to-vacuum matrix elements of local operators. This signals the breakdown of the operator expansion. The corresponding critical dimension is found. We propose a Borel improvement of the power series. On one hand, it makes the two-point functions less sensitive to high-order terms, and on the other hand, it transforms the standard dispersion representation into a certain integral representation with exponential weight functions. As a result we obtain a set of the sum rules for the observable spectral densities which correlate the resonance properties to a few vacuum-to-vacuum matrix elements. As the last bid to specify the sum rules we estimate the matrix elements involved and elaborate several techniques for this purpose.

1. Introduction

Quantum chromodynamics is widely believed nowadays to be a true theory of strong interactions. Because of the celebrated asymptotic freedom of QCD [1], it is especially simple when applied to the so-called hard processes. Indeed, at short distances the effective coupling constant of the quark-gluon interaction α_s becomes small and the interaction can be treated perturbatively. The simplicity of the theory seems to be in accord with the experimental observations such as an (approximate) scaling in deep inelastic scattering.

On the other hand, any comprehensive theory must include large-distance dynamics as well. In particular quark interaction within hadrons is strong by definition, since it binds quarks into unseparable pairs. At present there is no quantitative framework within QCD to deal with this strong interaction and such a fundamental

* Permanent address: Institute for Nuclear Physics, Novosibirsk 90, USSR.

problem as evaluation of the hadron spectrum is out of the reach of the theory yet.

Moreover, recent indication is that quark confinement is due to the non-Abelian nature of QCD and non-perturbative effects. There are two examples of such effects that attracted great attention: the Belavin-Polyakov-Tyupkin-Schwartz classical solutions (instantons) [2] and the Gribov gauge ambiguities for strong Yang-Mills fields [3]. Although the progress in understanding the structure of non-Abelian theories is impressive, the feeling is that it can hardly be translated into a computational scheme yet.

For this reason, resonance physics is approached nowadays on phenomenological grounds, by assuming some simple ansatz which will hopefully be justified by an ultimate development of the theory. A well-known example of this kind is the bag model which introduces an energy density inside hadrons.

Here we attempt to approach resonance physics from the "short-distance side". This has an advantage of basing the results on the first principles of the theory alone. The most straightforward derivation refers to integrals like

$$\int_0^\infty e^{-s/M^2} s\sigma_1(s)\, ds\,, \tag{1.1}$$

where $\sigma_1(s)$ is the cross section for e^+e^- annihilation into hadrons with isotopic spin $I = 1$, and M^2 is a variable.

To be sensitive to the resonance contribution, it is necessary to be able to evaluate integrals (1.1) at M^2 of order m_ρ^2 and our claim is that it is indeed possible. Then QCD clearly constrains the resonance properties in a severe way. In particular, we will get

$$g_\rho^2/4\pi \simeq 2\pi/e\,, \tag{1.2}$$

where e is the base of the natural logs and g_ρ determines the electronic decay width of the ρ, $\Gamma(\rho \to e^+e^-) = \frac{1}{3}\alpha^2\, m_\rho\, 4\pi/g_\rho^2$. Moreover, we are able to evaluate the ρ mass and find the result in agreement with the data.

Similar results are obtained for other resonances and mesons such as ω, φ, K^*, π, A_1. Thus, QCD fixes the properties of a single resonance.

Still, we do not claim a complete calculation of the spectrum. The reason is that not the whole interval of M^2 is available for an analysis. We can perform the computation at as low M^2 as m_ρ^2 but cannot penetrate to still lower values of M^2. An important piece of information about the $M^2 \to 0$ region is lacking and, as a result, our predictions are approximate. The accuracy is of order 5–10% and further improvements would require efforts beyond the scope of the present paper.

There is a long way to go before we can substantiate eq. (1.2) and its generalizations and we find it convenient to divide the whole material into two parts. In the first part we concentrate on theoretical foundations for the QCD sum rules which eventually lead to relations like (1.2). The applications are considered in the subsequent paper [4] (hereafter referred to as (II)).

The central object in our theoretical studies is the so-called power terms or corrections. The power corrections are due to non-perturbative effects of QCD. The simplest, although a bit misleading way to explain this is to remind the reader that, for example, the instanton density is proportional to $\exp(-\mathrm{const}/\alpha_s(M))$ where $\alpha_s(M)$ is the running coupling constant. Since $\alpha_s(M) \sim 1/\ln M$, we deal in fact with a power correction in M^{-2}.

The basic idea behind all the applications is that it is the power terms (not higher orders in the α_s series) that limit asymptotic freedom, if one tries to extend the short-distance approach to larger distances.

Phenomenologically, the power corrections are introduced *via* non-vanishing vacuum expectation values such as

$$\langle 0|\bar{q}q|0\rangle \neq 0 , \qquad \langle 0|G^a_{\mu\nu}G^a_{\mu\nu}|0\rangle \neq 0 , \tag{1.3}$$

where q is a quark field and $G^a_{\mu\nu}$ is the gluon field strength tensor. They vanish by definition in the standard perturbation theory.

We will argue that QCD relates the resonance properties to these vacuum expectation values and in this way resonance physics reflects the vacuum structure of QCD. (Note that the quark vacuum average, $\langle 0|\bar{q}q|0\rangle$, has been known for a long time [5] while the gluon condensate, $\langle 0|G^a_{\mu\nu}G^a_{\mu\nu}|0\rangle$, was discussed first in ref. [6].)

Our starting point is the T product of two currents and the Wilson operator expansion [7] for it; e.g., for the $I = 1$ piece of the electromagnetic current $j^{(\rho)}_\mu$ one can write

$$i\int \mathrm{d}x\, e^{iqx}\, \mathrm{T}\{j^{(\rho)}_\mu(x), j^{(\rho)}_\nu(0)\}$$
$$= (q_\mu q_\nu - q^2 g_{\mu\nu}) \sum_n C_n O_n , \tag{1.4}$$

where O_n are local operators. Since the operators O_n have various dimensions, at large Q^2, eq. (1.4) can be considered as an expansion in inverse powers of Q^2 ($Q^2 = -q^2$).

The validity of the operator expansion is by no means trivial since we include the non-perturbative effects. Indeed, the standard derivation of the operator expansion [8] relies in fact on an analysis of Feynman graphs and is nothing else but a (very convenient) computational device to evaluate the graphs at large Q^2.

We will argue that eq. (1.4) still holds in the presence of the non-perturbative effects as far as a few first terms are concerned. However, in higher orders in Q^{-2} the operator expansion becomes invalid. We find a critical dimension corresponding to the breakdown of the expansion. The advantage of knowing the explicit instanton solutions [2] is taken at this point so that the results are specific for QCD.

Taking the vacuum-to-vacuum matrix element of expansion (1.4) reveals another manifestation of non-perturbative effects. Namely, within the standard perturbation theory only the unit operator would survive. The non-perturbative effects induce non-vanishing vacuum expectation values for other operators as well.

The matrix elements like (1.3) can be introduced on purely phenomenological grounds. Another possibility is to use the present knowledge of the non-perturbative solutions to evaluate them. It is too poor and vague nowadays, however, and we rely mostly on phenomenology. Still, we will try instanton calculus [9] as well as some other tricks to explore the relations among various vacuum-to-vacuum matrix elements.

Expansion (1.4) along with the vacuum-to-vacuum matrix elements of the operators involved specify the QCD predictions for the corresponding polarization operators. An alternative form is provided by the general dispersion relations which give the polarization operators in terms of the observable cross sections. Equating the two representations we get QCD sum rules.

In fact, there is a variety of sum rules which correspond to different summation procedures for the power terms. We will show that the sum rules for the first Borel transform of the polarization operator are most suitable for our purposes. It is just at this point that integrals over the cross sections with an exponential weight arise (see eq. (1.1)).

Thus, our aim here is to develop all the machinery needed to extract the resonance properties by means of QCD (as was already mentioned, the concrete applications are considered in [4]). The paper is organized in the following way. In sect. 2 we present the basic ideas in an intuitive language. Sect. 3 deals with the status of the operator expansion taking account of the non-perturbative effects. Sect. 4 is devoted to computation of the operator expansion coefficients for the case of two-point functions of various currents. Sect. 5 consideres the Borel transforms of the polarization operators. The next step is the estimates of the vacuum-to-vacuum matrix elements (sect. 6).

Note that some of the results advertized above have already been published in letter form [6,10,11] while some of the preliminary considerations appeared first in ref. [12]. In a few recent papers of other authors, the importance of the power terms associated with non-perturbative effects of QCD is also argued for, see refs. [13–15]. However, the principal ingredients of our approach have not been overlapped so far. Moreover, we find it convenient to discuss the literature in a special section (sect. 6 of II) after presenting various applications of the technique developed.

2. General strategy

In this section we introduce the reader to the basic ideas formalized and developed in the subsequent sections. We concentrate on the power corrections to asymptotic freedom as they arise in the language of the Feynman graphs and argue for their relevance to resonance physics.

2.1. Space-time picture of quark graphs

Consider the polarization operator induced by the electromagnetic current of a heavy quark. There are two such quarks, c and b, known "experimentally" but we shall not specify the flavor. The only thing which counts is that the quark mass $m_{\rm h}$ is large in the mass scale of strong interactions.

Perturbatively, the polarization operator is given as a series of quark graphs, and we depict three of them in fig. 1. The perturbative sum is valuable as far as the effective coupling constant $\alpha_{\rm s}$ is small. According to QCD it is indeed small at short distances. To ensure that we deal with a short-distance process consider the external momentum q to be small as compared to the quark mass. Then the quark propagates a distance of order $1/2m_{\rm h}$ which is small. Therefore, the coupling is weak and we can retain one or two first terms in the $\alpha_{\rm s}$ expansion.

Phrased another way, the integrals corresponding to the diagrams in fig. 1 are dominated by

$$p^2,\, k^2 \sim -m_{\rm h}^2\,,$$

where p and k are the virtual quark and gluon momenta. If $m_{\rm h}^2 \gg \mu^2$, where μ is a hadronic scale, the standard asymptotic freedom formulas apply to the quark and gluon Green functions.

Thus, the point $q^2 = 0$ (real photon) and $m_{\rm h}$ large belongs to the region of asymptotic freedom: everything is simple and computable (at least as far as we are satisfied with a few terms in the $\alpha_{\rm s}$ expansion and do not put such sophisticated questions as "what does the whole series mean?").

We are interested in probing larger distances, however. The reason is that in this way we can come closer to understanding the nature of the resonances and quark confinement.

We can do that by increasing q^2 and approaching the quark threshold, $q^2 = 4m_{\rm h}^2$. We will choose an alternative procedure which is more convenient for practical purposes. Namely, let us start at $q^2 = 0$ and compute higher derivatives of the polarization operator.

It is rather clear that the dominant contribution to the nth derivative comes

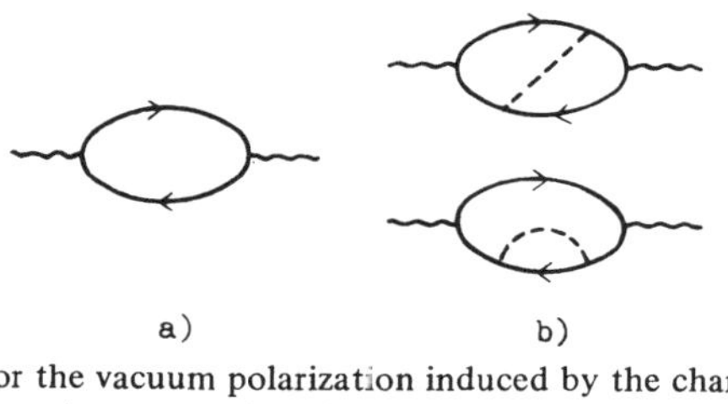

Fig. 1. Feynman graphs for the vacuum polarization induced by the charmed quark current. Solid, wavy and dashed lines denote quarks, photons and gluons respectively.

from the virtual momenta of order

$$p^2, k^2 \sim -m_h^2/n .$$

Indeed, the nth derivative is determined by integrals of the kind

$$\int \frac{d^4p}{(p^2 + m_h^2)^n} , \qquad \text{or} \qquad \int \frac{d^4p\, d^4k}{[(p+k)^2 + m_h^2]^n} ,$$

where we have performed the Wick rotation so that all the momenta are Euclidean.

For a fixed m_h^2 and n tending to infinity both p^2 and k^2 tend to zero. Nothing spectacular happens with vanishing p^2. Even at $p^2 = 0$, the heavy quark is highly virtual since it is off-mass-shell by m_h^2 and m_h is large. Therefore, its propagation is described by the standard perturbation theory.

On the other hand, if $k^2 \to 0$ the gluon in fig. 1b comes close to its would-be mass-shell. Due to confinement, the gluon propagator is strongly modified at low k^2 and perturbation theory becomes irrelevant.

Thus at high n, the gluon propagates a rather large distance and is sensitive to the confinement mechanism.

Most probably, confinement is due to non-perturbative effects of QCD which bring in a new mass scale, μ (in fact μ must be related to the distances where the coupling constant α_s reaches some critical value). We shall assume that for $k^2 >> \mu^2$ the non-perturbative corrections are negligible while for $k^2 \lesssim \mu^2$ they are most important.

It is clear then that the real expansion parameter for the power terms is $n\mu^2/m_h^2$ so that for $n \sim m_h^2/\mu^2$ the perturbative expansion is badly broken.

2.2. Power corrections and resonance physics

The argument of subsect. 2.1 demonstrates that at high n large distances come into the game. Taken alone, it does not provide convincing evidence in favor of the power terms, however. Therefore, it might be useful to indicate that there exist good reasons to believe in their importance, based on phenomenological observations.

Ther are two sources of large corrections at small k^2. First, according to QCD the coupling constant grows if the quark (gluon) virtuality decreases:

$$\alpha_s(Q) = \text{const}/\ln(Q/\Lambda) .$$

Formally, one approaches the infrared pole at $Q^2 = \Lambda^2$ and it can be the origin of large corrections.

Another source of corrections is non-perturbative terms which can be thought of as terms $\sim\exp(-\text{const}/\alpha_s(Q))$.

The strongest evidence in favor of relatively large power corrections is the observed difference between the mass spectra in the vector and axial-vector channels with isotopic spin $I = 1$. In the vector case there is a single prominent resonance, the ρ, while in the case of the axial-vector current there are two states one of which is much lighter than the ρ (the π meson) and the other is much heavier (the A_1 meson).

On the other hand, for massless quarks (and this is, beyond any doubt, a good approximation for the u and d quarks) the perturbative graphs do not differentiate between the vector and axial-vector currents. Thus, it is the spontaneous breaking of chiral symmetry that is responsible for the π-ρ-A_1 mass splittings. The symmetry breaking is signalled by the non-vanishing vacuum value of $\bar{\psi}\psi$. Thus we have an alternative:

either $\langle 0|\bar{\psi}\psi|0\rangle = 0$, $m_\rho = m_{A_1}$, no pion,

or $\langle 0|\bar{\psi}\psi|0\rangle \neq 0$, $m_\rho \neq m_{A_1}$, massless pion.

The alternative must be reflected in the polarization operators in some way. On purely dimensional grounds it is clear that the two possibilities can be distinguished only *via* the power corrections.

Analogously, the non-vanishing matrix element $\langle 0|G^a_{\mu\nu}G^a_{\mu\nu}|0\rangle$ signals the breaking of dilatation symmetry (we recall that $G^a_{\mu\nu}G^a_{\mu\nu}$ is proportional to the trace of the energy-momentum tensor). The "gluon condensate" $\langle 0|G^a_{\mu\nu}G^a_{\mu\nu}|0\rangle$ is in a sense connected with the emergence of mass parameters in QCD.

Other evidence in favor of the importance of power corrections is provided by the charmonium sum rules, i.e. by the QCD predictions for the leptonic widths in charmonium. Chronologically these sum rules were considered first [12,6]. We shall sketch the derivation in paper (II). A phenomenological estimate for $\langle 0|G^a_{\mu\nu}G^a_{\mu\nu}|0\rangle$ emerges as an outcome of the analysis.

Finally, let us mention another possibility, that both high orders in the α_s expansion and power corrections are equally important. The possibility cannot be ruled out *a priori*. Basing on independent estimates of the coupling constant [16], one might conclude that this is not the case and that is is the power corrections that play the major role. True, the independent estimates of α_s are not too conclusive (see a discussion in sect. 6 and paper (II)). However, the sum rules derived under the assumption that the coupling constant is small agree well with the data. The agreement observed justifies *a posteriori* the assumption that power corrections already become important at such virtualities that the coupling constant is still far from the infrared pole.

2.3. Basic idea

Now, that we hopefully have convinced the reader of the importance of power corrections we proceed to specify their notion in more detail and explain how one can parametrize them.

Qualitatively, we have already learned that to keep the power corrections small we must choose $\xi << 1$ where

$$\xi = \begin{cases} \mu^2 n/4m_h^2 \,, & \text{(heavy quarks)} \,, \\ \mu^2 n/Q^2 \,,\ Q^2 \equiv -q^2 \,, & \text{(light quarks)} \,. \end{cases}$$

Of course, we want to be much more quantitative and learn the precise meaning of μ, find coefficients, etc.; that is to construct a computational scheme and try to calculate hadrons. This is achieved through introducing new phenomenological parameters. (Attempts to extract these parameters theoretically are discussed in sect. 6.)

Now we explain the procedure in its gross features leaving aside all the reservations (which are, of course, essential) and technical details (which are practically important). Turn again to the graph 1b with a gluon exchange, but consider now the gluon Green function $\mathcal{D}_{\mu\nu}(k^2)$ as an exact one. Furthermore, let us split $\mathcal{D}_{\mu\nu}(k^2)$ into two parts

$$\mathcal{D}_{\mu\nu}(k^2) = \frac{g_{\mu\nu}}{k^2} + \left(\mathcal{D}_{\mu\nu} - \frac{g_{\mu\nu}}{k^2}\right) , \tag{2.1}$$

where we have chosen the Feynman gauge for the sake of definiteness. At large k^2 the Green function is given by the first term because of asymptotic freedom. Disregarding for the moment the calculable logarithmic corrections, we assume that the bracketed term in eq. (2.1) falls off as some power of k^2 at large k^2.

To get the answer for the graph we must collect all other factors and integrate over k^2. Then the first term in eq. (2.1) is absorbed into the standard perturbation theory while the second one represents something new. Since the difference $(\mathcal{D}_{\mu\nu}(k^2) - g_{\mu\nu}/k^2)$ is presumably large only at small k^2 we can expand the rest of the amplitude in k^2 and approximate $k^2 = 0$.

To be careful, we must first extract the gluon field strength tensor, $G^a_{\mu\nu}G^a_{\mu\nu}$, not to violate the gauge invariance, and put $k^2 = 0$ afterwards. (In other words, gauge invariance implies that modification of the propagator is accompanied by a change in the vertices.)

Integrating with $\mathcal{D}_{\mu\nu} - g_{\mu\nu}/k^2$ results in a number which is sensitive to the gluon dynamics at large distances. Once we have a theory of confinement we can evaluate it. In the absence of such a theory we are forced to introduce a new parameter which is equivalent to the vacuum expectation value

$$\langle 0 | G^a_{\mu\nu}G^a_{\mu\nu} | 0 \rangle \,. \tag{2.2}$$

It is important that we can study vacuum properties using simple Feynman graphs as a tool. The matter is that all other lines in the graph, (in this particular case the quark lines) are far off their would be mass shell and are known.

For high derivatives the diagram 1b is in fact a perfect probe, which detects gluon waves propagating through the vacuum. The probe is point-like while the gluon wave length is relatively large. The inner structure of the probe is known, so one can extract information about the gluon-wave intensity.

If better accuracy is desired one must keep further terms in the k^2 expansion of the amplitude. This introduces naturally further parameters, such as *

$$\langle 0|\, \mathcal{D}_\mu G^a_{\mu\alpha}\, \mathcal{D}_\nu G^a_{\nu\alpha}\, |0\rangle$$

and so on. Moreover, if we consider a graph with two or more gluon lines with low momenta, it cannot be calculated in terms of the parameters mentioned above and we need new ones such as

$$\langle 0|f^{abc} G^a_{\mu\nu} G^b_{\nu\sigma} G^c_{\sigma\mu}|0\rangle\,, \qquad \langle 0|G^a_{\mu\nu} G^a_{\nu\sigma} G^b_{\sigma\gamma} G^b_{\gamma\mu}|0\rangle\,, \ldots\,.$$

Thus, there arises a series of parameters, all of them being independent in the absence of a consistent theory. To keep the problem manageable we must cut the series in some way. Increasing the number of gluon fields or their derivatives implies increasing dimension of operators, and, therefore, introducing extra powers of m_h^2. Thus, if power corrections are small compared to unity, the first term in the series dominates, generally speaking. All others can be safely omitted.

Thus, we will keep only the power correction of the lowest dimension and go to such n that it becomes sizable but still rather small, say, 30%. Then we would expect that such n represent a boundary for asymptotic freedom: at higher n it breaks badly since the corrections blow up almost immediately. Rather arbitrarily, we estimate neglected power corrections as a square of the kept one. The sum rules themselves will show whether this assumption is reasonable.

Then for heavy quarks we are left with a single parameter $\langle G^2\rangle$ [6,17] which is determined from experimental data by fitting charmonium sum rules.

At first sight, we do not make much progress since we describe the data with a new free parameter in hand. Not quite so. First of all, we are able to check the self-consistency of the calculation by considering charmonium sum rules alone. What is more important is that the same parameter controls the asymptotic freedom breaking for light quarks. Indeed, if solid lines in fig. 1 denote light quarks, nothing is changed from a principal point of view. We must just substitute the heavy mass by a light one and ensure that large momentum Q flows through quark lines. (The effect of the change must still be non-trivial, see sect. 2 in paper (II).)

Thus, we are able to verify that the same force confines both heavy and light quarks.

As was already mentioned, light quarks result in a new vacuum average. They can also penetrate into large distances and this effect is important at high n. Phenomenologically, such effects are described by vacuum expectation values of

* Due to equations of motion this parameter can be expressed in terms of light-quark operators.

quark fields

$$\langle 0|\bar{\psi}\psi|0\rangle\,, \tag{2.3}$$

$$\langle 0|\bar{\psi}\Gamma_1\psi\bar{\psi}\Gamma_2\psi|0\rangle\,, \tag{2.4}$$

where $\Gamma_{1,2}$ are some matrices acting on color, flavour and Lorentz indices.

The matrix element (2.3) can be found by using the PCAC hypothesis (see, e.g., [5,18]). As for the matrix elements (2.4), we keep only the vacuum intermediate-state contribution. This approximation is substantiated in sect. 6. Then the expectation values do not introduce new free parameters.

An experienced reader has certainly recognized the operator expansion technique [7] in the procedure describe above. It is quite a standard procedure by now and may not need further justification. It is worth emphasizing, however, that usually the operator expansion is used within perturbation theory. In this case the operator expansion is well established [8] and is, in fact, a technical device. We are going to rely on the operator expansion beyond perturbation theory. Every step here is a new one and by no means evident.

In particular, one can worry whether the integration with $\mathcal{D}_{\mu\nu} - g_{\mu\nu}/k^2$ is dominated by low k^2, so that the approximation $k^2 \sim 0$ for the gluon emission amplitude is justified. Even if this integral is convergent, one can expect that further expansion in k^2 generates integrals which are ultraviolet unstable. Then the procedure becomes inconsistent.

In other words, we must show that the matrix elements introduced, like (2.2), (2.3), (2.4), are determined by large distances. Rather surprisingly, we can do that even now, in the absence of a complete theory of confinement. Indeed, there exist good reasons to believe that at short distances the leading non-perturbative corrections are generated by instantons. As was mentioned above, numerically the calculations are still uncertain. However, as far as problems of convergence of integrals are concerned they can be clarified. We will show that all the assumptions concerning the validity of the operator expansion which we are using turn out to be justified. It turns out that the operator expansion breaks down only at relatively high order in Q^{-2}. For pure gluonic fields a polarization operator can be represented as

$$\Pi(Q^2) = \left[(\text{perturbation theory}) + \sum_{k=2,\ldots,5}\left(\frac{\mu^2}{Q^2}\right)^k + O(Q^{-11})\right],$$

so that expansion in Q^{-2} is guaranteed as far as the few first terms are concerned. Since we keep power terms small we confine ourselves to the leading corrections and the use of the expansion is justified. If the existence of nearly massless quarks is accounted for, then the series can be extended up to terms $\sim Q^{-14}$ inclusively.

After this preliminary exposition of the approach used, we proceed to a more detailed and technical presentation of the results obtained.

3. Operator expansion and non-perturbative effects

We start the systematic derivation with a discussion of the operator expansion. For the sake of convenience it is divided into two parts: in the present section we consider general problems while computational details are referred to sect. 4. Subsect. 3.1 contains some definitions and generalities which are not specific, in fact, for non-perturbative effects. The principal subsections are 3.2 and 3.3.

3.1. General remarks

We start by introducing notations common to all the cases. Consider the T product of two currents $j^{\mathrm{A}}, j^{\mathrm{B}}$ which can be either light or heavy quark currents. The basic assumption is that at large external momentum q or for a large internal mass m_{h} the operator expansion [7] is valid:

$$i\int \mathrm{d}x\, \mathrm{e}^{iqx} \mathrm{T}\{j^{\mathrm{A}}(x), j^{\mathrm{B}}(0)\} = \sum_n C_n^{\mathrm{AB}}(q)\, O_n \,, \tag{3.1}$$

where C_n^{AB} are coefficients, O_n are local operators constructed from light quark (u,d,s) or gluon fields. To be more precise, we assume the validity of the expansion only in the few first terms (see subsect. 3.2 for more detail).

The operators O_n are conveniently classified according to their Lorentz spin and dimension d. We will consider only spin-zero operators since only these contribute to the vacuum expectation value. Naturally, the operators in (3.1) satisfy such general requirements as gauge invariance with respect to the gluon field. An important characteristic is operator dimension. An increase in dimension implies extra powers of μ^2/Q^2 or $\mu^2/4m_{\mathrm{h}}^2$ for the corresponding contribution, where μ is some typical hadronic mass entering through the matrix element of O_n. So we list all the operators with zero Lorentz spin and $d \leqslant 6$ *.

$$\begin{aligned}
&I \text{ (the unit operator)}\,, && (d=0)\,,\\
&O_M = \bar{\psi} M \psi\,, && (d=4)\,,\\
&O_G = G^a_{\mu\nu} G^a_{\mu\nu}\,, && (d=4)\,,\\
&O_\sigma = \bar{\psi}\sigma_{\mu\nu} t^a \tilde{M} \psi G^a_{\mu\nu}\,, && (d=6)\,,\\
&O_\Gamma = \bar{\psi}\Gamma_1 \psi \bar{\psi}\Gamma_2 \psi\,, && (d=6)\,,\\
&O_f = f^{abc} G^a_{\mu\nu} G^b_{\nu\gamma} G^c_{\gamma\mu}\,, && (d=6)\,,
\end{aligned} \tag{3.2}$$

* Other operators can be reduced to those, given in eq. (3.2) plus full derivatives, for example,

$$\bar{\psi}\gamma_\mu t^a \mathcal{D}_\nu \psi G^a_{\mu\nu} = -\tfrac{1}{2}\bar{\psi}\gamma_\mu t^a \psi \mathcal{D}_\nu G^a_{\mu\nu} - i\bar{\psi}\mathcal{D}^2 \gamma_\mu \mathcal{D}_\mu \psi - i\bar{\psi}\overleftarrow{\mathcal{D}}_\mu \gamma_\mu \mathcal{D}^2 \psi$$
$$+ \text{ full derivatives}\,,$$

and the right-hand side can be expressed in terms of O_M, O_σ by using the equations of motion.

where $G^a_{\mu\nu}$ is the gluon field strength tensor, t^a are the Gell-Mann SU(3) matrices acting in the color space and normalized by the condition $\mathrm{Tr}(t^a t^b) = 2\delta^{ab}$, M, $\tilde{M}$ are matrices in flavor (u,d,s) space whose elements are proportional to quark masses. (The dimension of the operators O_M, O_σ indicated accounts for this fact.) We reserve the conventional notation λ^a for the SU(3) matrices acting in the flavor space. We will use the notation $\bar\psi \ldots \psi$ when the summation over $SU(3)_{\rm flavor}$ is assumed, and $\bar{q} \ldots q$ in other cases, for example, $\bar\psi\psi$ means $\bar{u}u + \bar{d}d + \bar{s}s$, but $\bar{q}q$ means $\bar{u}u$ or $\bar{d}d$ or $\bar{s}s$. $\Gamma_{1,2}$ stand for some matrices acting on the color, flavor and spinor indices of the quark fields, and are specified below.

A remark is in order here concerning our convention on the interaction Lagrangian. Our definition of the quark-gluon coupling constant throughout the paper is $\alpha_s = g_s^2/4\pi$ and the interaction Lagrangian is of the form $\frac{1}{2}g_s\bar\psi t^a\gamma_\mu\psi b^a_\mu$, where b^a_μ is the gluon field.

Eq. (3.2) gives a complete set of operators which satisfy such general principles as Lorentz invariance, gauge invariance and having dimension $d \leqslant 6$. Note that we include in the list, not only the leading power correction ($d = 4$) but the next term ($d = 6$) as well. The reason is that in many cases the coefficients C_Γ (corresponding to O_Γ) are large numerically since they are associated with a Born series of graphs while, say, C_G comes from a loop graph. Choosing Q^2 (or $m_{\rm h}^2$) large enough we could still get rid of operators with $d = 6$, but for practical purposes they turn out to be important.

The coefficients C_n are determined by momenta of high virtuality, $p^2 \sim Q^2, m_{\rm h}^2$. Since QCD is asymptotically free, the calculation of the coefficients is reliable. As for the matrix elements, they will be treated phenomenologically in sect. 6.

The expansion coefficients in eq. (3.1) are calculated as a series in α_s. Naturally, in practice one is confined to one or two first terms in the α_s expansion. As an example of the relations emerging let us write out the answer for the current

$$j_\mu = \bar{q}\gamma_\mu q\ ,$$

in the imaginary world with a single light quark flavor. Assuming conventional $SU(3)_{\rm color}$ we find

$$i\int dx\, e^{iqx} T\{j_\mu(x)\, j_\nu(0)\} = (q_\mu q_\nu - q^2 g_{\mu\nu})$$

$$\times\left\{-\frac{1}{4\pi^2}(1+\alpha_s/\pi)\ln\frac{Q^2}{\mu^2} + \frac{2m_q}{Q^4}\bar{q}q\right.$$

$$+\frac{\alpha_s}{12\pi Q^4}G^a_{\mu\nu}G^a_{\mu\nu} - \frac{2\pi\alpha_s}{Q^6}\bar{q}\gamma_\alpha\gamma_5 t^a q\bar{q}\gamma_\alpha\gamma_5 t^a q$$

$$\left. -\frac{4\pi\alpha_s}{9Q^6}\bar{q}\gamma_\alpha t^a q\bar{q}\gamma_\alpha t^\alpha q + \ldots\right\}, \tag{3.3}$$

where $Q^2 \equiv -q^2$.

The derivation is, in fact, given in sect. 4, where the realistic case with many quark flavors is discussed.

3.2. Status of operator expansion

As was already mentioned in sect. 1, the validity of the operator expansion in our case is by no means obvious. The problem is that non-perturbative effects are included, while the standard derivation of the operator expansion relies heavily on the Feynman graph analysis [8]. Fortunately, recently considerable progress has been made in understanding non-perturbative terms in QCD [2,3,9]. This permits us to justify the operator expansion to the extent we really use it. The basic point is that for pure Yang-Mills theory the leading correction to the perturbative treatment at short distances is associated with the one-instanton solution. The effect of (nearly) massless quarks has not been fully incorporated into the theory yet, but it only extends the validity of the operator expansion.

Let us emphasize that the effect of non-perturbative terms in QCD is twofold:

(a) they induce non-vanishing vacuum expectation values, such as $\langle 0|G^a_{\mu\nu}G^a_{\mu\nu}|0\rangle$ which in standard perturnation theory vanish by definition;

(b) they break down the operator expansion itself, starting from some power $Q^{-d_{cr}}$.

The distinction between the two cases lies in the fact that the former effect is determined by the large-size instantons whose scale is of order of the confinement radius, $\rho \sim R_{conf}$. The latter effect is due to the small-size instantons, whose scale is controlled by the choice of the external parameter, $\rho \sim 1/Q$.

Let us give examples of the effects (a) and (b) above which arise within the instanton calculus [2,13,9]

(a) In the dilute-gas approximation one readily obtains

$$\langle 0|\frac{\alpha_s}{\pi}G^a_{\mu\nu}G^a_{\mu\nu}|0\rangle = \text{const} \times \int_0^{\rho_c} \frac{d\rho}{\rho^5}\, d(\rho)\,,$$

where ρ is the instanton scale size, $d(\rho)$ is the instanton density, $d(\rho) \sim \exp\{-2\pi/\alpha_s(\rho)\}$, and the cut-off ρ_c is introduced since the integral is divergent at the upper limit of integration. Indeed, at small ρ

$$d(\rho) \sim \rho^{\epsilon}\,, \qquad \epsilon \sim 11\,.$$

Thus we see, that $\langle 0|G^a_{\mu\nu}G^a_{\mu\nu}|0\rangle$ is contributed by instantons and the effect is controlled by the large-distance dynamics.

(b) Consider now the correlation function of two pseudoscalar densities

$$\Pi^{(P)} = i\int dx\, e^{iqx}\langle 0|T\{j^{(P)}(x), j^{(P)}(0)\}|0\rangle\,, \qquad j^{(P)} = \bar{u}i\gamma_5 u - \bar{d}i\gamma_5 d\,.$$

Then using the fermion zero-eigenmode solution found by 't Hooft [9] one can find for a one-instanton contribution (the anti-instanton gives the same)

$$\Pi^{(P)}_{\text{one inst}} = 2Q^2\int\frac{d\rho}{\rho}\, d(\rho)[K_{-1}(\sqrt{Q^2}\rho)]^2\,,$$

where K_{-1} is the McDonald function.

Now, the dominant contribution comes from $\rho \sim Q^{-1}$ and there is no need to introduce a cut-off by hand. Neglecting the log factors in $d(\rho)$, i.e., taking it to be

$$d(\rho) = \text{const} \times \rho^{\epsilon} ,$$

we find

$$\Pi^{(P)}_{\text{one inst}} = 2Q^2\, d\left(\rho = \frac{1}{Q}\right) \frac{2^{\epsilon-3}}{\Gamma(\epsilon)} [\Gamma(\tfrac{1}{2}\epsilon)]^4 \frac{\epsilon}{\epsilon - 2} ,$$

and at large ϵ the following numerical approximation works well:

$$d\left(\rho = \frac{1}{Q}\right) \frac{2^{\epsilon-3}}{\Gamma(\epsilon)} [\Gamma(\tfrac{1}{2}\epsilon)]^4 \frac{\epsilon}{\epsilon - 2} \approx d\left(\rho = \frac{\epsilon}{6Q}\right) .$$

This is an example of an effect which is induced by non-perturbative solutions and which breaks the operator expansion.

What is the physical meaning of the operator expansion? It assumes the possibility of separating short and large distance effects. Short distances are governed by asymptotic freedom and can be treated perturbatively. The corresponding contribution is reflected in the operator expansion coefficients. The large-distance contribution is accounted for phenomenologically, through various vacuum-to-vacuum matrix elements. It is clear then that the contribution of the large-size fluctuations, independent of Q^2, can be consistently kept within the framework of the operator expansion. As for the small-scale fluctuations with $\rho \sim 1/Q$, they modify asymptotic freedom itself and cannot be included into the operator expansion, at least in its present form.

Our central point is that for a pure Yang-Mills field it is easy to find explicitly the critical dimension up to which the Wilson expansion is valid. Really, in pure gluodynamics the leading correction is due to the BPST solution [2]

$$G^a_{\mu\nu}(x; x_0, \rho) = -\frac{4}{g_s} \frac{\eta_{\mu\nu\alpha}\rho^2}{[(x - x_0)^2 + \rho^2]^2} , \tag{3.4}$$

where x_0 is the instanton center and ρ is its scale. (Euclidean space-time is implied.) In the operator expansion for two colorless gluon densities only local products of the field strength tensors are involved:

$$\langle 0 | \underbrace{G^a_{\mu\nu}(0) \ldots G^b_{\alpha\beta}(0)}_{n \text{ factors}} | 0 \rangle . \tag{3.5}$$

(The particular form of contraction of both color and Lorentz indices is inessential here.) In the dilute-gas approximation (see ref. [13]) eq. (3.5) is reduced to the integral over x_0 and ρ. The x_0 integration is always convergent resulting in an expression which depends on n in the following way:

$$\langle 0 | \underbrace{G^a_{\mu\nu}(0) \ldots G^b_{\alpha\beta}(0)}_{n} | 0 \rangle \sim \int \mathrm{d}\rho\, \rho^{-2n-1}\, d(\rho) , \tag{3.6}$$

where $d(\rho)$ is the instanton density [9]

$$d(\rho) = \text{const}\left(\frac{2\pi}{\alpha_s(\rho)}\right)^6 \exp(-2\pi/\alpha_s(\rho)) ,$$

$$\frac{2\pi}{\alpha_s(\rho)} = \frac{2\pi}{\alpha_s(\rho_0)} + 11 \ln(\rho_0/\rho) , \tag{3.7}$$

and the numbers are given for $SU(3)_{color}$.

As is readily seen, the integral (3.6) is divergent at the upper limit for $n \leqslant 5$ and at the lower limit of integration for $n \geqslant 6$.

Thus, at $d \leqslant 10$, the vacuum expectation values (3.5) are determined by large-distance dynamics. (At large distances the dilute-gas approximation becomes invalid, and to avoid confusion we should emphasize that we do not use eq. (3.6) for numerical estimates. The one-instanton solution can help only to clarify the question of the integral convergence at small or large ρ.)

Two (and more) instanton contributions are proportional to even higher powers of ρ. For this reason the one-instanton contribution is dominant at short distances.

Starting from $d = 12$ the vacuum expectation values (3.5) become infrared stable which automatically means that the $\rho \sim 1/Q$ instantons come into play. Here the operator expansion must be forgotten.

Thus, the expansion in Q^{-2} cannot be extended to any power in fact. In pure gluon theory with $SU(3)_{color}$ symmetry it takes the form

$$i\int dx\, e^{iqx} \langle 0 | T\{j^A(x), j^B(0)\} | 0 \rangle$$

$$= (\text{perturbation theory}) \times [1 + \sum_{k=2,3,4,5} C_k^{AB} (\mu^2/Q^2)^k$$

$$+ O(Q^{-11})] . \tag{3.8}$$

However, for the sake of brevity we will still use the term "power-correction series".

Now, as to the light quarks. As is well-known, inclusion of the light quarks changes the theory drastically [9]. Let the quark mass vanish; in the real world m_u, $m_d \sim 5$ MeV [19,18] and it is clear that one can safely neglect $m_{u,d}$. In the limit $m_{u,d} \to 0$ the one-instanton contribution to the functional integrals as a rule vanishes [9]. The only exception is the polarization operators induced by scalar and pseudoscalar currents of the light quarks, i.e., chirality changing currents (thus, the correction evaluated in point (b) above does not vanish for massless quarks). Therefore, even a qualitative understanding of the instanton effects requires a knowledge of the effective quark mass generation mechanism. Needless to say, the present theory is far from providing it.

One can argue, however, that the presence of massless quarks affects the critical value d_{cr} but not the very fact of the operator expansion breaking. If one considers the strange quark to be heavy enough then the operator expansion is likely to be valid up to Q^{-16}. Indeed, for dimensional estimates one can invoke the following expression for the effective quark mass [20]:

$$m_{\text{eff}}(Q) = m_0(Q) + \frac{16\pi\alpha_s(Q)\langle 0|\bar{q}q(Q)|0\rangle}{Q^2}, \qquad (3.9)$$

where (Q) refers to the normalization point.

Then each quark results in an extra factor of the type

$$[\rho m_{\text{eff}}(\rho)](\rho\mu)^{-2/3}, \qquad (3.10)$$

in the instanton density. The factor $(\rho\mu)^{-2/3}$ in eq. (3.10) is associated with the modification of the logarithmic dependence of the effective coupling constant $\alpha_s(\rho)$ which induces, in turn, a change in the instanton density (see eq. (3.7)). In the theory with two massless (u, d) and one massive (s) quarks an extra damping factor Q^{-5} for the small-size instanton contribution emerges in this way. If instantons of a size smaller than the inverse mass of a heavy (say, charmed) quark are considered, then presence of these quarks must be accounted for as well.

At present, it is not clear whether eq. (3.9) can be taken literally but it seems to us quite safe as far as dimensional estimates are concerned. Still, let us mention that in the literature an even higher power of Q^{-2} for the small-size instanton contribution has been argued (see, e.g., ref. [13]).

An interesting question is how the instanton contribution is manifested in measurable cross sections.

3.3. Operator expansion and cross sections

The QCD results for the polarization operator discussed so far can be translated into the predictions for the corresponding cross sections. The well-known example of this kind is [21]

$$R(s) = \frac{\sigma(e^+e^- \to \text{hadrons})}{\sigma(e^+e^- \to \mu^+\mu^-)} = \sum_q 3Q_q^2 [1 + \alpha_s(s)/\pi], \qquad (3.11)$$

where Q_q are the quark charges and α_s is the coupling constant.

We would say that eq. (3.11) corresponds to asymptotic freedom. The series in α_s can be extended to higher powers and we denote by $R(s)_{\text{pert.th.}}$ the (symbolic) sum over α_s.

Now, turn to the power corrections. In the limit of extremely high energies, $s \to \infty$, the only correction to survive is due to the instantons of small size. The terms in the operator expansion which correspond to the instantons of large size do not modify the cross section (in accordance with intuition which seemingly says that the cross section is decided by short distance). For example, if we take it for granted that the operator expansion works up to the Q^{-14} terms while the small-size instantons show up in the Q^{-16} piece, then

$$R(s) = [R(s)]_{\text{pert.th.}} + (\mu^2/s)^8 + \dots , \qquad (s \to \infty) .$$

Changing the power of the small-size instanton contribution to $\Pi(Q^2)$ would change the approach of $R(s)$ to its asymptotic behaviour.

On the other hand, terms of lower order in Q^{-2} which are described by the operator expansion reflect the change in the cross section at relatively low energy.

These conclusions follow immediately from the equation

$$\Pi(Q^2) - \Pi(Q^2)_{\text{pert.th.}} = (12\pi^2 \sum_q Q_q^2)^{-1} \int_0^\infty \frac{R(s) - [R(s)]_{\text{pert.th.}}}{s + Q^2} \, ds .$$

Expanding in Q^{-2} we see that the convergence of the integral $\int (R - R_{\text{pert.th.}}) \times s^{n-1} \, ds$ at $s \to \infty$ is correlated with the validity of the operator expansion up to terms Q^{-2n}

Strictly speaking, the statements must be qualified taking account of the possibility of oscillating contributions to the cross section, but we are reluctant to consider such a possibility on physical grounds.

3.4. *Summary*

In this section we have substantiated the validity of the operator expansion up to some critical value of the operator dimension. The dimension is certainly quite high and the precise value of it can be reasonably guessed starting with the instanton solutions.

We have also established the connection between the asymptotic behaviour of the cross section and the polarization operator in the presence of non-perturbative power corrections.

4. Operator expansion for various currents

In this section we deal with two-point functions of various currents. The set of currents considered is motivated by the forthcoming applications.

There are three distinct mass parameters relevant to the problem under consideration. The first one is the inverse radius of confinement, $R_{\text{conf}}^{-1} \equiv \mu$, which is manifested through various vacuum expectation values, e.g., $\langle 0|(\alpha_s/\pi) \, G_{\mu\nu}^a G_{\mu\nu}^a|0\rangle \sim \mu^4$. The second parameter is the quark mass itself. And, finally, an external mass scale is introduced by the momentum q in the definition of the two-point function (3.1).

To apply the operator expansion, at least one of the last parameters has to be large as compared to μ. Thus, there are three possibilities:

(i) $m_q \lesssim \mu , \qquad Q^2 >> \mu^2 ,$

(ii) $m_q >> \mu$, $Q^2 << m_q^2$,

(iii) $m_q >> \mu$, $Q^2 \gtrsim m_q^2$.

(Here $Q^2 \equiv -q^2$.)

The first possibility implies an expansion in μ^2/Q^2 and m_q^2/Q^2. The corresponding technique will be referred to as the light-quark expansion. For heavy quarks it is convenient to consider possibility (ii) above and expand in Q^2/m_q^2, μ^2/m_q^2. It is just what we shall always do, exploiting consistently the so called heavy-quark expansion [22].

Consideration of the possibility (iii) is completely legitimate within the approach used, but we will never consider this for practical reasons.

For heavy quarks we always choose $Q^2 = 0$ and compute the derivatives with respect to Q^2. This can be considered as a substitution for a change in Q^2 in the polarization operator itself. In general, the two procedures are equivalent but in the case $Q^2 = 0$ all the equations simplify greatly.

It is worth noting that the bulk of the applications is devoted to the light quarks and the consideration of heavy quarks is partly auxiliary (see sect. 2 of paper II). The operator expansion is more tractable for heavy quarks since there is no quark vacuum expectation value. Thus, we start our exercises with heavy quarks and then proceed to the light ones.

4.1. Vector current of heavy quarks: the unit operator

The vector current of, say, charmed quarks has the form

$$j_\mu^{(c)} = \bar{c}\gamma_\mu c .$$

Note that we do not include the quark charge, $Q_c = \frac{2}{3}$ in this case, in the definition of the current. The operator expansion takes the form

$$i\int dx\, e^{iqx} T\{j_\mu^{(c)}(x), j_\nu^{(c)}(0)\} = (q_\mu q_\nu - q^2 g_{\mu\nu}) \times [C_I I + C_G O_G + ...] , \tag{4.1}$$

where the operator O_G is defined in eq. (3.2) and the dots stand for operators of higher dimension.

To the lowest order in α_s the T product (4.1) is given by a single graph of fig. 1a. The corresponding result for the coefficient C_I is conveniently represented in a dispersion form:

$$C_I^{(0)} = -\frac{Q^2}{\pi}\int \frac{\mathrm{Im}\, C_I^{(0)}(s)}{s(s+Q^2)}\, ds ,$$

$$\mathrm{Im}\, C_I^{(0)} = \frac{1}{4\pi}\frac{v(3-v^2)}{2}\,\theta(s-4m_c^2) , \qquad v = (1-4m_c^2/s)^{1/2} . \tag{4.2}$$

Instead of studying C_I as a function of Q^2 one can choose to calculate all the derivatives of C_I at $Q^2 = 0$ for which one gets (see, e.g., the review [12]):

$$\frac{1}{n!}\left(-\frac{\mathrm{d}}{\mathrm{d}Q^2}\right)^n C_I^{(0)}\bigg|_{Q^2=0} = \frac{3}{4\pi^2}\,\frac{2^n(n+1)(n-1)!}{(2n+3)!!}\,(4m_c^2)^{-n}\,.$$

The correction of the first order in α_s is given by fig. 1b. The corresponding imaginary part,

$$\mathrm{Im}\,C_I^{(1)}(s) = \mathrm{Im}\,C_I^{(0)}(s)\times\left\{1+\tfrac{4}{3}\alpha_s\left[\frac{\pi}{2v}-\frac{v+3}{4}\left(\frac{\pi}{2}-\frac{3}{4\pi}\right)\right]\right\}, \tag{4.3}$$

can be easily extracted from Schwinger, [23].

It follows from eqs. (4.2), (4.3) that the first-order corrections to the moments of $C_I^{(0)}$ are equal to

$$\left(-\frac{\mathrm{d}}{\mathrm{d}Q^2}\right)^n C_I^{(1)}\bigg/\left(-\frac{\mathrm{d}}{\mathrm{d}Q^2}\right)^n C_I^{(0)}\bigg|_{Q^2=0} = 1+\alpha_s\left[\frac{4\sqrt{\pi}}{3}\,\frac{\Gamma(n+\frac{3}{2})}{\Gamma(n+1)}\right.$$
$$\times\frac{1-1/(3n+3)}{1-1/(2n+3)}-\frac{\pi}{2}+\frac{3}{4\pi}-\frac{2}{3\sqrt{\pi}}\left(\frac{\pi}{2}-\frac{3}{4\pi}\right)\frac{\Gamma(n+\frac{3}{2})}{\Gamma(n+2)}\,\frac{1-2/(3n+6)}{1-1/(2n+3)}$$
$$\left.-\frac{4n\ln 2}{\pi}\right]. \tag{4.4}$$

The last term in the square brackets is due to the mass renormalization. (We normalize m_c at the Euclidean point $p^2 = -m_c^2$. For details see ref. [12]. Notice that the n asymptotics of the coefficients are always determined by the imaginary part in the non-relativistic region. Indeed, the weight factor in the dispersion integral for the nth moment is proportional to $(1-v^2)^{n-1}$ where v is the c-quark velocity, $v = (1-4m_c^2/s)^{1/2}$. Therefore, for high n only $v^2 \lesssim 1/n$ are essential.

This fact permits us to find, for high n, the whole series in α_s: in the non-relativistic limit the quark interaction reduces to the well-known Coulomb problem and the corresponding imaginary part can be computed exactly. The results are included in the review paper [12] and we will not dwell on them here.

4.2. *Vector current of heavy quarks; operator $G^a_{\mu\nu}G^a_{\mu\nu}$*

So far we have discussed ordinary perturbation theory which is absorbed into the coefficient C_I. Now we turn to computation of the coefficient C_G which is more specific.

To this end, let us consider formally matrix elements of expansion (4.1) over quark and gluon states. The idea is that expansion (4.1) is a general one and holds, in particular, in perturbation theory. To single out the operator $G^a_{\mu\nu}G^a_{\mu\nu}$, choose the gluon state. Then, to lowest order in the coupling constant, all the operators drop off except for the operator G^2 and we are left with the graphs of fig. 2.

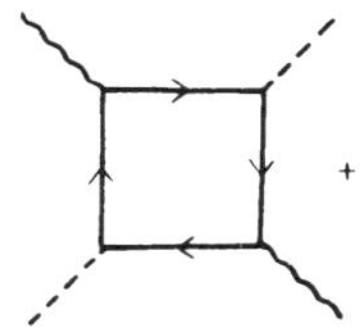

Fig. 2. Graphs giving rise to the operator $G^a_{\mu\nu}G^a_{\mu\nu}$ in the operator expansion. Notation is the same as in fig. 1.

There is some complication due to the fact that $G^a_{\mu\nu}G^a_{\mu\nu}$ vanishes for real gluons ($k^2 = 0, ek = 0$). However, one is free to choose the polarizations arbitrarily since the whole procedure can be considered as an evaluation of the matrix element of the T-product of four currents and our aim is just to find C_G in some way.

One more technical remark is in order. Straightforward calculation of the graph of fig. 2 yields not only the structure, $(q_\mu q_\nu - q^2 g_{\mu\nu})\, G^a_{\alpha\beta}G^a_{\alpha\beta}$, we are interested in but also $q_\alpha G^a_{\mu\alpha} q_\beta G^a_{\nu\beta}$.

The latter can be represented as

$$q_\alpha G^a_{\mu\alpha} q_\beta G^a_{\nu\beta} = q_\alpha q_\beta [G^a_{\mu\alpha}G^a_{\nu\beta} - \tfrac{1}{12}(g_{\alpha\beta}g_{\mu\nu} - g_{\mu\beta}g_{\nu\alpha})\, G^a_{\gamma\delta}G^a_{\gamma\delta}]$$

$$+ \tfrac{1}{12}(q^2 g_{\mu\nu} - q_\mu q_\nu)\, G^a_{\gamma\delta}G^a_{\gamma\delta}\ .$$

When averaged over vacuum state the first term vanishes since, by the symmetry argument,

$$\langle 0|G^a_{\mu\alpha}G^a_{\nu\beta}|0\rangle \sim (g_{\mu\nu}g_{\alpha\beta} - g_{\mu\beta}g_{\nu\alpha})\ .$$

Thus we are left with the second term alone, which is the structure needed. Vanishing of the term in the square brackets exemplifies the general rule according to which non-zero Lorentz spin operators can be safely omitted.

Keeping in mind the remarks made and performing an explicit calculation of fig. 2 one gets

$$C_G(Q^2) = \frac{\alpha_s}{12\pi}\frac{1}{4Q^4}\left\{\frac{3(a+1)(a-1)^2}{a^2}\frac{1}{2\sqrt{a}}\ln\frac{\sqrt{a}+1}{\sqrt{a}-1} - \frac{3a^2-2a+3}{a^2}\right\}, \qquad (4.5)$$

where

$$a = 1 + 4m_c^2/Q^2\ , \qquad (Q^2 \equiv -q^2)\ .$$

We have derived the same result in an alternative way as well. Namely, one can consider the graph of fig. 2 for slightly virtual gluons ($k^2 \neq 0$) and collect all terms of second order in the gluon momentum k. The calculation is simplified by taking the polarizations of the gluons to be the same and averaging both over this polarization, $\overline{e_\mu e_\nu} \to -\tfrac{1}{2}g_{\mu\nu}$ and over the gluon 4-momentum, $\overline{k_\mu k_\nu} \to \tfrac{1}{4}k^2 g_{\mu\nu}$.

In applications, we are interested first of all in the moments. They can be computed directly from eq. (4.5), expanding in powers of $Q^2/4m_c^2$. The simplest way, however, is to take one step back and not perform the last integration over the Feynman parameter, leaving the integral representation for C_G:

$$C_G = \frac{\alpha_s}{6\pi Q^2} \int_0^1 dx \left\{ -\frac{x(1-x)}{m_c^2 + x(1-x)Q^2} + \frac{m_c^2}{[m_c^2 + x(1-x)Q^2]^2} \times \left(\tfrac{1}{6} - \tfrac{4}{3}x(1-x)\right) + \frac{m_c^4}{[m_c^2 + x(1-x)Q^2]^3}\left(\tfrac{1}{6} + \tfrac{1}{3}x(1-x)\right) \right\} . \tag{4.6}$$

Expanding in Q^2 can now be trivially performed:

$$\left.\frac{(-d/dQ^2)^n C_G}{(-d/dQ^2)^n C_I^{(0)}}\right|_{Q^2=0} = -\frac{n(n+1)(n+2)(n+3)}{2n+5}\frac{4\pi\alpha_s}{9}(4m_c^2)^{-2} , \tag{4.7}$$

which completes our computation of C_G.

As for the coefficients C_M, C_f, C_Γ, C_σ (see eqs. (3.1), (3.2)) they appear only in higher orders in α_s. The relevant graphs are displayed in fig. 3. We do not undertake their calculation in this paper.

Notice, that we have introduced in fact an "external field", a fluctuating gluon field in the vacuum. It acts on quarks and is characterized by $\langle 0|G^a_{\mu\nu}G^a_{\mu\nu}|0\rangle$, $G^a_{\mu\nu}G^a_{\mu\nu}$ being the simplest function of the gluon field strength tensor, to which one can prescribe the non-vanishing vacuum expectation value without violating general principles. Therefore, the answer for C_G can be extracted in principle from the known results referring to QED calculations of the electron polarization operator in an external electromagnetic field. (See e.g. ref. [24].) The non-linearity of QCD does not manifest itself as far as the G^2 term is concerned. However, since we are going to include consideration of the vacuum averages of, say,

$$\bar{\psi}\psi \qquad \text{or} \qquad f^{abc}G^a_{\mu\nu}G^b_{\nu\sigma}G^c_{\sigma\mu}$$

(which are not encountered in QED), we prefer to perform all the calculations independently from the very beginning.

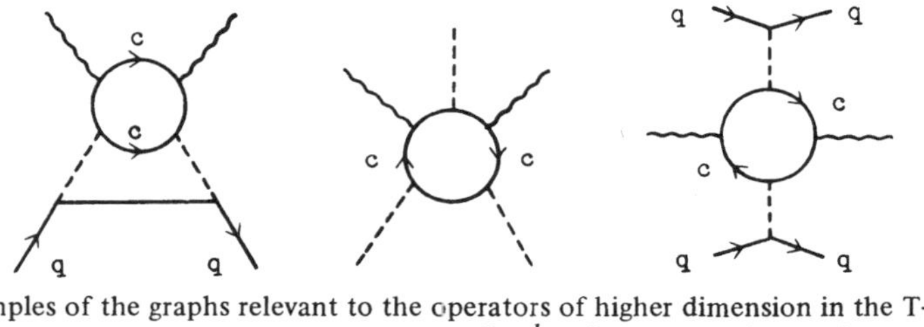

Fig. 3. Examples of the graphs relevant to the operators of higher dimension in the T-product of two heavy quark currents: (a) $m_q\bar{q}q$; (b) $f_{abc}G^a_{\mu\nu}G^b_{\nu\sigma}G^c_{\sigma\mu}$; (c) $\bar{q}\gamma_\mu t^a q\bar{q}\gamma_\mu t^a q$.

4.3. Pseudoscalar and scalar currents of heavy quarks

One can certainly construct the polarization operators induced by other currents as well, for example,

$$j^{(\mathrm{P})} = \bar{c}\, i\gamma_5 c\ , \quad j^{(\mathrm{S})} = \bar{c}c\ .$$

The corresponding calculations were performed in collaboration with M.B. Voloshin. The results can be useful for the consideration of the 0^- and 0^+ charmonium states. In particular, in ref. [17] the pseudoscalar charmonium (the so-called η_c state) was discussed in detail. For the sake of completeness, we give here the final answers for the expansion coefficients * (for definitions see eqs. (3.1), (3.2)).

$\bar{c}i\gamma_5 c$ current

$$\frac{1}{(n+1)!}\left(-\frac{\mathrm{d}}{\mathrm{d}Q^2}\right)^{n+1} C_I\bigg|_{Q^2=0} = \frac{3}{8\pi^2}\,\frac{1}{(4m_c^2)^n}\,\frac{2^n(n-1)!}{(2n+1)!!}\,[1+a_n^{(\mathrm{P})}\alpha_s]\ , \tag{4.8}$$

where

$$\tfrac{3}{4}a_n^{(\mathrm{P})} = \frac{(2n+1)!!}{2^{n+1}n!}\left[\pi - \frac{4}{3(n+1)} + \frac{0.1}{3(n+1)(n+2)}\right] - 1 + \frac{0.69}{2n+3}$$
$$+\frac{1}{\pi}\left[2 - \frac{3}{2n} - \frac{3}{n+1} - \frac{1}{2(n+2)}\right] - \frac{3n\ln 2}{\pi}\ , \tag{4.9}$$

Furthermore,

$$\frac{(-\mathrm{d}/\mathrm{d}Q^2)^{n+1}C_G}{(-\mathrm{d}/\mathrm{d}Q^2)^{n+1}C_I}\bigg|_{Q^2=0} = -\frac{n(n+1)(n+2)(n-3)}{2n+3}\,\frac{4\pi\alpha_s}{9}\,(4m_c^2)^{-2}\ . \tag{4.10}$$

$\bar{c}c$ current

$$\frac{1}{(n+1)!}\left(-\frac{\mathrm{d}}{\mathrm{d}Q^2}\right)^{n+1} C_I\bigg|_{Q^2=0} = \frac{3}{8\pi^2}\,\frac{1}{(4m_c^2)^n}\,\frac{3\cdot 2^n(n-1)!}{(2n+3)!!}\,[1+a_n^{(\mathrm{S})}\alpha_s]\ , \tag{4.11}$$

where

$$\tfrac{3}{4}a_n^{(\mathrm{S})} = \frac{(2n+3)!!}{3\cdot 2^{n+1}(n+1)!}\left[\pi - \frac{\pi - 6/\pi}{n+2}\right] - \tfrac{1}{2}\left(\frac{\pi}{2} - \frac{3}{\pi}\right)$$
$$+\frac{1}{\pi}\left[4 - \frac{2}{n} - \frac{3}{n+1} - \frac{4}{n+2} - \frac{1}{n+3}\right] - \frac{3n\ln 2}{\pi}\ , \tag{4.12}$$

and finally,

$$\frac{(-\mathrm{d}/\mathrm{d}Q^2)^{n+1}C_G}{(-\mathrm{d}/\mathrm{d}Q^2)^{n+1}C_I}\bigg|_{Q^2=0} = -\frac{n(n+1)(n+2)(3n+7)}{2n+5}\,\frac{4\pi\alpha_s}{9}\,(4m_c^2)^{-2}\ . \tag{4.13}$$

* We keep the same notation for the coefficients independently of the current considered, although $C_n(Q^2)$ are determined by the current structure, of course, and are calculated separately for each case. We hope that this makes no confusion.

Notice, that the high-n behaviour of C_I, C_G is determined by the non-relativistic expressions for the imaginary parts of the corresponding Feynman graphs. Therefore one almost immediately finds, that at high n

$$\frac{(-\mathrm{d}/\mathrm{d}Q^2)^{n+1}\, C_{I,G\,(\mathrm{scalar})}}{(-\mathrm{d}/\mathrm{d}Q^2)^{n+1}\, C_{I,G\,(\mathrm{pseudoscalar})}} \sim 1/n\ .$$

On the other hand, the contribution of the G^2 term relative to that of the unit operator is proportional to n^3 in both cases. Numerically, the power correction in the scalar channel is approximately 3 times as large as that in the pseudoscalar one.

4.4. *Vector current of light quarks* $(\bar{q}\gamma_\mu q)$

For definiteness let us consider the current with the ρ-meson quantum numbers

$$j_\mu^{(\rho)} = \tfrac{1}{2}(\bar{u}\gamma_\mu u - \bar{d}\gamma_\mu d)\ . \tag{4.14}$$

In the case of light quarks we introduce a large external momentum q $(-q^2 \equiv Q^2 >> \mu^2)$. The operator expansion has the form:

$$i\int e^{iqx}\mathrm{d}x\ \mathrm{T}\{j_\mu^{(\rho)}(x), j_\nu^{(\rho)}(0)\} = (q_\mu q_\nu - q^2 g_{\mu\nu})$$

$$\times\, [C_I I + C_G O_G + C_M O_M + C_\sigma O_\sigma + C_f O_f]\ , \tag{4.15}$$

where O_i are given in eq. (3.2) and we omitted terms of higher order in Q^{-2}.

The calculations are now slightly more complicated than for heavy quarks because new vacuum averages enter the game. In particular, in zeroth order in α_s the coefficients C_I and C_M are non-vanishing (see figs. 1a and 4a, respectively). The explicit result is:

$$C_I^{(0)} = -\frac{1}{8\pi^2}\ln\frac{Q^2}{\mu^2}\ , \qquad C_M^{(0)}O_M = \frac{1}{2Q^4}(m_u\bar{u}u + m_d\bar{d}d)\ . \tag{4.16}$$

Graphs of first order in α_s both induce corrections to these coefficients and give rise to further operators in the expansion. The former effect is illustrated in figs. 1b, 4b:

$$C_I = -\frac{1}{8\pi^2}\left[(1+\alpha_s/\pi)\ln\frac{Q^2}{\mu^2} + 3\,\frac{m_u^2 + m_d^2}{Q^2}\right]\ , \tag{4.17}$$

$$C_M/C_M^{(0)} = 1 + \alpha_s/3\pi\ . \tag{4.18}$$

Notice that the term proportional to $m_{u,d}^2$ and the α_s correction to C_M are numerically small and we will omit them in further applications.

The coefficient C_G can be evaluated through the same kind of graphs as represented in fig. 2, with a substitution of the heavy quark by the light one. There is

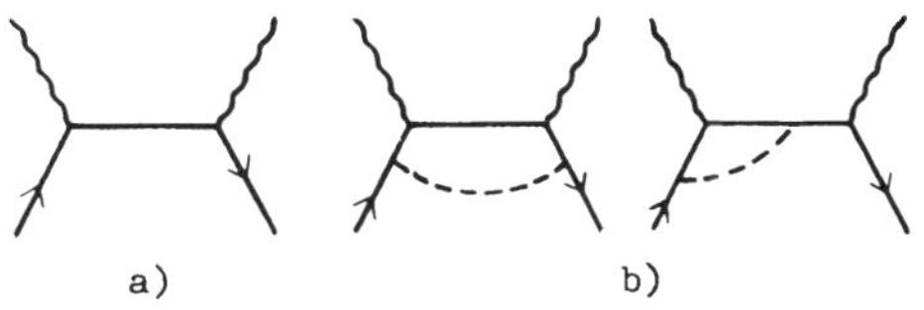

Fig. 4. Graphs giving rise to the operator $O_M = \bar{\psi}M\psi$: (a) the lowest order; (b) the one-loop correction.

an important computational difference between the two cases, however. The point is that taking the two-gluon matrix element no longer singles out the operator $G^a_{\mu\nu}G^a_{\mu\nu}$. The reason is that the operator $m_q\bar{q}q$ contributes to the two-gluon matrix element to the same order in α_s as well (see fig. 5), and these two effects must be separated.

An explicit calculation yields

$$C_G O_G = -\left[\frac{\alpha_s}{24\pi Q^4} - \frac{2\alpha_s}{24\pi Q^4}\right] G^a_{\mu\nu}G^a_{\mu\nu} \, . \tag{4.19}$$

Here, the first term corresponds to fig. 2 and can be readily obtained by evaluating the $1/Q^4$ asymptotics of the coefficient C_G obtained in subsect. 4.2 (see eq. (4.5)). The subtracted term can be immediately obtained by a straightforward calculation of the graph in fig. 5 and eq. (4.16).

The meaning of the subtraction procedure is, in fact, simple. Indeed, the $1/Q^4$ asymptotics of the diagram in fig. 2 received contributions from two distinct regions of the virtual momenta, $p^2 \sim Q^2$ and $p^2 \sim m_q^2$, respectively. Clearly enough, only the former region must be included into the coefficient C_G, while the integration over small p must be absorbed into the matrix element of another operator. The distinction is important for theories with confined quarks. The contribution of short distances, $p^2 \sim Q^2$ is computed reliably and is kept intact. As for the matrix element, it is drastically changed by the non-perturbative effects, which for example, make it very improbable to find a light quark with $p^2 \sim m_q^2$ (recall that, e.g., $m_d + m_u \simeq 11$ MeV [19,18]). Therefore, matrix elements must be treated separately.

One can readily check that the coefficient C_G given by eq. (4.19) does correspond to high virtualities, $p^2 \sim Q^2$.

Now, we come to a new kind of operators, $im_q\bar{q}\sigma_{\mu\nu}t^a qG^a_{\mu\nu}$ and $\bar{q}\Gamma q\bar{q}\Gamma q$. The

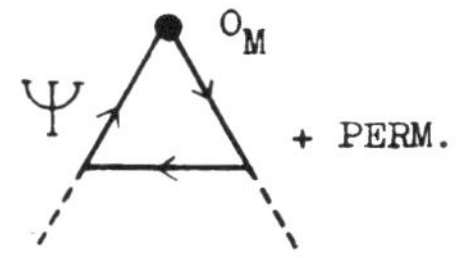

Fig. 5. The two-gluon matrix element of the operator O_M.

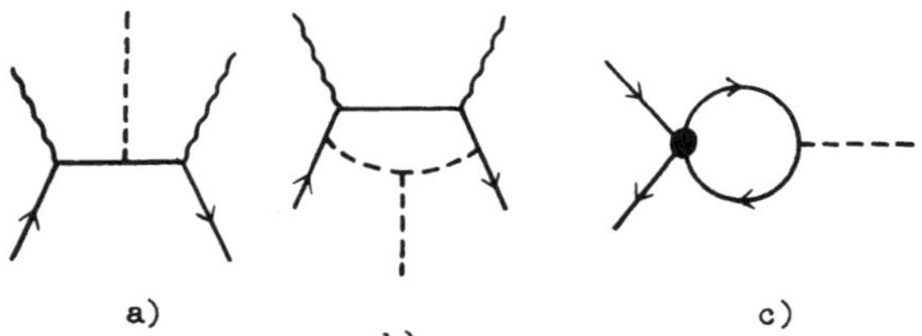

Fig. 6. Graphs relevant to the operator $O_\sigma = m\bar{\psi}\sigma_{\mu\nu}t^a\psi G^a_{\mu\nu}$: (a) the lowest order; (b) the one-loop correction; (c) the O_Γ-O_σ mixing. The closed circle denotes a four-fermion operator.

coefficient $C^{(0)}_\sigma$ can be found by computing the matrix element associated with the graph in fig. 6a:

$$C^{(0)}_\sigma O_\sigma = \frac{ig_s}{12Q^8}(m_u^3\bar{u}\sigma_{\mu\nu}t^a u + m_d^3\bar{d}\sigma_{\mu\nu}t^a d)\, G^a_{\mu\nu}\, . \tag{4.20}$$

Note that $C^{(0)}_\sigma$ contains an extra power of m_q^2/Q^2 and is severely suppressed in this way. On general grounds alone one asserts that the mass term must be inserted at least once. Further suppression is specific for the graph considered (fig. 6a). Even if in higher orders (fig. 6b) this suppression goes away, the numerical smallness surely persists. That is why the operator $\bar{\psi}\sigma_{\mu\nu}t^a\psi G^a_{\mu\nu}$ does not seem to play any important role here.

On the other hand, the four-fermion operators $\bar{\psi}\Gamma\psi\bar{\psi}\Gamma\psi$ are very important.

There exist two types of relevant diagrams. Indeed, large momentum q can flow either through an internal gluon or quark line (see figs. 7a and 7b, respectively). To find the coefficient C_Γ in the latter case, we must extract the k^2 factor from the quark-gluon vertex (k is the gluon momentum), so that the gluon propagator k^{-2} is cancelled out and a point-like operator arises. Straightforward calculation leads in this case to an operator $\bar{q}\gamma_\nu t^a q \mathcal{D}_\alpha G^a_{\alpha\nu}$ which is reduced to a four-fermion form by using the equations of motion:

$$\mathcal{D}_\mu G^a_{\mu\nu} + \tfrac{1}{2}g_s \sum_q \bar{q}\gamma_\nu t^a q = 0\, . \tag{4.21}$$

Fig. 7. Diagrams for four-fermion operators in the T product of two light quark currents.

From an explicit evaluation of the graphs in fig. 7a we find

$$-\frac{g_s^2}{8Q^6}(\bar{u}\gamma_\mu\gamma_5 t^a u - \bar{d}\gamma_\mu\gamma_5 t^a d)(\bar{u}\gamma_\mu\gamma_5 t^a u - \bar{d}\gamma_\mu\gamma_5 t^a d)\,, \tag{4.22}$$

while the graph 7b adds the following piece:

$$-\frac{g_s^2}{36Q^6}(\bar{u}\gamma_\mu t^a u + \bar{d}\gamma_\mu t^a d)\sum_{q=u,d,s}\bar{q}\gamma_\mu t^a q\,. \tag{4.23}$$

Collecting all the terms together we find for the operator expansion (4.15)

$$i\int dx\, e^{iqx}\mathrm{T}\{j_\mu^{(\rho)}(x), j_\nu^{(\rho)}(0)\} = (q_\mu q_\nu - q^2 g_{\mu\nu})\Bigg\{-\frac{1}{8\pi^2}\left(1+\frac{\alpha_s}{\pi}\right)\ln\frac{Q^2}{\mu^2}$$
$$+\frac{1}{2Q^4}(m_u\bar{u}u + m_d\bar{d}d) + \frac{\alpha_s}{24\pi Q^4}G^a_{\mu\nu}G^a_{\mu\nu} - \frac{\pi\alpha_s}{2Q^6}$$
$$\times(\bar{u}\gamma_\mu\gamma_5 t^a u - \bar{d}\gamma_\mu\gamma_5 t^a d)^2 - \frac{\pi\alpha_s}{9Q^6}(\bar{u}\gamma_\mu t^a u + \bar{d}\gamma_\mu t^a d)\sum_{q=u,d,s}\bar{q}\gamma_\mu t^a q\Bigg\}. \tag{4.24}$$

4.5. Axial and pseudoscalar currents of light quarks

All the calculations for the current with A_1 quantum numbers

$$j_\mu^{(A_1)} = \tfrac{1}{2}(\bar{u}\gamma_\mu\gamma_5 u - \bar{d}\gamma_\mu\gamma_5 d)\,, \tag{4.25}$$

run in parallel to those for a vector current and we will give only the final answer for the difference between the vector and the axial currents:

$$i\int dx\, e^{iqx}\mathrm{T}\{j_\mu^{(A_1)}(x), j_\nu^{(A_1)}(0) - j_\mu^{(\rho)}(x), j_\nu^{(\rho)}(0)\}$$
$$= -g_{\mu\nu}\frac{1}{Q^2}(m_u\bar{u}u + m_d\bar{d}d) - (q_\mu q_\nu - g_{\mu\nu}q^2)$$
$$\times\frac{2\pi\alpha_s}{Q^6}(\bar{u}_L\gamma_\mu t^a u_L - \bar{d}_L\gamma_\mu t^a d_L)(\bar{u}_R\gamma_\mu t^a u_R - \bar{d}_R\gamma_\mu t^a d_R)\,. \tag{4.26}$$

Here $q_{L,R} = \frac{1}{2}(1\pm\gamma_5)\,q$. As for the isoscalar current, there are some extra terms due to the so-called triangle anomaly [25]. We plan to discuss the question in detail in a separate publication.

In applications, we will also need the operator expansion for the pseudoscalar current with π-meson quantum numbers:

$$j^{(\pi)} = \tfrac{1}{2}i(\bar{u}\gamma_5 u - \bar{d}\gamma_5 d)\,.$$

We give, directly, the final result

$$i\int dx\, e^{iqx}\, T\{j^{(\pi)}(x), j^{(\pi)}(0)\} = \frac{3}{16\pi^2} Q^2 \ln\frac{Q^2}{\mu^2}$$

$$-\frac{1}{4Q^2}(m_u \bar{u}u + m_d \bar{d}d) + \frac{\alpha_s}{16\pi Q^2} G^a_{\mu\nu} G^a_{\mu\nu}$$

$$+\frac{\pi\alpha_s}{4Q^4}(\bar{u}\sigma_{\mu\nu}\gamma_5 t^a u - \bar{d}\sigma_{\mu\nu}\gamma_5 t^a d)^2$$

$$+\frac{\pi\alpha_s}{6Q^4}(\bar{u}\gamma_\mu t^a u + \bar{d}\gamma_\mu t^a d) \sum_{q=u,d,s} \bar{q}\gamma_\mu t^a q\,. \qquad (4.27)$$

4.6. Anomalous dimensions

Relations obtained so far are valid to the lowest order in the strong interaction coupling constant α_s. It is clear that the results stand as they are if higher-order corrections are included but both α_s and all the operators are normalized at Q^2 * (by the normalization point for an operator we mean here the standard convention according to which quark (gluon) matrix elements of the operator are equal to those of the free field theory at the normalization point).

Once we want to keep the Q^2 dependence explicit we must choose, however, an independent normalization point. Under the change of the normalization point the operators get factors $(\alpha_s(\mu)/\alpha_s(Q))^{\delta/b}$ where δ is the anomalous dimension ** and b is the coefficient in the Gell-Mann–Low β function, $b = 11 - \frac{2}{3}n_f$ (for our purposes we can take n_f to be equal to 3 since the effect of heavy virtual quarks turns to be negligible). This recipe corresponds to a summing of the log terms of order $(\alpha_s \ln(Q^2/\mu^2))^n$ which arise in perturbation theory. In this subsection we will give the values of δ for the operators introduced above.

For the unit operator the anomalous dimension vanishes and the summation of

* The statement is not quite accurate as far as the unit operator is concerned. It would be precise if we meant, say, the derivative dC_I/dQ^2. The coefficient C_I itself is logarithmic even to the zeroth order in α_s, so the renormalization-group effects are slightly more complicated here. Namely, $(1 + \alpha_s/\pi)\ln(Q^2/\mu^2)$ goes into [21]:

$$\left[1 + \left(\frac{4}{b}\ln\frac{\alpha_s(\mu)}{\alpha_s(Q)}\right)\left(\ln\frac{Q^2}{\mu^2}\right)^{-1}\right]\ln\frac{Q^2}{\mu^2}\,.$$

This peculiarity is inessential since we always work with the derivatives $(-d/dQ^2)^n\, C_I(Q^2)$, to which the general statement made above is applicable in full.

** To be more precise, if there are more than one operator of a given dimension the anomalous dimension δ must be substituted by an anomalous dimension matrix.

the log terms reduces * to a mere substitution of α_s by $\alpha_s(Q)$. Thus the effect of higher orders on $C_I(Q)$ is small in all the cases considered, except for the pseudoscalar current. (Indeed, one can obtain the polarization operator for pseudoscalar currents by multiplying that for the corresponding axial-vector currents by $q_\mu q_\nu/(2m_q)^2$. The anomalous dimension of mass ($\delta = -4$) is then converted into the power dependence on log Q^2 of the unit operator contribution.)

The anomalous dimension of the operator $\bar{q}q$ is equal to 4. However, it always enters the operator expansion multiplied by a quark mass m_q which also depends on the normalization point. The net effect is that the product $m_q\bar{q}q$ does not depend, in fact, on the normalization point. In other words, all the log factors are absorbed into the definition of the mass **.

Similar arguments hold for the operator $\alpha_s G^a_{\mu\nu}G^a_{\mu\nu}$. Independence of $\alpha_s G^a_{\mu\nu}G^a_{\mu\nu}$ on the normalization point can be asserted in a number of ways. In particular, one can express $\alpha_s G^a_{\mu\nu}G^a_{\mu\nu}$ in terms of the trace of the energy-momentum tensor [26]:

$$\theta_{\mu\mu} = -\frac{9\alpha_s}{8\pi} G^a_{\mu\nu}G^a_{\mu\nu} + \sum_q m_q\bar{q}q \,, \tag{4.28}$$

where corrections of higher orders in α_s are omitted. Since the anomalous dimension of the conserved quantity, $\theta_{\mu\nu}$, vanishes, the same is true for $\alpha_s G^a_{\mu\nu}G^a_{\mu\nu}$. This also can be checked by direct calculation.

The effect of higher orders is most drastic for the case of the coefficient C_σ. The reason is that to lowest order, C_σ is greatly suppressed (see eq. (4.20)) by an extra factor m_q^2/Q^2. This suppression is rather accidental and does not persist in higher orders. For this reason calculating the loop graphs is crucial to find C_σ.

Let us remark that a consistent approach would require a consideration of both one- and two-loop graphs (figs. 6b,c) to find C_σ. In fact, a very similar analysis has been performed in ref. [27] where the operators

$$\bar{s}\sigma_{\mu\nu}(1 \pm \gamma_5)\, t^a dG^a_{\mu\nu} \,, \qquad \bar{s}\sigma_{\mu\nu}(1 \pm \gamma_5)\, dF_{\mu\nu} \,, \qquad (F_{\mu\nu} \text{ is the photon field}) \,,$$

in the weak strangeness changing, effective Hamiltonian were treated. We will not go into details here since anyhow the resulting contribution of O_σ is extremely small.

For the four-fermion operators, the computation of the anomalous dimension matrix is rather standard. The details can be found in the appendix, and here we mention only some of the results.

The dominant contribution to the vacuum-to-vacuum matrix elements is associated with the operator

$$\bar{\psi}_L\lambda^a t^b\gamma_\mu\psi_L\bar{\psi}_R\lambda^a t^b\psi_R \,, \tag{4.29}$$

* See first footnote of this subsection.

** Taking account of the next orders in α_s, the renormalization-invariant quantity is $(1 + 2\alpha_s/\pi + ...)\, m_q\bar{q}q$.

where λ^a and t^a stand for the SU(3) matrices acting in flavor and color spaces respectively. It mixes with the operator

$$\bar{\psi}_L \lambda^a \gamma_\mu \psi_L \bar{\psi}_R \lambda^a \gamma_\mu \psi_R \,. \tag{4.30}$$

Within the framework of the factorization hypothesis (i.e., vacuum insertions in all the channels, see sect. 6), the vacuum expectation values of these operators are connected with each other:

$$\langle 0|\bar{\psi}_L \lambda^a t^b \gamma_\mu \psi_L \bar{\psi}_R \lambda^a t^b \gamma_\mu \psi_R|0\rangle = \tfrac{16}{3}\langle 0|\bar{\psi}_L \lambda^a \gamma_\mu \psi_L \bar{\psi}_R \lambda^a \gamma_\mu \psi_R|0\rangle \,. \tag{4.31}$$

Eq. (A.15) of the appendix and eq. (4.31) imply that the strong-interaction effects reduce to multiplication of (4.29) by a factor $(\alpha_s(\mu)/\alpha_s(Q))^{8/9}$, so that

$$\alpha_s(Q)\, \bar{\psi}_L \lambda^a t^b \gamma_\mu \psi_L \bar{\psi}_R \lambda^a t^b \gamma_\mu \psi_R|_Q \rightarrow$$
$$(\alpha_s(Q)/\alpha_s(\mu))^{1/9}\, \alpha_s(\mu)\, \bar{\psi}_L \lambda^a t^b \gamma_\mu \psi_L \bar{\psi}_R \lambda^a t^b \gamma_\mu \psi_R|_\mu \,. \tag{4.32}$$

Here Q and μ indicate the normalization points. Thus, all the Q dependence is manifested only *via* $(\alpha_s(Q)/\alpha_s(\mu))^{1/9}$ and is extremely weak.

Other four-fermion operators are encountered in the operator expansion with numerically small coefficients. Equations given in the appendix allow one to write out a full answer in every particular case. Let us give an example. For the ρ-meson current (4.14), the coefficient of the Q^{-6} term in the corresponding polarization operator is of the form:

$$-\tfrac{112}{81}\, \pi \langle \bar{q}q\rangle\, \alpha_s(Q)\, \kappa^{8/9} \eta^{(\rho)}(Q)\,, \tag{4.33}$$

$$\eta^{(\rho)}(Q) = 1.29 - 0.29\,\kappa^{-0.14} + 0.07\,\kappa^{-0.56} - 0.07\,\kappa^{-1.27}\,, \tag{4.34}$$

where $\langle \bar{q}q\rangle$ means $\langle 0|\bar{u}u(\mu)|0\rangle$ or $\langle 0|\bar{d}d(\mu)|0\rangle$ or $\langle 0|\bar{s}s(\mu)|0\rangle$, and

$$\kappa = \alpha_s(\mu)/\alpha_s(Q) = 1 + \alpha_s(\mu)\frac{b}{4\pi}\ln\frac{Q^2}{\mu^2}\,.$$

At $\kappa = 1$ the right-hand side of eq. (4.34) is normalized to unity so that averaging over the vacuum state we come back to the coefficient in front of the Q^{-6} term in the curly brackets of eq. (4.24). It is readily seen that the Q dependence implied by eq. (4.33) is very weak in fact. To illustrate the point let us note that for a realistic choice of κ, $\kappa = 3 - 5$ the parameter $\eta^{(\rho)}$ is given by

$$\eta^{(\rho)} = 1.06 \text{ to } 1.08\,,$$

and its deviation from unity serves as a measure of the operator mixing. Moreover, the residual Q dependence is partly cancelled by the multiplicative factor $\alpha_s(Q)\,\kappa^{8/9} \sim (\ln Q)^{-1/9}$.

To summarize, most of the operator expansion coefficients start with $\alpha_s(Q)$. One might expect, therefore, that at large Q^2 the coefficients are small since the quark-gluon coupling constant falls off logarithmically with Q^2. However, the analysis performed shows that the Q dependence of α_s is cancelled by the anomalous dimensions of the operators. The cancellation is not rigorous in all the cases but holds very well numerically.

4.7. Summary

In this section we have computed the operator expansion coefficients associated with the two-point function of the currents relevant to the forthcoming applications.

Explicit calculation has been performed for graphs of zero and first order in α_s. The computational recipe is simple: apart from evaluating the standard Feynman integrals it is necessary to cut the graphs in all possible ways over the gluon and light quark lines. The cut lines are then "annihilated" into vacuum. It is just these cut diagrams that determine the coefficients in front of the power terms.

The physical meaning of the procedure is that at low virtualities the gluon and quark propagators are modified drastically and this modification cannot be accounted for perturbatively. Large distances, therefore, are accounted for phenomenologically, through the vacuum expectation values of local operators.

In the approximation considered the expansion is especially simple for heavy quarks and is given by:

$$i\int \mathrm{d}x\, \mathrm{e}^{iqx} \mathrm{T}\{j^{\mathrm{A}}(x), j^{\mathrm{B}}(0)\} = C_I I + C_G G^a_{\mu\nu} G^a_{\mu\nu} ,$$

where we have found C_G as a function of $Q^2/4m_c^2$ for the following currents:

$$\bar{c}\gamma_\mu c , \qquad \bar{c} i\gamma_5 c , \qquad \bar{c}c .$$

(c stands for charmed, or more generally, any heavy quark field.)

For the light quarks the expansion takes the form

$$i\int \mathrm{d}x\, \mathrm{e}^{iqx} \mathrm{T}\{j^{\mathrm{A}}(x), j^{\mathrm{B}}(0)\} = C_I I + C_M \bar{\psi} M \psi + C_G G^a_{\mu\nu} G^a_{\mu\nu} + C_\Gamma \bar{\psi}\Gamma\psi\bar{\psi}\Gamma\psi ,$$

and we considered explicitly the following currents:

$$\bar{q}\gamma_\mu q , \qquad \bar{q}\gamma_\mu \gamma_5 q , \qquad \bar{q} i\gamma_5 q .$$

The coefficients depend in a non-trivial way on the current structure and the quark mass. In paper (II) we will show that the variety in the expansion coefficients results in a variety of resonance properties.

5. Sum rules

5.1. Introductory remarks

Taking the vacuum-to-vacuum matrix element of the operator expansion we get the QCD representation for the polarization operators. Say, for the current with the ρ-meson quantum numbers defined in eq. (4.14) we have

$$\Pi_{\rm QCD}(Q^2) = -\frac{1}{8\pi^2}\ln\frac{Q^2}{\mu^2} + \frac{1}{2Q^4}\langle 0|m_u\bar{u}u + m_d\bar{d}d|0\rangle$$

$$+\frac{1}{24Q^4}\langle 0|\frac{\alpha_s}{\pi}G^a_{\mu\nu}G^a_{\mu\nu}|0\rangle + \dots\,, \tag{5.1}$$

where we exhibit explicitly only the few first terms. (We recall that $Q^2 = -q^2$.)

On the other hand, the general dispersion relation gives

$$\Pi(Q^2) = \frac{1}{\pi}\int\frac{\mathrm{Im}\,\Pi_{\rm phys}(s)\,ds}{s+Q^2}\,, \tag{5.2}$$

where $\mathrm{Im}\,\Pi_{\rm phys}(s)$ is proportional to measurable cross sections such as that for e^+e^- annihilation into hadrons.

The sum rule is given by:

$$\Pi_{\rm QCD}(Q^2) = \Pi(Q^2) = \frac{1}{\pi}\int\frac{\mathrm{Im}\,\Pi_{\rm phys}(s)\,ds}{s+Q^2}\,. \tag{5.3}$$

It is useful only at large Q^2 since in this case the theory allows computation of $\Pi_{\rm QCD}(Q^2)$.

Expansions like (5.1) serve as a basis for the sum rules. In the present section we consider the next logical step, the derivation of the general form of the sum rules. Although the form exhibited in eq. (5.3) is the most conventional one we will show that it is not the most convenient to study resonances.

Equations like (5.3) lead to predictions which can be checked experimentally. The implications are especially simple at large s. The well-known result arising in this way is the prediction [21] for the e^+e^- annihilation total cross section:

$$\sigma(e^+e^- \to \text{hadrons}) = \frac{4\pi\alpha^2}{s}\sum_i Q_i^2\left(1+\frac{\alpha_s(s)}{\pi}\right)\,, \qquad s\to\infty\,, \tag{5.4}$$

where Q_i are the quark charges and α_s is the running coupling constant.

The novel feature of the sum rules considered in this paper is the inclusion of the power terms, $(\mu^2/Q^2)^k$. Certainly, the asymptotic region is not the best place to search for such terms and we turn to lower energies.

Our consideration in this section is addressed to the case of the light quarks, which is central in the applications. Moreover, it turns out that the mathematical procedure is most simple for the light quarks.

We will show that there exists a variety of alternative forms of the sum rules which corresponds to freedom in the summation procedure for the power terms. For example, one can consider, instead of the polarization operator, its Borel transform. Note that for the sake of brevity we shall use expressions like "summation of the Q^{-2} series". In fact the series is truncated, since the operator expansion breaks at some critical operator dimension (see sect. 2). In fact, all the results are general enough to cope with the real situation.

The choice of the summation prescription fixes the weight function in the integral over the spectral density which enters the sum rules. Thus, the factor $(s + Q^2)^{-1}$ in the integrand in the right-hand side of eq. (5.3) can be replaced by an exponential, $\exp(-s/Q^2)$, or a Bessel function, say $J_1(2\sqrt{s/Q^2})/\sqrt{s}$.

Since we are interested in resonance physics, we would like to have a weight function which enhances the low-energy contribution relative to the high-energy one. On the other hand, it is desirable to present the Q^{-2} series in a way that suppresses the high-order contributions since in practice we are confined to the first one or two terms in the Q^{-2} expansion.

There is no surprise that, in general, these two requirements are self-contradictory and making progress in one respect implies paying the price of a setback in the other.

Our main result is that a balance can still be reached to some extent and that there exists an optimal choice. It refers to the first Borel transform of the polarization operator. For that choice, QCD fixes such integrals as

$$\int e^{-s/M^2} \operatorname{Im} \Pi_{\text{phys}}(s)\, ds\,, \tag{5.5}$$

higher order in the M^{-2} expansion being factorially suppressed.

Apart from the choice of the most suitable form of the sum rules we discuss the possibility of determining both the coupling constant and mass of a low-lying resonance starting from the sum rules. We will argue that such a possibility does exist due to the gap in the dimensions in the operator expansion: there is the unit operator of zero dimension, and the leading power corrections come from dimension 4, with no terms of dimension 2.

The procedure is as follows. In subsect. 5.2 we consider "conventional" sum rules for the polarization operator. In subsect. 5.3 we transform them by taking the derivatives $(-d/dQ^2)^n\, \Pi(Q^2)$ with both Q^2 and the number of the derivative n, tending to infinity while their ratio Q^2/n is kept finite. In subsect. 5.4 it is shown that taking the limit of $Q^2 \to \infty$, $n \to \infty$, Q^2/n fixed, is equivalent to deriving sum rules for the Borel transform of the polarization operator. In subsect. 5.5 further Borel transforms are introduced and examined, while in subsect. 5.6 the final choice of the sum rules is substantiated and the advantages of the first Borel transform are discussed in detail.

5.2. Sum rules for the polarization operator

In sects. 3 and 4, the operator expansion for the T-product of two currents was discussed in detail. The polarization operator $\Pi(Q^2)$ is defined as the vacuum average of this product:

$$(q_\mu q_\nu - q^2 g_{\mu\nu})\, \Pi(Q^2) = i \int dx\, e^{iqx} \langle 0 | T\{j_\mu(x), j_\nu(0)\} | 0 \rangle\,, \tag{5.6}$$

where we take two vector currents just for the sake of definiteness. The general structure of $\Pi(Q^2)$ can be inferred from QCD and is given by *

$$Q^2\left(-\frac{\mathrm{d}}{\mathrm{d}Q^2}\right)\Pi(Q^2) = h_0 + \frac{h_2}{(Q^2)^2} + \frac{h_3}{(Q^2)^3} + \ldots , \tag{5.7}$$

where the coefficients h_i are dimensional (with the subscript i specifying the dimension, $h_i \sim (\mathrm{mass}^2)^i$) and are related to the vacuum expectation values of the relevant operators in the operator expansion and to the expansion coefficients. For example, in the case of the current $j_\mu^{(\rho)} = \frac{1}{2}(\bar{u}\gamma_\mu u - \bar{d}\gamma_\mu d)$ an explicit form of h_2 is

$$h_2 = \langle 0|(m_\mathrm{u}\bar{u}u + m_\mathrm{d}\bar{d}d) + \tfrac{1}{12}\frac{\alpha_\mathrm{s}}{\pi}G^a_{\mu\nu}G^a_{\mu\nu}|0\rangle .$$

The coefficients h_i can depend on Q^2 only weakly, *via* log factors.

On the other hand, the function $\Pi(Q^2)$ satisfies the general dispersion relation:

$$-\frac{\mathrm{d}}{\mathrm{d}Q^2}\Pi(Q^2) = \frac{1}{\pi}\int\frac{\mathrm{Im}\,\Pi(s)\,\mathrm{d}s}{(s+Q^2)^2} , \tag{5.8}$$

where $\mathrm{Im}\,\Pi(s)$ is subject, in principle, to direct experimental determination. For example, in the case of the current $j_\mu^{(\rho)}$ mentioned above, it is proportional to the cross section of $\mathrm{e}^+\mathrm{e}^-$ annihilation into hadrons with the isotopic spin equal to unity.

Equating the r.h.s. of eqs. (5.7) and (5.8) gives the sum rules which constrain the experimental cross section provided that QCD is the right theory of strong interactions. To apply QCD one must be sure, however, that Q^2 is large enough. In practice it is important to known which Q^2 can be considered as large. We will turn back to discussion of this point later on.

5.3. Differentiating the polarization operator: limit of Q^2, $n \to \infty$

QCD allows one to compute $(\mathrm{d}/\mathrm{d}Q^2)\,\Pi(Q^2)$ at any Q^2 provided that Q^2 is large. Thus, there exists a continuum family of sum rules. One can choose an alternative procedure: fix some large Q^2 and evaluate a number of derivatives with respect to Q^2. Intuitively, one feels that the two procedures are equivalent to each other: computing many derivatives at some Q^2 implies learning the function at lower Q^2 as well and, therefore, at some n we probe small Q^2.

Moreover, if Q^2 tends to infinity then the number of derivatives calculable in a reliable way is also arbitrarily large and we will consider the limit

$$Q^2 \to \infty , \qquad n \to \infty , \qquad Q^2/n \equiv M^2 \text{ fixed} . \tag{5.9}$$

In this way we introduce a new variable M^2 instead of Q^2. The meaning of this procedure is clarified in subsect. 5.4: it corresponds in fact to introducing the Borel transform of $\Pi(Q^2)$.

* We choose to work with the derivative of $\Pi(Q^2)$ since the constant term in $\Pi(Q^2)$ is not defined.

Let us rewrite the sum rules in terms of the new variable M^2. Introduce to this end an operator

$$\hat{L}_M = \lim_{Q^2\to\infty, n\to\infty, Q^2/n=M^2} \frac{1}{(n-1)!}(Q^2)^n\left(-\frac{d}{dQ^2}\right)^n , \tag{5.10}$$

and apply it to both the right- and left-hand sides of eq. (5.3). The result is

$$\frac{1}{\pi M^2}\int \mathrm{Im}\ \Pi(s)\, e^{-s/M^2} ds = h_0 + \frac{h_2}{2!(M^2)^2} + \frac{h_3}{3!(M^2)^3} + \dots , \tag{5.11}$$

where the coefficients h_i determine $\Pi_{QCD}(Q^2)$ and are defined in eq. (5.7). Note the appearance of the exponential factor in the integral over the imaginary part. Although we start with a conventional dispersion representation, the final result has no direct resemblance to the dispersion relations any longer.

Equations like (5.11) play a crucial role in our analysis and we will turn to a discussion of some of their properties.

5.4. Moments and the Borel improvement of the power series

Here we will show that the limiting procedure (5.10) which introduces the new variable, M^2, is equivalent to the Borel improvement of the Q^{-2} series.

Let us first recall some definitions *. Consider a function $f(x)$. Furthermore, introduce $\tilde{f}(\lambda)$ which is related to $f(x)$ *via* the following equation:

$$\tilde{f}(\lambda) = \frac{1}{2\pi i}\int_{c-i\infty}^{c+i\infty} e^{\lambda/x} f(x)\, x\, d\frac{1}{x} , \tag{5.12}$$

where the integration contour runs to the right of all the singularities of the function $f(x)$. The function $\tilde{f}(\lambda)$ is called the Borel transform of $f(x)$. The inverse transformation is given by:

$$f(x) = \int_0^\infty \tilde{f}(\lambda)\, e^{-\lambda/x}\, d\lambda/x . \tag{5.13}$$

To clarify the meaning of the Borel transform assume that $f(x)$ is given as an expansion in x (which can be asymptotic, however):

$$f(x) = a_0 + a_1 x + a_2 x^2 + \dots + a_k x^k + \dots . \tag{5.14}$$

Then the corresponding expansion of $\tilde{f}(\lambda)$ in λ has the form

$$\tilde{f}(\lambda) = \frac{a_0}{0!} + \frac{a_1\lambda}{1!} + \frac{a_2\lambda^2}{2!} + \dots + \frac{a_k\lambda^k}{k!} + \dots , \tag{5.15}$$

* A physical approach to the Borel summation technique is reviewed in ref. [28].

so that the coefficients are factorially suppressed as compared to the case (5.14).

This suppression of higher-order terms in the expansion implies that the approximation of the whole series by the few first terms is more reliable for $\tilde{f}(\lambda)$ than for $f(x)$.

It is readily seen that all these definitions are realized when one introduces M^2 instead of Q^2. Indeed, we have considered the limit

$$\tilde{\Pi}(M^2) = \lim_{\substack{Q^2, n \to \infty \\ Q^2/n = M^2}} \frac{1}{(n-1)!} (Q^2)^n \left(-\frac{\mathrm{d}}{\mathrm{d}Q^2}\right)^n \Pi(Q^2) ,$$

or using the notation (5.10)

$$\tilde{\Pi}(M^2) = \hat{L}_M \, \Pi(Q^2) . \tag{5.16}$$

The meaning of the operator $\hat{L}_M$ becomes clear if we turn again to a particular term in the power expansion. As is readily seen, application of $\hat{L}_M$ transforms the series in Q^{-2} into a series in M^{-2}:

$$\hat{L}_M \left(\frac{1}{Q^2}\right)^k = \frac{1}{(k-1)!} \left(\frac{1}{M^2}\right)^k . \tag{5.17}$$

Let us emphasize the appearance of the factorial suppression $1/(k-1)!$. This fact alone convinces us that we are dealing with the Borel transform of the polarization operator.

However, as mentioned in sect. 2, the polarization operator cannot be represented as an infinite series in Q^{-2}; the expansion may break at some place. Therefore, it is worth mentioning that the equivalence of the limiting procedure (5.10) to the Borel transform can be shown in a general form.

The only thing which is indeed important is that the polarization operator satisfies the standard dispersion relations. Indeed, let us write down an analog of the inverse transformation (5.13) for the function $\tilde{\Pi}(M^2)$:

$$\int_0^\infty \tilde{\Pi}(M^2) \, \mathrm{e}^{-Q^2/M^2} \, Q^2 \mathrm{d} \frac{1}{M^2} = \frac{Q^2}{\pi} \int_0^\infty \mathrm{d}s \, \mathrm{Im}\, \Pi(s) \int_0^\infty \frac{1}{M^2} \, e^{-(Q^2+s)/M^2} \, \mathrm{d} \frac{1}{M^2} , \tag{5.18}$$

where we have used the integral representation for $\tilde{\Pi}(M^2)$ in terms of the imaginary part $\mathrm{Im}\, \Pi(s)$. Furthermore, performing the integration over M^2 gives

$$\int_0^\infty \tilde{\Pi}(M^2) \, \mathrm{e}^{-Q^2/M^2} Q^2 \mathrm{d} \frac{1}{M^2} = \frac{Q^2}{\pi} \int_0^\infty \frac{\mathrm{Im}\, \Pi(s) \, \mathrm{d}s}{(s+Q^2)^2} = Q^2 \left(-\frac{\mathrm{d}}{\mathrm{d}Q^2}\right) \Pi(Q^2) , \tag{5.19}$$

which proves that $\tilde{\Pi}(M^2)$ does coincide with the Borel transform of the function $Q^2(-\mathrm{d}/\mathrm{d}Q^2)\, \Pi(Q^2)$.

Eq. (5.17) specifies the effect of the operator $\hat{L}_M$ on the power terms. As was mentioned above the expansion coefficients h_i may have some log dependence on Q^2. In fact, it can enter either through α_s corrections ($\alpha_s(Q) \sim 1/\ln Q$) or through anomalous dimension factors of the kind $1/(\ln Q)^\epsilon$. Therefore, we also need to know how such factors are transformed. We will show that under the procedure, $h_i(Q^2)$ are replaced by the same functions of M^2:

$$h_i(Q^2) \to h_i(M^2)\,. \tag{5.20}$$

To this end use the following representation:

$$\left(\frac{1}{Q^2}\right)^k \left(\frac{1}{\ln(Q^2/\mu^2)}\right)^\epsilon = \frac{1}{\epsilon\Gamma(\epsilon)} \left(\frac{1}{Q^2}\right)^k \int_0^\infty \mathrm{d}z\; \mathrm{e}^{-z^{1/\epsilon}\ln(Q^2/\mu^2)}\,. \tag{5.21}$$

Furthermore, applying the operator $\hat{L}_M$ and using eq. (5.17) yields

$$\hat{L}_M \left[\left(\frac{1}{Q^2}\right)^k \left(\frac{1}{\ln(Q^2/\mu^2)}\right)^\epsilon\right] = \frac{1}{\Gamma(k)} \left(\frac{1}{M^2}\right)^k \left(\frac{1}{\ln(M^2/\mu^2)}\right)^\epsilon \left[1 + \mathrm{O}\left(\frac{1}{\ln(Q^2/\mu^2)}\right)\right], \tag{5.22}$$

which is just what we wanted to prove.

Eqs. (5.15) and (5.20) exaust all the transformations needed in practical applications.

5.5. Further Borel transforms

The motivation to turn to the Borel transform is to improve the approximation of the whole series of the power corrections by the first few terms (in fact we will keep terms of order M^{-4}, M^{-6} and neglect the others). One can reiterate the procedure and introduce a new variable M'^2. This subsection deals with these further improvements of the series while the meaning of the results obtained is discussed in the next subsection.

Divide the left- and right-hand sides of eq. (5.11) by M^2 and apply the operator $\hat{L}_{M'}$ to the both. Then we get

$$\frac{1}{\pi M'} \int_0^\infty \mathrm{Im}\,\Pi(s)\, J_1\left(\frac{2\sqrt{s}}{M'}\right)(s)^{-1/2}\,\mathrm{d}s$$

$$= h_0 + \frac{h_2}{(2!)^2 (M'^2)^2} + \frac{h_3}{(3!)^2 (M'^2)^3} + \ldots\,, \tag{5.23}$$

where we have used that

$$\hat{L}_{M'} \frac{1}{M^4}\,\mathrm{e}^{-s/M^2} = \frac{1}{M'^3\sqrt{s}}\, J_1\left(\frac{2\sqrt{s}}{M'}\right),$$

and J_1 is the Bessel function.

One can easily generate the "daughter" sum rules as well. For example, differentiating eq. (5.23) with respect to $1/M'^2$ we come to the following relation:

$$\frac{1}{\pi M'^2}\int \mathrm{Im}\,\Pi(s)\,J_0(2\sqrt{s}/M')\,\mathrm{d}s = -M'^2(\mathrm{d}/\mathrm{d}M'^2)\,h_0 +$$

$$+\frac{h_2}{2!1!(M'^2)^2}+\frac{h_3}{3!2!(M'^2)^3}+\ldots\,. \tag{5.24}$$

Note that the right-hand side of this equation vanishes in the limit of asymptotic freedom. Indeed, the coefficient h_0 depends on M^2 only *via* $\alpha_s(M')$ and

$$-M'^2\frac{\mathrm{d}\alpha_s}{\mathrm{d}M'^2}=\frac{b\alpha_s^2}{4\pi}\,,$$

so that $M'^2(\mathrm{d}/\mathrm{d}M'^2)\,h_0$ can be neglected in the approximation considered.

The sequence of Borel transforms can be continued. It would introduce the hypergeometrical functions as a weight in the integral over the spectral density. We will not go into details here and conclude with a remark that in all these cases the weight function is not positive definite.

5.6. *On the choice of the form of sum rules*

Thus, starting with the dispersion relation for the polarization operator and applying the Borel transform we come to a variety of sum rules.

From a practical point of view some particular choice of sum rules may turn to be most helpful. Thus, it is mostly a matter of convenience as to which sum rules are used in the analysis of the experimental data. Indeed, all the dynamical information is confined to the knowledge of some of the expansion coefficients and of the corresponding matrix elements.

What are the qualities we would like to embody into the sum rules? To be sensitive to a single resonance the integrals over the cross sections must be concentrated in as narrow an energy region as possible.

On the other hand, all the evaluations of the polarization operator in QCD are confined to a few first terms in Q^{-2} and introducing a factorial-like suppression of higher orders is desirable from this point of view. In particular, let us remind the reader that starting from some rather high power of Q^{-2}, small-size instantons come into the game. They break the operator expansion and may bring a contribution which might be qualitatively different from those considered so far. It certainly would be nice to suppress this contribution numerically. Schematically, we have:

$$\hat{L}_M\begin{bmatrix}\text{small-size instanton}\\ \text{contribution in } Q^2\end{bmatrix}\sim\frac{1}{(\frac{1}{2}\epsilon)!}\begin{bmatrix}\text{small size instanton}\\ \text{contribution in } M^2\end{bmatrix},$$

where ϵ is the exponent encountered in the instanton density function, $\mathrm{d}(\rho)\sim\rho^{\epsilon}$. One expects that $\epsilon = 11-16$ [9].

In general, the two requirements for the sum rules show in opposite directions. Still, we will argue that the sum rules for the first Borel transform of the polarization operator represent the optimal choice. The sum rules are examplified by eq. (5.11) and we turn now to their discussion.

M^2 tending to infinity implies moving towards asymptotic freedom. Indeed, only the first perturbative term, h_0, survives in this limit in eq. (5.11). In other words, we deal with short-distance dynamics in this limit. On the other hand, taking low M^2 puts emphasis on the large distances.

It is remarkable that this purely theoretical distinction between short and large distances is directly manifested in the interplay between the resonance and high-energy contributions. Indeed, because of the exponential cut-off, only $s \lesssim M^2$ contributes to the integrals over the imaginary part. Thus, if M^2 is of the order of a resonance mass, say, m_ρ^2 then the integral is dominated by a single resonance.

At large M^2 the corrections to asymptotic freedom are small, and, as a reflection of this, the integral over the physical states is dominated by high energies. Diminishing M^2 enhances the resonance contribution, on one hand, and increases the power corrections to asymptotic freedom, on the other.

Power corrections specify the very notion of "high" and "low" M^2: for large M^2 the power corrections are small while for low M^2 they become dominant and it is necessary to sum up all the power terms to have a reliable answer for $\tilde{\Pi}(M^2)$. This sets a natural bound, $M^2_{\rm crit}$, on M^2, which can be used in our approximation (which keeps only first terms in the M^{-2} expansion).

For $M^2 > M^2_{\rm crit}$ one may hope that the power terms of lowest dimension which are kept explicit represent the leading corrections to asymptotic freedom while the higher orders can be safely neglected.

As was proclaimed many times above we aim at extracting the QCD predictions for a single resonance. To fulfill the task we are inclined to choose M^2 as low as possible. On the other hand, taking M^2 too low makes the whole calculation unreliable since the power corrections become large.

Our central point, which rests entirely on numerical estimates, is that it is still possible to make a balance between the low and high M^2 tendencies and find such M^2 that on the one hand, a resonance dominance is guaranteed, and on the other hand, the power corrections are still moderate and tractable.

In achieving this aim we are helped by using the sum rules for the first Borel transform of the polarization operator. Indeed, it introduces the exponential cut-off into the integral over the spectral density. Thus, if we take both Q^2 and M^2 to be the same and of order, say, m_ρ^2 then the low-lying meson dominance is much more prominent for the sum rules (5.11) than for (5.3). On the other hand, the approximation of the whole series by the few first terms is also better for the Borel transform than for the polarization operator. These are advantages of the sum rules for the first Borel transform which single them out among the other possibilities.

At first sight, further Borel improvements could do even better. Indeed, the process of improving the theoretical accuracy can be extended to any degree by repetitions of the Borel transform. By this we mean, that one can achieve an arbitrary accuracy of calculation at a fixed value of an external parameter such as Q^2, M^2, M'^2 and so on, by going to a Borel transform of a high order.

One feels, however, that the very possibility of an unlimited improvement of the accuracy implies that the value of the external parameter becomes non-representative of the energy scale needed to verify the sum rules. A closer examination reveals that this is indeed the case. The point is that for higher Borel transforms, the weight function in the sum rules is not positive definite. Examples of this kind have been already given in eqs. (5.23), (5.24). In these equations we have oscillating Bessel functions as a weight.

It is clear that in the limit of an infinite number of repetitions of the Borel transforms, the sum rules are entirely controlled by the high-energy contribution, independent of the value of the external parameter.

It is remarkable that the correspondence between the choice of the parameter Q^2, M^2, ... and the distances which are essential, dynamically works, strictly speaking only in the case of the polarization operator and its first Borel transform.

We find it difficult to analyse the sum rules with an oscillating weight function and choose to work with the first Borel transform.

Although we do not use further Borel transforms we do not rule out the possibility that they are instructive in some respects. Note, as an example, that the integral with J_0 vanishes in the limit of high M^2 while the integral with J_1 does not (see eqs. (5.23) and (5.24)). Since the Bessel functions are nearly periodical, an impression arises that the resonances are "tuned" to some wavelength in energy to make the difference between J_0 and J_1 so profound.

5.7. Sum rules for resonances masses

In conclusion of this section let us add one comment of more technical nature concerning the possibility of extracting from the sum rules both the coupling constant and mass of a resonance.

Assume that at some M^2 the integral $\int e^{-s/M^2}\,\mathrm{Im}\,\Pi(s)\,ds$ is saturated by a single resonance, and that we are still in a "safe" region so that the expansion (5.11) is an expansion in a small parameter.

It is convenient under this circumstance to consider a sum rule which is obtained by differentiating eq. (5.11) with respect to $1/M^2$ *:

$$\frac{1}{\pi M^4}\int \mathrm{Im}\,\Pi(s)\,e^{-s/M^2} s\,ds = h_0 - \frac{h_2}{2!(M^2)^2} - \frac{2h_3}{3!(M^2)^3} - \dots \,. \tag{5.25}$$

Certainly, this relation is not an independent one. Nevertheless, it is useful since it

* It is convenient prior to differentiating to multiply eq. (5.11) by M^2. Direct differentiation yields a linear combination of eqs. (5.11), (5.25).

allows control of the accuracy of the mass calculation.

Indeed, if both (5.11) and (5.25) were single-resonance dominated, then the ratio of the respective left-hand sides would produce an experimental value of the resonance mass squared. The possibility of predicting the number theoretically depends on whether power corrections to (5.11) and (5.25) are still tractable or not.

Repetition of differentiation with respect to $1/M^2$ would eliminate some of the power corrections. Thus, the second derivative eliminates the M^{-4} term and so on. It does not imply, of course, that the accuracy of the theoretical calculation gets improved in this way. On the contrary, elimination of the corrections which are calculated in some way implies loosing control over the accuracy of the entire procedure. We will push asymptotic freedom to the limit of its applicability where sum rules become sensitive to the resonance contribution. We cannot do that without learning the critical value of M^2.

Therefore, to make use of the sum rules for higher derivatives in $1/M^2$ we must introduce higher orders in the M^{-2} expansion, estimate the corresponding matrix elements and so on. This is out of the scope of the present paper and we will confine ourselves to the sum rules (5.11), (5.25).

It is amusing that the first differentiation with respect to $1/M^2$ does not introduce any new parameters and eq. (5.25) is as reliable as eq. (5.11). This is quite specific for QCD. Indeed, there is a gap in dimensions in the operator expansion. It starts with the unit operator which has vanishing dimension and proceeds directly to terms of dimension four ($m_q\, \bar{q}q,\ G^a_{\mu\nu}G^a_{\mu\nu}$). There are no operators of dimension two because of the gauge invariance of QCD. Indeed, for scalar gluons we would have an operator of dimension two bilinear in the boson field. The same is true for a vector gluon field but with no gauge invariance. In these cases the first differentiation of eq. (5.11) with respect to $1/M^2$ would eliminate the leading power correction and would require consideration of higher order in M^{-2}.

Thus, we can say that it is just the gauge invariance of QCD that ensures the possibility of learning from the sum rules both the coupling constant and mass of low-lying states.

5.8. Conclusions

To summarize, the first Borel transform realizes the optimal choice which ensures both resonance dominance in the sum rules and the suppression of higher-order power corrections. The sum rules are given by:

$$\frac{1}{\pi M^2}\int_0^\infty \mathrm{Im}\,\Pi(s)\,\mathrm{e}^{-s/M^2}\,\mathrm{d}s = h_0 + \frac{h_2}{2!(M^2)^2} + \frac{h_3}{3!(M^2)^3} + \dots ,$$

$$\frac{1}{\pi M^4}\int_0^\infty \mathrm{Im}\,\Pi(s)\,\mathrm{e}^{-s/M^2} s\,\mathrm{d}s = h_0 - \frac{h_2}{2!(M^2)^2} - \frac{2h_3}{3!(M^2)^3} - \dots .$$

In forthcoming publications we will show that the right-hand side is calculable at $M^2 \lesssim 1$ GeV^2 which implies severe constraints on the resonance properties.

If at $M^2 \lesssim$ GeV^2 the power corrections are relatively small in both of the above sum rules, then the mass and the coupling constant of the lowest state can be found (examples will be given in subsequent parts of the paper). This possibility is in fact due to the gap in the dimensions of the operators entering the operator expansion.

To avoid misstatement, let us emphasize that a complete theory would include the possibility of considering the limit $M^2 \to 0$ as well. In this limit all the contributions to the integral over the imaginary part die away exponentially but the lowest-lying state dominates over the others. Moreover, one can say for this reason that finding the spectrum means considering the limit of $M^2 \to 0$. Clearly enough, we cannot go to the limit $M^2 = 0$ with our sum rules. Our statement will be that the resonance properties are fixed to a great extent by studying moderate M^2.

Thus far, about the first Borel transform. Further Borel transforms introduce integrals like

$$\int \mathrm{Im}\,\Pi(s)\,J_1(2\sqrt{s}/M')(s)^{-1/2}\,\mathrm{d}s\,, \qquad \int \mathrm{Im}\,\Pi(s)\,J_0(2\sqrt{s}/M')\,\mathrm{d}s\,,$$

so that oscillating weight functions emerge. For this reason we will not consider the corresponding sum rules although they might be interesting.

6. Matrix elements

6.1. Introduction

To specify the sum rules we need the vacuum expectation values for various operators. So far, we have encountered the following matrix elements:

$$\langle 0|\bar{\psi}\psi|0\rangle\,, \qquad \langle 0|\bar{\psi}\Gamma_1\psi\bar{\psi}\Gamma_2\psi|0\rangle\,, \qquad \langle 0|G^a_{\mu\nu}G^a_{\mu\nu}|0\rangle\,,$$

and

$$\langle 0|\bar{\psi}\sigma_{\mu\nu}t^a\psi G^a_{\mu\nu}|0\rangle\,, \qquad \langle 0|f_{abc}G^a_{\mu\nu}G^b_{\nu\sigma}G^c_{\sigma\mu}|0\rangle\,,$$

where the division into two groups is purely pragmatic: in the former case the matrix elements govern the leading power corrections while in the latter case we deal with small terms which are actually suppressed in the numerical applications. To have a 10% accuracy in predicting the resonance properties we need an estimate valid within a factor of two for the leading terms and an order of magnitude evaluation for small corrections.

Unlike the case of the operator expansion coefficients there is no standard procedure to compute the matrix elements. Indeed, the matrix elements are sensitive to the large-distance dynamics, and the present understanding of it is far from complete. Therefore, it is rather clear that we must use some experimental information as an input.

An extreme attitude would be to determine all the matrix elements from the sum rules themselves. The sum rules are economical enough in the sense that the same matrix elements enter different sum rules. Thus, we can sacrifice some of them and still have independent and crucial tests of QCD. This would be a purely phenomenological approach.

Another way is to speculate on theoretical estimates. It might be difficult to invent a computational device but the stake is rather high: once the estimate turns to be successful a new understanding of the large-distance interactions is achieved.

We will choose a middle, or exploratory way. Namely we will rely on phenomenology as far as derivation of the principal results is concerned. On the other hand, we will try our best to estimate the matrix elements in an independent way using one or another approximation. Once we find support for a theoretical framework in some case, we will trust similar estimates for other matrix elements.

In particular, in subsects. 6.2, 6.3 we extract $\langle 0|\bar{q}q|0\rangle$ from the experimental data. But we do not use sum rules for this purpose. The point is that the non-vanishing $\langle 0|\bar{q}q|0\rangle$ is the simplest manifestation of spontaneous chiral symmetry breaking and has been discussed for quite a long time for this reason. Therefore, we use some of the earlier results to get estimates of $\langle 0|\bar{q}q|0\rangle$. The basic ingredients here are the $\pi \to \mu\nu$ decay coupling constant, $f_\pi \simeq 0.95\ m_\pi$, and an idea on the SU(3) breaking mass scale, $\Delta \sim 150$ MeV. At least at first sight, these data have no connection with the resonance physics which will be studied below and, we are happy to find one of the key parameters in a very independent way.

The rest of the section is devoted to efforts to compute all the matrix elements starting from $\langle 0|\bar{q}q|0\rangle$ on purely theoretical grounds. We use to this end three techniques:

(i) dominance of the vacuum intermediate state;
(ii) matching of the light and heavy quark expansions;
(iii) the dilute instanton-gas approximation.

Only point (ii) is fairly new while the two others have been occasionally used in the literature. In particular, the dilute-gas approximation is strongly advocated in ref. [13]. Still, our way of using it is somewhat different.

The techniques mentioned above are discussed in subsects. 6.5–6.8. We choose to demonstrate them with concrete examples rather than dwell on the general theory.

We find the assumption (i) to be reliable enough to use without further reservations. As for the approximations (ii) and (iii) above we would like to be more careful. They seem to be good for a rough estimate but inadequate for a more quantitative treatment. For this reason having estimated one of the key parameters, $\langle 0|G^a_{\mu\nu}G^a_{\mu\nu}|0\rangle$, by virtue of the approximations (ii) and (iii), we shall return to its discussion in the subsequent paper. There we determine $\langle 0|G^a_{\mu\nu}G^a_{\mu\nu}|0\rangle$ phenomenologically, by fitting the sum rules for heavy quarks (the charmonium sum rules).

Strictly speaking, all the estimates considered in this section are model dependent. To have control over the approximations used, it would be important to have some general results for the matrix elements. Unfortunately, little can be said on general grounds alone. Still, we are able to fix the signs of some of the matrix elements. The results are presented in subsect. 6.9. It turns out that all the estimates performed stand the test and give signs which agree with the general rules.

6.2. Matrix element $\langle 0|m_q\bar{q}q|0\rangle$

This is the simplest case in fact since the matrix element can be evaluated by the standard current algebra technique.

Start with the identity

$$\langle \pi^-|\varphi_\pi|0\rangle = 1 ,$$

where φ_π stands for the operator of the pion field. By virtue of the PCAC hypothesis it is related to the quark field and masses:

$$\varphi_\pi = \frac{i(m_\mathrm{u} + m_\mathrm{d})}{m_\pi^2 f_\pi}\bar{d}\gamma_5 u , \qquad (6.1)$$

where f_π is the $\pi \to \mu\nu$ decay constant, $f_\pi = 0.95\ m_\pi$.

Reducing the pion field and letting the pion momentum tend to zero one finds then, in a standard way:

$$\langle \pi^-|\bar{d}\gamma_5 u|0\rangle = \frac{i}{f_\pi}\langle 0|\bar{u}u + \bar{d}d|0\rangle . \qquad (6.2)$$

Combining eqs. (6.1) and (6.2) yields:

$$(m_\mathrm{u} + m_\mathrm{d})\langle 0|\bar{u}u + \bar{d}d|0\rangle = -m_\pi^2 f_\pi^2 , \qquad (6.3)$$

which fixes $\langle 0|(m_\mathrm{u} + m_\mathrm{d})(\bar{u}u + \bar{d}d)|0\rangle$. Note also that by virtue of the isotopic invariance

$$\langle 0|\bar{u}u|0\rangle = \langle 0|\bar{d}d|0\rangle , \qquad (6.4)$$

and this relation is expected to hold to within several per cent. If $SU(3)_\mathrm{flavor}$ is assumed then eq. (6.4) can be extended to the case of the strange quark, $\langle 0|\bar{s}s|0\rangle$. Further discussion of eq. (6.4) can be found in ref. [29].

6.3. Quark masses

Ratios of the quark masses can be extracted from the observed masses of the pseudoscalar mesons provided that the masses are small. Moreover, one can also include the electromagnetic contribution which turns out to be finite by virtue of the PCAC hypothesis. Explicitly, one obtains in the first order in the mechanical quark masses:

$$m^2(\mathrm{K}^+) = (m_\mathrm{u} + m_\mathrm{s})\,C + \gamma_{\mathrm{K}^+} ,$$

$$m^2(\mathrm{K}^0) = (m_\mathrm{d} + m_\mathrm{s})\,C + \gamma_{\mathrm{K}^0}\,,$$

$$m^2(\pi^+) = (m_\mathrm{u} + m_\mathrm{d})\,C + \gamma_{\pi^+}\,,$$

$$m^2(\pi^0) = (m_\mathrm{u} + m_\mathrm{d})\,C + \gamma_{\pi^0}\,. \tag{6.5}$$

Here C is an SU(3) invariant mass parameter which is related to the vacuum average of the quark density, and γ's denote the electromagnetic self-energies.

Moreover, using current algebra techniques, one can show that [30]

$$\gamma_{\pi^0} \approx \gamma_{\mathrm{K}^0} \approx 0\,, \qquad \gamma_{\pi^+} \approx \gamma_{\mathrm{K}^+}\,,$$

which implies in turn

$$\frac{m_\mathrm{d} - m_\mathrm{u}}{m_\mathrm{d} + m_\mathrm{u}} = \frac{m^2(\mathrm{K}^0) - m^2(\mathrm{K}^+) + m^2(\pi^+) - m^2(\pi^0)}{m^2(\pi^0)} \approx 0.29\,, \tag{6.6a}$$

$$\frac{m_\mathrm{s} + m_\mathrm{d}}{m_\mathrm{u} + m_\mathrm{d}} = \frac{m^2(\mathrm{K}^0)}{m^2(\pi^0)} \approx 14\,. \tag{6.6b}$$

These results are in no way new, of course (see, e.g. ref. [5] and recent reviews [18]).

Note that eq. (6.6a) indicates that the isotopic breaking in the u, d masses is quite strong. The result can be checked independently by studying the sum rules for $\rho\omega$ mixing. We shall publish the analysis separately [29].

Eqs. (6.5), (6.6) fix the products $m_\mathrm{q}\langle 0|\bar{q}q|0\rangle$ or the mass ratios but not the masses themselves. It is not incidental of course and the reason is that the quark masses are not invariant with respect to change in the normalization point while all the observables do not depend on the choice of the renormalization procedure. Thus, to specify the mass parameters we need further hypotheses and reasoning.

The masses and operators considered so far are normalized at the point of the ultraviolet cut-off, $\mu = \Lambda$. One can readily introduce a mass parameter normalized at some finite point as well. The commonly accepted convention relates the quark masses to the inverse propagator [31]:

$$G^{-1}(p) = \hat{p}A(p^2) - B(p^2)\,, \qquad m(\mu) = \left.\frac{B(p^2)}{A(p^2)}\right|_{p^2=-\mu^2}\,, \tag{6.7}$$

and we will follow this convention. Then one can express $m(\Lambda)$, $\bar{q}q(\Lambda)$ in terms of $m(\mu)$ and $\bar{q}q(\mu)$ normalized at $p^2 = -\mu^2$. For μ tending to infinity, the mass tends to zero. The ratio of the masses stands finite, however. Similarly, $m(\mu) \to 0$ and $\bar{q}q(\mu) \to \infty$ if $\mu \to \infty$ but their product is finite at any μ.

Note that the product $m_\mathrm{q}\bar{q}q$ is renormalization invariant in the leading log approximation. If corrections of order α_s, not only $(\alpha_\mathrm{s} \ln(\Lambda^2/\mu^2))^n$, are accounted for the renormalization invariant quantity is given by:

$$(1 + \gamma_m)[m_\mathrm{q}\bar{q}q]_\mu\,, \tag{6.8}$$

where the function $\gamma_m = -\mu(d/d\mu)\ln m(\mu)$ determines the dependence of the mass on the normalization point and is given by

$$\gamma_m = \frac{4\alpha_s(\mu)}{2\pi} + O(\alpha_s^2) .$$

In an asymptotically free theory the product (6.8) tends to $m_q\bar{q}q$ at $\mu \to \infty$ since $\alpha_s \to 0$.

All these redefinitions are useful only if one can relate $m(\mu)$ or $\bar{q}q(\mu)$ to observables in a reasonable way. Several hypotheses of this kind have been introduced in literature (with no explicit discussion of the choice of the normalization point, however).

In particular, Leutwyler [19] proposes to use the SU(6) relation

$$\langle 0|i\bar{u}\gamma_5 d|\pi\rangle = \sqrt{\tfrac{1}{2}} \sum_\mu \langle 0|\bar{u}\gamma_\mu d|\rho_\mu\rangle \qquad (6.9)$$

to fix the quark masses. Indeed, the matrix element for the current-ρ meson transition is known experimentally while the matrix element $\langle 0|i\bar{u}\gamma_5 d|\pi\rangle$ is reduced to the coupling constant f_π and quark masses by virtue of the equation of motion:

$$\langle 0|\partial_\mu \bar{u}\gamma_\mu\gamma_5 d|\pi\rangle = i(m_u + m_d)\langle 0|\bar{u}\gamma_5 d|\pi\rangle . \qquad (6.10)$$

Clearly, eq. (6.9) cannot hold for an arbitrary normalization point since $\bar{u}\gamma_5 d$ depends on μ, while the operator $\bar{u}\gamma_\mu d$ is not affected by a change in μ. Since SU(6) is a symmetry of constituent rather than current quarks it seems natural to assume that eq. (6.9) is valid at the normalization point of order of a typical hadronic mass $\mu \sim \kappa$ *.

Eqs. (6.9), (6.10) lead to the conclusion that light quarks are really light:

$$m_u + m_d \simeq 11 \text{ MeV} , \qquad (\mu = \kappa) . \qquad (6.11)$$

Moreover, eqs. (6.6) define then all the masses separately:

$$m_s \simeq 150 \text{ MeV} , \qquad m_d \simeq 7 \text{ MeV} , \qquad m_u \simeq 4 \text{ MeV} ,$$

which implies in turn:

$$\langle 0|\bar{u}u|0\rangle = \langle 0|\bar{d}d|0\rangle \simeq -(250 \text{ MeV})^3 . \qquad (6.12)$$

As mentioned above, these estimates refer to a low normalization point and are complemented, therefore, by the condition:

$$\alpha_s(\kappa) \sim 1 ,$$

which completes the set of parameters which are used throughout this paper to

* In the rest of the paper μ itself stands for a typical hadronic mass. But following a well-established tradition we denote here by μ the running normalization point, and use for this reason κ for the hadronic mass scale.

evaluate the resonance properties within QCD. (To be precise we use $\alpha_s = 0.7$ which corresponds to $\kappa = 0.2$ GeV if α_s is normalized as suggested by the J/ψ decays, $\alpha_s(2m_c) = 0.2$. The mass scale $\kappa \sim 200$ MeV emerges as a reasonable guess for the typical hadronic mass in a number of ways, see below.)

It would be important to check the validity of the parameters used in independent ways. Such possibilities do exist and provide further evidence in favor of the choice made.

In particular, $m_s = 150$ MeV seems reasonable as a scale of SU(3) symmetry breaking; the strange-quark mass is the only parameter in QCD which destroys the exact $SU(3)_{\text{flavor}}$. Namely, if one assumes that

$$\langle \Sigma | \bar{s}s | \Sigma \rangle \simeq \tfrac{1}{2} \langle \Xi | \bar{s}s | \Xi \rangle \,, \qquad (6.13)$$

$$\langle p | \bar{s}s | p \rangle \simeq 0 \,,$$

then $m_s = 150$ MeV gives a very reasonable fit to the observed SU(3) mass splittings [18].

Eqs. (6.13) amount in fact to identification of the constituent and current quarks at some low normalization point, i.e., it is close in spirit to the assumption (6.9).

Small quark masses serve first of all as a measure of the symmetry breaking and are not manifested, as a rule, in dynamical effects. The reason is that for confined quarks it is the quark virtuality, not mass, that counts. Thus, if the anomalous magnetic moment of the nucleon is of order e/m_N, for a light quark it is of order eR, where R is the radius of confinement.

It does not mean, however, that small quark masses are a purely conceptual device with no direct experimental consequences. The possibility of finding $m_{u,d}$ from the data rests on the fact that matrix elements of pseudoscalar and axial currents are related to each other through a quark mass:

$$\langle A | \partial_\mu j^5_\mu | B \rangle = m_q \langle A | j^5 | B \rangle$$

(see, e.g., eq. (6.10)). Unfortunately, there is no source of j_5 in nature. But the j_5 current emerges in theoretical studies in at least two ways:

(i) First, amplitudes governed by short-distance dynamics are reduced to matrix elements of various local operators. In particular, the operators

$$j^5 j^5 \qquad \text{or} \qquad j^s j^s \qquad (j^5 = \bar{\psi} i \gamma_5 \psi \,, j^s = \bar{\psi}\psi) \,,$$

can arise compensating in this way for the absence of an observable j_5. Any attempt to estimate the matrix elements in this case is crucially dependent on the quark masses.

Just such a situation is realized for weak non-leptonic decays and was discussed in detail in ref. [32]. The whole set of the parameters used in these papers coincides with that quoted above. The results turn out to be quite encouraging. Both the $\Delta I = \frac{1}{2}$ and $\Delta I = \frac{3}{2}$ amplitudes were calculated for K-meson and hyperon decays.

In all the cases the theoretical predictions agree with the data within a factor of 1.5. If one keeps in mind that this type of calculation includes matching of short- and large-distance techniques, the accuracy achieved goes far beyond any sceptic's guess.

Moreover, the $\Delta I = \frac{1}{2}$ matrix elements are extremely sensitive to the quark masses and their smallness provides a qualitative explanation for the $\Delta I = \frac{1}{2}$ selection rule.

Thus, weak non-leptonic interactions set a "precedent" for the accuracy one can expect using the parameters quoted above.

(ii) The sum rules themselves can serve as a tool to find the quark masses, e.g., one can introduce a pseudoscalar current and construct the sum rules for the corresponding spectral density. We shall discuss these sum rules in a subsequent publication. Here we note only that the sum rules do indicate the smallness of the quark masses. It is difficult, however, to achieve high accuracy in their determination.

6.4. Coupling constant α_s

It is worth emphasizing that we assume in fact that the coupling constant is rather small at almost all distances relevant to hadronic physics. To be more quantitative, our choice corresponds to normalizing the constant to unity at momenta comparable to the pion mass:

$$\alpha_s(0.2\ \mathrm{GeV}) \sim 1\ . \qquad (6.14)$$

(For those readers who are used to the parametrization $\alpha_s(Q^2) = 4\pi/(b \ln(Q^2/\Lambda^2))$, we note that eq. (6.14) corresponds to

$$\Lambda \sim 100\ \mathrm{MeV}\ .)$$

Most theoreticians would prefer a higher normalization point [33] * and therefore we sketch the argument in favor of our choice.

First, eq. (6.14) is consistent with the observed smallness of the J/ψ hadronic width and can be reconstructed in fact from $\alpha_s(2m_c) = 0.2$ which follows from the application of the Appelquist-Politzer recipe to J/ψ [16]. It also provides at least qualitative understanding of the observed smallness of the $\omega\varphi$ mixing. Moreover, the choice (6.14) is favored by successful calculation of the matrix elements for the weak non-leptonic decays mentioned above.

It is also worth noting that in most recent times the choice of a low value of α_s has got support from numerical exercises in instanton physics. Thus, according to ref. [13], power corrections become important for $\alpha_s \sim \frac{1}{3}$ so that asymptotic freedom cannot be extended beyond this point. We keep the power corrections explicit in our calculations (to avoid misunderstanding let us remark that the definition of α_s in our paper differs from that of ref. [13] by a factor $(2\pi)^{-1}$).

* Estimates of α_s obtained by fitting data on deep inelastic scattering can be found for example in ref. [34].

In principle the coupling constant α_s is subject to a direct experimental determination through measurements of the total cross section of e^+e^- annihilation into hadrons. According to the current data the cross section, say, just below the J/ψ production is rather high and this would imply rather large α_s [33], the corollary reached by most of the theoreticians. As is well-known, however, the present experimental data are qualified for possible systematic errors and this uncertainty is just crucial for the α_s determination. Thus, we would prefer to wait for better accuracy to make the final judgement on the value of α_s *.

If the choice (6.14) is correct, then the experimental cross section at energies of about 2 GeV is expected to exceed the simple-minded quark counting by less than 10%.

6.5. Matrix elements $\langle 0|\bar\psi\Gamma_1\psi\bar\psi\Gamma_2\psi|0\rangle$

To estimate these matrix elements we will reduce them to the square of the vacuum expectation value for $\bar\psi\psi$. This corresponds to retaining the vacuum intermediate state in all the channels and neglecting the contribution of all the other states (the assumption to be substantiated below).

In this approximation we have

$$\langle 0|\bar\psi\Gamma_1\psi\,\bar\psi\Gamma_2\psi|0\rangle = N^{-2}[(\mathrm{Tr}\,\Gamma_1\,\mathrm{Tr}\Gamma_2) - \mathrm{Tr}(\Gamma_1\Gamma_2)]\langle 0|\bar\psi\psi|0\rangle^2\,, \tag{6.15}$$

where the normalization factor N is defined as

$$\langle 0|\bar\psi_A\psi_B|0\rangle = \frac{\delta_{AB}}{N}\langle 0|\bar\psi\psi|0\rangle\,, \tag{6.16}$$

and the subscripts A,B include spin, color and flavor. For example, in the case of SU(3) symmetry $N = 36$ $(36 = 4\times 3\times 3)$ and $\langle 0|\bar\psi\psi|0\rangle = \langle 0|\bar uu + \bar dd + \bar ss|0\rangle$. If SU(3) breaking is taken into account explicitly, then the flavor indices are not included and

$$\langle 0|\bar q_A q_B|0\rangle = \frac{\delta_{AB}}{N}\langle 0|\bar qq|0\rangle\,,$$

with $N = 3\times 4 = 12$.

Let us give a few examples of eq. (6.15):

$$\langle 0|\bar q\gamma_\mu\gamma_5 t^a q\bar q\gamma_\mu\gamma_5 t^a q|0\rangle = -\langle 0|\bar q\gamma_\mu t^a q\bar q\gamma_\mu t^a q|0\rangle = \tfrac{16}{9}\langle 0|\bar qq|0\rangle^2\,,$$

$$\langle 0|\bar q\sigma_{\mu\nu}\gamma_5 t^a q\bar q\sigma_{\mu\nu}\gamma_5 t^a q|0\rangle = \tfrac{16}{3}\langle 0|\bar qq|0\rangle^2\,. \tag{6.17}$$

A few words are now in order on the validity of the approximation considered. It is worth noting first, that the vacuum state dominance is widely used in many-body physics [35] if a symmetry is spontaneously broken (e.g., the Cooper pair condensate).

* See note added in proof.

In our case the dominance of the vacuum intermediate state is due to:

(a) rather large value of $\langle 0|\bar{\psi}\psi|0\rangle$,

(b) duality between the quark and physical states.

To illustrate the point, we will discuss other contributions which we neglect in eq. (6.15), e.g., for the one-pion intermediate state the coupling constant is fixed while the integral over the pion momentum diverges quadratically. Introducing a cut-off, Λ_π, one can readily find that the pion contribution equals that of the vacuum if

$$\Lambda_\pi^2 \sim 2\ \mathrm{GeV}^2\ .$$

Clearly, such a choice of cut-off is hardly acceptable. Indeed, by definition only the effects that go beyond standard perturbation theory are included into the matrix elements $\langle 0|\bar{\psi}\Gamma_1\psi\bar{\psi}\Gamma_2\psi|0\rangle$. As for the quarks and gluons they are accounted for explicitly. From the experimental data on deep inelastic scattering we learn, however, that at $Q^2 = 2\ \mathrm{GeV}^2$ we are inside the "quark territory" and to keep the contribution of an "elementary" pion is quite senseless at such momenta.

Thus, we must introduce a lower cut-off. As we shall see below, the sum rules based on asymptotic freedom work down to $Q^2 \sim m_\rho^2$ and, therefore, m_ρ^2 constitutes a reasonable upper bound on Λ_π^2:

$$\Lambda_\pi^2 \lesssim 0.6\ \mathrm{GeV}^2\ .$$

Then the pion contribution is at least four times lower than that of the vacuum. Thus we see that the quantity $|\langle 0|\bar{q}q|0\rangle| \simeq (0.25\ \mathrm{GeV})^3$ is in fact quite large despite the apparent smallness of the scale involved (0.25 GeV). We shall come back to this point later on.

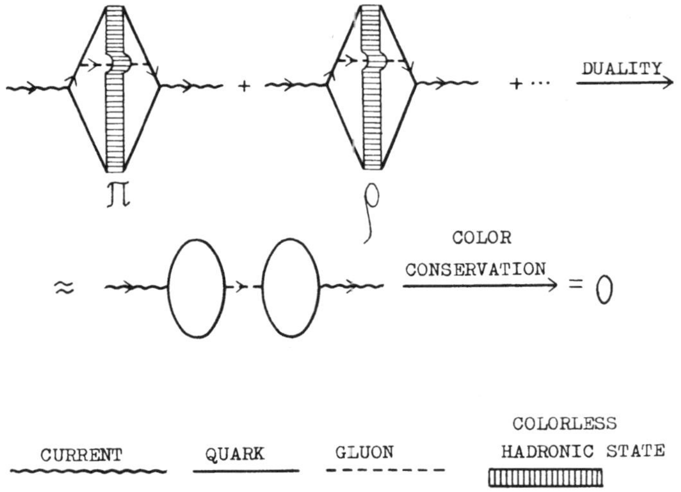

Fig. 8. The non-vacuum contribution to the $\bar{q}\Gamma q\bar{q}\Gamma q$ term in the operator expansion. Arrows mark the lines with a large virtual momentum.

Argument (b) above implies that the pion contribution is in fact suppressed somewhat further. Indeed, duality implies that counting both the quark and physical states may well become a double counting since they reproduce each other, if an averaging over some energy is performed [36].

The pion can readily be seen to be dual (along with other physical states) to a quark graph which vanishes by virtue of color conservation (see fig. 8).

Thus, we must keep in the matrix element only the contribution which violates duality. Most probably this implies a low cut-off of order 0.2 GeV. Indeed, we shall encounter this mass scale as typical for hadron physics many times and see no reason to reject it now. Moreover, as argued in (II), sum rules for $\omega\varphi$ mixing give an opportunity to check the guess in a direct way.

To summarize, the pion contribution is seemingly very much smaller that that of the vacuum state and constitutes, say, $\frac{1}{20}$ of the latter. Taking the safe upper bound $\Lambda_\pi^2 \sim m_\rho^2$ still suppresses the pion contribution by a factor of 4 relative to the vacuum state. This is sufficient for our purposes since we are aimed at an estimate which is correct within a factor of 2.

6.6. How large are the power corrections?

Thus, we have demonstrated that the vacuum expectation values are in some sense large: a high cut-off on other states is needed to make a comparable contribution. Since numerically, on the other hand, $\langle 0|\bar{q}q|0\rangle \simeq -(0.25\ \text{GeV})^3$, it might worth reiterating the argument.

The "largeness" of $\langle 0|\bar{q}q|0\rangle$ is due to hidden numerical factors like $(2\pi)^{-3}$ which enter the phase space for any state except for that of vacuum.

To be more quantitative, let us assume that the non-perturbative effects lead to a complete cancellation of the standard contribution coming from an ordinary Feynman graph up to some cut-off p_{int}, while at higher momenta perturbation theory stays untouched.

In other words, since the vacuum expectation values reflect the modification of ordinary graphs at low virtuality (see subsect. 2.3 and fig. 9), let us normalize the effect to perturbation theory.

One can easily find that the cut-off needed to imitate the power corrections due

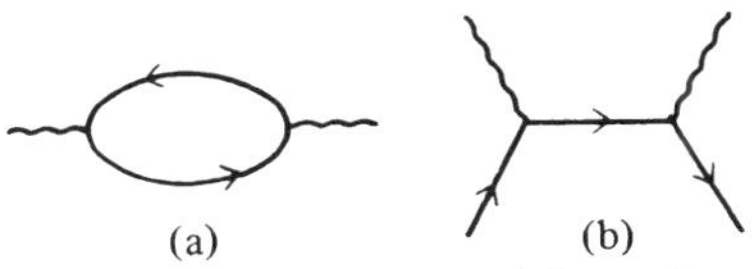

Fig. 9. (a) A loop graph of perturbation theory containing a phase-space factor $\sim\pi^{-2}$. (b) The same graph with a cut line determines the $m_q\bar{q}q$ term in the operator expansion. The π^{-2} factor does not appear explicitly.

to $\langle 0|\bar{q}q|0\rangle$, $\langle 0|G^2|0\rangle \neq 0$, is indeed high:

$$p_{\rm int}^2 \sim 3\ {\rm GeV}^2\ ,$$

which is a resemblance of the high cut-off discussed in subsect. 6.5 and non-acceptable for the same reasons.

Since the only place left for the quark and gluon propagators to deviate from asymptotic freedom is at low momenta, the deviation must be much more violent than a mere modification of perturbation theory by order unity.

We shall demonstrate in paper (II) that this conclusion, which is based on purely phenomenological analysis, is in amusing correspondence with the instanton-based picture of the non-perturbative effects.

In conclusion, let us notice that the hypothesis of the vacuum intermediate state dominance, like all the others which we introduce here, leads to some direct experimental consequences and is subject to an independent check in this way. In particular, it implies $\rho\omega$ degeneracy which is well-established experimentally (see (II) for details).

6.7. Dilute instanton-gas approximation: matrix elements $\langle 0|G^a_{\mu\nu}G^a_{\mu\nu}|0\rangle$, $\langle 0|f_{abc}G^a_{\mu\nu}G^b_{\nu\sigma}G^c_{\sigma\mu}|0\rangle$

So far we have discussed evaluation of matrix elements by saturating them by the intermediate vacuum state. We find the foundation of the approximation rather solid and will use the results for the four-fermion operators obtained in this way without further reservation. Unfortunately, the technique is not universally applicable. In particular, the gluonic operators $G^a_{\mu\nu}G^a_{\mu\nu}$, $f_{abc}G^a_{\mu\nu}G^b_{\nu\sigma}G^c_{\sigma\mu}$ cannot be treated in this way (f_{abc} here are the SU(3) color structure constants).

Here we utilize another approximation, that is the dilute instanton-gas approximation elaborated in ref. [13]. The use of the approximation is qualified for important theoretical uncertainties, and we shall turn back to their discussion at the end of the subsection. First let us demonstrate the technique with the example of the matrix elements specified in the title of the subsection. Start with the Belavin-Polyakov-Schwartz-Tyupkin solution (3.4).

As far as the instanton size is small, one can use the dilute instanton gas approximation [13]. In this approximation the matrix elements under consideration reduce to the one-instanton (+ anti-instanton) contribution. Using the explicit expression (3.4) and performing the integration over the instanton position we come to

$$\langle 0|g_s^2 G^a_{\mu\nu}G^a_{\mu\nu}|0\rangle_{\rm inst.+anti\text{-}inst.} = 2^6\pi^2 \int_0^{\rho_c} \frac{d\rho}{\rho^5}\, d(\rho)\ , \tag{6.18}$$

$$\langle 0|g_s^3 f_{abc}G^a_{\mu\nu}G^b_{\nu\sigma}G^c_{\sigma\mu}|0\rangle_{\rm inst.+anti\text{-}inst.}$$

$$= \frac{3 \cdot 2^8 \cdot \pi^2}{5} \int_0^{\rho_c} \frac{d\rho}{\rho^7} d(\rho) , \qquad (6.19)$$

where g_s is the quark-gluon coupling constant, $d(\rho)$ is the instanton density function [13]. The integrals in (6.18), (6.19) are infrared divergent and the cut-off ρ_c is introduced for this reason.

Now, $d(\rho)$ is given in eq. (3.7). The numerical constant entering this equation depends on the α_s renormalization procedure. It is typically of order 10^{-1} [13,14]. Note that all the numbers here correspond to pure gluodynamics with $SU(3)_{color}$.

Eqs. (6.19), (6.18) give

$$\langle 0 | g_s^3 f_{abc} G^a_{\mu\nu} G^b_{\nu\sigma} G^c_{\sigma\mu} | 0 \rangle = \tfrac{12}{5} \langle \rho^{-2} \rangle \langle 0 | g_s^2 G^a_{\mu\nu} G^a_{\mu\nu} | 0 \rangle , \qquad (6.20)$$

where $\langle \rho^{-2} \rangle$ is the mean value of ρ^{-2} as weighted with the instanton density function. Since the integrands are peaked at the upper limit of integration $\langle \rho^{-2} \rangle \simeq \rho_c^{-2}$.

Thus, the value of the cut-off, ρ_c, is a key parameter for all the instanton estimates. Moreover, the matrix elements in point are proportional to a high power of ρ_c. Therefore, any rough estimate of ρ_c is of little value since the corresponding uncertainty in the matrix elements is enormous.

For this reason we would prefer the other way: find one of the matrix elements from independent sources, fix ρ_c and evaluate further matrix elements.

In particular, one can find $\langle 0 | g_s^2 G^a_{\mu\nu} G^a_{\mu\nu} | 0 \rangle$ from the sum rules for charmonium decays

$$\left\langle 0 \left| \frac{\alpha_s}{\pi} G^a_{\mu\nu} G^a_{\mu\nu} \right| 0 \right\rangle \simeq 0.012 \text{ GeV}^4 , \qquad (6.21)$$

(see (II) and ref. [6]). The corresponding value of ρ_c is [10]

$$\rho_c \sim 1/200 \text{ MeV} , \qquad (6.22)$$

and

$$\langle 0 | g_s^3 f_{abc} G^a_{\mu\nu} G^b_{\nu\sigma} G^c_{\sigma\mu} | 0 \rangle \simeq 0.045 \text{ GeV}^6 \simeq (0.59 \text{ GeV})^6 . \qquad (6.23)$$

The reference to the experimental data to fix $\langle 0 | g_s^2 G^a_{\mu\nu} G^a_{\mu\nu} | 0 \rangle$ is not in fact consistent with the logic of the present section: our plan is to fix only $\langle 0 | \bar{q} q | 0 \rangle$ from the data and reduce all the other matrix elements to this quantity. In subsect. 6.8 we shall remove this inconsistency by computing $\langle 0 | g_s^2 G^2 | 0 \rangle$ theoretically in terms of $\langle 0 | \bar{q} q | 0 \rangle$. This is achieved at the price of another approximation, however, and we would like to isolate eq. (6.21) which is on firmer basis than theoretical speculations presented in subsect. 6.8.

There are apparent weaknesses of the one-instanton approximation. The need for a cut-off seems to be the most dangerous. Thus, the approximation is no better than, say, an evaluation of the weak interaction *via* a four-fermion theory with a cut-off. We will demonstrate in paper II that the similarity extends rather far and the result depends on the way the cut-off is introduced.

Another well-known disease is the lack of understanding of the effect of the light quarks. Formally, the instanton density vanishes in the presence of massless quarks. One may hope, however, that due to spontaneous symmetry breaking, quarks are not so important for the matrix elements controlled by instantons of large size.

Finally, let us notice that the instanton-based estimates can be tried for a limited number of matrix elements, since the matrix elements of the type $\langle 0|G^n|0\rangle$ with high n are ultraviolet divergent.

Still, the estimates can be useful. Indeed, the cut-off is quite high, $\rho_c \sim$ 1/200 MeV, and it is difficult to imagine that larger distances modify the answer drastically. Moreover, relations between the matrix elements like eq. (6.20) have a better chance of withstanding the modifications provided by a complete theory.

6.8. Matching of light and heavy quarks

Light and heavy quarks have been considered in different ways so far. For light quarks we expand in the quark mass. For heavy quarks we choose external momentum $Q^2 \sim 0$ and expand in the inverse quark mass. Imagine that there exists a quark of an intermediate mass which is neither light nor heavy. Strictly speaking we cannot use any of the techniques. An optimist, however, could try to reverse the statement and say that for a fictitious quark of an intermediate mass both approximations are approximately valid and lead to similar results.

The world of heavy quarks is simpler in the sense that there is no need to introduce independent vacuum expectation values of the type $\langle 0|\bar{h}h|0\rangle$, where h is a quark of heavy mass $m_{\rm h}$. Indeed, it is reduced to the matrix elements of gluonic operators. Explicitly, we have in the first approximation (see fig. 10):

$$\langle 0|\bar{h}h|0\rangle = -\frac{1}{12\,m_{\rm h}}\left\langle 0\left|\frac{\alpha_s}{\pi}G^a_{\mu\nu}G^a_{\mu\nu}\right|0\right\rangle + \ldots , \tag{6.24}$$

which is valid as far as higher orders in $m_{\rm h}^{-1}$ are negligible.

Now, let us try to extrapolate eq. (6.24) to lower mass. Then the matrix element $\langle 0|\bar{h}h|0\rangle$ increases until we come to the point where terms of higher order in $m_{\rm h}^{-1}$ neglected so far become important. At this point little can be said for certain. But we know that once we cross the boundary of the world of the light quarks, $\langle 0|\bar{h}h|0\rangle$ becomes mass independent.

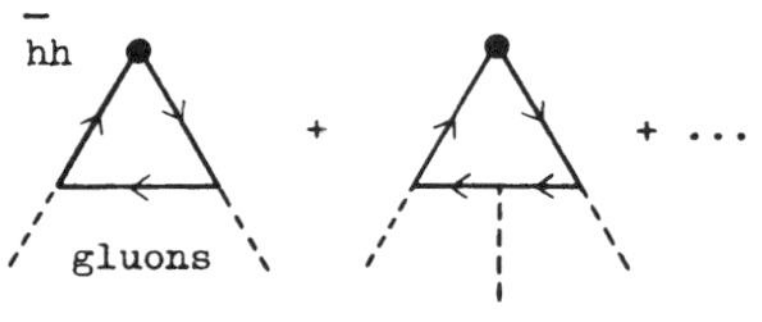

Fig. 10. The heavy quark expansion for $\bar{h}h$.

Moreover, the breaking of eq. (6.24) is due to power corrections so that the change in the regime is hopefully fast. Thus, we would like to speculate that the light and heavy quarks match smoothly. More specifically, the approximation amounts to saying that eq. (6.24) can serve for the purpose of an estimate, if the left-hand side is evaluated in fact for a light quark while the right-hand side is taken at some transition point $m_{\rm h}$ which marks the boundary of the world of heavy quarks.

We see that the guess is right in sign, to say the least. Indeed, the $\langle 0|G^a_{\mu\nu}G^a_{\mu\nu}|0\rangle$ is positive on general grounds (see subsect. 6.9) while $\langle 0|\bar{q}q|0\rangle$ is negative as follows from the phenomenological analysis presented in previous subsections.

Now, as to the absolute magnitude. The prediction for the expectation value of $G^a_{\mu\nu}G^a_{\mu\nu}$ depends on the critical value of $m_{\rm h}$ for which eq. (6.24) is still valid.

A consistent way of determining the critical value of $m_{\rm h}$ would be to compute next orders in $m_{\rm h}^{-1}$ and find the point where they become appreciable. This seems to be a tractable problem and the corresponding calculations are in progress (they are being performed by Novikov and the present authors).

Here we would like to rely more on intuition. The strange quark mass is about 150 MeV. The corresponding hadron scale is represented by $m_{\rm K}^2$. It is not too small, indeed, so that we are rather close to the boundary of heavy quarks. Moreover, all the numerical studies in the sum rules (see (II) for details) indicate that at m_ρ^2 the expansion in the inverse mass is valid and higher orders in the operator expansion can be neglected. Thus, the critical mass is somewhere between $m_{\rm K}^2$ and m_ρ^2.

Therefore, $m_{\rm h} \sim 0.2$ GeV seems to be a reasonable guess for the critical value of $m_{\rm h}$ which can be still treated as "heavy". In this way we come to

$$\left\langle 0\left|\frac{\alpha_s}{\pi}G^a_{\mu\nu}G^a_{\mu\nu}\right|0\right\rangle \sim -12\, m_{\rm h}\langle 0|\bar{h}h|0\rangle\Big|_{m_{\rm h}=200\,{\rm MeV}} \sim 0.03\ {\rm GeV}^4\ . \qquad (6.25)$$

In (II) we will determine the same matrix element phenomenologically and the corresponding value has already been mentioned above (see eq. (6.21)). It differs from the rough estimate by a factor of 2.5 which is indeed not bad.

Thus, we can rely on this technique as far as rough estimates of various matrix elements are concerned. To give perspective, let us mention that matching of light and heavy quarks allows one to estimate $\langle 0|\bar{q}\,\sigma_{\mu\nu}t^a qG^a_{\mu\nu}|0\rangle$. This matrix element is crucial for analysing the sum rules for charmed mesons. The analysis is being performed in collaboration with Novikov.

6.9. On the signs of the matrix elements

In the previous subsections we have proposed a number of techniques for estimating the vacuum-to-vacuum matrix elements. In particular, we have found that

$$\langle 0|G^a_{\mu\nu}G^a_{\mu\nu}|0\rangle > 0 \qquad \text{and} \qquad \langle 0|\bar{q}\gamma_\mu t^a q\bar{q}\gamma_\mu t^a q|0\rangle < 0\ .$$

(see eqs. (6.21) and (6.17), respectively).

However, all the estimates explored so far are model dependent. In this subsection we will argue that the signs just mentioned can be established in fact on very general grounds. The only assumption needed is the possibility to perform the averaging over the vacuum state by means of the functional integration in the Euclidean space-time. This is indeed a very plausible assumption. Moreover, the formulation of the theory in Euclidean space-time may be considered as primary nowadays.

Consider first $G^a_{\mu\nu}G^a_{\mu\nu}$. In Minkowski space-time it is not a positive definite quantity since

$$G^a_{\mu\nu}G^a_{\mu\nu} = \sum_{\substack{i,k=1,2,3\\ i>k}} (-2G^a_{0i}G^a_{0i} + 2G^a_{ik}G^a_{ik}) , \qquad \text{(Minkowski)} .$$

However, written in Euclidean space it does become positive definite:

$$G^a_{\mu\nu}G^a_{\mu\nu} = \sum_{\substack{i,k=1,2,3\\ i>k}} (2G^a_{0i}G^a_{0i} + 2G^a_{ik}G^a_{ik}) , \qquad \text{(Euclidean)} .$$

Therefore, representing the vacuum-to-vacuum matrix element as a functional integral we find that it is positive definite:

$$\langle 0|G^a_{\mu\nu}G^a_{\mu\nu}|0\rangle = \frac{\int \mathcal{D}A \ldots \mathrm{e}^{-S} \sum (2G^a_{0i}G^a_{0i} + G^a_{ik}G^a_{ik})}{\int \mathcal{D}A \ldots \mathrm{e}^{-S}} > 0 .$$

To consider the four-fermion operator mentioned above let us use first the equation of motion:

$$\bar{q}\gamma_\mu t^a q = -\frac{2}{g_s} \mathcal{D}_\mu G^a_{\mu\nu} ,$$

which reduces the four-fermion operator to the gluon fields:

$$(\bar{q}\gamma_\mu t^a q)(\bar{q}\gamma_\mu t^a q) = \frac{4}{g_s^2}(\mathcal{D}_\mu G^a_{\mu\nu})(\mathcal{D}_\alpha G^a_{\alpha\nu})$$

$$\xrightarrow{\text{Euclid}} -(\mathcal{D}_j G^a_{j0})^2 - (\mathcal{D}_0 G^a_{0i} + \mathcal{D}_j G^a_{ji})^2 < 0 .$$

When we saturate the matrix element of this operator by the vacuum intermediate state, the minus sign emerges by virtue of the Fierz transformation. Since we cannot introduce the color intermediate states the agreement in signs between the model estimates and general equations does not seem to be trivial and is gratifying, therefore.

6.10. Conclusions

In the first part of the present section we have argued in favor of light quarks, $m_{u,d} \sim 5$ MeV, $m_s \sim 150$. These low masses become more and more conventional nowadays, so that presentation is partly of a review kind. Still we hope that the discussion of the value of the coupling constant which complements the choice of the mass may be new to some of the readers. The same is possibly true for the reference to the non-leptonic decays as a support for the choice of parameters favored in this paper.

In sect. 6 we proposed several techniques for rough estimates of the matrix elements relevant to the sum rules. It might worth emphasizing that even an estimate within a factor of two is very valuable for our purposes because the corresponding uncertainty in predicting the resonance properties turn out to be 10–15% (see II)). Our feeling is now that it is possible to get an estimate of this kind practically for any matrix element.

From a more general point of view, this section completes the discussion of the general problems of the approach proposed. Indeed, we started with a proof that it is possible to use the Wilson operator expansion as far as a few first terms in the Q^{-2} expansion are concerned. Then we evaluated explicitly the coefficients in the two-point functions of various currents. Then we substantiated the use of the first Borel transform of the polarization operator as most convenient to study resonances within QCD. Finally, in this section we estimated the vacuum-to-vacuum matrix elements which specify the sum rules completely. Now everything is ready for applications.

Note added in proof

In subsect. 6.4 we have called for better accuracy in measuring the e^+e^- annihilation cross section at relatively low energies. Due to the efforts of the Novosibirsk, Orsay and Frascati groups such data do exist now. The data have been analyzed by Eidelman et al. (to be published), and the analysis confirms our guess that asymptotic freedom is violated by the power corrections rather than by higher orders in α_s.

Appendix

Four-fermion operators

This appendix is devoted to summation of the $(\alpha_s \ln(Q^2/\mu^2))^n$ terms in the operator expansion coefficients corresponding to the four-fermion operators of the type $O_\Gamma = \bar{\psi}\Gamma\psi\bar{\psi}\Gamma\psi$. Many of the results are already known but are scattered in numerous publications [37], so we find it helpful to collect them all in one place. The whole procedure is rather standard so we only sketch the derivation. In subsect. A.1. all independent structures are listed and classified. In subsect. A.2 we write out anomalous dimension matrices and solve the corresponding renormalization-group equations. Subsect. A.3 presents the final results.

A.1. Operators

First of all it is necessary to give proper names to various four-fermion opera- tors which were previously all called O_Γ. In considering the T-product of vector and axial-vector currents the following independent operators are encountered:

$$\left.\begin{aligned}
P_1 &= \bar\psi_L\gamma_\mu\psi_L\bar\psi_R\gamma_\mu\psi_R\,,\\
P_2 &= \bar\psi_L\gamma_\mu t^a\psi_L\bar\psi_R\gamma_\mu t^a\psi_R\,,\\
P_3 &= \bar\psi_L\gamma_\mu\psi_L\bar\psi_L\gamma_\mu\psi_L + (L\to R)\,,
\end{aligned}\right\}(1.1)$$
$$P_4 = \bar\psi_L\gamma_\mu t^a\psi_L\bar\psi_L\gamma_\mu t^a\psi_L + (L\to R)\,, \qquad \text{(A.1)}$$
$$\left.\begin{aligned}
P_5 &= \bar\psi_L\lambda^a t^b\gamma_\mu\psi_L\bar\psi_R\lambda^a t^b\gamma_\mu\psi_R\,,\\
P_6 &= \bar\psi_L\lambda^a\gamma_\mu\psi_L\bar\psi_R\lambda^a\gamma_\mu\psi_R\,.
\end{aligned}\right\}(8.8)$$

Below we will discuss them in detail. For completeness we also write out here the operators involved in the expansion for scalar and pseudoscalar densities:

$$\begin{aligned}
P_7 &= \bar\psi_R\psi_L\bar\psi_R\psi_L + (L\leftrightarrow R)\,,\\
P_8 &= \bar\psi_R t^a\psi_L\bar\psi_R t^a\psi_L + (L\leftrightarrow R)\,,\\
P_9 &= \bar\psi_R\lambda^a\psi_L\bar\psi_R\lambda^a\psi_L + (L\leftrightarrow R)\,,\\
P_{10} &= \bar\psi_R\lambda^a t^b\psi_L\bar\psi_R\lambda^a t^b\psi_L + (L\leftrightarrow R)\,.
\end{aligned} \qquad \text{(A.2)}$$

In the above equations ψ denotes the fermion field carrying both color ($i, j, k, ... =$ 1, 2, 3) and SU(3) flavor ($\alpha, \beta, \gamma, ... = 1, 2, 3$) indices. The summation is implicit; thus for example

$$\bar\psi\gamma_\mu\psi = \bar u_i\gamma_\mu u^i + \bar d_i\gamma_\mu d^i + \bar s_i\gamma_\mu s^i\,.$$

Furthermore t^a and λ^a stand for the Gell-Mann SU(3) matrices acting in color and flavor spaces respectively. (They are normalized by the condition $\mathrm{Tr}(\lambda^a\lambda^b) = \mathrm{Tr}(t^at^b) = 2\delta^{ab}$.) The subscript L (R) labels the left- (right-) handed spinors,

$$\psi_L = \tfrac12(1+\gamma_5)\,\psi\,, \qquad \psi_R = \tfrac12(1-\gamma_5)\,\psi\,.$$

We indicated in parenthesis in eq. (A.1) the properties of the operators with respect to the $SU(3)_R \otimes SU(3)_L$ symmetry. We included in the list only the $SU(3)_{\text{flavor}}$ singlet operators since only these have non-vanishing vacuum expectation values (under the assumption of the vacuum $SU(3)_{\text{flavor}}$ symmetry). For completeness we write down a few formulae often used in working with four-fermion operators.

The Fierz transformations

$$\begin{aligned}
\bar\psi_{1L}\gamma_\mu\psi_{2L}\bar\psi_{3L}\gamma_\mu\psi_{4L} &= \bar\psi_{1L}\gamma_\mu\psi_{4L}\bar\psi_{3L}\gamma_\mu\psi_{2L}\,,\\
\bar\psi_{1L}\gamma_\mu\psi_{2L}\bar\psi_{3R}\gamma_\mu\psi_{4R} &= -2\bar\psi_{1L}\psi_{4R}\bar\psi_{3R}\psi_{2L}\,.
\end{aligned} \qquad \text{(A.3)}$$

(Here the subscripts 1, 2, ... in ψ_{1L}, ψ_{2L}, ... stand symbolically for both color and flavor, e.g. $1 \equiv \{i_1, \alpha_1\}$, etc.).

Relation for the Gell-Mann matrices

$$t^a_{ij} t^a_{mn} = 2\delta_{in}\delta_{jm} - \tfrac{2}{3}\delta_{ij}\delta_{mn} \,. \tag{A.4}$$

A.2. Mixing matrices

In the operator expansion there emerge operators normalized at a (running) point Q. The problem is to express them in terms of operators normalized at a (fixed) point μ, where μ is of the order of the inverse confinement radius. First, our knowledge of matrix elements, if any, refers just to this point. Second, the entire Q dependence must be exhibited explicitly.

In the leading logarithmic approximation the standard renormalization-group technique allows one to change easily one normalization point to another. All we have to know is the mixing matrix in the one-loop approximation.

In a slightly different language our procedure is as follows. When operators (A.1) are "dressed" with gluons, there arise logarithmic corrections of the type $(\alpha_s \ln(Q^2/\mu^2))^n$ which contain an explicit Q dependence and call for summation. The summation can be performed either diagrammatically, after the corresponding diagram selection, or by means of the renormalization group.

Below we will use the anomalous dimension language which seems more compact.

Notice that the operators P_{1-4} and $P_{5,6}$ possess different selection rules with respect to $SU(3)_R \otimes SU(3)_L$; therefore under the mixing they split into two distinct groups, P_1, P_2, P_3, P_4 and P_5, P_6.

In principle, the operators P_1–P_4 with the vacuum quantum numbers can mix also with a pure gluon operator O_f:

$$O_f = g_s^3 f^{abc} G^a_{\mu\nu} G^b_{\nu\sigma} G^c_{\sigma\mu} \,, \qquad (d = 6) \,.$$

However, the effect is important only for the O_f coefficient since the O_f-O_Γ feedback is weak; it is suppressed by an overall factor α_s^2 which enters along with the conventional $(\alpha_s \ln(Q^2/\mu^2))^n$. An example of the diagram responsible for the O_Γ-O_f mixing is given in fig. 11. The three-loop structure of the diagram results in a numerical smallness which makes the effect completely inessential.

A few technical remarks are in order now. There are two completely distinct types of one-loop graph, call them conventional (fig. 12a) and annihilational (fig. 12b). In fig. 12 we denoted the four-fermion operator by a closed circle and slightly split the fermion lines in order to indicate explicitly which fermion line corresponds to this or that fermion bracket. (Each fermion bracket $(\bar\psi\gamma_\mu\psi)$, or $(\bar\psi\gamma_\mu\lambda^a\psi)$ or $(\bar\psi\gamma_\mu\lambda^a t^b\psi)$ is denoted by a single solid line without cuts.)

The gluon propagator $1/k^2$ in the annihilational graphs is cancelled out since the

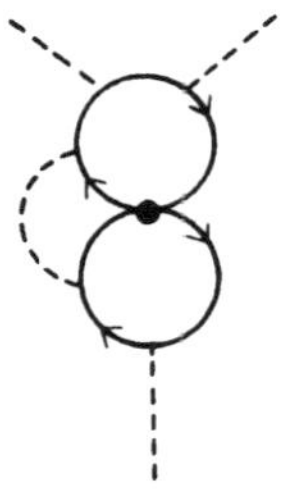

Fig. 11. An example of the diagram responsible for the mixing of the four-fermion operators O_Γ (closed circle) with the three-gluon operator $O_f = g_s^3 f_{abc} G^a_{\mu\nu} G^b_{\nu\sigma} G^c_{\sigma\mu}$.

adjacent quark loop is proportional to k^2. In other words, if a four-fermion operator is inserted into the annihilational diagram it yields the operator $\bar{\psi}\gamma_\mu t^a \psi \mathcal{D}_\nu G^a_{\mu\nu}$ where $\mathcal{D}_\nu$ is the covariant derivative and $G^a_{\mu\nu}$ is the gluon field strength tensor. The latter operator in turn reduces to a four-fermion form by virtue of the equation of motion:

$$\mathcal{D}_\mu G^a_{\mu\nu} = -\tfrac{1}{2} g_s \bar{\psi}\gamma_\nu t^a \psi \ .$$

Calculations are conveniently performed in the Landau gauge for a gluon field. In this gauge the anomalous dimension of a fermion field vanishes and one is left with the anomalous dimension of the local operators.

In the Landau gauge the following rules take place:

(i) a gluon line "dressing" a solid quark line (one fermion bracket) gives zero;

(ii) a gluon line attached to two distinct brackets gives zero for operators of the type $j_{\mu L} j_{\mu L}, j_{\mu R} j_{\mu R}$, if it connects the ingoing and the outgoing lines, and for operators of the type $j_{\mu L} j_{\mu R}$, if it connects two ingoing or two outgoing lines;

(iii) the annihilational graphs exist only for the operators P_2, P_3, P_4.

After these preliminary remarks we give the answer for the anomalous dimension

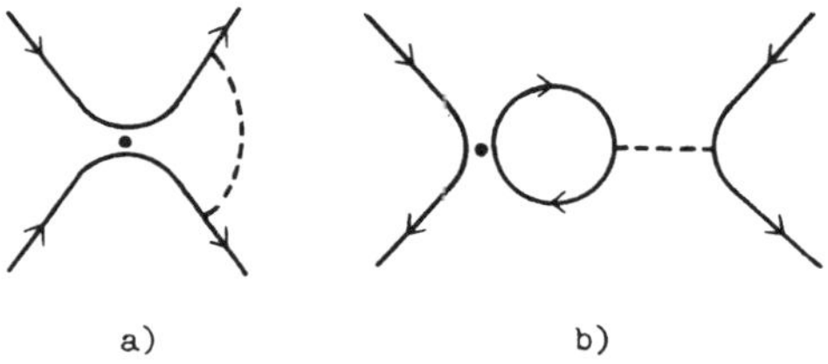

Fig. 12. Graphs determining the anomalous dimension matrix for the four-fermion operators: (a) a conventional graph; (b) annihilational graphs. The four-fermion operator is denoted by the closed circle.

matrix

$$\begin{bmatrix} P_1 \\ P_2 \\ P_3 \\ P_4 \end{bmatrix}_\Lambda = \left\{ \mathbf{1} + \frac{\alpha_s}{4\pi} \ln \frac{\Lambda^2}{\mu^2} \begin{bmatrix} 0 & \frac{3}{2} & 0 & 0 \\ \frac{16}{3} & 5 & 0 & -1 \\ 0 & -\frac{2}{3} & 0 & -\frac{11}{6} \\ 0 & -\frac{32}{9} & -\frac{16}{3} & \frac{2}{9} \end{bmatrix} \right\} \begin{bmatrix} P_1 \\ P_2 \\ P_3 \\ P_4 \end{bmatrix}_\mu , \tag{A.5}$$

where the subscripts μ, Λ stand for the normalization point of the operators.

The following combinations are diagonal:

$$\begin{aligned} D_1 &= P_1 - 0.213P_2 + 0.138P_3 + 0.029P_4 , && (\delta_1 = -1.134) , \\ D_2 &= 0.787P_1 + P_2 + 0.154P_3 - 0.196P_4 , && (\delta_2 = 6.774) , \\ D_3 &= -0.505P_1 - 0.238P_2 + P_3 - 0.560P_4 , && (\delta_3 = 2.989) , \\ D_4 &= -1.188P_1 + 0.759P_2 + 1.565P_3 + P_4 , && (\delta_4 = -3.407) . \end{aligned} \tag{A.6}$$

In parenthesis the corresponding eigenvalues of the mixing matrix are indicated, which just determine the anomalous dimensions. For example,

$$D_1(Q) = \kappa^{-1.134/b} D_1(\mu) . \tag{A.7}$$

Here

$$\kappa = \frac{\alpha_s(\mu)}{\alpha_s(Q)} = 1 + \frac{\alpha_s(\mu)}{4\pi} b \ln \frac{Q^2}{\mu^2} ; \tag{A.8}$$

b is the coefficient in the Gell-Mann-Low function $\beta(\alpha_s)$ for the effective charge,

$$\beta(\alpha_s) = -\frac{b\alpha_s^2}{2\pi} + O(\alpha_s^3) . \tag{A.9}$$

For three quark flavors (and we always work in the region where only three flavors are essential)

$$b = 9 . \tag{A.10}$$

The procedure of converting the operators $P_i(Q)$, normalized at Q, into the operators $P_i(\mu)$, normalized at μ, includes in fact two steps. At first one rewrites the operators P_i in terms of D_i. At this point it is convenient to introduce matrix notation. For example, eq. (A.6) can be rewritten as

$$[D] = [X]\,[P] ,$$

where $[D]$ and $[P]$ are the operator columns, and $[X]$ is the 4×4 matrix which can be easily read off eq. (A.6).

Then

$$[P] = [X^{-1}]\,[D] ,$$

where

$$[X^{-1}] = \begin{bmatrix} 0.784 & 0.170 & -0.080 & -0.035 \\ -0.593 & 0.769 & -0.160 & 0.078 \\ 0.534 & 0.048 & 0.495 & 0.271 \\ 0.546 & -0.457 & -0.749 & 0.475 \end{bmatrix} . \tag{A.11}$$

It is a trivial matter for the operators D_i to proceed from a normalization point Q to a point μ, since they are simply multiplied by $\kappa^{\delta_i/b}$ (see eq. (A.7)). The corresponding values of δ_i are given in eq. (A.6). In matrix form

$$[D(Q)] = [\kappa^{\delta/b}]\,[D(\mu)] \;,$$

where

$$[\kappa^{\delta/b}] = \begin{bmatrix} \kappa^{-1.134/b} & & & 0 \\ & \kappa^{6.774/b} & & \\ & & \kappa^{2.989/b} & \\ 0 & & & \kappa^{-3.407/b} \end{bmatrix} \tag{A.12}$$

Finally, the operators $D_i(\mu)$ are transformed into the operators $P_i(\mu)$ with the help of the $[X]$ matrix:

$$[P(Q)] = [X^{-1}]\,[\kappa^{\delta/b}]\,[X]\,[P(\mu)] \;. \tag{A.13}$$

This relation (together with eqs. (A.6), (A.11), (A.12)) is our final result for the Q dependence of the operators P_{1-4}. As for the operators $P_{5,6}$ the answer here is even much simpler. The mixing matrix looks like

$$\begin{bmatrix} P_5 \\ P_6 \end{bmatrix}_\Lambda = \left\{ \mathbf{1} + \frac{\alpha_s}{4\pi} \ln \frac{\Lambda^2}{\mu^2} \begin{bmatrix} 7 & \frac{16}{3} \\ \frac{3}{2} & 0 \end{bmatrix} \right\} \begin{bmatrix} P_5 \\ P_6 \end{bmatrix}_\mu , \tag{A.14}$$

and the diagonal combinations are:

$$D_5 = P_5 + \tfrac{2}{3} P_6 \;, \qquad (\delta_5 = 8) \;,$$

$$D_6 = -\tfrac{3}{16} P_5 + P_6 \;, \qquad (\delta_6 = -1) \;.$$

Thus

$$\begin{bmatrix} P_5 \\ P_6 \end{bmatrix}_Q = \begin{bmatrix} \frac{8}{9} & -\frac{16}{27} \\ \frac{1}{6} & \frac{8}{9} \end{bmatrix} \begin{bmatrix} \kappa^{8/b} & 0 \\ 0 & \kappa^{-1/b} \end{bmatrix} \begin{bmatrix} 1 & \frac{2}{3} \\ -\frac{3}{16} & 1 \end{bmatrix} \begin{bmatrix} P_5 \\ P_6 \end{bmatrix}_\mu . \tag{A.15}$$

A.3. Vacuum expectation values and final result

For the vacuum-to-vacuum matrix elements of the operators $P_{1-6}(\mu)$ we use the factroization hypothesis discussed in detail in sect. 4. The effective recipe is to insert the vacuum intermediate state in all possible channels, then $\langle 0|P_i|0\rangle$ are expressed in terms $\langle 0|\bar{\psi}\psi|0\rangle^2$. Namely,

$$\langle P_2(\mu)\rangle = \tfrac{16}{3}\langle P_1(\mu)\rangle = -\tfrac{8}{27}\langle\bar{\psi}\psi\rangle^2 ,$$

$$\langle P_3(\mu)\rangle = \langle P_4(\mu)\rangle = 0 ,$$

$$\langle P_5(\mu)\rangle = \tfrac{16}{3}\langle P_6(\mu)\rangle = -\tfrac{128}{81}\langle\bar{\psi}\psi\rangle^2 . \qquad \text{(A.16)}$$

We recall that $\langle\bar{\psi}\psi\rangle$ means $\langle 0|\bar{u}u + \bar{d}d + \bar{s}s|0\rangle$. Using (A.13), (A.15) and (A.16) one can finally find

$$\langle P_1(Q)\rangle = (-0.020\kappa^{-1.134/b} + 0.195\kappa^{6.774/b} + 0.030\kappa^{2.989/b} - 0.019\kappa^{-3.407/b})\langle P_2(\mu)\rangle ,$$

$$\langle P_2(Q)\rangle = (0.015\kappa^{-1.134/b} + 0.883\kappa^{6.774/b} + 0.060\kappa^{2.989/b} + 0.042\kappa^{-3.407/b})\langle P_2(\mu)\rangle ,$$

$$\langle P_3(Q)\rangle = (-0.014\kappa^{-1.134/b} + 0.055\kappa^{6.774/b} - 0.187\kappa^{2.989/b} + 0.145\kappa^{-3.407/b})\langle P_2(\mu)\rangle ,$$

$$\langle P_4(Q)\rangle = (-0.014\kappa^{-1.134/b} - 0.525\kappa^{6.774/b} + 0.283\kappa^{2.989/b} + 0.255\kappa^{-3.407/b})\langle P_2(\mu)\rangle ,$$

$$\langle P_5(Q)\rangle = \kappa^{8/b}\langle P_5(\mu)\rangle ,$$

$$\langle P_6(Q)\rangle = \tfrac{3}{16}\kappa^{8/b}\langle P_5(\mu)\rangle .$$

References

[1] D.J. Gross and F. Wilczek, Phys. Rev. Lett. 30 (1973) 1343;
H.D. Politzer, Phys. Rev. Lett. 30 (1973) 1346.

[2] A. Belavin, A. Polyakov, A. Schwartz and Yu. Tyupkin, Phys. Lett. 59B (1975) 85.

[3] V.N. Gribov, *in* Proc. 12th LINP Winter School of Physics, Leningrad, 1977, vol. 1, page 147 (English translation: SLAC-TRANS-176 (1977)); Nucl. Phys. B139 (1978) 1.

[4] M. Shifman, A. Vainshtein and V. Zakharov, Nucl. Phys. B147 (1979) 448.

[5] M. Gell-Mann, R. Oakes and B. Renner, Phys. Rev. 175 (1968) 2195.

[6] A. Vainshtein, V. Zakharov and M. Shifman, Pisma ZhETF (USSR) 27 (1978) 60.

[7] K. Wilson, Phys. Rev. 179 (1969) 1499.

[8] K. Symanzik, Comm. Math. Phys. 23 (1971) 49;
C. Callan, Phys. Rev. D5 (1972) 3302;
R. Brandt, Fortschr. Phys. 18 (1970) 249.

[9] C.G. Callan, R. Dashen and D.J. Gross, Phys. Rev. D17 (1978) 2717;
A. Polyakov, Nucl. Phys. B120 (1977) 429;
A. Belavin and A. Polyakov, Nucl. Phys. B123 (1977) 429;
G. 't Hooft, Phys. Rev. Lett. 37 (1976) 8; Phys. Rev. D14 (1976) 3432;
C.G. Callan et al., Phys. Lett. 63B (1976) 334; 66B (1977) 375;
R. Jackiw and C. Rebbi, Phys. Rev. Lett. 37 (1976) 172.

[10] M. Shifman, A. Vainshtein and V. Zakharov, Phys. Lett. 76B (1978) 471.
[11] M. Shifman, A. Vainshtein and V. Zakharov, Moscow preprint ITEP-54 (1978) Phys. Rev. Lett., submitted.
[12] V. Novikov, L. Okun, M. Shifman, A. Vainshtein, M. Voloshin and V. Zakharov, Moscow preprint ITEP-83 (1977); Phys. Reports 41 (1978) 1.
[13] C.G. Callan, R. Dashen and D.J. Gross, see ref. [9].
[14] N. Andrei and D.J. Gross, Phys. Rev. D18 (1978) 468.
[15] L. Baullieu, J. Ellis, M.K. Gaillard and W.J. Zakrzewski, Phys. Lett. 77B (1978) 290.
[16] T. Appelquist and H. Politzer, Phys. Rev. Lett. 34 (1975) 43.
[17] M. Shifman, A. Vainshtein, M. Voloshin and V. Zakharov, Phys. Lett. 77B (1978) 80;
A. Vainshtein et al., Yad. Fiz. 28 (1978) 465.
[18] S. Weinberg, Harvard preprints HUTP-77/AO57, HUTP-77/AO61 (1977).
[19] H. Leutwyler, Phys. Lett. 48B (1974) 45; Nucl. Phys. B76 (1974) 413.
[20] H.D. Politzer, Nucl. Phys. B117 (1976) 397.
[21] T. Appelquist and H. Georgi, Phys. Rev. D8 (1973) 4000;
A. Zee, Phys. Rev. D8 (1973) 4038.
[22] A. Vainshtein, V. Zakharov and M. Shifman, JETP Lett. 22 (1975) 55;
E. Witten, Nucl. Phys. B104 (1976) 445; B122 (1977) 109;
V. Novikov et al., Phys. Reports 41 (1978) 1.
[23] J. Schwinger, Particles, sources and fields, (Addison-Wesley, 1973) v.2.
[24] V.N. Baier, V.M. Katkov and V.M. Strakhovenko, ZhETF (USSR) 67 (1974) 453, and references therein.
[25] S. Adler, Phys. Rev. 177 (1969) 2426;
J.S. Bell and R. Jackiw, Nuovo Cim. 60A (1969) 47.
[26] R. Crewther, Phys. Rev. Lett. 28 (1972) 1421;
M. Chanowitz and J. Ellis, Phys. Lett. 40B (1972) 397; Phys. Rev. D7 (1973) 2490;
J. Collins, L. Duncan and S. Joglekar, Phys. Rev. D16 (1977) 438.
[27] A. Vainshtein, V. Zakharov and M. Shifman. JETP Lett. 23 (1976) 602; JETP (Sov. Phys.) 45 (1977) 670;
M. Shifman, A. Vainshtein and V. Zakharov, Phys. Rev. D, in press.
[28] V.S. Popov, V.L. Eletsky and A.V. Turbiner, ZhETF (USSR) 74 (1978) 445.
[29] M. Shifman, A. Vainshtein and V. Zakharov, Nucl. Phys. B147 (1979) 519.
[30] R. Dashen, Phys. Rev. 183 (1969) 1245.
[31] H. Georgi and H. Politzer, Phys. Rev. D14 (1976) 1829.
[32] A. Vainshtein, V. Zakharov and M. Shifman, JETP Lett. 22 (1975) 55; JETP (Sov. Phys.) 45 (1977) 670.
[33] R. Shankar, Phys. Rev. D15 (1977) 755.
[34] A. DeRújula, H. Georgi and H. Politzer, Ann. of Phys. 103 (1977) 315;
M. Gluck and E. Reya, Phys. Rev. D16 (1977) 3242;
G.C. Fox, Nucl. Phys. B131 (1977) 107;
A. Buras and K. Gaemers, Nucl. Phys. B132 (1978) 249.
[35] A.A. Abrikosov, L.P. Gor'kov and I.E. Dzyaloshinski, Methods of quantum field theory in statistical physics, (Fizmatgiz, Moscow, 1962).
[36] J.J. Sakurai, Phys. Lett. 46B (1973) 207.
[37] M.K. Gaillard and B.W. Lee, Phys. Rev. Lett. 33 (1974) 108;
G. Altarelli and L. Maiani, Phys. Lett. 52B (1974) 351;
M. Shifman, A. Vainshtein and Zakharov, Nucl. Phys. B120 (1977) 316;
F. Wilczek, S.B. Treiman, A. Zee and R.L. Kingsley, Phys. Rev. D12 (1975) 2768.

Nuclear Physics B147 (1979) 448–518

QCD AND RESONANCE PHYSICS. APPLICATIONS

M.A. SHIFMAN, A.I. VAINSHTEIN * and V.I. ZAKHAROV
Institute of Theoretical and Experimental Physics, Moscow, 117259, USSR

Received 24 July 1978

Resonance properties are investigated within the QCD-based approach to resonance physics developed earlier. We extend first the dispersion charmonium theory to include power terms due to the non-perturbative effects of QCD. As a byproduct, an estimate for the gluonic vacuum expectation value, $\langle 0|G^a_{\mu\nu}G^a_{\mu\nu}|0\rangle$, emerges. The main emphasis is made on the analysis of the ρ, ω, φ, K^* mesons. Predictions are formulated for integrals of the type $\int \mathrm{Im}\,\Pi\, e^{-s/M^2}\, ds$ where Im Π is an appropriate spectral density. It is shown that there exist such M^2 that the integrals are dominated by a single resonance, on one hand, and are calculable in a reliable way, on the other. As a result we are able to calculate the resonance coupling constants and masses. The typical accuracy achieved is about 10%. The power terms considered explain both the π-ρ-A_1 mass splittings and the observed pattern of the SU(3) symmetry breaking in the vector nonet. We discuss, also, the relation between our approach and more traditional ones. A few original remarks concerning the MIT bag model, instanton calculus, etc. are included.

1. Introduction

In the preceeding paper [1] (hereafter referred to as(I)) we have developed the general technique for deriving the QCD sum rules. Here we use them to evaluate the resonance properties.

The sum rules for the light quarks look like

$$\frac{1}{\pi M^2}\int_0^\infty e^{-s/M^2}\,\mathrm{Im}\,\Pi(s)\,ds = h_0 + \frac{h_2}{2!(M^2)^2} + \frac{h_3}{3!(M^2)^3} + \dots , \qquad (1.1)$$

where M^2 is a variable, the coefficients h_i are calculable theoretically while the imaginary part of the polarization operator, Im Π(s) is directly related to the cross section for hadron production in lepton-lepton collisions. The sum rules are expected to be valid as long as the power corrections (terms proportional to $h_{2,3,\dots}$) do not dominate. Otherwise, inclusion of higher orders in M^{-2} is necessary.

Thus, M^2 is bounded from below and this limits the approach discussed. However,

* Permanent address: Institute for Nuclear Physics, Novosibirsk 90, USSR.

we will show that the theoretical calculation is reliable down to such low M^2 that the integral over the cross section is dominated by a single resonance. This is our principal result.

Moreover, we will try various currents and therefore various polarization operators. Then the right-hand sides modify in a well defined way, through the coefficients h_i. The left-hand sides are determined by the masses and couplings of the resonances with the corresponding quantum numbers. Thus, there is the exciting possibility of observing a correlation between the properties of the resonances and the coefficients h_i which store the information on both simple quark graphs and vacuum-to-vacuum matrix elements.

The main emphasis is made on mesons constructed from light quarks. More specifically, we will consider the ρ, ω, φ, K^*, π, A_1, their masses and leptonic decays.

In sect. 2 we start, however, with sum rules for the charmonium decays (the J/ψ meson). The purpose is twofold. First, the statement above on the calculability of the coefficients h_i is to be taken with reservation. The point is that they include both the operator expansion coefficients (which can be found beyond any doubt) and the matrix elements, evaluation of which is a matter of an inspiration rather than a standard technique. Thus far, we are left with no reliable estimate of one of the basic matrix elements, $\langle 0|G^a_{\mu\nu}G^a_{\mu\nu}|0\rangle$ ($G^a_{\mu\nu}$ being the gluon field strength tensor.) Therefore, we use the charmonium sum rules to fix this single parameter so that all the subsequent fits are parameter free. In fact, we not only extract $\langle 0|G^a_{\mu\nu}G^a_{\mu\nu}|0\rangle$ but also check immediately the self-consistency of the whole approach by analyzing various charmonium sum rules. The theory stands the check well and gives quite unambigous predictions, say, for the pseudoscalar ground state, the η_c meson.

We open this section with a review of the so-called dispersion theory of charmonium [2] which is chronologically the first example of a successful application of QCD to compute the electronic width of a single resonance. In its original formulation [2] it makes no use of the power corrections and is based on asymptotic freedom [3] alone. Therefore, we use the opportunity to discuss not only the advantages but also the limitations of such sum rules. The discussion serves as a fresh impetus to introduce the power corrections due to non-perturbative effects.

Thus, one can say that the real applications start with sect. 3 where we proceed to light mesons. As a crucial test we choose the π-ρ-A_1 system (sects. 3 and 4). Indeed, the ρ meson can be called the most typical of all the resonances and it is natural to start with it (sect. 3). Furthermore, the difference in the observed spectra of the low-lying states in the vector and axial-vector channels (π-ρ-A_1) is a challenge to any resonance theory. Perturbation series of QCD with massless quarks make no distinction between the two cases and it is entirely up to the power corrections to provide an explanation (see sect. 4). We argue that the sum rules give a very clear qualitative explanation of the π-ρ-A_1 mass splittings.

Apart from qualitative predictions the sum rules lead to well-defined quantitative results as well. In particular, we are able to compute $g_\rho^2/4\pi$, m_ρ^2, $g_{A_1}^2/4\pi$, f_π^2, the computation depending heavily on the power terms. The accuracy of the results obtained is about ten per cent and in all cases they agree with the data.

In sect. 5, the results obtained are extended to other vector mesons. It is a natural step again since the ρ, ω, φ are produced directly in e^+e^- annihilation and the experimental information is rich enough to provide a crucial test of the sum rules. We find that the sum rules lead to a specific pattern of SU(3) violations which is in accord with the observations. Moreover, we indulge in discussion of the (approximate) ρ-ω degeneracy and the effect of the ω-φ mixing. The theory turns to explain the data in a very natural way. Moreover quite a refined treatment is involved so that both the ρ-ω mass splitting and the ω-φ mixing are sensitive to rather intimate properties of the theory which are difficult to check otherwise.

Finally, sect. 6 is devoted to a discussion of literature. In fact, it complements not only the present paper but (I) as well. Apart from using the opportunity to pay tribute to earlier papers on the same or similar subjects we concentrate on specific features of the QCD sum rules which distinguish them from the approaches proposed earlier.

2. Charmonium sum rules

2.1. Introduction

Here we study sum rules for heavy quarks. More specifically, we will concentrate on the J/ψ meson which is relevant to the charm-anticharm production by the electromagnetic current. As explained in the introduction to (I), heavy quarks are sensitive to the gluon condensate, or gluon fields in vacuum. In the first approximation the condensate is characterized by $\langle 0|G^a_{\mu\nu}G^a_{\mu\nu}|0\rangle$ and we extract this matrix element in subsect. 2.4.

The matrix element enters the sum rules as a free parameter and we will show that choosing it in the proper way makes the sum rules agree with the data. The agreement between the theory and experiment is not trivial: first, the sign of $\langle 0|G^a_{\mu\nu}G^a_{\mu\nu}|0\rangle$ coincides with what is expected on general grounds (see subsect. 6.9 of (I)) and, second, we have a variety of sum rules corresponding to various derivatives of the polarization operator. We show that the theory reproduces well the n dependence (n is the number of the derivative) as long as calculations are reliable theoretically.

Apart from the determination of $\langle 0|G^a_{\mu\nu}G^a_{\mu\nu}|0\rangle$ the principal result of the present section is that the charmonium sum rules support the idea of the crucial rôle of the power corrections in breaking asymptotic freedom.

In subsect. 2.2 we give some background to the sum rules and review to this end the dispersion theory of charmonium. The theory is based on asymptotic freedom and is quite successful in predicting the basic charmonium properties. The main lesson which we would like to draw from the consideration is that asymptotic freedom can work even for a single resonance, an idea which may well be foreign to many readers.

Any attempt to extend the dispersion theory of charmonium in a simple-minded way meets serious problems, however. We list some of them in subsect. 2.3. The answers to the questions arising can be obtained, in fact, only within a theory which does include power corrections. We proceed to this theory starting from subsection 2.4 and stick to it until the end of the paper.

2.2. Dispersion theory of charmonium

In this subsection we review the calculation of the J/ψ electronic width by means of the asymptotic freedom sum rules [2].

Consider the vacuum polarization induced by the electromagnetic current of charmed quarks:

$$i\langle 0|\int dx\, e^{iqx}\, T\{j_\mu^{(c)}(x), j_\nu^{(c)}(0)\}|0\rangle$$

$$= (q_\mu q_\nu - q^2 g_{\mu\nu})\, \Pi^{(c)}(Q^2)\,, \tag{2.1}$$

where

$$j_\mu^{(c)} = \bar{c}\gamma_\mu c\,, \qquad Q^2 = -q^2\,.$$

The function $\Pi^{(c)}$ satisfies standard dispersion relations:

$$-\frac{d}{dQ^2}\Pi^{(c)} = \frac{1}{12\pi^2 Q_c^2}\int \frac{R_c(s)\, ds}{(s+Q^2)^2}\,, \qquad R_c = \frac{3s\sigma_c}{4\pi\alpha^2}\,. \tag{2.2}$$

Here σ_c stands for the cross section of charm production in e^+e^- annihilation which includes both mesons with hidden charm (J/ψ, ψ', ...) and pairs of particles with naked charm ($D\bar{D}$ mesons and so on); $Q_c = \frac{2}{3}$ is the c-quark charge.

Eq. (2.2) defines $\Pi^{(c)}(Q^2)$ in terms of the observable cross section for any Q^2. In fact, we have made an additional assumption here that the current $j_\mu^{(c)}$ is responsible for the entire cross section of charm production, i.e., we have neglected the electromagnetic current of light quarks. This is justified by virtue of the Zweig-Iizuka rule [4].

An alternative way of evaluating $\Pi^{(c)}$ in deep Euclidean region is provided by asymptotic freedom. The point was discussed in detail in paper (I) and we just mention that for heavy quarks it is possible to apply asymptotic freedom at, say, $Q^2 = 0$. The only condition on Q^2 is that $Q^2 + 4m_c^2 \gg \mu^2$ where μ is a typical hadronic mass. The choice $Q^2 = 0$ simplifies all the expressions greatly and we use it throughout the paper.

Furthermore, introduce the moments $\mathcal{M}_n$ defined as

$$\mathcal{M}_n = \frac{1}{12\pi^2 Q_c^2}\int \frac{R_c(s)\, ds}{s^{n+1}} = \frac{1}{n!}\left(-\frac{d}{dQ^2}\right)^n \Pi^{(c)}\Big|_{Q^2=0}\,. \tag{2.3}$$

The simplest quark loop with no gluons at all gives

$$\mathcal{M}_n^{(0)} = \frac{3}{4\pi^2} \frac{2^n(n+1)(n-1)!}{(2n+3)!!} \frac{1}{(4m_c^2)^n} , \tag{2.4}$$

and asymptotic freedom ensures that this is a good approximation at least for the first few moments.

In subsect. 2.4 we shall present expressions for $\mathcal{M}_n$ which follow from a more sophisticated framework, but here concentrate on this simple result.

The same moments can be evaluated in terms of the physical cross section:

$$\sigma(e^+e^- \to \text{charm}) = \sum_{V=J/\psi,\psi',\ldots} 12\pi^2 \delta(s - m_V^2)\, \Gamma(V \to e^+e^-)\, m_V^{-1} + \text{cont.} .$$

Equating both representations one finds the charmed quark mass:

$$m_c = 1.26 \text{ GeV} , \tag{2.5}$$

which is one of our key parameters. (The notion of a mass of a confined quark can be introduced only with reservations. It is convenient to define the mass at a Euclidean point [5] and hereafter we choose $p^2 = -m_c^2$ (and the Landau gauge for the gluon field). In fact, the difference between various normalization points becomes apparent only if higher orders in α_s are included. The terms of order α_s have actually been included in the analysis [2] and eq. (2.5) accounts for these corrections.)

Eq. (2.5) implies that at $Q^2 = 0$ we are off the singularity of quark graphs by $\sim$6 GeV^2 (6 GeV^2 = $4m_c^2$). On the other hand, the scaling behaviour in deep inelastic scattering sets in at about 1 GeV^2. Therefore, the few first derivatives of $\Pi^{(c)}$ are calculable in a reliable way.

The guess turns to be true. The ratios of the first four moments given by eq. (2.4) to their experimental counterparts * can be found in table 1. We see that agreement between theory and experiment is excellent.

What is most remarkable is that the contribution of a single resonance, the J/ψ meson, to the experimental values of $\mathcal{M}_n$ is very substantial. It gives about 50% of the total for $n = 1$ and about 90% for $n = 4$. Therefore, the experimental and theoretical values of $\mathcal{M}_n$ can agree with each other only if the properties of the J/ψ are correlated with those of the simplest graph. In particular, one can find the electronic width of the J/ψ in terms of its mass. To this end one eliminates the quark mass by considering the ratio of appropriate powers of $\mathcal{M}_{3,4}$:

$$\Gamma(J/\psi \to e^+e^-) \simeq \frac{2^{11} \cdot 11^3}{3^8 \cdot 5^4 \cdot 7} \frac{\alpha^2}{\pi} M_{J/\psi} \simeq 5 \text{ keV} . \tag{2.6}$$

This result is in good agreement with the data although it somewhat oversimplifies

* Because of the poor knowledge of the continuum cross section, the "experimental" numbers reflect our own prejudices as well. Eq. (2.10) is used to compute them. For the discussion of the uncertainties in $(\mathcal{M}_n)_{\text{exp}}$ see subsect. 2.4.

Table 1
Comparison of the theoretical and experimental numbers for the moments of the ratio

$$R_c = \frac{\sigma(e^+e^- \to \text{charm})}{\sigma(e^+e^- \to \mu^+\mu^-)}\,, \qquad \mathcal{M}_n = \text{const} \int \frac{R_c(s)}{s^{n+1}}\,ds$$

n	1	2	3	4
$\frac{(\mathcal{M}_n)_{th}}{(\mathcal{M}_n)_{exp}}$	1.0	1.0	1.0	1.1

the situation: playing with high powers of the moments is dangerous since even small corrections can become large.

Thus we can say that the J/ψ meson belongs to asymptotic freedom in the sense that its properties are well understood on the basis of perturbative calculation of vacuum polarization induced by the electromagnetic current of charmed quarks.

For higher moments agreement between theory and experiment worsens rapidly. Phenomenologically, it is clear that the perturbative and experimental values of $\mathcal{M}_n$ must be different for high n. Indeed, the physical spectrum contains resonances while there is no structure in the quark cross section. Increasing n means increasing the weight function in the dispersion integral at relatively low s, where the physical and quark cross sections are different. If imaginary parts are different the moments must deviate also.

We will discuss in detail the theoretical correction responsible for this phenomenon later on. Now let us only say that the indication is that the correction unaccounted for is a sharp function of n. While it is negligible for $n \leqslant 4$ it constitutes about 15% of the experimental value of $\mathcal{M}_5$ and amounts to 50% at $n = 8$.

To summarize, there is some kind of duality between the bare quark graph and the J/ψ meson. The meson is dual to the quark cross section over the energy interval starting at the unphysical quark threshold, ~6 GeV^2, and ending at the physical threshold of charmed particle production $s_{th} \simeq 16$ GeV^2. The weight function in the dispersion integral which gives the precise meaning to the notion of duality is of the kind s^{-n}, where $n = 3, 4$ work excellently while higher n are hardly acceptable. The c-quark mass (defined in the Euclidean region) enters all the relations in an essential way. It can be determined from the sum rules themselves.

2.3. The key questions

According to quantum chromodynamics, a quark-gluon interaction does not depend on the quark flavor. It is natural to assume, therefore, that QCD universally applies to charmonium and ordinary resonances constructed from light quarks, say, to the ρ meson. We will argue that this is indeed the case. Any generalizations to the case of light quarks meet serious problems, however. We begin by listing some questions which serve as a starting point for a further investigation.

First of all, (i) what is the ρ meson needed for? Indeed, according to the approach considered the very existence of the J/ψ is intimately related to the difference between the bare quark mass, m_c, and that of the lightest charmed partible, the D-meson. There is some kind of duality between the J/ψ and quark graph *below* the physical threshold. On the other hand, the ρ meson is well above threshold, and asymptotic freedom sum rules could be perfectly satisfied by a more or less smooth cross section starting at $s = 4m_\pi^2$.

Even if the existence of the ρ is taken for granted, there arises another question which seems to be even more difficult to answer. (ii) Why is the ρ meson dual to $\sim$1.5 GeV^2 of the bare quark graph while the J/ψ is dual to approximately 10 GeV^2? Duality is understood here in a standard (although a bit vague) way: namely, we compare the smeared cross section of resonance production with the quark cross section. For the ρ meson the duality interval is set by the ρ' mass, while for the J/ψ it is given by $4m_D^2 - 4m_c^2 \sim 10$ GeV^2. To avoid confusion, let us emphasize that it is no problem to understand the difference discussed on *phenomenological* grounds. Mass splittings between, say, the ρ, ρ' and J/ψ, ψ' are approximately the same. Since the particle mass grows, the duality interval measured in masses squared, Q^2, increases.

However, we are trying to understand the resonance properties without referring to experimental data, just by considering quark graphs. The corresponding amplitudes depend on Q^2 for light quarks or on $Q^2 + 4m_h^2$ for heavy quarks.

The first impression, therefore, is that the cross sections must look similar if viewed from the same distance, say, $4m_c^2$ (for light quarks it implies $Q^2 = 4m_c^2$ while for heavy quarks $Q^2 = 0$; the distance to the singularity is then the same in both cases). The impression is in conflict with the data which indicate that the structure inherent to the charm production cross section has a much larger scale. The difference reveals itself in that asymptotic freedom breaks at a lower moment number n for heavy quarks. Any consistent theory must explain the observation. Below, we will show that this question has a nice answer within QCD and is related to the resolution of the next problem.

(iii) What is the reason for breaking of asymptotic freedom in high n moments? Phenomenologically, it is due to the existence of the resonance. Theoretically it is due to some kind of correction. The origin of the correction is the real key problem.

Our guess has been proclaimed many times above: the breaking of asymptotic freedom is due to the power corrections. The fast n dependence of the correction observed phenomenologically supports the idea qualitatively. We turn now to a thorough quantitative analysis which includes power terms, deliberately ignored so far.

2.4. Matrix element $\langle 0|G^a_{\mu\nu}G^a_{\mu\nu}|0\rangle$

Let us try to introduce non-perturbative effects into the dispersion charmonium theory discussed so far. The leading non-perturbative effect can be parametrized in terms of the vacuum expectation value $\langle 0|G^a_{\mu\nu}G^a_{\mu\nu}|0\rangle$ which is one of the key parameters for all the subsequent applications. The charmonium sum rules allow one to find it and provide the first opportunity to put on trial the very idea of the rôle of power corrections.

Taking advantage of the calculations performed in sect. 4 of (I), we are able to find the moments defined in eq. (2.3) to much better accuracy than specified in eq. (2.4). Including both the $\alpha_s(m_c)$ term and the leading power correction we get:

$$\mathcal{M}_n = \frac{3\cdot 2^n(n+1)(n-1)!}{4\pi^2(2n+3)!!}\,\frac{1}{(4m_c^2)^n}\left[1 + a_n^{(V)}\alpha_s - \frac{n(n+1)(n+2)(n+3)}{2n+5}\,\phi\right], \tag{2.7}$$

where

$$a_n^{(V)} = \frac{4\sqrt{\pi}}{3}\,\frac{\Gamma(n+\frac{3}{2})}{\Gamma(n+1)}\,\frac{1-(3n+3)^{-1}}{1-(2n+3)^{-1}} - \tfrac{1}{2}\pi + \frac{3}{4\pi}$$

$$- \frac{2}{3\sqrt{\pi}}\left(\tfrac{1}{2}\pi - \frac{3}{4\pi}\right)\frac{\Gamma(n+\frac{3}{2})}{\Gamma(n+2)}\,\frac{1-2(3n+6)^{-1}}{1-(2n+3)^{-1}} - \frac{4n\ln 2}{\pi}\,,$$

$$\phi = \tfrac{4}{9}\pi^2\langle 0|(\alpha_s/\pi)\,G^a_{\mu\nu}G^a_{\mu\nu}|0\rangle\,(4m_c^2)^{-2}\,.$$

On the other hand, moments can be found from the experimental data using eq. (2.2). Unlike the simplified analysis of subsect. 2.2 we deal now with three parameters: the quark-gluon coupling constant α_s, the charmed quark mass m_c and ϕ which we are interested in.

The problem of the choice of the coupling constant has been discussed in detail above (see sect. 6 of (I)). In essence, we rely on the determination of α_s which follows from the total hadronic width of the J/ψ [6]:

$$\alpha_s(m_c) = 0.2\,.$$

Once α_s is fixed we are left with two parameters: ϕ and m_c. In principle, there is no difficulty in finding them both. Indeed, the power correction is a steep function of n and can be just neglected at low n. Then the sum rules depend on m_c alone and the result of the fit is given by eq. (2.5). Since it follows from the moments with low n, it stays untouched by more recent progress.

Thus, the sum rules for higher n, where asymptotic freedom does not work any longer, are free to determine ϕ without shifting any other parameter (see fig. 1.).

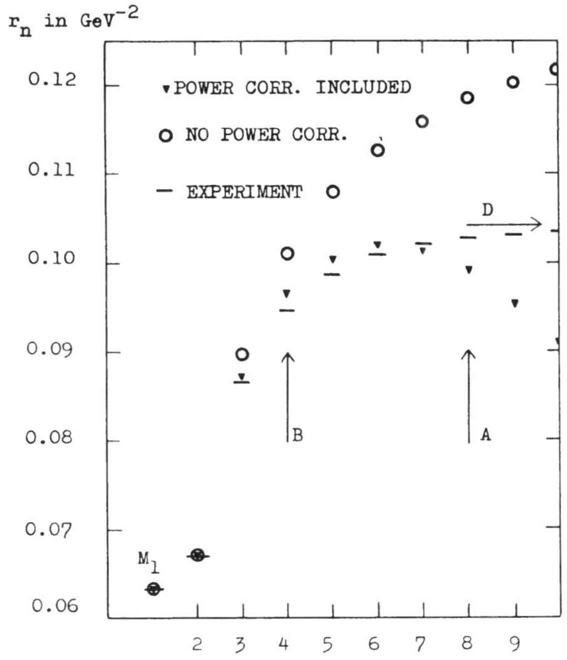

Fig. 1. The ratio of the moments $r_n = \mathcal{M}_n/\mathcal{M}_{n-1}$ *versus n*. (For definitions see eqs. (2.3), (2.9)). Arrow A marks the 20% level for the power correction. Arrow B separates the regions of small and large experimental uncertainties: to the right of this arrow the uncertainty is $\lesssim 1\%$. Arrow D shows the asymptotic value of $(r_n)_{\rm exp}$, $1/m^2_{J/\psi}$. The point M_1 serves for a normalization; it gives $(4\pi/3)^2\mathcal{M}_1$.

In this way we come to an estimate:

$$\phi \simeq 1.35 \cdot 10^{-3}\ , \qquad \left\langle 0 \left| \frac{\alpha_s}{\pi} G^a_{\mu\nu} G^a_{\mu\nu} \right| 0 \right\rangle \simeq 0.012\ \mathrm{GeV}^4\ , \tag{2.8}$$

which completes our race for the parameters.

Note that the result (2.8) agrees with the general statement on the sign of the vacuum expectation value in point (see subsect. 6.9 of (I)). Thus, the deviations from asymptotic freedom can be ascribed to the non-vanishing of $\langle 0|G^a_{\mu\nu}G^a_{\mu\nu}|0\rangle$ with no contradiction with general principles. If otherwise, the whole scheme would have been ruined.

A few comments are now in order concerning the above estimates of the parameters. To extract the value of ϕ we considered the ratio

$$r_n = \frac{\mathcal{M}_n}{\mathcal{M}_{n-1}} = \frac{n^2 - 1}{n^2 + \frac{3}{2}n} \frac{1}{4m_c^2}$$

$$\times \left[1 + (a_n^{(V)} - a_{n-1}^{(V)})\, \alpha_s - \frac{(6n+14)\, n\, (n+1)(n+2)}{(2n+3)(2n+5)} \phi \right], \tag{2.9}$$

where the coefficients $a_n^{(V)}$ are given in eq. (2.7). Unlike the moments themselves the ratios r_n do not contain high powers of m_c and for this reason are stable against a possible error in the determination of m_c. Note also that eq. (2.9) is expected to hold only as long as the G^2 term can be really treated as a correction, i.e., does not become too large. Thus n is bounded by $n \lesssim 10$.

Using the experimental data we have computed r_n for $n = 2$ to 10 *. The corresponding numbers are represented by the horizontal lines in fig. 1. Unfortunately it is impossible to indicate the experimental uncertainties by the conventional error bars. The problem is that the errors in $r_2, \ldots r_{10}$ are correlated and any parametrization of the experimental cross section generates a set of the numbers. The set displayed in fig. 1 corresponds to the following resonance widths and continuum model:

$$\begin{aligned}
&\Gamma(J/\psi(3095) \to e^+e^-) = 4.50\ \text{keV}, \\
&\Gamma(\psi'(3684) \to e^+e^-) = 1.95\ \text{keV}, \\
&\Gamma(\psi''(3772) \to e^+e^-) = 360\ \text{eV}, \\
&\Gamma(\psi(4033) \to e^+e^-) = 360\ \text{eV}, \\
&\Gamma(\psi(4116) \to e^+e^-) = 390\ \text{eV}, \\
&\Gamma(\psi(4400) \to e^+e^-) = 460\ \text{eV}, \\
&(R_c)_{\text{continuum}} = \tfrac{4}{3}\left(1 + \frac{\alpha_s(s)}{\pi}\right) \theta(s - (4.2\ \text{GeV})^2).
\end{aligned} \tag{2.10}$$

The model is consistent with the data within the experimental uncertainties.

Another choice of the widths would shift all the horizontal lines concertedly. It is important, however, that at $n \geqslant 4$ the values of r_n are stable: any variation in the cross section within the experimental uncertainties shifts r_4, r_5 ... by less than ~1%. The reason is that at $n \geqslant 4$ the J/ψ almost saturates the ratios and r_n with $n \geqslant 4$ are expressed in terms of the J/ψ mass alone. The latter is known to high precision.

Fig. 1 also displays the theoretical predictions for r_n. Open circles correspond to neglect of the power correction, $\phi = 0$. They show a deviation from the data which grows fast with n. Choosing ϕ as indicated by eq. (2.8) improves the fit considerably (closed triangles in fig. 1). At $n \geqslant 8$ the G^2 correction exceeds 20% and higher corrections show up.

* We are grateful to M. Polikarpov and M. Vysotsky who performed computer calculations for various fits.

A simple-minded estimate of the neglected power corrections is that they are of the order of the square of the leading power correction which is kept explicit (at least, it appears to be true as far as the n dependence is concerned). By comparing predictions (2.9) with the data one can convince oneself that eq. (2.9) does hold within the uncertainty guessed for all n considered.

Thus, the situation with the charmonium sum rules turns out to be the best one could hope for: the leading power correction improves the fit substantially and shows the way to higher orders when it becomes too strong.

Although the success seems to be impressive it is worth emphasizing that the present numerical analysis is qualified by some uncertainties of both experimental and theoretical origin.

First, the low-n sum rules are rather sensitive to the continuum contribution (let us remind the reader that it is most sizable for $n = 1$ and constitutes about 50% of the total in this case). However, the experimental cross section in the continuum region is measured rather poorly and we rely mostly on an idea of how the data could look like rather than on the results of measurements. The sum rules with $n \geqslant 5$ are saturated by the J/ψ and the accuracy in evaluation of the integral over the experimental cross section is better. However, if we stick to these sum rules alone, then both m_c and ϕ can be somewhat varied in a correlated way without making the fit much worse.

Second, the coupling constant α_s itself can depend on n since the characteristic distance relevant to the problem varies, generally speaking, with n. At present we are able to make only an educated guess on this dependence and cannot give a decisive answer.

In the worst case, the uncertainty mentioned above can shift the vacuum expectation value (2.8) by a factor of 2.

2.5. *Sum rules for the* η_c

Sum rules for the pseudoscalar charmonium can be derived in a similar way. It was done in collaboration with Voloshin and is published separately [7]. Here we just summarize the results for the sake of completeness.

The main point is that the J/ψ-η_c mass splitting can be expressed in terms of the same parameter ϕ (or, more exactly, in terms of a certain combination of m_c and ϕ which is stable against possible variations in each of them (see subsect. 2.4)).

If the η_c plays the same role in the spectrum of the pseudoscalar states as the J/ψ does in the vector channel then

$$m(\eta_c) = (3.00 \pm 0.03)\ \text{GeV}\ ,$$

and it cannot be identified with X(2.83) detected experimentally [8].

If the spectrum of the pseudoscalar states (for reasons not understood so far) looks more complicated and if there are two nearly degenerate states, then the X(2.83) could be one of these states. However, the other one is predicted to have the mass

$$m(\eta_c) = (3.0 - 3.2) \text{ GeV}$$

and is coupled to the $\bar{c}\gamma_5 c$ current much stronger than X(2.83).

It is worth noting that a high accuracy is required on the part of the sum rules to distinguish between the masses 2.83 and 3.0 GeV but the sum rules do posess such an accuracy, at the present level of understanding.

Existence of a charmonium state with mass ~ 3.0 GeV and the properties expected on theoretical grounds is in no way rules out experimentally and further search for it seems fully justified.

2.6. Conclusions

It seems fair to say that the first application of the sum rules with power corrections included turns to be encouraging. First, we have fixed $\langle 0|(\alpha_s/\pi)\, G^a_{\mu\nu}G^a_{\mu\nu}|0\rangle$ which is a necessary ingredient in further applications. What is more important, the sum rules look very reasonable as far as comparison with the data is concerned. In particular, we keep in mind that the power correction reproduces the deviation from asymptotic freedom as it starts at $n = 4, 5$ (n is the number of the derivative of the polarization operator at $Q^2 = 0$) and grows for higher n.

In fact, the world of heavy quarks deserves further consideration. In particular, we did not try sum rules for P-wave states of charmonium. The first impression is that the power correction is too large here to probe a single level by means of the sum rules. But we have not been too deep into the problem.

Application to the Υ family seems to be most interesting. In principle we have developed all the machinery to suggest the fit to the Υ family as an exercise for the reader. Anyhow, the sum rules are sensitive enough to isolate a single meson in this case, as follows immediately from the analysis of the power corrections. We do not rule out, however, that some specific problems arise, and have in fact one particular problem in mind: growth of the Coulomb-like interaction which is quite unimportant for the J/ψ (see the review [9] for details) but can become appreciable for the Υ. We hope to come back to these problems in a future publication.

3. Mass and electronic width of the ρ meson

Starting with this section we consider mesons constructed from light quarks. The whole framework is assumed to be standard in the sense that changing the current and resonances under consideration reduces to a mere recomputation of the operator expansion coefficients. All the other ingredients and, in particular, the relevant matrix elements, do not vary, and we have fixed all the parameters in the preceeding sections.

In this section we will study the ρ meson. We will try to set the pattern for further applications and to be rather detailed here to avoid unnecessary repetitions in the subsequent sections.

The finding of this section is that there exists a region of the variable M^2 such that:

(a) Integrals like $\int \exp(-s/M^2)\,\mathrm{Im}\,\Pi(s)\,\mathrm{d}s$ are given by asymptotic freedom and are calculable in a reliable way;

(b) the same integrals over the experimental cross section are saturated by a single resonance, say, the ρ meson.

This allows one to evaluate the ρ meson mass and electronic width.

The procedure is as follows. In subsect. 3.1 the sum rules are written down explicitly. In subsect. 3.2 the ρ meson properties are evaluated by considering M^2 in the region just mentioned. In subsects. 3.3, 3.4 the region of M^2 available for the analysis is somewhat extended by using a simple but plausible model for the continuum cross section. In subsects. 3.5, 3.6 we discuss in detail the power corrections. Subsect. 3.7 is devoted to a discussion of the uncertainties in the predictions obtained.

3.1. Sum rules

Consider the $j_\mu^{(\rho)}$ current with the ρ meson quantum numbers:

$$j_\mu^{(\rho)} = \tfrac{1}{2}(\bar{u}\gamma_\mu u - \bar{d}\gamma_\mu d)\,. \tag{3.1}$$

The operator expansion for the T-product of two such currents was constructed in sect. 4 of (I). Taking the vacuum-to-vacuum matrix element gives, by definition, the polarization operator $\Pi^{(\rho)}$:

$$(q_\mu q_\nu - q^2 g_{\mu\nu})\,\Pi^{(\rho)}(q^2) = i\int \mathrm{d}x\, \mathrm{e}^{iqx}\langle 0|\mathrm{T}\{j_\mu^{(\rho)}(x)\, j_\nu^{(\rho)}(0)\}|0\rangle\,, \tag{3.2}$$

and *via* the operator expansion it is related to the matrix elements of the relevant operators ($q^2 \equiv -Q^2$ is assumed to be large and negative).

On the other hand, $\Pi^{(\rho)}(q^2)$ satisfies the standard dispersion relation

$$\Pi^{(\rho)}(Q^2) = \Pi^{(\rho)}(0) - \frac{Q^2}{12\pi^2}\int \frac{\mathrm{d}s\, R^{I=1}(s)}{s(s+Q^2)}\,, \tag{3.3}$$

where

$$R^{I=1} = \frac{\sigma(\mathrm{e}^+\mathrm{e}^- \to \text{hadrons}\,, I=1)}{\sigma(\mathrm{e}^+\mathrm{e}^- \to \mu^+\mu^-)}\,.$$

Indeed, the $j_\mu^{(\rho)}$ coincides with the isovector part of the electromagnetic current and is responsible, therefore, for the production of the $I = 1$ hadrons in $\mathrm{e}^+\mathrm{e}^-$ annihilation.

It is worth noting that for our purposes $\sigma(\mathrm{e}^+\mathrm{e}^- \to \text{hadrons}, I = 1)$ can be identified with the cross section for production of an even number of pions so that there is no principal difficulty with extracting it from the data.

Following the general procedure substantiated in sect. 5 of (I), we apply the $\hat{L}_M$

procedure,

$$\hat{L}_M = \lim_{\substack{Q^2\to\infty, n\to\infty \\ Q^2/n\equiv M^2\,\text{fixed}}} \frac{1}{(n-1)!}(Q^2)^n\left(-\frac{d}{dQ^2}\right)^n$$

to the both representations of the polarization operator introduced above. As a result the following sum rules arise:

$$\int ds\, e^{-s/M^2} R^{I=1}(s) = \tfrac{3}{2}M^2\Bigg[1 + \frac{\alpha_s(M)}{\pi} + \frac{4\pi^2\langle 0|m_u\bar{u}u + m_d\bar{d}d|0\rangle}{M^4} + \tfrac{1}{3}\pi^2\frac{\left\langle 0\left|\frac{\alpha_s}{\pi}G^a_{\mu\nu}G^a_{\mu\nu}\right|0\right\rangle}{M^4} - 2\pi^3\frac{\langle 0|\alpha_s(\bar{u}\gamma_\alpha\gamma_5 t^a u - \bar{d}\gamma_\alpha\gamma_5 t^a d)^2|0\rangle}{M^6}$$
$$- \tfrac{4}{9}\pi^3\frac{\langle 0|\alpha_s(\bar{u}\gamma_\alpha t^a u + \bar{d}\gamma_\alpha t^a d)\sum_{q=u,d,s}\bar{q}\gamma_\alpha t^a q|0\rangle}{M^6}\Bigg], \qquad (3.4)$$

which are a starting point for all further considerations. (Let us remind the reader that the sum rules are valid as far as the power corrections in eq. (3.4) do not become dominant).

The matrix elements entering the right-hand side of eq. (3.4) can be extracted from independent experimental data (+ some theoretical ideas). We will use the set motivated in sect. 6 of paper I (see also the previous section):

$$\langle 0|m_u\bar{u}u + m_d\bar{d}d|0\rangle = -\tfrac{1}{2}f_\pi^2 m_\pi^2 = -1.7\cdot 10^{-4}\ \text{GeV}^4,$$

$$\langle 0|\frac{\alpha_s}{\pi}G^a_{\mu\nu}G^a_{\mu\nu}|0\rangle = 1.2\cdot 10^{-2}\ \text{GeV}^4, \qquad (3.5)$$

$$\langle 0|\alpha_s(\bar{u}\gamma_\alpha\gamma_5 t^a u - \bar{d}\gamma_\alpha\gamma_5 t^a d)^2|0\rangle = \tfrac{32}{9}\,\alpha_s\langle 0|\bar{q}q|0\rangle^2 \simeq 6.5\cdot 10^{-4}\ \text{GeV}^4,$$

$$\langle 0|\alpha_s(\bar{u}\gamma_\alpha t^a u + \bar{d}\gamma_\alpha t^a d)\sum_{q=u,d,s}\bar{q}\gamma_\alpha t^a q|0\rangle \simeq -\tfrac{32}{9}\,\alpha_s\langle 0|\bar{q}q|0\rangle^2 \simeq -6.5\cdot 10^{-4}\ \text{GeV}^4.$$

The first two operators here do not depend on the normalization point. The last two (with account of α_s) are practically independent of it as well (see sect. 4 of (I)). However, in estimating their vacuum-to-vacuum matrix element we used the factorization hypothesis which seems to be valid for a low normalization point, $\mu \sim R_{\text{conf}}^{-1}$, $\alpha_s(\mu) \sim 1$.

If one takes $\alpha_s(2.5\ \text{GeV}) = 0.2$ as implied by the analysis of the J/ψ decays [6] and $\mu = 0.2$ GeV, then $\alpha_s(\mu) = 0.7$. Using this number literally and

$$|\langle 0|\bar{q}q|0\rangle| = \frac{f_\pi^2 m_\pi^2}{2(m_u + m_d)} \simeq (0.25 \text{ GeV})^3 , \qquad (m_u + m_d \simeq 11 \text{ MeV})$$

(see sect. 6 in paper (I)), we obtain the last two estimates in eq. (3.5).

The uncertainty in the parameters used is seemingly bounded by a factor of 2. The effect of this uncertainty will be discussed in subsect. 3.7. Up to this point we consistently use estimates (3.5) in all the cases.

Substituting eq. (3.5) into (3.4) gives

$$\int e^{-s/M^2} R^{I=1}(s)\, ds$$
$$= \tfrac{3}{2}M^2 \left[1 + \frac{\alpha_s(M)}{\pi} + 0.1\left(\frac{0.6 \text{ GeV}^2}{M^2}\right)^2 - 0.14\left(\frac{0.6 \text{ GeV}^2}{M^2}\right)^3\right], \tag{3.6}$$

so that the sum rules are completely specified numerically. (For α_s we use α_s (2.5 GeV) = 0.2 and extrapolate it to lower M by means of the standard asymptotic freedom formula).

Sometimes it is helpful to consider the second sum rule which can be obtained by differentiating eq. (3.6) with respect to $1/M^2$ and looks as

$$\int e^{-s/M^2} R^{I=1}(s)\, s\, ds = \tfrac{3}{2}M^4 \left[1 + \frac{\alpha_s(M)}{\pi} - 0.1\left(\frac{0.6 \text{ GeV}^2}{M^2}\right)^2 + 0.28\left(\frac{0.6 \text{ GeV}^2}{M^2}\right)^3\right]. \tag{3.7}$$

Note that the M^{-4} and M^{-6} terms are comparable to each other at $M^2 = m_\rho^2 = 0.6$ GeV^2. The reader might conclude that this signals the breaking of the whole expansion in M^{-2} at $M^2 = m_\rho^2$. In fact, this is not so. The point is that the coefficient h_3 is anomalously large as compared to h_2. The reason is readily traced to the fact that h_3 comes from the Born graphs while h_2 is due to the loop graphs (see sect. 4 in (I)). This "anomaly" is not reiterated in higher orders. Moreover, the M^{-4} term is proportional to $\langle 0|G^a_{\mu\nu}G^a_{\mu\nu}|0\rangle$ and represents in a way the effect of the gluon confinement. The M^{-6} correction is proportional to $\langle 0|\bar{\psi}\Gamma\psi\bar{\psi}\Gamma\psi|0\rangle$ and can be thought of as a result of quark confinement. It is amusing that both share control over the resonance properties equally although their respective estimates come from very different sources.

3.2. Expansion of asymptotic freedom

What is most remarkable about eqs. (3.6), (3.7) is that even at $M^2 = m_\rho^2 = 0.6$ GeV^2 the power corrections to the unit term are relatively small:

$$\int e^{-s/m_\rho^2} R^{I=1}(s)\, ds = \tfrac{3}{2}m_\rho^2 [1 + 0.1 + 0.1 - 0.14] ,$$

$$\int e^{-s/m_\rho^2} R^{I=1}(s)\, s\, ds = \tfrac{3}{2}m_\rho^4 [1 + 0.1 - 0.1 + 0.28] . \tag{3.8}$$

On the other hand, at such M^2 the integral over the physical cross section is dominated by a single ρ.

As a rough approximation, we neglect both (a) the corrections to asymptotic freedom (delegated by the unit term in eqs. (3.8)) and (b) the continuum contribution as compared to that of the ρ. In the limit of vanishing ρ-meson width the latter appears as:

$$R_\rho^{I=1} = \frac{12\pi^2 m_\rho^2}{g_\rho^2}\,\delta(s - m_\rho^2)\,.$$

Then we get

$$\int e^{-s/m_\rho^2} R_\rho^{I=1}\, ds|_{\rm exp} \simeq \frac{12\pi^2 e^{-1}}{g_\rho^2}\, m_\rho^2\,, \tag{3.9}$$

and comparing it with eq. (3.8) (with all the corrections suppressed) we come to

$$\frac{g_\rho^2}{4\pi} \simeq \frac{2\pi}{e} \simeq 2.3\,, \tag{3.10}$$

which is one of our main results.

Thus, asymptotic freedom severely constraints the properties of a single resonance; we would call this phenomenon: *expansion of asymptotic freedom.*

The prediction (3.10) can be confronted with the experimental number [10]

$$\left(\frac{g_\rho^2}{4\pi}\right)_{\rm exp} = 2.36 \pm 0.18\,.$$

We can try one step further and evaluate the ρ-meson mass. To this end it is convenient to consider eq. (3.7). Keeping the ρ-meson contribution alone gives

$$\int e^{-s/M^2} R^{I=1}(s)\, s\, ds \simeq \frac{m_\rho^4}{g_\rho^2}\, 12\pi^2 e^{-m_\rho^2/M^2}\,. \tag{3.11}$$

Eq. (3.11) is valid as far as M^2 is small enough so that the continuum is negligible; we will turn back to the discussion of this point later on.

Whether it is possible to evaluate the integral (3.11) within QCD or not depends on the power corrections: if they are large, then the prediction is non-reliable, while if they do not dominate, then it is possible to find the mass. Numerically we have:

$$\frac{\int s\, e^{-s/M^2} R^{I=1}(s)\, ds}{\int e^{-s/m^2} R^{I=1}(s)\, ds} = M^2\, \frac{1 + \frac{\alpha_s(M)}{\pi} - 0.1\left(\frac{0.6\ {\rm GeV}^2}{M^2}\right)^2 + 0.28\left(\frac{0.6\ {\rm GeV}^2}{M^2}\right)^3}{1 + \frac{\alpha_s(M)}{\pi} + 0.1\left(\frac{0.6\ {\rm GeV}^2}{M^2}\right)^2 - 0.14\left(\frac{0.6\ {\rm GeV}^2}{M^2}\right)^3}$$

Now, if we take $M^2 = m_\rho^2$ and saturate the integrals over the cross section by the ρ

contribution alone we have for the ratio in the l.h.s., m_ρ^2. If we had the same m_ρ^2 in the r.h.s., then the ρ-meson solution to the sum rules would be self-consistent. Instead, we get

$$(m_\rho^2)_{\text{th}} \simeq 1.2\,(m_\rho^2)_{\text{exp}} \tag{3.12}$$

which is still quite satisfactory by itself. However, we would add in haste that eq. (3.12) is *not* our final result for the ρ-meson mass (see subsect. 3.4.).

3.3. A rough model of experimental data

Up to now we have considered a fixed value of M^2, $M^2 = m_\rho^2$. In fact, there is a continuum set of sum rules and the crucial question is whether the predictions for the ρ-meson mass and width do not depend on the choice of M^2.

The choice of $M^2 = m_\rho^2$ is specific since both the theoretical corrections and continuum contribution are negligible at this point. To be able to vary M^2, we need some model for the continuum on one hand and for higher power corrections, on the other. Then we can probe higher M^2 where ρ dominance is not so prominent and try lower M^2 where power corrections become more important.

An elaborated analysis of the sum rules would assume a smooth curve for the cross section which reproduces both the threshold and asymptotic behaviour and includes peaks corresponding to resonances. Then, the positions and widths of the peaks could be found, along with other possible fit parameters by considering the sum rules for a relatively wide range of M^2.

As a first step, however, it seems appropriate to exploit a rough model just to see whether the sum rules work or not. Indeed, it is in fact for the first time that we can test QCD beyond mere perturbation theory, and a rough model is preferable for a qualitative analysis.

Therefore, we will assume that the cross section can be approximated by one or two resonances + continuum which starts at s_0 and coincides at higher energies with the quark cross section:

$$R_{\text{cont}} = R_{\text{pert th}}\,\theta(s - s_0)\,. \tag{3.13}$$

We leave s_0 as a fit parameter.

Moreover, we will confine ourselves here to the approximation of a vanishing total width which renders the integration over a resonance trivial. Thus, for the ρ meson we use *

$$R_\rho^{I=1} = \frac{12\pi^2}{g_\rho^2}\, m_\rho^2\, \delta(s - m_\rho^2)\,.$$

* Note that corrections linear in the ratio Γ_ρ/m_ρ are present (the first impression might be that only terms of the order Γ_ρ^2/m_ρ^2 arise). Nevertheless, the error introduced through the $\Gamma_\rho/m_\rho = 0$ approximation is unimportant numerically.

(g_ρ is defined in a standard way: $\langle 0|j_\mu^{\text{e.m.}}|\rho\rangle = m_\rho^2 g_\rho^{-1} \epsilon_\mu$.)

With a model of experimental data in hand we are in position to extend the region of M^2 used in the sum rules. It is worth noting, however, that it would be much better not to rely on any model but to evaluate the integrals (3.6), (3.7) in a straightforward way by integrating the experimental cross section and confronting the result with the theoretical predictions for the same integrals. One could hope to extract in this way the power corrections from the experimental data.

We need the measurements of the cross section with high accuracy, however, and could not find them in the literature *. This is one of the reasons whey we try to extract predictions for a single resonance for which the experimental accuracy is satisfactory.

It seems to be a common belief nowadays that to probe short distances and check QCD one must go to high energies. We see that in fact measurements at very moderate (by present standards) energies $s \sim 2$ GeV^2 can provide not only a test of asymptotic freedom but information on the mechanism of its breaking as well. Unfortunately, many accelerators for such energies have already been shut down. By this remark we do not intend of course to disregard the possibility of encountering something completely new at high energies. But for the present theories, measurements at $s \sim 2$ GeV^2 are quite crucial.

3.4. Evaluation of the ρ-meson mass and electronic width

In subsect. 3.2 we found the ρ-meson mass and electronic width neglecting the corrections due to the power terms and continuum contribution. Here we would like to include both.

As explained above we assume the following rough model:

$$R^{I=1}(s) = \frac{12\pi^2 m_\rho^2}{g_\rho^2}\,\delta(s - m_\rho^2) + \tfrac{3}{2}\left(1 + \frac{\alpha_s(s)}{\pi}\right)\theta(s - 1.5\ \text{GeV}^2)\,, \qquad (3.14)$$

where the chosen value of s_0, $s_0 = 1.5$ GeV^2, is suggested by the experimental data.

The model is intended to convey only the gross features of the experimental data. In particular, we ignore some structures in the 1–1.5 GeV region.

Since we are dealing with a smeared cross section we feel that the continuum model adopted reproduces the data with an accuracy not worse than 30% (locally, at some particular energy the discrepancy can be larger).

We would like to keep the accuracy of our calculations at a 10% level, and therefore will not consider the sum rules in which the continuum contribution exceeds 30%.

Neither do we allow the theoretical correction to be large. Namely, we assume that unaccounted power terms are of the order of the square of the power corrections which are kept explicit in eqs. (3.6), (3.7). Again, the calculation is stopped

* See, however, note added in proof in ref. [1].

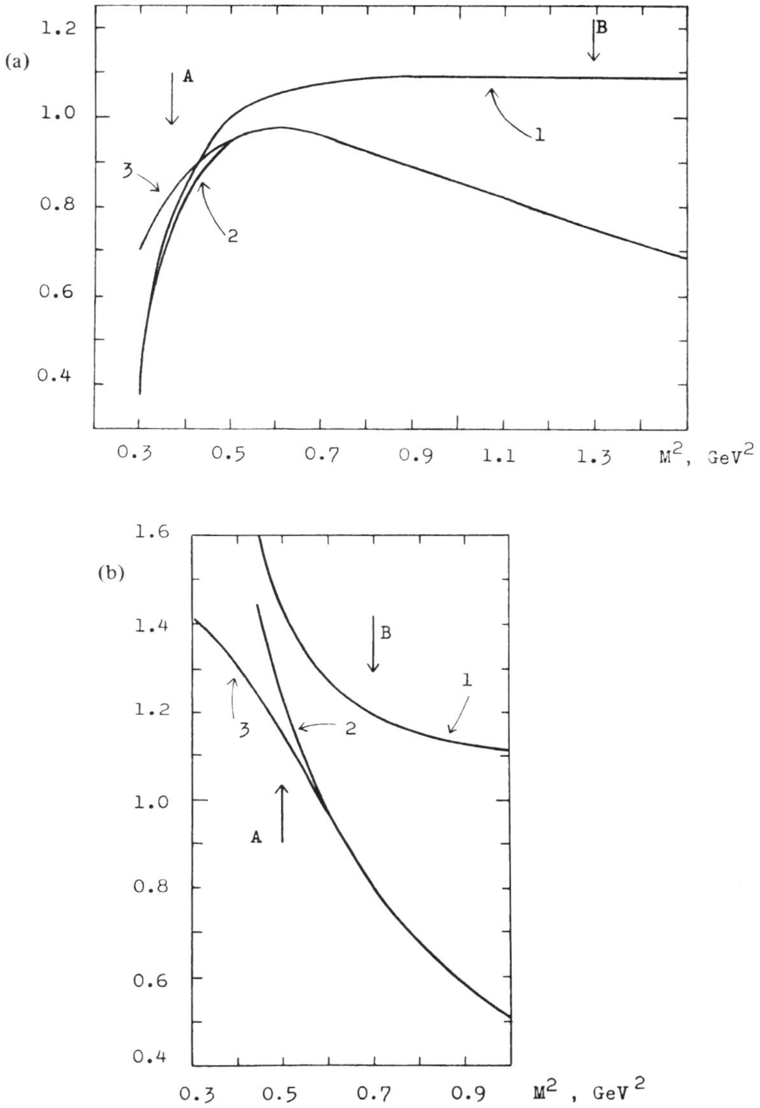

Fig. 2. (a) The function $\frac{2}{3}M^{-2}\int e^{-s/M^2}R^{I=1}(s)\mathrm{d}s$ *versus* M^2. (b). The function $\frac{2}{3}M^{-4}\int e^{-s/M^2}$ $R^{I=1}(s)\, s\mathrm{d}s$ *versus* M^2. Curve 1, the theoretical prediction. Curve 2, the theoretical prediction with the continuum contribution subtracted. The following continuum model is used

$$R^{I=1}_{\mathrm{cont}}(s) = 1.5(1 + \alpha_s(s)/\pi)\, \theta(s - 1.5\ \mathrm{GeV}^2)\ .$$

Curve 3, the rho-meson contribution with $m_\rho^2 = 0.6\ \mathrm{GeV}^2$, $4\pi/g_\rho^2 = 0.414$. Arrows A and B indicate M^2 for which the power correction and the continuum contribution reach 30% of the total, respectively. The continuum contribution grows for larger M^2, while the power corrections become more important with diminishing M^2. The region between the arrows is most sensitive to the resonance contribution and reliable from a theoretical point of view.

once the corresponding uncertainty reaches 10% of the total.

Thus, the experimental uncertainty due to the continuum contribution and the theoretical uncertainty due to the higher order power terms bracket the interval of M^2 suitable for determination of m_ρ^2, g_ρ^2: choosing M^2 too high makes the continuum contribution unacceptably large while taking M^2 too low emphasizes unaccounted power corrections.

Still, we can now evaluate m_ρ^2, g_ρ^2 for a range of M^2 and see whether our predictions are stable against a change in the choice of M^2.

The results are summarized in figs. 2a,b which correspond to the sum rules (3.6) and (3.7). Curves (1) here represent the theoretical predictions with inclusion of the power corrections. Curves (2) are the same theoretical predictions but with the continuum contribution subtracted. Therefore, curves (2) must coincide with the resonance contribution, if the theory and the continuum model are correct.

Let us concentrate first on fig. 2a. Then the shape of the curve (3) is determined by two parameters g_ρ^2, m_ρ^2 (in the limit $\Gamma_\rho = 0$). The former is correlated with the height of the curve and the latter is correlated with the position of its maximum. Our fit looks like

$$\left(\frac{4\pi}{g_\rho^2}\right)_{\rm th} \simeq 0.414\,, \qquad (m_\rho^2)_{\rm th} \simeq 0.6\ {\rm GeV}^2\,, \tag{3.15}$$

which is our final result for the ρ meson.

Arrows A and B indicate the "region of confidence" in the sense explained above. We see that, within these limits, curves (2) and (3) practically coincide with each other. For $M^2 \lesssim 0.5$ there is a hint of the deviations which can be readily attributed to higher order power corrections, however. For this interpretation to be true the series of power corrections must be sign alternating. Note that without inclusion of the power corrections the curves (1) and (2) in fig. 2a flatten out at the level of 1.1 at $0.3 < M^2 < 0.6$ and deviate from the real ρ contribution (curve (3)) by 20–30%, even for M^2 belonging to the "confidence region".

Note that the position and the height of the maximum of the theoretical curve (2) in fig. 2a control the predictions for m_ρ^2 and $g_\rho^2/4\pi$, respectively. The "confidence region" between the arrows in fig. 2b is much narrower than that in fig. 2a. Since the sum rules (3.7) are most sensitive to the ρ-meson mass, this implies in fact that the prediction for the mass is less accurate than the evaluation of the coupling constant (for further discussion see subsect. 3.7).

We could perform the analysis in an alternative way and find $m_\rho^2, g_\rho^2/4\pi$ from eqs. (3.6), (3.7) as a function of M^2. The results are stable against a change in M^2 as long as we do not cross to the left of the arrow "A" where higher power terms must have been included (see also subsect. 3.5). This indicates that the model of the cross section and the sum rules are self-consistent. Moreover the theoretical and experimental values of mass and electronic width coincide with each other within the errors. We interpret the agreement between theory and experiment as a strong argument in favour of the approach developed.

3.5. Limits to the asymptotic freedom

In subsect. 3.2 we argued that asymptotic freedom can be expanded into the resonance region. In particular, the predictions for the ρ-meson mass and electronic width follow directly from equating the bare polarization operator with the physical one, with seemingly no parameter left.

Now we would like to emphasize that asymptotic freedom alone is of course helpless to fix the resonance properties. Indeed, there is no dimensional parameter (the quark masses are presumably small), at least as far as we do not consider the sum of the whole series in α_s which is not a well-defined notion. Therefore, it is just the power corrections that set the limits to standard perturbation theory and introduce a mass scale for the resonances in this way.

To substantiate the point turn back to, say, the ρ-meson mass. Our prediction for the mass reads as

$$m_\rho^2 = M^2 f_{\text{cont}}(M^2)\, f_{\text{th corr}}(M^2)\,, \tag{3.16}$$

where M^2 stands for the ratio of the integrals

$$\int e^{-s/M^2} R^{I=1}(s)\, s\, ds\,, \qquad \int e^{-s/M^2} R^{I=1}(s)\, ds\,,$$

as evaluated in an asymptotically free field theory with a small effective coupling constant, while the correction factors $f_{\text{cont}}(M^2)$ and $f_{\text{th corr}}(M^2)$ account for the continuum contribution and non-perturbative terms, respectively. Stripped of the correction factors, asymptotic freedom would lead to a prediction $m_\rho^2 = M^2$ with an arbitrary M^2 which is senseless.

It is just the factors f_{cont} and $f_{\text{th corr}}$ which bring in the dimensional parameters. In one case it is the beginning of the continuum *, $s_0 = 1.5\ \text{GeV}^2$, while in the other case the mass scale is associated with the non-vanishing vacuum expectation $\langle 0|G^a_{\mu\nu}G^a_{\mu\nu}|0\rangle$, $\langle 0|\bar{q}q|0\rangle$.

In more detail, the correction factors look as follows:

$$f_{\text{th corr}} = \frac{1 + \alpha_s/\pi - 0.1\left(\dfrac{0.6\ \text{GeV}^2}{M^2}\right)^2 + 0.28\left(\dfrac{0.6\ \text{GeV}^2}{M^2}\right)^3}{1 + \alpha_s/\pi + 0.1\left(\dfrac{0.6\ \text{GeV}^2}{M^2}\right)^2 - 0.14\left(\dfrac{0.6\ \text{GeV}^2}{M^2}\right)^3}\,,$$

$$f_{\text{cont}} = \frac{[\text{total}/(\text{total} - \text{continuum})]_0}{[\text{total}/(\text{total} - \text{continuum})]_1}\,. \tag{3.17}$$

* For further comments see to the end of this subsection.

Here the subscripts "0" and "1" refer to the integrals

$$\int e^{-s/M^2} R^{I=1}(s)\, ds\,, \qquad \int e^{-s/M^2} R^{I=1}(s)\, s\, ds\,,$$

respectively; "total" and "continuum" stand for the total theoretical value and the continuum contribution to the integrals, respectively. (The latter is specified by eq. (3.13).)

Corrections $f_{\rm cont}$ and $f_{\rm th\ corr}$ can compensate for the M^2 factor in eq. (3.16) and stabilize the prediction at a fixed mass. Indeed, the factors $f_{\rm cont}$ and $f_{\rm th\ corr}$ *versus* M^2 are depicted in fig. 3. (The A and B arrows indicate, as usual, M^2 sensitive to the resonance contribution, and reliable from the theoretical point of view). The former factor is close to unity at small M^2, $M^2 \lesssim 0.4$ GeV^2 and is substantial at high M^2, $M^2 \gtrsim 0.8$. For the theoretical correction the situation is just reversed.

Moreover, as seen from fig. 3, the product $M^2 f_{\rm cont} f_{\rm th\ corr}$ does not vary for M^2 between the A and B arrows and coincides with $(m_\rho^2)_{\rm exp}$. Of course, this is not a new result but a simple repetition of our argument in subsect. 3.4. We just wanted to make more explicit the rôle of the power corrections.

It is worth noting that introducing the dimensional parameter $s_0 \simeq 1.5$ GeV^2 through the continuum model is in fact unsatisfactory. It might make a false impression that the sum rules just relate the ρ mass to s_0 introduced "by hand" and that

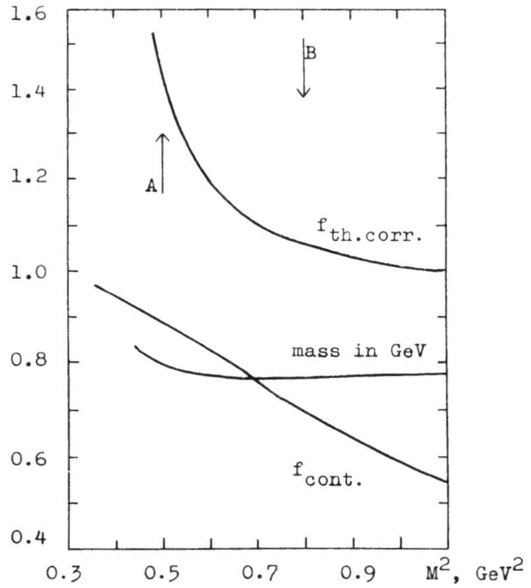

Fig. 3. Theoretical prediction for the rho meson mass. The resonance mass is defined as $(M^2 f_{\rm th\ corr} f_{\rm cont})^{1/2}$. Also shown are the factors $f_{\rm th\ corr}$ and $f_{\rm cont}$ given by eq. (3.17).

is all. In fact, we can consider s_0 in eq. (3.13) as a fit parameter and find it from the sum rules themselves. Then the only mass scale is manifestly due to the theoretical corrections associated with the power terms. The fit gives

$$s_0 \approx 2.5\ m_\rho^2\ ,$$

so that all the conclusions are untouched.

3.6. More on the power corrections

Since the computation of the ρ mass and electronic width is one of our central points we pause here to add a few further comments on the power corrections.

In principle, one can imagine three distinct cases to realize:

(a) the correction factor $(f_{\text{th corr}} - 1)$ becomes rather large when $(f_{\text{cont}} - 1)$ is also appreciable;

(b) the factor f_{cont} approaches unity just at M^2 where $f_{\text{th corr}}$ deviates from it;

(c) there exists a region of M^2 where both f_{cont} and $f_{\text{th corr}}$ are well approximated by unity.

The case (c) would imply that the theory was wrong. Indeed, the prediction $m_\rho^2 = M^2$ would be self-contradictory.

The case (a) would imply that the cross section is structured in a rather complicated way: it would require the smooth cross section at high energy to match the resonances *via* an extended intermediate region which is neither asymptotic nor belongs to resonance physics. Still the theory could be correct and tested numerically for a cross section smeared over many resonances. (It would be two-parameter theory in a way; apart from the mass of the lowest lying state there would be something else.)

The case (b) implies the theory to be simple: "high" and "low" energies are not gapped, and the resonances conspire to bring the asymptotic value of the smeared cross section to as low energies as possible. The simplicity of the theory is manifested in the simplicity of its testing experimentally: the theory is correct only if f_{cont} approaches unity just at $M^2 = m_\rho^2$ and theoretical corrections start to increase at the same point as well. We saw that this is just what happens.

Qualitatively, case (b) does correspond to the violation of asymptotic freedom by power terms. Indeed, the change of the asymptotic to the resonance behaviour is then fast and there is no place for an extended intermediate region. If the log corrections were to be blamed for the breaking of asymptotic freedom then one can imagine that the case (a) would materialize. Indeed, log M^2 varies much more slowly than, say, M^{-6}.

Although we do think that the results obtained indicate the validity of the underlying field theory let us emphasize once more that we assume a certain model for the cross section. A complete theory would make this last assumption superfluous. The summation of the series of the power corrections would, hopefully, produce a resonance structure explicitly. We cannot prove this, however, and cannot claim, therefore, evaluation of the spectrum of the resonances. The sum rules as they stay now allow only a check on the existence of a resonance solution. In the case of the ρ it works well. In the case of an axial current there are two resonances, such as π, A_1 and we demonstrate in sect. 4 that the sum rules do indicate the difference.

All the model assumptions aside, one can say that the power corrections fix the sign of deviations from asymptotic freedom. Indeed, at large M^2 the physical cross section must approach the quark one. At lower M^2, the corrections enter the game. In particular, we predict that the corrections are negative in the case of the sum rules (3.6) and positive in the case of the sum rules (3.7). If our model of the continuum is worthy enough to reproduce at least the signs (and we do feel so) then figs. 2a, 2b demonstrate that the data respect the theory. The region of M^2 where the power corrections become appreciable but still manageable is very narrow in fig. 2b. First the correction improves the agreement between experiment and theory and then, when it becomes too large to be trusted, worsens it. Therefore, the neighboring terms in the power expansion must be of the opposite sign.

To summarize our lengthy discussion, the power corrections to the sum rules set the mass scale and fix in this way the ρ-meson mass. The model-independent test of the power corrections is provided by the signs of the deviations from asymptotic freedom. So far, QCD stands the tests well.

3.7. On the accuracy of the theoretical predictions

So far we have taken the favored set of parameters (3.5) at its face value without trying to vary it. As explained in sect. 6 of (I), the parameters are fixed from independent sources such as weak non-leptonic decays [11]. In sect. 2, we also used the charmonium sum rules to find $\langle 0|G^a_{\mu\nu}G^a_{\mu\nu}|0\rangle$. The success achieved in these calculations encourages further use of the parameters.

Still, there is some uncertainty and we feel that the estimates of $\langle 0|\alpha_s G^a_{\mu\nu}G^a_{\mu\nu}|0\rangle$, $\langle 0|\alpha_s \bar\psi\Gamma\psi\bar\psi\Gamma\psi|0\rangle$ can be changed within a factor of two. The question is how the theoretical predictions for the resonances are changed under such a variation.

To find the answer turn again to figs. 2 which were used to extract the predictions. Changing the vacuum expectation values roughly speaking, shifts curves (1) and (2), and the A arrow as a whole in the horizontal direction.

It is rather clear that the prediction for the electronic width is affected only slightly by this shift. Indeed, the prediction comes from the height of the curve (2) in fig. 2a at its maximum, and the change in the position is of little importance.

In other words the stability of the prediction for the electronic width can be understood in the following way. Eq. (3.10) follows from a consideration of the function

$$x\,\mathrm{e}^{-x}$$

near its extremum point $x = 1$. Here, x stands for the ratio m_ρ^2/M^2.

Thus the computation of g_ρ^2 seems to be very reliable. The major uncertainty, which amounts to 10% at most, is associated with, say the zero-width approximation and can be readily diminished.

The prediction for the ρ-meson mass is correlated with the position of the maximum of the curve (2) in fig. 2a and is more sensitive to the variation in the parameters. Indeed, power corrections are the only source of the mass scale in our analysis. Thus, multiplying $\langle 0|\alpha_s \bar{\psi}\Gamma\psi\bar{\psi}\Gamma\psi|0\rangle$ by a factor of 2 implies a change in the scale by a factor of $\sqrt[6]{2}$. Therefore, the uncertainty in computation of m_ρ^2 amounts to 30%.

In other words, the variation of the parameters to their extreme would cause the arrows A and B in fig. 2b to coincide, so that we are left with a single value of M^2 to find the ρ mass.

Since the accuracy of the mass computation is not so high and since some dimensional parameters have been introduced into the theory "by hand" (*via* the vacuum expectation values) one may get disappointed and say that the theoretical predictions are rather trivial.

It is worth emphasizing , therefore, that despite the uncertainty involved the whole calculation is far from being a simple dimensional estimate.

Indeed we start with equation

$$\langle 0|\bar{q}q|0\rangle = -m_\pi^2 f_\pi^2/2(m_\mathrm{u} + m_\mathrm{d}) \, ,$$

which is one of the parameters introduced "by hand". If one would take naïvly $|\langle 0|\bar{q}q\,|0\rangle| \sim m_\pi^3$ which corresponds to $(m_\mathrm{u} + m_\mathrm{d}) \sim m_\pi$, then the ratio $\langle 0|\bar{q}q|0\rangle^2/m_\rho^6$ would have been of order 10^{-4} and one could conclude that the corresponding power correction can be safely neglected.

Moreover, even if one is aware that the quark mass can be as low as ~ 5 MeV, still $\langle 0|\bar{q}q|0\rangle^2/m_\rho^6 \sim 10^{-3}$ and the power corrections do not seem to play any appreciable rôle. The real calculation, however, boosts such an estimate by a numerical factor of about 200 ($\sim 6\pi^3$) and only for this reason do the power corrections become important. The sign of the correction is fixed and agrees with the data. Moreover, it is remarkable that an omission of the term proportional to $\langle 0|\bar{\psi}\Gamma\psi\bar{\psi}\Gamma\psi|0\rangle$ would leave us with the G^2 correction which has a "wrong" sign. Both signs cannot be changed and we see that the balance is quite delicate.

As for the second dimensional parameter, $\langle 0|\alpha_s G_{\mu\nu}^a G_{\mu\nu}^a|0\rangle$, it is borrowed from the analysis of the sum rules for charm production. However, as explained in sect. 2, any translation of the sum rules for heavy particles into the language of light mesons is far from being trivial, since the "duality interval" depends not only on the virtuality considered but on the quark mass as well. The theory also stood this test.

3.8. Questions answered: problems ahead

In subsect. 2.3 we posed some questions which seemed to be crucial in applying QCD to resonances. Now we have got answers to these questions.

First of all, the existence of the ρ is required to some extent by the theory which includes power corrections. Indeed, if we had asymptotic freedom alone, then the physical cross section most probably would have been a smooth function up to very low values of s. When power correction is included, a simple cross section which coincides with the bare one starting from the threshold is no longer a solution to the sum rules. Some structure is required and a resonance, with specific mass and electronic width, fits the sum rules well.

Moreover, we explained theoretically the difference in the duality intervals for the J/ψ and the ρ mesons. If we study the sum rules for heavy particles at $Q^2 = 0$ and choose $Q^2 = 4m_c^2$ for the light quarks so that the distance from the threshold is the same, the power correction for heavy particle grows as an extra power of n (n is the number of the derivative). It implies that in the s scale, the cross section for charm is more structured than that for light particles.

Thus, we have some hints that the framework developed has something to do with reality. To confirm it, it would be important to explain in the same way other salient features of the hadronic spectrum. Without trying to be exhaustive, let us list some qualitative features of the spectrum to be explained by any theory of hadrons.

(a) The independence of the mass spacing on the quark flavor (e.g., $m_{\psi'} - m_{J/\psi} \simeq m_{\rho'} - m_{\rho}$).

(b) The nearly massless pion.

(c) The difference between the spectra of vector and axial-vector states.

(d) The growth of the mass of low-lying resonances with their spin.

(e) The pattern of the $SU(3)_{\text{flavor}}$ breaking.

(f) The status of the gluon bound states (the so-called gluonium).

(g) The absence of mesons with exotic quantum numbers.

Only problem (a) has been treated so far. In the subsequent sections we will address ourselves to some other problems but not to all of them.

4. Axial vector current

So far we have discussed vector currents which are a special case because of the possibility of measuring the corresponding leptonic widths directly in e^+e^- annihilation. From the purely theoretical point of view other currents such as scalar, pseudoscalar and so on are not worse but the experimental information is limited in most cases to the masses of the low-lying states. The corresponding current-induced widths are not measurable as a rule and this hampers any detailed comparison of the theory with experiment.

Still, the axial vector current with $I = 1$ occupies the better position among the others. First, the current coupling to the π meson, f_π, is very well known. Moreover, the recent discovery of a heavy lepton [12] allows one to measure through the decays, $\tau \to \nu_\tau + X$, the coupling of a hadronic state X to the axial current. (We assume at this point the standard picture according to which the τ is a spin-$\frac{1}{2}$ lepton with its own neutrino ν_τ, and the weak interaction Hamiltonian is given by the product of the currents.)

In this section we will consider the QCD sum rules for the spectral density induced by the $I = 1$ axial vector current. The central point is the computation of the constant f_π in terms of the ρ-meson mass (subsect. 4.4). Other numerical predictions concern the integral over the spectral density in the region associated with the A_1 meson.

We will concentrate also on one of the basic problems as to how the QCD sum rules distinguish between the vector and axial vector cases. In particular, in subsect. 4.1 we prove the existence of a massless pion in the chiral limit. The result is very well known [13]. There is a unique possibility, however, to identify a certain term in the operator expansion with the contribution of a single physical state and we will dwell upon this. In subsect. 4.5 we discuss the difference between the vector and axial vector channels.

4.1. Massless pion

The common belief is that the pion mass vanishes in the limit of exact chiral symmetry so that the pion is a Goldstone particle. A vanishing pion mass is a central point for any theory of spontaneous chiral symmetry breaking. However, we approach the resonance region from high Q^2 and cannot probe too low Q^2. Therefore, we are not sensitive, in general, to the mass scale m_π^2. The first impression is that we can set the pion mass equal to zero "by hand" but cannot prove $m_\pi^2 = 0$ starting from the operator expansion. We will argue that this is not so and that the operator expansion in this particular case is powerful enough to indicate the presence of a massless particle in the limit $m_q \to 0$.

Let us first formulate the result and discuss its implications. Consider the T-product of axial vector currents

$$\Pi^a_{\mu\nu} = i \int e^{iqx} dx \langle 0 | T \{ a^+_\mu(x), a^-_\nu(0) \} | 0 \rangle \tag{4.1}$$

$$= -\Pi_1(Q^2)\, g_{\mu\nu} + \Pi_2(Q^2)\, q_\mu q_\nu ,$$

$$a_\mu(x) = \bar{u}(x)\, \gamma_\mu \gamma_5 d(x) . \tag{4.2}$$

Since the axial vector current is not conserved in the real world there are now two independent functions $\Pi_1(Q^2)$ and $\Pi_2(Q^2)$. In the limit of vanishing quark mass the function $\Pi_1(Q^2) + Q^2\Pi_2(Q^2) \equiv \Pi_\parallel$ becomes a polynomial since its imaginary part vanishes, $\mathrm{Im}(\Pi_1 - s\Pi_2) = 0, m_q \to 0$. Now, switch on a small quark mass and keep terms linear in this mass. The central point is that for large Q^2 the function $\Pi_1 + Q^2\Pi_2$ is exactly calculable in this approximation:

$$\Pi_1 + Q^2\Pi_2 = \frac{(m_u + m_d)\langle 0 | \bar{u}u + \bar{d}d | 0 \rangle}{Q^2} . \tag{4.3}$$

It is remarkable that there are no terms of higher orders in Q^{-2}. On dimensional grounds alone one can imagine having such terms as $(m_u + m_d)\langle 0|(\bar{u}u + \bar{d}d)^3|0\rangle$ Q^{-8} but our statement is that they do not actually appear (see below). Let us show that eq. (4.3) implies the existence of a nearly massless pion. Compare to this end eq. (4.3) with the general dispersion representation for the polarization operator:

$$\Pi_1 = C_1 + C_2 Q^2 + \frac{1}{\pi}\int \frac{\text{Im}\,\Pi_1(s)\,\text{d}s}{s+Q^2} ,$$

$$\Pi_2 = C + \frac{1}{\pi}\int \frac{\text{Im}\,\Pi_2(s)\,\text{d}s}{s+Q^2} , \tag{4.4}$$

$$\Pi_1 + Q^2\Pi_2 = \tilde{C}_1 + \tilde{C}_2 Q^2 + \frac{1}{\pi}\int \frac{\text{Im}\,\Pi_1 - s\,\text{Im}\,\Pi_2}{s+Q^2}\,\text{d}s ,$$

where C, $C_{1,2}$, $\tilde{C}_{1,2}$ are subtraction constants (note that we wrote for simplicity the integral over the imaginary part as if there were no subtractions. It does not matter: say, by differentiating twice we remove any memory of subtractions).

Moreover, the spectral densities Im $\Pi_{1,2}$ can be expressed in terms of the physical state contributions:

$$\text{Im}\,\Pi_1 = \sum_{\text{A}} \pi h_{\text{A}}^2 \delta(s - m_{\text{A}}^2) ,$$

$$\text{Im}\,\Pi_2 = \sum_{\text{A}} \pi h_{\text{A}}^2 m_{\text{A}}^{-2} \delta(s - m_{\text{A}}^2) + \sum_{\text{P}} \pi h_{\text{P}}^2 \delta(s - m_{\text{P}}^2) , \tag{4.5}$$

where the indices P and A refer to the pseudoscalar (spin-0) and pseudovector (spin-1) states, respectively, and $m_{\text{A,P}}$, $h_{\text{A,P}}$ denote the corresponding masses and residue constants. From eqs. (4.3)–(4.5) we find:

$$\frac{(m_u + m_d)\langle 0|\bar{u}u + \bar{d}d\,|0\rangle}{Q^2} + \text{O}(m_q^2) = -\sum_{\text{P}} h_{\text{P}}^2 m_{\text{P}}^2 (Q^2 + m_{\text{P}}^2)^{-1} , \tag{4.6}$$

which holds only if there exists a pseudoscalar state satisfying the conditions

$$m_{\text{P}}^2 = \text{O}(m_q^1) , \qquad h_{\text{P}} = \text{O}(m_q^0) , \tag{4.7}$$

while all the states with a non-vanishing mass decouple in the chiral limit:

$$h_{\text{P}} = \text{O}(m_q^1) \quad \text{if} \quad m_{\text{P}} = \text{O}(m_q^0) .$$

Indeed, any other (non-singular) solution for h_{P}, m_{P} either gives no linear in m_q term or generates, along with it, an infinite series of the kind $m_q Q^{-2n}$, $n = 2,3,4$ ... .

The state (4.7) is naturally identified with the pion so that we rederive the well-known relation [14] (see sect. 6 of (I)):

$$f_\pi^2 m_\pi^2 = -(m_u + m_d)\langle 0|\bar{u}u + \bar{d}d|0\rangle\,, \tag{4.8}$$

which manifestly demonstrates the vanishing of the pion mass in the chiral limit (provided that $\langle 0|\bar{u}u + \bar{d}d|0\rangle \neq 0$).

Thus, existence of a nearly massless pion is implied by eq. (4.3). To derive in turn eq. (4.3), consider $q_\mu q_\nu \Pi^a_{\mu\nu}$ where $\Pi^a_{\mu\nu}$ is defined in eq. (4.1). By virtue of the equations of motion it is related to the T-product of the pseudoscalar densities:

$$q_\mu q_\nu \Pi^a_{\mu\nu} = Q^2(\Pi_1 + Q^2\Pi_2) = -i(m_u + m_d)^2$$

$$\times \int dx\, e^{iqx}\langle 0|T\{\bar{d}(x)\,\gamma_5 u(x)\,,\bar{u}(0)\,\gamma_5 d(0)\}|0\rangle + \text{const}\,. \tag{4.9}$$

The constant on the right-hand side accounts for possible contact terms which usually arise if one differentiates a T-product. This constant corresponds to the Q^{-2} term in the combination $\Pi_1 + Q^2\Pi_2$.

Thus, this combination is proportional to $(m_u + m_d)^2$ except for a possible Q^{-2} term which calls for special consideration. The consideration can be given either by evaluating the commutator or the corresponding Feynman graphs. In particular, the graph depicted in fig. 4 contributes to $q_\mu q_\nu \Pi^a_{\mu\nu}$ in first order in $m_{u,d}$ and gives rise to eq. (4.3). Clearly enough, the same graph corresponds to the contact term arising due to the differentiation of the T-product.

Derivation of eq. (4.3) completes the proof. We have shown that the operator expansion requires a massless pion in the limit of $m_{u,d} = 0$, if $\langle 0|\bar{q}q|0\rangle \neq 0$.

Although the result is trivial by itself it is amusing to have a particular term in the operator expansion identified with a contribution of a single physical state.

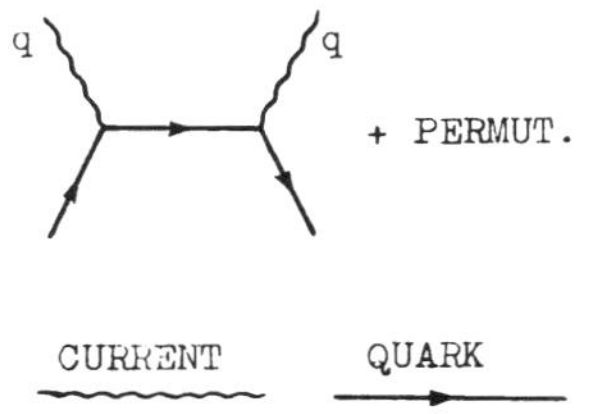

Fig. 4. The graph giving rise to a contact term in the product $q_\mu q_\nu \Pi^a_{\mu\nu}$, $\Pi^a_{\mu\nu}$ being the polarization operator induced by an axial-vector current (for the definition of $\Pi^a_{\mu\nu}$ see eq. (4.1)). The momentum carried by the current is denoted by q.

4.2. U(1) problem

The presence of a massless particle can be demonstrated in the $I = 1$ channel since in this case the axial vector current is conserved. As for the current

$$\sqrt{\tfrac{1}{3}}(\bar{u}\gamma_\mu\gamma_5 u + \bar{d}\gamma_\mu\gamma_5 d + \bar{s}\gamma_\mu\gamma_5 s)\,,$$

the triangle anomaly [15] invalidates the proof. This can resolve, in our language, the so-called U(1) problem * i.e., the absence of a nearly massless pseudoscalar particle with $I = 0$. To be quantitative we must consider the triangle graphs and see whether they are important numerically. Moreover, instantons of small size can contribute an anomalous amount in this case [17] because of the 't Hooft multifermion effective interaction [18].

We feel that a careful analysis of this kind would introduce too much of a new element and to make the consideration uniform we postpone the computations in the $I = 0$ channel until a future publication. By doing so we in no way intend to draw a veil over the U(1) problem. Its constructive resolution, i.e., an evaluation or at least a rough estimate of the η' mass, starting from QCD dynamics seems to be crucial for the whole framework.

4.3. Sum rules

After the general remarks on massless particles we proceed now to a regular derivation of the results mentioned in the introduction to the present section.

The operator expansion of the two axial vector currents was obtained in sect. 4 of (I). Taking the vacuum-to-vacuum matrix element produces the polarization operator $\Pi^a_{\mu\nu}$ defined in eq. (4.1). For simplicity we neglect the u- and d-quark masses, i.e., choose to work in the limit $m_\pi^2 = 0$. The error introduced in this way is of order $m_\pi^2/\mathrm{GeV}^2 \sim 0.02$, and negligible. In this limit the current (4.2) is conserved and $\Pi_1(Q^2) + Q^2\Pi_2(Q^2) = 0$ so that there is only one independent structure function.

Starting with the dispersion representation (4.4), using the operator expansion and applying the $\hat{L}_M$ procedure (see (I)) we find sum rules for, say, Π_2:

$$\int \mathrm{e}^{-s/M^2}\,\mathrm{Im}\,\Pi_2(s)\,\mathrm{d}s = \frac{M^2}{4\pi}\Bigg[1 + \frac{\alpha_s(M)}{\pi} + \tfrac{1}{3}\pi^2\cdot\frac{\left\langle 0\left|\frac{\alpha_s}{\pi}G^a_{\mu\nu}G^a_{\mu\nu}\right|0\right\rangle}{M^4}$$
$$+\frac{4\pi^3\alpha_s\langle 0|\bar{u}\gamma_\alpha\gamma_5 t^a d\,\bar{d}\gamma_\alpha\gamma_5 t^a u|0\rangle}{M^6} \tag{4.11}$$
$$-\tfrac{4}{9}\pi^3\alpha_s\frac{\langle 0|(\bar{u}\gamma_\alpha t^a u + \bar{d}\gamma_\alpha t^a d)\sum_{q=u,d,s}\bar{q}\gamma_\alpha t^a q|\rangle}{M^6}\Bigg].$$

* An exhaustive discussion is given in ref. [16].

Substituting the matrix elements as given by eqs. (3.5) we finally get

$$\int e^{-s/M^2}\ \mathrm{Im}\ \Pi_2(s)\ \mathrm{d}s = \frac{M^2}{4\pi}\left\{1 + \frac{\alpha_s(M)}{\pi} + 0.1\left(\frac{0.6\ \mathrm{GeV}^2}{M^2}\right)^2 + 0.22\left(\frac{0.6\ \mathrm{GeV}^2}{M^2}\right)^3\right\}, \tag{4.12}$$

while the sum rule for Π_1 takes the form:

$$\int e^{-s/M^2}\ \mathrm{Im}\ \Pi_1(s)\ \mathrm{d}s = \frac{M^4}{4\pi}\left\{1 + \frac{\alpha_s(M)}{\pi} - 0.1\left(\frac{0.6\ \mathrm{GeV}^2}{M^2}\right)^2 - 0.44\left(\frac{0.6\ \mathrm{GeV}^2}{M^2}\right)^3\right\}. \tag{4.13}$$

As in the case of the ρ meson, we assume a simplified version of the cross section:

$$\mathrm{Im}\ \Pi_2 = \pi f_\pi^2 \delta(s) + \pi m_{A_1}^2 g_{A_1}^{-2} \delta(s - m_{A_1}^2) + \frac{1}{4\pi}\left(1 + \frac{\alpha_s(s)}{\pi}\right)\theta(s - s_0),$$

$$\mathrm{Im}\ \Pi_1 = \pi m_{A_1}^4 g_{A_1}^{-2} \delta(s - m_{A_1}^2) + \frac{s}{4\pi}\left(1 + \frac{\alpha_s(s)}{\pi}\right)\theta(s - s_0), \tag{4.14}$$

where the constants g_{A_1}, f_π are defined in the standard way:

$$\langle 0|\bar{u}\gamma_\mu\gamma_5 d|\pi\rangle = i f_\pi P_\mu\ ; \qquad \langle 0|\bar{u}\gamma_\mu\gamma_5 d|A_1\rangle = g_{A_1}^{-1} m_{A_1}^2 \epsilon_\mu\ .$$

Thus, all the preliminary work is done and we are in position to determine the coupling constants starting from the sum rules.

4.4. Computation of f_π

Saturating the sum rules at relatively low M^2 by a single pion fixes its coupling constant. Basically, the derivation is the same as for the ρ meson (see sect. 3) and we just sketch it.

Choosing $M^2 = m_\rho^2$ apparently ensures the π-meson dominance since there is no other state of low mass in the channel considered. On the other hand, the corrections to asymptotic freedom represented by the unit term in the right-hand side of eq. (4.12) are still moderate at such M^2.

Neglecting for the moment these corrections as well as contributions of states of higher mass to Im Π_2, we find

$$f_\pi = m_\rho/2\pi \simeq 125\ \mathrm{MeV}\ , \tag{4.15}$$

which is to be compared with the experimental value

$$f_\pi|_{\mathrm{exper}} \simeq 133\ \mathrm{MeV}\ . \tag{4.16}$$

Eq. (4.15) is an analog of the prediction (3.10) for the ρ coupling constant. Both follow from applying asymptotic freedom at $M^2 = m_\rho^2$. Their phenomenological success indicates that the coupling constant $\alpha_s(m_\rho^2)$ is indeed relatively small.

Eq. (4.15) holds for a particular choice of M^2, $M^2 = m_\rho^2$. As the next step we must check stability against variations in M^2. At this stage the role of the power corrections becomes manifest. Moreover, considering the sum rules (4.12) as a function of M^2 we will derive certain predictions for A_1.

4.5. Power corrections

Before proceeding to the numerical estimates of the A_1 coupling constant, we turn to the qualitative side of the problem. Chiral symmetry is known to be almost exact in nature. This implies the nearly vanishing of u- and d-quark masses. Setting $m_{u,d} = 0$ means in turn that standard perturbation theory cannot distinguish between vector and axial vector currents.

On the other hand, the difference in the resonance spectra in the two channels is quite striking. Instead of a single prominent resonance, the ρ in the vector case, we have widely split π, A_1 in the axial vector current density. The two cases are distinguished by the vacuum structure which is manifested in the non-vanishing vacuum expectation value, $\langle 0|\bar{q}q|0\rangle \neq 0$. The vacuum expectation values enter sum rules through the power corrections. Therefore, if we are right in our guess that first terms in the M^{-2} expansion already reveal the structure of the spectrum, there is an exciting possibility to watch the correlations between the simple quark graphs and the resonance masses.

To clarify the role of the power corrections it is instructive to consider the difference between the vector and axial vector densities. Within the accuracy of our computation we are free to consider three sum rules

$$\int e^{-s/M^2} [\mathrm{Im}\,\Pi_2 - \mathrm{Im}\,\Pi^{(\rho)}]\, ds = \frac{M^2}{4\pi}\, 0.36 \left(\frac{0.6\ \mathrm{GeV}^2}{M^2}\right)^3 , \tag{4.17a}$$

$$\int s\, e^{-s/M^2} [\mathrm{Im}\,\Pi_2 - \mathrm{Im}\,\Pi^{(\rho)}]\, ds = \frac{M^4}{4\pi} (-0.72) \left(\frac{0.6\ \mathrm{GeV}^2}{M^2}\right)^3 , \tag{4.17b}$$

$$\int s^2 e^{-s/M^2} [\mathrm{Im}\,\Pi_2 - \mathrm{Im}\,\Pi^{(\rho)}]\, ds = \frac{M^6}{4\pi} (0.72) \left(\frac{0.6\ \mathrm{GeV}^2}{M^2}\right)^3 . \tag{4.17c}$$

As usual, the last two equations follow immediately from the first one (4.17a) by differentiating with respect to $1/M^2$. It is important, however, that there is no need to introduce further parameters to derive the right-hand sides. Any extra differentiation would require introducing higher-order power corrections. Finally, note that in the limit $M^2 \to \infty$, eqs. (4.17a), (4.17b) coincide with the celebrated Weinberg sum rules [19].

It is important that the sign of the right-hand side of eqs. (4.17) is changed at each step. The phenomenological manifestations of these signs are readily identified in the first two cases. Indeed, the lightest hadron, the π, contributes to Π_2. With diminishing M^2 its contribution grows relative to all the other states due to the exponential weight factor. In compliance with it, the right-hand side of eq. (4.17a) is positive.

Furthermore, multiplying Π_2 by s cancels the π contribution. Indeed, the pole term $1/s$ then becomes a subtraction constant and drops off under differentiation. Then the ρ meson becomes the lightest state contributing to the sum rule. In full accordance with it, the r.h.s. of eq. (4.17b) is negative.

Multiplying by one more power of s shifts the weight to even higher masses and the excess of the axial vector density over the vector one is expected here.

At present, there is no data to check the third prediction but the interpretation of the first two signs seems to be indeed remarkable.

Turn now to the power corrections to the sum rules in the vector and axial vector channels taken separately. There is an important difference between the two cases. In the vector channel the M^{-4} and M^{-6} terms partly cancel each other while in the axial vector case they build up to each other. Qualitatively, this again looks satisfactory. Indeed, the energy gap between the π and A_1, contributing to Π_2, is wider than between the ρ and ρ' in the vector case. One can say that the cross section is more smoothened for the vector current than for the axial vector one. The power corrections clearly signal this difference.

Thus, even without any detailed analysis one is inclined to say that the signs of the power corrections are in amusing accord with the observed spectra.

4.6. *Estimates of the A_1 coupling constant*

The estimates of the A_1 coupling constant are more sensitive to the power corrections and the continuum model. Moreover we will see that the sum rules can hardly distinguish the A_1 from the continuum which is expected to start rather close to the A_1. Thus, it is better to say that the sum rules constrain the integrated cross section in the region covered by the A_1.

To be more precise, the sum rules cannot be reconciled with a wide gap between the A_1 and the continuum (by "continuum" we understand the region of multiparticle production with the smeared total cross section close to that predicted by the parton model). There is no similarity with the case of ρ in this respect; the reason is that the A_1 is quite heavy. Moreover the total A_1 width is rather large and it is no surprise that the sum rules are not sensitive to the leakage of the resonance cross section.

Therefore, it is better to discuss integrated cross sections. We have tried a number of models satisfying the sum rules and convinced ourselves that in all the cases

$$\int_{s_1}^{s_2} \tilde{\rho}_A \, ds = (1.1 - 1.3)\ \mathrm{GeV}^2 \,, \qquad (4.18a)$$

$$\int_{s_1}^{s_2} \tilde{\rho}_A \, ds/s = 0.8 - 1.0 \,, \tag{4.18b}$$

where $s_1 = (0.85 \text{ GeV})^2$, $s_2 = (1.35 \text{ GeV})^2$ and $\tilde{\rho}_A$ is the reduced spectral density defined as

$$\tilde{\rho}_A = [\text{Im}\, \Pi_2]_{\text{exp}}/[\text{Im}\, \Pi_2]_{\text{pert th}} \simeq \frac{4\pi}{1.08} [\text{Im}\, \Pi_2]_{\text{exp}} \,.$$

Thus, the net contribution of the A_1 is no larger than that of the bare quark cross section smeared over the same energy interval. Therefore, a conspicuous A_1 would contradict the sum rules.

The picture is in qualitative accord (or rather, is in no contradiction) with the experimental observations [20]. Indeed, according to ref. [20] the decay channel $\tau \to \nu_\tau + A_1$ does not dominate over other modes of the type $\tau \to \nu_\tau + (\geqslant 3$ charged hadrons). The absence of a conspicuous peak in the total cross section does not rule out the possibility, of course, of observing the A_1 in some particular channel, such as

$$(\text{axial vector current}) \to A_1 \to \rho\pi \,.$$

Experimentally it was found just in this way [20].

Now, as to the numerical considerations. Fig. 5 confronts various models of the experimental cross section with the sum rules. Fig. 5a corresponds to the sum rule (4.12). The dominant contribution here comes from the π. To illustrate the sensitivity to the continuum contribution we looked at three simple models: two extreme models with a strong A_1 and no A_1 at all, and an intermediate model:

Model 1 (π + A_1 + continuum)

$f_\pi = 133$ MeV , $m_{A_1}^2 = 1.53\ \text{GeV}^2$, $4\pi/g_{A_1}^2 = 0.227$, $s_0 = 1.53\ \text{GeV}^2$;

Model 2 (π + A_1 + continuum)

$f_\pi = 133$ MeV , $m_{A_1}^2 = 1.21\ \text{GeV}^2$, $4\pi/g_{A_1}^2 = 0.14$, $s_0 = 1.21\ \text{GeV}^2$;

Model 3 (π + continuum)

$f_\pi = 133$ MeV , $4\pi/g_{A_1}^2 = 0$, $s_0 = 0.75\ \text{GeV}^2$.

For all the models the parameters specified are substituted into the following expression for the imaginary part:

$$\text{Im}\, \Pi_2 = \pi f_\pi^2 \delta(s) + \pi m_{A_1}^2 g_{A_1}^{-2} \delta(s - m_{A_1}^2) + \frac{1}{4\pi}\left(1 + \frac{\alpha_s(s)}{\pi}\right) \theta(s - s_0) \,. \tag{4.19}$$

As seen from fig. 5, all the models reproduce well the gross features of the sum rules and deviate from the theoretical predictions only in minor details. The agreement improves if the cross section is made more realistic: a smooth structure in the beginning instead of the theta function; a broad A_1 peak instead of the delta function; a few small bumps here and there. Introducing these minor modifications to the models there is no difficulty in making the fit absolute.

The need for introducing these modifications is better seen in fig. 5b which summarizes the fits to the sum rule (4.13). An important new point here is that the pion gives no contribution and the sum rule is more sensitive to the cross section in the GeV region. The variation in the models amounts to 5–10% and becomes, in principle, detectable. Keeping in mind the possibility of introducing the minor modifications we still cannot say at the moment which model is better.

Note that all the models satisfy eq. (4.18) and the check of this prediction seems most crucial for the theory. As for the detailed structure it can vary.

4.7. Conclusions

The consideration of the π and A_1 mesons seems to bring a new success to the QCD sum rules. The mass splittings between π and ρ, ρ and A_1 are well-understood;

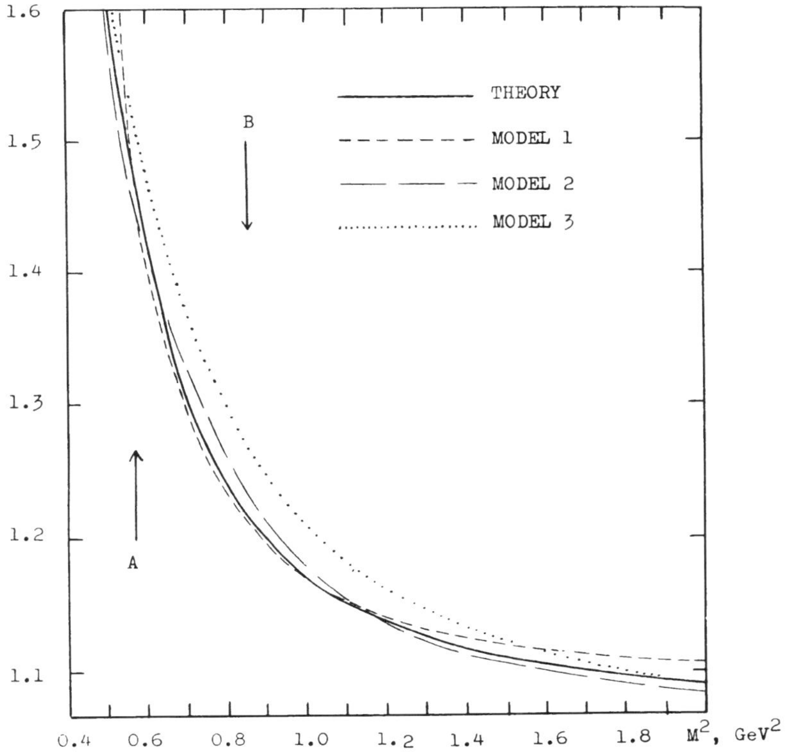

Fig. 5a.

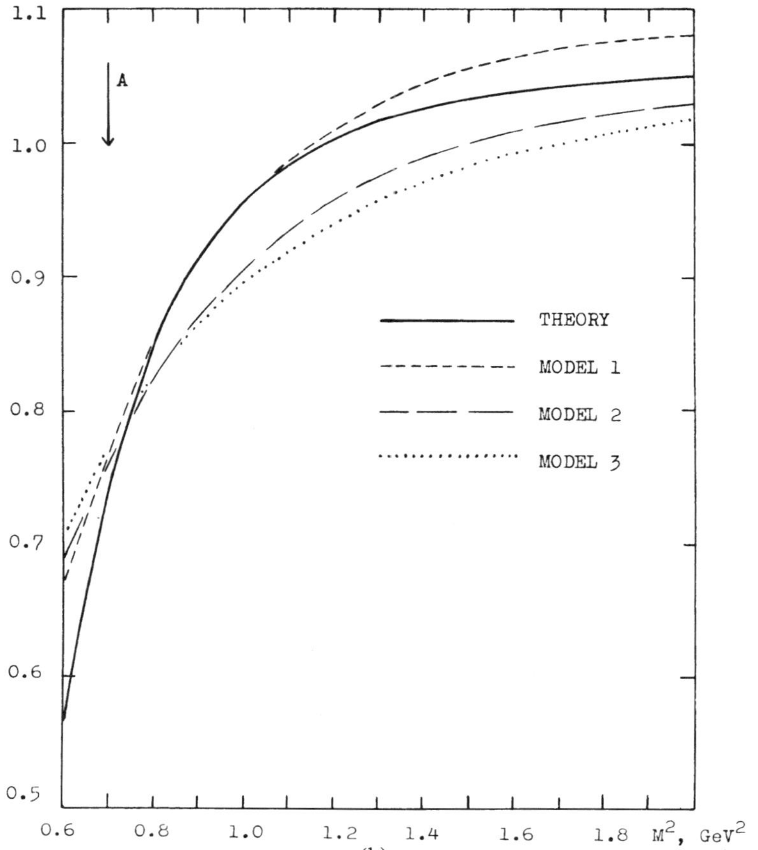

(b)

Fig. 5. (a) The function $4\pi M^{-2}\int e^{-s/M^2}$ Im $\Pi_2 ds$ *versus* M^2. (b) The function $4\pi M^{-4}\int e^{-s/M^2}$ Im $\Pi_1 ds$ *versus* M^2. For all the explanations see subsection 4.6. Arrow A indicates M^2 for which the power corrections give 30% of the total theoretical prediction. Arrow B in (a) marks M^2 for which the pion contribution gives 70% of the total.

the coupling constant f_π is computable in a reliable way. Further tests can be provided by a careful study of the τ decays.

5. Vector nonet

Since the sum rules are successful in describing the π, ρ, A_1 mesons it is natural to ask whether the results generalize to other mesons as well. As one of the extensions we consider here the nonet of vector mesons. As usual we are interested in masses and leptonic decays.

Saying that we consider the vector mesons almost automatically means that we study the SU(3) breaking. Indeed, in the limit of an exact $SU(3)_{flavor}$ the only dynamical problem is the ω-φ mixing. If it is known (and it *is* known to be small *), the masses and coupling constants are reduced to those of the ρ.

Confining ourselves to the deviations from exact SU(3) does not necessarily mean that the effects in point are small. Thus, the symmetry is violated rather heavily in masses, say, $m_\varphi^2 \simeq 1.7\ m_\rho^2$. Most of the theorists are seemingly inclined to assume that the same breaking occurs in the dimensionless coupling constants [22,23]. Moreover, there are some proposals to normalize the coupling constants to the fourth power of mass.

Within the dynamical framework developed, the coupling constants g_ρ^2, g_ω^2, g_φ^2, $g_{K^*}^2$ are fixed within, say, 10%. Therefore, we are able to distinguish between various possibilities of the symmetry breaking mentioned above. Roughly speaking the violation in the square of the dimensionless coupling constant is linear in mass. For example, we get

$$\frac{g_\varphi^2 : g_\rho^2}{(g_\varphi^2 : g_\rho^2)_{\text{exact SU(3)}}} \simeq 1.3\ , \quad (\text{compare } m_\varphi/m_\rho \simeq 1.3)\ .$$

True, the uncertainty in predicting the SU(3) violations in the ratios g_φ^2/g_ρ^2, $g_{K^*}^2/g_\rho^2$, etc. is rather large since the effects are of order 30% while the constants themselves are known with 10% accuracy. But for certain, the signs of the violations are fixed unambiguously. Moreover, the calculated values of m_φ, m_{K^*}, m_ω fall close to the experimental ones.

It is remarkable that the SU(3) violating effects are associated essentially with the vacuum expectation value $\langle 0|m_s\bar{s}s|0\rangle$ alone. This expectation value is more or less known theoretically **, a happy obstacle which permits one to start working without lengthy speculation.

The φ meson is dealt with in subsects. 7.1, 7.2 while the K^* is studied in subsect. 7.3.

Consideration of the vector mesons would be incomplete without computing the ω-φ mixing. Experimentally the mixing is known to be small [21]. We shall demonstrate that the QCD sum rules do imply a very reasonable value for the mixing parameter. Thus, at least in this case, QCD explains the success of the Zweig rule [4] on a more fundamental level. Moreover, unlike the standard three-gluon picture [6] which was claimed to give a wrong sign [25], the sum rules seem to reproduce the sign as well.

Without going into details here, let us indicate that the ω-φ mixing is sensitive to non-vacuum intermediate state contributions into the matrix elements $\langle 0|\bar{q}\Gamma q\bar{q}\Gamma q|0\rangle$, veiled so far. The reason is that the leading, vacuum, contribution is suppressed

* Experimental data can be found in ref. [21].
** For a recent review, see ref. [24].

strongly and there arises an interesting possibility to check the accuracy of the vacuum state saturation. The accuracy turns out to be rather high, a non-vacuum contribution turns out to be about 6% of the vacuum one (subsect. 7.4). Rather surprisingly we can put the estimate on trial immediately, by considering the ρ-ω mass splitting (subsect. 7.5).

5.1. Sum rules for the $j_\mu^{(\varphi)}$ current

Introduce the $j_\mu^{(\varphi)}$ current with the φ-meson quantum numbers:

$$j_\mu^{(\varphi)} = -\tfrac{1}{3}\bar{s}\gamma_\mu s\,, \tag{5.1}$$

and define the corresponding polarization operator in a standard way:

$$(q_\mu q_\nu - q^2 g_{\mu\nu})\,\Pi^{(\varphi)}(Q^2) = i\int e^{iqx}dx\,\langle 0|\mathrm{T}\{j_\mu^{(\varphi)}(x)\,,\ j_\nu^{(\varphi)}(x)\}|0\rangle\,,$$

$$(Q^2 \equiv -q^2)\,.$$

The operator expansion can be read off eq. (3.3) of ref. [1]:

$$9(\Pi^{(\varphi)} - \Pi^{(\varphi)}_{\text{pert. th}}) = \left\langle 0\left|\frac{2m_s\bar{s}s}{Q^4} + \frac{\alpha_s}{12\pi}\frac{G^a_{\mu\nu}G^a_{\mu\nu}}{Q^4}\right.\right.$$

$$-\frac{2\pi\alpha_s}{Q^6}\,\bar{s}\gamma_\mu\gamma_5 t^a s\,\bar{s}\gamma_\mu\gamma_5 t^a s \tag{5.2}$$

$$\left.\left.-\frac{4\pi\alpha_s}{9Q^6}\,\bar{s}\gamma_\mu t^a s\sum_{q=u,d,s}\bar{q}\gamma_\mu t^a q\right|0\right\rangle,$$

where we neglect, as usual, higher powers in Q^{-2}.

All the operators entering the r.h.s. of eq. (5.2) are normalized at the (Euclidean) point Q. On the other hand, the estimates of the matrix elements given in sect. 6 of (I) refer to a low normalization point, $\mu \sim R^{-1}$ where R is the confinement radius.

Therefore, to make use of the estimates of the matrix elements we must change the normalization point for the operators. The first two of them $m_s\bar{s}s$ and $\alpha_s G^a_{\mu\nu}G^a_{\mu\nu}$ are renormalization invariant in the leading log approximation which we accept here. The procedure is more involved in the case of the four-fermion operators; it is considered in detail in the appendix to ref. [1]. Combining all the factors gives

$$\langle 0|[\alpha_s(\bar{s}\gamma_\mu\gamma_5 t^a s\,\bar{s}\gamma_\mu\gamma_5 t^a s + \tfrac{2}{9}\bar{s}\gamma_\mu t^a s\sum_{q=u,d,s}\bar{q}\gamma_\mu t^a q)]_Q|0\rangle$$

$$= \alpha_s(\mu)\,\tfrac{112}{81}\,\langle 0|(\bar{q}q)_\mu|0\rangle^2\kappa^{-1/9}\eta^{(\varphi)}(\kappa)\,, \tag{5.3}$$

where

$$\langle 0|(\bar{q}q)_\mu|0\rangle \equiv \langle 0|(\bar{u}u)_\mu|0\rangle = \langle 0|(\bar{d}d)_\mu|0\rangle = \langle 0|(\bar{s}s)_\mu|0\rangle \,, \tag{5.4}$$

$$\eta^{(\varphi)}(\kappa) = 0.86 + 0.24\,\kappa^{-0.14} - 0.10\,\kappa^{-1.27} \,, \tag{5.5}$$

$$\kappa = \alpha_s(\mu)/\alpha_s(Q) \,.$$

The relations hold under two assumptions: (a) the $SU(3)_{flavor}$ symmetry for the quark vacuum expectation values (see eq. (5.4)), and (b) the intermediate vacuum state dominance in the matrix elements $\langle 0|(\bar{q}\Gamma q\bar{q}\Gamma q)_\mu|0\rangle$ (for a discussion see subsect. 6.5 in (I)).

Note that the Q^2 dependence enters only through the factor $\kappa^{-1/9}\eta^{(\varphi)}(\kappa)$ and is very weak in fact. Thus,

$$\eta^{(\varphi)} = 1.03 - 1.04 \,, \qquad \kappa = \text{several units} \,,$$

and $\kappa^{-1/9}\eta^{(\varphi)}(\kappa)$ equals unity within a few per cent.

It is worth emphasizing that the closeness of $\eta^{(\varphi)}$ to unity cannot be conjectured on general grounds alone. Any simple-minded approximation would overestimate the Q^2 dependence.

Although $\eta^{(\varphi)} \simeq 1$ cannot be proved in advance it could be well forseen. We have learnt such things from experience and get used to them. Indeed, the same is true for the ρ sum rules. Earlier we observed the weak Q dependence of the four-fermion operators entering the effective Hamiltonian of weak non-leptonic interactions [11]. Thus, we can guess the right answer in advance but cannot rely on the guess until the computation is performed thoroughly.

Eqs. (5.2)–(5.5) specify the sum rules completely. Since we are interested primarily in the SU(3) violations, let us make a few remarks on manifestations of the non-vanishing mass of the strange quark.

The leading contribution to the sum rules comes from $\Pi^{(\varphi)}_{\text{pert th}}$. There is no term linear in m_s here, and we keep the correction quadratic in m_s. As for the non-perturbative effects, we keep the correction linear in m_s to the leading Q^{-4} order while the term of order Q^{-6} is treated in the limit of exact SU(3). The reason is that the Q^{-6} term constitutes only a small fraction of the total sum.

Thus, in all the cases we keep the corrections O(10%) and neglect terms O(1%). To be more specific let us display the contributions of various terms into the sum rules at $M^2 = 1\ \text{GeV}^2$:

$$\Pi^{(\varphi)}_{\text{pert. th}} \sim 100\% \,, \qquad m_s^2 \text{ correction} \sim 5\% \,,$$

$$M^{-4}\begin{cases} m_s\bar{s}s \text{ term} \sim 30\% \,, \\ G^2 \text{ term} \sim 6\% \,, \end{cases}$$

$$M^{-6}\,\{(\bar{q}\Gamma q)(\bar{q}\Gamma q) \text{ term} \sim 8\% \,.$$

The final form of the sum rules arises after applying the operator $\hat{L}_M$ introduced in sect. 5 of (I). In this way we come to the sum rules for the first Borel transform of $\Pi^{(\varphi)}$:

$$\frac{9}{\pi M^2}\int e^{-s/M^2}\,\mathrm{Im}\,\Pi^{(\varphi)}(s)\,ds$$

$$=\frac{1}{4\pi^2}\left[1+\frac{\alpha_s(M)}{\pi}-\frac{6m_s^2(M)}{M^2}+\frac{8\pi^2\langle 0|m_s\bar{s}s|0\rangle}{M^4}\right. \qquad (5.7)$$

$$\left.+\frac{\pi^2}{3}\frac{\left\langle 0\left|\frac{\alpha_s}{\pi}G^a_{\mu\nu}G^a_{\mu\nu}\right|0\right\rangle}{M^4}-\frac{448}{81}\pi^3\alpha_s(\mu)\frac{\langle 0|\bar{q}q|0\rangle^2}{M^6}\right].$$

The matrix elements entering eq. (5.7) are discussed in detail in sect. 6 of (I). Note that when evaluating $\langle 0|m_s\bar{s}s|0\rangle$ one can use the SU(3) symmetry to fix $\langle 0|\bar{s}s|0\rangle$ since the mass m_s enters explicitly. Thus, we use eqs. (3.5) and

$$m_s = 150\ \mathrm{MeV}\,, \qquad \langle 0|\bar{s}s|0\rangle = -(0.25\ \mathrm{GeV})^3\,.$$

The values above correspond to the point $\alpha_s(\mu) = 0.7$.

Thus, numerically eq. (5.7) becomes:

$$9\int e^{-s/M^2}\,\mathrm{Im}\,\Pi^{(\varphi)}(s)\,ds=\frac{M^2}{4\pi}\left[1+\frac{\alpha_s(M)}{\pi}\right.$$

$$\left.-0.225\,\frac{0.6\ \mathrm{GeV}^2}{M^2}\left(\frac{\alpha_s(M)}{\alpha_s(\mu)}\right)^{8/9}-0.4\left(\frac{0.6\ \mathrm{GeV}^2}{M^2}\right)^2-0.14\left(\frac{0.6\ \mathrm{GeV}^2}{M^2}\right)^3\right]. \qquad (5.8a)$$

Differentiating this equation with respect to M^2 gives the second sum rule:

$$9\int e^{-s/M^2}\,\mathrm{Im}\,\Pi^{(\varphi)}(s)\,s\,ds=\frac{M^4}{4\pi}\left[1+\frac{\alpha_s(M)}{\pi}\right.$$

$$\left.+0.4\left(\frac{0.6\ \mathrm{GeV}^2}{M^2}\right)^2+0.28\left(\frac{0.6\ \mathrm{GeV}^2}{M^2}\right)^3\right]. \qquad (5.8b)$$

Note that the correction proportional to m_s^2/M^2 drops out of this sum rule.

Eqs. (5.8) are a starting point for evaluating the mass and electronic width of the φ.

5.2. *Electronic width and mass of the φ*

Proceeding in a standard way we accept a simplified model for the cross section associated with the $j_\mu^{(\varphi)}$ current:

$$\mathrm{Im}\,\Pi^{(\varphi)}=\frac{\pi m_\varphi^2}{g_\varphi^2}\delta(s-m_\varphi^2)+\frac{1}{36\pi}(1+\alpha_s(s)/\pi)\,\theta(s-s_0)\,,$$

where m_φ, g_φ, s_0 are the fit parameters to be determined from the sum rules. Keeping a single resonance in the spectral density, the φ, we rely on the phenomenological observation that the φ is an $\bar{s}s$ bound state to an accuracy of several per cent *.

As a first guess we choose s_0 as

$$s_0 = 1.95 \text{ GeV}^2 , \tag{5.9}$$

i.e., assume the gap between the meson and the beginning of the continuum to be the same for the φ as for the ρ. In fact, the value (5.9) emerges from the sum rules as well.

We would like to use the sum rules only as far as the continuum contribution does not exceed 30% of the total theoretical prediction. Thus, we need only a rough model for the continuum. Likewise, we do not want to go to such M^2 that the power corrections become too large.

Prior to proceeding to detailed computations let us make a few qualitative remarks. Even a rough estimate reveals immediately that the power corrections are much larger in the φ-channel than in the ρ one. This is indeed remarkable since an increase in the mass of the lowest-lying state is signalled in this way: since the φ is substantially heavier than the ρ, the structure in the cross section is of larger scale and the asymptotic freedom must be broken at higher M^2. Otherwise, the theory would be inconsistent with the observations.

Roughly speaking, the mass of a lowest-lying state is equal to M^2_{cr} which constitutes a boundary for asymptotic freedom. To be more specific, consider the ratio of the sum rule for the spectral density and that multiplied by s; this is the simplest way to compute the mass. In the case of the φ the sum rules are given by eqs. (5.8) while the ρ-channel is considered in sect. 3. Then one can readily check that the correction factor due to the power terms reaches 1.4 at $M^2 \sim 1$ GeV^2 in the case of the φ and at $M^2 \sim 0.6$ GeV^2 in the case of the ρ. Thus, the power corrections reproduce, in a way, the observed mass splitting.

As for the coupling constant the SU(3) breaking effects are not so drastic. Indeed, to extract g^2_φ, g^2_ρ one can consider equivalent values of M^2, i.e., such M^2 that the power corrections look similar. Thus, taking $M^2 \sim 1$ GeV^2 we see that the power corrections in eq. (5.8a) give ~20% of the total. The continuum contribution is about the same. Neglecting them both, we get as a first rough approximation:

$$g^2_\varphi/4\pi \simeq 9\pi/e , \tag{5.10}$$

which is an evident SU(3) generalization of eq. (3.10) for the ρ. Actually, there are some violations of SU(3) because of the differences in the continuum contributions in the φ and ρ channels for M^2 defined as "equivalent" above.

If we extend the analysis to include the corrections both due to the power terms

* For a detailed discussion of the Zweig rule in φ decays, see ref. [26].

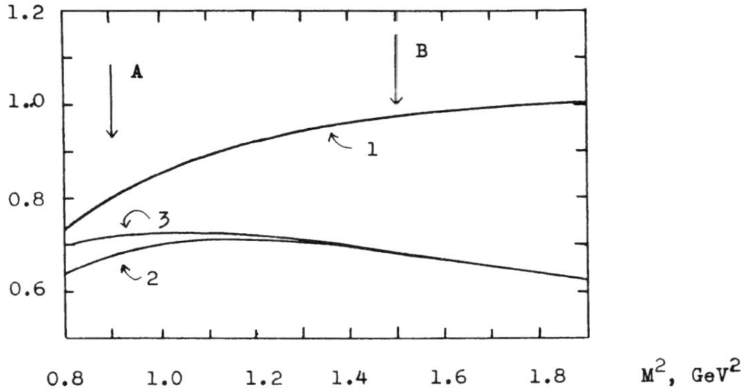

Fig. 6. The function $36\pi M^{-2}\int e^{-s/M^2}\,\mathrm{Im}\,\Pi^{(\varphi)}(s)\mathrm{d}s$ *versus* M^2. Curve 1, the theoretical prediction. Curve 2, the theoretical prediction with the continuum contribution subtracted. The following continuum model is used:

$$\mathrm{Im}\,\Pi^{(\varphi)}_{\mathrm{cont}} = (36\pi)^{-1}(1 + \alpha_s(s)/\pi)\,\theta(s - 1.95\ \mathrm{GeV}^2)\,.$$

Curve 3, the φ meson contribution with $m^2_\varphi = 1.04\ \mathrm{GeV}^2$, $4\pi/g^2_\varphi = 0.07$. Arrows A and B indicate M^2 for which the power correction and the continuum contribution reach 30% of the total respectively. The continuum contribution grows for larger M^2 while the power corrections become important with diminishing M^2.

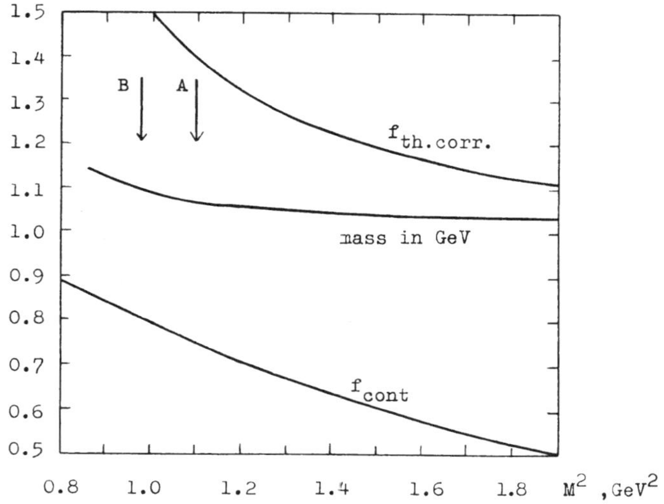

Fig. 7. Theoretical prediction for the φ meson mass. The resonance mass is defined as $(M^2 f_{\mathrm{th\ corr}}\, f_{\mathrm{cont}})^{1/2}$. Also shown are the factors $f_{\mathrm{th\ corr}}$ and f_{cont}.

and continuum we get *:

$$g_\varphi^2/4\pi \simeq 14\ , \qquad m_\varphi \simeq 1.07\ \mathrm{GeV}\ , \qquad s_0 \simeq 1.95\ \mathrm{GeV}^2\ , \tag{5.11}$$

which is our final result here.

The validity of the fit obtained is illustrated in figs. 6 and 7. The first one refers to the sum rule (5.8a). Curve 1 gives the theoretical prediction; curve 2 represents the same prediction with the continuum subtracted; curve 3 refers to the resonance contribution. If curves 2 and 3 coincide with each other then the fit is perfect.

We see that the fit (5.11) works well. Moreover, the experimental value of the coupling constant [21],

$$(g_\varphi^2/4\pi)_{\mathrm{exp}} = 11.7 \pm 0.9\ ,$$

falls rather close to the theoretical prediction.

Fig. 7 displays the result for the φ mass. The simplest way to fix the mass is to divide eq. (5.8b) by (5.8a). If the spectral density contained a single φ peak then the l.h.s. would yield m_φ^2. The presence of other contributions is reflected in the correction factor f_{cont}. The detailed discussion of the procedure can be found in subsect. 3.5. We are pleased to find $(m_\varphi)_{\mathrm{th}}$ only 50 MeV larger than $(m_\varphi)_{\mathrm{exp}}$.

Eq. (5.11) gives our best fit to the sum rules but the result can certainly be varied to some extent and we proceed now to a discussion of the corresponding uncertainty in the predictions.

It is instructive to this end to compare the sum rules for the φ and for the ρ. Even a first glance at figs. 6 and 2a reveals the difference between the sum rules. First, the power corrections start earlier in the φ-channel, as mentioned above. What is more important is that, say, $M^{-2}\int \exp(-s/M^2)\ \mathrm{Im}\ \Pi(s)\ \mathrm{d}s$ is a much smoother function of M^2 in the case of the φ. The reason is that the corrections proportional to M^{-2} and M^{-4} dominate in the φ sum rules while in the case of the ρ the leading correction comes from the M^{-6} term. This distinction is due to the m_s and m_s^2 effects.

As a result the whole picture for the φ becomes a bit more ambiguous. In particular, the maximum of the theoretical curve (2) becomes less conspicuous and it is just the position and the height of the maximum that determine the mass and width of a resonance. Moreover at M^2 around the maximum, the continuum contribution is about 10% of the total in the ρ sum rules and twice as much in the φ case.

Clearly enough, the net effect is that the uncertainty in determining the φ parameters is larger. To be more quantitative, note that both s_0 and g_φ^2 can be varied together. The meaning of such a variation is rather obvious: one can strengthen the resonance contribution and weaken the continuum one without shifting the total

* Notice, that the analysis of the present paper implies $g_\varphi^2/g_\rho^2 \simeq 5.8$, while in our earlier paper [27] there was obtained a somewhat larger number for this ratio. In the latter paper we overestimated the $m_s\bar{s}s$ contribution and as a result overestimated the SU(3) breaking effect.

considerably. One can lower the coupling constant g_φ^2 by up to 10% in this way.

Turn now to fig. 7 which summarizes the situation with the mass computation. The arrows A and B mark here the boundaries of the confidence region. Note, however, that they have changed their usual order so that arrow A is now to the right of the arrow B. Thus, we are left with no M^2 for which both the continuum and power corrections are safely small. Therefore, a detailed version of the continuum model can affect the mass computation.

The overall conclusion is that the uncertainty in g_φ^2 and m_φ^2 is about 10%.

It is worth emphasizing that we discuss here the uncertainty in the fit parameters under the condition that the power corrections are well-known. Thus, it cannot be contrasted directly with the discussion of possible variations in the parameters given in subsect. 3.7, since in this subsection we allowed for an uncertainty in the power corrections. Now we do not include such an uncertainty. The reason is that the experience with the ρ sum rules convinces us that the power corrections are reproduced correctly.

5.3. *The K^* meson*

Consider the current with the K^*-meson quantum numbers:

$$j_\mu^{(K^*)} = \bar{u}\gamma_\mu s \,. \tag{5.12}$$

Then the sum rules look the same as for the ρ, except for the operator $m_q\bar{q}q$ which violates SU(3) explicitly. (In fact the statement is qualified for neglect of the SU(3) violation in the contribution of the unit operator. The violation is proportional to m_s^2 and is present both in the transverse and longitudinal parts of the polarization operator. Numerically it is (roughly speaking) 4 time smaller than the m_s^2 contribution to the φ sum rules and can be safely omitted.)

Thus, defining the polarization operator in the standard way

$$\Pi_{\mu\nu}^{(K^*)} = i\int \mathrm{d}x\, e^{iqx}\langle 0|\mathrm{T}\,\{\bar{u}(x)\,\gamma_\mu s(x)\,,\,\bar{s}(0)\,\gamma_\nu u(0)\}\,|0\rangle\,,$$

we get for the change:

$$\Delta\Pi_{\mu\nu} = \left\langle 0\left| g_{\mu\nu}\frac{(m_u - m_s)(\bar{s}s - \bar{u}u)}{Q^2} + (q_\mu q_\nu - q^2 g_{\mu\nu})\frac{m_u\bar{u}u + m_s\bar{s}s}{Q^4}\right|0\right\rangle . \tag{5.13}$$

Since $m_u \ll m_s$, the second term on the right-hand side is just one half of the m_s correction in the φ sum rules. As for the first term it is a newcomer and is associated with current non-conservation. It is of second order in the SU(3) breaking, however, and can be safely neglected.

After these preliminary remarks we are in a position to write down the sum rules

$$\int e^{-s/M^2} \operatorname{Im} \Pi^{(K^*)}(s)\, ds = \frac{M^2}{4\pi} \left\{ 1 + \frac{\alpha_s(M)}{\pi} - 0.15\left(\frac{0.6\ \mathrm{GeV}^2}{M^2}\right)^2 - 0.14\left(\frac{0.6\ \mathrm{GeV}^2}{M^2}\right)^3 \right\}, \tag{5.14a}$$

$$\int e^{-s/M^2} \operatorname{Im} \Pi^{(K^*)}(s)\, s\, ds = \frac{M^4}{4\pi} \left\{ 1 + \frac{\alpha_s(M)}{\pi} + 0.15\left(\frac{0.6\ \mathrm{GeV}^2}{M^2}\right)^2 + 0.28\left(\frac{0.6\ \mathrm{GeV}^2}{M^2}\right)^3 \right\}, \tag{5.14b}$$

where the numbers arise from the use of $m_s = 150$ MeV, $\langle 0|\bar{s}s|0\rangle = -(0.25\ \mathrm{GeV})^3$ and eqs. (3.5).

Our toy model for the experimental data, to substitute in the left-hand side becomes:

$$\operatorname{Im} \Pi^{(K^*)} = \frac{\pi m_{K^*}^2}{g_{K^*}^2}\, \delta(s - m_{K^*}^2) + \frac{1}{4\pi}\left(1 + \frac{\alpha_s(s)}{\pi}\right)\theta(s - s_0)\,, \tag{5.15}$$

where g_{K^*} is defined in a standard way, $\langle 0|\bar{u}\gamma_\mu s|K^*\rangle = m_{K^*}^2 g_{K^*}^{-1} \epsilon_\mu$, ϵ_μ being the polarization vector. As for the continuum, s_0, it is natural to assume within the quark model that

$$s_0 = 1.7\ \mathrm{GeV}^2\,, \tag{5.16}$$

which is just in the middle of the corresponding s_0 in the ρ and φ sum rules. The validity of eq. (5.16) is confirmed by the sum rules as well.

Here we deviate from the straightforward road we have followed in working with the ρ and φ, and start with considering the K* mass rather than its coupling to the $\bar{u}\gamma_\mu s$ current. The reason is that it is just the K* mass that is known experimentally, not its coupling constant.

The procedure has become routine by now. Divide eq. (5.14b) by eq. (5.14a). Then

$$m_{K^*} = (M^2 f_{\text{th corr}} f_{\text{cont}})^{1/2}\,,$$

where

$$f_{\text{th corr}} = \frac{1 + \dfrac{\alpha_s(M)}{\pi} + 0.15\left(\dfrac{0.6\ \mathrm{GeV}^2}{M^2}\right)^2 + 0.28\left(\dfrac{0.6\ \mathrm{GeV}^2}{M^2}\right)^3}{1 + \dfrac{\alpha_s(M)}{\pi} - 0.15\left(\dfrac{0.6\ \mathrm{GeV}^2}{M^2}\right)^2 - 0.14\left(\dfrac{0.6\ \mathrm{GeV}^2}{M^2}\right)^3}\,, \tag{5.17}$$

$$f_{\text{cont}} = \frac{[\text{total}/(\text{total}-\text{continuum})]_0}{[\text{total}/(\text{total}-\text{continuum})]_1} \,.$$

Here the scripts "0, 1" refer to the integrals (5.14a), (5.14b), respectively.

The results are displayed in fig. 8. The arrows A and B mark, as usual, the boundaries of the "confidence region sensitive to a single resonance". We see that the arrows' positions practically coincide with each other so that there is not so much confidence as one might like. Still, we compute the mass to be

$$(m_{\text{K}^*})_{\text{th}} = 0.93 \text{ GeV} \,,$$

which is only 4% higher than the experimental mass. Thus, the sum rules turn to be reliable.

An estimate of the K^* current transition constant is also not without interest. (The only reliable way to measure g_{K^*} known to us is to observe the corresponding decay of the heavy lepton, $\tau \to \nu_\tau \text{K}^*$. It seems to be a very difficult experiment to perform at the moment but we hope that progress will be fast.)

Fig. 9 displays the sum rule (5.14a). Curve 1 represents the right-hand side. Subtracting the continuum contribution we come to curve 2 which must, therefore, coincide with the K^* contribution. The latter is given by curve 3 provided the following

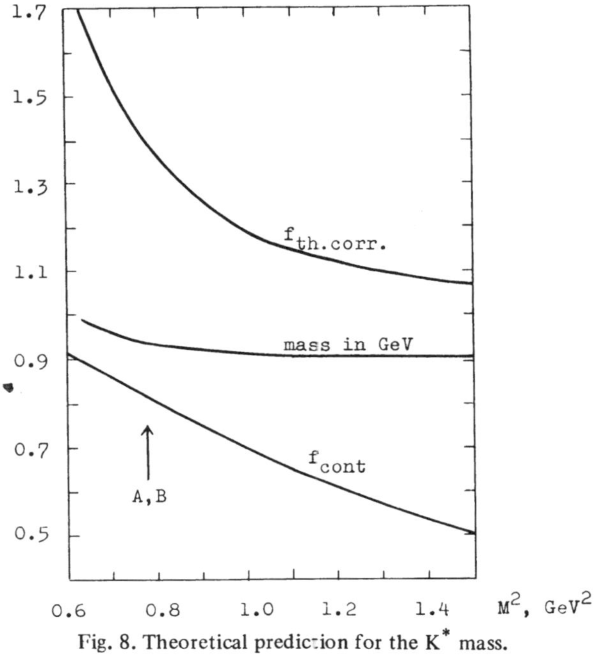

Fig. 8. Theoretical prediction for the K^* mass.

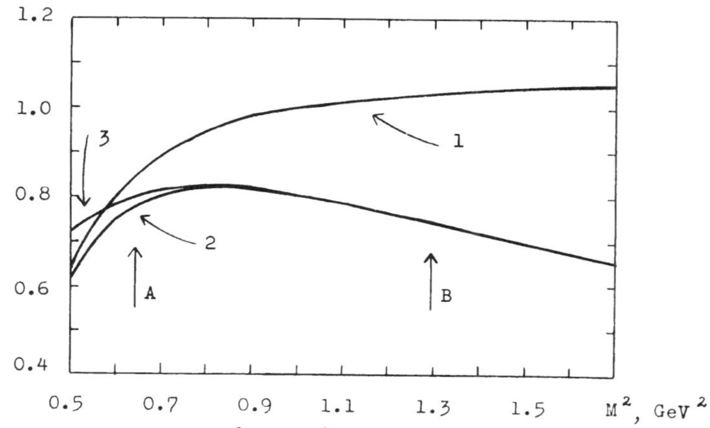

Fig. 9. The function $4\pi M^{-2} \int e^{-s/M^2}$ Im $\Pi^{(K^*)}(s)$ ds *versus* M^2. Curve 1, the theoretical prediction. Curve 2, the theoretical prediction with the continuum contribution subtracted. The following continuum model is used:

$$\text{Im}\, \Pi^{(K^*)}_{\text{cont}} = (4\pi)^{-1}(1 + \alpha_s(s)/\pi)\, \theta(s - 1.7\ \text{GeV}^2)\ .$$

Curve 3, the K* meson contribution with $m^2_{K^*} = 0.8$ GeV², $4\pi/g^2_{K^*} = 0.72$.

values of the parameters are accepted:

$$4\pi/g^2_{K^*} = 0.72\ , \qquad m^2_{K^*} = 0.8\ \text{GeV}^2\ . \tag{5.18}$$

The exact SU(3)$_{\text{flavor}}$ symmetry would imply $g^2_{K^*}/g^2_\rho = 0.5$, while the sum rules give

$$g^2_{K^*}/g^2_\rho \simeq 0.56\ , \tag{5.19}$$

where we use the values of g_{K^*}, g_ρ as specified by eqs. (5.18) and (3.15), respectively.

We see that the SU(3) violating effect in the ratio of the coupling constants (5.19) is ~2.7 times weaker than that in the ratio of the masses, $m^2_{K^*}/m^2_\rho$. Just the same thing happened with the φ (see subsect. 5.2).

5.4. ω-φ mixing

As noted in sect. 1, the novel feature of the ω-φ mixing sum rules is that they are sensitive to a non-vacuum contribution in the matrix elements $\langle 0|\bar{q}\Gamma q \bar{q}\Gamma q|0\rangle$. Therefore, they extend our understanding of the non-perturbative effects. To demonstrate the point, we will first show that the contribution of the vacuum intermediate state is suppressed in the case considered, and that this is the reason behind the observed smallness of the mixing.

For the relevant polarization operator,

$$\Pi^{(\omega\varphi)}_{\alpha\beta} = i \int e^{iqx} dx\, \langle 0|T\{j^{(\varphi)}_\alpha(x), j^{(\omega)}_\beta(0)\}|0\rangle\ ,$$

$$\Pi_{\alpha\beta}^{(\omega\varphi)} \equiv (q_\alpha q_\beta - q^2 g_{\alpha\beta})\, \Pi^{(\omega\varphi)} \,,$$

we get

$$\Pi^{(\omega\varphi)} = \left\langle 0 \left| \left[\frac{\pi\alpha_s}{9Q^6} \bar{s}\gamma_\alpha\gamma_5 t^a s (\bar{u}\gamma_\alpha\gamma_5 t^a u + \bar{d}\gamma_\alpha\gamma_5 t^a d) \right]_Q \right| 0 \right\rangle , \qquad (5.20)$$

where the subscript Q indicates the normalization point, the $j_\alpha^{(\varphi)}$ current is defined in eq. (5.1), while $j_\alpha^{(\omega)}$ is normalized in the following way:

$$j_\alpha^{(\omega)} = \tfrac{1}{6}(\bar{u}\gamma_\alpha u + \bar{d}\gamma_\alpha d) \,. \qquad (5.21)$$

Eq. (5.20) corresponds to the graphs in fig. 10a.

It is worth emphasizing that the polarization operator starts with a power "correction" this time. Of course, if we turn Q^2 to infinity then we would be left with the graph of fig. 10b; but this is a four-loop graph, and is seemingly small for M^2 of order of the resonance masses.

The salient feature of eq. (5.20) is that it is not possible to sandwich the matrix element with a vacuum intermediate state in any way. This is the formal reason for $\Pi^{(\omega\varphi)}$ to be small.

Thus, we must include contributions which come from two sources. (a) Mixing with the operators whose matrix elements allow for a vacuum intermediate state. Indeed, the hypothesis on the vacuum state dominance applies at a low normalization

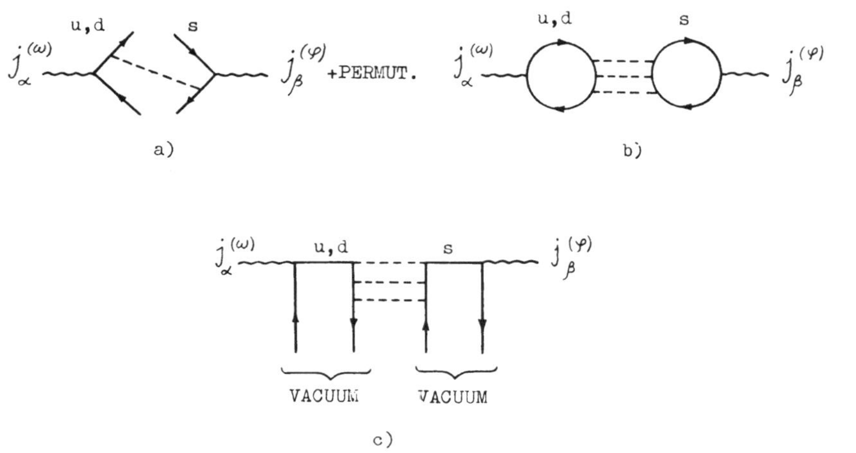

Fig. 10. The ω-φ mixing in the operator expansion language. (a). A graph pertinent to the operator

$$[\bar{s}\gamma_\alpha\gamma_5 t^a s (\bar{u}\gamma_\alpha\gamma_5 t^a u + \bar{d}\gamma_\alpha\gamma_5 t^a d)]_Q \,.$$

(b). Perturbative contribution to $\Pi^{(\omega\varphi)}$. (c) A simplest graph contributing to eq. (5.22) in the leading log approximation.

point, $\mu \sim R_{\text{conf}}^{-1}$ (see sect. 6 in (I)) while the straightforward evaluation of the operator expansion coefficient refers to the point $\mu \sim Q$ (see eq. (5.20)).

(b) Non-vacuum intermediate states.

The first effect can be treated within the standard renormalization group technique (see the appendix of ref. [1]):

$$[\Pi^{(\omega\varphi)}]_a = \frac{\pi\alpha_s(\mu)}{9Q^6}\left(-\frac{32}{27}\right)\langle 0|(\bar{q}q)_\mu|0\rangle^2 \kappa^{-1/9}\eta^{(\omega\varphi)}(\kappa)\,, \tag{5.22}$$

$$\eta^{(\omega\varphi)} = 1 - 0.02\,\kappa^{-1.02} - 1.23\,\kappa^{-0.14} + 0.18\,\kappa^{-0.56} + 0.07\,\kappa^{-1.27}\,,$$

where κ is defined in eq. (5.5). Note that the factor $\eta^{(\omega\varphi)}$ is normalized to zero at $\kappa = 1$, which just corresponds to the vacuum state dominance at the point μ. Then eq. (5.22) extrapolates $\eta^{(\omega\varphi)}$ to any κ.

A priori, $\eta^{(\omega\varphi)}$ could be of order unity for κ = several units. In fact it is fairly small for all reasonable κ:

$$\eta^{(\omega\varphi)} \simeq 0.05\,, \quad \kappa = 3\,; \qquad \eta^{(\omega\varphi)} \simeq 0.09\,, \quad \kappa = 5\,.$$

This is one of our central points: QCD explains the smallness of the observed ω-φ mixing *.

Choosing $\kappa = 3$ gives numerically:

$$[\Pi^{(\omega\varphi)}]_\alpha = -1.5 \cdot 10^{-5}\left(\frac{0.6\ \text{GeV}^2}{Q^2}\right)^3\,. \tag{5.23}$$

Since the vacuum intermediate state is highly suppressed we must be more careful about non-vacuum contributions which are otherwise unimportant (point "b" above).

We will estimate the non-vacuum contribution to the matrix element

$$\langle 0|\bar{d}\gamma_\alpha\gamma_5 t^a d\,\bar{u}\gamma_\alpha\gamma_5 t^a u|0\rangle\,,$$

which determines in fact the ω-ρ mass splitting (see subsect. 5.5) and can be SU(3) rotated to what we need for the ω-φ mixing estimates.

Keeping the contribution of colorless intermediate states in all possible channels implies the following general representation for the matrix element:

$$\langle 0|(\bar{d}\gamma_\alpha\gamma_5 t^a d\bar{u}\gamma_\alpha\gamma_5 t^a u)_\mu|0\rangle = \tfrac{16}{9}\sum_{i=\text{S,P,V,A}}\int_0^{Q_0^2}\Pi^{(i)}(Q^2)$$
$$\times Q^2\text{d}Q^2/16\pi^2\,, \tag{5.24}$$

* Note that the α_s expansion for $\eta^{(\omega\varphi)}$ starts with the $\alpha_s^2\ln^2(Q^2/\mu^2)$ term: $\eta^{(\omega\varphi)} = 0.03\,\alpha_s^2\ln^2(Q^2/\mu^2) + \ldots$, which corresponds to the two-loop graph of fig. 10c evaluated in the leading log approximation.

where $\Pi^{(i)}$ are polarization operators, e.g.,

$$\Pi^{(\mathrm{P})} = i \int \mathrm{d}x \, \mathrm{e}^{iqx} \langle 0 | \mathrm{T} \{ \bar{d}(x) \, i\gamma_5 u(x) \, , \ \bar{u}(0) \, i\gamma_5 d(0) \} | 0 \rangle \, ,$$

$\frac{16}{9}$ is due to the Fierz transformation and color matrices, and we introduced a cut-off Q_0^2 to avoid the double-counting (see below). Note also that the normalization point in eq. (5.24) equals μ, while the original operators in eq. (5.20) are normalized at Q. Using the appendix in ref. [1], one can easily show that

$$[\alpha_s \bar{s} \gamma_\alpha \gamma_5 t^a s (\bar{u} \gamma_\alpha \gamma_5 t^a u + \bar{d} \gamma_\alpha \gamma_5 t^a d)]_Q$$

$$\approx [\alpha_s \bar{s} \gamma_\alpha \gamma_5 t^a s (\bar{u} \gamma_\alpha \gamma_5 t^a u + \bar{d} \gamma_\alpha \gamma_5 t^a d)]_\mu \, ,$$

and the same is true of course for the operator

$$\alpha_s \bar{d} \gamma_\alpha \gamma_5 t^a d \, \bar{u} \gamma_\alpha \gamma_5 t^a u \, .$$

Our guess is that the cut-off Q_0 is quite low, $Q_0 \sim 0.2$ GeV (see sect. 6 in (I)). Indeed, all the contributions representable in the standard form:

$$\Pi^{(i)} = \Pi^{(i)}_{\mathrm{pert,\,th}} + \sum_k C_k (\mu^2/Q^2)^k \, , \tag{5.25}$$

are already accounted for in the initial operator expansion for $\Pi^{(\omega\varphi)}$. In particular, quark graphs are absorbed into $\Pi^{(\omega\varphi)}_{\mathrm{pert.\,th}}$, etc. Thus the integration in eq. (5.24) covers only such Q^2 for which the representation (5.25) is senseless. This presumably happens for $Q^2 \lesssim R^{-2}_{\mathrm{conf}}$. We would like to speculate that higher virtualities, $Q^2 \gtrsim R^{-2}_{\mathrm{conf}}$, must not be included in eq. (5.24). Moreover, it is reasonable to approximate the sum in (5.24) by the pseudoscalar term alone since it is contributed by the lightest hadronic state, the pion. Then we have

$$\Pi^{(\mathrm{P})} = [f_\pi m_\pi^2/(m_\mathrm{u} + m_\mathrm{d})]^2 Q^{-2} \, , \qquad (Q^2 \sim 0) \, . \tag{5.26}$$

Substituting eq. (5.26) into eq. (5.24) and integrating over Q^2 up to $Q_0^2 = 0.04$ GeV^2 we finally get:

$$\langle 0 | (\bar{d} \gamma_\alpha \gamma_5 t^a d \bar{u} \gamma_\alpha \gamma_5 t^a u)_\mu | 0 \rangle \simeq \frac{1}{9\pi^2} \left(\frac{f_\pi m_\pi^2}{m_\mathrm{u} + m_\mathrm{d}} \right)^2 Q_0^2 \simeq 0.1 \, (250 \text{ MeV})^6 \tag{5.27}$$

and, by virtue of the SU(3) symmetry:

$$\langle 0 | [\bar{s} \gamma_\alpha \gamma_5 t^a s (\bar{u} \gamma_\alpha \gamma_5 t^a u + \bar{d} \gamma_\alpha \gamma_5 t^a d)]_\mu | 0 \rangle \simeq 0.2 \, (250 \text{ MeV})^6 \, .$$

Combining all the factors we find for the non-vacuum contribution to the $\Pi^{(\omega\varphi)}$:

$$[\Pi^{(\omega\varphi)}]_b \simeq 5.7 \cdot 10^{-5} \left(\frac{0.6 \text{ GeV}^2}{Q^2} \right)^3 \, .$$

We realize of course that the estimates used are rather crude, but these are best we can do. So let us go ahead and see what happens.

Now that we have all the numbers specified, the sum rules become

$$\frac{1}{\pi M^2}\int e^{-s/M^2}\ \mathrm{Im}\ \Pi^{(\omega\varphi)}(s)\ ds \simeq 2\cdot 10^{-5}\left(\frac{0.6\ \mathrm{GeV}^2}{M^2}\right)^3 , \qquad (5.28)$$

$$\frac{1}{\pi M^4}\int e^{-s/M^2}\ \mathrm{Im}\ \Pi^{(\omega\varphi)}(s)\ s\ ds \simeq -4\cdot 10^{-5}\left(\frac{0.6\ \mathrm{GeV}^2}{M^2}\right)^3 .$$

We would like to saturate the integral over the physical spectral density Im $\Pi^{(\omega\varphi)}$ with the ω, φ mesons. Certainly it is not reasonable for all the M^2, and M^2 of order m_ω^2, m_φ^2 is the best candidate. Indeed, from the experience with the ρ, φ sum rules we learn that for lower M^2, higher power corrections are important while for higher M^2, heavier states also contribute. To motivate the choice of M^2 thoroughly we should have gone further and considered next power corrections and so on, but we just speculate that at $M^2 = \frac{1}{2}(m_\omega^2 + m_\varphi^2)$ both higher power corrections and continuum contribution can be neglected.

Now we are in a position to find the ω-φ mixing and a few words are in order on the phenomenology. The standard definition of the mixing parameter ϵ is

$$\varphi = \varphi_0 + \epsilon\omega_0 , \qquad \omega = \omega_0 - \epsilon\varphi_0 ,$$

where ω_0 and φ_0 are made of the light and strange quarks, respectively. Experimental data imply [21]

$$\epsilon_{\mathrm{exp}} = 0.04 - 0.05 .$$

Theoretically we must introduce into the sum rules two parameters, $\epsilon(m_\varphi^2)$ and $\epsilon(m_\omega^2)$ where

$$\langle 0|j_\alpha^{(\varphi)}|\omega\rangle = -\frac{m_\varphi^2}{g_\varphi}\ \epsilon(m_\omega^2)\ \omega_\alpha ,$$

$$\langle 0|j_\alpha^{(\omega)}|\varphi\rangle = \frac{m_\omega^2}{g_\omega}\ \epsilon(m_\varphi^2)\ \varphi_\alpha ,$$

where ω_α, φ_α are the polarization vectors of the mesons. Experimentally one cannot distinguish between $\epsilon(m_\omega^2)$ and $\epsilon(m_\varphi^2)$, since ω and φ have masses close to each other. The sum rules are sensitive, however, to two parameters:

$$\epsilon = \tfrac{1}{2}\left[\epsilon(m_\omega^2) + \epsilon(m_\varphi^2)\right] , \qquad \epsilon' = \frac{\epsilon(m_\varphi^2) - \epsilon(m_\omega^2)}{m_\varphi^2 - m_\omega^2} .$$

The reason is that experiments are performed for particles on their mass shell while we view the ω-φ system from a distance of order $2m_{\omega,\varphi}^2$. Then the contributions proportional to $(m_\varphi^2 - m_\omega^2)^{-1}$ in the ω, φ peaks cancel each other and we are more sensitive to details.

Using

$$\mathrm{Im}\,\Pi^{(\omega\varphi)} = \frac{\pi m_\omega^2}{g_\omega g_\varphi}\,\epsilon(m_\varphi^2)\,\delta(s-m_\varphi^2) - \frac{\pi m_\varphi^2}{g_\omega g_\varphi}\,\epsilon(m_\omega^2)\,\delta(s-m_\omega^2)\,,$$

and $g_\varphi = -\sqrt{\frac{1}{2}}g_\omega$, we find

$$\epsilon_{\mathrm{th}} = 5\cdot 10^{-2}\,.$$

Note the amusing coincidence between the theoretical and experimental values of ϵ. Although it seems rather fortuitous, the explanation of the observed mixing by an order of magnitude seems to be remarkable.

It is worth emphasizing that leaving the vacuum contribution alone (term "a" above) would lead to the mixing parameter of the opposite sign, in accordance with the general observation of ref. [25].

The reader who finds our estimates too speculative and unmotivated may be appeased to some extent by the fact that the ω-ρ mass splitting allows for an independent estimate of the non-vacuum contribution.

5.5. The ω-ρ degeneracy

There is no general ground for the ω, ρ to be degenerate. The nearly coinciding masses of the ρ and ω are a consequence of strong-interaction dynamics. It is amusing that the degeneracy has a natural explanation within the framework developed. Indeed, as is readily seen, the polarization operators corresponding to the $j_\alpha^{(\rho)}$ and $j_\alpha^{(\omega)}$ currents are identical to each other as far as dominance of the intermediate vacuum state holds. Therefore, our prediction is

$$m_\omega^2 \simeq m_\rho^2\,, \qquad g_\omega^2 \simeq 9g_\rho^2\,.$$

The relation between the coupling constants holds within the experimental errors which are rather large, however. Further measurements of $g_{\rho,\omega}^2$ seem to be interesting for this reason.

We can take one step further and include the non-vacuum contribution discussed in detail in the subsect. 5.4. First, since the ratio

$$\frac{\langle 0|(\bar{d}\gamma_\alpha\gamma_5 t^a d\,\bar{u}\gamma_\alpha\gamma_5 t^a u)_\mu|0\rangle}{\langle 0|(\bar{u}\gamma_\alpha\gamma_5 t^a u\,\bar{u}\gamma_\alpha\gamma_5 t^a u)_\mu\,|0\rangle} \simeq 0.06 \tag{5.29}$$

is positive, the $(\bar{q}\Gamma q)^2$ correction gets slightly enhanced in the ω-channel and diminished in the ρ one. This indicates immediately that the ρ becomes lighter and ω heavier, an effect which is welcomed. Quantitatively one can get

$$m_\omega^2/m_\rho^2 = (f_{\mathrm{cont}}^{(\omega)}/f_{\mathrm{cont}}^{(\rho)})(f_{\mathrm{th\ corr}}^{(\omega)}/f_{\mathrm{th\ corr}}^{(\rho)})\,,$$

where the ratio $f_{\mathrm{th\ corr}}^{(\omega)}/f_{\mathrm{th\ corr}}^{(\rho)}$ is calculable under the assumption specified above:

$$\frac{f^{(\omega)}_{\text{th corr}}}{f^{(\rho)}_{\text{th corr}}} = \frac{1 + 0.28 \cdot 0.15(0.6\ \text{GeV}^2/M^2)^3}{1 - 0.14 \cdot 0.15(0.6\ \text{GeV}^2/M^2)^3} \simeq 1.06 \qquad \text{at} \qquad M^2 = 0.6\ \text{GeV}^2 .$$

Here the factor "0.15" is a reminiscence of the "0.06" entering eq. (5.29): if it were not for the two types of the four-fermion operators (see eq. (3.4)), this factor would be equal to $0.12 = 0.06 \cdot 2$.

The ratio $f^{(\omega)}_{\text{cont}}/f^{(\rho)}_{\text{cont}}$ accounts for the difference in the continuum which cannot be found in a reliable way and introduces an element of uncertainty for this reason. The point $M^2 = m_\rho^2$ seems to be most appropriate to apply the sum rules. If $f^{(\omega)}_{\text{cont}}/f^{(\rho)}_{\text{cont}} = 1$ then

$$(m_\omega - m_\rho)_{\text{th}} \sim 20\ \text{MeV} ,$$

which is correct in sign and is very reasonable in magnitude.

Even keeping in mind the uncertainty due to the continuum, we still can say that the estimate of the non-vacuum contribution gets new support. Note that the sign of the mass splitting does not correspond to the degenerate "bare" ω and ρ with a subsequent mixing of ω and φ. The ω is heavier than the ρ in the imaginary world with no strange quarks at all.

5.6. Conclusions

It is convenient to divide the problems considered into two groups: SU(3) violations in the ρ, K^*, φ and the ω-φ mixing + the ρ-ω degeneracy. In the former case we exploit the fact that the leading SU(3) violating term in the sum rules is associated with the operator $m_s\bar{s}s$ and its contribution is well-defined numerically. Therefore, we are able to evaluate all the masses and coupling constants. The results agree with the data within 5–10% in all cases when the checks are possible.

We found the following approximate law for SU(3) violation in the residue constants (normalized to their SU(3) values):

$$f^{(\mathrm{V})} \simeq \text{const}\ m_{\mathrm{V}} , \qquad (\mathrm{V} = \rho, \mathrm{K}^*, \varphi)$$

where

$$f^{(\rho)} = m_\rho^2/g_\rho^2 , \qquad f^{(\varphi)} = \tfrac{9}{2} m_\varphi^2/g_\varphi^2 , \qquad \text{etc} .$$

The ω-φ mixing is a more subtle problem, that is more interesting for a theoretical analysis. Because of the strong suppression of the intermediate vacuum state in the matrix elements $\langle 0|\bar{q}\Gamma q\bar{q}\Gamma q|0\rangle$ involved, the result is sensitive to non-vacuum contributions. The rough estimates of the latter lead to surprisingly good agreement with the data. Moreover, both the near degeneracy and the magnitude of the ρ-ω mass splitting are well-understood within the same framework.

6. Comments on literature *

In this paper we have undertaken the task of evaluating the meson leptonic widths and masses by means of the QCD sum rules. We realize that resonance physics has been a subject of an intense theoretical study for the last twenty years or so and we certainly cannot aim at compiling a full list of references. Moreover, the history of studies is so rich that once you get a relation which is successful there are good chances that something similar has been mentioned in literature. It is not necessarily fruitful to extend the discussion and compare all the numerical predictions. Instead, we would like to concentrate on QCD and, through confronting it with other approaches, to clarify its specific features.

We include in this section some original remarks as well and, certainly, use the opportunity to pay tribute to earlier works which show the same direction.

We choose to discuss other approaches in more or less chronological order and present them in the following sequence. (i) Algebraic SU(3) breaking. (ii) Duality. (iii) The Weinberg sum rules. (iv) Bag model. (v) Polarization operator at complex Q^2. (vi) The Migdal program. (vii) Instantons. (viii) More on the dilute gas approximation. (ix) Latest papers.

6.1. Algebraic versus dynamical approach to SU(3) breaking

One of the traditional issues in resonance physics is $SU(3)_{flavor}$ breaking. The common approach is to assume certain SU(3) transformation properties for the SU(3) breaking term and derive algebraic relations among the amplitudes [31]. We have both setbacks and advantages, as compared to this procedure.

Our main problem is that the SU(3) breaking parameter is given, roughly spea-

* Before proceeding to a discussion it might be useful to provide a short guide to our own work. It consists of a detailed version and letter-type publications [27,7,28,29]. The former is divided in turn into two parts (ref. [1] plus the present paper), mostly for technical reasons. Some preliminary results are published first, however, in the theorized part of the ITEP review on charmonium, ref. [30] (see also ref. [9]). In particular, it is argued here that power terms, not higher orders in α_s, are responsible for the breaking of asymptotic freedom; a non-vanishing vacuum expectation value $\langle 0|G^a_{\mu\nu}G^a_{\mu\nu}|0\rangle$ is introduced and shown that it is associated with the Q^{-4} correction to the polarization operator, with no place left for the Q^{-2} terms. Here we present the original elaborated paper on which these remarks are based. Thus, it took about a year to complete the study. Explicit computation of the operator expansion coefficients with account of the G^2 term is published first in ref. [27] along with evaluation of the $\rho \to e^+e^-$ decay width. The first extension of the results obtained to the case of the η_c is performed in collaboration with Voloshin (ref. [7]). Ref. [28] summarizes the implications of the results obtained for instanton physics, and finally ref. [29] can be considered as an extended abstract to the whole series. Until after submitting the paper [27], we were unaware of any other attempt to study phenomenological implications of the non-perturbative effects of QCD. In fact, the instanton-induced power corrections were introduced meantime in ref. [17] where the behaviour of the effective coupling is considered. These corrections vary in many respects with what we do (see subsect. 6.7).

king, by m_K^2/m_ρ^2 and is quite large. For this reason the sum rules do not fix the SU(3) breaking effects on a single resonance to a high degree of accuracy. For example, it is not simple to derive

$$m_\varphi^2 - m_\rho^2 = 2(m_{K^*}^2 - m_\rho^2) ,$$

which is rather obvious within the quark model.

On the other hand, the dynamical framework we rely upon in many cases goes far beyond a mere algebraic approach. Thus, we are able to reproduce the observed pattern of SU(3) breaking for the vector mesons: the effects are large on masses and moderate on the coupling constants. The sum rules indicate the ω-ρ degeneracy, a conclusion which can be reached only on dynamical grounds.

Thus, the sum rules lead to a deeper understanding of the mechanism of the symmetry breaking. The main lesson is that chiral symmetry breaking is essential for the mass splittings within SU(3) and SU(2) multiplets as well. For example, the effect of the ω-ρ mixing is [32]

$$m_{\rho\omega}^2 \sim (m_d - m_u)\langle 0|\bar{u}u + \bar{d}d\,|0\rangle . \tag{6.1}$$

Note that similar relations for Goldstone particles are well-known [14,24]

$$m_\pi^2 = -f_\pi^{-2}(m_u + m_d)\langle 0|\bar{u}u + \bar{d}d\,|0\rangle , \tag{6.2}$$

and follow from symmetry considerations alone.

If one tries to think in terms of quarks as light as 5 MeV the problem arises as to how such quarks transfer their mass to mass differences of hadrons. Indeed, quarks are seemingly ultrarelativistic and their masses can be neglected. The masses become essential only if quarks are somehow slowed down within hadrons. An appealing, although intuitive proposal is that the effective quark mass appears as [33] $m(p) = m_0 + 16\pi\alpha_s\langle 0|\bar{q}q\,|0\rangle/p^2$. Equations like (6.1), (6.2) qualitatively agree with such a picture since they demonstrate that both quark mass difference and chiral symmetry breaking are necessary to make apparent SU(3) and SU(2) violations in low-energy hadron physics.

6.2. *Duality*

A close connection between resonance and continuum cross sections has been discussed for many years. In strong-interaction physics, it was argued that averaging over resonances extrapolates the Regge asymptotics to low energies [34,35]. Duality sum rules for the current densities were introduced first by Sakurai [36] (see also [37]) while the most general formulation of the resonance-background correspondence is given by Bjorken and Kogut [38]. Duality is an indispensible part of our intuition now, to say nothing of the fact that it eventually developed into very sophisticated schemes.

As demonstrated above, QCD also leads to certain dispersion sum rules which correlate the resonance properties with high-energy asymptotic behaviour. Thus, there is a certain similarity between the two approaches. The differences are also great, however. First of all, the QCD dispersion sum rules follow directly from the assumed structure of the Lagrangian. Therefore, they are much less phenomenological than the standard duality relations. This conceptual difference manifests itself in many practical matters as well. For example, there is intrinsic uncertainty of the standard duality as to which polarization operator, $\Pi(q^2), q^2\Pi(q^2), ...$, applies the idea of smooth extrapolation of the asymptotic curve into the resonance region. The numerical results are quite different in each the case, however. Thus, the accuracy of duality relations is no better than, say, a factor of 2. As we have seen, the QCD prescription is far from being trivial in this respect. QCD sets up quite a peculiar exponential weight in the dispersion sum rules, i.e., the predictions arise for such integrals as

$$\int e^{-s/M^2}(s\sigma(e^+e^- \to X))\, ds \, .$$

Therefore, duality in, say the π- and ρ- channels has rather different meanings.

Moreover, for heavy quarks we expand in the inverse quark mass, which is completely foreign to the standard duality approach, and still are able to relate the properties of the ρ and J/ψ mesons.

If we turn to practical applications, the differences become even deeper. The point is that duality has been used to guess the asymptotic behaviour of the cross section starting from the leptonic widths of resonances, while QCD takes just the opposite way. Thus, in the review lectures [36] there is a variety of asymptotic values for the cross section for charm production in e^+e^- annihilation depending on the quark mass. In QCD it is certainly not possible to change the high-energy behaviour by varying the mass.

To summarize, the QCD sum rules can be considered as a justification and refinement of the duality relations. They seem to be much more precise in their meaning and quantitative in their predictions although sharing the spirit of earlier works.

6.3. The Weinberg sum rules

The well-known Weinberg relations [19]

$$\int (\rho_V - \rho_A)\, ds = 0 \, , \qquad \int (\rho_V - \rho_A)\, s ds = 0 \, , \tag{6.3}$$

(where ρ_V and ρ_A are dimensionless spectral functions) are in fact forerunners to the QCD dispersion sum rules. Although eqs. (6.3) were originally derived within the current algebra framework, QCD incorporates and justifies them. The QCD analysis of the Weinberg sum rules was performed first in refs. [39,40] which also set up a more general framework for deriving further sum rules of the same type.

Our starting point is close to that of ref. [40]. Still, the general strategy and details are different. The authors of ref. [40] concentrated themselves on extending eqs. (6.3) to other densities and combined the currents in such a way as to cancel the

$m_q\bar{q}q$ power corrections which otherwise could contribute to the second sum rule [39].

We are mostly interested in resonance physics and welcome power terms. We try to introduce such a weight function which enhances the low-energy contribution to the sum rules. As a result we get the following relations as an analog of sum rules (6.3):

$$\int e^{-s/M^2}(\rho_V - \rho_A)\, ds \sim C_0/M^4 ,$$

$$\int e^{-s/M^2}(\rho_V - \rho_A)\, s ds \sim C_1/M^2 . \tag{6.4}$$

We expect the low-lying resonances to saturate the sum rules at $M^2 \sim m_\rho^2$. At this point the weight factor is equal to 1 for the pion and e^{-1}, e^{-2} for the ρ and A_1 mesons, respectively.

We keep not only the $m_q\bar{q}q$ terms discussed in ref. [39] but introduce non-vanishing expectation values for the gluon operators, such as $\langle 0|G^a_{\mu\nu}G^a_{\mu\nu}|0\rangle$. As a result we hope to follow closely the pattern of violations of asymptotic freedom by power corrections in the resonance region. Moreover, we have studied separately integrals over ρ_V and ρ_A and get predictions for the resonance masses in terms of vacuum-to-vacuum matrix elements.

It might be worth noting that we do not rule out the possibility that the sum rules (6.4) are saturated by resonances even in the $M^2 \to \infty$ limit so that the original analysis of relations (6.3) is not modified. Moreover, it looks quite probable from the phenomenological point of view. We just wanted to derive such sum rules which are saturated by resonances beyond any doubt.

To summarize, the QCD sum rules coincide with the Weinberg sum rules for a particular choice of $M^2 (M^2 \to \infty)$. There are further sum rules in QCD which are sensitive to low-energy contributions, and have dimension parameters built in. Their study allows for crucial checks of QCD.

6.4. Bag model

The basic assumption of the bag model [41] is that quarks are confined to a finite space volume. The stability of the bag is due to the positive volume energy density which does not allow the bag to expand, on one hand, and quark pressure which does not allow it to collapse, on the other. The model is manifestly phenomenological in nature and does not involve the QCD Lagrangian directly. However, the expectation is that the net effect of the QCD confinement mechanism is reproduced by introducing the energy density (for a review see, e.g., refs. [42,43]).

The considerations of the present paper may shed some light on the origin of the volume energy introduced by the bag model and lead to a a change in the conceptual structure of the model.

The point is that a non-vanishing $\langle 0|G^a_{\mu\nu}G^a_{\mu\nu}|0\rangle$ implies a negative vacuum energy density. Indeed, according to QCD the operator $G^a_{\mu\nu}G^a_{\mu\nu}$ is proportional to the trace of the energy momentum tensor [44]:

$$\theta_{\mu\mu} = \frac{\beta(\alpha_s)}{4\alpha_s} G^a_{\mu\nu}G^a_{\mu\nu} ,$$

$$\beta(\alpha_s) = \frac{-b\alpha_s^2}{2\pi} + O(\alpha_s^3) , \qquad b = 9 , \tag{6.5}$$

where we put the mechanical mass of light quarks equal to zero while heavy quarks decouple and can be omitted in the estimates given below (for a detailed discussion see ref. [45]). The non-vanishing r.h.s. of eq. (6.5) is due to the so-called triangle anomaly [44], while the naïve use of the equations of motion leads to a traceless $\theta_{\mu\nu}$ (note also that $\beta(\alpha_s)$ in eq. (6.5) stands for the Gell-Mann-Low function; its appearence can be traced back to the renormalization invariance of $\theta_{\mu\mu}$).

On the other hand, by virtue of Lorentz invariance

$$\langle 0|\theta_{\mu\mu}|0\rangle = 4\epsilon ,$$

where ϵ is the vacuum energy density. Thus, the phenomenological analysis of the present paper implies

$$\epsilon \simeq -\tfrac{9}{32} \left\langle 0 \left| \frac{\alpha_s}{\pi} G^a_{\mu\nu}G^a_{\mu\nu} \right| 0 \right\rangle \simeq -0.0035 \text{ GeV}^4 . \tag{6.6}$$

We see that the vacuum energy is negative. Although the conclusion is drawn on phenomenological grounds it is, in fact, of very general nature. The only assumption needed to show $\epsilon < 0$ is the possibility to perform the calculation of ϵ in Euclidean space-time so that $G^a_{\mu\nu}G^a_{\mu\nu}$ is positive definite.

Bringing in quarks disturbs the gluon vacuum fields and requires some energy. Thus, a bag can be made out of vacuum: usually one assumes that positive energy is inside the bag while we would like to say that there is negative energy outside the bag. Apart from purely verbal difference this could explain some paradoxes of the bag model. We have in mind the difficulty in coupling the bag model with the quark counting rule [46]. If the success of the quark model is not coincidental, it implies that there are no extra degrees of freedom except those of quarks. On the other hand the bag apparently introduces new degrees of freedom. If the bag is made out of vacuum as speculated above, and vacuum is "everywhere", then it might help to understand better the quark counting rule.

Our argument is purely qualitative of course and there is a long way before one can say something more definite. Nevertheless, we are pleased to find that the vacuum energy density introduced here, $|\epsilon| = (240 \text{ MeV})^4$ is close numerically to the volume energy density, $B \sim (200 \text{ MeV})^4$ determined phenomenologically [41].

6.5. Polarization operator at complex Q^2

The polarization operator far off the real Q^2 axis, $Q^2 = Q_0^2 + i\Delta$ (where Δ is of order several GeV) has been evaluated within QCD in ref. [47]. As a result the sum rules for a smeared cross section of e^+e^- annihilation,

$$\int \frac{\sigma(e^+e^- \to \text{hadrons})}{(s+Q_0^2)^2 + \Delta^2}\, s\, ds$$

emerge *. The sum rules were used in an attempt to extract the number of quarks flavors from the data.

It is argued in ref. [47] for the first time and in very clean terms, that the only assumption needed to apply QCD is to keep far away from the singularities of the quark graphs. The remark is crucial, to our mind, for the whole framework. However, at present we are at variance with the authors of ref. [47] both in some principal and practical matters.

First, when evaluating the smeared cross section (see above) the authors of ref. [47] suppress the J/ψ contribution with no comments. It might be due to a belief that a single resonance cannot dominate QCD sum rules. We are positive that there exist sum rules saturated by a resonance. The suppression of the J/ψ in ref. [47] is of little practical importance since the primary interest of the paper is higher energies where the existence of the J/ψ is inessential. Thus, ref. [47] does not apply to the resonance region discussed here.

Second, we do not think now that the distance to the singularity is the only parameter that counts. The idea can be true for the α_s expansion discussed in ref. [47] but certainly fails once the power corrections are switched on. Indeed, it has been mentioned many times above that the power corrections distinguish, say, between light and heavy quarks: even if we stay at equal distances to the singularity, the number of derivatives calculable within QCD crucially depends on the quark mass.

Moreover, approaching a singularity from different directions can produce different effects as well. Indeed, for negative q^2 the power series seems to be sign alternating (see subsects. 3.4, 3.6). Therefore, for positive q^2 all the power corrections build up to each other. Thus, the accuracy of approximating the whole series by the first terms can depend rather drastically on the position in the complex Q^2 plane. The statement is specific for the power corrections. The α_s expansion which is an expansion in $1/\ln|Q^2|$ complies with the argument of ref. [47].

To improve the convergence of the power series we introduce the Borel trans-

* Some alternative versions of the sum rules were obtained in ref. [48,49] within an approach close in spirit to that of Poggio, Quinn and Weinberg [47].

form of the polarization operator. Still, the global properties of the series of the power corrections remain rather obscure at the moment and further study may bring excitement.

6.6. The Migdal program

Migdal has recently argued [50] that asymptotic freedom of strong interactions fixes the masses of all the resonances in terms of, say, the ρ meson mass. The idea is to approximate polarization operators by a sum over poles (corresponding to resonances) and require that this sum reproduces the QCD perturbative result for the polarization operators at space-like Q^2 to the best accuracy possible with a minimal number of poles *.

Although it might seem that our method is close in spirit to that of Migdal, the only common point in fact is that they attack one and the same problem. This does not mean, of course, that we could not borrow much from the papers [50,51] and hopefully pay something back. We would like just to emphasize that the realizations of the programs are rather complementary than identical to each other.

Thus, if successful, the Migdal program goes far beyond. Indeed, we think that as $s \to \infty$ the bare quark cross section coincides with the physical one and that this exhausts the consequences of QCD for high energies. Migdal tries to extract the Regge trajectories as a whole, i.e., to determine the masses even in the limit $M^2 \to \infty$. This can be done, to our mind, only at the price of additional assumptions, which may still be interesting and, what is most important, true. We are interested in the low-lying resonances, or in the energy region where the structure of cross sections becomes prominent. For such resonances we feel on firmer ground.

What is most important, we introduce power corrections and this brings in a mass parameter. The only input information in the Migdal program is the α_s series for the polarization operators with various Lorentz structure. Thus, the way in which a mass scale arises is completely different. The power corrections depend drastically on the channel under consideration and we think that the observed variety of the resonance properties cannot be understood without power terms. A good example is the π-ρ-A_1 system: the vector and axial vector spectral densities are quite different while the α_s expansion is the same in the limit of massless quarks. One may hope that a kind of unification of the Migdal approach and that developed here may turn fruitful.

In conclusion, we would like to thank A.A. Migdal for interesting and stimulating discussions.

6.7. Instantons

We have already discussed the rôle of instantons in sect. 3 of (I). To summarize, instantons allow one to compute the critical dimension starting from which there is

* We take the freedom to formulate the Migdal program in rather loose terms. For precise wordings see the original papers [50,51].

no operator expansion. The critical dimension turns out to be quite high. Instantons of small size would control the first correction which goes beyond the operator expansion, but for the applications considered so far there is no need to introduce this correction. Indeed, high powers of Q^{-2} seem to be overshadowed by the leading power terms such as Q^{-4}, Q^{-6} at least for such Q^2 that the whole expansion does not collapse.

6.7.1. Large scale instantons. Instantons of large size do contribute to, say, $\langle 0|G^a_{\mu\nu}G^a_{\mu\nu}|0\rangle$ which governs the Q^{-4} correction. But the corresponding integral over the instanton size is infrared divergent so that the dilute-gas approximation fails and other field configurations become equally important. Therefore, we treat $\langle 0|G^a_{\mu\nu}G^a_{\mu\nu}|0\rangle$ on purely phenomenological grounds without relying on the instanton-borne estimates.

The reader might get confused since sometimes we say something like "the one-instanton approximation is inadequate for our purposes" but on other occasions we argue "let us use the one-instanton approximation...". The reason is of course that at present the instanton calculus [52,18,53,17] is a theory with a cut-off. An interpretation of such a theory is, as always, dubious. An optimist may hope that introducing a cut-off would give numerical results with resonable accuracy. A pessimist could say that introducing a cut-off implies that something else, not instantons, is important and instantons are irrelevant to the power corrections we study here.

We would prefer an alternate point of view: to be on firm phenomenological ground as far as basic predictions are concerned on the one hand, and to explore one-instanton estimates, on the other.

In practical applications we cannot evaluate, say, $\langle 0|G^a_{\mu\nu}G^a_{\mu\nu}|0\rangle$ by using a preconceived value of the cut-off ρ_c. The reason is that the dependence on ρ_c is so strong that an uncertainty in ρ_c of the order of a factor two makes the estimate much too unstable. Therefore, we take the other way and determine ρ_c from the value of $\langle 0|G^a_{\mu\nu}G^a_{\mu\nu}|0\rangle$ extracted in a phenomenological way. The result is published separately [28] and indicates that the dilute instanton-gas approximation can be relevant for a wide range of distances, i.e., the value of ρ_c is quite high:

$$\rho_c \sim 1/200 \text{ MeV} . \tag{6.7}$$

Then we use this value of ρ_c to estimate further matrix elements such as $\langle 0|f_{abc}G^a_{\mu\nu}G^b_{\nu\sigma}G^c_{\sigma\mu}|0\rangle$.

6.7.2. Small-size instantons. A numerical estimate of the small-size instanton contribution is, in fact, also very uncertain. The reason is that the mechanism of the effective quark mass generation is far from being clear and that the effective coupling constant is poorly known. Still, let us present some estimates which are in line with the choice of the parameters used throughout this paper.

Concentrate on a chirality changing current, say,

$$j^{(\mathrm{P})} = \bar{u}i\gamma_5 u - \bar{d}i\gamma_5 d .$$

Then there is no problem with the mass generation since the result is independent of the quark masses. Moreover, all the necessary equations are already given in ref. [1] so that we just substitute the numbers here.

The instanton (+ anti-instanton) contribution into the polarization operator induced by the $j^{(P)}$ current is

$$\Pi^{(P)}_{\text{inst + ant-inst}} = 4Q^2 \int [K_{-1}(Q\rho)]^2 m_s d(\rho)\, d\rho \tag{6.8}$$

$$= 4Q^2 [m_s \rho\, d(\rho)]_{\rho=1/Q} \frac{2^{\epsilon-3}}{\Gamma(\epsilon)} \left[\Gamma\left(\frac{\epsilon}{2}\right)\right]^4 \frac{\epsilon}{\epsilon-2},$$

where $d(\rho)$ is the density of instantons of size ρ:

$$d(\rho) = C\left(\frac{2\pi}{\alpha_s(\rho)}\right)^6 e^{-2\pi/\alpha_s(\rho)} = C\left(\frac{2\pi}{\alpha_s(\rho)}\right)^6 (\Lambda\rho)^b , \tag{6.9}$$

and the parameter Λ determines the effective coupling constant: $\alpha_s(Q) = 2\pi \times [b \ln(Q/\Lambda)]^{-1}$. Finally,

$$b = 9 , \qquad \epsilon = b + 1 .$$

Our choice of the coupling constant, $\alpha_s = 0.2$ at $Q^2 = 6\ \text{GeV}^2$, corresponds to

$$\Lambda = 0.075\ \text{GeV} .$$

We realize of course that such value of Λ is much lower than that favored by fits to deep inelastic cross section but we would prefer to stick to it for the reasons explained in subsect. 6.4 of (I).

Apart from uncertainty in Λ there are some other numerical factors which are difficult to specify. In particular, the constant C entering the definition of $d(\rho)$ (see eq. (6.9)) depends on the renormalization procedure. (Note that we include in C the factor due to the integration over the fermion fields. If the Pauli-Villars regularization is used then it is equal to 1.3^{N_f}).

It is impossible to identify the renormalization procedure which corresponds to $\alpha_s(6\ \text{GeV}^2) = 0.2$. Therefore, let us just mention that in all the cases considered in the literature the constant C turns to be around 0.1 or several times larger.

It is convenient to normalize the instanton contribution to ordinary perturbation theory which gives

$$\Pi^{(P)}_{\text{pert th}} = \frac{3}{4\pi^2} Q^2 \ln(Q^2/\mu^2) .$$

Moreover, we use the sum rules for the first Borel transform of the polarization operator and consider, therefore,

$$\hat{L}_M \Pi^{(P)}_{\text{inst+anti-inst}} = 4M^2 [m_s \rho d(\rho)]_{\rho=1/M}$$

$$\times \frac{2^{\epsilon-3}}{\Gamma(\epsilon)} \left[\Gamma\left(\frac{\epsilon}{2}\right)\right]^3 \frac{\epsilon}{2} \approx 4M^2 \left[m_s \rho d(\rho)\right]_{\rho=(\epsilon/2e)^{1/2}M^{-1}} , \tag{6.10}$$

where e is the base of natural logs, and

$$\hat{L}_M \Pi^{(P)}_{\text{pert th}} = \frac{3M^2}{4\pi^2} . \tag{6.11}$$

Note that eq. (6.10) uses a numerical approximation which works well for the realistic value of ϵ, $\epsilon \sim 10$.

Finally, we get for the ratio of the instanton and standard leading contributions:

$$\frac{\text{inst + anti-inst}}{\text{pert th}} \approx \tfrac{16}{3} \pi^2 \left[m_s \rho d(\rho)\right]_{\rho=(\epsilon/2e)^{1/2}M^{-1}} . \tag{6.12}$$

The ratio is of order unity for $M^2 = m_\rho^2$. Indeed, according to our choice $\alpha_s(m_\rho) = 0.3$ and taking $m_s = 150$ MeV, $C = 0.1$ we get:

$$\left[m_s \rho d(\rho)\right]_{\rho=(\epsilon/2e)^{1/2}m_\rho^{-1}} \approx \left(\frac{0.29\ \text{GeV}^2}{m_\rho^2}\right)^5 ;$$

$$\frac{\text{inst + anti-inst}}{\text{pert th}} = \frac{\text{eq. (6.10)}}{\text{eq. (6.11)}} \approx \left(\frac{0.64\ \text{GeV}^2}{m_\rho^2}\right)^5 \sim 1 , \quad \text{at } M^2 = m_\rho^2 .$$

Thus, in the case of pseudoscalar current small-size instantons are likely to show up at $M^2 \sim m_\rho^2$. For vector and axial vector currents which we considered so far, the contribution is further suppressed because of the smallness of the quark masses (see for a discussion sect. 3 of (I)). The suppression could be of the order $(m_{u,d}/m_\rho)^2$ or $(\langle 0|\bar{q}q|0\rangle/m_\rho^3)^2$ which amounts to factor $\lesssim 10^{-3}$ but, in fact, we cannot present at the moment a motivated estimate of this suppression factor.

To summarize, the estimate of the small-size-instanton contribution is too uncertain to be convincing. Still, we would like to conclude that it is not necessarily inconsistent to neglect this contribution to the sum rules considered so far. We did neglect it and observed that the sum rules agree with the data. For us this agreement is the very last and best argument in favour of the approximation used.

6.7.3. Comments on the literature. The instanton-induced corrections were introduced first by Callan, Dashen and Gross [17] who studied the Q^2 dependence of the effective coupling constant. They discovered that the power corrections dominate as far as asymptotic freedom breaking is concerned. The same conclusion can be reached starting from the charmonium sum rules (see refs. [30] and [1]). It is amusing that such different approaches lead to the same conclusion.

A closer examination reveals, however, that the power corrections introduced in ref. [17] and in our paper differ from each other. Indeed, the authors of ref. [17] argue that the power corrections to the coupling constant can be consistently treated within the one-instanton approximation since they come from instantons of a variable size, $\rho \sim 1/Q$. On the other hand, as demonstrated in our papers the leading correction to the two-point functions considered is controlled by the large scale field fluctuations which does not vary with Q^2. Thus, we cannot use the one-instanton approximation literally. The polarization operators considered are, in a sense, directly measurable. The authors of ref. [17] do not indicate in which way, at least in principle, their results can be confronted with experiment.

Moreover, turning to the effect of instantons on a heavy quark-antiquark pair interaction the authors of ref. [17] * insist again that the instantons of a size larger than the distance between the quarks are irrelevant since their fields cancel between the quark and antiquark. According to our point of view, the leading effect is due to the non-vanishing vacuum expectation value $\langle 0|G^a_{\mu\nu}G^a_{\mu\nu}|0\rangle$. As explained in the introduction to our paper [1] (see also sects. 2, 3 in this paper) quarks detect long wave gluon fields flowing through vacuum.

Thus, we see that as far as details are concerned, the differences are great. Still it might be premature to go into details because the very subject is quite new. The similarity of the qualitative pictures developed seems more important and indeed encouraging.

6.8. More on the dilute instanton-gas approximation

The dilute instanton-gas approximation as a quasi-quantitative framework to study non-perturbative effects within QCD was developed in ref. [17]. It implies a cut-off on the integrals over the instanton size. The approximation is seemingly very useful for the purpose of orientation and we rely on it in some of our estimates (see sects. 3, 6 of (I)). Still one must keep in mind that, as any model with a cut-off, the one-instanton approximation cannot be taken too literally, and introducing a certain cut-off in one integral does not imply that it is the same in another case. We will demonstrate explicitly an example of this kind. By doing so we in no way want to disgrace the use of the approximation for rough estimates.

Consider the vacuum expectation value $\langle 0|G^a_{\mu\nu}G^a_{\mu\nu}|0\rangle$. For a pure Yang-Mills field it is given by the following functional integral

$$\int \mathcal{D}A_\mu \mathrm{e}^{-S(A_\mu)} G^a_{\mu\nu}G^a_{\mu\nu} \Big/ \int \mathcal{D}A_\mu \mathrm{e}^{-S(A_\mu)} ,$$

where the Wick rotation to Euclidean space-time is implied. The instanton (anti-instanton) solution contributes as a configuration with an action equal to $2\pi/\alpha_s$. Using the standard saddle point technique one comes to

$$\left\langle 0\left|\frac{\alpha_s}{\pi} G^a_{\mu\nu}G^a_{\mu\nu}\right|0\right\rangle_{\text{inst + anti-inst}} = 2\int [\mathrm{d}\rho\, d(\rho)/\rho^5] \int \mathrm{d}^4 x_0 \frac{\alpha_s}{\pi} G^a_{\mu\nu}(0;x_0,\rho)\, G^a_{\mu\nu}(0;x_0,\rho) , \qquad (6.13)$$

* See also ref. [54].

where ρ is the instanton size, x_0 is its center and $d(\rho)$ is the instanton density which for an SU(3) color group is equal to [17]

$$d(\rho) = \text{const} \left[\frac{2\pi}{\alpha_s}\right]^6 \exp\left(-\frac{2\pi}{\alpha_s(\rho)}\right).$$

Integrating over x_0, we are left with an integral over ρ which diverges at large ρ. Furthermore, introducing a cut-off at $\rho = \widetilde{\rho}_c$ we get:

$$\left\langle 0 \left| \frac{\alpha_s}{\pi} G^a_{\mu\nu} G^a_{\mu\nu} \right| 0 \right\rangle = 16 \int_0^{\widetilde{\rho}_c} \frac{d\rho}{\rho^5} d(\rho) . \tag{6.14}$$

A cut-off on the instanton size was introduced originally in the course of an evaluation of the vacuum energy density associated with instantons [17]. The energy density can be represented in the form [17]:

$$\epsilon = -2 \int_0^{\rho_c} \frac{d\rho}{\rho^5} d(\rho) , \tag{6.15}$$

where we reserved as a possibility, that ρ_c and $\widetilde{\rho}_c$ do not coincide.

Our point is that a naïve identification $\rho_c = \widetilde{\rho}_c$ would lead to a contradiction. Indeed, $\langle 0|(\alpha_s/\pi)\, G^2|0\rangle$ can be related to ϵ in a model-independent way since $\alpha_s G^a_{\mu\nu} G^a_{\mu\nu}$ is proportional to the trace of the energy-momentum tensor (see subsect. 6.4) Thus, we come to

$$\left\langle 0 \left| \frac{\alpha_s}{\pi} G^a_{\mu\nu} G^a_{\mu\nu} \right| 0 \right\rangle = -\frac{32}{b}\, \epsilon, \qquad \text{(no quarks)} . \tag{6.16}$$

Now, if the cut-offs ρ_c and $\widetilde{\rho}_c$ are the same, the general relation (6.16) is violated. There is not much surprise in that. Indeed, a general proof of eq. (6.16) (and it can be readily given) includes a change of variables in the functional integral. It is meaningful and correct for a finite integral, while introducing a cut-off in an arbitrary way does not respect the general relations.

Thus, the only way to recoincile eqs. (6.14)–(6.16) with each other is to assume that

$$\int_0^{\rho_c} d\rho\, \rho^{-5} d(\rho) = \tfrac{1}{4} b \int_0^{\widetilde{\rho}_c} d\rho' \rho^{-5} d(\rho) . \tag{6.17}$$

In the case considered the inconsistency of the approximation affects the computation of a numerical coefficient. It is not ruled out that in some other cases functions are also affected.

6.9. Latest papers

Recently, there appeared a number of papers dealing with phenomenological manifestations of instantons. In particular, Andrei and Gross [55] evaluate the instanton contribution to the polarization operator associated with the electromagnetic current (neglecting the effect of nearly massless quarks on instantons). They show explicitly that the corrections of order Q^{-2} do not arise and find the integral over instanton sizes which gives the Q^{-4} term. To our mind, the absence of the Q^{-2} correction is of a very general nature [30,27]. Moreover, if we combine our general operator expansion [27] with evaluation of $\langle 0|G^a_{\mu\nu}G^a_{\mu\nu}|0\rangle$ within the dilute-gas approximation (see above and ref. [28]) then we get a number which differs from the result of ref. [55] by a factor $\frac{3}{2}$. As explained above there is no need to worry about such factors within the dilute-gas approximation because of the inconsistencies of the cut-off procedure.

There is some overlap with the paper of Baullieu et al. [56] as well. Here the correction to the cross section for e^+e^- annihilation due to the instantons of small size, $\rho \sim 1/Q$, is evaluated. We agree as far as general relations are concerned. Numerically our estimates differ, however (see also subsect. 6.7). According to our point of view the leading non-perturbative correction is due to the large-scale vacuum fluctuations and these are still moderate at $Q^2 \sim m_\rho^2$, while Baullieu et al. [56] prefer the contribution of the small-size instantons to be of order unity already at $Q^2 = (1-2)\ \mathrm{GeV}^2$. The difference is mostly due to different values of α_s used in the analysis. We think that our estimates are *a posteriori* supported by the success of the sum rules. The authors of ref. [56] also come to the conclusion that large scale instantons do not affect the cross section at high energies (compare with subsect. 3.3 of our paper [1]).

It is common to all the papers that we have seen so far, that the emphasis is made on the instanton contribution. As repeatedly explained above, more general field configurations are seemingly crucial to break asymptotic freedom, confine quarks and make resonances out of quarks.

7. Conclusions

What new did we learn in constructing the QCD-based approach to resonance physics (refs. [1,32] + the present paper)?

Since the work covers many questions any selection of the "main" results for a short summary may turn rather arbitrary. Still let us try to sketch the argument and to this end group the results as follows: (a) theoretical foundations; (b) vacuum-to-vacuum matrix elements; (c) consequences for resonances; (d) general picture.

(a) We concentrated on the power corrections to asymptotic freedom which are due to non-perturbative effects of QCD. It is argued that one can rely on the standard Wilson expansion when dealing with a first few terms in the Q^{-2} expansion for a two-point function of two external currents. If it were not for the non-perturbative effects, only the unit operator in the expansion would survive. It just incorporates the whole series in the effective quark-gluon coupling constant α_s.

The non-perturbative effects are represented by the vacuum expectation values such as $\langle 0|G^a_{\mu\nu}G^a_{\mu\nu}|0\rangle$ which vanish by definition within the standard perturbation theory.

For higher powers of Q^{-2}, the very operator expansion is not valid and small-size non-perturbative vacuum fluctuations show up. The corresponding "critical" power of Q^{-2} can be readily found by means of the instanton calculus.

Instead of a polarization operator one can introduce an infinite series of its Borel transforms and the Q^{-2} expansion for the polarization operator is explicitly rewritten in terms of the corresponding expansions for the Borel transforms.

The first Borel transform turns to be of special importance. First, it sharpens the integral over physical spectral densities still keeping it positive definite and, second, suppresses in a factorial-like way higher-power corrections. The corresponding sum rules are, therefore, most suitable to study resonances within QCD.

(b) We would like to introduce into the theoretical use a variety of the vacuum-to-vacuum matrix elements. We discussed how to extract them from experimental data and proposed independent theoretical estimates.

The most interesting result seems to be

$$\left\langle 0\left|\frac{\alpha_s}{\pi}G^a_{\mu\nu}G^a_{\mu\nu}\right|0\right\rangle \sim 10^{-2}\ \mathrm{GeV}^4\ ,$$

where $G^a_{\mu\nu}$ is the gluon strength tensor. It is remarkable that this matrix element can be related to the vacuum energy density ϵ. The latter turns to be negative, $\epsilon = -(3\text{–}4)\cdot 10^{-3}\ \mathrm{GeV}^4$. The value of ϵ is close in absolute magnitude to the positive volume energy density inside hadrons postulated by the bag model. Thus, it is possible to make a bag out of the vacuum.

Another potentially important relation is

$$\langle 0|\bar{h}h|0\rangle|_{m_\mathrm{h}\sim 200\ \mathrm{MeV}} \sim -\frac{1}{12m_\mathrm{h}}\left\langle 0\left|\frac{\alpha_s}{\pi}G^a_{\mu\nu}G^a_{\mu\nu}\right|0\right\rangle,$$

where m_h is the mass of fictitious quark h. The right-hand side represents the first term in the heavy quark expansion for $\langle 0|\bar{h}h|0\rangle$. The relation implies a smooth matching of the light and heavy quark techniques at intermediate masses.

Another valuable approximation is the dominance of the vacuum intermediate state which is substantiated on phenomenological and theoretical grounds. Applying the approximation one gets, e.g.,

$$\langle 0|\bar{q}\Gamma q\bar{q}\Gamma q|0\rangle = \frac{1}{N^2}\left[(\mathrm{Tr}\ \Gamma)^2 - (\mathrm{Tr}\ \Gamma^2)\right]\langle 0|\bar{q}q|0\rangle^2\ ,$$

where q is a quark field and N is a normalization factor.

Another approximation involves a cut-off ρ_c in integrals over instanton size ($\rho_c \sim (0.2\ \text{GeV})^{-1}$). It gives, e.g.,

$$\langle 0|g_s^3 f_{abc} G^a_{\mu\nu} G^b_{\nu\sigma} G^a_{\sigma\mu}|0\rangle \simeq \tfrac{12}{5}\, \rho_c^{-2} \langle 0|g_s^2 G^a_{\mu\nu} G^a_{\mu\nu}|0\rangle\,.$$

Applying one or another technique we are able to estimate more or less any vacuum-to-vacuum matrix element in terms of a single number, say, $\langle 0|G^a_{\mu\nu}G^a_{\mu\nu}|0\rangle$. (However, in the most important applications we prefer to rely on phenomenological grounds, and not on any estimate.)

(c) The QCD sum rules relate the vacuum-to-vacuum matrix elements to the observable cross sections. We claim both qualitative and quantitative successes for QCD in predicting the properties of a single resonance. In particular the signs of the integrals

$$\int (\rho_V - \rho_A)\, e^{-s/M^2} ds < 0\,,$$

$$\int (\rho_V - \rho_A)\, e^{-s/M^2} s ds > 0\,,$$

which are fixed by the power terms reflect the inequalities

$$m_\pi < m_\rho\,, \qquad m_\rho < m_{A_1}$$

(ρ_V and ρ_A denote the vector and axial vector dimensionless spectral densities).

Moreover, the first power corrections add to each other in the axial vector case and subtract from each other in the integrals over ρ_V. This implies

$$m_{\rho'} - m_\rho < m_{A_1} - m_\pi\,.$$

We have considered the whole vector-meson nonet: $\varphi, \rho, \omega, K^*$. The result is that the SU(3) violations are large in masses and are considerably smaller in the dimensionless coupling constants.

Quantitatively, we have computed

$$f_\pi, g^2_{A_1}\,, \quad g^2_V/4\pi\,, \quad m^2_V\,, \quad \text{where} \quad V = \rho, \omega, \varphi, K^*\,.$$

All the numbers depend crucially on the mass scale set by the power corrections. Although we borrow, in a way, the power correction from independent experimental data the relations obtained are far from being a simple dimensional estimate and involve large, non-trivial coefficients (which, say, transform, $(2m_\pi/m_\rho)^6$ into a quantity of order unity).

It is worth emphasizing that we did not try an elaborate mathematical analysis of the sum rules. It seemed more important to us to reach a qualitative understanding and we are satisfied with the 10% accuracy achieved in most cases. Indeed, since the very approach is new it seemed crucial to probe its validity. Therefore, we prefer not to deepen into the debris of computer calculations, sophisticated fit programs, arranging error bars here and there ... Instead we tried simple models which satisfied our intuition.

Still, let us note that in many cases we found that the well-established information on low-energy interactions is too poor and we look forward to further experimental efforts to provide checks of QCD.

(d) The phenomenological success of the sum rules up to as low-mass parameter M^2 as $M^2 = m_\rho^2$ seems to confirm the conjectured smallness of $\alpha_s(m_\rho)$. On the other hand, the power corrections are large in a sense. Namely, if one normalizes them to perturbation theory, then power corrections correspond to sweeping off the perturbative integrals over momenta p^2 bounded by

$$p^2 \lesssim \Lambda^2_{\text{p.t.}} \simeq 3\ \text{GeV}^2\ .$$

To be rather small at m_ρ^2 and have the net effect corresponding to high $\Lambda^2_{\text{p.t.}}$ implies violent vacuum fluctuations at larger distances.

Thus, apart from m_ρ there is another mass scale, say, 0.2 GeV. Although the difference between the m_ρ and 0.2 GeV may seem rather symbolic at the first sight, the high powers of mass ratios encountered in the sum rules make it very important in fact.

The picture is in accord with the instanton-based estimates. Indeed, the instanton density is a very steep function of the instanton size. It can be small at $\rho \sim 1/m_\rho$ and large at $\rho \sim (0.2\ \text{GeV})^{-1}$. As noted by V.N. Gribov the same is seemingly true for other vacuum fluctuations as well. Indeed, the growth of the fluctuation density is mostly due to the phase space.

The success of the QCD sum rules indicates also that the small-size instantons have little bearing, if any, on the resonance properties. Although large-scale instantons may be important along with other vacuum fluctuations, the dilute instanton-gas approximation is not consistent within the context of our program. (The statement is qualified by the fact that we do not consider here the $I = 0$ axial vector case.)

Asymptotic freedom is violated by power corrections which are a fast function of Q^2 and take the asymptotic cross section "by surprise" and this is the origin of the observed duality between the resonance and continuum cross sections.

We see that the papers include statements of various kinds which are not directly related to each other. For us, everything can be summarized in a few words:

QCD sum rules fix properties of a single resonance.

It might be too much to swallow at once but we hope that a well-disposed reader can at least accept some of the results obtained.

We are grateful to V.N. Gribov, B.L. Ioffe, I.Yu. Kobzarev, M.S. Marinov, A.A. Migdal, V.A. Novikov, L.B. Okun, M.I. Polikarpov, A.M. Polyakov, E.V. Shuryak,

M.B. Voloshin, M.I. Vysotski and Alexander Zamolodchikov for valuable discussions.

References

[1] M. Shifman, A. Vainshtein and V. Zakharov, Nucl. Phys. B147 (1979) 385.
[2] V.A. Novikov et al., Phys. Rev. Lett. 38 (1977) 626; Phys. Lett. 67B (1977) 409.
[3] D.J. Gross and F. Wilczek, Phys. Rev. Lett. 30 (1973) 1343;
H. Politzer, Phys. Rev. Lett. 30 (1973) 1346.
[4] S. Okubo, Phys. Lett. 5 (1963) 165;
G. Zweig, CERN preprints TH 401, TH 412 (1964);
J. Iizuka, Prog. Theor. Phys. Suppl. 37-38 (1966) 21.
[5] H. Georgi and H. Politzer, Phys. Rev. D14 (1976) 1829.
[6] T. Appelquist and H. Politzer, Phys. Rev. Lett. 34 (1975) 43; Phys. Rev. D12 (1975) 1404.
[7] M. Shifman, A. Vainshtein, M. Voloshin and V. Zakharov, Phys. Lett. 77B (1978) 80; Yad. Fiz. 28 (1978) 465.
[8] DASP group, W. Braunschweig et al., Phys. Lett. 67B (1977) 243.
[9] V. Novikov, L. Okun, M. Shifman, A. Vainshtein, M. Voloshin and V. Zakharov, Phys. Reports 41 (1978) 1.
[10] D. Benaksas et al., Phys. Lett. 39B (1972) 289.
[11] A. Vainshtein, V. Zakharov and M. Shifman, JETP (Sov. Phys.) 45 (1977) 670.
[12] M. Perl et al., Phys. Rev. Lett. 38 (1977) 83.
[13] Y. Nambu, Phys. Rev. Lett. 4 (1960) 380;
H. Pagels, Phys. Reports 16 (1975) 220.
[14] M. Gell-Mann, R. Oakes and B. Renner, Phys. Rev. 175 (1968) 2195.
[15] S. Adler, Phys. Rev. 177 (1969) 2426;
J.S. Bell and R. Jackiw, Nuovo Cim. 60A (1969) 47.
[16] S. Weinberg, Phys. Rev. D11 (1975) 3583.
[17] C. Callan, R. Dashen and D. Gross, Phys. Rev. D17 (1978) 2717.
[18] G. 't Hooft, Phys. Rev. Lett. 37 (1976) 8; Phys. Rev. D14 (1976) 3432.
[19] S. Weinberg, Phys. Rev. Lett. 18 (1967) 507.
[20] G. Alexander et al., Phys. Lett. 73B (1978) 99.
[21] J. Parrour et al., Phys. Lett. 63B (1976) 357;
A.D. Bukin et al., Yad, Fiz. 27 (1978) 976.
[22] T. Das, V.S. Mathur and S. Okubo, Phys. Rev. Lett. 19 (1967) 470;
V.S. Mathur and S. Okubo, Phys. Rev. 181 (1969) 2148.
[23] R.P. Feynman, Photon-hadron interactions (Benjamin, Mass., 1972), lecture 16.
[24] S. Weinberg, Harvard preprints HUPT-77/A057, HUPT-77/A061 (1977).
[25] J. Arafune, M. Fukugita and Y. Oyanagi, Phys. Lett. 70B (1977) 221.
[26] S. Okubo, Phys. Rev. D16 (1977) 2336.
[27] A. Vainshtein, V. Zakharov and M. Shifman, ZhETF Pisma 27 (1978) 60.
[28] M. Shifman, A. Vainshtein and V. Zakharov, Phys. Lett. 76B (1978) 471.
[29] M. Shifman, A. Vainshtein and V. Zakharov, Moscow preprint ITEP-54 (1978), Phys. Rev. Lett., submitted.
[30] V. Novikov, L. Okun, M. Shifman, A. Vainshtein, M. Voloshin and V. Zakharov, Moscow preprint ITEP-83 (March 1977).

[31] R.H. Dalitz, *in* Elementary particle physics, ed. G. Takeda and A. Fujii, Tokio-New-York (1967);
J.J. Kokkedee, The quark model, (Benjamin, N.Y., Amsterdam, 1969).
[32] M. Shifman, A. Vainshtein and V. Zakharov, Nucl. Phys. B147 (1979) 519.
[33] H. Politzer, Nucl. Phys. B117 (1976) 397.
[34] A.A. Logunov, L.D. Soloviev and A.N. Tavkhelidze, Phys. Lett. 24B (1967) 181.
[35] R. Dolen, D. Horn and C. Schmidt, Phys. Rev. 166 (1968) 1768.
[36] J.J. Sakurai, Phys. Lett. 46B (1973) 207; New duality in electromagnetic interactions, Erice Lectures (1973).
[37] A. Bramon and M. Greco, Nuovo Cim. Lett. 3 (1972) 693;
M. Greco, Nucl. Phys. B63 (1973) 398.
[38] J.D. Bjorken and J. Kogut, Phys. Rev. D8 (1973) 1341.
[39] K. Wilson, Phys. Rev. 179 (1969) 1499.
[40] L. Bernard, H. Dunkan, J. LoSecco and S. Weinberg, Phys. Rev. D12 (1976) 792.
[41] A. Chodos et al., Phys. Rev. D9 (1974) 3471; D10 (1974) 2599;
T. DeGrand et al., Phys. Rev. D12 (1975) 2060.
[42] V. Weisskopf, CERN preprint TH-2068 (1975).
[43] P. Hasenfratz and F. Kuti, Phys. Reports 40 (1978) 45.
[44] R. Crewther, Phys. Rev. Lett. 28 (1972) 1421;
M. Chanowitz and J. Ellis, Phys. Lett. 40B (1972) 397;
J. Collins, L. Dunkan and S. Joglekar, Phys. Rev. D16 (1977) 438.
[45] M. Shifman, A. Vainshtein and V. Zakharov, Phys. Lett. 78B (1978) 443.
[46] V. Matveev, R. Muradyan and A. Tavkhelidze, Nuovo Cim. Lett. 7 (1973) 719;
S. Brodsky and G. Farrar, Phys. Rev. Lett. 31 (1973) 1153; Phys. Rev. D11 (1975) 1309.
[47] E. Poggio, H. Quinn and S. Weinberg, Phys. Rev. D13 (1976) 1958.
[48] R. Shankar, Phys. Rev. D15 (1977) 755.
[49] V. Kuzmin, A. Tavkhelidze and K. Chetyrkin, ZhETF Pisma 25 (1977) 456;
K. Chetyrkin and N. Krasnikov, Nucl. Phys. B119 (1977) 174.
[50] A.A. Migdal, Ann. of Phys. 110 (1978) 46.
[51] A.A. Migdal, Batavia preprint FERMILAB-PUB 77/74 (1977);
H.G. Dosch, J. Kripfganz and M.G. Schmidt, Phys. Lett. 70B (1977) 337.
[52] A. Belavin, A. Polyakov, A. Schwartz and Yu. Tyupkin, Phys. Lett. 59B (1975) 85.
[53] A. Polyakov, Nucl. Phys. B120 (1977) 429;
A. Belavin and A. Polyakov, Nucl. Phys. B123 (1977) 429;
C.G. Callan et al., Phys. Lett. 63B (1976) 334; Phys. Lett. 66B (1977) 375;
R. Jackiw and C. Rebbi, Phys. Rev. Lett. 37 (1976) 172.
[54] F. Wilczek and A. Zee, Phys. Rev. Lett. 40 (1978) 83.
[55] N. Andrei and D.J. Gross, Phys. Rev. D18 (1978) 468.
[56] L. Baullieu, J. Ellis, M.K. Gaillard and W.J. Zakrzewski, Phys. Lett. 77B (1978) 290.

3. Coefficients in the Operator Product Expansion

Introduction

Any application of the sum rule method to real problems starts with the construction of the operator product expansion. As was explained above, in practice the coefficients in front of the various operators are calculated using perturbation theory. In the first works devoted to QCD sum rules the standard computational technique based on Feynman graphs has been used. Soon it became clear, however, that this way of calculation is extremely inconvenient in gauge theories, especially for gluon operators of high dimension. Indeed, in order to get the final gauge-invariant answer from the Feynman diagrams one has to deal with a large number of intermediate terms which are not gauge-invariant and cancel each other only at the very end of the calculation.

Practical needs called for a new method of calculation which would ensure explicit gauge invariance at every step. Gradually such a method has been developed. More exactly, two related but not identical variants have been proposed and used in the literature. The first variant extends to non-Abelian theories [1,2] the general operator formalism invented by Schwinger many years ago [3] for calculations in external electromagnetic fields. The second variant concretizes one of the aspects of the operator formalism, namely, the use of the so-called Fock–Schwinger gauge for an external gluon field,

$$(x - x_0)_\mu A^a_\mu = 0,$$

where A^a_μ is the gluon field and $x_{0\mu}$ is an arbitrary point in coordinate space. Often x_0 is chosen to coincide with the origin, $x_0 = 0$.

The Fock–Schwinger gauge produced a "technological revolution" which resulted in a real breakthrough in the field of OPE. It was invented in the classical era of QED [3–5] and then rediscovered many times in QCD [6–8], under different names (the coordinate gauge, the fixed-point gauge, etc.). A similar gauge known as "normal" [9] is exploited in the theory of (super) gravity where it allows one to express the metric $g_{\mu\nu}(x)$ in terms of the Riemann curvature tensor $R_{\mu\nu\alpha\beta}(x)$.

The computational scheme based on the Fock–Schwinger gauge transformed the tiresome procedure of determination of the OPE coefficients into an "industrial process". The scheme has emerged from the wonderful work [8] done by Dubovikov and Smilga.

Various concrete elements of the construction have been worked out in refs. [10–13]. (Ref. [13] represents a more or less independent evolution line.) The approach proved to be so efficient that at present it is applied not only in QCD sum rules but in many other problems. The reader will find a few examples (anomalies, two-loop Gell-Mann–Low functions, etc.) in this chapter.

Both variants – the Schwinger operator technique and the Fock–Schwinger gauge – have certain virtues and nicely supplement each other. Sometimes the former turns out to be more convenient, while in other problems the latter leads to the final answer in a more straightforward way. In general, we recommend to the reader to master both approaches. A detailed pedagogical presentation of the material is given by Novikov et al. [14].

A part of this review is reproduced at the end of this chapter with a few insertions which I have written additionally. A special exercise – calculation of the two-loop β function in QED – will help to understand, I hope, all subtleties. Actually, I strongly recommend to those readers who are unfamiliar with this subject to start directly with this review and then, later, return to the original papers reproduced in the beginning of this chapter.

The chapter starts with the paper [15], which discusses the general status of the operator product expansion. Then we proceed with the practical aspects of the procedure – calculation of the the coefficients functions. First we reproduce the works [7,8], which introduce the Fock–Schwinger gauge in the context of QCD. The main issue discussed in ref. [8] – the analytic properties of correlation functions in self-dual fields – is only indirectly connected with the topic under consideration. It is very interesting by itself, however. Moreover, the results of ref. [8] allow one to prove the vanishing of certain OPE coefficients without explicit calculations. The best known example of this type is the coefficient of G^3 in the two-point functions of the currents $\bar{\psi}\Gamma\psi$, where ψ is the massless-quark field and Γ stands for an arbitrary combination of γ matrices. This fact, the absence of the G^3 term in the operator expansion for massless quarks, is explicitly checked in ref. [16], which is also presented here, basically for pedagogical reasons.

The details of the computational scheme based on the Fock–Schwinger gauge were developed in a series of works [10–13]. For this reprint volume that one is chosen, ref. [12], which puts special emphasis on the method *per se*, not on particular applications. It is worth noting that the Schwinger operator formalism is less familiar, even to QCD experts, and attracts less attention than it deserves. In particular, I am not aware of any clear and sufficiently complete pedagogical discussion of this formalism in the literature. There are only a few scattered remarks and rather technical papers. For the reader who is not afraid of technical difficulties I may recommend refs. [1,2,17].

A mini-review on the Schwinger operator formalism is given at the end of this chapter in my commentary preceeding the review by Novikov et al. [14].

Nuclear Physics B249 (1985) 445-471

WILSON'S OPERATOR EXPANSION: CAN IT FAIL?

V.A. NOVIKOV, M.A. SHIFMAN, A.I. VAINSHTEIN and V.I. ZAKHAROV

Institute of Theoretical and Experimental Physics, Moscow 117259, USSR

Received 13 June 1984

Within the framework of some simple models we discuss the status of the operator product expansion (OPE) in the presence of nonperturbative effects. We consider, in particular, the 4d Higgs model, 2d sigma model and the Schwinger model. The general formulation of OPE is presented and it is demonstrated that there exists a consistent procedure allowing one to define unambiguously both coefficient functions and matrix elements of composite operators. One of the key elements of the procedure is the introduction of an auxiliary parameter, the normalization point μ. For the simplest T-products discussed in the literature earlier we construct the corresponding OPE explicitly. Then we check its validity by comparing the results for the two-point functions with independent direct calculations of the same correlators. Although the general procedure is standard and does not vary from one theory to another, numerically the relative role of perturbative and nonperturbative contributions in vacuum condensates is different in different theories. The two extremes considered are the $\lambda\varphi^4$ theory with no spontaneous breaking of the symmetry and the O(N) sigma model in the limit $N\to\infty$. In the former case there is only perturbative contribution to $\langle\varphi^2\rangle$, while in the latter case the perturbative pieces are suppressed by $1/N$ factors. Numerically QCD is much closer to the O(N) sigma model in the large-N limit. Comments on specific features of QCD are presented.

The operator product expansion (OPE) at short distances was proposed by Wilson 15 years ago [1]. At that time asymptotically free field theories [2] had not yet been discovered and the operator expansion was considered as an alternative to the traditional lagrangian field theory [1, 3]. Applications of OPE appeared immediately [1, 4]. Suffice it to refer to numerous papers on deep inelastic scattering. Moreover, practically all works on QCD in its beginning period were developed within the same framework.

The very status of the OPE has also been examined, although the effort invested in this issue is much less intensive.

A rigorous formulation of the OPE within perturbation theory has been worked out [5]. In other words, the famous Wilson relation

$$i\int e^{iqx}\,dx\,T\{j_A(x)j_B(0)\}\xrightarrow[q\to\infty]{}\sum_n D_n^{AB}(q)O_n(0) \tag{1}$$

can be justified order by order in perturbation theory, and there exists a well-defined algorithm allowing one to calculate the coefficient function $C_n^{AB}(q)$ for any local operator O_n (for a recent discussion see ref. [6]). It is well known however that quantum chromodynamics (QCD) - and this theory is most important in applications - is far from being exhausted by perturbation theory. Thus one naturally comes to the problem of extending the operator expansion to the case of nonperturbative dynamics.

The problem became even more acute in connection with the development of the so-called QCD sum rules [7], which utilize both the operator product expansion and the notion of large nonperturbative effects in QCD. Various two-point functions induced by colourless currents are calculated with the help of expansion (1), while the nonperturbative effects enter through vacuum matrix elements of local operators, such as $\langle 0|\alpha_s G^a_{\mu\nu} G^a_{\mu\nu}|0\rangle$, $\langle 0|\bar{q}q|0\rangle$, where $G^a_{\mu\nu}$ and q are the gluon field strength tensor and light quark field, respectively. Moreover, in practice one relies on eq. (1) at "intermediate" values of q^2 and evaluates the coefficient functions to leading order in $\alpha_s(q)$, where $\alpha_s(q)$ is the running coupling constant.

Is it legitimate to use such a procedure? Can one give to it a solid foundation or does it bear an approximate character? Is it universal or specific of QCD? These questions naturally arise once the phenomenological success of the sum rules is established.

Our purpose here is to answer these questions. To this end we will both formulate the procedure in general terms and illustrate it by simple examples.

There is nothing particularly new about what we are going to say. Moreover, we have had the chance to present our views earlier (see refs. [7, 8], especially sect. 6 of ref. [8]). However we feel that our discussion was somewhat sporadic and not as clear and consistent as it should have been. This is perhaps one of the reasons why there appeared recently a number of papers [9-15] which checked the procedure within some simple models and, moreover, claimed the OPE to be invalid at least in some cases. Reading these papers has been the primary stimulus for us to reconsider the problem in a more systematic way.

We believe that the difficulties which some authors have in establishing the validity of the OPE (see, e.g., refs. [9, 12]) are in essence of a common origin. As a rule, they forget about one and the same element of the routine procedure, namely introduction of the normalization point μ for the operators considered. True, the notion of the normalization point is an auxiliary one; however, it cannot be avoided by a lucky choice of the normalization scheme since it has a well-defined physical meaning (see below). The dependence of some operators on μ is well known to anyone who has ever tried to calculate the anomalous dimensions of operators in perturbation theory; in the leading log approximation the μ-dependence reduces to factors of the type $[\alpha_s(Q)/\alpha_s(\mu)]^{\gamma/b}$, where the coefficients b and γ appear as expansion coefficients of the Gell-Mann-Low function and the anomalous dimension [16]. Certainly, in many cases this rather weak μ-dependence is of little interest, and moreover, one can pick up a few operators which have vanishing anomalous dimension. Thus, it is not of great importance, as far as practical applications are concerned. The situation changes, however, once the theoretical status of OPE is considered. In the presence of nonperturbative effects the parameter μ begins to play a key role - only after its introduction does the procedure become theoretically meaningful.

The physical meaning of the OPE in asymptotically free theories is especially transparent. One starts by specifying the theory at short distances. Apart from the bare lagrangian $\mathscr{L}_0$ one can include various terms, $\sum_i \eta_i(x)O_i(x)$, where $O_i(x)$ are some local operators. This allows one to incorporate in the analysis various correlation functions of the type $\langle T\{O_A(x)O_B(O)\}\rangle$. At momenta of order of the ultraviolet cut-off M this lagrangian provides a full description of any process. If external momenta are much smaller than M one integrates first over all field configurations ranging from M down to some μ. The asymptotic freedom guarantees that it is possible to perform such a calculation. The result of the integration is a new *effective lagrangian*

$$\mathscr{L}_{\text{eff}} = \tilde{\mathscr{L}}_0(x, M, \mu) + \sum_i \eta_i(x) C_i(M, \mu) O_i(x, \mu) , \tag{2}$$

which is constructed from the operators normalized at the point μ (in particular, $\tilde{\mathscr{L}}_0$ depends on the "running" coupling constant $\alpha_s(\mu)$). The high-frequency fields determine the coefficient functions in front of the operators $O_i(x, \mu)$. The effective lagrangian (2) is to be considered as an ordinary lagrangian with respect to all fluctuations whose frequencies are less than μ. As a next step one can reiterate the procedure going down to some lower value of μ, and so on; the only limitation comes from the condition $\alpha_s(\mu) \ll 1$.

Integrating over high-frequency fluctuations does not imply that we are accounting for perturbation theory alone. For example, one can (and has to) account explicitly for small-size instantons (see ref. [8]), and, at least in principle, for other nonperturbative contributions. The accuracy in calculating the coefficient functions is determined by our knowledge of the theory at short distances ($x < \mu^{-1}$), and is under control since the coupling constant $\alpha_s(\mu)$ is small.

Such a procedure for constructing the OPE was actually proposed first by Wilson (see, e.g., ref. [17]) in connection with his formulation of the renormalization group. Within this framework the very question of whether the OPE is valid is absurd. Indeed, if the contribution of the domain between M and μ is accounted for correctly, what can be wrong?

There are, however, other questions which are reasonable:

(i) Is it possible to determine - even to some approximation - the coefficient functions $C_i(\mu)$, keeping only a perturbative series? As has already been mentioned, in QCD one evaluates coefficients $C_i(\mu)$ keeping only the simplest Feynman graphs. Is it a good approximation? Can it be extended to other models?

(ii) How strong is the dependence of the operators on the normalization point μ, and can it be neglected, and if so, when?

(iii) If vacuum condensates are unknown theoretically is it possible to extract them phenomenologically from the analysis of power corrections in the sum rules? And so on.

As we shall argue below, the answers to these questions are actually not universal and vary from one model to another, or even from one particular problem to another within the same model.

Although the questions to which we address ourselves are quite general, we will discuss them in detail for some simple models highlighted in earlier papers on similar subjects [9-15]: the 4-dimensional Higgs model, 2-dimensional σ-models at large N, and finally, 2-dimensional Schwinger electrodynamics. In all these cases we introduce the normalization point μ, construct explicitly the OPE for some Green functions, and compare the results with the solutions obtained by other methods. In all these cases the validity of the OPE is confirmed.

Our conclusions on the OPE in the Higgs model differ from those of refs. [9, 10], which claim the OPE to be invalid (see also ref. [18]). We do not find any problems with the OPE in 2-dimensional σ-models either. This assertion is at variance with that of David [12], who experiences certain difficulties in the next-to-leading order in $1/N_c$. According to ref. [12] there exists no infrared stable definition of the matrix elements of composite operators. (To avoid confusion let us mention that in an earlier paper by the same author [11] treating the O(N) sigma model in the leading approximation, the OPE was found to be perfectly consistent.) As for the Schwinger model our response to the criticism in the literature seems to be somewhat slow: the objections with respect to the OPE in this model - which prompted to some extent the present investigation - are cancelled by their authors. Moreover, there appeared a detailed check of the OPE by Danilov and Dyatlov [14], who confirmed its validity in this model. However, in spite of the similarity of the final conclusions of all the papers devoted to the subject which are circulating at the moment, we feel that some aspects of the problem call for further consideration, and we do not cut short the corresponding section of our paper.

Along with analysing some particular models we will try to outline a general view on the OPE in QCD to which we adhere when constructing and using the sum rules. Namely, we will try to emphasize those elements which distinguish QCD from, say, the 2-dim σ-model, despite the fact that both theories share such basic properties as asymptotic freedom and dimensional transmutation. We keep in mind now some numericals which turn out to be crucial in practical applications.

In our opinion, attempts to prove failure of the OPE are reminiscent of attempts to work out a *perpetuum mobile.* Although everybody knows that such a device does not exist, sometimes it is not easy to establish the falsehood of a given project. Similarly, to unveil the weak points in the criticism of the OPE we must go through a lot of technical details first.

The procedure is as follows. In sect. 2 we consider the $\lambda\varphi^4$ in four dimensions. Sect. 3 is devoted to sigma models. Schwinger electrodynamics are discussed in sect.

4. Finally, sect. 5 contains conclusions and remarks on some specific features of the OPE within QCD.

2. Real scalar field (the Higgs model)

In this section we will consider, following refs. [9, 10], the simplest example, namely $\lambda\varphi^4$ theory in four-dimensional space-time. It is well known that the "wrong" sign of the mass term in this model spurs the vacuum instability and condensation of the φ-field. Moreover, the vacuum expectation value $\langle 0|\varphi|0\rangle$ is nonanalytical in the coupling constant. Thus, in spite of its triviality the model mimics in a sense nonperturbative effects. The problem is formulated as follows: one considers some correlation function which is known at large (euclidean) q^2 and, alternatively, constructs the OPE for the same function. Then the results are confronted with each other.

The actual procedure of constructing the OPE in refs. [9, 10] relies on a simplified recipe according to which the coefficient functions C_i are determined exclusively by perturbation theory over the mathematical vacuum (i.e. the vacuum with $\langle\varphi\rangle = 0$), while the vacuum condensates are entirely due to nonperturbative effects. This procedure (which is approximately valid in QCD, see below) has lead to a paradoxical conclusion on the failure of the OPE.

We will argue that once the regular OPE is used there is no contradiction whatsoever. The key point is that the vacuum expectation values, such as $\langle\varphi\rangle$, do receive contributions from perturbation theory. Moreover, numerically the nonperturbative effects are not superior in the case at hand. It is just the latter point that makes this model drastically different from QCD.

To sharpen the problem we will consider first even a more trivial example: the "right" sign of the mass term; nonperturbative effects are absent altogether. The lagrangian then takes the form

$$\mathcal{L} = \tfrac{1}{2}(\partial_\mu\varphi)^2 - \tfrac{1}{2}m_0^2\varphi^2 - \frac{\lambda_0}{4!}\varphi^4 , \tag{3}$$

where m_0 and λ_0 are the bare mass and coupling constant, respectively, and φ is a real scalar field. Moreover, the theory is assumed to be regularized in some way, say, by means of the Pauli–Villars regulator fields. All observable (i.e. renormalized) quantities can be expressed then in terms of m_0, λ_0 and the ultraviolet cut-off M.

If $m_0^2 > 0$, symmetry with respect to the transformation $\varphi \to -\varphi$ is not broken, and one can use the standard perturbative expansion in λ_0. In particular, the Green function $D(q)$ of the field φ in the one-loop approximation is determined by the graph in fig. 1:

$$D(q) = -i\int \mathrm{d}^4x\, \mathrm{e}^{iqx}\langle 0|\mathrm{T}\{\varphi(x)\varphi(0)\}|0\rangle$$
$$= \frac{1}{q^2 - m_0^2}\left[1 + \frac{\lambda_0}{32\pi^2}\,\frac{M^2 - m_0^2\ln(M^2/m_0^2)}{q^2 - m_0^2}\right]. \tag{4}$$

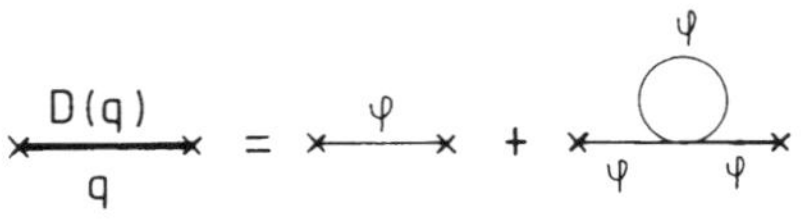

Fig. 1. Propagator of the φ-field in $\lambda\varphi^4$ theory ($m_0^2 > 0$) in the one-loop approximation.

Now, let us establish the correspondence between this expression for the Green function and the OPE for the T-product

$$-i\int e^{iqx}\,d^4x\,T\{\varphi(x)\varphi(0)\} \tag{5}$$

at large euclidean momenta ($q^2 \equiv -Q^2$, $Q^2 \gg m^2$). Keeping only the first two terms in $1/Q^2$ one gets for the asymptotics of $D(q)$

$$D(q) \to -\frac{1}{Q^2} + \frac{1}{Q^4}\left[m_0^2 + \frac{\lambda_0}{32\pi^2}\left(M^2 - m_0^2 \ln\frac{M^2}{m_0^2}\right)\right]. \tag{6}$$

Respectively, in the OPE one must keep the first two terms, namely, operators of dimension 0 and 2 – the unit operator $\mathbb{1}$ and the operator φ^2.

The coefficient C_φ in front of the operator φ^2 in the leading (Born) approximation is determined by the graph of fig. 2, and is equal to

$$C_\varphi = \frac{\lambda_0}{2Q^4} \neq 0\,. \tag{7}$$

Moreover, the coefficient $C_{\mathbb{1}}$ in the approximation considered is determined by the same graphs in fig. 1 which were used above for fixing $D(q)$. Superficially, the perturbative contribution as a whole is included already in $C_{\mathbb{1}}$, and to ensure the correspondence between eq. (6) and the OPE, one has to put $\langle 0|\varphi^2|0\rangle = 0$. Actually this is wrong. In particular, one can readily convince oneself that proceeding in this way it would be impossible to use the renormalization group equations to sum up the log corrections of the type $(\lambda \ln M^2)^n$.

The genuine recipe of the OPE prescribes the inclusion in the coefficient functions only contributions of virtual momenta $p > \mu$, where μ is the normalization point. The contributions corresponding to $p < \mu$ are to be referred to as the matrix elements of the operators. On the other hand, in evaluating the graphs in fig. 1 one integrates over all momenta p.

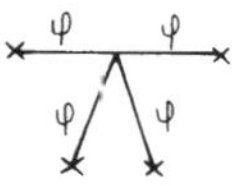

Fig. 2. $C_\varphi\varphi^2$ in the T-product $-i\int d^4x\, e^{iqx}T\{\varphi(x)\varphi(0)\}$.

Thus, to get $C_{\mathbb{1}}$ one has to remove from the perturbative graphs the domain $p<\mu$. In the case considered such a subtraction is readily achieved. Namely, one has to subtract from the term proportional to λ_0 in eq. (4) just the same expression with the cut-off M substituted by the normalization point μ. Then one gets

$$C_{\mathbb{1}}=-\frac{1}{Q^2}+\frac{1}{Q^4}\left[m_0^2+\frac{\lambda_0}{32\pi^2}\left(M^2-\mu^2-m_0^2\ln\frac{M^2}{\mu^2}\right)\right]; \tag{8}$$

it is assumed here that $\mu^2 \gg m_0^2$.

A few explanatory remarks are now in order. First, as far as large euclidean Q^2 are considered, one can say that there are two independent coupling constants. The first one is the dimensionless coupling constant λ_0, while the second one is the dimensionless ratio m_0^2/μ^2. If the normalization point $\mu^2 \gg m_0^2$, then the expansion in m_0^2/μ^2 is justified. The coefficient functions $C_{\mathbb{1}}(Q^2, \mu^2)$, $C_\varphi(Q^2, \mu^2)$ are analytical in both "coupling constants", as they should be, while the Green function (4) is nonanalytical in m_0^2/μ^2 (contains $\ln m_0^2$).

Second, the power terms in eq. (8) proportional to M^2, μ^2 depend on the regularization scheme. Eq. (8) refers actually to the case of the Pauli–Villars regularization with the minimal set of the regulator fields. Introducing extra regulator fields one can get any coefficient in front of the power terms. In particular, one can ensure their vanishing. If one uses dimensional regularization these power terms are absent from the very beginning (see below). On the other hand, the log term in eq. (8) does not depend on the details of the renormalization procedure.

The observation that terms $\sim M^2$, μ^2 are actually not fixed does not imply any physical ambiguity. One just must keep in mind that all calculations are to be performed within one and the same regularization scheme. For example, if we used dimensional regularization then the term proportional to M^2 would be absent both in eq. (4) and (8). As for the normalization point μ, this parameter is auxiliary and must drop off from the final answer for the correlation functions in any case.

Finally, let us check that the Green function defined in eq. (4) does coincide in the approximation considered with the OPE for the T-product $-i\langle 0|\mathrm{T}\{\varphi(x)\varphi(0)\}|0\rangle$. In other words, we have to check the validity of the following equation:

$$D(q)=\langle 0|C_{\mathbb{1}}(q,\mu)+C_\varphi(q,\mu)\varphi^2(\mu)|0\rangle\,. \tag{9}$$

As has been mentioned already there are no nonperturbative effects in this version of the theory, and a nonvanishing vacuum expectation value $\langle 0|\varphi^2(\mu)|0\rangle$ arises within perturbation theory thanks to the contribution of momenta $p<\mu$. The corresponding tadpole graph is given in fig. 3. Cutting off the integration over the virtual momenta $p_{\max}=\mu$, we get

$$\langle 0|\varphi^2(\mu)|0\rangle=\frac{1}{16\pi^2}\left(\mu^2-m_0^2\ln\frac{\mu^2}{m_0^2}\right). \tag{10}$$

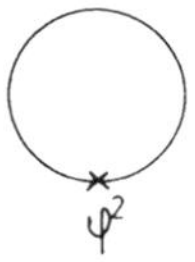

Fig. 3. Vacuum expectation value $\langle 0|\varphi^2|0\rangle$ in perturbation theory.

Combining this expression with eqs. (7) and (8) we find that the OPE (see eq. (9)) is, of course, valid.

Some readers may perhaps think at this stage, that the introduction of the normalization point μ and the corresponding discrimination of virtual momenta is an awkward manoeuvre, an unnecessary element. Below, we proceed to the summation of the log terms by means of the renormalization group equations. Our purpose is to demonstrate that with no normalization point the very notion of the renormalization group makes no sense.

2.1. MASS RENORMALIZATION. SUMMATION OF THE LEADING LOGARITHMS

Expression (8) for the coefficient C_1 contains both the bare mass m_0 and the ultraviolet cut-off M. Introducing the running mass $m(\mu)$,

$$m^2(\mu) = m_0^2 + \frac{\lambda_0}{32\pi^2}\left(M^2 - \mu^2 - m_0^2 \ln\frac{M^2}{\mu^2}\right)$$

$$= m^2(\mu_0)\left(1 + \frac{\lambda_0}{16\pi^2}\ln\frac{\mu}{\mu_0}\right) + \frac{\lambda_0}{32\pi^2}(\mu_0^2 - \mu^2)\,, \tag{11}$$

normalized at μ removes the dependence on the cut-off M. Furthermore, in the next order in λ_0 there arises renormalization of the coupling constant λ.

Using the renormalization group equations, there is no difficulty in "improving" the results obtained above by summing up all terms of the kind $(\lambda \ln \mu^2)^k$, or the leading logs. For simplicity let us assume that the normalization procedure is chosen in such a way that power divergences are absent.

Then the renormalization group equations read

$$\mu\frac{\mathrm{d}}{\mathrm{d}\mu}C_\varphi = \frac{\lambda(\mu)}{16\pi^2}C_\varphi\,, \tag{12a}$$

$$\mu\frac{\mathrm{d}}{\mathrm{d}\mu}C_1 = \frac{m^2(\mu)}{8\pi^2}C_\varphi\,, \tag{12b}$$

$$\mu\frac{\mathrm{d}}{\mathrm{d}\mu}m^2(\mu) = \frac{\lambda(\mu)}{16\pi^2}m^2(\mu)\,, \tag{12c}$$

$$\mu\frac{\mathrm{d}}{\mathrm{d}\mu}\lambda(\mu) = 3\frac{\lambda^2(\mu)}{16\pi^2}\,. \tag{12d}$$

Fig. 4. The simplest graph determining the anomalous dimension of φ^2, and the corresponding renormalization group equation for the coefficient C_φ.

Eq. (12a) reflects a nonvanishing anomalous dimension of operator φ^2 (see fig. 4), while eqs. (12c) and (12d) are the standard equations for the running mass and coupling constant. The only subtle point is eq. (12b), which describes mixing of the unit operator with the operator φ^2.

Eqs. (12) are readily integrable and the solutions are as follows:

$$\lambda(\mu)=\frac{\lambda(\mu_0)}{1-3(\lambda(\mu_0)/16\pi^2)\ln(\mu/\mu_0)},$$

$$m^2(\mu)=\left[\frac{\lambda(\mu)}{\lambda(\mu_0)}\right]^{1/3} m^2(\mu_0),$$

$$C_\varphi(\mu)=\left[\frac{\lambda(\mu)}{\lambda(\mu_0)}\right]^{1/3} C_\varphi(\mu_0),$$

$$C_1(\mu)=C_1(\mu_0)-2\frac{m^2(\mu_0)C_\varphi(\mu_0)}{\lambda(\mu_0)}\left(\left[\frac{\lambda(\mu)}{\lambda(\mu_0)}\right]^{-1/3}-1\right), \tag{13}$$

where $\lambda(\mu_0)$, $m^2(\mu_0)$, $C_\varphi(\mu_0)$, $C_1(\mu_0)$ are initial conditions. It is convenient to choose $\mu_0=Q$. Then the coefficients $C_1(\mu_0=Q)$ and $C_\varphi(\mu_0=Q)$ contain no logarithms if expressed in terms of $\lambda(Q)$, $m^2(Q)$, and reduce to the corresponding Born results:

$$C_1(\mu_0=Q)=-\frac{1}{Q^2}\left(1-\frac{m^2(Q)}{Q^2}\right),$$

$$C_\varphi(\mu_0=Q)=\frac{\lambda(Q)}{2Q^4}. \tag{14}$$

Finding in this way the coefficients $C_1(\mu)$ and $C_\varphi(\mu)$, we get an explicit expression for OPE:

$$\begin{aligned} D(q) &= \langle 0|C_1(\mu)+C_\varphi(\mu)\varphi^2(\mu)|0\rangle \\ &= -\frac{1}{Q^2}+\frac{1}{Q^4}\left\{\left[\frac{\lambda(Q)}{\lambda(\mu)}\right]^{1/3} 2m^2(\mu)\right. \\ &\quad \left.+\left[\frac{\lambda(Q)}{\lambda(\mu)}\right]^{2/3}\left(\tfrac{1}{2}\lambda(\mu)\langle\varphi^2(\mu)\rangle-m^2(\mu)\right)\right\}. \end{aligned} \tag{15}$$

Here $\langle\varphi^2(\mu)\rangle$ is the vacuum expectation value of the operator φ^2 normalized at μ. If $\mu \sim m$ then the vacuum expectation value vanishes to the accuracy considered (see eq. (10) with the power term μ^2 dropped off). Therefore the final result for $D(q)$ reduces to

$$D(q) = -\frac{1}{Q^2} + \frac{1}{Q^4} m^2(m) \left\{ 2\left[\frac{\lambda(Q)}{\lambda(m)}\right]^{1/3} - \left[\frac{\lambda(Q)}{\lambda(m)}\right]^{2/3} \right\}. \tag{16}$$

To compare this result with eq. (6) one has to expand eq. (16) in $\lambda(m)\ln(Q/m)$. Terms of the first order in this parameter are cancelled out, in full accordance with eq. (6).

Note that so far we have considered the theory with the "right" sign of the mass term and no spontaneous symmetry breaking.

2.2. THE HIGGS MODEL (SPONTANEOUS SYMMETRY BREAKING)

Turn now to the Higgs version of the model when m_0^2 in eq. (3) is negative. Then the condensate of the φ-field develops and it is nonanalytical in the coupling constant. We will demonstrate that the operator expansion (15) results in the correct expression for the Green function $D(q)$ in this case as well.

Note first of all that $\langle\varphi^2\rangle$ is given by its classical value if the normalization point μ is of the order of the physical mass

$$\mu^2 \sim m_{\rm phys}^2 = -2m^2(m_{\rm phys}) . \tag{17}$$

Indeed, under such a choice

$$\tfrac{1}{2}\lambda(m_{\rm phys})\langle\varphi^2(m_{\rm phys})\rangle = -3m^2(m_{\rm phys}) = \tfrac{3}{2}m_{\rm phys}^2 . \tag{18}$$

Taking this fact into account, we get for the operator expansion (15)

$$D(q) = -\frac{1}{Q^2} + \frac{m_{\rm phys}^2}{Q^4} \left\{ 2\left(\frac{\lambda(Q)}{\lambda(m_{\rm phys})}\right)^{2/3} - \left(\frac{\lambda(Q)}{\lambda(m_{\rm phys})}\right)^{1/3} \right\}. \tag{19}$$

Let us compare this expression with that arising from perturbation theory over the physical vacuum (i.e. the vacuum with the shifted value of φ and quantum fluctuations $\varphi - \langle\varphi\rangle$). In the one-loop approximation there exists a single graph depending on Q (see fig. 5a). Adding the contribution of the Born graph of fig. 5b, we find

$$D(q) = -\frac{1}{Q^2} + \frac{1}{Q^4} m_{\rm phys}^2 \left(1 + \frac{3\lambda}{16\pi^2} \ln\frac{Q}{m_{\rm phys}}\right) + \mathrm{O}(\lambda^2) . \tag{20}$$

The subtractions here are performed at euclidean virtual momenta of order $p \sim m_{\rm phys}$. In particular, Q-independent terms coming from other diagrams are absorbed into the mass renormalization.

Expanding eq. (19) in λ we reproduce eq. (20).

Fig. 5. The Green function $D(q)$ in perturbation theory over the "physical" vacuum in the version with spontaneous symmetry breaking, $\langle\varphi\rangle \neq 0$.

It is worth emphasizing that the summation of the log factors performed above resulted in two terms (see the factor in braces in eq. (19)) and does not reduce to a mere anomalous dimension of mass. (Compare with ref. [9] which assumes that the effect of log corrections is exhausted by the anomalous dimension of mass in the example at hand.)

The validity of the OPE in the models with spontaneous symmetry breaking has been discussed in refs. [9, 10]. We hope that the consideration given above is sufficient to elucidate the reasons for the "difficulties and paradoxes" in these works.

Indeed, for the consistent use of the OPE it is imperative to divide - by introducing the normalization point - the integration domain in the Feynman graphs into two parts corresponding to coefficient functions and matrix elements. In particular, just in this way does one introduce the running coupling constant and mass - both quantities emerge as coefficients in the effective action. The subtraction procedure used in ref. [9] does not account for this. The same criticism applies to paper [18], which deals with the case of negative m^2. Although such a distinction between low and high p is not very essential in the leading log approximation, it is indispensable conceptually. Only the introduction of μ allows one to fix the coefficient functions independently of the sign of m^2.

3. Two-dimensional O(N) Sigma Model

The lagrangian has the form

$$\mathscr{L} = \tfrac{1}{2}(\partial_\mu \sigma^a(x))(\partial_\mu \sigma^a(x)) ,$$

$$\sigma^a \sigma^a = \frac{N}{f} , \qquad (21)$$

where a is the "colour" index, $a = 1, \dots, N$. At large N the model can be solved exactly by means of the $1/N$ expansion. We do not describe the model here in any detail, and refer the reader to the original papers [19, 20] or to a recent review [21]. Let us note only that, like QCD, the two-dimensional sigma model is asymptotically free [22]:

$$f(\mu) = \frac{4\pi}{\ln(\mu^2/m^2)} . \qquad (22)$$

Moreover, although the original lagrangian (21) has no mass parameter there is dynamical mass generation:

$$m^2 = \mu^2 \exp(-4\pi/f(\mu)), \tag{23}$$

so that the σ-particle mass depends nonanalytically on the coupling constant.

The "physical" vacuum in this model respects O(N) invariance and differs rather drastically from the "mathematical" vacuum of perturbation theory. In particular, vacuum condensates of the type $\langle 0|f(\partial_\mu \sigma^a)^2|0\rangle$ develop. In leading $1/N$ order one has

$$\langle 0|f(\partial_\mu \sigma^a)^2|0\rangle = N\mu^2 \exp\left(-\frac{4\pi}{f(\mu)}\right) = Nm^2, \tag{24}$$

while vacuum expectation values of other composite operators factorize (see, e.g., ref. [21] for details):

$$\langle 0|[f(\partial_\mu \sigma^a)^2]^k|0\rangle = [Nm^2]^k. \tag{25}$$

Thus, the physical vacuum is nontrivial and imitates, to some extent, the properties of the QCD vacuum. On the other hand, the model is simple enough to check the validity of the OPE thoroughly.

The validity of the OPE in this model has been discussed in the literature. First, in two independent papers [11, 23] the following Green function has been considered:

$$S(q^2) = i \int d^2x\, e^{iqx} \langle 0|T\{j_S(x) j_S(0)\}|0\rangle,$$

$$j_S = f(\partial_\mu \sigma^a)^2, \tag{26}$$

in leading $1/N$ approximation. The reason for choosing this particular operator, j_S, is that the product $f(\partial_\mu \sigma^a)^2$ is renormalization group invariant.

The choice (26) turned out to be lucky in one respect, and simultaneously unlucky in another respect. The point is that the OPE basis for the T-product of $j_S(x) j_S(0)$ consists of operators of the type $[f(\partial_\mu \sigma^a)^2]^k$. As mentioned above, the matrix elements of these operators in leading order just factorize (see eq. (25)), and therefore the OPE for this Green function reduces, after all, to a combination of the coefficient functions and powers of the condensate $\langle 0|f(\partial_\mu \sigma^a)^2|0\rangle$. In the approximation considered in refs. [11, 23] neither the former nor the latter depend on the normalization point μ. For example, naively one gets

$$\langle 0|f(\partial_\rho \sigma^a)^2|_\mu|0\rangle = -f(\mu) N \int_{p<\mu} \frac{d^2p}{(2\pi)^2} \frac{p^2}{p^2+m^2}$$

$$= \frac{Nf(\mu)}{4\pi}\left(-\mu^2 + m^2 \ln\frac{\mu^2}{m^2}\right). \tag{27}$$

A more careful analysis reveals (see refs. [23, 21] for details) that there is no power term $(-\mu^2)$, and the correct answer is

$$\langle 0|f(\partial_\rho \sigma^a)^2|_\mu|0\rangle = \frac{Nf(\mu)}{4\pi} m^2 \ln\frac{\mu^2}{m^2} = \text{const} \tag{28}$$

(compare with eq. (24)). Thus, there is no need to introduce the normalization point explicitly. As a result, if $N \to \infty$ the general OPE simplifies considerably and reduces to the naive recipe according to which the coefficient functions are determined entirely by perturbation theory (p.t.) over the mathematical vacuum (and only p.t.), while the matrix elements absorb all nonperturbative effects, and moreover, receive no contribution from perturbation theory. Just this version of the OPE is often used in practical applications of QCD sum rules. Unlike QCD where this naive formulation is manifestly approximate, in the large-N limit in the sigma model it has a sound theoretical basis - it becomes exact if $N = \infty$.

Thus, all complications associated with the necessity of introducing the normalization point can be ignored in the limit $N \to \infty$. As we have already seen, however, such a situation is rather an exception than a rule. In particular, there is no reason to believe that a similar procedure holds in the next-to-leading order. From this point of view the analysis of refs. [11, 23] does not touch upon the problems which are important in other applications of the OPE.

As a consequence, ref. [12] has appeared presenting an attempt (doomed to failure from the very beginning) to extend straightforwardly the naive procedure of the leading N approximation to the next-to-leading order. According to this work there exist some basic ambiguities in formulation of the OPE (comprising nonperturbative effects) which are related to infrared properties of the theory, or renormalons.

In this section we reconsider the two-point function (26) with inclusion of $1/N$ corrections and reiterate some points which we think are important in any analysis of the status of the OPE in field theory.

The operator basis which we deal with in the product $\mathrm{T}\{j_S(x) j_S(0)\}$ includes operators of the type α^n where

$$\alpha = \frac{1}{\sqrt{N}} f(\partial_\mu \sigma^a)^2 \tag{29}$$

(the normalization factor, $1/\sqrt{N}$, arises in a natural way, see, e.g., ref. [21]). Moreover, following ref. [12] let us consider the first composite operator in this series, namely $\alpha^2(\mu)$. Since it depends on the normalization point we introduce μ from the very beginning. In this respect it differs from α which is renormalization group invariant.

The vacuum expectation value of the operator α^2 normalized at μ can be defined in the following way:

$$\langle 0|\alpha^2(\mu)|0\rangle = \int_{\text{Eucl.}\, p<\mu} \frac{d^2 p}{(2\pi)^2} D^{(\alpha)}(p)\,, \tag{30}$$

where $D^{(\alpha)}(p)$ is the propagator of the α-field which is known explicitly to the leading order:

$$D^{(\alpha)}(p) = i \int d^2x\, e^{ipx} \langle 0|T\{\alpha(x)\alpha(0)\}|0\rangle$$
$$= (2\pi)^2 \delta^{(2)}(p) \langle \alpha \rangle^2 - \frac{4\pi\sqrt{p^2(p^2+4m^2)}}{\ln \dfrac{\sqrt{p^2+4m^2}+\sqrt{p^2}}{\sqrt{p^2+4m^2}-\sqrt{p^2}}} . \tag{31}$$

The first term in eq. (31) proportional to the delta function corresponds to the factorizable piece of the matrix element $\langle \alpha^2 \rangle$. To leading order $\langle \alpha \rangle = \sqrt{N} m^2$, and there also exist some corrections of order $1/N$. However, the factorizable part should not concern us here.

As for the connected part of the T-product (31) it is of order $1/N$, as compared to the disconnected one. Thus, in order to evaluate it in the next-to-leading approximation it is sufficient to know $D^{(\alpha)}$ in the leading order in $1/N$. Eqs. (30), (31) account for this observation.

Performing the obvious substitution

$$x = \left[\sqrt{1+\frac{p^2}{4m^2}} + \sqrt{\frac{p^2}{4m^2}} \right]^4 , \tag{32}$$

we easily get

$$\langle 0|\alpha^2(\mu)|0\rangle_{\text{nonfactorizable}} = -m^4 \int_1^{A(\mu)} \frac{dx}{x^2} (x-1)^2 (\ln x)^{-1} , \tag{33}$$

where

$$A(\mu) = \left[\sqrt{1+\frac{\mu^2}{4m^2}} + \sqrt{\frac{\mu^2}{4m^2}} \right]^4 .$$

Moreover, expressing the result of integration in (33) in terms of the standard functions we find

$$\langle 0|\alpha^2(\mu)|0\rangle_{\text{nonfactorizable}}$$
$$= -m^4 \{ \mathrm{Ei}(\ln A(\mu)) + \mathrm{Ei}(-\ln A(\mu)) - 2 \ln \ln A(\mu) - 2C \} , \tag{34}$$

where $C = 0.5772$ is the Euler constant.

What lesson can be learnt about the OPE from these exercises? To answer this question we have to study the limit $\mu^2 \gg m^2$. Indeed, only under such a choice of μ can the coefficient functions of the OPE be found perturbatively (see eq. (22)). It goes without saying that the operators entering the OPE are normalized at the same point μ.

For $\mu^2 \gg m^2$ expression (34) simplifies to

$$\langle 0|\alpha^2(\mu)|0\rangle_{\text{nonfactorizable}}$$
$$= -2\mu^4[e^{-L}\text{Ei}\,L] - \frac{4}{L}\mu^2 m^2 + 2m^4\left[C + \ln L - \frac{1}{L} + \frac{4}{L^2}\right] + O\left(\frac{m^6}{\mu^2}\right), \qquad (35)$$

where

$$L = \ln\frac{\mu^4}{m^4} \gg 1\,, \quad e^{-L}\text{Ei}\,L \underset{L\to\infty}{\approx} \sum_{n=0}^{\infty}\frac{n!}{L^{n+1}}\,.$$

To interpret various terms in eq. (35), compare it to the general operator expansion for a T-product of two renormalization group-invariant currents j_S:

$$i\int d^2x\, e^{ipx}\text{T}\{j_S(x)j_S(0)\}$$
$$= C_0(p^2,\mu^2)\mathbb{1} + C_1(p^2,\mu^2)\alpha(0)|_\mu + C_2(p^2,\mu^2)\alpha^2(0)|_\mu + \cdots. \qquad (36)$$

As explained above, in the leading order in $1/N$ neither the coefficients nor the operators depend actually on the normalization point μ. Eq. (35) demonstrates, however, that this is not so in the next order, and the condensate $\langle 0|\alpha^2(\mu)|0\rangle$ does depend on μ in a nontrivial way. Furthermore, it is obvious that this dependence is to be cancelled out by other terms in the r.h.s. of eq. (36), since the l.h.s. of the same equation has no parameter μ at all.

Let us trace this cancellation in more detail. Consider first the power terms $\sim\mu^4$, μ^2. In sect. 2 we have mentioned that terms of this type depend actually on details of the renormalization procedure. Eq. (30) uses a sharp cut-off in the integration over virtual momenta p and actually assumes the integrand to contain a θ-function. Introducing the cut-off in a more soft way one can, in principle, make the power terms vanishing. Both dimensional regularization and the Pauli-Villars procedure with extra regulator fields are suitable for this purpose. In these cases the power dependence on μ is eliminated from coefficients C_i as well.

If we still stick for simplicity to the θ-function cut-off then the power terms μ^4 in eq. (35) cancel against a similar term coming from $C_0(p^2,\mu^2)\langle 0|\mathbb{1}|0\rangle$. Indeed, it is quite obvious that just such a dependence of C_0 on μ arises in p.t. graphs subject to the constraint that only the region of $p^2_{\text{Euclid.}} > \mu^2$ is integrated over.

Moreover, the term proportional to $\mu^2 m^2$ in eq. (35) is cancelled against a similar term associated with $C_1(p^2,\mu^2)\times\langle 0|\alpha(\mu)|0\rangle$. In this case the factor μ^2 reflects subtraction from p.t. graphs determining C_1 the domain $p^2 < \mu^2$, while the factor m^2 represents the nonperturbative condensate $\langle 0|\alpha(\mu)|0\rangle$.

Finally, consider the terms proportional to m^4 which are uniquely determined and do not depend on the regularization procedure. It is perfectly clear - and we would like to emphasize this point - that terms $m^4/\ln(\mu^4/m^4)$, $m^4/\ln^2(\mu^4/m^4)$, $m^4\ln\ln(\mu^4/m^4)$ are a manifestation of the anomalous dimension of the operator

α^2 which does not vanish in the next-to-leading $1/N$ approximation. Indeed, if some operator O acquires a nonvanishing anomalous dimension of order $1/N$, then

$$O(\mu) = C(\mu, \mu_0)O(\mu_0) ,$$

where

$$\begin{aligned} C(\mu, \mu_0) &= \exp \int_{f(\mu_0)}^{f(\mu)} \mathrm{d}f \frac{\gamma(f)}{\beta(f)} = \exp \left\{ \frac{1}{N} \int_{f(\mu_0)}^{f(\mu)} \mathrm{d}f \left[\frac{a_{-1}}{f} + a_0 + a_1 f + \cdots \right] \right\} \\ &= (\text{a } \mu\text{-independent factor}) \\ &\quad \times \exp \left\{ \frac{1}{N} (a_{-1} \ln f(\mu) + a_0 f(\mu) + \tfrac{1}{2} a_1 f^2(\mu) + \cdots) \right\} \\ &\approx (\text{a } \mu\text{-independent factor}) \\ &\quad \times \left\{ 1 + \frac{1}{N} (a_{-1} \ln f(\mu) + a_0 f(\mu) + \tfrac{1}{2} a_1 f^2(\mu) + \cdots) \right\} . \end{aligned} \tag{37}$$

Substituting eq. (22) and comparing the general expression (37) with our result we readily identify the coefficients

$$a_{-1} = -2 , \qquad a_0 = -\frac{1}{4\pi} , \qquad a_1 = \frac{1}{4\pi^2} , \ldots .$$

As usual, the logarithmic dependence on μ^2 of the matrix element is cancelled out by the analogous log dependence of the coefficient $C_2(p^2, \mu^2)$.

Thus, to the next-to-leading order both the coefficient functions and the operators entering the OPE for $\mathrm{T}\{j_S(x) j_S(0)\}$ do depend on μ^2, and this dependence cannot be eliminated by any "improvement" of the regularization procedure. On the other hand, the T-product itself does not depend on μ^2 of course.

Note also that subtraction of the region of low p^2 in the graphs representing $C_0(p^2, \mu^2)$ results in the following perturbative series:

$$\frac{1}{N} \sum_k \mu^4 k! \left(\frac{f(\mu)}{4\pi} \right)^k .$$

Thus, unlike the case of the leading $1/N$ approximation the perturbative expansion contains an infinite number of terms and, moreover, the series is factorially divergent. The latter is typical of any field theory.

We hope that the considerations above have convinced the reader that introduction of the normalization point is indeed necessary if we are going to deal with $1/N$ corrections. Moreover, generally speaking, v.e.v.'s of composite operators are contributed both by truly nonperturbative fluctuations and p.t. graphs. If a similar thing happened in the leading order the study of nonperturbative effects via the OPE would be a more difficult task although the validity of the OPE would not be questioned in any case.

It is worth emphasizing that the O(N) sigma model with $N \to \infty$ imitates QCD to a great extent. In both cases introduction of the auxiliary point μ is unavoidable as far as the general procedure is concerned. In both cases, however, the μ-dependence is not important numerically. In the sigma model this latter point can be demonstrated taking the $N \to \infty$ limit. In the case of QCD there also exists a domain of μ where, on the one hand, the coupling constant $\alpha_s(\mu)$ is small (so that the expansion of the coefficient functions in $\alpha_s(\mu)$ is justified), and on the other hand, the μ-dependence of the matrix elements can be safely neglected. In other words, possible perturbative contributions like μ^4 are negligible in comparison with the nonperturbative piece m^4. The existence of such a μ was demonstrated by simple estimates in the very first papers on QCD sum rules [7] and is confirmed by all further experience. It remains to be clarified what theoretical parameter, if any, commands such a hierarchy of mass scales in QCD.

In conclusion of this section let us compare our results with those of ref. [12]. According to ref. [12] there exists no definition of the vacuum condensate $\langle 0|\alpha^2|0\rangle$ which is free of ambiguities in the next-to-leading $1/N$ order. As a result, at the best the status of the OPE is unclear, according to the author of ref. [12]. Our assertions are quite the opposite.

For the reader's convenience it might be worth going into more technical detail at this point since the technique of ref. [12] looks superficially very different from what we are using here. First, the dimensional regularization is chosen in ref. [12] so that the power terms do not arise from the very beginning and the whole business reduces to terms proportional to m^4 (see eq. (35)). As explained above the μ-dependence of these terms is due to the anomalous dimension of the operator $\alpha^2(\mu)$ ($\gamma(\alpha^2) \sim 1/N \neq 0$). Although there is no mention of the normalization point μ in ref. [12] it is quite clear that the parameter $1/\epsilon$ introduced therein plays the same role as $\ln(\mu^2/m^2)$ in the routine procedure. Then the limit $\epsilon \to 0$ considered in ref. [12] is equivalent to tending $\ln(\mu^2/m^2) \to \infty$ in our notation. In this limit the terms $m^4/\ln(\mu^2/m^2)$, $m^4/\ln^2(\mu^2/m^2)$ are negligible, and the only way the anomalous dimension shows up in this case is through the terms of order $m^4 \ln\ln(\mu^2/m^2)$. Indeed, it is straightforward to see that the factor $\ln\ln(\mu^2/m^2)$ corresponds to the factor $\ln(\epsilon^{-1})$ figuring in the dimensional regularization (DR) scheme:

$$\ln\ln\frac{\mu^2}{m^2} = \int^{\mu^2}\frac{dp^2}{p^2}\frac{1}{\ln(p^2/m^2)} \to \int^{\infty}\frac{dp^2}{p^2}(p^2)^{-\epsilon}\frac{1}{\ln(p^2/m^2)}$$

$$\xrightarrow[\epsilon\to 0]{} m^{2\epsilon}[-\ln\epsilon - C + O(\epsilon)]. \tag{38}$$

Following the standard prescription of DR the author of ref. [12] then tries to put $\epsilon \to 0$ and finds that the result for $\langle 0|\alpha^2|0\rangle$ depends on the details of the limiting procedure. This observation served as a basis for the negative conclusion – the impossibility of defining the vacuum expectation value of the composite operator α^2.

The existence of a singularity in ϵ at $\epsilon=0$ and the fact that $\ln(\epsilon+i0)\neq\ln(\epsilon-i0)$ are undoubtedly true. These have nothing to do, however, with the validity or invalidity of the OPE, since almost any operator has a μ-dependence associated with the anomalous dimension. Only the products of the type $C_2\langle\alpha^2\rangle$ have physical meaning and do not depend on μ. If one still wishes to get rid of the $\ln\ln\mu^2$ (or $\ln(1/\epsilon)$) factors, one has to multiply the operator α^2 by $\exp(-\int \mathrm{d}f\,\gamma(f)/\beta(f))$. After this has been done the effect of the anomalous dimension is eliminated and the limit $\mu\to\infty$ ($\epsilon\to 0$) becomes meaningful and unambiguous.

4. The Schwinger model

The 2-dim Schwinger model [24] is well suited for studying the status of the OPE since an exact operator solution exists in this case (see, e.g., ref. [25]). Thus, one can confront an exact answer for some Green function with its OPE counter-partner. On the other hand, no asymptotic freedom is not logarithmic in the Schwinger model. Moreover, some other features are specific of the model and have no parallel in QCD so that the lessons for the latter are limited. Nevertheless, the problem of checking OPE in the Schwinger model can be formulated and successfully solved.

In standard notation the lagrangian looks as follows:

$$\mathcal{L}=-\tfrac{1}{4}F_{\mu\nu}F_{\mu\nu}+\bar{\psi}i\gamma_\mu D_\mu\psi\,, \tag{39}$$

where

$$iD_\mu=i\partial_\mu+eA_\mu\,,$$

$$\gamma_0=\begin{pmatrix}0&1\\1&0\end{pmatrix},\quad \gamma_1=\begin{pmatrix}0&-1\\1&0\end{pmatrix},\quad \gamma_5=\gamma_0\gamma_1=\begin{pmatrix}1&0\\0&-1\end{pmatrix}.$$

Let us confine ourselves to considering the currents which are bilinear in the fermion fields. Moreover, two-point functions for the vector and axial-vector currents immediately reduce to a triviality, and for this reason we will consider only the two-point functions induced by the scalar currents:

$$j^+=\bar{\psi}_R\psi_L\,,\qquad j^-=\bar{\psi}_L\psi_R\,,$$

$$\psi_{L,R}=\tfrac{1}{2}(1\pm\gamma_5)\psi\,. \tag{40}$$

There exist two basically different two-point functions induced by these scalar currents:

$$\Pi_{++}(x)=\langle 0|\mathrm{T}\{j^+(x)j^+(0)\}|0\rangle\,, \tag{41}$$

$$\Pi_{+-}(x)=\langle 0|\mathrm{T}\{j^+(x)j^-(0)\}|0\rangle\,. \tag{42}$$

The former vanishes identically in perturbation theory because of chirality conservation and, consequently, is proportional to the vacuum condensates, while $\Pi_{+-}(x)$ is contributed by ordinary Feynman graphs.

Starting from the exact solution of the model [25] one can derive

$$\Pi_{+\pm}(x) = \tfrac{1}{4}A^2\, \mathrm{e}^{\pm 4\pi i\Delta(x,m^2)} = \tfrac{1}{4}A^2\, \mathrm{e}^{\mp 2K_0(mx)}, \tag{43}$$

where $\Delta(x, m^2)$ is the propagator of a scalar field with mass m:

$$m^2 = \frac{e^2}{\pi}, \tag{44}$$

while the constant A equals

$$A = \frac{\gamma m}{2\pi}, \quad \gamma = \mathrm{e}^C \approx \mathrm{e}^{0.577} \simeq 1.78\,.$$

Moreover, K_0 is the McDonald function of the following argument:

$$m(-x^2)^{1/2} = m(x^2_{\text{Euclid}})^{1/2} \equiv mx\,.$$

Expanding the exact solution (43) for $x \to 0$ we find

$$\Pi_{++} = \frac{\gamma^2 m^2}{16\pi^2}(\tfrac{1}{2}\gamma z)^2[1 + 2(\tfrac{1}{2}z)^2(\ln(\tfrac{1}{2}\gamma z) - 1) + \mathrm{O}(z^4)]\,,$$

$$\Pi_{+-} = \frac{\gamma^2 m^2}{16\pi^2}\left(\frac{2}{\gamma z}\right)^2[1 - 2(\tfrac{1}{2}z)^2(\ln(\tfrac{1}{2}\gamma z) - 1)$$

$$+ 2(\tfrac{1}{2}z)^4(\ln^2(\tfrac{1}{2}\gamma z) - \tfrac{9}{4}\ln(\tfrac{1}{2}\gamma z) + \tfrac{11}{8}) + \mathrm{O}(z^6)]\,, \tag{45}$$

where $z = mx$.

Note that the coupling constant $e = m\sqrt{\pi}$ enters the log functions in eq. (45) as their argument. Thus, the small-x expansion for the correlators Π_{+-}, Π_{++} are manifestly nonanalytical in e. This does not imply, however, any breaking of the OPE or the impossibility of calculating the coefficient functions perturbatively. From the previous exercises the reader may immediately conclude that expansion (45) is merely an expansion of functions Π_{+-}, Π_{++} near the singularity but *not* the OPE. Comparing (45) to the actual OPE calls for additional efforts.

In the momentum representation (at euclidean momenta $Q^2=-q^2$) correlation functions (45) convert to

$$\Pi_{++}(Q)=-\frac{\gamma^4 m^6}{\pi Q^6}+O\left(\frac{m^8}{Q^8}\right),$$

$$\Pi_{+-}(Q)=\frac{1}{4\pi}\left\{\ln\frac{M^2}{Q^2}+\frac{m^2}{Q^2}-\frac{m^4}{Q^4}\left(\ln\frac{Q^2}{m^2}+\tfrac{1}{4}\right)+O\left(\frac{m^6}{Q^6}\right)\right\}, \qquad (46)$$

where M is the ultraviolet cut-off*.

With the exact solution in hand we can attack the problem from another direction, namely, let us construct the OPE for the T-products

$$T\{j^+(x)j^+(0)\}, \quad T\{j^+(x)j^-(0)\}.$$

We start from the correlation function which is identically zero in perturbation theory.

Since in this case the chirality is changed by four units the operator of the lowest dimension must contain a product of four ψ-fields (two $\bar\psi_R$ and two ψ_L). However, the product $(\bar\psi_R\psi_L)^2=0$ thanks to the Grassmann nature of spinors. Therefore, the relevant operator of the lowest dimension looks as follows:

$$\bar\psi_R\psi_L\partial^2\bar\psi_R\psi_L.$$

It could be written in another form, $\bar\psi_R\partial_\mu\psi_L\bar\psi_R\partial_\mu\psi_L$, of course.

* While performing the Fourier transformation we have used the relations

$$\Pi(q)=i\int d^2x\, e^{-iqx}\Pi(x),$$

which in euclidean notation becomes

$$\Pi(Q)=\int d^2x\, e^{iQx}\Pi(x)=2\pi\int dx\, xJ_0(Qx)\Pi(x);$$

$$\int_0^\infty dx\, x^\epsilon J_0(x)=2^\epsilon\frac{\Gamma(\frac{1}{2}+\frac{1}{2}\epsilon)}{\Gamma(\frac{1}{2}-\frac{1}{2}\epsilon)}.$$

Substituting $\epsilon=n+\delta$ and expanding in δ we find

$$\int_0^\infty dy\, y^k J_0(y)=0 \qquad (k=1,3,5),$$

$$\int_0^\infty dy\, y\ln(y)J_0(y)=-1, \qquad \int_0^\infty dy\, y^3\ln(y)J_0(y)=4,$$

$$\int_0^\infty dy\, y^3\ln^2(y)J_0(y)=8(1-\ln\tfrac{1}{2}\gamma),$$

$$\int_0^\infty dy\, y^5\ln(y)J_0(y)=-64.$$

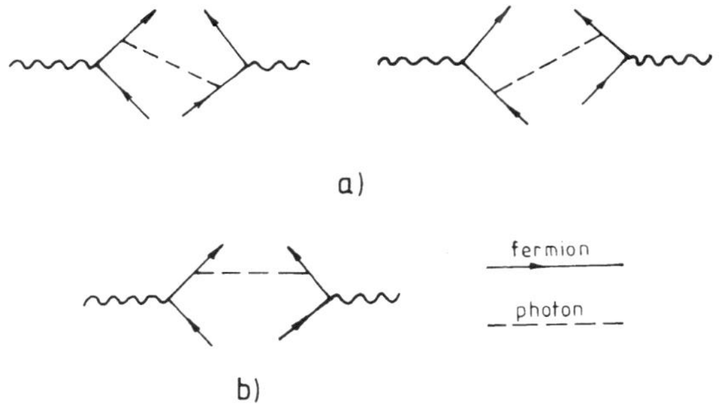

Fig. 6. Feynman graphs determining the coefficient function in front of the operator $\bar{\psi}_R\psi_L\partial^2\bar{\psi}_R\psi_L$ in the T-product $T\{j^+(x)j^+(0)\}$ in the Schwinger model. Graph 6b vanishes in the Feynman gauge.

The corresponding coefficient function is determined by the graphs in fig. 6a. (Note that in the Feynman gauge the graph in fig. 6b vanishes). The evaluation is readily performed and in the leading order in Q^{-2} the OPE for $\Pi_{++}(Q)$ takes the form

$$\Pi_{++} = \frac{16e^2}{Q^6}\langle 0|\bar{\psi}_R\psi_L\partial^2\bar{\psi}_R\psi_L|0\rangle\,. \tag{47}$$

To compare this equation with the exact answer one has to fix the vacuum condensate first.

In the accepted approximation the operator $\bar{\psi}_R\psi_L\partial^2\bar{\psi}_R\psi_L$ does not depend on the normalization point μ (see below) and we can use the following definition of the vacuum expectation value:

$$\begin{aligned}\langle 0|\bar{\psi}_R\psi_L\partial^2\bar{\psi}_R\psi_L|0\rangle &= \lim_{x\to 0}\left(\frac{\partial}{\partial x_\mu}\right)^2\langle T\{\bar{\psi}_R\psi_L(x),\,\bar{\psi}_R\psi_L(0)\}\rangle \\ &= \lim_{x_{\rm Euclid}\to 0} -\left(\frac{\partial}{\partial x_\mu^{\rm Euclid}}\right)^2 \Pi_{++}(x_{\rm Euclid}) \\ &= -\frac{\gamma^4 m^4}{16\pi^2}\,,\end{aligned} \tag{48}$$

where the Wick rotation in the x-space is performed and expansion (45) is used.

Combination of eqs. (47) and (48) yields the complete coincidence of the OPE and the exact solution:

$$\Pi_{++}(Q^2\to\infty) = -\frac{\gamma^4 m^6}{\pi Q^6}\,.$$

Thus, the OPE is fully confirmed at this point.

Proceed now to the case of the two-point function Π_{+-} which turns out to be less trivial.

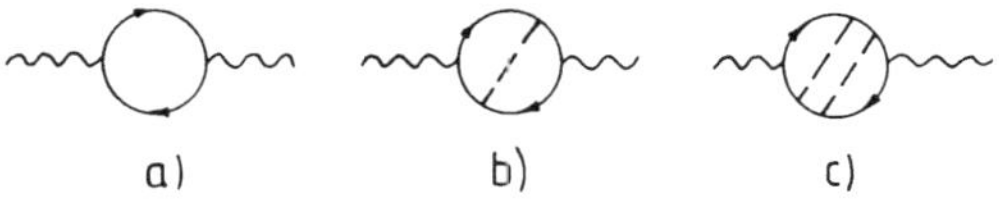

Fig. 7. The graphs determining the coefficient $C_{\mathbb{1}}$ corresponding to the unit operator in $\mathrm{T}\{j^+(x)j^-(0)\}$.

In this problem the OPE starts from the unit operator. The coefficient $C_{\mathbb{1}}$ is determined by the graph in fig. 7. The contribution of the graph in Fig. 7a is of order e^0 and reduces to

$$C_{\mathbb{1}}(\text{fig. 7a}) = \frac{1}{4\pi} \ln \frac{M^2}{Q^2}, \tag{49}$$

while the other graphs are actually of higher order in e, and have not been computed by us.

Furthermore, the coefficient function in front of the operator $F_{\mu\nu}^2$ is determined by the graph in fig. 8; explicit calculation shows that the corresponding coefficient vanishes.

This result could be predicted. Indeed, the operator $F_{\mu\nu}^2$ has Lorentz spin zero. It is convenient therefore to average the T-product

$$\mathrm{d}^2x\, \mathrm{e}^{iqx} \mathrm{T}\{j^+(x)j^-(0)\}$$

over directions of vector q from the very beginning, with fixed q^2. Furthermore, the graphs of fig. 8 reduce to products of fermion propagators in an external field. Using the Fock–Schwinger gauge for the external field (for details see review [26]) one finds, upon averaging over the direction of q, expressions of the type

$$\gamma_\mu \not{A}_{\text{ext}} \gamma_\mu ,$$

which vanish identically in two dimensions ($\gamma_\mu \gamma_{\alpha_1} \gamma_{\alpha_2} \cdots \gamma_{\alpha_{2k+1}} \gamma_\mu \equiv 0$).

For the same reason one can disregard the graph in fig. 9a in evaluating the contribution of the four-fermion operator $\bar{\psi}_R \psi_L \bar{\psi}_L \psi_R$. As a result, the corresponding coefficient is determined entirely by fig. 9b and is readily found:

$$-\frac{4e^2}{Q^4} \bar{\psi}_R \psi_L \bar{\psi}_L \psi_R . \tag{50}$$

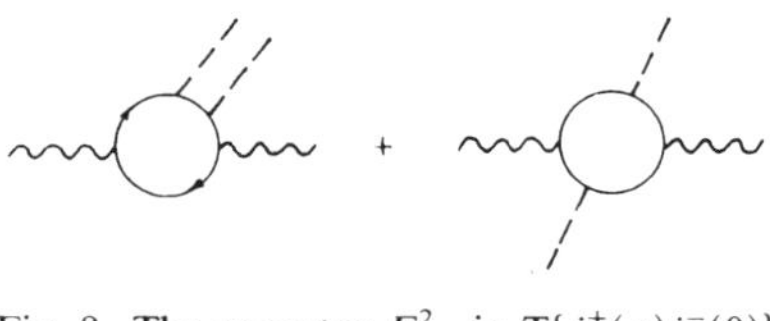

Fig. 8. The operator $F_{\mu\nu}^2$ in $\mathrm{T}\{j^+(x)j^-(0)\}$.

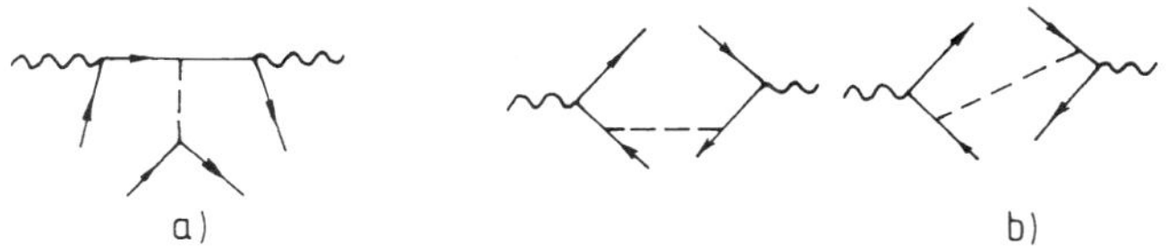

Fig. 9. The operator $\bar{\psi}_R\psi_L\bar{\psi}_L\psi_R$ in the T-product of $j^+(x)$ and $j^-(0)$. The contribution of graph 9a vanishes.

Combining eqs. (49) and (50) gives the following OPE for Π_{+-}:

$$\Pi_{+-}(Q) = \frac{1}{4\pi}\ln\frac{M^2}{Q^2} + C_{\mathbb{1}}(\text{fig. 7b}) + C_{\mathbb{1}}(\text{fig. 7c}) - \frac{4e^2}{Q^4}\langle 0|\bar{\psi}_R\psi_L\bar{\psi}_L\psi_R|0\rangle + O(1/Q^6)\,, \tag{51}$$

where $C_{\mathbb{1}}$(fig. 7b) and $C_{\mathbb{1}}$(fig. 7c) actually represent radiative corrections to the leading term in $C_{\mathbb{1}}$ and are of order e^2 and e^4 (i.e. m^2/Q^2 and m^4/Q^4) respectively.

To compare the OPE with the exact answer one needs again to know the corresponding vacuum condensate, this time

$$\langle 0|\bar{\psi}_R\psi_L\bar{\psi}_L\psi_R|0\rangle\,.$$

Let us try first the following natural regularization for the vacuum matrix element;

$$\langle 0|\bar{\psi}_R\psi_L\bar{\psi}_L\psi_R|0\rangle = \lim_{x\to 0}\langle 0|\mathrm{T}\{\bar{\psi}_R(x)\psi_L(x),\, \bar{\psi}_L(0)\psi_R(0)\}|0\rangle = \lim_{x\to 0}\Pi_{+-}(x)\,. \tag{52}$$

We immediately convince ourselves that it is impossible to put $x=0$, since in this limit

$$\Pi_{+-}(x) = \frac{1}{4\pi^2x^2} + \frac{m^2}{8\pi^2}\left(\ln\frac{2}{\gamma m x} + 1\right) + O(x^2)\,.$$

Hopefully, this observation does not discourage the reader. The reason for the singularity can be traced again to the nonvanishing anomalous dimension. The four-fermion operator at hand does depend on the normalization point μ, and the latter should be introduced explicitly. To logarithmic accuracy $x = 1/\mu$ and

$$\langle 0|\bar{\psi}_R\psi_L\bar{\psi}_L\psi_R|_\mu|0\rangle = \frac{\mu^2}{4\pi^2} + \frac{m^2}{16\pi^2}\ln\frac{\mu^2}{m^2} + \text{const} + O(m^2/\mu^2)\,. \tag{53}$$

Upon substituting this expression in eq. (51) the μ-dependence must cancel out. More specifically, the cancellation occurs between the contribution of the operator $\bar{\psi}_R\psi_L\bar{\psi}_L\psi_R$ and that of $C_{\mathbb{1}}$(fig. 7b), $C_{\mathbb{1}}$(fig. 7c). The latter must contain, in particular, a $\log(Q^2/\mu^2)$ which, combined with $\log(\mu^2/m^2)$ from $\langle(\bar{\psi}\psi)^2\rangle$ gives rise to $\log(Q^2/m^2)$, which is μ-independent.

This chain of arguments allows us to fix the log piece to order Q^{-4}:

$$\Delta\Pi_{+-} = -\frac{4e^2}{Q^4}\frac{m^2}{16\pi^2}\ln\frac{Q^2}{m^2} = -\frac{m^4}{4\pi Q^4}\ln\frac{Q^2}{m^2},$$

which does coincide with what follows from the exact answer (cf. eq. (46)). Note that comparison of $O(Q^{-2})$ terms and nonlogarithmic $O(Q^{-4})$ terms would require explicit evaluation of C_1(fig. 7b) and C_1(fig. 7c) - which task we have not undertaken.

The examples considered are sufficient we think to illustrate the OPE within the Schwinger model. It might be worth emphasizing that in calculating the OPE coefficients within perturbation theory we have used the operator basis of perturbation theory. The latter must not be confused with the operator basis exploited in the exact operator *solution.* It is perfectly clear that the perturbative C_i's have nothing to do with the expansion coefficient appearing in the operator solution.

Moreover, the observation that the exact operator *solution* is strongly nonanalytical in e^2 has nothing to do with the problem of the OPE. Indeed, comparison of the operator *solution* with the OPE is legitimate only upon averaging over the only common state - the vacuum. Some of the singularities of the operator *solution* drop off then. The remaining nonanalyticity is absorbed into matrix elements of composite operators (after introducing the normalization point μ).

5. More on OPE in QCD

It might be worth reiterating our viewpoint on the status and applications of the OPE in QCD. Indeed, a good portion of applications of the OPE refers to this theory. OPE (with inclusion of nonperturbative effects) is exploited in a most intensive way in QCD sum rules [7]. Phenomenologically, these applications are quite successful (see, e.g., the review talks [27]). In practice one accepts a simplified version of OPE. Namely, the coefficients are calculated in perturbation theory, while nonperturbative effects are manifested through various vacuum expectation values, like $\langle \alpha_s G^a_{\mu\nu} G^a_{\mu\nu} \rangle$; it is assumed that the p.t. contribution to vacuum condensates is absent.

This procedure is justified numerically in most cases (see below). However, any attempt to stick to this procedure literally results in a conflict with the general principles as amply illustrated by refs. [10-12].

Indeed, the very notion of perturbative and nonperturbative fluctuations is rather foreign to the OPE as discussed above. Instead, the fluctuations are distinguished by their frequencies (or sizes). While high-frequency fields are accounted for in the coefficient functions, the matrix elements include low frequencies. The definition assumes introduction of the normalization point μ. All this being said, it would be unwise, however, to overlook the simple fact that in asymptotically free theories high frequencies are described by perturbation theory to a good approximation, while low frequencies can well be dominated by nonperturbative fluctuations.

This, the division between high and low frequencies and the division between perturbative and nonperturbative parts can be approximately the same. Certainly, the correspondence is not absolute. Thus, the presence of small-size instantons results in some exponential terms, $\exp(-2\pi/\alpha_s(\mu)) = (\Lambda/\mu)^b$, in the coefficient functions [8]. Moreover, at least in some regularization schemes (see sect. 3, the θ-function cut-off) there arise in the C_i's powers of μ of a more trivial nature, associated with the subtraction of the $p<\mu$ domain from the p.t. graphs.

Within such a subtraction scheme the analogous power terms automatically appear in the vacuum expectation values of composite operators. Say that the following definition of the gluon condensate is conceivable:

$$\langle 0|\alpha_s G^a_{\mu\nu} G^a_{\mu\nu}|0\rangle = \mu^4 \sum_k c_k \left(\frac{\alpha_s(\mu)}{\pi}\right)^k + m^4 ,$$

$$c_k \neq 0 , \quad m = \text{const}\, M \exp(-2\pi/b\alpha_s(M))$$

(compare with eq. (35)). Then the μ^4 piece cancels out similar corrections in the coefficient functions.

However, all these subtleties of definitions and procedures can safely be disregarded everywhere, as far as practical applications of QCD are concerned, except for the logarithmic factors due to the anomalous dimensions of composite operators. To be more cautious, as far as we are aware there are no counter examples. So, what is specific of QCD numerically?

There exist two trends in choosing the normalization point μ. Superficially, these trends are contradictory to each other. First, one would like to choose μ large so that $\alpha_s(\mu)/\pi \ll 1$. This ensures theoretical control over the region $p>\mu$. On the other hand, one would like to choose μ as low as possible. The purpose of this is to diminish possible perturbative contributions to the vacuum condensates so that one would be dealing with genuine nonperturbative effects. It is highly nontrivial - and could not be proved in advance - that there exists some "window" where both requirements on μ are met.

This property of the QCD vacuum has been established only numerically so far. The physical reality behind this is that the nonperturbative effects become dominating when the effective coupling constant $\alpha_s(\mu)/\pi = 2/b\ln(\mu/\Lambda)$ is considerably lower than unity.

Let us exemplify this passage by a simple estimate. Take $\mu = (3-4)\Lambda$. Then $a_s(\mu)/\pi \leq 0.2$, and calculating the coefficient functions C_i as a series (perhaps asymptotic) in $\alpha_s(\mu)$ seems justified (and can be confirmed by explicit evaluation of, say, the first two terms). On the other hand, the power μ-dependence of $\langle 0|O_i(\mu)|0\rangle$ (fixing this dependence requires explicit and strict regularization) is inessential and can be neglected. The corresponding redefinition of $\langle(\alpha_s/\pi)G^a_{\mu\nu}G^a_{\mu\nu}\rangle$ would be of order

$$\frac{\alpha_s}{\pi} 6(N_c^2-1) \int_0^{\mu^2} \frac{dp^2\, p^2}{16\pi^2} \sim \frac{3\alpha_s(\mu)}{2\pi^3}\mu^4 .$$

Even if $\Lambda = 150$ MeV this contribution is of the order of a few per cent (it does not exceed 20 per cent anyhow) of the phenomenological number

$$\langle 0|\frac{\alpha_s}{\pi} G^a_{\mu\nu} G^a_{\mu\nu}|0\rangle_{\exp} \sim 0.012\ \mathrm{GeV}^4 .$$

This dominance of nonperturbative effects simplifies the analysis of sum rules greatly, and allows us to determine sufficiently accurately various vacuum condensates by the sum rule technique in spite of the presence of perturbative "backgrounds". If the situation in QCD was the same as, say, in $O(N)$ sigma model with low N the analysis of sum rules would be more cumbersome. Nevertheless, the OPE would not be questioned in any case. One would have had just to indicate each time carefully which procedure of dealing with perturbative terms is chosen. In reality, however, QCD looks much more like the $O(N)$ σ-model with $N \to \infty$, where all perturbative contributions to the condensates can be disregarded.

It might be worth mentioning that the estimate of possible perturbative contributions to $\langle \alpha_s G^2/\pi \rangle$ given above actually repeats similar arguments of the original papers on sum rules [7, 8]. Just this kind of estimate, not general theorems concerning lagrangian field theories, motivated the use of what we call the simplified version of the OPE: the coefficient functions are found perturbatively while vacuum condensates are nonperturbative. Certainly, even this simplified recipe has never ignored the effect of the log factors due to anomalous dimensions.

To summarize, there is no room for breaking of the OPE in its general form. We have explicitly checked it on a number of examples and the ultimate success was evident from the very beginning. However, there is a variety of possibilities so far as the relative role of perturbative and nonperturbative contributions to vacuum condensates is concerned. The large-N sigma model falls into the same class as QCD: perturbative corrections to vacuum condensates can be forgotten in the first approximation. Consideration of the most trivial tadpole graph in $\lambda\varphi^4$ theory (see the beginning of sect. 2) gives an example of just the opposite kind: there is no nonperturbative contribution to the $\langle \varphi^2 \rangle$ condensate. Identification of the "class" to which some particular theory belongs seems to be a real problem for theoretical investigation rather than checking of OPE validity.

One of the authors (A.V.) is grateful to Prof. I. Dyatlov for valuable discussions of the OPE in the Schwinger model.

References

[1] K. Wilson, Phys. Rev. 179 (1969) 1499
[2] H. Fritzsch and M. Gell-Mann, in Proc. XVI Int. Conf. on High energy physics, vol. 2 (Batavia, 1972), p. 135;
H.D. Politzer, Phys. Rev. Lett. 30 (1973) 1346;
D.J. Gross and F. Wilczek, Phys. Rev. Lett. 30 (1973) 1343

[3] K. Wilson, Phys. Rev. D3 (1971) 1818
[4] K. Wilson, in Proc. 1971 Int. Symp. on Electron and photon interactions at high energies, ed. N. Mistry (New York, 1972) p 115
[5] W. Zimmermann, in Lectures on elementary particles and quantum field theory, ed. S. Deser, M. Grisaru and H. Pendelton (MIT Press, Cambridge, Mass., 1971) vol. 1
[6] K.G. Chetyrkin, S.G. Gorishny and F.V. Tkachov, Phys. Lett. 119B (1983) 407;
K.G. Chetyrkin, Phys. Lett. 126B (1983) 371
[7] M.A. Shifman, A.I. Vainshtein and V.I. Zakharov, Nucl. Phys. B147 (1979) 385
[8] V.A. Novikov, M.A. Shifman, A.I. Vainshtein and V.I. Zakharov, Nucl. Phys. B174 (1980) 378
[9] H.R. Quinn and S. Gupta, Phys. Rev. D26 (1982) 499
[10] H.R. Quinn and S. Gupta, Phys. Rev. D27 (1983) 980
[11] F. David, Nucl. Phys. B209 (1982) 433
[12] F. David, Nucl. Phys. B234 (1984) 237
[13] M. Soldate, Stanford preprint SLAC-PUB-3054 (1983); Ann. of Phys., in press
[14] G.S. Danilov and I.T. Dyatlov, Yad. Fiz., in press
[15] A.N. Tavkhelidze and V.F. Tokarev, preprint П-0314 (1983)
[16] H.D. Politzer, Phys. Reports 14 (1974) 130
[17] K. Wilson and J. Kogut, Phys. Reports 12 (1974) 75
[18] C. Taylor and B. McClain, Phys. Rev. D28 (1983) 1364
[19] W. Bardeen, B. Lee and R. Shrock, Phys. Rev. D14 (1976) 985
[20] E. Brezin and J. Zinn-Justin, Phys. Rev. B14 (1976) 3110
[21] V. Novikov, M. Shifman, A. Vainshtein and V. Zakharov, Phys. Reports, in press
[22] A.M. Polyakov, Phys. Lett. 59B (1975) 79
[23] A.I. Vainshtein, V.I. Zakharov, V.A. Novikov and M.A. Shifman, Moscow preprint ITEP-64 (1983)
[24] J. Schwinger, Phys. Rev. 128 (1962) 2425
[25] J.H. Lowenstein and J.A. Swieca, Ann. of Phys. 68 (1971) 172
[26] V.A. Novikov, M.A. Shifman, A.I. Vainshtein and V.I. Zakharov, Moscow preprint ITEP-140 (1983); Fortschr. der Phys., in press
[27] H. Rubinstein, Talk at the XXI Int. Conf. on High energy physics, Paris, 1982;
B.L. Ioffe, Talk at the XXII Int. Conf. on High energy physics, Leipzig, 1984

A SIMPLE AND COMPLETE LORENTZ-COVARIANT GAUGE CONDITION

C. CRONSTRÖM [1]
Nordita, DK-2100 Copenhagen Ø, Denmark

Received 16 October 1979

We discuss a very simple gauge condition, for abelian or non-abelian gauge theories, which is both Lorentz-covariant and complete. Using this gauge a straightforward solution to the problem of expressing the vector potential A in terms of the field strength G is obtained.

1. Introduction. In this note a very simple Lorentz-covariant and complete gauge condition is proposed and discussed. The gauge condition in question fixes the vector potential uniquely, apart from global gauge transformations. Furhtermore, using the gauge condition in question, a straightforward solution is obtained to the inversion problem, i.e. the problem of expressing the vector potential A in terms of the field tensor G.

We consider a general theory of the Yang–Mills type [1], with a semi-simple compact gauge group G. A hermitian matrix description is used for the vector potential A,

$$A_\mu(x) = A_\mu^a(x) T_a \,, \tag{1}$$

where the T_a's are hermitian matrix representatives of the generators of the group G.

The gauge condition we propose is simply the following,

$$x^\mu A_\mu(x) = 0 \,. \tag{2}$$

The gauge condition (2) is deceptively simple, but has not, to the best of my knowledge, been discussed previously in the literature. In fact, the only previously given gauge condition which is complete in non-abelian gauge theories, is the following superaxial gauge condition,

$$\begin{aligned} A_0(x^0,0,0,0) &= 0\,,\\ A_1(x^0,x^1,0,0) &= 0\,,\\ A_2(x^0,x^1,x^2,0) &= 0\,,\\ A_3(x^0,x^1,x^2,x^3) &= 0\,. \end{aligned} \tag{3}$$

To the best of my knowledge, the condition (3) was first given by Coleman [2].

In eq. (2) the point $x^\mu = 0$ occupies a somewhat special position in that eq. (2) implies that

$$A_\mu(0,0,0,0) = 0\,. \tag{4}$$

However, instead of eq. (2) we could equally well take

$$(x^\mu - z^\mu) A_\mu(z + (x - z)) = 0\,, \tag{5}$$

where z is a fixed point in Minkowski space. Taking $z = 0$ simplifies the notation, and implies no essential losss of generality, whence we restrict our consideration to $z = 0$ in what follows.

Besides the usefulness of any particular gauge condition, there are two questions which arise about any gauge condition [3].

The first question is whether the gauge condition is attainable, i.e. given an arbitrary potential A_μ not satisfying the gauge-condition, can one find a gauge transformation Ω such that the gauge-transformed A_μ,

$$A_\mu \overset{\Omega}{\to} A_\mu^\Omega = \Omega^{-1} A_\mu \Omega - (\mathrm{i}/g)\Omega^{-1}\partial_\mu \Omega\,, \tag{6}$$

satisfies the gauge condition.

The second question concerns the completeness

[1] On leave from Department of Theoretical Physics, University of Helsinki. Supported in part by the Academy of Finland.

(or uniqueness) of the gauge condition, i.e. if one has one potential A_μ satisfying the gauge condition, can one find a gauge transform of that potential which also satisfies the gauge condition in question. If not, then the gauge condition is complete.

2. *Attainability and completeness of the gauge condition.* We assume from the outset that space–time is simply connected and that the vector potentials we consider are regular, i.e. are continuously differentiable up to second order. Let us consider an arbitrary regular vector potential A_μ, not satisfying eq. (2). Let

$$B_\mu = \omega^{-1} A_\mu \omega - (i/g)\omega^{-1}(\partial_\mu \omega) . \qquad (7)$$

Imposing the condition (2) on B_μ leads to the following differential equation for ω,

$$x^\mu \partial_\mu \omega = -igx^\mu A_\mu(x)\omega . \qquad (8)$$

The differential equation (8) has the solution

$$\omega(x) = P \exp\left[-ig \int_0^1 dt\, x^\mu A_\mu(tx)\right]\omega_0 , \qquad (9)$$

where ω_0 is an arbitrary initial value for ω, and P stands for "time-ordering" in the variable t. Thus, the gauge condition (2) is attainable by means of the gauge transformation $\omega(x)$ given in eq. (9).

We then consider the question of completeness. Assume that there is one vector potential A_μ satisfying (2) and also a gauge transform A_μ^Ω of that potential which also satisfies the condition (2). Using the relation (6) we obtain then a differential equation for Ω, namely

$$x^\mu \partial_\mu \Omega = 0 . \qquad (10)$$

However, according to a classical theorem (Euler's theorem), eq. (10) implies that Ω is a homogeneous function of zeroth degree,

$$\Omega(x) = \Omega(tx) , \qquad (11)$$

where t is arbitrary. Since $\Omega(x)$ is regular, in particular at the origin, we can take $t = 0$ in eq. (11) and obtain

$$\Omega(x^0, x^1, x^2, x^3) = \Omega(0) = \text{const} . \qquad (12)$$

The gauge condition (2) is thus a complete gauge condition, apart from global (space–time independent) gauge transformations.

The function $\omega(x)$ defined by eq. (9) transforms as follows under a gauge transformation Ω,

$$\omega(x) \xrightarrow{\Omega} \omega^\Omega(x) = \Omega^{-1}(x)\omega(x)\omega_0^{-1}\Omega(0)\omega_0^\Omega . \qquad (13)$$

Thus, by requiring that the initial value ω_0 in eq. (9) transforms as

$$\omega_0^\Omega = \Omega^{-1}(0)\omega_0 , \qquad (14)$$

we remove the global gauge transformation $\Omega(0)$ of B_μ defined by eq. (7), which otherwise would be induced by a local gauge transformation $\Omega(x)$ of A_μ.

3. *The inversion problem.* The field tensor $G_{\mu\nu}(x)$ is defined as follows,

$$G_{\mu\nu}(x) = \partial_\nu A_\mu(x) - \partial_\mu A_\nu(x) - ig[A_\mu(x), A_\nu(x)] . \qquad (15)$$

The field-tensor therefore satisfies the Bianchi identity,

$$\nabla_\alpha(A)G_{\beta\gamma} + \nabla_\beta(A)G_{\gamma\alpha} + \nabla_\gamma(A)G_{\alpha\beta} = 0 , \qquad (16)$$

where ∇_α denotes the covariant derivative,

$$\nabla_\alpha(A)\phi \equiv \partial_\alpha \phi + ig[A_\alpha, \phi] . \qquad (17)$$

Assume now that the vector potential A_μ in eq. (15) satisfies the gauge condition (2). Then we have

$$x^\nu G_{\mu\nu}(x) = x^\nu \partial_\nu A_\mu(x) + A_\mu(x) . \qquad (18)$$

Replacing x by tx in eq. (18), where t is an arbitrary parameter, we obtain,

$$(d/dt)(tA_\mu(tx)) = tx^\nu G_{\mu\nu}(tx) . \qquad (19)$$

Integrating eq. (19), we then obtain,

$$A_\mu(x) = \int_0^1 dt\, tx^\nu G_{\mu\nu}(tx) . \qquad (20)$$

The solution (20) to the inversion problem $G_{\mu\nu}(x) \to A_\mu(x)$ is very simple compared to the solution given by Halpern [4], who uses a variant of the super-axial gauge (3) in his construction of A_μ in terms of $G_{\mu\nu}$.

One still has to check that the solution (20) actually solves eq. (15), i.e. that one actually obtains the field tensor $G_{\mu\nu}(x)$ by inserting A_μ given by eq. (20) in eq. (15). It is not difficult to show that the Bianchi-identities (16), with A_μ given by eq. (20), are necessary and sufficient conditions for A_μ to be the solution to eq. (15), in the gauge specified by eq. (2). The Bianchi identities are thus constraint equations of the $G_{\mu\nu}(x)$ in the gauge given by eq. (2).

4. Discussion and symmary. The gauge condition we have considered here, eq. (2), is a very simple, Lorentz-covariant gauge condition, which has been shown to give a unique (apart from global gauge transformations) specification of the vector potential A_μ in any non-abelian (or abelian) gauge theory. We have restricted our considerations to regular potentials A_μ (and therefore regular gauge transformations) i.e. twice continuously differentable potentials in a simply connected Minkowski space. The gauge condition considered has been shown to lead to a very simple solution to the inversion problem of obtaining the vector potential from the field-strength. Quantising gauge theory in the complete covariant gauge given here, is an interesting question which will be dealt with in future communications.

I am indebted to the staff of NORDITA for hospitality, and to P. Cvitanović, H.B. Nielsen and H. Römer for useful discussions.

References

[1] R.L. Mills and C.N. Yang, Phys. Rev. 96 (1954) 191.

[2] S. Coleman, Erice Lectures 1975, in: New Phenomena in subnuclear physics, Part A, ed. Z. Zichichi; see also: C.W. Bernard, N.H. Christ, A.H. Guth and E.J. Weinberg, Phys. Rev. D16 (1977) 2967.

[3] R.J. Jackiw, Gauge-specification in a non-abelian gauge theory, talk given at Orbis Scientia (Coral Gables, FL, January 1978), CTP-preprint 697 (February 1978).

[4] M.B. Halpern, Phys. Rev. D19 (1979) 517.

Nuclear Physics B185 (1981) 109–132

ANALYTICAL PROPERTIES OF THE QUARK POLARIZATION OPERATOR IN AN EXTERNAL SELF-DUAL FIELD

M.S. DUBOVIKOV

Moscow Physical Technical Institute, Moscow, USSR

A.V. SMILGA

Institute of Oceanology, Academy of Sciences, USSR

Received 25 February 1980
(Revised 21 October 1980)

It is shown that in the lowest orders in α_s, light quark polarization operators in an arbitrary multi-instanton external field have the asymptotic form $\Pi(q^2) \sim \Pi^{(0)} + C/q^2 +$ exponential terms and the terms $\sim 1/q^4$, $\sim 1/q^6$, etc., are absent. We use the operator expansion language that corresponds to the vanishing of the contributions of higher-dimension operators to the polarization operator in the lowest order in α_s in a self-dual external field. It is shown that in a self-dual field quark and gluon vacuum condensate densities are strictly connected:

$$m\langle\bar{\psi}\psi\rangle_0 = -\frac{1}{8\pi}\langle\alpha_s G^a_{\mu\nu} G^a_{\mu\nu}\rangle_0 .$$

This relation is strongly violated in experiment, displaying the inapplicability of the saddle approximation in calculating continual integrals. The abelian case is also discussed. The physical pictures in the abelian and non-abelian cases are very similar. It is shown, in particular, that in the constant electromagnetic self-dual field, zero-mode solutions are present.

1. Introduction

The most striking feature of QCD distinguishing it sharply from the less sophisticated theories is its complicated vacuum structure. There are an infinite number of vacuum states differing by the value of the topological charge [1–3]. The field configurations describing the tunneling transitions between those states contribute to the continual integrals corresponding to the amplitudes of various physical processes. The contributions of vacuum transitions of non-zero topological charge are non-perturbative in their origin and fall rapidly with increasing energy. Thus, in hard processes where typical energies are large, the effects due to the non-trivial vacuum structure can be neglected and standard perturbation theory works well.

Low-energy physics is a different story.

The non-perturbative effects due to vortical vacuum transitions are of crucial importance here. These effects determine the masses and widths of resonances, low-energy scattering amplitudes and are probably responsible for confinement.

In the last two years non-perturbative effects in QCD have become a subject of intense study. In most of the works, various continual integrals were estimated in the saddle approximation where only the lowest non-trivial saddle points corresponding to the instanton and anti-instanton transitions were taken into consideration.

Such an approximation (the so-called dilute-gas approximation [4]) cannot be regarded as a satisfactory one.

The integrals over instanton size ρ arising in this technique have the form [5]

$$\int \frac{d\rho}{\rho^5} e^{-2\pi/\alpha_s(\rho)} .$$

If substituting $\alpha_s(\rho)$ for the asymptotic freedom expression

$$\alpha_s(\rho) = 2\pi/b \ln(1/\mu\rho) , \qquad b = 11 - \tfrac{2}{3}N_f ,$$

the integrand grows rapidly ($\sim\rho^4$) with increase of ρ, so that contributions to the integral mainly come from the large ρ region where the coupling constant together with the instanton density is large and the asymptotic freedom expression as well as the dilute-gas approximation is no longer valid. One has no idea of how to calculate the total continual integral in the absence of any small parameter and so the exact value of non-perturbative effects remains unknown.

A hope was expressed in the literature that the situation can be improved if multi-instanton configurations were accounted for along with one-instanton configurations. One of our results is that in QCD this hope is groundless: if continual integrals are saturated by self-dual configurations, a certain relation between quark and gluon vacuum condensate densities exists which is sharply violated in experiment.

There is a phenomenological approach, however, which enables one to obtain various quantitative predictions for non-perturbative effects without calculating continual integrals. This approach was suggested in ref. [6]. The main idea of this work consisted in applying the operator expansion technique in the non-perturbative region. Since the present work is devoted to settling a question arising in the framework of the operator expansion approach, we start with a brief reminder of its principal features. In sect. 2 (which may be regarded as part of the introduction) the question to be settled is formulated and the plan of the remainder of the paper is specified.

2. Operator expansion versus one-instanton approximation

Consider the polarization operator of, say, the vector currents:

$$\Pi_{\mu\nu}(Q) = \int e^{iQx} \langle 0|T\{j_\mu(x) j_\nu(0)\}|0\rangle \, d^4x , \tag{1}$$

where Q and x are euclidean 4-vectors. All the following considerations will be taken in euclidean space.

According to the operator expansion hypothesis [7], expression (1) at large Q can be expanded in the infinite series

$$\Pi_{\mu\nu}(Q) = \sum_n c_n(Q)\langle 0|O_n|0\rangle(Q_\mu Q_\nu - \delta_{\mu\nu}Q^2)\,, \tag{2}$$

where O_n are the local spin-zero gauge-invariant operators: $O_n = 1$, four-dimensional $\alpha_s G^a_{\mu\nu}G^a_{\mu\nu}$, $m_q\bar{\psi}\psi$, six-dimensional $g_s^3 f^{abc}G^a_{\mu\nu}G^b_{\nu\sigma}G^c_{\sigma\mu}$, $m_q g_s\bar{\psi}\sigma^{\mu\nu}t^a\psi G^a_{\mu\nu}$, etc. Here and below we limit ourselves to the one-flavour case for the sake of brevity.

Because of the complicated vacuum structure, the operators acquire non-zero vacuum expectation values. These values could in principle be determined if the corresponding total continual integral could be calculated. As this is not the case, the vacuum expectation values should be introduced as phenomenological parameters. The point is that one can obtain a lot of non-trivial physical predictions by the use of just two such parameters, $\langle\alpha_s G^a_{\mu\nu}G^a_{\mu\nu}\rangle_0$ and $m\langle\bar{\psi}\psi\rangle_0$, the densities of the gluon and the quark vacuum condensate, respectively. The coefficients $C_{\alpha_s G^a_{\mu\nu}G^a_{\mu\nu}}(q)$ and $C_{m\bar{\psi}\psi}(q)$ in (1) can be calculated so that one obtains [6]

$$\begin{aligned}&\int \mathrm{d}^4x\, \mathrm{e}^{iQx}\langle \mathrm{T}\{j_\mu(x)j_\nu(0)\}\rangle_0 \\ &= (Q_\mu Q_\nu - \delta_{\mu\nu}Q^2)\left\{-\frac{1}{4\pi^2}\ln\frac{Q^2}{\mu^2} + \frac{2m_q}{Q^4}\langle\bar{q}q\rangle_0\right. \\ &\left. + \frac{1}{12\pi Q^4}\langle\alpha_s G^a_{\mu\nu}G^a_{\mu\nu}\rangle_0 + \mathrm{O}(1/Q^6)\right\}\,,\end{aligned} \tag{3}$$

where $j_\mu = \bar{q}\gamma_\mu q$. We see that non-perturbative effects display themselves as power corrections to perturbation theory results.

On the other hand, there are a number of works where the vector polarization operator was obtained in the one-instanton approximation by calculating the simplest quark loop (fig. 1) in the external instanton (anti-instanton) field [8–11] and averaging over instanton positions. The answer is*

$$\begin{aligned}\Pi_{\mu\nu}(Q) &= \Pi^{(0)}_{\mu\nu}(Q) - \tfrac{4}{3}(Q_\mu Q_\nu - \delta_{\mu\nu}Q^2)\int n_0(\rho)\,\mathrm{d}\rho \\ &\times\left\{\frac{1}{Q^4} - \frac{3\rho^2}{Q^2}\int_0^1 \mathrm{d}x\, K_2(2\rho Q/(1-x^2)^{1/2})\right\}\,,\end{aligned} \tag{4}$$

where $n_0(\rho) = D(\rho)/\rho^5$ is the instanton density.

Now we compare expressions (3) and (4). It can easily be checked that the lowest-dimensional power terms in expressions (3) and (4) coincide. Note that in the instanton field the terms in (3) connected with the vacuum expectation values

* Note the disagreement with the expression of ref. [9]. Our result for the second term in eq. (4) is three times smaller than that of ref. [9].

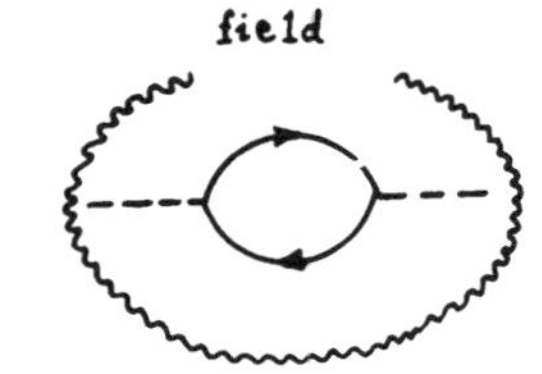

Fig. 1. Quark polarization operator in an external field.

$\langle\alpha_s G^a_{\mu\nu} G^a_{\mu\nu}\rangle_0$ and $m_q\langle\bar{\psi}\psi\rangle_0$ are of the same order. This is due to the fact that in the instanton field $\langle\bar{\psi}\psi\rangle_0$, arising from the zero-mode contribution to the quark Green function, is proportional to $1/m_q$, so that the suppression $\sim m_q$ in the expression $2m_q\langle\bar{q}q\rangle_0/Q^4$ is fictitious. We shall return to the discussion of this point in sect. 5. We see that in eq. (4) an additional contribution is present besides the power term. This contribution falls exponentially with increasing Q^2 (this is so if the instanton size ρ is *fixed*; the integration over ρ results in an effective power falloff $\Pi^{\text{inst}}_{\mu\nu}(Q) \sim Q^2 Q^{-11+2N_f/3}$ because of the contribution of small instanton sizes $\rho \sim 1/Q$[6] and cannot be determined in the framework of the operator expansion technique).

The presence of the exponential term in (4) is natural and could be expected in advance. A much more surprising fact is the absence of the power terms $\sim 1/Q^4$, $\sim 1/Q^6$ etc., in eq. (4). These terms arise naturally in the operator expansion language and it is not clear why the contribution of higher-dimensional operators should vanish.

In this paper we show that the vanishing of the higher-dimensional operators contribution in eq. (4) is not causal, being a natural consequence of self-duality of the instanton field. We show, moreover, that the same phenomenon occurs in any (anti-) self-dual field, i.e. in any multi-instanton field configuration corresponding to a saddle point of the euclidean continual integral.

To avoid misunderstanding, note that the terms $\sim 1/Q^4$ are actually present in the operator expansion (3). In ref. [6] the coefficients of the four-fermion operators $\bar{\psi}\Gamma\psi\bar{\psi}\Gamma\psi$ in the expansion of $\Pi_{\mu\nu}(Q)$ were calculated. These coefficients are proportional to α_s, however, which is beyond our approximation (the terms $\sim\alpha_s$ arise, for example, due to gluon exchange in the quark loop (fig. 2), the sort of graph which we do not consider here).

One may wonder why the operator $\alpha_s\bar{\psi}\Gamma\psi\bar{\psi}\Gamma\psi$ is stated to be of first order in α_s, whereas the operators $\alpha_s G^a_{\mu\nu} G^a_{\mu\nu}$ and $g_s^3 f^{abc} G^a_{\mu\nu} G^b_{\nu\sigma} G^c_{\sigma\mu}$ are of zero order. The point is that typical gluon fields contributing to continual integrals have the magnitude $A_\mu \sim 1/g$, so that the expressions $\alpha_s G^a_{\mu\nu} G^a_{\mu\nu}$ and $g_s^3 f^{abc} G^a_{\mu\nu} G^b_{\nu\sigma} G^c_{\sigma\mu}$ (as well as $m\bar{\psi}\psi$ and $m g_s \bar{\psi}\sigma^{\mu\nu} t^a \psi G^a_{\mu\nu}$) are renormalization-invariant values.

The main purpose of this paper is the verification of the validity of the non-perturbative operator expansion for multi-instanton field configurations. In perturbative QCD the operator expansion technique was accomodated about 7 years ago [12]. However, its justification based on the direct calculation of Feynman graphs

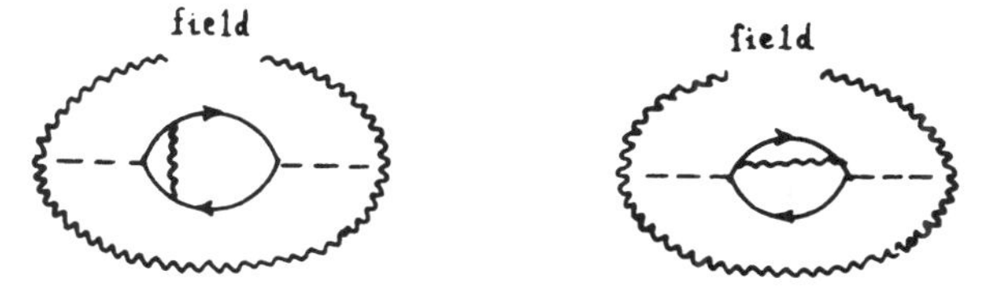

(a) (b)

Fig. 2. Radiative corrections to the quark polarization operator in an external field.

came comparatively recently [13]. Operator expansion in the non-perturbative region is not a habitual technique yet and doubts about its applicability can be found in the literature [9]. Under these circumstances a detailed investigation of how the non-perturbative operator expansion works does not seem to be superfluous*.

We show that operator expansion works perfectly well and that relation (3) holds for any self-dual field, with higher-order terms being absent. As an important byproduct, a definite relation between $m_q\langle\bar{q}q\rangle_0$ and $\langle\alpha_s G^a_{\mu\nu}G^a_{\mu\nu}\rangle_0$ is obtained. As has already been mentioned and will be discussed in more detail in the final section, this relation contradicts experiment. Thus, we conclude that the saddle approximation is a bad one in estimating continual integrals.

The rest of the paper is organized as follows. In sect. 3 we briefly recall the technique of the calculation of the quark Green function [15] and the polarization operator [8–11] in the external self-dual field and explain where the vanishing of higher-dimensional power terms comes from (namely, it follows from meromorphness of the trace of the polarization operator in the coordinate space $\Pi_{\mu\nu}(x-y)$). In sect. 4 we discuss the analytical properties of the fermion Green function and polarization operator in a self-dual gauge field and show how the operator expansion for the polarization operator works. To make our arguments more transparent we first discuss the more simple case of a constant self-dual electromagnetic field which preserves all the principal physical features of the non-abelian case. In sect. 5 we discuss some physical implications of the results obtained and draw conclusions.

3. Quark Green function and polarization operator in a self-dual field

Our analysis of the analytical properties of the quark Green function is based on the technique developed in ref. [15], where the quark Green functions in the instanton field and the 't Hooft multi-instanton field were found. We repeat the main results of this paper without proof.

The Green function of a light fermion in a multi-instanton field can be presented in the form

$$G(x, y) = -\frac{1}{m}P_0(x, y) + S(x, y) + m\Delta(x, y) + O(m^2)\,, \tag{5}$$

* The same problem was investigated in ref. [14] in another example. In this work the validity of the operator expansion for the tensor gluon polarization operator in an arbitrary self-dual field has been demonstrated explicitly.

where $\Delta(x, y)$ is the Green function of a scalar colour-triplet particle in the field and

$$P_0(x, y) = \sum_n \psi_n^{(0)}(x)\psi_n^{(0)+}(y)\,, \tag{6}$$

where $\psi_n^{(0)}(x)$ are the zero-mode solutions. The m-independent part of the fermion Green function $S(x, y)$ can be expressed through the Green function of a scalar particle according to

$$S(x, y) = i\gamma_\mu \frac{1+\gamma^5}{2}\left(\frac{\partial}{\partial x_\mu} - it^a A_\mu^a(x)\right)\Delta(x, y) - i\gamma_\mu \frac{1-\gamma^5}{2}\Delta(x, y)\left(\frac{\overleftarrow{\partial}}{\partial y_\mu} + it^a A_\mu^a(y)\right). \tag{7}$$

The polarization operator $\Pi_{\mu\nu}(x, y)$ is given by

$$\Pi_{\mu\nu}(x, y) = -\mathrm{Sp}\,\{\gamma^\mu G(x, y)\gamma^\nu G(y, x)\}\,. \tag{8}$$

Substituting expression (5) for $G(x, y)$ in eq. (8), we find that the terms $\sim 1/m^2$ and $\sim 1/m$ in $\Pi_{\mu\nu}(x, y)$ vanish (which follows from the chirality properties of zero-mode solutions) and only the constant terms survive:

$$\Pi_{\mu\nu}(x, y) = -\mathrm{Sp}\,\{\gamma^\mu S(x, y)\gamma^\nu S(y, x)\} + \mathrm{Sp}\,\{\gamma^\mu \Delta(x, y) \times \gamma^\nu P_0(y, x)\} + \mathrm{Sp}\,\{P_0(x, y)\gamma^\nu \Delta(y, x)\gamma^\mu\} + \mathrm{O}(m)\,. \tag{9}$$

Expression (9) should be averaged over positions of the field configuration as a whole. After this, translational invariance is restored and the averaged $\Pi_{\mu\nu}$ depends only on $x - y$.

Averaging consists in integrating over field positions and multiplying over the density of the field configurations with a given shape dn_0. For self-dual fields constituting saddle points of continual integrals, dn_0 coincides with the value of the 't Hooft determinant accounting for small fluctuations around the given field. To obtain a finite, physically sensible answer the integration must be performed after the free polarization operator has been subtracted. Technically it is more convenient to integrate over $\mathrm{d}^4x\, \mathrm{d}^4y$ keeping $x - y$ fixed, leaving the field in place. Thus, we define

$$r_{\mu\nu}^A(X) = dn_0 \int \mathrm{d}^4x\, \mathrm{d}^4y\, \delta(x - y - X)[\Pi_{\mu\nu}^A(x, y) - \Pi_{\mu\nu}^0(x - y)]\,. \tag{10}$$

The total polarization operator is then given by the sum of $\Pi_{\mu\nu}^0(x-y)$ and $r_{\mu\nu}^A(x-y)$ integrated over the parameters determining the shape of the field configuration. The Fourier transform of eq. (10) is the contribution of a multi-instanton field of given shape to the usual momentum-space polarization operator.

The power asymptotics of $r_{\mu\nu}(Q)$ at $Q^2 \to \infty$ is determined by the singularities of $r_{\mu\nu}(X)$ at $X = 0$, so if $r_{\mu\nu}(X)$ has no singularities except a pole $\sim 1/X^2$ (and we shall show it to be the case), $r_{\mu\mu}(Q) \sim 1/Q^2 +$ exponentially falling terms at $Q^2 \to \infty$. This statement is rigorously proved in appendix A (see also [11]) and here we illustrate its validity by a clear example.

Suppose $r_{\mu\mu}(Q)$ has a term $\sim 1/Q^4$ at $Q^2 \to \infty$. Then its Fourier transform,

$$r_{\mu\mu}(x-y) \sim \int e^{iQ(x-y)} d^4Q/Q^4 \sim \ln(x-y)^2 ,$$

has a branching point at $(x-y)^2 = 0$, which we supposed to be absent. The Fourier transform of $1/Q^6$ is $\sim (x-y)^2 \ln(x-y)^2$ and also has a branching point. Thus the terms $\sim 1/Q^4$, $\sim 1/Q^6$ etc., must be absent in the Fourier transform of a meromorphic function. The Fourier transform of the pole contribution $1/(x-y)^2$ results in a term $\sim 1/Q^2$.

Thus, in order to prove the vanishing of higher-dimensional operator contributions to $\Pi_{\mu\nu}(Q)$, the analytical properties of $r^A_{\mu\nu}(X)$ should be investigated. It follows from eqs. (5), (7), (9) and (10) that $r^A_{\mu\nu}(x-y)$ is expressed through the functions $P_0(x, y)$ given by eq. (6) and describing the zero-mode influence and $\Delta(x, y)$, the Green function of a scalar massless particle in the field under consideration. The analytical properties of $r^A_{\mu\nu}(x-y)$ are determined, therefore, by the analytical properties of $P_0(x, y)$ and $\Delta(x, y)$. $P_0(x, y)$ is an analytical function as zero-mode solutions are analytical functions. The scalar Green function $\Delta(x, y)$ in the multi-instanton field can be found in semi-explicit form [16–17],

$$\Delta(x, y) = \frac{N^+(x)N(y)}{4\pi^2(x-y)^2} , \tag{11}$$

where $N(x)$ is a quaternion $k+1$-dimensional row (k is the topological charge of the solution). $N(x)$ satisfies the condition $N^+(x)N(x) = 1$ and the multi-instanton solution is expressed through $N(x)$ as

$$A_\mu(x) = N^+(x)\partial_\mu N(x) . \tag{12}$$

(Here $\Delta(x, y)$ and $A_\mu(x)$ are written in the quaternion representation. It may be connected with the conventional matrix form: for any quaternion $A = A^0 + A^1\hat{\imath} + A^2\hat{\jmath} + A^3\hat{k}$ a matrix $\hat{A} = \sum_{\alpha=0}^3 \sigma_\alpha A^\alpha$, where $\sigma_0 = 1$, $\sigma_j = -i\tau_j$ are the Pauli matrices, may be constructed.) We see from eq. (11) that the analytical properties of $\Delta(x, y)$ at the point $x - y = 0$ coincide with those of the free particle Green function. The same statement is true, of course, for the one-instanton (and also the 't Hooft multi-instanton) Green function found in ref. [15]. So the integrand in eq. (10) has no logarithmic singularities at the point $x - y = 0$. The integral (10) over field positions is convergent (which follows from the absence of infinities in the physical polarization operator in the real vacuum) and so do not give rise to new singularities.

Thus, the singular structure at $X = 0$ of the averaged trace of the polarization operator $\langle \Pi_{\mu\mu} \rangle$ depending only on $X^2 = (x-y)^2$ amounts to poles $\sim 1/X^6$, $\sim 1/X^4$, $\sim 1/X^2$. The singularities in $\langle \Pi_{\mu\nu}(X) \rangle$ are of the same order because the traceless zero-order singular structure $\delta_{\mu\nu} - 4X_\mu X_\nu/X^2 = (\frac{1}{4}\delta_{\mu\nu}\partial^2 - \partial_\mu\partial_\nu)X^2 \ln X^2$ is precluded by vector current conservation. The above discussion and the rigorous analysis of appendix A display the absence of terms $\sim 1/Q^4$, $\sim 1/Q^6$, etc., in $\Pi_{\mu\nu}(Q)$.

Now we are going to show that the non-vanishing power contribution $\sim 1/Q^2$ to $\Pi_{\mu\nu}(Q)$ in any self-dual field is given exactly by the operator expansion result (3). To do this, more detailed information on $\Delta(x, y)$ than just the absence of a branching at the point $x - y = 0$ is required. Namely, we need to know the expansion of the numerator of eq. (11) at the point $x = y$ up to $(x - y)^4$. In sect. 4, such an expansion is found. Our analysis is based only on self-duality of the field, without any use of the intricate expressions (11), (12) for the multi-instanton solution and the scalar Green function.

4. Operator expansion for the fermion polarization operator in a self-dual gauge field

4.1. ABELIAN CASE

Consider the polarization operator of the vector electron current in a constant electromagnetic self-dual field. For the electromagnetic field self-duality means $\boldsymbol{E} = \boldsymbol{H}$ in euclidean space, i.e. $\boldsymbol{E} = i\boldsymbol{H}$ in Minkowski space. This condition cannot be satisfied for non-zero physical fields and the problem is rather of academic importance. We regard it as a preliminary step to the much more complicated non-abelian case. It is remarkable, however, that the simple case possesses all the distinguishing features of the multi-instanton case. The expression for the polarization operator can be found here in closed form.

Note first of all that in the constant electromagnetic (anti-) self-dual field zero modes do appear. We state, in other words, that in this field the euclidean equation

$$\gamma_\mu D_\mu \psi = \gamma_\mu(\partial_\mu - ieA_\mu)\psi = 0\,,$$

where ψ is a massless fermion field, has a number of non-trivial solutions $\psi_z(x)$ localized near the point $x = z$. We give the expressions for these solutions, omitting intermediate steps, in the gauge $A_\mu(x) = -\frac{1}{2}x_\nu F_{\mu\nu}(x)$:

$$\psi_z(x) = \frac{eH}{2\pi\sqrt{2}}\, \mathrm{e}^{iez_\nu x_\mu F_{\mu\nu}/2}\, \mathrm{e}^{-eH(x-z)^2/4} \begin{pmatrix} \overset{+}{\varphi}_{\boldsymbol{n}} \\ \pm \overset{+}{\varphi}_{\boldsymbol{n}} \end{pmatrix}, \tag{13}$$

where $\overset{+}{\varphi}_{\boldsymbol{n}}$ is the positive spin projection unit eigenvector of the operator $\boldsymbol{\sigma n} = \boldsymbol{\sigma H}/H$: $(\boldsymbol{\sigma n})\overset{+}{\varphi}_{\boldsymbol{n}} = \overset{+}{\varphi}_{\boldsymbol{n}}$, $\overset{+}{\varphi}_{\boldsymbol{n}}(\overset{+}{\varphi}_{\boldsymbol{n}})^* = 1$. The positive sign on the lower bispinor components corresponds to the case of self-dual field $\boldsymbol{E} = \boldsymbol{H}$ (in euclidean space) and the negative sign to the anti-self-dual field $\boldsymbol{E} = -\boldsymbol{H}$. In strict analogy with the instanton case, the zero-mode solution (13) has the definite chirality: $\gamma^5\psi_z = -\psi_z$ for the self-dual field and $\gamma^5\psi_z = \psi_z$ for the anti-self-dual field. The presence of the infinite number of zero modes in the constant self-dual field is well understood in the framework of the Atiyah–Singer theorem as the topological charge,

$$Q = \frac{e^2}{16\pi^2} \int F_{\mu\nu} F^d_{\mu\nu} \, d^4 x \,,$$

of such a field is infinite [18–19]. The problem is very closely related to the case of two-dimensional motion in the presence of a constant magnetic field where an infinite number of zero modes also exists.

The zero modes (13) produce the non-zero fermion vacuum expectation value

$$m\langle\bar{\psi}\psi\rangle_0 = -\sum_n \psi_n^{(0)}(0)\psi_n^{(0)+}(0) \,, \tag{14}$$

where $\psi_n^{(0)}(x)$ constitute the orthogonal set of zero-mode solutions. The functions $\psi_z(x)$ given by eq. (13) are not orthogonal but can be orthogonalized by the use of a standard procedure. If we choose $\psi_1^{(0)}(x) = \psi_0(x)$, the functions $\psi_2^{(0)}(x)$, $\psi_3^{(0)}(x)$, etc., orthogonal to $\psi_1^{(0)}(x)$ are expressed through linear combinations of the functions $\psi_z(x)$:

$$\psi_n^{(0)}(x) = \tilde{\psi}_z(x) = \psi_z(x) - \psi_0(x) \int \psi_z^+(x')\psi_0(x') \, d^4 x' \,.$$

It can easily be seen that $\tilde{\psi}_z(0) = 0$, so that only the first mode $\psi_1^{(0)}(x) = \psi_0(x)$ contributes to the vacuum expectation value (14). Substituting it into eq. (14), we get $m\langle\bar{\psi}\psi\rangle_0 = -e^2H^2/4\pi$. Remembering that $\langle F_{\alpha\beta}F_{\alpha\beta}\rangle_0 = F_{\alpha\beta}F_{\alpha\beta} = 4H^2$, we get the relation

$$\frac{\alpha}{4\pi}\langle F_{\alpha\beta}F_{\alpha\beta}\rangle_0 = -m\langle\bar{\psi}\psi\rangle_0 \,. \tag{15}$$

The non-abelian analog of eq. (15) derived in subsect. 4.2 is very important for us and we discuss it in detail later.

In the presence of zero modes, the fermion Green function is expressed through the Green function of a scalar particle in just the same way as in the non-abelian case [cf. eq. (7)]:

$$\begin{aligned} S(x, y) = {} & \gamma_\mu \frac{1+\gamma^5}{2}\left(\frac{\partial}{\partial x_\mu} - ieA_\mu(x)\right)\Delta(x, y) \\ & - \gamma_\mu \frac{1-\gamma^5}{2}\left(\frac{\partial}{\partial y_\mu} + ieA_\mu(y)\right)\Delta(x, y) \,. \end{aligned} \tag{16}$$

Just as in the non-abelian case, the polarization operator is given by eq. (9). We see again that knowledge of the scalar Green function $\Delta(x, y)$ is of crucial importance here. In this simple case it may be found explicitly.

Define the function $F(z, x)$ according to

$$\Delta(z, z+x) = \frac{1}{4\pi^2 x^2} F(z, x) \,. \tag{17}$$

From the identity $D_y^2\Delta(x,y)=\delta(x-y)$, following from the definition of $\Delta(x,y)$, we get for $F(z,x)$

$$D_\mu D_\mu F - 4x_\mu/x^2 D_\mu F = 0\,, \tag{18}$$

where

$$D_\mu = \partial/\partial x_\mu - ieA_\mu(z+x)\,.$$

The boundary conditions are $F(z,0)=1$, $\Delta(z,\infty)=0$. We seek the solution of eq. (18) in the form

$$F(z,x)=\phi(z,x)\exp\left[ie\int_0^1 x_\mu A_\mu(z+\alpha x)\,\mathrm{d}\alpha\right]\equiv \phi E\,. \tag{19}$$

The action of the operator D_μ on the exponential E amounts to

$$D_\mu E = ieR_\mu E\,, \tag{20}$$

where

$$\begin{aligned} R_\mu &= \frac{\partial}{\partial x_\mu}\int_0^1 \mathrm{d}\alpha\, x_\beta A_\beta(z+\alpha x) - A_\mu(z+x) \\ &= \int_0^1 x_\nu F_{\mu\nu}(z+\alpha x)\alpha\,\mathrm{d}\alpha \\ &= \tfrac{1}{2}x_\nu F_{\mu\nu}(z+x) - \tfrac{1}{6}x_\nu x_\sigma \partial_\nu F_{\mu\sigma}(z+x)+\cdots \\ &\quad + \frac{(-1)^n x_\nu x_{\rho_1}\cdots x_{\rho_n}}{(n+2)!}\partial_{\rho_1}\cdots\partial_{\rho_n}F_{\mu\nu}(z+x)+\cdots\,. \end{aligned} \tag{21}$$

This result can be checked directly by expanding

$$\frac{\partial}{\partial x_\mu}\int_0^1 \mathrm{d}\alpha\, x_\beta A_\beta(z+\alpha x) - A_\mu(z+x)$$

and comparing it with r.h.s. of eq. (21). It may also be obtained by noting that

$$\Delta x_\mu\left[\frac{\partial}{\partial x_\mu}\int_0^1 x_\sigma A_\sigma(z+\alpha x)\,\mathrm{d}\alpha - A_\mu(z+x)\right] = \oint_C A_\mu(\xi)\,\mathrm{d}\xi_\mu\,, \tag{22}$$

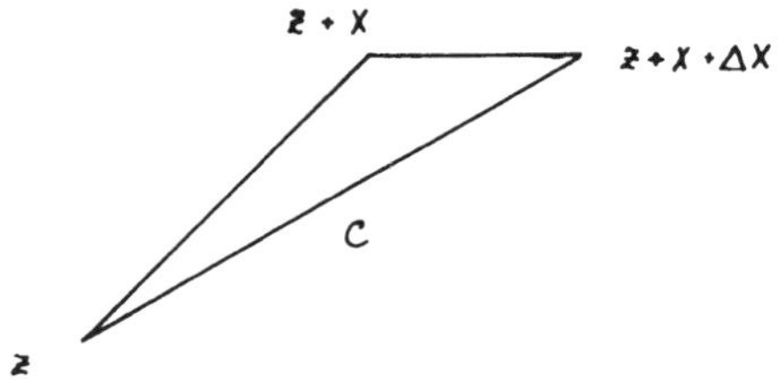

Fig. 3. The integration contour in eq. (22).

where the contour C is depicted in fig. 3. The integral of A_μ over a closed contour is equal to the flow of the field $F_{\mu\nu}$ through that contour which amounts to $x_\nu \Delta x_\mu \int_0^1 \alpha \, d\alpha \, F_{\mu\nu}(z+\alpha x)$ in our case. So we are led to eq. (21) (cf. [20, 21]).

So far we have not specified the properties of the field. Now we substitute eq. (19) into eq. (18), make use of eq. (20) and note that $\partial_\mu R_\mu = 0$ because of equations of motion $\partial_\mu F_{\mu\nu} = 0$. (Note that any self-dual field $F_{\mu\nu} = \pm F^d_{\mu\nu}$ necessarily satisfies the equations of motion $\partial_\mu F_{\mu\nu} = 0$. This follows from the second pair of Maxwell equations $\partial_\mu F^d_{\mu\nu} = 0$, valid for any field $F_{\mu\nu} = \partial_\mu A_\nu - \partial_\nu A_\mu$. The generalization for the non-abelian case is straightforward; any self-dual field $\hat{G}_{\mu\nu} = \pm \hat{G}^d_{\mu\nu}$ satisfies the equations of motion $[\hat{D}_\mu, \hat{G}_{\mu\nu}] = 0$, as follows from the identity $[\hat{D}_\mu, \hat{G}^d_{\mu\nu}] = 0$, valid for any field $\hat{G}_{\mu\nu} = i[\hat{D}_\mu, \hat{D}_\nu]$ [15].) We get

$$4x^2\partial^2\phi - 4x_\mu/x^2\partial_\mu\phi + ieR_\mu\partial_\mu\phi - e^2R_\mu R_\mu \, . \tag{23}$$

This equation can be further simplified if we suppose ϕ to be function of x^2 only. In the constant field, $R_\mu = \frac{1}{2}F_{\mu\nu}x_\nu$ and we get

$$4x^2\phi''(x^2) = e^2\frac{x_\nu x_\sigma}{4}F_{\mu\nu}F_{\mu\sigma}\phi(x^2) \, . \tag{24}$$

Eq. (24) has non-zero solutions (and we know that $\phi(0) = 1$) if and only if the field satisfies the condition

$$F_{\mu\nu}F_{\mu\sigma} = \tfrac{1}{4}\delta_{\nu\sigma}F_{\alpha\beta}F_{\alpha\beta} \, . \tag{25}$$

(Here and below we stick to the euclidean space description.) The requirement (25) can be regarded as the system of equations on the components of the tensor $F_{\mu\nu}$. The solution of this system is

$$\boldsymbol{E} = \pm\boldsymbol{H} = \boldsymbol{n}\sqrt{\tfrac{1}{4}F_{\alpha\beta}F_{\alpha\beta}} \, , \tag{26}$$

where $\boldsymbol{n}$ is an arbitrary 3-space unit vector, i.e. the field must be either self-dual or anti-self-dual. The property (25) for (anti-) self-dual fields may also be checked directly if one substitutes into eq. (25) the identities

$$F_{\mu\nu} = \pm\tfrac{1}{2}\varepsilon_{\mu\nu\rho\varepsilon}F_{\rho\varepsilon} \, , \qquad F_{\mu\sigma} = \pm\tfrac{1}{2}\varepsilon_{\mu\sigma\kappa\lambda}F_{\kappa\lambda} \, .$$

After substituting eq. (25) into eq. (24) and making use of eq. (26) we get the trivial equation

$$\phi''(x^2) = \tfrac{1}{16}e^2H^2\phi(x^2) \, , \tag{27}$$

which has a general solution

$$\phi(x^2) = A\,e^{eHx^2/4} + B\,e^{-eHx^2/4} \, . \tag{28}$$

From the boundary conditions $\phi(0) = 1$ and $\Delta(\infty) = 0$ we get $\phi(x^2) = e^{-eHx^2/4}$ and hence

$$F(z,x) = e^{-eHx^2/4}\,e^{ie\int_0^1 x_\mu A_\mu(z+\alpha x)\,d\alpha} \, . \tag{29}$$

In the gauge $A_\mu = -\frac{1}{2}x_\nu F_{\mu\nu}$, where the relation $x_\mu A_\mu = 0$ holds, the exponential E vanishes and we finally arrive at the result

$$\Delta(x, y) = \frac{1}{4\pi^2(x-y)^2}\, e^{-eH(x-y)^2/4}\,. \tag{30}$$

It can be shown that if the field is not self-dual and hence does not satisfy the condition (25), the analytical solutions of eq. (23) are absent.

Now we are to substitute eq. (30) into eq. (16) to find the electron Green function $S(x, y)$ and then calculate the polarization operator by use of eq. (9). The second term in eq. (9) comprizes the zero-mode projection operator $P_0(x, y)$ given by the infinite sum, eq. (6). Fortunately it need not be calculated: its value may be reconstructed from vector current conservation dictating the transverse structure $\sim(\delta_{\mu\nu}Q^2 - Q_\mu Q_\nu)$ of the final answer. Omitting intermediate steps we present here the final answer for the polarization operator of a massless fermion in a constant self-dual electromagnetic field:

$$\Pi_{\mu\nu}(Q) = (Q_\mu Q_\nu - Q^2\delta_{\mu\nu})\left\{-\frac{1}{12\pi^2}\ln\frac{Q^2}{\mu^2} - \frac{e^2H^2}{3\pi^2 Q^4}\right.$$
$$\left.\times\left[1 - e^{-Q^2/2eH} + \frac{Q^2}{4eH}\, e^{-Q^2/2eH} + \frac{Q^4}{8e^2H^2}\, Ei\left(\frac{-Q^2}{2eH}\right)\right]\right\}. \tag{31}$$

This result is to be compared with the abelian analog of the operator expansion result, eq. (3):

$$i\int d^4x\, e^{iQx}\langle T\{j_\mu(x)j_\nu(0)\}\rangle_0$$
$$= (Q_\mu Q_\nu - \delta_{\mu\nu}Q^2)\left\{-\frac{1}{12\pi^2}\ln(Q^2/\mu^2) + 2m\langle\bar\psi\psi\rangle_0/Q^4\right.$$
$$\left. + (\alpha/6\pi Q^4)\langle F_{\mu\nu}F_{\mu\nu}\rangle_0 + O(1/Q^6)\right\}. \tag{32}$$

Making use of the relation between $\langle\bar\psi\psi\rangle_0$ and $\langle F_{\mu\nu}F_{\mu\nu}\rangle_0$, eq. (15), the coincidence of the lowest-order power terms $\sim 1/Q^2$ in eqs. (31) and (32) can be immediately demonstrated. The higher-order power terms $\sim 1/Q^4$, etc., are absent in the polarization operator eq. (31) due to the simple analytical structure of the scalar Green function, eq. (30).

We see that besides the lowest-dimensional operator expansion term $\sim e^2H^2/Q^2$, only the terms falling exponentially with increasing Q^2 are present. These terms are non-analytical in H, having an essential singularity at the point $H = 0$.

The last observation is in accordance with the general results of ref. [22] where the electron polarization operator in an arbitrary constant electromagnetic field was found. The expression in the general case is extremely complicated, containing an intricate double integral. This integral can be expanded in the field and the expansion coefficients at higher than second powers of the field can be shown to turn into infinities in the limit $m \to 0$, so that the massless fermion polarization operator in a constant electromagnetic field is not analytical at the point $H = 0$.

4.2. NON-ABELIAN CASE

This is a central section of the article. We give a simple proof of the analyticity of $(x-y)^2 \hat{\Delta}(x, y)$ at the point $x = y$ ($\hat{\Delta}(x, y)$ is the scalar Green function in the multi-instanton field; the hat indicates matricity), show how the operator expansion works and obtain the relation between gluon and quark vacuum expectation values.

As earlier we rewrite $\hat{\Delta}(z, z+x)$ in the form (17) and obtain for $\hat{F}(z, x)$ the equation

$$\partial^2 \hat{F} - \frac{4x_\mu}{x^2} \partial_\mu \hat{F} - 2i\hat{A}_\mu \partial_\mu \hat{F} - (i\partial_\mu \hat{A}_\mu + \hat{A}_\mu \hat{A}_\mu)\hat{F} = 0\,, \tag{33}$$

with the boundary conditions $\hat{F}(z, 0) = 1$, $\hat{\Delta}(z, \infty) = 0$. The coupling constant g_s is included in the definition of fields so that the covariant derivative is $\hat{D}_\mu = \partial_\mu - i\hat{A}_\mu$ and the field strength is

$$\hat{G}_{\mu\nu} = \partial_\mu \hat{A}_\nu - \partial_\nu \hat{A}_\mu + i[\hat{A}_\mu, \hat{A}_\nu] = i[\hat{D}_\mu, \hat{D}_\nu]\,.$$

When analysing the analytical properties of $\hat{F}(z, x)$ it is convenient to choose the gauge*

$$x_\mu A_\mu(z+x) = 0 \tag{34}$$

(the Lorentz-gauge analysis is presented in appendix B). We seek the solution of eq. (33) in the form

$$\hat{F}(z, x) = 1 + x^2 \hat{w}(z, x)\,.$$

The equation for $\hat{w}(z, x)$ in the fixed-point gauge (34) is

$$x^2 \partial^2 \hat{w} - 2ix^2 \hat{A}_\mu \partial_\mu \hat{w} = (\hat{A}_\mu \hat{A}_\mu + i\partial_\mu \hat{A}_\mu)(1 + x^2 \hat{w})\,. \tag{35}$$

We shall now prove that in the multi-instanton field $\hat{A}_\mu \hat{A}_\mu + i\partial_\mu \hat{A}_\mu \sim x^2 \hat{f}(z, x)$ is an analytical function at the point $x = 0$. If so, we can divide eq. (35) over x^2 to obtain for $\hat{w}$ a linear differential equation with analytical coefficients, the coefficient at the highest derivative $\partial^2 \hat{w}$ not turning to zero at the point $x = 0$. The well-known theorem of differential equation theory states that the solution of such an equation must be analytical at the point $x = 0$. (Remember that the high-energy power asymptotics of $\Pi_{\mu\nu}(Q)$ is determined completely by the analytical properties of $\hat{\Delta}(z, z+x)$ at this point). The sense of the theorem is rather transparent. If expanding

* This gauge was first suggested in ref. [23] and has recently been reinvented independently in ref. [24]. Later we shall refer to it as the "fixed-point gauge".

$\hat{w}$ in x, inserting this series into the equation and equating the coefficients at the same powers of x we find the equation system for the expansion coefficients to be self-consistent if and only if the conditions of the theorem are satisfied.

In the fixed-point gauge the potential $\hat{A}_\mu$ is represented through the field according to

$$\hat{A}_\mu(x) = x_\nu \int_0^1 \alpha \, d\alpha \, \hat{G}_{\mu\nu}(\alpha x) \tag{36}$$

(we do not display the dependence on z for the sake of brevity). One of the ways to prove it is to use the identity (valid in the fixed-point gauge)

$$\begin{aligned} \hat{A}_\mu(y) &= \frac{\partial}{\partial y_\mu}(y_\rho A_\rho(y)) - y_\rho \partial \hat{A}_\rho(y)/\partial y_\mu \\ &= y_\rho \hat{G}_{\rho\mu}(y) - y_\rho \partial \hat{A}_\mu(y)/\partial y_\rho \,. \end{aligned} \tag{37}$$

Substitute y for αx and rewrite this in the form

$$\alpha x_\rho \hat{G}_{\mu\rho}(\alpha x) = \hat{A}_\mu(\alpha x) + x_\rho \frac{\partial \hat{A}_\mu(\alpha x)}{\partial x_\rho} = \frac{\partial}{\partial \alpha}[\alpha \hat{A}_\mu(\alpha x)] \,. \tag{38}$$

The integration of eq. (38) over α results in eq. (36). Thus we can write

$$\hat{A}_\mu(x)\hat{A}_\mu(x) = x_\nu x_\sigma \int_0^1 \alpha \, d\alpha \, \hat{G}_{\mu\nu}(\alpha x) \int_0^1 \beta \, d\beta \, \hat{G}_{\mu\sigma}(\beta x) \,. \tag{39}$$

For self-dual fields the identity

$$\hat{G}_{\mu\nu}(x)\hat{G}_{\mu\sigma}(y) + \hat{G}_{\mu\sigma}(x)\hat{G}_{\mu\nu}(y) = \tfrac{1}{2}\delta_{\nu\sigma}\hat{G}_{\alpha\beta}(x)\hat{G}_{\alpha\beta}(y)$$

holds. Substituting it into eq. (39) we find that

$$\hat{A}_\mu \hat{A}_\mu \sim x^2 \,.$$

As for $\partial_\mu \hat{A}_\mu$, the use of the *equations of motion* $[\hat{D}_\mu, \hat{G}_{\mu\nu}] = 0$ leads to

$$\begin{aligned} \partial_\mu \hat{A}_\mu &= i x_\nu \int_0^1 \alpha^2 \, d\alpha \, [\hat{A}_\mu(\alpha x), \hat{G}_{\mu\nu}(\alpha x)] \\ &= i x_\nu x_\sigma \int_0^1 \alpha \, d\alpha \int_0^\alpha \beta \, d\beta \, [\hat{G}_{\mu\sigma}(\beta x), \hat{G}_{\mu\nu}(\alpha x)] \,. \end{aligned} \tag{40}$$

Again, in a self-dual field it can be shown to be proportional to x^2 and thus the property $\hat{A}_\mu \hat{A}_\mu + i\partial_\mu \hat{A}_\mu = x^2 f(x)$ and, hence the analyticity of $\hat{w}$ and $\hat{F}$ follows just from self-duality of the field and the knowledge of the explicit multi-instanton solution, eq. (12), is not necessary. As we have already learned, the simple analytical properties of $\hat{\Delta}(x, y)$ provide the absence of the higher-order power terms $\sim 1/Q^4$, $\sim 1/Q^6$, etc., in $\Pi_{\mu\nu}(Q)$. Now consider the terms contributing to the leading power asymptotics $\sim 1/Q^2$ more intently.

The zero-mode contribution to the polarization operator (10) averaged over positions of the field is

$$q^A_{\mu\nu}(x)=\frac{\delta_{\mu\nu}}{2\pi^2x^2}\,dn_0\int d^4z\,P_0(z,z)+\text{reg. terms}\,. \tag{41a}$$

The value

$$-\frac{1}{m}\,dn_0\int d^4z\,P_0(z,z)=-\frac{dn_0}{m}\int d^4z\sum_n\psi_n^{(0)}(z)\psi_n^{(0)+}(z)$$

has the sense of the vacuum fermion density $\langle\bar\psi\psi\rangle_0$ produced by the multi-instanton field. Taking the Fourier transform of eq. (41a) we get

$$q^A_{\mu\nu}(Q)=-\frac{2\delta_{\mu\nu}}{Q^2}\,m\langle\bar\psi\psi\rangle_0+\text{exp. falling terms}\,. \tag{41b}$$

Consider now the first term $p_{\mu\nu}(x,y)=-\mathrm{Sp}\,\{\gamma_\mu\hat S(x,y)\gamma_\nu\hat S(y,x)\}$ in eq. (9). The quark Green function $\hat S(x,y)$ is expressed through the scalar Green function $\hat\Delta(x,y)$ according to eq. (7). Recall that

$$\hat\Delta(z,z+x)=\frac{1}{4\pi^2x^2}[1+x^2\hat w(z,x)]\,, \tag{42}$$

where the function $\hat w(z,x)$ is the solution of eq. (35). In order to find exactly the singular part of $p^A_{\mu\nu}(x)$ as we did for $q^A_{\mu\nu}(x)$, we need to know the expansion coefficients of $\hat w(z,x)$ up to the x^2 term. Higher powers of x fail to produce structure $\sim 1/x^2$ in the polarization operator. If substituting the expansion for $\hat w(z,x)$ into eq. (35) and equating the coefficients of the lowest powers to zero, we find that the solution of the system obtained is not unique. One can freely choose the values of the first two coefficients, $\hat w(z,x)=\hat A(z)+x_\mu\hat B_\mu(z)+\cdots$, and also add a structure $\sim x_\mu x_\nu\hat D_{\mu\nu}(z)$ to the third term, where $\hat D_{\mu\nu}(z)$ is an arbitrary traceless tensor. Then the coefficients at higher powers of x will depend on the parameters $\hat A(z)$, $\hat B_\mu(z)$ and $\hat D_{\mu\nu}(z)$. The values of $\hat A(z)$, $\hat B_\mu(z)$ and $\hat D_{\mu\nu}(z)$ are determined by the boundary condition at infinity—we must require the function $\hat\Delta(z,z+x)$ to die out as $x\to\infty$ [cf. the abelian case; eqs. (28), (29)].

The values of $\hat A(z)$, $\hat B_\mu(z)$ and $\hat D_{\mu\nu}(z)$ can be extracted, in principle, from the exact solution, eq. (11), but we do not know how to do it in the general case. Fortunately, we do not need to know them as we shall see that the terms $\hat A(z)+x_\mu\hat B_\mu(z)+x_\nu x_\mu\hat D_{\mu\nu}(z)$ in $\hat w(z,x)$ do not contribute to the singular part of $p^A_{\mu\nu}(x)$.

Substituting eqs. (39) and (40) into eq. (35), we find the general solution of eq. (35) near the point $x=0$ in the form

$$\hat w(z,x)=\hat A(z)+\hat B_\mu(z)x_\mu+\hat D_{\mu\nu}(z)x_\mu x_\nu$$
$$+\tfrac{1}{512}G^a_{\mu\nu}(z)G^a_{\mu\nu}(z)+\mathrm{O}(x^3)\,, \tag{43}$$

and hence

$$\hat{\Delta}(z, z+x) = \frac{1}{4\pi^2 x^2}[1 + x^2\hat{A}(z) + x^2 x_\mu \hat{B}_\mu(z)$$
$$+ x^2 x_\mu x_\nu \hat{D}_{\mu\nu}(z) + \tfrac{1}{512} x^4 G^a_{\mu\nu}(z) G^a_{\mu\nu}(z) + O(x^5)], \tag{44}$$

where $D_{\mu\mu}(z) = 0$. The quark Green function is determined through the scalar Green function according to eq. (7) and is given by

$$\hat{S}(z, z+x) = -\frac{i\gamma_\mu x_\mu}{2\pi^2 x^4}(1 - \tfrac{1}{512} x^4 G^a_{\mu\nu}(z) G^a_{\mu\nu}(z) - x^2 x_\mu x_\nu \hat{D}_{\mu\nu}(z))$$
$$- \frac{(1-\gamma^5)\gamma_\mu x_\nu}{8\pi^2 x^2} \int_0^1 \hat{G}_{\mu\nu}(z + \alpha x)\alpha \, \mathrm{d}\alpha (1 + x^2 \hat{A}(z))$$
$$+ i\gamma_\mu \hat{B}_\mu(z)/4\pi^2 - \frac{i(1+\gamma^5)\gamma_\mu}{8\pi^2}\left(\frac{\partial \hat{A}}{\partial z_\mu} + x_\nu \frac{\partial \hat{B}_\nu}{\partial z_\mu}\right) + O(x^2). \tag{45}$$

Substituting this into the first term in eq. (9) and integrating over field positions according prescription (10) we get eventually,

$$p^A_{\mu\nu}(x) = \frac{x_\mu x_\nu - x^2 \delta_{\mu\nu}}{96\pi^4 x^4} \, dn_0 \int \mathrm{d}^4 z \, G^a_{\alpha\beta}(z) G^a_{\alpha\beta}(z) + \text{reg. terms}. \tag{46a}$$

We see that unknown parameters $\hat{A}(z)$, $\hat{B}_\mu(z)$ and $\hat{D}_{\mu\nu}(z)$ do not contribute to the singular part of the polarization operator. Note that the last statement is valid *only if the integration over field positions is performed.* The singular terms in $p^A_{\mu\nu}(z, x)$ are affected by $\hat{A}(z)$, $\hat{B}_\mu(z)$.

The Fourier transform of eq. (46a) is

$$p^A_{\mu\nu}(Q) = -\frac{2Q_\mu Q_\nu + \delta_{\mu\nu} Q^2}{12\pi Q^4} \langle \alpha_s G_{\alpha\beta} G_{\alpha\beta} \rangle_0$$
$$+ \text{exp. falling terms} \tag{46b}$$

(the coupling constant $\alpha_s = g_s^2/4\pi$ is restored). The tensor structure of the total polarization operator given by the sum of expressions (41b) and (46b) must be transverse, $\Pi_{\mu\nu}(Q) \sim Q^2\delta_{\mu\nu} - Q_\mu Q_\nu$, because of vector current conservation. This condition can be satisfied only if the gluon and fermion vacuum densities are related according to

$$\frac{1}{8\pi}\langle \alpha_s G^a_{\alpha\beta} G^a_{\alpha\beta} \rangle_0 = -m\langle \bar{\psi}\psi \rangle_0 . \tag{47}$$

This is the non-abelian analog of eq. (15).

The identity (47) can be checked directly for the one-instanton case (as was done, in fact, in ref. [8] where the transverse tensor structure for $\Pi_{\mu\nu}(Q)$ was obtained after adding the exact expressions for $p^{\text{inst.}}_{\mu\nu}(x, y)$ and $q^{\text{inst.}}_{\mu\nu}(x, y)$). The generalization of

(47) to the multi-instanton case is rather trivial and could be expected in advance. The physical content of eq. (47) amounts to the statement that both the total action and the number of zero modes in the multi-instanton field are proportional to the topological charge of the field [1, 18, 19].

5. Discussion and conclusions

We have shown that the quark polarization operator in a multi-instanton field has no contribution from operators of higher than four dimensions in lowest order in α_s. What are the physical implications of this fact?

Consider first the 6-dimensional operators, namely, $g_s^3 f^{abc} G^a_{\mu\nu} G^b_{\nu\sigma} G^c_{\sigma\mu}$ and $m\alpha_s \bar{\psi}\sigma^{\mu\nu} t^a \psi G^a_{\mu\nu}$. In the instanton field these operators acquire non-zero expectation values, but we know that they do not contribute to the polarization operator asymptotics. The two statements are compatible only if the coefficients of the 6-dimensional operators in expansion (2) are equal to zero. (Note that the contribution of the operator $m\alpha_s \bar{\psi}\sigma^{\mu\nu} t^a \psi G^a_{\mu\nu}$ is connected to the zero-mode term $\sim \mathrm{Sp}\,\{\gamma_\mu \hat{P}_0(x, y)\gamma^\nu \hat{\Delta}(y, x)\}$, whereas the operator $g_s^3 f^{abc} G^a_{\mu\nu} G^b_{\nu\sigma} G^c_{\sigma\mu}$ contributes to the term $\sim \mathrm{Sp}\,\{\gamma_\mu \hat{S}(x, y)\gamma_\nu \hat{S}(y, x)\}$ [see eq. (9)]. The analyticity arguments of sect. 3 work for the zero-mode and non-zero-mode contributions separately, so that the coefficients of both operators must vanish.)

The contribution of 6-dimensional operators to the vector current polarization operator was analysed in ref. [6]. The main contribution was found to be given by the operators $\bar{\psi}\Gamma\psi\bar{\psi}\Gamma\psi$. As mentioned in sect. 2, the contribution of these operators is suppressed $\sim\alpha_s(Q)$ and is not sensible in our approach.

The coefficient of the operator $m\alpha_s \bar{\psi}\sigma^{\mu\nu} t^a \psi G^a_{\mu\nu}$ was calculated and turned out to include the additional damping factor $\sim m^2$ so that it vanishes in the massless quark limit.

In ref. [6] the coefficients in the operator expansion were calculated by taking the matrix elements between plane waves. Say, the coefficients of the operators $\bar{\psi}\Gamma\psi\bar{\psi}\Gamma\psi$ were calculated by use of figs. 4a, b, the coefficient of the operator $m\alpha_s \bar{\psi}\sigma^{\mu\nu} t^a \psi G^a_{\mu\nu}$ was calculated by means of fig. 5a and the coefficient of the operator $\sim G^3$ can be determined in principle by figs. 5a, b.

In the framework of the plane-wave approach, the vanishing of the coefficient of $m\alpha_s \bar{\psi}\sigma^{\mu\nu} t^a \psi G^a_{\mu\nu}$ can be observed as an "experimental fact" but has no clear

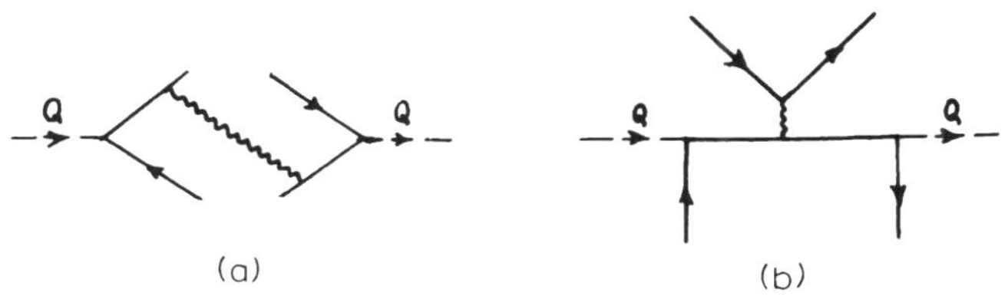

Fig. 4. Calculation of the coefficients in the operator expansion for the operators $\bar{\psi}\Gamma\psi\bar{\psi}\Gamma\psi$.

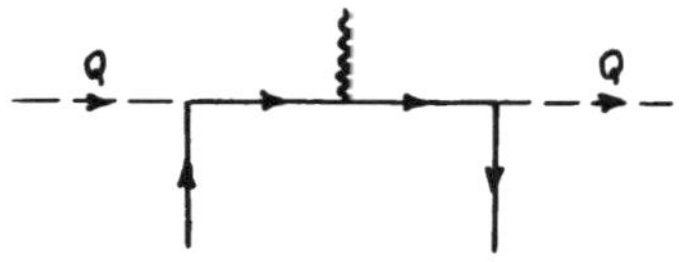

Fig. 5. Calculation of the coefficient of the operator $\alpha_s G^a_{\mu\nu}\bar{\psi}\sigma^{\mu\nu}t^a\psi$.

explanation. In our approach it is a natural consequence of the simple analytical properties of the polarization operator in a self-dual field.

The coefficient of $g_s^3 f^{abc} G^a_{\mu\nu} G^b_{\nu\sigma} G^c_{\sigma\mu}$ was not calculated in ref. [6] and its contribution was neglected on the grounds that the one-loop graphs of fig. 6a, b must include the small factor $\sim 1/4\pi^2$ as compared to the tree graphs of fig. 4. This argument is quite sound, but our results make it superfluous as we proved rigorously that the coefficient of the operator $\sim G^3$ is equal to zero in the massless quark limit.

On the level $\sim G^4$ there are several possible 8-dimensional operators built out of the field strength tensor $\hat{G}_{\mu\nu}$. Some of them, such as the operator $(G^a_{\mu\alpha}G^a_{\mu\beta} - \frac{1}{4}\delta_{\alpha\beta}G^a_{\mu\nu}G^a_{\mu\nu})^2$, have zero expectation value in any self-dual field and we cannot say anything about the coefficient of this operator. The other operators acquire non-zero expectation values in a self-dual field and their coefficients must then vanish.

Another important observation follows from relation (47), valid in any multi-instanton field. The true vacuum expectation values $\langle \alpha_s G^a_{\mu\nu} G^a_{\mu\nu} \rangle_0$ and $m\langle \bar{\psi}\psi \rangle_0$ are given by the continual integrals

$$\langle G^a_{\mu\nu} G^a_{\mu\nu} \rangle_0 = \int \prod \mathrm{d}A\, \mathrm{d}\psi\, \mathrm{d}\bar{\psi}\, G^a_{\mu\nu}(x) G^a_{\mu\nu}(x)\, \mathrm{e}^{-\int \mathcal{L}(A,\psi,\bar{\psi})\,\mathrm{d}^4x}, \tag{48a}$$

$$m\langle \bar{\psi}\psi \rangle_0 = \int \prod \mathrm{d}A\, \mathrm{d}\psi\, \mathrm{d}\bar{\psi}\, m\bar{\psi}(x)\psi(x)\, \mathrm{e}^{-\int \mathcal{L}(A,\psi,\bar{\psi})\,\mathrm{d}^4x}. \tag{48b}$$

Nowadays the only way of evaluating continual integrals we know is the integrand expansion near the saddle points; that is, near the *multi-instanton solutions*. But in multi-instanton fields relation (47) holds and if the saddle evaluation of the continual integrals (48) gave fairly correct results, one could expect the true vacuum expectation values to be related according to eq. (47) too. We know, however, that this is

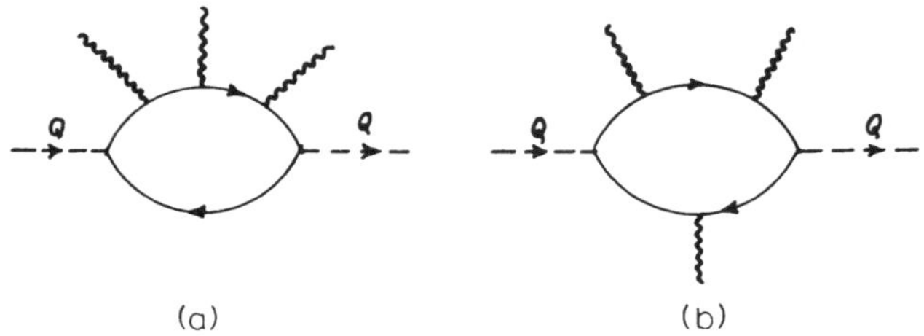

Fig. 6. Calculation of the coefficient of the operator $\sim G^3$.

not the case. The gluon vacuum expectation value was estimated in ref. [6] to have the value

$$\left\langle \frac{\alpha_s}{8\pi} G^a_{\mu\nu} G^a_{\mu\nu} \right\rangle_0 \approx 1.5 \times 10^{-3}\,\mathrm{GeV}^4 ,$$

which is ~10 times larger than

$$-m\langle \bar{\psi}\psi \rangle_0 = f_\pi^2 m_\pi^2 \approx 1.7 \times 10^{-4}\,\mathrm{GeV}^4 ,$$

so that the contribution of the operator $m\bar{\psi}\psi$ to the polarization operator is negligible compared to that of the operator $\alpha_s G^a_{\mu\nu} G^a_{\mu\nu}$ [6].

The breaking of relation (47) unequivocally displays the non-applicability of the saddle approximation in evaluating continual integrals. We conclude, therefore, in full accordance with [6] (see also [25, 26]), that until progress in calculating continual integrals more complicated than gaussian ones is achieved, the operator expansion approach with vacuum expectation values $\langle \alpha_s G^a_{\mu\nu} G^a_{\mu\nu} \rangle_0$, etc., appearing as phenomenological parameters is the only way to obtain reliable numerical results.

So far, we have only discussed the vector current polarization operator. All the analyticity arguments of sect. 3 can equally be applied to the polarization operators of other light quark currents.

We remark, however, that our arguments do not work in the heavy quark case. The simple analytical properties of the quark Green function (5) are spoiled just in the next order in m, where branching at the point $x = y$ appears [27]. Thus, there is nothing precluding the presence of contributions from higher-dimensional operators in the operator expansion for polarization operators of heavy quark currents and one can expect them to appear.

Our arguments do not work also for polarization operators of gluon currents (such as $j_5^G(x) \sim \alpha_s G^a_{\mu\nu}(x) G^{ad}_{\mu\nu}(x)$) where the simplest gluon loops in external fields should be considered. The gluon Green function in an external self-dual field is expressed through convolutions of the scalar Green function which lead to logarithmically divergent integrals [15] and the branching point at $x = y$ appears.

The gluon polarization operators may be handled by the use of quite another sort of argument. It turned out that the higher power terms in the tensor gluon current operator vanish and the situation is very similar to that in light quark case [14].

In conclusion we briefly repeat our main results.

(i) We showed that in lowest order in α_s polarization operators of light quark currents in an external multi-instanton field receive no contribution from higher than four-dimensional operators in expansion (2) and that they are presented in the form $\Pi(Q^2) \sim \Pi^{(0)}(Q) + C/Q^2 +$ exp. falling terms. This result follows from the simple analytical properties of the polarization operator in the coordinate representation, namely, from the absence of branching points at $(x-y)^2 = 0$.

The physical implications of this result are rather scarce, unfortunately. We can just show the absence of some small corrections to the contributions from the operators $\alpha_s G^a_{\mu\nu} G^a_{\mu\nu}$, $m\bar{\psi}\psi$ and $\bar{\psi}\Gamma\psi\bar{\psi}\Gamma\psi$ discussed earlier [6]. However, the result is interesting from the methodological viewpoint. It settles puzzling questions which arise after comparing the operator expansion (3) with the one-instanton result (4) and strengthens our confidence in the operator expansion method.

(ii) We note also that, in an arbitrary multi-instanton field corresponding to a saddle point of the continual integral, relation (47) between quark and gluon vacuum densities holds. This relation is strongly broken in experiment and thus we have *experimental evidence* for the fact that the saddle approximation is a very bad one for calculating continual integrals.

We wish to thank B.L. Ioffe, V.A. Novikov, A.S. Schwars, M.A. Shifman, M.B. Voloshin and V.I. Zakharov for helpful discussions.

Appendix A

We discuss here the connection between the asymptotics of $\tilde{\Pi}_{\mu\nu}(Q)$, $Q^2 \to \infty$ and the singularities of $\Pi_{\mu\nu}(x)$ at the point $x = 0$. This connection is due to the fact that the Fourier integral expressing $\tilde{\Pi}_{\mu\nu}(Q)$ through $\Pi_{\mu\nu}(x)$ amounts to the complex contour integral.

Consider first the high-energy asymptotics of the Fourier transform of a function $f(x)$ independent of the direction of euclidean 4-vector x. The Fourier transform of such a function is given by

$$\tilde{f}(Q) = \int e^{iQx} f(x)\, d^4x = \frac{4\pi^2}{Q^2} \int_0^\infty J_1(Qr) f(r) r^2\, dr, \tag{A.1}$$

where $r = \sqrt{x^2}$, $J_1(z)$ is the Bessel function. Making use of the well-known relation between cylindrical functions,

$$J_1(z) = \tfrac{1}{2}\{H_1^{(1)}(z + i0) + H_1^{(1)}(-z + i0)\}, \qquad \operatorname{Im} z = 0, \qquad \operatorname{Re} z > 0,$$

the integral (A.1) is rewritten through the contour integral in the complex r-plane

$$\tilde{f}(Q) = \frac{2\pi^2}{Q^2} \int_C H_1^{(1)}(Qr) f(r) r^2\, dr, \tag{A.2}$$

where the contour C is depicted in fig. 7, the solid line corresponding to the cut of the Hankel function $H_1^{(1)}(z)$. The asymptotic relation

$$H_1^{(1)}(z) = -i\sqrt{2/i\pi z}\, e^{iz}, \qquad |z| \to \infty \tag{A.3}$$

enables us to shift the contour in the upper half-plane, bending the singularities of $f(r)$ as shown in fig. 7. The dominating contribution to the asymptotics of $\tilde{f}(Q)$ is due to the nearest singularity of $f(r)$ in the upper half-plane. If, e.g., the nearest singularity lies at the point $r = ia$ as shown in fig. 7, $\tilde{f}(Q) \sim e^{-aQ}$ in the large Q region.

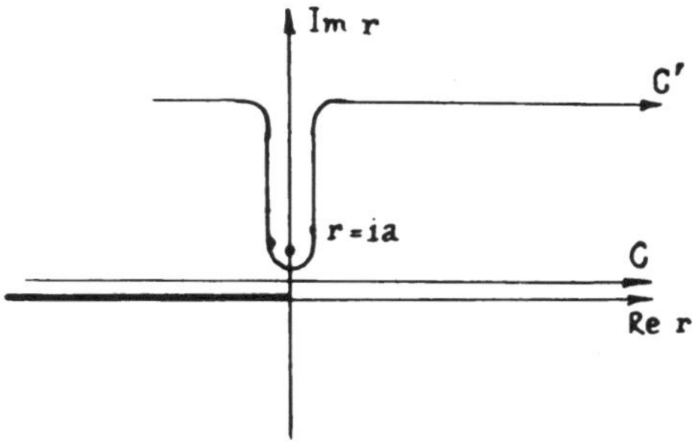

Fig. 7. Shifting the contour in the complex r-plane in eq. (A.2).

If the singularities of $f(r)$ at the point $r=0$ are present, particular consideration is required. In this case our arguments are applicable if substituting $f(r)$ (for $f(\sqrt{r^2+a^2})$ and letting $a \to 0$ afterwards. It does not change the value of the integral (A.1) if it converges at the point $r=0$.

If the integral diverges at small r, this procedure is, in fact, one of the possible ways of ultraviolet regularization required necessarily in that case.

Consider an example: let $f(r)=A(r)/r^2$, where $A(r)$ is analytical at the point $r=0$ [it is the form of the function $r^A_{\mu\mu}(x)$ of eq. (10)]. The function $A(r)/(r^2+a^2)$ [which is to be substituted for $f(r)=A(r)/r^2$ in the integral (A.2) according to our prescription] has a pole in the upper half-plane at $r=ia$ and the calculation of the asymptotics of (A.2) amounts to taking the residue at this pole in the limit $a \to 0$. We have

$$\tilde{f}(Q)=\lim_{a\to 0}\frac{2\pi^2}{Q^2}2\pi i H_1^{(1)}(iQa)\left(-\frac{a^2}{2ia}\right)A(ia)=\frac{4\pi^2}{Q^2}A(0)\,. \tag{A.4}$$

So the pole at $x^2=0$ results in power asymptotics of $\tilde{f}(Q)$. The singularities at non-zero r result in exponential corrections to eq. (A.4). Our arguments are easily generalized to an arbitrary function $f(x)$. Indeed, in the general case, the Fourier image of the tensor functions $\sim x_{\alpha_1}\cdots x_{\alpha_n}\varphi(x^2)$ must be found. A function of this kind can be represented as $p_n^{\alpha_1\cdots\alpha_n}(\partial_\alpha)\chi(x^2)$ (where $p_n^{\alpha_1\cdots\alpha_n}$ is the nth order tensor linear differential operator made out of ∂_{α_i}). The Fourier image of it is equal to $p_n(iQ_\alpha)\tilde{\chi}(Q^2)$ where $\tilde{\chi}(Q^2)$ is the Fourier image of $\chi(x^2)$ and the problem is reduced to the case discussed above.

Appendix B

We repeat here the analysis of the analytical properties of the polarization operator in a multi-instanton field performed in subsect. 4.2 in the Lorentz gauge $\partial_\mu\hat{A}_\mu=0$. The result coincides with that obtained earlier in the fixed-point gauge $x_\mu\hat{A}_\mu=0$.

The Lorentz-gauge analysis is more complicated and comprises some technical tricks which may be useful elsewhere.

As earlier, we represent $\hat{\Delta}(x, y)$ in the form (11) and obtain for $\hat{F}(z, z+x)$ the equation

$$\hat{D}_\mu \hat{D}_\mu \hat{F} - 4x_\mu \hat{D}_\mu \hat{F}/x^2 = 0 . \tag{B.1}$$

We seek the solution of eq. (B.1) in the form

$$\hat{F}(x) = \hat{\phi}(x)\hat{E}(x) , \tag{B.2}$$

where

$$\begin{aligned} \hat{E}(x) &= P_\alpha \exp\left[i\int_0^1 x_\mu \hat{A}_\mu(\alpha x)\,\mathrm{d}\alpha \right] \\ &= 1 + ix_\mu \int_0^1 \hat{A}_\mu(\alpha x)\,\mathrm{d}\alpha - x_\mu x_\nu \int_0^1 \mathrm{d}\alpha \int_0^\alpha \mathrm{d}\beta \hat{A}_\mu(\alpha x)\hat{A}_\nu(\beta x) + \cdots \\ &= 1 + \hat{E}^{(1)}(x) + \hat{E}^{(2)}(x) + \cdots \end{aligned} \tag{B.3}$$

is the path-ordered exponential (the dependence on z is not displayed). Now, quite analogous to the abelian case [see eqs. (18), (19)] we want to obtain the equation for the new unknown function $\hat{\phi}$ by substituting eq. (B.2) into eq. (B.1). To this end, the result of the action of operator $\hat{D}_\mu$ on the path-ordered exponential must be found. We are going to show that

$$\hat{D}_\mu(x)\hat{E}(x) = \hat{K}_\mu(x)\hat{E}(x) , \tag{B.4a}$$

where

$$\begin{aligned} \hat{K}_\mu(x) = ix_\nu \Big\{ &\tfrac{1}{2}\hat{G}_{\mu\nu}(x) - \tfrac{1}{6}x_\eta[\hat{D}_\eta, \hat{G}_{\mu\nu}(x)] + \cdots \\ &+ \frac{(-1)^n x_{\eta_1} \cdots x_{\eta_n}}{(n+2)!}[\hat{D}_{\eta_1}, \cdots, [\hat{D}_{\eta_n}, \hat{G}_{\mu\nu}(x)] \cdots] + \cdots \Big\} . \end{aligned} \tag{B.4b}$$

The proof of this beautiful result consists of several steps which we briefly outline here.

On the first step we learn to differentiate the path-ordered exponential. We state that $\partial_\mu \hat{E} = \hat{R}_\mu \hat{E}$, where

$$\begin{aligned} \hat{R}_\mu(x) = &\int_0^1 \mathrm{d}\alpha \hat{T}_{,\mu}(\alpha x) + \int_0^1 \frac{\mathrm{d}\alpha}{\alpha}\int_0^\alpha \mathrm{d}\beta [\hat{T}(\alpha x), \hat{T}_{,\mu}(\beta x)] \\ &+ \int_0^1 \frac{\mathrm{d}\alpha}{\alpha}\int_0^\alpha \frac{\mathrm{d}\beta}{\beta}\int_0^\beta \mathrm{d}\gamma [\hat{T}(\alpha x), [\hat{T}(\beta x), \hat{T}_{,\mu}(\gamma x)]] + \cdots \end{aligned} \tag{B.5}$$

(we define $\hat{T}(\xi) = i\xi_\mu \hat{A}_\mu(\xi)$; $\hat{T}_{,\mu}(\xi) = \partial\hat{T}(\xi)/\partial\xi_\mu$). This property can be proved by the use of induction, comparing the terms with equal powers of $\hat{T}$ on both sides of $\partial_\mu \hat{E} = R_\mu \hat{E}$. The function $\hat{R}_\mu$ given by eq. (B.5) satisfies the integral equation

$$\hat{R}_\mu(x) = \int_0^1 \mathrm{d}\alpha \{\hat{T}_{,\mu}(\alpha x) - [\hat{R}_\mu(\alpha x), \hat{T}(\alpha x)]\} . \tag{B.6}$$

Then

$$\hat{K}_\mu(x) = \hat{R}_\mu(x) - i\hat{A}_\mu(x)$$

satisfies the equation

$$\hat{K}_\mu(x) = \int_0^1 \hat{T}_{,\mu}(\alpha x)\, \mathrm{d}\alpha + \int_0^1 \mathrm{d}\alpha \Big[\hat{T}(\alpha x), \hat{K}_\mu(\alpha x) - i\hat{A}_\mu(x) + i\int_0^1 [\hat{T}(\alpha x), \hat{A}_\mu(\alpha x)\Big] \mathrm{d}\alpha\,. \tag{B.7}$$

We now note that

$$\int_0^1 \hat{T}_{,\mu}(\alpha x)\, \mathrm{d}\alpha - i\hat{A}_\mu(x) = -ix_\nu \int_0^1 \alpha\, \mathrm{d}\alpha (\partial_\mu \hat{A}_\nu - \partial_\nu \hat{A}_\mu)(\alpha x)\,, \tag{B.8}$$

which can be proved by expanding eq. (B.8) in a series in x again. Note that in the fixed-point gauge eq. (B.8) is reduced to eq. (36) discussed earlier. Combining the result (B.8) with the last term in eq. (B.7) we get

$$\hat{K}_\mu(\alpha x) = ix_\nu \int_0^1 a\, \mathrm{d}\alpha \hat{G}_{\mu\nu}(\alpha x) + i\int_0^1 \alpha\, \mathrm{d}\alpha x_\nu [\hat{A}_\nu(\alpha x), \hat{K}_\mu(\alpha x)]\,. \tag{B.9}$$

If we substitute $i\hat{A}_\nu = \partial_\nu - \hat{D}_\nu$ and note that $x_\nu \partial \hat{K}_\mu(\alpha x)/\partial(\alpha x_\nu) = \partial \hat{K}_\mu(\alpha x)/\partial\alpha$, we get

$$\int_0^1 \hat{K}_\mu(\alpha x)\, \mathrm{d}\alpha = ix_\nu \int_0^1 \alpha\, \mathrm{d}\alpha \hat{G}_{\mu\nu}(\alpha x) - x_\nu \int_0^1 \alpha\, \mathrm{d}\alpha [\hat{D}_\nu, \hat{K}_\mu](\alpha x)\,,$$

and hence

$$\hat{K}_\mu(x) = ix_\nu \hat{G}_{\mu\nu}(x) - x_\nu [\hat{D}_\nu(x), \hat{K}_\mu(x)]\,. \tag{B.10}$$

One can be convinced that the solution of eq. (B.10) is given by the infinite series (B.4).

Now we are ready to substitute $\hat{F}(x) = \hat{\phi}(x)\hat{E}(x)$ into eq. (B.1) to obtain the equation for $\hat{\phi}$. As in the fixed-point gauge, we represent $\hat{\phi}$ as $\hat{\phi}(x) = 1 + x^2 \hat{w}(x)$ and arrive at

$$x^2[\hat{D}_\mu, [\hat{D}_\mu, \hat{w}]] + 2x^2[\hat{D}_\mu, \hat{w}]\hat{K}_\mu + (1 + x^2\hat{w})\{\hat{K}_\mu^2 + [\hat{D}_\mu, \hat{K}_\mu]\} = 0\,. \tag{B.11}$$

The solution of this equation is analytical at the point $x = 0$ provided $\hat{K}_\mu^2 + [\hat{D}_\mu, \hat{K}_\mu] \sim x^2$. The property $\hat{K}_\mu^2 \sim x^2$ is easily obtained from eq. (B.4) and self-duality of the field. When proving $[\hat{D}_\mu, \hat{K}_\mu] \sim x^2$, the equations of motion should be used too. Then the commutator of $\hat{D}_\mu$ with the first term of eq. (B.4) becomes zero, the commutator with the second term is $\sim x_\nu x_\eta [\hat{G}_{\mu\nu}, \hat{G}_{\mu\eta}]$ (the Jacobi identity was used) and also becomes zero. The subsequent terms can be transformed by use of the Jacobi identity so that two field strength tensors appear:

$$[\hat{D}_\mu, \hat{K}_\mu^{(n)}] \sim \cdots x_\nu \hat{G}_{\mu\nu} \cdots x_\eta \hat{G}_{\mu\eta} \cdots,$$

proportional to x^2 because of self-duality.

Thus $\hat{K}_\mu^2 + [\hat{D}_\mu, \hat{K}_\mu] \sim x^2$ in the multi-instanton field, the solution of eq. (B.11) is regular at the point $x = 0$, having exactly the same form as in the fixed-point gauge case, eq. (43).

The transfer from the scalar Green function to the polarization operator is a little more complicated than in the fixed-point gauge case due to the presence of the path-ordered exponential. It can be performed, however, by making use of relations (B.3)–(B.4) to obtain expression (46a) which does not depend on the gauge, of course.

References

[1] A. Belavin et al., Phys. Lett. 59B (1975) 85
[2] C.G. Callan, R. Dashen and D.J. Gross, Phys. Lett. 63B (1976) 334
[3] R. Jackiw and C. Rebbi, Phys. Rev. Lett. 37 (1976) 172
[4] C.G. Callan, R. Dashen and D.J. Gross, Phys. Rev. D17 (1978) 2717
[5] G. 't Hooft, Phys. Rev. D14 (1976) 3432
[6] M.A. Shifman, A.I. Vainstein and V.I. Zakharov, Nucl. Phys. B147 (1979) 385
[7] K. Wilson, Phys. Rev. 179 (1969) 1499
[8] R. Carlitz and Ch. Lee, Phys. Rev. D17 (1978) 3238
[9] N. Andrei and D.J. Gross, Phys. Rev. D18 (1978) 469
[10] C. Baulieu et al., Phys. Lett. 77B (1978) 290
[11] T. Appelquist and R. Shankar, Phys. Rev. D18 (1978) 2952
[12] H.D. Politzer, Phys. Reports 14 (1974) 129
[13] Yu.L. Dokshitzer, D.I. Dyakonov and S.I. Troyan, Lectures at 13th LINR Winter School, Leningrad, 1978;
D. Amati, R. Petronzio and G. Veneziano, Nucl. Phys. B146 (1978) 29
[14] V.A. Novikov et al., Preprint ITEP-27 (1980)
[15] L.S. Brown et al., Phys. Rev. D17 (1978) 1583
[16] N.H. Christ, E.J. Weinberg and N.K. Stanton, Phys. Rev. D18 (1978) 2013
[17] E. Corrigan et al., Nucl. Phys. B140 (1978) 31
[18] M.F. Atiyah and I.M. Singer, Ann. Math. 87 (1968) 484
[19] J. Kiskis, Phys. Rev. D15 (1977) 2329
[20] S. Mandelstam, Ann. of Phys. 19 (1962) 1
[21] E. Corrigan and B. Hasslacher, Phys. Lett. 81B (1979) 181
[22] V.N. Baier, V.M. Katkov and V.M. Strakhovenko, ZhETF (USSR), 68 (1975) 405
[23] V.A. Fateev, A.S. Schwars and Yu.S. Tyupkin, Lebedev Phys. Institute preprint no. 155 (1976)
[24] C. Cronström, Phys. Lett. 90B (1980) 267
[25] M.A. Shifman, A.I. Vainstein and V.I. Zakharov, Preprint ITEP-57 (1979)
[26] M.A. Shifman, A.I. Vainstein and V.I. Zakharov, Preprint ITEP-81 (1979)
[27] A.M. Din, J. Finjord and W.J. Zakrzewski, Nucl. Phys. B153 (1979) 46

The calculation of power corrections in the fixed-point gauge

A. V. Smilga

Institute of Theoretical and Experimental Physics, USSR Academy of Sciences
(Submitted 21 May 1981)
Yad. Fiz. **35**, 473–484 (February 1982)

We propose a simple method of calculating nonperturbative power corrections in QCD. We use the fixed-point gauge $(x-x_0)_\mu A_\mu(x)=0$ which has the advantage that the potential $A_\mu(x)$ is directly expressed in terms of the field strength tensor $G_{\mu\nu}(x)$ and its covariant derivatives. This makes it possible to specify a clear algorithm for selecting the coefficients of gauge-invariant structures in the polarization operators of various currents in the vacuum field. This method is used to calculate the coefficient of the structure $\langle G^a_{\mu\nu} G^a_{\mu\nu}\rangle_0$ in the polarization operator of vector currents for heavy quarks. Our result coincides with previous calculations. We have also done an original calculation of the coefficients of the structure $\langle G^a_{\mu\nu} G^a_{\mu\nu}\rangle_0$ for the baryon currents corresponding to the proton and the Δ isobar. The predictions for the mass of the proton and the Δ isobar obtained recently by B. L. Ioffe from a sum-rule analysis are found to be insensitive to the effect of these power corrections, which he had not taken into account.

PACS numbers: 12.35.Cn, 11.15.Bt, 12.70. + q

1. INTRODUCTION

Recently in Ref. 1 a new approach has been proposed for investigating hadronic physics at low energies in the framework of QCD. It is well known that in the energy region ~1 GeV ordinary perturbation theory in QCD based on an expansion in the coupling constant α_s does not work anymore, and that nonperturbative effects which determine the properties of hadronic resonances and are responsible for confinement become important.

According to present ideas these effects are due to the complicated structure of the QCD vacuum: apart from ordinary perturbative vacuum fluctuations which are present in simpler field theories, nonperturbative fluctuations which are not analytic in the coupling constant are characteristic for QCD [the simplest example of such a fluctuation in the BPTSh instanton (Ref. 2)]. At present there is little one can say about the wave function of the vacuum which describes the detailed structure of these fluctuations. The calculations which are based on the dilute-instanton-gas hypothesis assuming that the vacuum structure is mainly determined by isolated one-instanton fluctuations are intrinsically inconsistent since in the answer the following integral with respect to the size of the instanton enters:

$$\int d\rho/\rho^5 d(\rho)\, e^{-2\pi/\alpha_s(\rho)},$$

which is determined by the region of large ρ where the coupling constant α_s is large and the original assumption does not apply. Moreover, the hypothesis that instanton fluctuations dominate (and even the more general hypothesis of the dominance of multi-instanton field configurations which correspond to saddle points in the contour integral) is in direct contradiction with experiment.[3]

It turns out, however, that a rather large amount of information on nonperturbative contributions to physical quantities can be obtained without a detailed knowledge of the vacuum structure. More precisely, it is sufficient to have a rough knowledge of this structure, which reduces to the knowledge of a small number of definite parameters. The most important of these parameters are the vacuum averages $\langle\bar\psi\psi\rangle_0$ and $\langle G^a_{\mu\nu}G^a_{\mu\nu}\rangle_0$ from first principles. However, in Ref. 1 it is shown that these values determine a large number of observable quantities, so that, having determined the values of $\langle\bar\psi\psi\rangle_0$ and $\langle G^a_{\mu\nu}G^a_{\mu\nu}\rangle$ in some way, one can compare nontrivial predictions with experiment.

The main technique used in Ref. 1 is the operator-expansion technique. Let us consider, for instance, the polarization operator of quark vector currents

$$\Pi_{\mu\nu}(p)=i\int e^{ipx}\langle j_\mu(x)j_\nu^+(0)\rangle_0 d^4x,$$

$j_\mu(x)=\bar q(x)\gamma_\mu q(x)$, in the nonphysical region below threshold $p^2<4m_q^2$. If p^2 is sufficiently far from the threshold, small values of x play the main role in the integral, and the operator product $j_\mu(x)j_\nu^+(0)$ can be expanded in powers of x according to the hypothesis of the Wilson operator expansion. Then, the polarization operator takes the form

$$\Pi_{\mu\nu}(p)=(p_\mu p_\nu-g_{\mu\nu}p^2)\sum_n c_n(p^2,m^2)\langle O_n\rangle_0,$$

where m is the quark mass, and the O_n are various gauge-invariant operators, $O_n=1$, $G^a_{\mu\nu}G^a_{\mu\nu}$, $\bar\psi\psi$, $f^{abc}G^a_{\mu\nu}\times G^b_{\nu\sigma}G^c_{\sigma\mu}$, etc. At a sufficiently large distance from the threshold the sum is determined by the first few terms and the contribution of higher-dimensional operators is small. The polarization operator in the nonphysical region is subsequently connected with observable quantities by dispersion relations. The main aim of our work is to propose a new convenient method of calculating the coefficients c_n for the various operators O_n in an aribtrary n-point function with colorless legs. In our opinion this method has a definite advantage compared to the methods proposed earlier to calculate c_n. The plan of the paper is as follows. In Sec. 2 we describe the fixed-point gauge $(x-x_0)_\mu A_\mu(x)=0$, which we use, and describe the algorithm for calculating the power corrections to an arbitrary n-point function with colorless legs in the nonphysical kinematicel region. In Sec. 3 we present the results of the calculations of the coefficient $c_G(q^2, m^2)$ of the operator $G^a_{\mu\nu}G^a_{\mu\nu}$ and compare our calculations with the results of Ref. 1. In Sec. 4 we calculate the coefficients of the operator $G^a_{\mu\nu}G^a_{\mu\nu}$ in the polarization operators for the baryon currents with the quantum numbers of the nucleon and the Δ isobar.

The polarization operators of the baryon currents and their sum rules, which make definite predictions possible on the mass of low-lying baryon resonances, have been studied previously in Ref. 4. In that paper the coefficient of the operator $G^a_{\mu\nu}G^a_{\mu\nu}$ was not calculated and only estimates were made, from which it could be expected that these power corrections have little effect on the sum rules. Our concrete calculation shows that this is indeed the case, and taking account of terms $\sim\langle G^2\rangle_0$ in the sum rules changes the predictions for the resonance masses only by a few per cent.

In Sec. 5 we summarize the main results of our work and discuss the possibilities for further applications of our method.

2. VACUUM FIELDS IN THE FIXED-POINT GAUGE

The fixed-point gauge implies imposing the condition

$$(x-x_0)_\mu A^a_\mu(x)=0, \tag{1}$$

where x_0 is an arbitrary point in space. This gauge was first proposed by Schwinger,[5] and subsequently repeatedly and independently rediscovered by a number of authors.[3,6,7] The first-point gauge is convenient since the potential $A^a_\mu(x)$ is expressed directly in terms of the field-strength tensor $G^a_{\mu\nu}(x)$:

$$A^a_\mu(x)=-\int_0^1 \alpha d\alpha G^a_{\mu\rho}(\alpha x+(1-\alpha)x_0)(x-x_0)_\rho.$$

It is clear that in this gauge the potential $A^a_\mu(x)$ depends significantly on the choice of the point x_0 and that the gauge condition violates translation invariance (hence the name fixed-point gauge). It is convenient to choose the origin at the point x_0 and write

$$A^a_\mu(x)=\int_0^1 \alpha d\alpha G^a_{\rho\mu}(\alpha x)x_\rho. \tag{2}$$

For our purposes another representation of $A^a_\mu(x)$ is more convenient:

$$A_\mu(x)={}^1/_2 x_\rho G_{\rho\mu}(0)+{}^1/_3 x_\alpha x_\rho D_\alpha G_{\rho\mu}(0)+\frac{1}{4\cdot 2!}x_\alpha x_\beta x_\rho D_\alpha D_\beta G_{\rho\mu}(0)+\dots \tag{3}$$

A proof of equation (2) is given in Refs. 3 and 5–7, and relation (3) is obtained in Ref. 8 where power corrections to the vacuum average for the Wilson loop have been calculated in the fixed-point gauge:

$$W(C)=\left\langle \exp\left[i\oint_C A_\mu(\xi)d\xi_\mu\right]\right\rangle_0,\quad A_\mu(\xi)=A^a_\mu(\xi)t^a.$$

In equation (3) the color index a has been omitted the following notation has been used

$$(D_\alpha R)^a=\partial_\alpha R^a-gf^{abc}A^b_\alpha R^c.$$

Let us now describe the procedure for calculating power corrections to the polarization operators in the fixed-point gauge. Let us consider, for instance, the polarization operator of vector currents in the external

FIG. 1.

field of some specific gluon vacuum fluctuation $A^a_\mu(x)$. It is described by the infinite set of diagrams which have been presented schematically in Fig. 1. The first term of the set describes the contribution of lowest-order perturbation theory, the second term corresponds to scattering once on the vacuum fluctuation, the third term to scattering twice, etc. (for simplicity all corrections of ordinary perturbation theory have been neglected; to take them into account does not pose a principal difficulty).

Every external field line has to be contracted with the Fourier transform of the potential $A^a_\mu(q)$, which we will write in the fixed-point gauge. Using equation (3) we get

$$A^a_\mu(q)=-\frac{i(2\pi)^4}{2}G^a_{\rho\mu}(0)\frac{\partial}{\partial q_\rho}\delta(q)-\frac{(2\pi)^4}{3}(D_\alpha G_{\rho\mu}(0))^a\frac{\partial^2}{\partial q_\rho\partial q_\alpha}\delta(q)+\dots \tag{4}$$

The potential $A^a_\mu(q)$ in this form has to be substituted into the Feynman integrals which correspond to the diagrams of Fig. 1 and one has to average over all possible vacuum fluctuations. From the symmetry of the wave function of the vacuum it follows immediately that a diagram with one vacuum gluon leg gives zero upon averaging since $\langle G_{\rho\mu}\rangle_0=\langle D_{\alpha_1}\cdots D_{\alpha_n}G_{\rho\mu}\rangle_0=0$.

Diagrams with two vacuum legs do not go to zero since they contain the terms

$$\sim\langle G^a_{\mu\nu}G^b_{\alpha\beta}\rangle_0={}^1/_{96}\delta^{ab}(g_{\mu\alpha}g_{\nu\beta}-g_{\mu\beta}g_{\nu\alpha})\langle G^a_{\mu\nu}G^a_{\mu\nu}\rangle_0 \tag{5}$$

(the tensor structure is determined from symmetry considerations). Terms of higher order also appear, proportional to averages of the type $\langle DGDG\rangle_0$, $\langle GDDG\rangle_0$, etc. (neglecting terms of higher order in α_s, one can show that the vacuum averages of all gluonic operators of dimension 6 can be expressed in terms of the average of the operator $f^{abc}G^a_{\mu\nu}G^b_{\nu\sigma}G^c_{\sigma\mu}$).[9]

Now the algorithm for calculating the coefficients of the structures $\sim\langle G^2\rangle_0$, $\langle fG^3\rangle_0$, etc. in the operator expansion is clear. To find the coefficient of the operator G^2 one should taken the diagrams in Fig. 1 with two gluon legs, use the expression for the external field in the form (4), restricting onself to the first term of the expansion, take the average according to the rule (5), and calculate the corresponding Feynman integrals (here one should make sure that large virtual momenta flow through all internal lines in a diagram; see the discussion in Sec. 3). To find the coefficient of the operator fG^3 one should also keep the next terms of the expansion in (4) in the diagrams with two gluon legs and select the structures $\sim(DG)(DG)$, $G(DDG)$, and $(DDG)G$ in the product $A^a_\mu(q)A^b_\nu(q')$. It is also necessary to calculate the diagrams with three gluon legs, in which case it is sufficient to restrict oneself to the first term of the expansion in (4). The higher the dimension of the operator of interest, the larger the number of legs of the diagrams one has to take into account and/or the more terms in the expansion for the potential $A^a_\mu(q)$ of the external field one has to take into account.

This method of calculating power corrections is close in spirit to the technique of Schwinger in an external field used in Ref. 9. There and here the gauge-invariant structures $\sim G^2$, fG^3, etc. emerge automatically and do not have to be constructed by hand as one has to do in the plane-wave method which was originally proposed.[1] In our view, however, the fixed-point gauge technique has the advantage compared to the Schwinger technique that it contains a simple formal algorithm for calculating the coefficients of operators of any dimension. It is, for instance, easy to use this algorithm for analytic computer calculations.

3. CALCULATION OF THE COEFFICIENT OF THE OPERATOR G^2 IN THE POLARIZATION OPERATOR OF VECTOR CURRENTS

In this section we will present the results of the calculation by our method of power corrections to the polarization operator of vector currents. The results of the calculation coincide with the results of Ref. 1: a quantity which makes sense physically, as the coefficient $c_G(q^2, m^2)$ undoubtedly does, is of course independent of the way it is calculated. Consequently, this section has an illustrative character.

According to the algorithm which we have described in the previous section, to calculate the coefficient of the operator $G^a_{\mu\nu}G^a_{\mu\nu}$ in lowest order in α_s we have to calculate the diagrams of Figs. 2a, b, and c, while the potential of the external field should be taken in the form

$$A^a_\alpha(u) = -\frac{i(2\pi)^4}{2}G^a_{\rho\alpha}\frac{\partial}{\partial u_\rho}\delta(u),$$

$$A^b_\beta(v) = -\frac{i(2\pi)^4}{2}G^b_{\sigma\beta}\frac{\partial}{\partial v_\sigma}\delta(v).$$

Taking the average according to the rule (5) and performing the integration with respect to $d^4u\, d^4v/(2\pi)^8$, we get

$$[\Pi_{\mu\nu}(q)]_{\text{Fig. 2a}} = -\frac{ig^2\langle G^a_{\mu\nu}G^a_{\mu\nu}\rangle_0}{96}(g_{\alpha\beta}g_{\rho\sigma} - g_{\alpha\sigma}g_{\beta\rho})\int\frac{d^4p}{(2\pi)^4}\frac{\partial^2}{\partial u_\rho\,\partial v_\sigma}$$
$$\times \mathrm{Sp}\left\{\gamma^\mu\frac{1}{\hat p - m}\gamma^\beta\frac{1}{\hat p + \hat v - m}\gamma^\nu\frac{1}{\hat p - \hat u - m}\gamma^\alpha\frac{1}{\hat p' - m}\right\}\Big|_{u=v=0}, \qquad (6a)$$

$$[\Pi_{\mu\nu}(q)]_{\text{Fig. 2b, c}} = -\frac{ig^2\langle G^a_{\mu\nu}G^a_{\mu\nu}\rangle_0}{96}(g_{\alpha\beta}g_{\rho\sigma} - g_{\alpha\sigma}g_{\beta\rho})\int\frac{d^4p}{(2\pi)^4}\frac{\partial^2}{\partial u_\rho\,\partial v_\sigma}$$
$$\times \mathrm{Sp}\left\{\gamma^\mu\frac{1}{\hat p - m}\gamma^\beta\frac{1}{\hat p + \hat v - m}\gamma^\alpha\frac{1}{\hat p + \hat u + \hat v - m}\gamma^\nu\frac{1}{\hat p' - m}\right\}\Big|_{u=v=0}, \qquad (6b)$$

where $p' = p + q$. We note here the small delicate point that formally in the vertex connected with the current j_ν in the diagrams of Fig. 2 momentum is not conserved This is due to the fact that the fixed-point gauge violates translation invariance. Generally speaking, we therefore have to put $q' \neq q$ and calculate

$$\Pi_{\mu\nu}(q, q') = \Pi_{\mu\nu}(q)\delta(q - q') + A_{\mu\nu}(q)\delta'(q - q') + \ldots$$

It is not difficult to convince oneself, however, that in a gauge-invariant quantity like the polarization operator all translation non-invariant terms $A_{\mu\nu}(q)$ etc. go to zero.

Contracting the indices μ and ν in expression (6) [we do not lose any information in that case since $\Pi_{\mu\nu}(q) \sim (q_\mu q_\nu - g_{\mu\nu}q^2)$ because of conservation of the vector current] and calculating the traces, we get

$$[\Pi_{\mu\mu}(q^2)]_{\text{Fig. 2a}} = -ig^2\langle G^2\rangle_0\int\frac{d^4p}{(2\pi)^4}\frac{pp'}{(p^2 - m^2)^2(p'^2 - m^2)^2}, \qquad (7a)$$

$$[\Pi_{\mu\mu}(q^2)]_{\text{Fig. 2b+Fig. 2c}} = -4im^2g^2\langle G^2\rangle_0\int\frac{d^4p}{(2\pi)^4}\frac{pp' - 2p^2}{(p^2 - m^2)^4(p'^2 - m^2)}. \qquad (7b)$$

We see that the second Feynman integral, which corresponds to the diagram in which the two gluon lines are emitted from a single quark line, has a factor m^2. In the limit $m \to 0$, however, expression (7b) does not go to zero since in this limit there occurs a strong infrared divergence in the integral with respect to d^4p and the integral itself goes as $\sim 1/m^2$ for $m \to 0$.

It is not correct, however, to add the asymptotic expressions of (7a) and (7b) for $m \to 0$ and to say that this result is the contribution of the structure $\langle G^a_{\mu\nu}G^a_{\mu\nu}\rangle_0$ to the polarization operator of massless quarks. It is only legitimate to calculate the coefficient functions by Feynman diagrams in the operator-expansion method if large virtual momenta flow through all internal lines in a diagram. In the diagrams of Figs. 2b and c the integrals are determined by the region $p^2 \sim m^2$ and one therefore has to say that these diagrams do not give a contribution to the coefficient of the operator $G^a_{\mu\nu}G^a_{\mu\nu}$ in the polarization operator of massless quarks. For massless quarks on the other hand the graph with an opened quark loop (Fig. 3) gives a contribution.

The situation here is completely analogous to the calculation of the coefficient $c_G(q^2, m^2)$ by the plane wave method,[1] where in the transition to the limit $m \to 0$ in determining the coefficient $c_G(q^2)$ for massless quarks the contribution to the Feynman integral from the region of small virtual momenta was also subtracted. The advantage of our method is that here all garbage is swept into one corner, as it were: the diagrams of Figs. 2b and c. In the plane-wave method there exists a region of small virtual momenta in all graphs and the separation is not so clear.

An elementary calculation of the integral (7a) in the limit $m \to 0$ gives a simple answer which coincides with the result in Ref. 1:

$$[c_G(Q^2)]_{m=0} = 1/12\pi Q^4, \quad Q^2 = -q^2,$$

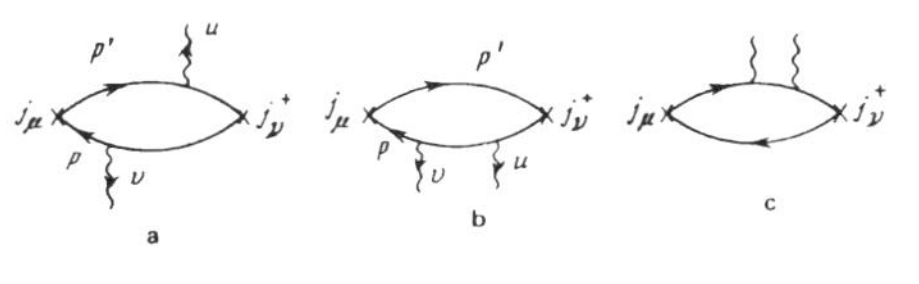

FIG. 2.

FIG. 3.

where the coefficient $[c_G(Q^2)]_{m=0} \neq c_G(Q^2, 0)$ (!) is defined according to

$$\Pi_{\mu\nu}(q) = (q_\mu q_\nu - g_{\mu\nu} q^2)[c_I + c_G \langle \alpha_s G^a_{\mu\nu} G^a_{\mu\nu} \rangle_0 + \ldots]. \quad (8)$$

Here α_s is included in the definition of the vacuum average to guarantee remornalization-group invariance.

If the mass of the quark is large ($m \gg \mu$, where μ is the characteristic hadronic scale), the contribution of expression (7b) has to be included. The calculation gives

$$c_G(Q^2, m^2) = \frac{1}{12\pi Q^2} \int_0^1 dx \left\{ \frac{1}{3m^2} + \frac{2m^2 x(1-x)}{[Q^2 x(1-x)+m^2]^2} - \frac{m^2 Q^2}{3[Q^2 x(1-x)+m^2]^3} - \frac{m^4[1+6x(1-x)]}{3[Q^2 x(1-x)+m^2]^3} \right\}$$
$$= \frac{1}{48\pi Q^4} \left\{ \frac{3(a+1)(a-1)^2}{a^2} \frac{1}{2\sqrt{a}} \ln \frac{1+\sqrt{a}}{\sqrt{a}-1} - \frac{3a^2-2a+3}{a^2} \right\}, \quad (9)$$

where $a = 1 + 4m^2/Q^2$. The intermediate integral with respect to the Feynman parameter x is given mainly in order that the interested reader can compare it with the corresponding integral in Ref. 1 [formula (4.6) on page 405] and can convince himself that the integrands have nothing in common. After the integration with respect to x the expressions for the coefficients coincide, which was of course expected.

4. THE CALCULATION OF THE CORRECTIONS $\sim\langle G^2\rangle_0$ TO THE POLARIZATION OPERATOR OF BARYON CURRENTS

In this section we will calculate the coefficients of the operator $G_{\mu\nu}G_{\mu\nu}$ in the polarization operators of baryon currents with the quantum numbers of the nucleon and the Δ isobar. The polarization operators of baryon currents have been studied previously in Ref. 4, where, in addition to the leading graph of perturbation theory, Fig. 4a, power corrections due to the effect of the quark condensate, Figs. 4b and c, have been taken into account. By means of the technique of Borel-transformed sum rules,[1] which connect the polarization operator in the Euclidean region with observable quantities, the author of Ref. 4 succeeded in obtaining impressive predictions for the mass of the nucleon and the Δ isobar: $m_N \approx 0.9$ GeV and $m_\Delta \approx 1.3$ GeV.

Here we will calculate the power corrections to the polarization operators of baryon currents described by the diagrams of Fig. 5 [diagrams in which the two gluon legs are emitted from the same quark line are not taken into account, since the power correction $\sim\langle G^2\rangle_0$ to the Green function of a massless quark goes to zero: compare the factor $\sim m^2$ of the integral (7b)]. In Ref. 4 arguments were given that the contribution of these diagrams should be suppressed compared to the contribution of the diagrams of Fig. 4 which were calculated earlier (the arguments come down to the point that the diagrams of Fig. 5 contain a larger number of loops, and for every loop a factor $\sim 1/4\pi^2$ appears in the Feynman integral). Our concrete calculation confirms this hypothesis; it turns out that the calculation of the correction $\sim\langle G^2\rangle_0$ shifts the predictions for the masses by a quantity of of the order of a per cent, and that they have practically no influence on the analysis performed in Ref. 4.

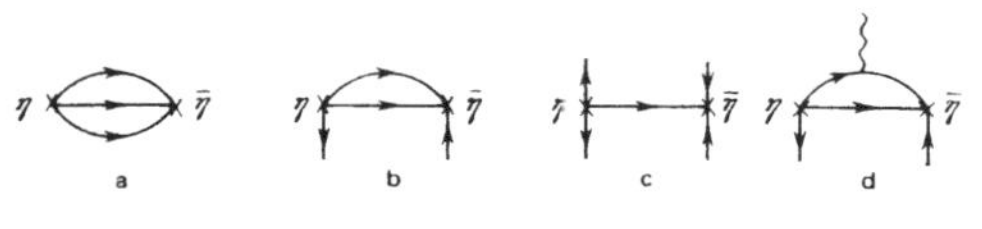

FIG. 4.

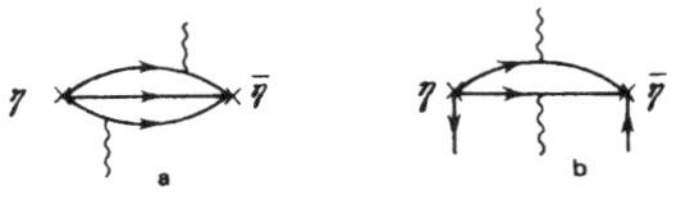

FIG. 5.

Let us recall the main ideas of Ref. 4. We consider the polarization operator

$$\hat{\Pi}(q) = i \int d^4x e^{iqx} \langle T\{\eta(x)\bar{\eta}(0)\}\rangle_0,$$

where

$$\eta(x) = \varepsilon^{abc}[u^a(x) C\gamma_\mu u^b(x)]\gamma^5\gamma^\mu d^c(x)$$

is the baryon current with the quantum numbers of the proton, and C is the Dirac charge-conjugation matrix. [The quantum numbers $T = \frac{1}{2}$ and $J^P = \frac{1}{2}^+$ do not uniquely define the structure of the current $\eta(x)$. Arguments in favor of choosing precisely this structure are given in Ref. 4.] We calculate the polarization operator in the Euclidean region $q^2 = -Q^2 < 0$. If Q^2 is sufficiently large, one can restrict oneself to calculating the leading graph of perturbation theory Fig. 4a, and also the first power corrections Fig. 4b–d and Fig. 5.

We select the coefficients of the structures $\hat{q}$ and 1 (the structure $\hat{q}$ corresponds to the diagrams of Figs. 4a, 4c, and 5a, and the structure 1 to the diagrams of Figs. 4b, 4d, and Fig. 5b). We now write a dispersion relation for these coefficients. In the right-hand part of the dispersion relation enter the contributions of the low-lying resonances (in this case the nucleon and its radial excitations) and the continuum which starts for instance at an energy $\sqrt{q^2} = 2$ GeV (at energies larger than 2 GeV the imaginary part of the polarization operator is assumed to coincide with the imaginary part of the simplest perturbative graph, Fig. 4a). Now we apply on both sides of the equation the Borel transformation

$$\hat{L}_M f(q^2) = \frac{1}{\pi} \int_0^\infty e^{-s/M^2} \operatorname{Im} f(s)\, ds. \quad (10)$$

The Borel transformation is convenient since it effectively enhances the contribution of the lowest-lying resonance compared to the contributions of the radially excited states and the continuum. The smaller the parameter M, the larger is the contribution of the ground state in the right-hand part and, on the other hand, the larger is the role which the power corrections from the lowest-dimensional operators play in the left-hand part.

It turns out that there exists a region of M in which one can restrict oneself to the leading power corrections in the left-hand part, and at the same time the right-hand part will be mainly saturated by a single resonance.

In this way, the sum rules make it possible to extract nontrivial information on the properties of the lightest resonance with given quantum numbers; one can for instance find its mass.

The simplest way to determine the mass of the nucleon is to put $M=m_N$. Then, the sum rules for the coefficients of the structures $\hat{q}$ and 1 give two equations with two unknowns: the mass of the nucleon and its coupling λ_N to the baryon current $\eta(x)$ (Ref. 4):

$$m_N^6+\frac{4}{3}a^2=2(2\pi)^4\lambda_N^2/e,\qquad 2am_N^3=2(2\pi)^4\lambda_N^2/e. \tag{11}$$

[Here only contributions from the lowest-dimensional operators, calculated by the graphs of Fig. 4, have been taken into account; $a=-(2\pi)^2\langle\bar{\psi}\psi\rangle_0\approx 0.55\ \mathrm{GeV}^3$]. In principle, a contribution from the vacuum average of the five-dimensional operator $ig\bar{\psi}\sigma_{\mu\nu}t^a\psi G^a_{\mu\nu}$ could appear in the second equation of the system (11). The calculation shows, however, that the coefficient of this operator is equal to zero in this case.

It can be seen that the system (11) leads to a quadratic equation with negative discriminant, i.e., it has no solution. However, the minimum of this quadratic trinomial is relatively close to zero and, as we will see later, taking account of the terms $\sim\langle G^2\rangle_0$ results in the appearance of a root of the equation. Without taking these terms into account the best estimate for the mass is $m_N=\sqrt[3]{a}\approx 0.82$ GeV. Taking account of the anomalous dimensions, i.e., of the dependence of the vacuum average $\langle\bar{\psi}\psi\rangle_0$ on the normalization point, increases the effective value of a, and the mass of the nucleon becomes almost exactly equal to the experimental value.

We will now calculate the power corrections $\sim\langle G^2\rangle_0$. Let us consider in more detail the technical side of the calculations. As in the previous section, we will use the fixed-point gauge in which $A^a_\mu(x)=\frac{1}{2}x_\rho G^a_{\mu\rho}(0)+\ldots$. For the problem in question it is convenient not to go over to the momentum representation, but to calculate the Feynman diagrams directly in the coordinate representation. In this case it is necessary to perform the integration with respect to the virtual momenta in the loops of the graphs in Figs. 5a and b, and the coordinate Green function of a massless fermion has the extremely simple form

$$\langle T\{\psi_i^a(x)\bar{\psi}_j^b(0)\}\rangle_{\rm pert.}=-\frac{i\delta^{ab}(\hat{x})_{ij}}{2\pi^2x^4} \tag{12}$$

(for massive particles, as is well known, the expression is more complicated and includes Bessel functions).

If a line of a vacuum gluon field (in the fixed-point gauge) is attached to the quark line, then it is easy to show that

$$\langle T\{\psi_i^a(x)\bar{\psi}_j^b(0)\}\rangle_0(G)=\frac{g(t^c)^{ab}(\hat{x}\sigma_{\alpha\beta}+\sigma_{\alpha\beta}\hat{x})_{ij}}{32\pi^2x^2}G^c_{\alpha\beta}+\ldots \tag{13}$$

[here $\sigma_{\alpha\beta}$ is defined as $\frac{1}{2}(\gamma_\alpha\gamma_\beta-\gamma_\beta\gamma_\alpha)$; all calculations are performed in Euclidean space]. If the quark line goes directly into the vacuum, the Green function takes the form

$$\langle T\{\psi_i^a(x)\bar{\psi}_j^b(0)\}\rangle_0(\psi)=-{}^1/{}_{12}\delta^{ab}\delta_{ij}\langle\bar{\psi}\psi\rangle_0. \tag{14}$$

Now, to calculate the contribution of the diagram in Fig. 5a for instance, we have to contract the fermion fields in the expression for the polarization operator $\langle T\{\eta(x)\bar{\eta}(0)\}\rangle_0$ in all possible ways [two pairs according to the rule (13) and one pair according to the rule (12)], take the vacuum average according to the rule (5), go over to the momentum representation by a Fourier transformation, and return to Minkowsky space. The fiarly simple calculations lead to the answer

$$[\hat{\Pi}_N(q)]_{\rm Fig.\ 5a}=-\frac{\langle\alpha_s G^a_{\mu\nu}G^a_{\mu\nu}\rangle_0\hat{q}\ln(\Lambda^2/-q^2)}{32\pi^3}, \tag{15a}$$

$$[\hat{\Pi}_N(q)]_{\rm Fig.\ 5b}=-\frac{\langle\alpha_s G^a_{\mu\nu}G^a_{\mu\nu}\rangle_0\langle\bar{d}d\rangle_0}{24\pi(q^2+i0)}, \tag{15b}$$

where Λ is the ultraviolet-cutoff parameter.

Not only the diagram of Fig. 5b, however, gives a contribution to the coefficient of the operator $\bar{\psi}\psi G^a_{\mu\nu}G^a_{\mu\nu}$, but also the old diagrams of Figs. 4b and d [one must take into account the next terms of the expansion in x in expression (14)]. The calculation of this coefficient is, however, primarily of academic interest; apart from the operator $\bar{\psi}\psi G^a_{\mu\nu}G^a_{\mu\nu}$ there exist also other operators of the same dimension of the type $f^{abc}\bar{\psi}\sigma_{\mu\nu}t^a\times\psi G^b_{\nu\sigma}G^c_{\sigma\mu}$ whose vacuum average introduces new unknown parameters. The calculation of the coefficient of $\langle\bar{\psi}\psi\rangle_0\langle G^2\rangle_0$ is, however, not without interest from a technical point of view and is described in detail in the Appendix. We will not take contributions from six-dimensional operators into account in the sum rules.

The same program can be performed for the polarization operator $\langle T\{\eta_\mu(x)\bar{\eta}_\nu(0)\}\rangle_0$ for the current

$$\eta_\mu(x)=\varepsilon^{abc}[u^a(x)C\gamma_\mu u^b(x)]u^c(x)$$

with the quantum numbers of the Δ isobar. The answer has the form

$$[\hat{\Pi}^\Delta_{\mu\nu}]_{\rm Fig.\ 5a}=\frac{5\langle\alpha_s G^a_{\mu\nu}G^a_{\mu\nu}\rangle_0}{9\cdot 32\pi^3}\Big\{\ln\frac{\Lambda^2}{-q^2}\Big[\hat{q}g_{\mu\nu}-\frac{3}{8}\gamma^\mu\gamma^\nu\hat{q}+\frac{3}{8}(q^\nu\gamma^\mu-q^\mu\gamma^\nu)-\frac{1}{8}(q^\nu\gamma^\mu+q^\mu\gamma^\nu)\Big]-q_\mu q_\nu\hat{q}/5(q^2+i0)\Big\}. \tag{16}$$

Let us now examine what the effect of the power corrections (15) and (16) is on the estimate of the mass of the nucleon and the Δ isobar. The sum rules for the structures $\hat{q}$ and 1 in the polarization operator for the nucleon current at $M=m_N$ now take the form

$$m_N^6+4a^2/3-\pi^2m_N^2\Big\langle\frac{\alpha_s}{\pi}G^2\Big\rangle_0=\frac{2(2\pi)^4}{e}\lambda_N^2,\qquad 2am_N^4=2(2\pi)^4\lambda_N^2/e. \tag{17}$$

The equation for the mass has the form

$$(m_N^3-a)^2+a^2/3-\pi^2m_N^2\Big\langle\frac{\alpha_s}{\pi}G^2\Big\rangle_0=0. \tag{18}$$

It can be seen that if $\langle\alpha_s/\pi G^2\rangle_0\gtrsim a^{4/3}/\pi^2\approx 0.015\ \mathrm{GeV}^4$ the equation has a root which does practically not differ from the naive estimate $m_N^3=a$. (In Ref. 1 the estimate $\langle\alpha_s/\pi G^2\rangle_0\approx 0.012\ \mathrm{GeV}^4$ was obtained. It is possibly somewhat underestimated since large Coulombic corrections to the high moments of the polarization operator of c-quark vector currents have not been taken into account. We think that the estimate $\langle\alpha_s/\pi G^2\rangle_0\approx 0.017\ \mathrm{GeV}^4$, obtained from a QCD sum-rule analysis for the Υ resonance,[11] is more reliable.) Thus, the power corrections $\sim\langle G^2\rangle_0$ improve the consistency of the sum rules, but have no effect on the estimate for the resonance mass. Our estimate for the nucleon coupling constant

$$\lambda_N^2\approx 1.05\cdot 10^{-3}\ \mathrm{GeV}^6 \tag{19}$$

differs also very little from the estimate of Ref. 4.

Similar calculations can be carried out for the Δ isobar. The sum rules for the coefficients of the structures $g_{\mu\nu}\hat{q}$ and $g_{\mu\nu}$ take the form

$$\begin{gathered} m_\Delta^6+20a^2/3-25\pi^2 m_\Delta^2\left\langle\frac{\alpha_s}{\pi}G^2\right\rangle_0\Big/18=\frac{5(2\pi)^4}{e}\lambda_\Delta^2,\\ \frac{20am_\Delta^4}{3}\left(1-\frac{m_0^2}{2m_\Delta^2}\right)=5(2\pi)^4\lambda_\Delta^2 m_\Delta/e, \end{gathered} \tag{20}$$

where m_0 is a parameter connected with the vacuum average

$$\langle ig\bar\psi\sigma_{\mu\nu}t^a\psi G^a_{\mu\nu}\rangle_0=m_0^2\langle\bar\psi\psi\rangle_0.$$

From an analysis of the sum rules for D mesons the estimate $m_0^2\gtrsim 0.5\ \mathrm{GeV}^2$ (Ref. 10) follows. Our terms $\sim\langle G^2\rangle_0$ shift the estimate for m_Δ upwards by about 10–15 MeV. A one per cent correction to the mass is far less than the uncertainties connected with the unknown parameter m_0 and with the unknown structure of the imaginary part of the polarization operator in the energy region from 1 to 2 GeV.

5. DISCUSSION AND CONCLUSIONS

In conclusion we can say that in our opinion this method for calculating power corrections is simpler and more clear than the methods used previously. We directly calculate the polarization operators in an external vacuum field of a general form. The fixed-point gauge (3) makes it possible to present them in the form of a series of gauge-invariant quantities which correspond to the vacuum averages of the gluonic operators O_n in the Wilson expansion, by means of a minimal number of technical operations which give a well defined algorithm. If a diagram we are interested in contains massless quark loops, it is convenient to perform the calculations directly in the coordinate representation, and go over to the momentum representation only at the very last stage, as was done in Sec. 4.

In this paper two examples of calculations in the fixed-point gauge were considered: an instructure one in which an old result was reproduced by a new method (section 3), and an original calculation in which the coefficients of the structure $\langle G^a_{\mu\nu}G^a_{\mu\nu}\rangle_0$ in the polarization operators of baryon currents were calculated. This calculation did not give any essentially new physical information. However, the investigation carried out in section 4 of the stability of the predictions of Ref. 4 for the baryon masses, with respect to the effects of higher orders which were unaccounted for, seems fairly useful to us.

The advantage of our method should manifest itself especially clearly in calculating the coefficients of higher-dimensional operators. Thus, in the fixed-point gauge the coefficients of the operators $\sim G^3$ and $\sim G^4$ in the Wilson loop were found.[8] Perhaps, the algorithm proposed in Sec. 2 is the only sensible way to find the coefficient of the operator $f^{abc}G^a_{\mu\nu}G^b_{\nu\sigma}G^c_{\sigma\mu}$ in the polarization operator of heavy quarks.

In conclusion I gladly express my gratitude to M. S. Dubvikov and B. L. Ioffe for fruitful discussions on a series of questions touched upon in this paper.

APPENDIX

Here we will describe the technique of calculating the coefficients of operators with a mixed quark-gluon character of the type $\bar\psi\psi G^2$ by the example of the polarization operator of nucleon currents. Also here, the calculations are most conveniently carried out in the fixed-point gauge. One can show that in the fixed-point gauge the following relation is satisfied:

$$\psi(x)=\psi(0)+x_\alpha\nabla_\alpha\psi(0)+{}^1/_2x_\alpha x_\beta\nabla_\alpha\nabla_\beta\psi(0)+\dots \tag{A.1}$$

Equation (A.1) differs from an ordinary Taylor series for $\psi(x)$ in that here the ordinary derivatives have been replaced by covariant ones (in Euclidean space $\nabla_\alpha=\partial_\alpha+igA^a_\alpha t^a$). The result (A.1) is completely analogous to the representation (3) for $A^a_\mu(x)$ given in Ref. 8.

To find the contribution of the diagram in Fig. 4b to the coefficient of $\langle\psi\psi\rangle_0\langle\alpha_s/\pi G^2\rangle_0$, terms up to x^4 in the expansion in x must be taken into account in the expression for $\langle T\{\psi^a_i(x)\psi^b_j(0)\}\rangle_0(\mathrm{vac})$.

Substituting (A.1) into (14) and using the property $[\nabla_\alpha,\nabla_\beta]=igG^a_{\alpha\beta}t^a$ and the equations of motion $\hat\nabla\psi=0$, it is not difficult to arrive at the answer

$$\begin{gathered}\langle T\{\psi^a_i(x)\bar\psi^b_j(0)\}\rangle_0(\mathrm{vac})=-\frac{1}{12}\delta^{ab}\delta_{ij}\langle\bar\psi\psi\rangle_0\\ \times\left[1+\frac{m_0^2x^2}{16}+\frac{\pi^2x^4}{288}\left\langle\frac{\alpha_s}{\pi}G^2\right\rangle_0+\dots\right]\end{gathered} \tag{A.2}$$

(here we have included the contributions of the other six-dimensional operators besides $\bar\psi\psi G^a_{\mu\nu}G^a_{\mu\nu}$). The required contribution is then equal to

$$[\hat\Pi_N(q)]_{\mathrm{Fig.\,4b}}(\bar\psi\psi G^2)=\frac{\langle\alpha_sG^a_{\mu\nu}G^a_{\mu\nu}\rangle_0\langle\bar dd\rangle_0}{36\pi(q^2+io)}. \tag{A.3}$$

The diagram of Fig. 4d also makes a contribution to the coefficient of $\bar\psi\psi G^2$ in $\hat\Pi_N(q)$. In complete analogy to (A.2) one can get

$$\begin{gathered}ig\langle T\{\bar\psi(0)\sigma_{\mu\nu}t^a\psi(x)G^a_{\mu\nu}(0)\}\rangle_0=\langle ig\bar\psi\sigma_{\mu\nu}t^a\psi G^a_{\mu\nu}\rangle_0\\ -\frac{\pi^2\langle\bar\psi\psi\rangle_0\left\langle\frac{\alpha_s}{\pi}G^2\right\rangle_0x^2}{192}\end{gathered} \tag{A.4}$$

and the corresponding contribution to $\hat\Pi_N(q)$ is

$$[\hat\Pi_N(q)]_{\mathrm{Fig.\,4d}}(\bar\psi\psi G^2)=-\frac{\langle\alpha_sG^a_{\mu\nu}G^a_{\mu\nu}\rangle_0\langle\bar dd\rangle_0}{384\pi(q^2+io)}. \tag{A.5}$$

Summing the contributions (A.3), (A.5), and (15b), we get

$$[\hat\Pi_N(q)](\bar\psi\psi G^2)=-\frac{19\langle\alpha_sG^a_{\mu\nu}G^a_{\mu\nu}\rangle_0\langle\bar dd\rangle_0}{9\cdot 2^7\pi(q^2+io)}. \tag{A.6}$$

Expression (A.6) can be used to estimate the contribution of seven-dimensional operators to the sum rules. Numerically it turns out to be extremely small and can be completely neglected in all physical applications.

[1]M. A. Shifman, A. I. Vainshtein, and V. I. Zacharov, Nucl. Phys. **B147**, 385, 448 (1979).
[2]A. Belavin *et al.*, Phys. Lett. **59B**, 85 (1975).
[3]M. S. Dubovikov and A. V. Smilga, Nucl. Phys. **B185**, 109 (1981).
[4]B. L. Ioffe, Nucl. Phys. **B188**, 317 (1981).
[5]J. Schwinger, Particles, Sources and Fields, vol I, § 11, Addison Wesley (1970) [Russian Transl., Mir (1973)].
[6]V. A. Fateev, A. S. Schwartz, and Yu. S. Tyupkin, Preprint Lebedev Phys. Inst. 155 (1976).
[7]C. Cronström, Phys. Lett. **90B**, 267 (1980).
[8]M. A. Shifman, Nucl. Phys. **B173**, 13 (1980).
[9]V. A. Novikov *et al.*, Preprint ITEP-27 (1980).
[10]V. A. Novikov *et al.*, Proc. Intern. Conf. Neutrino-78, Lafayette, page C278 (1978).
[11]M. B. Voloshin, Preprint ITEP-21 (1980).

Translated by L. J. Reinders

Nuclear Physics B207 (1982) 29–42

OPERATOR EXPANSION AT SHORT DISTANCE IN QCD

W. HUBSCHMID[1]

Institut für Theoretische Physik, Universität Bern, Sidlerstrasse 5, CH-3012 Bern, Switzerland

S. MALLIK[2]

Institut für Theoretische Kernphysik, Universität Karlsruhe, Kaiserstrasse 12, D-7500 Karlsruhe, Germany

Received 2 June 1982
(Corrected version received 23 July 1982)

We present a method of calculating coefficients of gluon operators in the operator product expansion of two-point functions at short distance. It is based on a short-distance expansion of the singular part of the quark propagator in the gluon field, the latter being treated as external We verify in full generality that the spin zero, gluon operator of dimension six does not contribute to the two-point functions of quark bilinears.

1. Introduction

Recently there has been a renewed interest in the operator product expansion (OPE) at short distance [1]. The OPE forms the basis of a remarkable type of sum rules in QCD derived by Shifman, Vainshtein and Zakharov [2]. Here one considers the OPE of two-point functions of composite operators. Because of non-perturbative vacuum fluctuations, the vacuum expectation values (v.e.v.s) of the higher dimensional operators are non-zero in QCD theory. The sum rules thus provide a means of studying the effect of the non-perturbative vacuum on the observed hadron spectrum, for example.

The OPE has been proven only in perturbation theory [3]. In the non-perturbative domain, where the sum rules assume the validity of OPE, one can at present only hope to check it in the context of specific soluble examples, for instance, the two-point functions in the background of an instanton [4, 5] or in an (anti-) self-dual constant abelian field [6, 7]. Indeed, the former example was discussed in ref. [8] where the authors noted the absence of terms with inverse power of Q^2 beyond $(Q^2)^{-2}$ in the expression for the vector polarisation operator, Q being the momentum carried by the vector current. This suggests definite relationships among coefficients of higher dimensional operators; in particular, the coefficient of the 3-gluon operator should vanish. Subsequently, this observation was extended to the polarisation operator in the background of any (anti-) self-dual field by showing

[1] Supported by Schweizerischer Nationalfonds.
[2] Supported by Deutsche Forschungsgemeinschaft.

the absence of the singularity structure required to give rise to higher inverse powers of Q^2 in the momentum space [5].

To carry out such verifications as well as for the more practical purpose of writing down the QCD sum rules, the first step consists in calculating the coefficients of higher dimensional operators appearing in the short-distance OPE. The original method [2] based on conventional Feynman diagrams does not seem to be the most convenient one for this purpose: It is not only laborious but also technically complicated if the dimension of the operator is high enough. It involves integration over loop momenta and one has to separate the contributions from regions of high and low momenta, the former alone contributing to the coefficient to be calculated and the latter being absorbed in the matrix element of some other operator.

In the present work we develop a direct and simpler method of calculating the coefficients of gluon operators in the short distance OPE*. Since we aim at extracting the expansion coefficient to lowest order in g_s, the strong coupling constant, we can treat the gluon field as classical (external). We then solve the differential equation satisfied by the quark propagator in this external field as a series expansion in powers of z_μ, thus working out its short-distance expansion directly in configuration space. This in turn gives the OPE for the different two-point functions at short distance. We emphasize that large distance dynamics does not enter our calculations at any stage. It thus avoids altogether the necessity of separating contributions from regions of short and long distances, a problem inherent in any method of calculating the coefficients based on Feynman diagrams.

In sect. 2 we show that only the singular part of the quark propagator which does not depend on the boundary conditions is required for obtaining the coefficients of OPE of different two point functions. We also discuss a gauge suitable for our calculation. The short-distance expansion of the propagator is worked out in sect. 3. In sect. 4 we describe how to extract Lorentz scalar operators from the tensor operators that can be found in general. In sect. 5 we extract the coefficients of gluon operators in the OPE at short distance. In particular, we find the spin zero gluon operator of dimension 6 occur with zero coefficient in all the two-point functions. We thus verify in full generality the conjecture [8, 5] based on instanton and (anti-) self-dual background field calculations.

2. Formulation of the problem

Let $q(z)$ denote the quark field of any flavour of mass m [the SU(3) colour index being suppressed]. Its equation of motion in an external (c-number) potential $A_\mu(z)$ is

$$(-i\nabla\!\!\!/ + m)q(z) = 0\,, \qquad \nabla_\mu = \partial_\mu - iA_\mu\,, \qquad A_\mu = g_s A_\mu^a \frac{\lambda^a}{2}\,, \tag{2.1}$$

* As far as we know, the first few terms of the expansion of the quark propagator in the external gluon field appear first in ref. [9]. Recently a number of authors [10] have also realised the simplicity and efficacy of this method. We thank the referee for the later references.

λ^a being the SU(3) Gell-Mann matrices. Then the quark propagator in the external potential satisfies

$$(-i\nabla\!\!\!/ + m)S(z,0) = \delta^4(z)\,. \tag{2.2}$$

This Green function or propagator defines our field theory of quarks in the external potential in the sense that all other higher Green functions can be constructed from it.

The complete solution of (2.2) may be written as

$$S(z,0|A) = S^{\text{sing}}(z,0|A) + S^{\text{reg}}(z,0|A)\,. \tag{2.3}$$

Here S^{sing} denotes the singular piece, which can be calculated without any knowledge of the boundary conditions at large distances. (Note, however, that it may be necessary to include in S^{sing} some regular terms containing local functions, in order that S^{sing} by itself satisfies eq. (2.2).) On the other hand, the regular piece S^{reg}, satisfying the homogeneous equation, depends on such boundary conditions; in particular, it may involve the external potential in a non-local way. Clearly this decomposition is not unique.

We shall also require $S(0,z|A)$. Considering the differential equation satisfied by $S(0,z|A)$, we easily get

$$S(0,z|A) = CS^{\text{T}}(z,0|-A^{\text{T}})C^{-1}\,, \tag{2.4}$$

where C is the charge conjugation matrix $C\gamma_\mu^{\text{T}}C^{-1} = -\gamma_\mu$. Here the superscript T denotes transpose of both Dirac and colour matrices.

We are interested in the OPE of

$$\Pi_{12}(z,0) = T{:}\bar{q}(z)\Gamma_1 q(z){:}\,{:}\bar{q}(0)\Gamma_2 q(0){:} \tag{2.5}$$

as $z_\mu \to 0$. Here Γ_1 and Γ_2 are any two Dirac matrices. Note, however, that, in the limit of massless quarks, the Wilson coefficients of pure gluon operators can be non-zero only for scalar ($\Gamma_1 = \Gamma_2 = 1$), vector ($\Gamma_1 = \gamma_\mu, \Gamma_2 = \gamma_\nu$) and tensor ($\Gamma_1 = \sigma_{\alpha\beta}, \Gamma_2 = \sigma_{\mu\nu}$) amplitudes. (The cases for the pseudoscalar and axial vector amplitudes are identical to that for scalar and vector amplitudes, respectively.) The mixed amplitudes like scalar–vector ($\Gamma_1 = 1, \Gamma_2 = \gamma_\mu$), vector–tensor ($\Gamma_1 = \gamma_\mu, \Gamma_2 = \sigma_{\alpha\beta}$) etc. have odd chirality and cannot receive a contribution from pure gluon operators.

We define normal ordering of two quark fields by

$$T\{q_\alpha(z)\bar{q}_\beta(0)\} = {:}q_\alpha(z)\bar{q}_\beta(0){:} - iS^{\text{sing}}_{\alpha\beta}(z,0|A)\,. \tag{2.6}$$

Different definitions of S^{sing}, differing from each other by a regular piece, amount only to redefinitions of the normal ordering of the quark bilinear above. Now we can expand Π_{12} using the familiar Wick's theorem

$$\begin{aligned}\Pi_{12}(z,0) = {}& \text{Tr}\,\{\Gamma_1 S^{\text{sing}}(z,0|A)\Gamma_2 S^{\text{sing}}(0,z|A)\} \\ & + i\{\Gamma_2 S^{\text{sing}}(0,z|A)\Gamma_1\}_{\alpha\beta}{:}q_\beta(z)\bar{q}_\alpha(0){:} + i\{\Gamma_1 S^{\text{sing}}(z,0|A)\Gamma_2\}_{\alpha\beta}{:}q_\beta(0)\bar{q}_\alpha(z){:} \\ & + \text{term regular as } z_\mu \to 0\,. \end{aligned} \tag{2.7}$$

where Tr indicates trace over both Dirac and colour matrices. For the coefficient of gluon operators, we thus need to calculate the trace term only. The main task is thus to find a short distance expansion of S.

All our calculations are carried out in the gauge [11, 5] where

$$z^{\mu}A_{\mu}(z)=0\,. \tag{2.8}$$

In this gauge the equation defining the field strength

$$G_{\mu\nu}(z)=\partial_{\mu}A_{\nu}(z)-\partial_{\nu}A_{\mu}(z)-i[A_{\mu}(z),A_{\nu}(z)] \tag{2.9}$$

can be easily integrated to give

$$A_{\mu}(z)=\int_0^1 \mathrm{d}t\, tz^{\nu}G_{\nu\mu}(tz)\,. \tag{2.10}$$

We expand $G_{\nu\mu}$ in the integrand around $z_{\alpha}=0$ obtaining ordinary derivatives of the field strength at this point. Now from the equation defining the gauge (2.8) we see that at $z_{\alpha}=0$ the potential as well as the derivatives of the potential, symmetrized in all the indices, are zero. Then the ordinary derivatives of the field strength are equal to the corresponding covariant derivatives. We thus get an expansion of the potential in terms of the field strength and its covariant derivatives at $z_{\alpha}=0$:

$$A_{\mu}(z)=\tfrac{1}{2}z^{\alpha}G_{\alpha\mu}+\tfrac{1}{3}z^{\alpha}z^{\beta}\nabla_{\alpha}G_{\beta\mu}+\tfrac{1}{8}z^{\alpha}z^{\beta}z^{\gamma}\nabla_{\alpha}\nabla_{\beta}G_{\gamma\mu}$$
$$+\tfrac{1}{30}z^{\alpha}z^{\beta}z^{\gamma}z^{\delta}\nabla_{\alpha}\nabla_{\beta}\nabla_{\gamma}G_{\delta\mu}+\cdots\,. \tag{2.11}$$

Note that it is only the part of the covariant derivative, $\nabla_{\alpha}\nabla_{\beta}\cdots G_{\lambda\mu}$, which is symmetric in all the indices except μ, which enters this expression. This symmetrisation will be understood in all expressions below where such derivatives may occur without the symmetrizing z_{α} factors.

3. Quark propagator

We are interested in the OPE of (2.5) in the limit of zero quark mass. So we ignore the quark mass term in the Dirac equation (2.2). It is convenient to put it in a Klein–Gordon form. Let

$$S(z,0|A)=i(\partial\!\!\!/-iA\!\!\!/(z))D(z,0|A)\,. \tag{3.1}$$

Then D satisfies

$$(\Box-P_{\mu}\partial^{\mu}-Q)D(z,0|A)=\delta^4(z)\,, \tag{3.2}$$

where

$$P_{\mu}=iz^{\alpha}G_{\alpha\mu}+\tfrac{2}{3}iz^{\alpha}z^{\beta}\nabla_{\alpha}G_{\beta\mu}+\tfrac{1}{4}iz^{\alpha}z^{\beta}z^{\gamma}\nabla_{\alpha}\nabla_{\beta}G_{\gamma\mu}+\cdots\,, \tag{3.3}$$

$$Q=\tfrac{1}{2}i\gamma G\gamma+\tfrac{2}{3}iz^{\alpha}\gamma^{\mu}\gamma^{\nu}\nabla_{\alpha}G_{\mu\nu}+z^{\alpha}z^{\beta}a_{\alpha\beta}+z^{\alpha}z^{\beta}z^{\gamma}b_{\alpha\beta\gamma}+z^{\alpha}z^{\beta}z^{\gamma}z^{\delta}c_{\alpha\beta\gamma\delta}+\cdots\,, \tag{3.4}$$

with

$$a_{\alpha\beta} = \tfrac{1}{4}\gamma^{\mu}\gamma^{\nu}(G_{\alpha\mu}G_{\beta\nu} + \tfrac{3}{2}i\nabla_{\alpha}\nabla_{\beta}G_{\mu\nu}) , \tag{3.5}$$

$$c_{\alpha\beta\gamma\delta} = \gamma^{\mu}\gamma^{\nu}(\tfrac{1}{16}G_{\alpha\mu}\nabla_{\beta}\nabla_{\gamma}G_{\delta\nu} + \tfrac{1}{16}\nabla_{\alpha}\nabla_{\beta}G_{\gamma\mu}G_{\delta\nu} + \tfrac{1}{9}\nabla_{\alpha}G_{\beta\mu}\nabla_{\gamma}G_{\delta\nu}) . \tag{3.6}$$

The expression for $b_{\alpha\beta\gamma}$ will not be required in our calculations. (We employ matrix notation for quantities with two indices, e.g. $\gamma G\gamma = \gamma^{\mu}\gamma^{\nu}G_{\mu\nu}$, $zG\gamma = z^{\mu}\gamma^{\nu}G_{\mu\nu}$, etc.).

To solve (3.2) we use an ansatz for D due to Hadamard [12],

$$D(z, 0|A) = -\frac{i}{4\pi^2}\left\{\frac{1}{z^2} + V(z)\ln z^2 + W(z)\right\} , \tag{3.7}$$

where z^2 is to be taken as $z^2 - i\varepsilon$ in the singular terms. V and W are assumed to be analytic functions of z, regular in the neighbourhood of $z = 0$. Inserting it in (3.2), the elimination of $\ln z^2$ and $1/z^2$ singularities in the resulting equation requires that V and W satisfy (note that $P_{\mu}z^{\mu} = 0$)

$$4z_{\mu}\partial^{\mu}V + 4V - Q = z^2h(z) , \tag{3.8}$$

$$(\Box - P_{\mu}\partial^{\mu} - Q)V = 0 , \tag{3.9}$$

$$(\Box - P_{\mu}\partial^{\mu} - Q)W = -h(z) , \tag{3.10}$$

where $h(z)$ is an unknown function, regular in z, to be determined by the two equations for V. Thus the original problem of finding D reduces to solving the above three partial differential equations.

Writing a power series expansion for V and h,

$$V(z) = v_0 + z^{\alpha}v_{1\alpha} + z^{\alpha}z^{\beta}v_{2\alpha\beta} + z^{\alpha}z^{\beta}z^{\gamma}v_{3\alpha\beta\gamma} + z^{\alpha}z^{\beta}z^{\gamma}z^{\delta}v_{4\alpha\beta\gamma\delta} + \cdots , \tag{3.11}$$

$$h(z) = h_0 + z^{\alpha}h_{1\alpha} + z^{\alpha}z^{\beta}h_{2\alpha\beta} + \cdots , \tag{3.12}$$

the first equation, (3.8), determines the expansion coefficients in V in terms of those of Q and h. The second equation for V determines the traces (in any two indices when the number of indices exceed two) of the coefficients in V, which may be used to fix the unknown coefficients in h. With a similar expansion for W,

$$W(z) = w_0 + z^{\alpha}w_{1\alpha} + z^{\alpha}z^{\beta}w_{2\alpha\beta} + z^{\alpha}z^{\beta}z^{\gamma}w_{3\alpha\beta\gamma} + z^{\alpha}z^{\beta}z^{\gamma}z^{\delta}w_{4\alpha\beta\gamma\delta} + \cdots , \tag{3.13}$$

the equation for W, (3.10), again determines only the trace of each of the expansion coefficients, the traceless parts being left arbitrary. Because of this feature of eqs. (3.9) and (3.10), it is useful to separate the expansion coefficients $v_{i\alpha\beta\ldots}$, $w_{i\alpha\beta\ldots}$ into a part which is traceless in any two indices and a part containing the trace. It proves useful to carry out this separation also in the remaining uncontracted indices. Such a decomposition for $v_{4\alpha\beta\gamma\delta}$, for example, reads

$$\begin{aligned} z^{\alpha}z^{\beta}z^{\gamma}z^{\delta}(v_{4\alpha\beta\gamma\delta} - \tilde{v}_{4\alpha\beta\gamma\delta}) &= \tfrac{3}{4}z^2z^{\alpha}z^{\beta}v_{4}{}^{\sigma}{}_{\sigma\alpha\beta} - \tfrac{1}{16}(z^2)^2v_{4}{}^{\sigma}{}_{\sigma}{}^{\lambda}{}_{\lambda} \\ &= \tfrac{3}{4}z^2z^{\alpha}z^{\beta}\tilde{v}_{4}{}^{\sigma}{}_{\sigma\alpha\beta} + \tfrac{1}{8}(z^2)^2v_{4}{}^{\sigma}{}_{\sigma}{}^{\lambda}{}_{\lambda} , \end{aligned} \tag{3.14}$$

Here and in the following a (˜) on a tensor indicates tracelessness in any two of its (uncontracted) indices.

Having described the method of solution in general, we now restrict ourselves to calculating the expansion coefficients of V and W sufficient for evaluating coefficients of gluon operators up to dimension six in the short-distance OPE of (2.5). Then dimension counting shows that we need only calculate the series for V and W up to z^4. (The series for h will enter only up to z^2.) One of the simplifications of working with the gauge (2.8) is that the lowest dimension of the operators we have to deal with is two. Also note that there is no spin zero gluon operator of dimensions five and non-zero spin ones of this dimension can, by multiplication with other operators, only give rise to operators of dimension higher than six. Thus terms with z^3 in V and W (and z in h), containing operators of dimension five, may be ignored for our purpose.

The solution for the coefficients of V are

$$v_0 = \tfrac{1}{8} i\gamma G\gamma\,, \qquad v_{1\alpha} = \tfrac{1}{12} i\gamma^\mu\gamma^\nu \nabla_\alpha G_{\mu\nu}\,,$$

$$v_{2\alpha\beta} = \tfrac{1}{12}\tilde{a}_{\alpha\beta} - \tfrac{1}{128} g_{\alpha\beta}(\gamma G\gamma)^2\,, \qquad \tilde{v}_{4\alpha\beta\gamma\delta} = \tfrac{1}{20}\tilde{c}_{\alpha\beta\gamma\delta}\,, \tag{3.15}$$

$$v_4{}^\sigma{}_{\sigma\alpha\beta} = \tfrac{1}{12} i\{2(Gv_2)_{\alpha\beta} + \tfrac{1}{2}(\gamma G\gamma)v_{2\alpha\beta} + \tfrac{2}{3}\nabla_\alpha G_{\beta\mu} v_1{}^\mu + \tfrac{2}{3}\gamma^\mu\gamma^\nu\nabla_\alpha G_{\mu\nu} v_{1\beta} - i a_{\alpha\beta} v_0\}\,.$$

Also we get

$$h_0 = -\tfrac{1}{4}\{\tfrac{3}{8}(\gamma G\gamma)^2 + a^\sigma{}_\sigma\}\,,$$
$$h_{2\alpha\beta} = 15(v_4{}^\sigma{}_{\sigma\alpha\beta} - \tfrac{1}{12} g_{\alpha\beta} v_4{}^\sigma{}_\sigma{}^\lambda{}_\lambda) - \tfrac{3}{4}(c^\sigma{}_{\sigma\alpha\beta} - \tfrac{1}{12} g_{\alpha\beta} c^\sigma{}_\sigma{}^\lambda{}_\lambda)\,. \tag{3.16}$$

The solution for the coefficients of W are

$$w_{2\alpha\beta} = \tilde{\omega}_{2\alpha\beta} + \tfrac{1}{4} g_{\alpha\beta} w_2{}^\sigma{}_\sigma\,,$$
$$w_2{}^\sigma{}_\sigma = \tfrac{1}{4}\{i(\gamma G\gamma)w_0 + \tfrac{3}{16}(\gamma G\gamma)^2 + \tfrac{1}{2} a^\sigma{}_\sigma\}\,, \tag{3.17}$$
$$w_4{}^\sigma{}_{\sigma\alpha\beta} = \tfrac{1}{12} i\{2(Gw_2)_{\alpha\beta} + \tfrac{1}{2}(\gamma G\gamma)w_{2\alpha\beta} + \tfrac{2}{3}\nabla_\alpha G_{\beta\mu} w_1{}^\mu$$
$$+ \tfrac{2}{3}\gamma^\mu\gamma^\nu\nabla_\alpha G_{\mu\nu} w_1{}^\mu - i a_{\alpha\beta} w_0 + i h_{2\alpha\beta}\}\,.$$

Here w_0, $w_{1\alpha}$, $\tilde{w}_{2\alpha\beta}$, $\tilde{w}_{4\alpha\beta\gamma\delta}$ are arbitrary.

With these expressions for V and W we can construct $S(z, 0|A)$ from eqs. (3.7) and (3.1):

$$4\pi^2 S(z, 0|A) = -\frac{2z}{(z^2)^2} + R + S \ln z^2 + T\,, \tag{3.18}$$

where

$$R = -\frac{i}{4z^2}\gamma^\mu \not{z}\gamma^\nu G_{\mu\nu} - \frac{i}{6z^2} z^\alpha \gamma^\beta \not{z}\gamma^\nu \nabla_\alpha G_{\beta\nu} - \frac{i}{8z^2} z^\alpha z^\beta z^\gamma \gamma^\nu \nabla_\alpha \nabla_\beta G_{\gamma\nu}$$

$$+ \frac{2}{z^2}\not{z} z\tilde{v}_2 z + \tfrac{1}{2}\not{z} v_2{}^\sigma{}_\sigma$$

$$+ \frac{1}{10z^2}\not{z} z^\alpha z^\beta z^\gamma z^\delta c_{\alpha\beta\gamma\delta} + \tfrac{3}{2}\not{z} z^\alpha z^\beta \tilde{v}_4{}^\sigma{}_{\sigma\alpha\beta} + \tfrac{1}{4}\not{z} z^2 v_4{}^\sigma{}_\sigma{}^\lambda{}_\lambda\,, \tag{3.19}$$

$$S = \gamma^\mu v_{1\mu} + 2\gamma\tilde{v}_2 z + \tfrac{1}{2}\not{z} v_2{}^\sigma{}_\sigma + \tfrac{1}{5}\gamma^\alpha z^\beta z^\gamma z^\delta c_{\alpha\beta\gamma\delta} + \tfrac{3}{2}(\not{z} z^\alpha z^\beta + z^2\gamma^\alpha z^\beta)\tilde{v}_4{}^\sigma{}_{\sigma\alpha\beta}$$

$$+ \tfrac{1}{2} z^2 \not{z} v_4{}^\sigma{}_\sigma{}^\lambda{}_\lambda$$

$$- \tfrac{1}{2} i z G\gamma(v_0 + z\tilde{v}_2 z + \tfrac{1}{4}z^2 v_2{}^\sigma{}_\sigma) - \tfrac{1}{3} i z^\alpha z^\beta z^\gamma \nabla_\alpha G_{\beta\gamma} v_{1\lambda} z^\lambda$$

$$- \tfrac{1}{8} i z^\alpha z^\beta z^\gamma \gamma^\delta \nabla_\alpha \nabla_\beta G_{\gamma\delta} v_0\,. \tag{3.20}$$

T is obtainable from S by replacing v_i's by w_i's and $c_{\alpha\beta\gamma\delta}$ by $20\tilde{w}_{4\alpha\beta\gamma\delta}$. It is straightforward to verify that the solution (3.18) satisfies (2.2).

In the following we shall choose $S^{\text{sing}}(z, 0)$ to be the one given by eq. (3.18) with the arbitrary functions of potentials, w_0, $w_{1\alpha}$, $\tilde{w}_{2\alpha\beta}$, $\tilde{w}_{4\alpha\beta\gamma\delta}$ set equal to zero.

4. Independent scalar operators

The expression for S contains operators * which are, in general, reducible Lorentz tensors. We are interested here only in operators which are Lorentz scalars. (They have also to be scalar in colour space, which is accomplished by taking trace over colour, denoted by the superscript c in the following.) We therefore project these reducible tensors onto an independent set of scalars for each operator dimension we are interested in.

Consider only gluon field strength and covariant derivatives in constructing operators. In dimension four, only one tensor is possible, $\text{tr}^c\, G_{\alpha\beta}G_{\mu\nu}$ ($\text{tr}^c\, \nabla_\alpha \nabla_\beta G_{\mu\nu} = 0$). It clearly contains only one scalar operator, which we choose to be $G^a_{\mu\nu}G^{a\mu\nu}$,

$$\langle 0|\text{tr}^c\, G_{\alpha\beta}G_{\mu\nu}|0\rangle = \tfrac{1}{24}\langle 0|g_s^2 G^a_{\rho\sigma}G^{a\rho\sigma}|0\rangle(g_{\alpha\mu}g_{\beta\nu} - g_{\alpha\nu}g_{\beta\mu})\,. \tag{4.1}$$

In dimension six, one has two tensors, $T_{1\alpha\beta\mu\nu\lambda\sigma} \sim \text{tr}^c(G_{\alpha\beta}G_{\mu\nu}G_{\lambda\sigma})$ and $T_{2\alpha\beta\mu\nu\lambda\sigma} \sim \text{tr}^c\,(\nabla_\alpha G_{\beta\gamma}\nabla_\mu G_{\nu\lambda})$. (Again $\text{tr}^c\,(\nabla_\alpha \nabla_\beta \nabla_\gamma \nabla_\delta G_{\mu\nu}) = 0$. Also the tensors $\text{tr}^c\,(\nabla_\alpha \nabla_\beta G_{\gamma\delta} G_{\mu\nu})$ and $\text{tr}^c\,(G_{\alpha\beta}\nabla_\mu \nabla_\nu G_{\lambda\sigma})$ can be reduced to T_2 and a total derivative by partial integration.) Analysis of spin content by Clebsch–Gordan decomposition ($G_{\alpha\beta} \sim D^{(1,0)} + D^{(0,1)}$, $\nabla_\mu \sim D^{(1/2,1/2)}$) shows that the tensors T_1 and T_2 contain one and three scalar operators, respectively. We choose these to be

$$J_0 = f^{abc}G^a_{\mu\rho}G^b_{\rho\sigma}G^c_{\sigma\mu}\,, \qquad J_1 = (\nabla_\mu G_{\rho\sigma})^a(\nabla_\mu G_{\rho\sigma})^a\,,$$
$$J_2 = (\nabla_\mu G_{\rho\sigma})^a(\nabla_\rho G_{\mu\sigma})^a\,, \qquad J_3 = (\nabla_\mu G_{\mu\sigma})^a(\nabla_\nu G_{\nu\sigma})^a\,. \tag{4.2}$$

* From here on we refer to the local functions of gauge fields as operators.

These four scalars are not independent, however; using Jacobi identity, we get

$$J_1 = -2J_0 + 2J_3\,, \qquad J_2 = -J_0 + J_3\,, \tag{4.3}$$

ignoring total derivatives. We choose J_0 and J_3 as the independent ones. The trace over colour in J_0 actually gives rise to two colourless operators with f and d type couplings, but the latter vanishes identically. The operator J_3 is not relevant in our calculation. In fact, we can remove it by imposing the condition, $\nabla^\mu G_{\mu\nu} = 0$, on the external field. (Of course, in QCD theory it gives rise to a four-quark operator, which does not occur in the lowest order of perturbation theory we are considering here.) Thus we need to project the tensors T_1 and T_2 onto the f type J_0 only:

$$\begin{aligned}&\langle 0|\mathrm{tr}^c\,(iG_{\alpha\beta}G_{\mu\nu}G_{\lambda\sigma})|0\rangle \\ &\quad = A[\{(g_{\alpha\mu}g_{\beta\lambda} - g_{\beta\mu}g_{\alpha\lambda})g_{\nu\sigma} - (g_{\alpha\nu}g_{\beta\lambda} - g_{\beta\nu}g_{\alpha\lambda})g_{\mu\sigma}\} - (\lambda \leftrightarrow \sigma)]\,,\end{aligned} \tag{4.4}$$

$$\begin{aligned}&\langle 0|\mathrm{tr}^c\,(\nabla_\mu G_{\rho\sigma}\nabla_\alpha G_{\beta\gamma})|0\rangle \\ &\quad = A[2(g_{\rho\beta}g_{\sigma\gamma} - g_{\sigma\beta}g_{\rho\gamma})g_{\mu\alpha} + (g_{\mu\rho}g_{\sigma\beta} - g_{\mu\sigma}g_{\rho\beta})g_{\alpha\gamma} - (g_{\mu\rho}g_{\sigma\gamma} - g_{\mu\sigma}g_{\rho\gamma})g_{\alpha\beta} \\ &\qquad + (g_{\rho\alpha}g_{\sigma\gamma} - g_{\sigma\alpha}g_{\rho\gamma})g_{\mu\beta} - (g_{\rho\alpha}g_{\sigma\beta} - g_{\sigma\alpha}g_{\rho\beta})g_{\mu\gamma}]\,,\end{aligned} \tag{4.5}$$

where $A = -\frac{1}{96}\langle 0|f^{abc}G^a_{\alpha\beta}G^b_{\beta\gamma}G^c_{\gamma\alpha}|0\rangle$. For later use we note

$$\langle 0|\tilde{v}_{2\alpha\beta}iG_{\mu\nu}|0\rangle = \langle 0|\{zv^{\sigma}_{2\sigma} - i(zG\gamma)v_0\}G_{\mu\nu}|0\rangle = 0\,. \tag{4.6}$$

In the above discussion we have only considered the independent gluon operators. Now it is instructive to enumerate also the other operators which involve quark fields. In the massless theory, a complete independent set of operators up to and including dimension six are given by

$$\begin{aligned}&O_0 = 1 \text{ (the unit operator)}\,, && O_1 = G^a_{\mu\nu}G^{\mu\nu a}\,, \\ &O_2 = f^{abc}G^a_{\mu\nu}G^b_{\nu\lambda}G^c_{\lambda\mu}\,, && O_3 = \bar{q}q\,, \\ &O_4 = \bar{q}\sigma_{\mu\nu}G^{\mu\nu}q\,, && O_5 = \bar{q}\Gamma_1 q\bar{q}\Gamma_2 q\end{aligned} \tag{4.7}$$

where Γ_1 and Γ_2 are appropriate matrices in the Dirac and colour space. For the even chirality correlators Π (scalar–scalar, vector–vector, etc.) the OPE is of the form

$$\Pi(z,0) = \sum_{i=0,1,2,5} C_i(z^2)O_i \tag{4.8}$$

while for the odd chirality correlators Π' (scalar-vector, vector-tensor, etc.) it is

$$\Pi'(z,0) = \sum_{i=3,4,5} C_i(z^2)O_i\,. \tag{4.9}$$

The coefficient of the unit operator is given by the usual perturbation theory.

The coefficients of the two odd-chirality quark operators, O_3 and O_4 in (4.9) can be read off from the second and third terms of eq. (2.7). The coefficient of the four-quark operator, O_5, cannot be calculated in the present framework; it involves a gluon propagator which is outside the external field formalism.

We are then left with the two gluon operators, O_1 and O_2 whose coefficients in (4.8) we calculate below to lowest order. The calculations are carried out with S^{sing} as defined below eq. (3.20). As already stated a regular piece, S^{reg}, can be absorbed in the definition of normal ordering of $\bar{q}(z)q(0)$. In fact, such a regular piece generates in eq. (2.7) extra terms which are exactly of the same structure as the second and third terms in it. In the present context, there is no operator involving $:\bar{q}q:$ which can contribute to (4.8) to the lowest order. This in turn implies that a regular piece cannot change the coefficients of O_1 and O_2 in (4.8). This is verified explicitly in the appendix.

For our purpose we can simplify the expression for S considerably by replacing in its terms the operators of highest dimension, we are interested in, by their v.e.v.'s. This we shall carry out in the next section.

5. Coefficients of operator expansion

We first evaluate the coefficients of $g_s^2 G^a_{\alpha\beta} G^{a\alpha\beta}$ in the short distance OPE of different two-point functions. Here we only need terms with operators up to dimension four in the quark propagator. Using the v.e.v. (4.1), it turns out that all contributions of dimension four operators drop out from the expression for S. Further, operators of dimension three are of no interest to us here. Thus we are left with only two terms in the propagator,

$$4\pi^2 S^{\text{sing}}(z, 0|A) = -\frac{2\not{z}}{(z^2)^2} - \frac{i}{4z^2}\gamma^\alpha \not{z} \gamma^\beta G_{\alpha\beta}\,, \tag{5.1}$$

We shall state the results for the coefficients as they occur in the invariant amplitudes for the Fourier transform of different two-point functions. Denoting the quark bilinears by

$$(S, V, T)(z) = \bar{q}(z)(1, \gamma_\mu, \sigma_{\mu\nu})q(z)\,,$$

we define the invariant amplitudes in the two point functions

$$i\int \mathrm{d}x\, \mathrm{e}^{iqx}\langle 0|\mathrm{T}\{S(x)S^+(0)\}|0\rangle = T_{SS}(q^2)\,, \tag{5.2}$$

$$i\int \mathrm{d}x\, \mathrm{e}^{iqx}\langle 0|\mathrm{T}\{V_\mu(x)V_\nu^+(0)\}|0\rangle = (q_\mu q_\nu - q^2 g_{\mu\nu})T_{VV}(q^2)\,, \tag{5.3}$$

$$i\int dx\, e^{iqx}\langle 0|T\{T_{\alpha\beta}(x)T^{\dagger}_{\mu\nu}(0)\}|0\rangle = (g_{\alpha\mu}g_{\beta\nu}-g_{\alpha\nu}g_{\beta\mu})T^{(1)}_{TT}(q^2)$$
$$+(g_{\alpha\mu}q_\beta q_\nu - g_{\beta\mu}q_\alpha q_\nu - g_{\alpha\nu}q_\beta q_\mu$$
$$+g_{\beta\nu}q_\alpha q_\mu)T^{(2)}_{TT}(q^2)\,. \qquad (5.4)$$

The contributions of $g_s^2 G^a_{\mu\nu}G^{a\mu\nu}$ to these invariant amplitudes, denoted by the superscript (G) are given by $(Q^2=-q^2)$

$$T^{(G)}_{SS}=\frac{M_1^4}{8Q^2}\,,\qquad T^{(G)}_{VV}=\frac{M_1^4}{12Q^4}\,,\qquad T^{(1)(G)}_{TT}=\frac{M_1^4}{24Q^2}\,,\qquad T^{(2)(G)}_{TT}=\frac{M_1^4}{12Q^4}\,, \qquad (5.5)$$

where

$$M_1^4=\langle 0|\frac{g_s^2}{4\pi^2}G^a_{\mu\nu}G^{a\mu\nu}|0\rangle\,.$$

We now turn to the case of the coefficients of $g_s^3 f^{abc}G^a_{\alpha\beta}G^b_{\beta\mu}G^c_{\mu\alpha}$ in two-point functions. Following our earlier discussion we consider only the singular part of the propagator S^{sing} [defined below eq. (3.20)] in evaluating these coefficients. That the regular piece does not change our results is shown in the appendix. Inserting v.e.v.'s (4.4), (4.5) we find that $\ln z^2$ terms in S coming from operators of dimensions six vanish. The expression for S^{sing} simplifies to

$$4\pi^2 S^{\text{sing}}(z,0|A) = -\frac{2\not{z}}{(z^2)^2}-\frac{i}{4z^2}\gamma^\mu\not{z}\gamma^\nu G_{\mu\nu}-\frac{i}{6z^2}z^\alpha\gamma^\mu\not{z}\gamma^\nu\nabla_\alpha G_{\mu\nu}$$
$$+\frac{i}{12}\gamma^\mu\nabla_\theta G_{\theta\mu}\ln z^2$$
$$+\frac{2}{z^2}\not{z}z\tilde{v}_2 z-\frac{i}{8z^2}z^\alpha z^\beta z^\gamma\gamma^\delta\nabla_\alpha\nabla_\beta G_{\gamma\delta}+\tfrac{1}{128}\not{z}(\gamma G\gamma)^2+\tfrac{1}{16}\not{z}a^\sigma_\sigma$$
$$+\ln z^2(2\gamma\tilde{v}_2 z+\tfrac{1}{2}\not{z}v^\sigma_{2\sigma}-\tfrac{1}{2}i z G\gamma v_0)-\tfrac{1}{288}Az^2\not{z}\,. \qquad (5.6)$$

Noting (2.4), the trace term in (2.7) can be evaluated. We pick out terms containing operators of dimension six and extract the scalar operator using the v.e.v.'s (4.4)–(4.6). We see, in particular, that the fourth and fifth terms and also the terms with $\ln z^2$ in (5.6) cannot contribute. The contribution of the sixth term also vanishes:

$$\langle 0|\nabla_\alpha\nabla_\beta G_{\lambda\sigma}G_{\mu\nu}|0\rangle z^\alpha z^\beta z^\lambda\Gamma_1\gamma^\sigma\Gamma_2\gamma^\mu\not{z}\gamma^\nu$$
$$\sim z^2\Gamma_1(z_\mu\gamma_\nu - z_\nu\gamma_\mu)\Gamma_2\gamma^\mu\not{z}\gamma^\nu=0\,. \qquad (5.7)$$

Thus we have

$$-(4\pi^2)^2 \operatorname{Tr}\{\Gamma_1 S^{\text{sing}}(z,0|A)\Gamma_2 S^{\text{sing}}(0,z|A)\}|_{\text{dim. 6 operators}}$$

$$=\frac{A}{72z^2}\operatorname{Tr}(\Gamma_1 \not{z}\Gamma_2 \not{z})+\frac{1}{36z^4}\operatorname{Tr}(\Gamma_1 z^\alpha\gamma^\beta \not{z}\nabla_\alpha G_{\beta\lambda}\gamma^\lambda\Gamma_2 z^\mu\nabla_\mu G_{\nu\sigma}\gamma^\sigma \not{z}\gamma^\nu)$$

$$-\frac{1}{32z^2}\operatorname{Tr}[\Gamma_1 \not{z}\{\tfrac{1}{8}(\gamma G\gamma)^2+a^\sigma_\sigma\}\Gamma_2\gamma^\alpha \not{z}\gamma^\beta iG_{\alpha\beta}]\,. \tag{5.8}$$

Working out the traces*, we find that the configuration space analogue of the invariant amplitudes (5.2)–(5.4) does not contain any singular term. We thus get the remarkable result that in the short-distance OPE of the two-point functions, the operator $g_s^3 f^{abc} G^a_{\alpha\beta} G^b_{\beta\gamma} G^c_{\gamma\alpha}$ appers in none of the invariant amplitudes.

It is really not necessary to evaluate the traces in (5.8) to see the null result for the amplitudes T_{SS}, T_{VV} and $T^{(1)}_{TT}$; it suffices to note that there are no $\ln^2 z^2$ and $\ln z^2$ terms in it. (For T_{VV}, the evaluation shows that there are no singular terms like $z_\mu z_\nu/z^2$, which is ensured by current conservation, once terms with $\ln z^2$ are absent.) However, for $T^{(2)}_{TT}$ it is necessary to check that the traces in (5.8) indeed do not give rise to a non-zero result.

6. Conclusion

We have presented a simple configuration space method of calculating the Wilson coefficients of gluon operators in the OPE of two-point functions at short distance. The method owes its simplicity over the one based on Feynman diagrams [2] in that here it is possible to carry out computations without being involved in large distances at any stage.

The quark propagator in configuration space consists of a singular part determined by short-distance dynamics alone and a regular part containing behaviour of the theory at large distances and specific boundary conditions. From eq. (2.6) it is clear that the regular part can be absorbed in the definition of normal ordering of the quark–antiquark operator. Then the coefficient of other operators of the same dimension appearing in the OPE would, in general, depend on the specific definition of this normal ordering. However, in the present calculation, where we consider operators of dimension four and six only, the situation is simple: In the massless theory there is only one gluon operator in each of these dimensions, O_1 in dimension 4 and O_2 in dimension 6. Because of the absence of any operator involving $:\bar{q}q:$ in these dimensions, there is no possibility of absorbing the regular piece in the definition of such an operator. It follows that the regular part of the propagator cannot change the coefficients of O_1 and O_2, calculated from the singular part alone [the first term in eq. (2.7)]. That it is indeed the case, is verified explicitly in the appendix.

* The traces were evaluated by using the program Reduce.

Our result that the spin-zero, gluon operator of dimension six does not show up in the OPE of two-point functions of quark bilinears agrees with an earlier anticipation based on explicit calculation of the vector two point functions in the background field of an instanton [8] and of any (anti-) self-dual field [5]. Whether it occurs in other two-point functions, e.g. of energy-momentum tensor, remains to be seen. The occurrence of the operator in the expansion of two-point functions of heavy quark bilinears has already been reported [13].

We are indebted to H. Leutwyler for explaining to us some aspects of the calculation presented here and for helpful discussions. We also thank J. Gasser and L. Jones for help during our computations and for comments. One of us (S.M.) appreciates the hospitality extended to him at the Institute for Theoretical Physics, University of Bern.

Appendix

Here we demonstrate that a regular piece of the quark propagator has no effect on our calculation of the coefficients of gluon operators, O_1 and O_2. It suffices to show that

$$A \equiv \langle 0|\Gamma_1 S^{\text{sing}}(z, 0|A)\Gamma_2 S^{\text{reg}}(0, z|A)|0\rangle = 0, \tag{A.1}$$

where $S^{\text{reg}}(0, z|A)$ is any solution of the homogeneous Dirac equation,

$$S^{\text{reg}}(0, z|A)(\overleftarrow{\not\partial} + i\not A) = 0\,. \tag{A.2}$$

We have already found the general solution for $S^{\text{reg}}(z, 0|A)$ (from which $S^{\text{reg}}(0, z|A)$ can be obtained by using (2.4)); it consists of terms in (3.18) which are proportional to w_0, $w_{1\alpha}$, $\tilde{w}_{2\alpha\beta}$ and $\tilde{w}_{4\alpha\beta\gamma\delta}$. However all we need here are the recursion relations among the expansion coefficients, which may be worked out directly. Let

$$S(0, z|A) = d_0 + d_{1\mu}z^\mu + d_{2\mu\nu}z^\mu z^\nu + d_{3\mu\nu\lambda}z^\mu z^\nu z^\lambda + \cdots\,. \tag{A.3}$$

Then eq. (A.2) gives

$$\begin{gathered} d_{1\mu}\gamma^\mu = 0\,, \\ 2d_{2\mu\nu}z^\mu\gamma^\nu + \tfrac{1}{2}id_0 z^\alpha G_{\alpha\beta}\gamma^\beta = 0\,, \\ 3d_{3\mu\nu\lambda}z^\mu z^\nu z^\lambda + \tfrac{1}{3}id_0 z^\alpha z^\beta \nabla_\alpha G_{\beta\mu}\gamma^\mu + id_{1\mu}z^\mu \frac{z^\alpha}{2} G_{\alpha\beta}\gamma^\beta = 0\,. \end{gathered} \tag{A.4}$$

Considering terms up to and including six, we have

$$A = -\frac{2}{z^4}\langle 0|\Gamma_1 z\Gamma_2 d_0 + \Gamma_1 z\Gamma_2 d_{1\mu} z^{\mu}$$

$$+\Gamma_1 z\Gamma_2 d_{2\mu\nu} z^{\mu} z^{\nu} + \Gamma_1 z\Gamma_2 d_{3\mu\nu\lambda} z^{\mu} z^{\nu} z^{\lambda}|0\rangle$$

$$-\frac{i}{4z^2}\langle 0|\Gamma_1\gamma^{\mu} z\gamma^{\nu} G_{\mu\nu}\Gamma_2 d_0 + \Gamma_1\gamma^{\mu} z\gamma^{\nu} G_{\mu\nu}\Gamma_2 d_{1\alpha} z^{\alpha}|0\rangle$$

$$-\frac{i}{6z^2}\langle 0|\Gamma_1 z^{\alpha}\gamma^{\beta} z\nabla_{\alpha} G_{\beta\nu}\gamma^{\nu}\Gamma_2 d_0|0\rangle$$

$$+\frac{1}{12} i \ln z^2 \langle 0|\Gamma_1\gamma^{\nu}\nabla_{\theta} G_{\theta\nu}\Gamma_2 d_0|0\rangle . \tag{A.5}$$

Noting the tensor structure of the different quantities and that d_i's are odd in γ_{μ}, the v.e.v.'s we meet in (A.5) may be written as

$$\begin{aligned}
&\langle 0|d_0|0\rangle = 0 , \qquad \langle 0|d_{1\mu}|0\rangle = a\gamma_{\mu} , \\
&\langle 0|d_{2\mu\nu}|0\rangle = 0 , \\
&\langle 0|d_{3\mu\nu\lambda}|0\rangle = b(g_{\mu\nu}\gamma_{\lambda} + g_{\mu\lambda}\gamma_{\nu} + g_{\nu\lambda}\gamma_{\mu}), \\
&\langle 0|G_{\mu\nu} d_0|0\rangle = 0 , \\
&\langle 0|G_{\mu\nu} d_{1\alpha}|0\rangle = c(g_{\mu\alpha}\gamma_{\nu} - g_{\nu\alpha}\gamma_{\mu}) + d\varepsilon_{\mu\nu\alpha\rho}\gamma^{\rho}\gamma_5 , \\
&\langle 0|\nabla_{\alpha} G_{\beta\mu} d_0|0\rangle = e(g_{\alpha\beta}\gamma_{\mu} - g_{\alpha\mu}\gamma_{\beta}) + f\varepsilon_{\alpha\beta\mu\rho}\gamma^{\rho}\gamma_5 ,
\end{aligned} \tag{A.6}$$

where a, b, c, d, e, f are scalar functions of gauge fields. Using further the translation invariance of the propagator, the condition $\nabla^{\mu} G_{\mu\nu} = 0$ on the gauge field and contractions with the totally antisymmetric tensor one can show that all the scalars in eqs. (A.6) vanish. Hence we prove the result (A.1).

References

[1] K.G. Wilson, Phys. Rev. 179 (1969) 1499
[2] M.A. Shifman, A.I. Vainshtein and V.I. Zakharov, Nucl. Phys. B147 (1979) 385
[3] W. Zimmermann, Ann. of Phys. 77 (1973) 536, 570
[4] N. Andrei and D. Gross, Phys. Rev. D18 (1978) 468
[5] M.S. Dubovikov and A.V. Smilga, Nucl. Phys. B185 (1981) 109
[6] H. Leutwyler, Phys. Lett. 96B (1980) 154; Nucl. Phys. B179 (1981) 129
[7] J. Finjord, Nucl. Phys. B194 (1982) 77
[8] V.A. Novikov, M.A. Shifman, A.I. Vainshtein and V.I. Zakharov, Nucl. Phys. B174 (1980) 378
[9] H. Fritzsch and H. Leutwyler, Phys. Rev. D10 (1974) 1624
[10] A.V. Smilga, Yad. Fiz. 35 (1982) 473;
E.V. Shuryak and A.I. Vainsthtein, Nucl. Phys. B199 (1982) 451; B201 (1982) 141;
T.M. Aliev, M.A. Shifman, Phys. Lett. 112B (1982) 401;
B.L. Ioffe, M.A. Shifman, Moscow preprint ITEP-134 (1981);
M.A. Shifman, Moscow preprint ITEP-49 (1982)

[11] J. Schwinger, Particles, sources and fields (Addison-Wesley, Reading, 1979) p. 271;
C. Cronström, Phys. Lett. 90B (1980) 267
[12] J. Hadamard, Le problème de Cauchy et les équations aux dérivées partielles linéaires hyperboliques (Paris, 1932) 143
[13] S.N. Nikolaev and A.V. Radyushkin, Phys. Lett. 110B (1982) 476

3.1. Calculations in external fields in quantum chromodynamics. Technical review

3.1.1. General formalism

This section gives an idea of a very convenient formalism proposed by Schwinger for the description of the motion of particles in arbitrary external fields. Chapter 1 of the review paper by Novikov et al. [14] (this paper is reproduced below) pursues the same aim but, unfortunately, it is rather lengthy and contains too many technicalities. Therefore, I think it is reasonable to substitute it by the following simple remarks. Then we give only chapter 2 of the above-mentioned review.

The Schwinger operator approach is formulated in coordinate space. Let us introduce a formal set of states $|x\rangle$, which are the eigenvalues of the coordinate operator X_μ:

$$X_\mu |x\rangle = x_\mu |x\rangle. \tag{3.1}$$

We also introduce the momentum operator P_μ which satisfies the following commutation relations

$$[P_\mu, X_\mu] = \mathrm{i} g_{\mu\nu}, \tag{3.2}$$

$$[P_\mu, P_\nu] = \mathrm{i} g T^a G^a_{\mu\nu}, \tag{3.3}$$

where T^a are the generators of the color group in the relevant representation of $SU(3)_c$. Thus, for quarks $T^a = \frac{1}{2}\lambda^a$, where λ^a stands for the Gell-Mann matrices.

The algebra (3.2), (3.3) is the basic instrument of the Schwinger formalism. It is rather obvious that in the coordinate basis P_μ acts as a covariant derivative. If

$$\langle y | x \rangle = \delta(x - y), \tag{3.4}$$

then

$$\langle y | P_\mu | x \rangle = \left(\mathrm{i} \frac{\partial}{\partial x_\mu} + g T^a A^a_\mu(x) \right) \delta(x - y). \tag{3.5}$$

For practical purposes sometimes it will be convinient to represent P_μ as the sum of two terms

$$P_\mu = p_\mu + g A_\mu, \tag{3.6}$$

where the action of p_μ in the coordinate basis reduces to the (ordinary) derivative, $p_\mu \to \mathrm{i}\, \partial_\mu$, and $A_\mu = T^a A^a_\mu$.

Now we can write down the formal expressions for the Green functions. For instance, the quark Green function describing the propagation from point 0 to point x is

$$S(x, 0) = \langle x | (\not{P} - m)^{-1} | 0 \rangle. \tag{3.7}$$

Equation (3.7), rather evident by itself, can be verified by applying the Dirac operator to both sides of eq. (3.7). Notice that in the presence of an external field there is no translational invariance, and $S(x, y)$ does not coincide with $S(x - y, 0)$.

We pause here to give a simple example of the application of the Schwinger formalism. A more sophisticated problem is considered, e.g., in chapter 1 of ref. [14] (see also references therein).

Let us try to calculate the matrix element of $\langle i\bar{Q}\gamma_5 Q\rangle$ in an external gluon field as an expansion in powers of m_Q^{-1} (here Q is the heavy-quark field and m_Q denotes the mass of the heavy quark. The reader familiar with the previous material will immediately recognize the so-called heavy-quark expansion, which was introduced in the given context by Shifman et al. (chapter 2 of this volume).

First of all,

$$\left\langle i\bar{Q}\gamma_5 Q\right\rangle = \langle x|\ \mathrm{Tr}\left\{\gamma_5(\not{P}-m)^{-1}\right\}|x\rangle,$$

where $|x\rangle$ is the eigenstate of the coordinate operator and Tr denotes the trace over both color and Lorentz indices. Inserting a unity in the braces,

$$1 = (\not{P}+m)^{-1}(\not{P}+m)$$

and taking into account the fact that $\gamma_\mu\gamma_\gamma = g_{\mu\nu} + \sigma_{\mu\nu}$ we get

$$\begin{aligned}\left\langle i\bar{Q}\gamma_5 Q\right\rangle &= \langle x|\ \mathrm{Tr}\left\{\gamma_5[(\not{P}+m)(\not{P}-m)]^{-1}(\not{P}+m)\right\}|x\rangle \\ &= \langle x|\ \mathrm{Tr}\left\{\gamma_5\left(P^2-m^2+\tfrac{1}{2}igT^aG^a_{\mu\nu}\sigma_{\mu\nu}\right)^{-1}(\not{P}+m)\right\}|x\rangle \qquad (3.8)\end{aligned}$$

(see eq. (3.3)). Moreover, since the trace of any odd number of γ matrices vanishes, $(\not{P}+m)$ in the last line reduces to m, and we arrive at

$$\begin{aligned}\left\langle i\bar{Q}\gamma_5 Q\right\rangle = m\langle x|\ \mathrm{Tr}\ \gamma_5\Bigg\{&\frac{1}{P^2-m^2}+\frac{1}{P^2-m^2}\frac{-ig}{2}(G\sigma)\frac{1}{P^2-m^2} \\ &+\frac{1}{P^2-m^2}\frac{-ig}{2}(G\sigma)\frac{1}{P^2-m^2}\frac{-ig}{2}(G\sigma)\frac{1}{P^2-m^2} \\ &+\cdots\Bigg\}|x\rangle,\end{aligned}$$

$$G\sigma \equiv G^a_{\mu\nu}T^a\sigma_{\mu\nu}. \qquad (3.9)$$

The first two terms on the right-hand side are obviously irrelevant because $\mathrm{Tr}\ \gamma_5 = \mathrm{Tr}(\gamma_5\sigma_{\mu\nu}) = 0$, and the first nontrivial contribution comes from the third term. The latter is explicitly proportional to the second power of $G_{\mu\nu}$; therefore to the leading order in m_Q^{-1} we can substitute the operator P_μ by p_μ in eq. (3.9), thus neglecting all terms in the external field except the quadratic one. If we also disregard the derivatives of $G_{\mu\nu}$ the operators $G_{\mu\nu}$ and $(p^2-m^2)^{-1}$ become commutative. Then eq. (3.9) reduces to

$$\left\langle i\bar{Q}\gamma_5 Q\right\rangle = -\frac{mg^2}{4}\langle x|\ \mathrm{Tr}\left\{\gamma_5\frac{1}{(p^2-m^2)^3}(G\sigma)(G\sigma)\right\}|x\rangle + \text{higher orders in } m^{-1}.$$

The operator G acting on the eigenstate $|x\rangle$ yields

$$G^a_{\mu\nu}|x\rangle = G^a_{\mu\nu}(x)|x\rangle,$$

which leads in turn to the following expression

$$\left\langle i\bar{Q}\gamma_5 Q\right\rangle = -\frac{mg^2}{4}\mathrm{Tr}_{\mathrm{color}}\left(G_{\mu\nu}(x)G_{\alpha\beta}(x)\right)\ \mathrm{Tr}_{\mathrm{Lorentz}}(\gamma_5\sigma_{\alpha\beta}\sigma_{\mu\nu})\int\frac{d^4p}{(2\pi)^4}\frac{1}{(p^2-m^2)^3}.$$

After taking the traces and performing the integration we, finally, get

$$\langle i\bar{Q}\gamma_5 Q\rangle = -\frac{1}{4m}\frac{\alpha_s}{4\pi}G^a_{\mu\nu}G^a_{\alpha\beta}\epsilon^{\mu\nu\alpha\beta} \equiv -\frac{1}{2m}\frac{\alpha_s}{4\pi}G^a_{\mu\nu}\widetilde{G^a_{\mu\nu}}. \tag{3.10}$$

This $O(m_Q^{-1})$ term in the heavy-quark expansion for $\langle i\bar{Q}\gamma_5 Q\rangle$ agrees with the famous triangle anomaly in the axial current.

Here we leave the general Schwinger approach and show how it can be used in the procedure based on the Fock–Schwinger gauge. The basic blocks of this procedure are the quark and gluon Green functions in a background field obeying the gauge condition

$$x_\mu\big(A_\mu(x)\big)_{\text{ext}} = 0.$$

Under this condition

$$A_\mu(x) = \frac{1}{2\cdot 0!}x_\rho G_{\rho\mu}(0) + \frac{1}{3\cdot 1!}x_\alpha x_\rho\big(\mathscr{D}_\alpha G_{\rho\mu}(0)\big) + \frac{1}{4\cdot 2!}x_\alpha x_\beta x_\rho\big(\mathscr{D}_\alpha\mathscr{D}_\beta\mathscr{D}_\rho G_{\rho\mu}(0)\big) + \cdots. \tag{3.11}$$

Assume we would like to construct the operator product expansion for the Green function describing the propagation of the massless quark in the background gluon field,

$$S(x,0) = \langle x|\not{P}^{-1}|0\rangle. \tag{3.12}$$

The Fourier transform of eq. (3.12) is

$$S(q) \equiv \int d^4x\ e^{iqx}\ \langle x|\not{P}^{-1}|0\rangle = \int d^4x\ \langle x|\frac{1}{\not{P}+\not{q}}|0\rangle, \tag{3.13}$$

where we have taken the fact into account that

$$e^{iqX}\ P_\mu = (P_\mu + q_\mu)\ e^{iqX},\quad e^{iqX}|x\rangle = e^{iqx}|x\rangle,\quad e^{iqX}|0\rangle = |0\rangle.$$

The expansion in $(G_{\mu\nu})_{\text{ext}}$ for $S(q)$ is equivalent to the expansion in powers of q^{-1}. At large q

$$S(q) = \int d^4x\ \langle x|\frac{1}{\not{q}} + \frac{1}{\not{q}}(-\not{P})\frac{1}{\not{q}} + \frac{1}{\not{q}}(-\not{P})\frac{1}{\not{q}}(-\not{P})\frac{1}{\not{q}} + \cdots|0\rangle. \tag{3.14}$$

Thus, basically all we have to do (apart from the γ algebra) is to find the matrix elements of the type

$$\int d^4x\ \langle x|P_{\mu_1}P_{\mu_2}\cdots P_{\mu_n}|0\rangle. \tag{3.15}$$

Recall that $P_\mu = p_\mu + gA_\mu$, and if A_μ occupies the right-most position we immediately get zero,

$$\langle x|(\cdots A_\mu)|0\rangle = \langle x|(\cdots A_\mu(0))|0\rangle = 0, \tag{3.16}$$

since $A_\mu(0) = 0$ (cf. eq. (3.11)). By the same token if p_μ occupies the left-most position, the matrix element also vanishes:

$$\int d^4x\ \langle x|(p_\mu\cdots)|0\rangle = \int d^4x\ d^4y\ \langle x|p_\mu|y\rangle\ \langle y|(\cdots)|0\rangle = -i\int d^4x\ d^4y\left(\frac{\partial}{\partial x_\mu}\right)\delta(x-y)\ \langle y|(\cdots)|0\rangle = 0. \tag{3.17}$$

Thus transposing all A's to the right and all p's to the left we eventually reduce the matrix elements (3.15) to commutators of the type $[p_\mu, A_\nu]|0\rangle = i\,\partial_\mu A_\nu|_{x=0}$. The latter expression, in turn, reduces to $\frac{1}{2}G_{\mu\nu}$ with the aid of eq. (3.11). For example,

$$\int d^4x \langle x | P_\mu P_\nu | 0\rangle = g \int d^4x \langle x | A_\mu p_\nu | 0\rangle = g \int d^4x \langle x | [A_\mu, p_\nu] | 0\rangle$$

$$= -ig \int d^4x \langle x | \frac{\partial}{\partial x_\nu} A_\mu | 0\rangle = -\frac{ig}{2} G_{\nu\mu}(0);$$

$$\int d^4x \langle x | P_\mu P_\nu P_\alpha | 0\rangle = \int d^4x\, g \langle x | A_\mu (p_\nu + gA_\nu) p_\alpha | 0\rangle$$

$$= \int d^4x \langle x | g[A_\mu, p_\nu] p_\alpha + g^2 A_\mu [A_\nu, p_\alpha] | 0\rangle$$

$$= \int d^4x \langle x | -ig[\partial_\nu A_\mu, p_\alpha] - ig^2 A_\mu \, \partial_\alpha A_\nu | 0\rangle$$

$$= \int d^4x \langle x | -g \, \partial_\alpha \, \partial_\nu A_\mu | 0\rangle = -\tfrac{1}{3} g \left(\mathscr{D}_\alpha G_{\nu\mu}(0) + \mathscr{D}_\nu G_{\alpha\mu}(0) \right).$$

In this way the construction of OPE reduces to a set of purely algebraic operations. This seems to be the easiest method for calculating the quark and gluon propagators in the Fock–Schwinger gauge. The corresponding explicit expressions including all terms up to $O(q^{-6})$ are presented below (see eqs. (2.34) and (2.35) in ref. [14]). In many cases it is convenient to return to coordinate space (cf. eq. (2.31) in ref. [14]). To this end one can use the formulae for the inverse Fourier transform collected in Appendix D in ref. [14].

Chapter II. Method Based on the Fock-Schwinger Gauge

2.1. The Fock-Schwinger Gauge

Correlation functions of colourless sources are gauge invariant. Any gauge condition imposed on the vacuum gluon field (considered as given) will eventually lead to one and the same final answer. Hence the choice of the gauge is, first of all, the question of convenience. The gauge proposed in electrodynamics by FOCK and then, independently, by SCHWINGER [7, 4] is as follows

$$(x - x_0)\, A_\mu{}^a(x) = 0 \tag{2.1}$$

where $A_\mu{}^a$ is the four-potential and x_0 is an arbitrary point in the space playing the role of a gauge parameter. This gauge was rediscovered several times in chromodynamics [20], and its virtues were revealed in subsequent works.

Let us notice from the very beginning that condition (2.1) is invariant under the scale transformation and inversion of coordinates, but breaks the translational symmetry. The latter restores itself in gauge invariant quantities. In other words, parameter x_0 should cancel out in all correlation functions induced by colourless sources. This fact — cancellation of x_0 dependence — may serve as an additional check of correctness of calculations (see, e.g. [21]). We shall not exploit this useful property, however, and shall put $x_0 = 0$ in all following considerations,

$$x_\mu A_\mu{}^a(x) = 0 \tag{2.2}$$

getting, as a reward, compact formulae for quark and gluon propagators.

The most important consequence from eq. (2.2) is the following: potential $A_\mu{}^a(x)$ is expressable directly in terms of the gluon field strength tensor [20]. Namely,

$$A_\mu{}^a(x) = \int\limits_0^1 \alpha\, d\alpha G^a_{\varrho\mu}(\alpha x)\, x_\varrho\,. \tag{2.3}$$

The derivation of this relation is quite straightforward. Start from identity

$$A_\mu(y) = \frac{\partial}{\partial y_\mu}\left(A_\varrho(y)\, y_\varrho\right) - y_\varrho \frac{\partial A_\varrho(y)}{\partial y_\mu}, \tag{2.4}$$

where, due to eq. (2.2) the first term in the right-hand side vanishes. The second term can be, evidently, rewritten as follows

$$-y_\varrho G_{\mu\varrho} - y_\varrho \frac{\partial A_\mu(y)}{\partial y_\varrho}\,. \tag{2.5}$$

Combining eqs. (2.4), (2.5) we get

$$A_\mu(y) + y_\varrho \frac{\partial A_\mu(y)}{\partial y_\varrho} = y_\varrho G_{\varrho\mu}(y). \tag{2.6}$$

Substitute now $y = \alpha x$. One immediately sees that the left-hand side of eq. (2.6) reduces to a full derivative $(d/d\alpha)\,\big(\alpha A_\mu(\alpha x)\big)$. Integrating over α from 0 to 1 we arrive at eq. (2.3).

For our purposes — and we are interested here in the expansion in local operators $O_n(0)$ — it is much more convenient to deal with another representation for $A_\mu{}^a(x)$ which stems [22] from eq. (2.3):

$$A_\mu(x) = \frac{1}{2\cdot 0!} x_\varrho G_{\varrho\mu}(0) + \frac{1}{3\cdot 1!} x_\alpha x_\varrho \big(D_\alpha G_{\varrho\mu}(0)\big) + \frac{1}{4\cdot 2!} x_\alpha x_\beta x_\varrho \big(D_\alpha D_\beta G_{\varrho\mu}(0)\big) + \cdots, \tag{2.7}$$

where D as usual denotes the covariant derivative. Eq. (2.7) shows that $A_\mu{}^a(x)$ is expressed directly in terms of the field strength tensor at the origin and its covariant derivatives.

Let us outline the proof of eq. (2.7). The key point is the following observation:

$$x_{\alpha_1} \dots x_{\alpha_n} [\partial_{\alpha_1} \dots \partial_{\alpha_n} G_{\varrho\mu}]_{x=0} = x_{\alpha_1} \dots x_{\alpha_n} [D_{\alpha_1} \dots D_{\alpha_n} G_{\varrho\mu}]_{x=0}, \tag{2.8}$$

or, in other words, the ordinary derivatives at the origin may be substituted by covariant derivatives.

Indeed, expanding four-potential in eq. (2.2) we notice that for any x

$$x_\mu \left[A_\mu{}^a(0) + x_{\alpha_1}\, \partial_{\alpha_1} A_\mu{}^a(0) + \frac{1}{2}\, x_{\alpha_1} x_{\alpha_2}\, \partial_{\alpha_1}\, \partial_{\alpha_2} A_\mu{}^a(0) + \cdots \right] = 0, \tag{2.9}$$

and hence

$$\begin{aligned} &x_\mu A_\mu{}^a(0) = 0, \\ &x_\mu x_{\alpha_1} \partial_{\alpha_1} A_\mu{}^a(0) = 0, \\ &x_\mu x_{\alpha_1} x_{\alpha_2} \partial_{\alpha_1} \partial_{\alpha_2} A_\mu{}^a(0) = 0, \dots \end{aligned} \tag{2.10}$$

Now one sees that

$$x_{\alpha_1}\, \partial_{\alpha_1} G^a_{\varrho\mu}(0) = x_{\alpha_1} D_{\alpha_1} G^a_{\varrho\mu}(0),$$

$$x_{\alpha_1} x_{\alpha_2}\, \partial_{\alpha_1}\, \partial_{\alpha_2} G^a_{\varrho\mu}(0) = x_{\alpha_1} x_{\alpha_2}\, \partial_{\alpha_1} D_{\alpha_2} G^a_{\varrho\mu}(0) = x_{\alpha_1} x_{\alpha_2} D_{\alpha_1} D_{\alpha_2} G^a_{\varrho\mu}(0),$$

etc. Further details are given in ref. [22] where the proof is based on the method of mathematical induction.

In order to obtain eq. (2.7) we expand $G(\alpha x)$ in eq. (2.3) in αx, apply eq. (2.8) and integrate over α.

Quite analogous representation is valid for the fermion field

$$\psi(x) = \psi(0) + x_{\alpha_1} D_{\alpha_1} \psi(0) + \frac{1}{2}\, x_{\alpha_1} x_{\alpha_2} D_{\alpha_1} D_{\alpha_2} \psi(0) + \cdots. \tag{2.11}$$

It is worth emphasizing one more that expansions (2.7), (2.11) contain only gauge covariant quantities. Just this fact ensures advantages of the given computational scheme over others.

2.2. The Quark Propagator. General Expression

Correlation functions of currents are determined by quark and gluon propagators, however, unlike the ordinary Feynman diagrams, these propagators describe propagation in external gluon (quark) fields. Concentrate at first on the gluon fields leaving aside for a while effects of quark condensate.

Let the quark be emitted at the point y and annihilated at the point x. Propagator

$$S(x, y) = \langle x| \frac{1}{\not{P} - m} |y\rangle = -i\langle T\{\psi(x), \bar{\psi}(y)\}\rangle \tag{2.12}$$

satisfies the standard equation

$$\left(i \frac{\partial}{\partial x_\mu} \gamma_\mu + gA_\mu(x) \gamma_\mu - m\right) S(x, y) = \delta^{(4)}(x - y), \tag{2.13}$$

where $A_\mu(x)$ is the vacuum field (2.7), $A_\mu = 1/2\, \lambda^a A_\mu{}^a$. If the field $A_\mu(x)$ is small in comparison with characteristic distances $(x - y)$ — and we always work only in these conditions — then $S(x, y)$ is representable in the form of the standard series:

$$\begin{aligned} iS(x, y) &= iS^{(0)}(x - y) + g \int d^4z iS^{(0)}(x - z)\, i\not{A}(z)\, iS^{(0)}(z - y) \\ &+ g^2 \int d^4z'\, d^4z iS^{(0)}(x - z')\, i\not{A}(z')\, iS^{(0)}(z' - z)\, i\not{A}(z)\, iS^{(0)}(z - y) \\ &+ \cdots, \end{aligned} \tag{2.14}$$

where $S^{(0)}(x - y)$ is the free quark propagator. The meaning of this expansion is quite transparent. The quark emitted at the point y and annihilated at the point x scatters 0, 1, 2, ... times off the vacuum field (2.7).

It would be in order here to draw the reader's attention to a subtlety. The propagator (2.14) is translationally non-invariant due to two reasons. First of all, the field A_μ depends on coordinate. In a sense this dependence is fictitious since the vacuum field, after averaging, is clearly invariant under translations. The true source of the symmetry breaking is the gauge condition (2.2) according to which the origin plays a special role. When one works with gauge-dependent quantities one should be careful since, for instance, $S(x, y)$ and $S(x - y, 0)$ are generally speaking absolutely different functions. After fixing the reference frame one should consistently perform all computations just in this frame. Only in final results for n-point functions of colourless currents we return to original situation — the answer is one and the same for any choice of the reference frame.

We will not dwell any more on general relations, although one could go rather far along these lines (Chapter I). Instead, we turn to particular examples and consider all cases one may encounter with in applications. Recall that for light quarks (u, d, s) the mass is a small parameter, $m_q \ll \mu$ (μ stands for characteristic hadronic scale). Starting in the zeroth approximation from massless quarks, we, then, account for corrections proportional to m_q perturbatively. On the contrary, for c, b and t quarks $m_Q \gg \mu$, which also results in simplifications. Two different limiting cases, $m_q \ll \mu$ and $m_Q \gg \mu$ are treated separately.

2.3. Heavy Quarks

If the quark mass m_Q is sufficiently large the corresponding quark condensate is not an independent parameter, and the problem completely reduces to a motion in external gluon field. Assume that we are interested in the two-point function of vector currents $\langle J_\mu(x), J_\nu(y)\rangle$ where

$$J_\mu(x) = \bar{Q}(x)\,\gamma_\mu Q(x). \tag{2.15}$$

The current is defined in such a way that it contains no derivatives — an important fact allowing one to choose from the very beginning the origin, say, at the point y. This reference frame will be called below standard. Then

$$\begin{aligned}\Pi_{\mu\nu}(q) &= i \int e^{iqx}\, d^4x \langle T\{J_\mu(x)\, J_\nu(0)\}\rangle \\ &= i \int e^{iqx}\, d^4x \operatorname{Tr}\{\gamma_\mu S(x, 0)\, \gamma_\nu S(0, x)\}.\end{aligned} \tag{2.16}$$

In this Chapter the symbol Tr means the trace operation with respect *to Lorentz and colour indices only.*

As is well known, if $m_Q \neq 0$ it is much more convenient to work in the momentum space. Since $S(x, 0)$ depends on a single variable, x, one can perform the following Fourier transformation

$$\begin{aligned}S(p) &= \int S(x, 0)\, e^{ipx}\, d^4x, \\ \tilde{S}(p) &= \int S(0, x)\, e^{-ipx}\, d^4x,\end{aligned} \tag{2.17}$$

and in terms of these functions

$$\Pi_{\mu\nu}(q) = i \int \operatorname{Tr}\{\gamma_\mu S(p)\, \gamma_\nu \tilde{S}(p - q)\}\, \frac{d^4p}{(2\pi)^4} \tag{2.18}$$

(see fig. 3). Notice that $S(p)$ and $\tilde{S}(p)$, generally speaking, do not coincide. If it were not for this subtlety, the expression would be superficially just the same as for the loop of free quarks.

The general expansion (2.14) after Fourier transformation turns into the following graphic series

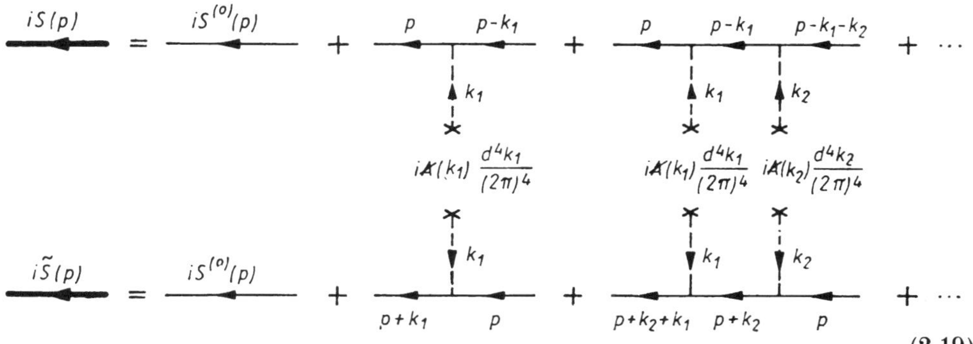

(2.19)

2*

where $A_\mu(k)$ is the Fourier transform of the vacuum field,

$$A_\mu(k) = \int A_\mu(z)\, e^{ikz}\, d^4z = \frac{-i(2\pi)^4}{2}\, G_{\varrho\mu}(0)\, \frac{\partial}{\partial k_\varrho}\, \delta^{(4)}(k)$$

$$+ \frac{(-i)^2\,(2\pi)^4}{3}\, \big(D_\alpha G_{\varrho\mu}(0)\big) \frac{\partial^2}{\partial k_\varrho\, \partial k_\alpha}\, \delta^{(4)}(k) + \cdots. \quad (2.20)$$

Each gluon line is accompanied by integration. Since $A_\mu(k)$ reduces to derivatives of the delta function the integration is performed trivially. Actually, it is equivalent to differentiation of all propagators preceding the given vertex (in the case of $\tilde{S}(p)$) or following it (in the case of $S(p)$).

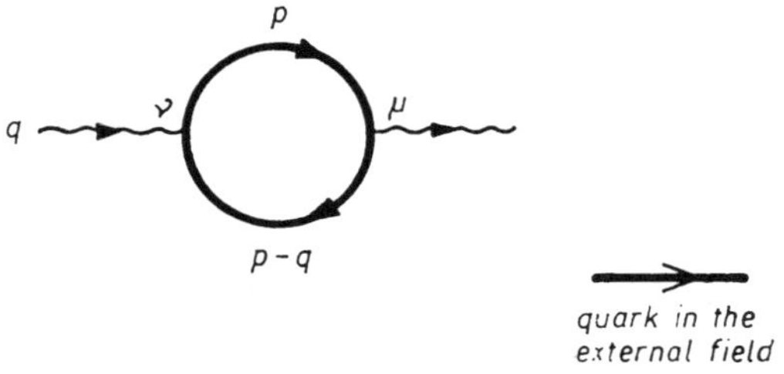

Fig. 3. Correlation function of quark currents in the external field

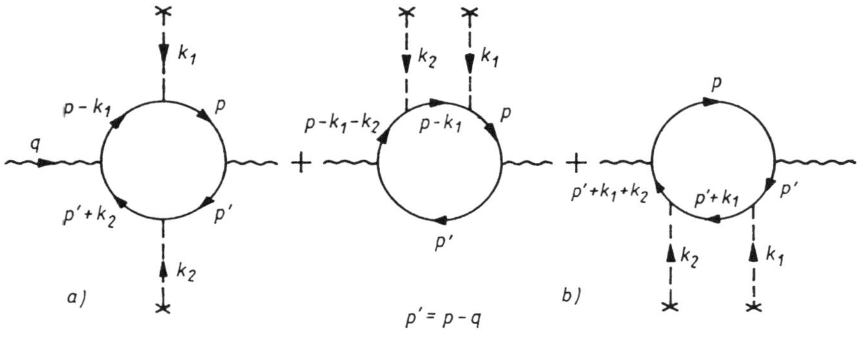

Fig. 4. G^2 correction in the correlation function of quark currents (heavy quarks)

To demonstrate in detail how the procedure works we will calculate, following ref. [6c], the G^2 correction on the two-point function (2.18). Operator G^2, evidently, can appear only in the diagrams where the external field $A(k)$ is inserted twice (fig. 4). Moreover, it is sufficient to keep only the first term of expansion for $A(k)$,

$$A_\alpha(k_1) = -\frac{i(2\pi)^4}{2}\, G_{\varrho\alpha}(0)\, \frac{\partial}{\partial k_{1\varrho}}\, \delta^{(4)}(k_1),$$

$$A_\beta(k_2) = -\frac{i(2\pi)^4}{2}\, G_{\sigma\beta}(0)\, \frac{\partial}{\partial k_{2\sigma}}\, \delta^{(4)}(k_2).$$

Using the fact that

$$\langle G^a_{\mu\nu}(0)\, G^b_{\alpha\beta}(0)\rangle = \frac{1}{96}\, \delta^{ab}(g_{\mu\alpha}g_{\nu\beta} - g_{\mu\beta}g_{\nu\alpha})\, \langle G^c_{\varrho\sigma} G^c_{\varrho\sigma}\rangle \quad (2.21)$$

and taking the colour trace we get

$$[\Pi_{\mu\nu}(q)]_a = -\frac{i}{96}\langle g^2 G^a_{\mu\nu} G^a_{\mu\nu}\rangle (g_{\varrho\sigma}g_{\alpha\beta} - g_{\varrho\beta}g_{\sigma\alpha}) \int \frac{d^4p}{(2\pi)^4} \frac{\partial}{\partial k_{1\varrho}} \frac{\partial}{\partial k_{2\sigma}} \mathrm{Tr}_L$$
$$\times \left\{\gamma_\mu \frac{1}{\not p - m} \gamma_\alpha \frac{1}{\not p - \not k_1 - m} \gamma_\nu \frac{1}{\hat{\not p}' + \not k_2 - m} \gamma_\beta \frac{1}{\hat{\not p}' - m}\right\}_{k_1=k_2=0},$$
$$[\Pi_{\mu\nu}(q)]_b = -\frac{i}{48}\langle g^2 G^a_{\mu\nu} G^a_{\mu\nu}\rangle (g_{\varrho\sigma}g_{\alpha\beta} - g_{\varrho\beta}g_{\sigma\alpha}) \int \frac{d^4p}{(2\pi)^4} \frac{\partial}{\partial k_{1\varrho}} \frac{\partial}{\partial k_{2\sigma}} \mathrm{Tr}_L$$
$$\times \left\{\gamma_\mu \frac{1}{\not p - m} \gamma_\alpha \frac{1}{\not p - \not k_1 - m} \gamma_\beta\right.$$
$$\left.\times \frac{1}{\not p - \not k_1 - \not k_2 - m} \gamma_\nu \frac{1}{\not p' - m}\right\}_{k_1=k_2=0},$$

where we have accounted for the fact that two diagrams of fig. 4b give equal contributions.

Due to current conservation the two-point function (2.18) is transversal,

$$\Pi_{\mu\nu}(q) = (q_\mu q_\nu - q^2 g_{\mu\nu})\, \Pi(q^2),$$

and, therefore, with no loss of information one can contract indices μ and ν which results in further simplifications:

$$[\Pi_{\mu\mu}]_a = -i\langle g^2 G^2\rangle \int \frac{d^4p}{(2\pi)^4} \frac{pp'}{(p^2 - m^2)^2\,(p'^2 - m^2)^2}, \tag{2.22}$$

$$[\Pi_{\mu\mu}]_b = -i4m^2\langle g^2 G^2\rangle \int \frac{d^4p}{(2\pi)^4} \frac{pp' - 2p^2}{(p^2 - m^2)^4\,(p'^2 - m^2)}. \tag{2.23}$$

As a result, the problem is reduced to the simplest Feynman integrals. Their computation is performed by standard methods and yields

$$\Pi^{(G)}_{\mu\nu}|_{\text{heavy quarks},1^-} = (q_\mu q_\nu - q^2 g_{\mu\nu}) \frac{1}{48} \left\langle \frac{\alpha_s}{\pi} G^2 \right\rangle \frac{1}{Q^4}$$
$$\times \left\{\frac{3(a+1)(a-1)^2}{a^2} \frac{1}{2\sqrt{a}} \ln\frac{\sqrt{a}+1}{\sqrt{a}-1} - \frac{3a^2 - 2a + 3}{a^2}\right\}, \tag{2.24}$$
$$a = 1 + (4m_Q^2/Q^2)$$

— the expression used in [1] in analysis of charmonium.

It is evident that the algorithm described above is easily generalized on gluon operators of higher dimension. Thus, in recent works [16, 21] coefficients in front of G^3 and G^4 in the two-point function (2.18) have been found by this technique. Certainly, the procedure is not less effective for other two-point functions. For instance, the G^2 correction in the correlation function of pseudoscalar currents $\bar{Q} i\gamma_5 Q$ is as follows

$$\Pi^{(G)}|_{\text{heavy quarks},0^-} = \frac{1}{48} \left\langle \frac{\alpha_s}{\pi} G^2 \right\rangle \frac{1}{4m_Q^2}$$
$$\times \left\{\frac{3(3a+1)(a-1)^2}{a^2} \frac{1}{2\sqrt{a}} \ln\frac{\sqrt{a}+1}{\sqrt{a}-1} - \frac{9a^2 + 4a + 3}{a^2}\right\}. \tag{2.25}$$

The careful reader, probably, has noticed that one of the simplifying devices is non-universal. We bear in mind the choice of the origin coinciding with coordinate of one of the currents. If the current contains derivatives, say, $\bar{Q}\overleftrightarrow{D}_\mu\gamma_5 Q$, it is impossible to put x or y equal to zero from the very beginning since the quark loop (in external field) is determined now by derivatives of the type $(\partial/\partial x_\alpha)\, S(x, y)$ and $(\partial/\partial y_\alpha)\, S(x, y)$. First of all, we must perform differentiation. This causes no difficulties, though. Instead of eq. (2.14), we have now

$$\begin{aligned}\frac{\partial}{\partial y_\alpha} iS(x, y) &= \frac{\partial}{\partial y_\alpha} iS^{(0)}(x-y) + g\int d^4z iS^{(0)}(x-z)\, i\not{A}(z) \frac{\partial}{\partial y_\alpha} iS^{(0)}(z-y) + \cdots \\ &= -\frac{\partial}{\partial x_\alpha} iS^{(0)}(x-y) + g\int d^4z iS^{(0)}(x-z)\, i\not{A}(z)\left(-\frac{\partial}{\partial z_\alpha}\right) \\ &\quad \times iS^{(0)}(z-y) + \cdots, \end{aligned} \tag{2.26}$$

and this last expression already permits $y = 0$. Moreover, Fourier transformation of eq. (2.26) is well-defined,

$$\begin{aligned}&\int d^4x e^{ipx}\left[\frac{\partial}{\partial y_\alpha} iS(x, y)\right]_{y=0} \\ &= ip_\alpha iS^{(0)}(p) + g\int iS^{(0)}(p)\, i\not{A}(k) \frac{d^4k}{(2\pi)^4}\, i(p-k)_\alpha\, iS^{(0)}(p-k) + \cdots;\end{aligned} \tag{2.27}$$

all further manipulations go along the lines sketched above.

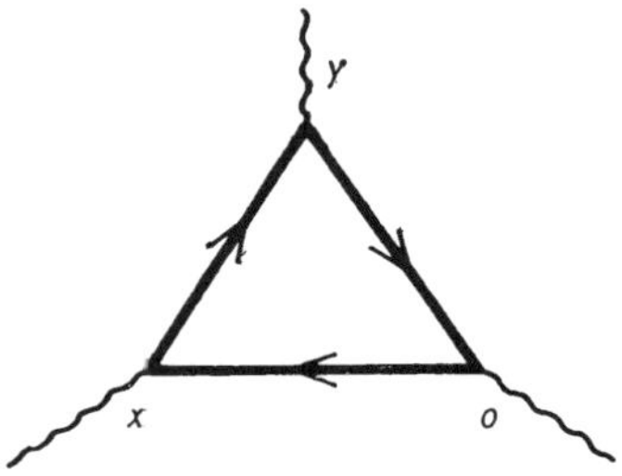

Fig. 5. Three-point function in the external field. The standard choice of the reference frame

Analysis of three-point functions would require additional, although rather trivial, modification of the algorithm. Even under the standard choice of the reference frame (fig. 5) in one of the propagators both coordinates are different from zero. We leave the problem of transition to the momentum space in this case to the reader as a simple exercise. The answer is illustrated in fig. 6. Let us especially call the reader's attention to the fact that momentum k_2 penetrated on the "foreign" line — the line corresponding to the first propagator, which never happens in two-point functions under the standard choice of the reference frame (i.e. $y = 0$).

Concluding this section let us make one more remark concerning the dependence on the quark mass. Varying this parameter — passing from heavy quarks to light ones — we eventually find ourselves in the domain $m \ll \mu$ where some diagrams become infrared — unstable even if all external momenta are euclidean and large. Consider, for instance, the graphs depicted in fig. 4b; the corresponding contribution to $\Pi_{\mu\mu}$ is given in eq. (2.23). At first sight it is proportional to m^2. One can easily convince oneself, however, that at $Q^2 \gg m^2$ characteristic virtual momenta in integral (2.23) $p^2 \sim m^2$ (and not $p^2 \sim Q^2$!), the integral is of order $1/m^2$, and the answer turns out to be of order m^0 — the situation not dangerous for heavy quarks since $p^2 \sim m^2 \gg \mu^2$ is the domain where calculations are reliable. On the contrary, for light quarks, $m^2 \ll \mu^2$, the result obtained in this way, evidently, has nothing to do with correct one. Indeed, when deriving eq. (2.23) we have assumed that the interval between consequent scatterings off the vacuum field is small, and in between the scatterings the quarks move as if they are free. If $p^2 \sim m^2 \ll \mu^2$ both assumptions become wrong. Propagation of the soft quark in these conditions is determined entirely by confinement dynamics. The graphs of the type of fig. 4b are absolutely senseless, and, instead of them, we draw the diagram of fig. 1 where the crosses mark soft lines. The corresponding effect is parametrized as $\langle \bar{u}u \rangle$ or $\langle \bar{d}d \rangle$.

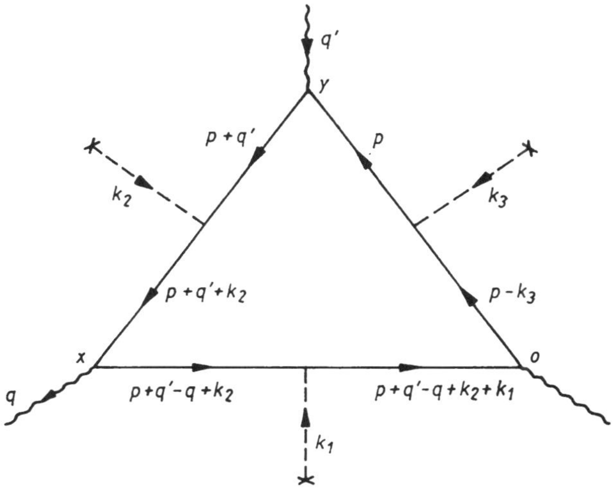

Fig. 6. One of the graphs determining G^3 correction to three-point function

Actually, the diagrams of fig. 4b represent an attempt to calculate quark condensate in terms of $\langle G^2 \rangle$, an attempt doomed to failure in the case at hand since $\langle \bar{u}u \rangle$ and $\langle \bar{d}d \rangle$ certainly do not reduce to a few rescatterings of a massless free quark. The following fact nicely demonstrates how far this model is from reality. According to eq. (2.23) the diagrams 4b give non-vanishing contribution to $\pi_{\mu\mu}(q)$ in the limit $m \to 0$. On the other hand, in terms of quark condensate the corresponding contribution is equal to $6m\langle \bar{\psi}\psi \rangle / Q^2$ (fig. 1). It tends to zero with $m \to 0$ provided that the vacuum expectation value $\langle \bar{\psi}\psi \rangle$ is non-singular in the chiral limit (and we known that this is indeed the case).

Thus, passing from heavy quarks to light ones it is impossible to mechanically put $m = 0$ in expressions for n-point functions. Such a procedure will generally speaking give a wrong answer for coefficients in front of the gluon operators. The defect can be eliminated by separating a piece associated with large distances and reparametrizing it in terms of quark operators.

This requires a few additional steps, actually quite superfluous, if one uses the virtues of the Fock-Schwinger gauge in full measure. The technique adequate for massless quarks, automatically yielding the correct operator expansion, is described below.

2.4. Massless Quarks. Gluon Operators

For $m \ll \mu$, when the quark mass can be treated perturbatively, the most economic procedure is as follows. First of all, practically up to the very last step there is no need in transition to the momentum space since the free quark propagator in the x representation looks just as simple as in the p representation,

$$S^{(0)}(x-y) = \frac{1}{2\pi^2} \frac{(\not{x} - \not{y})}{\left((x-y)^2\right)^2}. \tag{2.28}$$

This simple observation essentially lowers the number of intermediate integrations. Moreover, there exists a compact expression for quark propagator in external field which, being once found, serves in all further computations.

To begin with we consider a problem most important in the practical respect — the G^2 correction — and shall return to discussion of other operators somewhat later.

The starting point is expansion (2.14) for quark propagator. The terms containing A^3, A^4 etc. will, evidently, give operators of higher dimension, 6 and higher. Therefore, concentrate on the first two non-trivial terms.

If we are interested only in the operator $G^a_{\mu\nu}G^a_{\mu\nu}$ then the external field (2.7) can be taken in the form

$$A_\mu(z) = \frac{1}{2} z_\varrho G_{\varrho\mu}(0). \tag{2.29}$$

Then iteration of this expression will give G^2. One can easily convince oneself that gluon operators with derivatives in eq. (2.7) cannot induce operator $G^a_{\mu\nu}G^a_{\mu\nu}$ in the correlation function of colourless currents.

Armed with this information we write

$$\begin{aligned} S(x, y) = {} & S^{(0)}(x-y) - \frac{1}{(2\pi^2)^2}\frac{1}{2}\int \frac{\not{x}-\not{z}}{(x-z)^4} z_\varrho G_{\varrho\mu}(0)\, \gamma_\mu \frac{\not{z}-\not{y}}{(z-y)^4} d^4z \\ & + \frac{1}{(2\pi^2)^3}\frac{1}{2^2}\iint \frac{\not{x}-\not{z}'}{(x-z')^4} z_\varrho' G_{\varrho\mu}(0)\, \gamma_\mu \frac{\not{z}'-\not{z}}{(z'-z)^4} z_\alpha G_{\alpha\beta}(0) \\ & \times \gamma_\beta \frac{\not{z}-\not{y}}{(z-y)^4} d^4z\, d^4z' + \cdots, \end{aligned} \tag{2.30}$$

where the first integral is trivially calculable and generates G^2 in products of the type $S(x, y)\, S(y, x)$. The second integral contains G^2 from the very beginning and essentially simplifies after extracting the piece singlet with respect to Lorentz and colour indices (following eq. (2.21)). Just the singlet part determines corrections proportional to $\langle G^a_{\mu\nu}G^a_{\mu\nu}\rangle$.

It is easy to check that [23]

$$\begin{aligned} S(x,y) = {} & \frac{1}{2\pi^2}\frac{\not{r}}{(r^2)^2} - \frac{1}{8\pi^2}\frac{r_\alpha}{r^2}\tilde{G}_{\alpha\varphi}(0)\,\gamma_\varphi\gamma_5 \\ & + \left\{\frac{i}{4\pi^2}\frac{\not{r}}{(r^2)^2} y_\varrho x_\mu G_{\varrho\mu}(0) - \frac{1}{192\pi^2}\frac{\not{r}}{(r^2)^2}\left(x^2y^2 - (xy)^2\right) G_{\varphi\chi}(0)\, G_{\varphi\chi}(0)\right\} \\ & + \text{operators of higher dimension} \end{aligned} \tag{2.31}$$

where

$$r = x - y, \qquad G_{\alpha\beta} = \frac{g}{2} \lambda^a G^a_{\alpha\beta}, \qquad \tilde{G}_{\alpha\beta} = \frac{1}{2} \varepsilon_{\alpha\beta\mu\nu} G_{\mu\nu}.$$

Notice that under the standard choice of the origin, $y = 0$, the expression in braces vanishes at all. Therefore, in two-point functions induced by currents with no derivatives[3]) the G^2 correction is determined entirely by the first two terms in eq. (2.31).

For instance, computation for the vector current of massless quarks, $j_\mu = \bar{q}\gamma_\mu q$, proceeds as follows. We start from the general expression (2.16) and substitute eq. (2.31) for the quark propagator. Then we obtain

$$\begin{aligned} &\operatorname{Tr} \{\gamma_\mu S(x, 0)\, \gamma_\nu S(0, x)\} \\ &= \operatorname{Tr} \left\{\gamma_\mu \left(\frac{1}{2\pi^2} \frac{\not{x}}{x^4} - \frac{1}{8\pi^2} \frac{x_\alpha}{x^2} \tilde{G}_{\alpha\beta}\gamma_\beta\gamma_5\right) \gamma_\nu \left(-\frac{1}{2\pi^2} \frac{\not{x}}{x^4} + \frac{1}{8\pi^2} \frac{x_\gamma}{x^2} \tilde{G}_{\gamma\delta}\gamma_\delta\gamma_5\right)\right\} \\ &= -\frac{3}{\pi^4} \frac{2x_\mu x_\nu - x^2 g_{\mu\nu}}{x^8} - \frac{\langle g^2G^2\rangle}{384\pi^4} \frac{2x_\mu x_\nu + x^2 g_{\mu\nu}}{x^4}, \end{aligned} \tag{2.32}$$

where we have used also eq. (2.21). Having checked that the result (2.32) is transversal, as it should be, we contract indices μ and ν and go to the momentum space (the formulae collected in Appendix D will help us). Finally,

$$\Pi_{\mu\mu}(q)|_{\text{light quarks}, 1^-} = \frac{3}{4\pi^2} q^2 \ln Q^2 - \frac{1}{16\pi^2} \frac{1}{q^2} \langle g^2G^2\rangle, \tag{2.33}$$

which exactly coincides with the answer obtained first in ref. [1]. The whole computation takes no more than 10 minutes.

It is worth considering one more pedagogical example, only indirectly connected with the subject of the present review — the triangle anomaly in divergence of the axial current. Here the advantages of the proposed computational scheme are demonstrated from another point of view. If we regularize the axial current, following Schwinger, by virtue of ε splitting, then

$$\begin{aligned} \partial_\mu j^{\text{reg}}_{\mu 5} &= \frac{\partial}{\partial x_\mu} \left\{\bar{\psi}(x + \varepsilon)\, \gamma_\mu\gamma_5 \left(\exp \int_{x-\varepsilon}^{x+\varepsilon} igA_\varrho(y)\, dy_\varrho\right) \psi(x - \varepsilon)\right\} \\ &= \bar{\psi}(x + \varepsilon) \{-ig\not{A}(x + \varepsilon)\, \gamma_5 - \gamma_5 ig\not{A}(x - \varepsilon) + ig\gamma_\mu\gamma_5\varepsilon_\beta G_{\mu\beta}(0)\}\, \psi(x - \varepsilon), \end{aligned}$$

where the third term has emerged from differentiation of the exponential and, as usual, $A_\mu = 1/2\, \lambda^a A_\mu{}^a$.

This first step is more or less standard. (Notice, however, that we have used expansion (2.7) having substituted $A_\mu(y) = 1/2\, y_\varrho G_{\varrho\mu}(0)$). The further derivation is essentially shortened as compared to the normal route by use of expansion (2.31) for quark propagator in the Fock-Schwinger gauge. Only the second term of the expansion is operative. Namely,

$$\begin{aligned} \partial_\mu j^{\text{reg}}_{\mu 5} &= -ig \operatorname{Tr} [-2i\varepsilon_\varrho G_{\varrho\mu}(0)\, \gamma_\mu\gamma_5 S(x - \varepsilon, x + \varepsilon)] \\ &= -\frac{g^2}{2} G^a_{\varrho\mu}(0)\, \tilde{G}^a_{\alpha\varphi}(0)\, \frac{\varepsilon_\varrho\varepsilon_\alpha}{\varepsilon^2} \frac{1}{8\pi^2} \operatorname{Tr}_L \gamma_\mu\gamma_5\gamma_\varphi\gamma_5 = \frac{\alpha_s}{4\pi} G^a_{\alpha\beta}\tilde{G}^a_{\alpha\beta}. \end{aligned}$$

[3]) The expression in braces in eq. (2.31) is essential and cannot be omitted if one considers three-point functions provided that the current contain derivatives.

Of course, we have just rederived the well-known result but the whole business is the matter of a few minutes.

This short excursion seems to be sufficient for teaching the reader how he/she can calculate G^2 corrections in any correlation functions.

For currents with derivatives or in three-point Green functions, generally speaking, all four terms quoted in eq. (2.31) are operative. The procedure stays as simple and algorithmic as it was explained above; true, the number of elementary arithmetical operations somewhat increases.

There is no difficulty in extending the result to include gluon operators of higher dimension. Leaving aside all details we give here the complete expression for quark propagator (in p space) up to term $O(p^{-5})$ inclusively [6b]:

$$S(p) = \int e^{ipx}\, d^4x S(x, 0) = \frac{1}{\not p} - \frac{p_\alpha}{p^4} g\tilde{G}_{\alpha\beta}\gamma_\beta\gamma_5$$
$$+ \frac{2}{3} g \frac{1}{p^6} [p^2 D_\alpha G_{\alpha\beta}\gamma_\beta - \not p D_\alpha G_{\alpha\beta} p_\beta - p_\gamma D_\gamma p_\alpha G_{\alpha\beta}\gamma_\beta - 3ip_\gamma D_\gamma p_\alpha \tilde{G}_{\alpha\beta}\gamma_\beta\gamma_5]$$
$$+ \frac{2}{p^8} g \Big[ip^2 p_\gamma D_\gamma D_\alpha G_{\alpha\beta}\gamma_\beta - i\not p p_\gamma D_\gamma D_\alpha G_{\alpha\beta} p_\beta - i(p_\gamma D_\gamma)^2\, p_\alpha G_{\alpha\beta}\gamma_\beta$$
$$+ 2(p_\gamma D_\gamma)^2\, p_\alpha \tilde{G}_{\alpha\beta}\gamma_\beta\gamma_5 - \frac{1}{2} D_\gamma D_\gamma p^2 p_\alpha \tilde{G}_{\alpha\beta}\gamma_\beta\gamma_5 \Big]$$
$$+ \frac{1}{p^8} g^2[-2\not p p_\alpha G_{\alpha\beta} G_{\beta\gamma} p_\gamma + p^2 p_\alpha (G_{\alpha\beta} G_{\beta\delta} + G_{\beta\delta} G_{\alpha\beta})\, \gamma_\delta$$
$$- p^2 p_\alpha (G_{\alpha\beta} G_{\beta\delta} - G_{\beta\delta} G_{\alpha\beta})\, \gamma_\delta] + \cdots. \tag{2.34}$$

It is assumed that all operators are taken at the origin, for instance, $G_{\alpha\beta} = 1/2\, \lambda^a G^a_{\alpha\beta}(0)$.

The coefficient of gluon operator of (normal) dimension d is proportional to $(p)^{-d+1}$. In closed loops only several first terms of expansion (2.34) are meaningful — namely, those which do not lead to infrared divergences under integration over p. If the integrals over p are saturated in the momentum domain $p^2 \sim q^2$ where q is a large external moment then the expansion in terms of gluon operators is legitimate. Passing to operators of higher dimension we, sooner or later, encounter with an integral with a power singularity at $p^2 = 0$. The corresponding quark line becomes soft and must be parametrized in terms of quark operators. In this case we refuse to integrate over p and say that the quark line is "cut" (fig. 1 and, in more detail, Sec. 2.5).

At what particular step of the expansion (2.34) does the infrared instability emerge? This depends on the form of the vertices. If there are no derivatives in the vertices the power infrared divergence appears for operators of dimension 6. Derivatives in the vertices shift the boundary towards higher dimensions. The general rule is universal — coefficient in the operator expansions are always determined by short distances. If the calculation becomes sensitive to large distances, this means that, instead of the coefficient, we are "calculating" the gluon matrix element of some operator missed at the previous step — the action contradicting our original intensions. Actually, a mistake. We should return back by a step and include the missed operator in the analysis.

And what if the divergence is logarithmic? In this case one should not hurry throwing away the result as irrelevant. The infrared logarithm does not mean a mistake. Log of q^2/μ^2 in the coefficient of the gluon operator simply signals that there is another (quark or gluon) operator of the same dimension present in the expansion, and the conventional logarithmic mixing between them takes place. Clearly, it must be accounted for. An example of such a situation is presented in fig. 7.

Concluding this Section we quote for the sake of reference the expression for gluon propagator [6b] up to terms $O(q^{-6})$:

$$\begin{aligned} D_{\mu\nu}(q) &= \int d^4x\, e^{iqx}\, D_{\mu\nu}(x, 0) \\ &= \frac{g_{\mu\nu}}{q^2} + g\,\frac{2}{q^4}\, G_{\mu\nu} + g\,\frac{4i}{q^6}\,(qD)\, G_{\mu\nu} - g\,\frac{2i}{3q^6}\, g_{\mu\nu} D_\alpha G_{\alpha\beta} q_\beta \\ &\quad + g\,\frac{2}{q^8}\,(qD)\, D_\alpha G_{\alpha\beta} q_\beta g_{\mu\nu} + g\,\frac{2}{q^8}\left(q^2 D^2 G_{\mu\nu} - 4(qD)^2\, G_{\mu\nu}\right) \\ &\quad + g^2\,\frac{1}{2q^8}\, g_{\mu\nu}\left(q^2 G_{\alpha\beta}^2 - 4(q_\alpha G_{\alpha\beta})^2\right) + g^2\,\frac{4}{q^6}\, G_{\mu\alpha} G_{\alpha\nu}, \end{aligned} \tag{2.35}$$

where we use the following matrix notation

$$G_{\mu\nu} \equiv G_{\mu\nu}^{ab}(0) = f^{acb} G_{\mu\nu}^c(0). \tag{2.36}$$

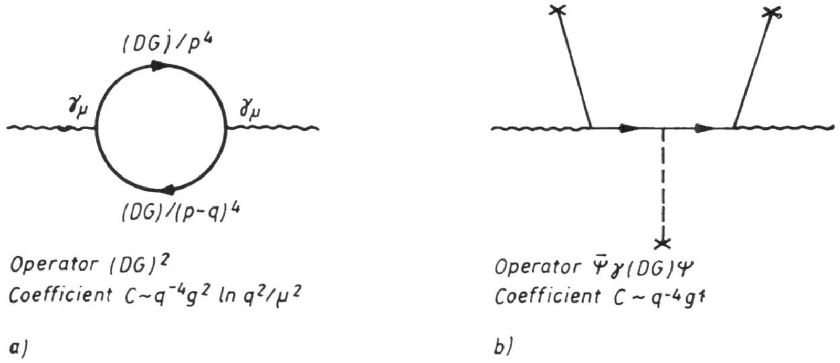

Fig. 7. Operators $(DG)^2$ and $\bar{\psi}\gamma(DG)\,\psi$ in the correlation function of vector currents. Diagram a gives an infrared logarithm reflecting mixing of the operators

Let us draw the reader's attention to one curious fact. In the quark propagator (2.34) it is impossible to separate the structure singlet with respect to Lorentz and colour indices, $G_{\mu\nu}^a G_{\mu\nu}^a$. Strictly speaking, this structure can be separated with zero coefficient as it is seen from eq. (2.31) with $y = 0$. At the same time, in gluon propagator it is explicitly present (see the last term in eq. (2.35). Not accidentally. Analytical properties of the propagators are such that $\overline{S(x, 0)}$ does not admit singularities of the type $\ln x^2$ while $\overline{D_{\mu\mu}(x, 0)}$ does admit it (here the bar denotes the Lorentz and colour averaging). One can easily prove the latter assertion following the line of reasoning of ref. [6a]. In this work a related problem was discussed: analytical properties of quark two-point functions in self-dual fields.

2.5. Quark Operators

Effects due to quark condensate are essential almost in all applications. They play the dominant role, for instance, in the sum rules for ϱ mesons, D mesons and, especially, baryons. As a rule, calculation of the corresponding coefficients is much simpler technically than for gluon operators. Almost always, after drawing the relevant graph, we immediately read off the answer. Still, even here there exist some useful devices which will be demonstrated below in a few examples.

As in other training exercises consider the correlation function (2.16) of vector currents, $j_\mu = \bar{\psi}\gamma_\mu\psi$, where ψ is the massless quark field. Let one of the quark line be soft. Then, cutting it (fig. 7), we get

$$\Pi_{\mu\nu}(q) = i \int e^{iqx}\, d^4x \langle T\{j_\mu(x)\, j_\nu(0)\}\rangle$$
$$\to -\int e^{iqx}\, d^4x \{\bar{\psi}(x)\, \gamma_\mu S(x, 0)\, \gamma_\nu\psi(0) + \bar{\psi}(0)\, \gamma_\nu S(0, x)\, \gamma_\mu\psi(x)\}\,, \tag{2.37}$$

where the cut lines are represented by the Heisenberg operators $\psi(x)$ and $\psi(0)$.

For large q, to the leading order in q^{-1}, the field $\psi(x)$ can be substituted by $\psi(0)$. Accounting for the fact that $\langle\bar{\psi}_\alpha{}^i(0)\, \psi_{j\beta}(0)\rangle = 1/3\, \delta_j{}^i\; 1/4\, \delta_{\alpha\beta}$ we arrive at the following expression

$$\Pi_{\mu\nu}(q) = -\frac{1}{12} \langle\bar{\psi}\psi\rangle \operatorname{Tr} \{\gamma_\mu S(q)\, \gamma_\nu + \gamma_\nu \tilde{S}(-q)\, \gamma_\mu\}, \tag{2.38}$$

which vanishes if the quark mass is neglected. The zero result in the chiral limit is obvious beforehand since the structure $\bar{\psi}\psi$ is chirally non-invariant and appears in the correlation function (2.37) only being multiplied by the quark mass m. Thus, at least this general property is fulfilled and there is no error.

Feeling complete satisfaction from the internal selfconsistency we now put the problem of determining the coefficient in front of $m\bar{\psi}\psi$. For this problem the approximation $\psi(x) = \psi(0)$ in eq. (2.37) is not sufficient, and it is necessary to account for the next term of expansion in (2.9)

$$\psi(x) = \psi(0) + x_\varrho D_\varrho \psi(0)$$

bearing in mind that $D_\varrho\psi$ will eventually become $\not{D}\psi = -im\psi$.

It is no surprise that $x_\varrho D_\varrho\psi(0)$ induces a new term in $\Pi_{\mu\nu}(q)$, namely,

$$-\int e^{iqx}\, d^4x \{\bar{\psi}(0)\, x_\varrho \overleftarrow{D}_\varrho \gamma_\mu S(x, 0)\, \gamma_\nu\psi(0) + \bar{\psi}(0)\, \gamma_\nu S(0, x)\, \gamma_\mu x_\varrho \vec{D}_\varrho \psi(0)\}\,.$$

Observing that

$$\langle\bar{\psi}_\alpha{}^i \vec{D}_\varrho \psi_{j\beta}\rangle = \frac{1}{3}\, \delta_j{}^i\, \frac{1}{16}\, (\gamma_\varrho)_{\beta\alpha} \langle\bar{\psi}\not{D}\psi\rangle = \frac{1}{48i}\, \delta_j{}^i (\gamma_\varrho)_{\beta\alpha}\, m\langle\bar{\psi}\psi\rangle$$

we reduce this extra term to

$$\frac{1}{48}\, m\langle\bar{\psi}\psi\rangle \frac{\partial}{\partial q_\varrho} \operatorname{Tr} \{-\gamma_\mu S(q)\, \gamma_\nu\gamma_\varrho + \gamma_\nu \tilde{S}(-q)\, \gamma_\mu\gamma_\varrho\}\,. \tag{2.39}$$

Combining now eqs. (2.38) and (2.39) we obtain

$$\Pi_{\mu\nu}(q) = 2m\langle\bar{\psi}\psi\rangle \left(\frac{q_\mu q_\nu}{q^4} - \frac{g_{\mu\nu}}{q^2}\right). \tag{2.40}$$

This result was exploited many times in ref. [1].

One more example referring to the operator of dimension 6,

$$\bar{\psi}\gamma_\nu\lambda^a\psi D_\mu G^a_{\mu\nu}\,,$$

appearing in eq. (2.16) already in the chiral limit. Correction to $\Pi_{\mu\mu}(q)$ associated with this operator is of order q^{-4}. The corresponding diagram is depicted in fig. 7b. The consistent procedure requires, generally speaking, expanding $\psi(x)$ up to terms $O(x^3)$, so that in $\Pi_{\mu\mu}(q)$ we deal with three different structures,

$$\Pi_{\mu\mu} \to -2 \int e^{iqx}\, d^4x \bar{\psi}(0) \{\gamma_\mu S^{(3)}(x, 0)\, \gamma_\mu + \overleftarrow{D}_\varrho \gamma_\mu S^{(2)}(x, 0)\, \gamma_\mu x_\varrho + \frac{1}{6} \overleftarrow{D}_\varrho \overleftarrow{D}_\gamma \overleftarrow{D}_\delta \gamma_\mu S^{(0)}(x, 0)\, \gamma_\mu x_\varrho x_\gamma x_\delta\}\, \psi(0), \tag{2.41}$$

where it is taken into account that both terms in (2.37) give one and the same contribution, and the subscript (0, 2, 3) marks the dimension of gluon operator in $S(x, 0)$ (for instance, $S^{(0)} = \text{const}\, \not{x}/x^4$ is the free quark propagator, $S^{(2)} = \text{const}\, x_\varrho \tilde{G}_{\varrho\varphi}\gamma_\varphi\gamma_5$, etc.).

Averaging over the vacuum state — and this is equivalent to separation of Lorentz-singlet operators — we convince ourselves that the second and the third terms in (2.41) are independent of x at all. Hence their Fourier transformation is proportional to $\delta^{(4)}(q)$, and one can forget about these pieces. We are left with the first term,

$$\Pi_{\mu\mu} = -2\bar{\psi}(0)\, \gamma_\mu S^{(3)}(q)\, \gamma_\mu \psi(0),$$

and will now again taste of the fruits of the work done beforehand. We bear in mind expansion (2.34) for $S(q)$ which implies

$$\Pi_{\mu\mu} = (-2)\,(-2)\, \frac{2}{3} \left(1 - \frac{1}{4} - \frac{1}{4}\right) g\langle \bar{\psi}(D_\alpha G_{\alpha\beta})\, \gamma_\beta \psi\rangle\, q^{-4} = \frac{2}{3} \frac{1}{q^4}\, g\langle \bar{\psi}\gamma_\beta \lambda^a \psi D_\alpha G^a_{\alpha\beta}\rangle. \tag{2.42}$$

The factor (-2) reflects the convolution of γ matrices, $\gamma_\mu(\ldots)\,\gamma_\mu$, the factor $(1 - 1/4 - 1/4)$ is a remant of three structures of dimension 3 which are averaged with respect to Lorentz indices in different ways.

2.6. Correlation Functions of Gluon Currents

We proceed in the same vein as with massless quarks. There is some subtlety, though, due to the fact that the four-potential A_μ plays a two-fold role,

$$A_\mu{}^a = (A_\mu{}^a)_{\text{ext}} + a_\mu{}^a, \tag{2.43}$$

where $(A_\mu{}^a)_{\text{ext}}$ is the vacuum field while $a_\mu{}^a$ describes deviations from the vacuum field. In other words, $a_\mu{}^a$ is a quantum fluctuation; gluon propagator in external field is defined in the following way:

$$D^{ab}_{\mu\nu} = \langle T\{a_\mu{}^a(x), a_\nu{}^b(0)\}\rangle. \tag{2.44}$$

The Fock-Schwinger gauge condition (2.2) is nonlocal in the momentum space, and one should work very hardly in order to write Green function $D^a_{\mu\nu}$ satisfying this condition. Even having overcome all difficulties one would certainly get an expression useless from the practical point of view. However, there is absolutely no need in this work. Indeed, let us assume that the Fock-Schwinger condition is imposed on external field

$$x_\mu\big(A_\mu{}^a(x)\big)_{\text{ext}} = 0. \tag{2.45}$$

Substituting eq. (2.43) into standard QCD Lagrangian we get

$$L = -\frac{1}{4} (G^a_{\mu\nu} G^a_{\mu\nu})_{\rm ext} - \frac{1}{2} (D_\mu{}^{\rm ext} a_\nu{}^a)^2 + \frac{1}{2} (D_\mu{}^{\rm ext} a_\nu{}^a)(D_\nu{}^{\rm ext} a_\mu{}^a) + \frac{1}{2} g a_\mu{}^a (G^b_{\mu\nu})_{\rm ext}\, a_\nu{}^c f^{abc} + \cdots,$$
$$D_\mu{}^{\rm ext} a_\nu{}^a = \partial_\mu a_\nu{}^a + g f^{abc} (A_\mu{}^b)_{\rm ext}\, a_\nu{}^c . \qquad (2.46)$$

This Lagrangian describes dynamics of field $a_\mu{}^a$ in the background field $(A_\mu{}^a)_{\rm ext}$ and still possesses gauge freedom; namely it is invariant under the transformations

$$\big(A_\mu(x)\big)_{\rm ext} \to (A_\mu{}')_{\rm ext} = U^{-1}(x) \big(A_\mu(x)\big)_{\rm ext}\, U(x),$$
$$a_\mu(x) \to a_\mu{}' = U^{-1}(x)\, a_\mu(x)\, U(x) + \frac{i}{g} U^{-1}(x)\, \partial_\mu U(x),$$

with arbitrary unitary matrix $U(x)$. It is evident that the first of these transformations does not violate condition (2.45): $x_\mu\big(A_\mu{}'(x)\big)_{\rm ext}$ automatically vanishes provided $x_\mu\big(A_\mu(x)\big)_{\rm ext} = 0$.

To extract the maximal profit from this gauge freedom we fix the gauge of a_μ as follows. We add to Lagrangian (2.46) the term

$$-\frac{1}{2} (D_\mu{}^{\rm ext} a_\mu{}^a)^2 \qquad (2.47)$$

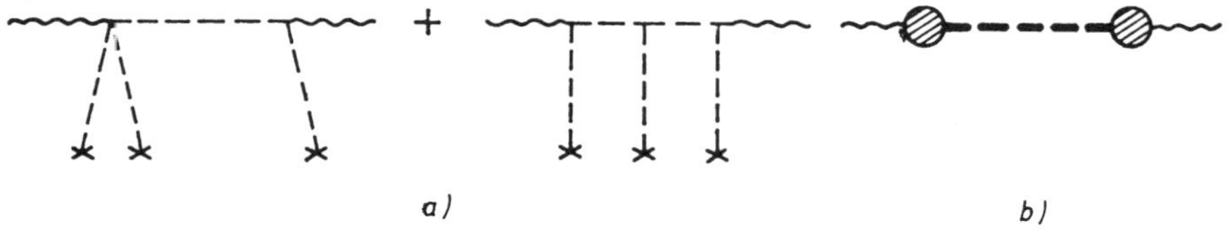

Fig. 8. Operator G^3 in the two-point function of gluon currents. Crosses mark soft lines annihilated by the vacuum (a); b — symbolic picture of the same two-point function in the external field. Gluon propagator (▬ ▬ ▬ ▬) and vertices (circles) are taken in the external field

and, certainly, the ghosts according to standard rules. This is the so called background gauge extensively exploited by DE WITT [24], 'T HOOFT [18] and in many recent works, both purely theoretical, and of more pragmatic character [25—27]. Gluon propagator, eq. (2.35), is given just in this gauge.

The exercise we concentrate on in this Section is the calculation of the G^3 correction in the two-point function of scalar gluonic currents,

$$j_s = \alpha_s G^a_{\mu\nu} G^a_{\mu\nu} . \qquad (2.48)$$

The coefficient of G^3 to the leading in α_s approximation is determined by the Born graphs of fig. 8a. The same diagrams in a symbolical form are depicted in fig. 8b; the external vacuum field is not shown here but it is implied that the gluon transferring the large momentum q propagates in a background field.

The next step is rewriting current (2.47) in terms of the background and quantum fields; as it is seen from fig. 8b it is sufficient to keep only linear in $a_\mu{}^a$ terms,

$$j_s(x) = 4\alpha_s (G^a_{\mu\nu})_{\text{ext}} \left(D_\mu{}^{\text{ext}} a_\nu{}^a(x)\right). \tag{2.49}$$

$O(G^3)$ terms arise in the current correlation function from two sources: from expansion of $G_{\mu\nu}(x)$ and from correction to the gluon propagator:

$$\begin{aligned} &i \int d^4x e^{iqx} \langle T\{j_s(x), j_s(0)\}\rangle \\ &\to 16\alpha_s{}^2 i \int d^4x e^{iqx} \left(G^a_{\mu\nu}(0) \frac{1}{2} \overleftarrow{D}_\alpha \overleftarrow{D}_\beta x_\alpha x_\beta \langle T\{\partial_\mu a_\nu(x), \partial_\gamma a_\delta(0)\}\rangle^{(0)}_{ab} G^b_{\gamma\delta}(0)\right. \\ &\quad \left. + G^a_{\mu\nu}(0) \langle T\{D_\mu a_\nu(x), D_\alpha a_\beta(0)\}\rangle^{(2)}_{ab} G^b_{\alpha\beta}(0)\right) \end{aligned} \tag{2.50}$$

where subscripts (0, 2) refer to dimension of gluon operators and subscript (ext) is omitted. In particular, (0) marks the free propagator

$$\langle T\{\partial_\mu a_\nu(x), \partial_\gamma a_\delta(0)\rangle^{(0)}_{ab} = \frac{1}{4\pi^2} \frac{\partial}{\partial x_\mu} \left(-\frac{\partial}{\partial x_\gamma}\right) \frac{1}{x^2} g_{\nu\delta} \delta_{ab} \,.$$

One can easily convince oneself that after separating Lorentz-singlet structure the first term in eq. (2.50) vanishes. The second term, in turn, reduces to

$$\begin{aligned} &16\alpha_s{}^2 i \int d^4x e^{iqx} G^a_{\mu\nu}(0) \langle T\{(\partial_\mu + gA_\mu) a_\nu(x), \partial_\alpha a_\beta(0)\}\rangle^{(2)}_{ab} G^b_{\alpha\beta}(0) \\ &= 16\alpha_s{}^2 i \int d^4x e^{iqx} G^a_{\mu\nu}(0) \left\langle T\left\{\left(-iq_\mu + g \frac{1}{2} x_\varrho G_{\varrho\mu}(0)\right) a_\nu(x), \partial_\alpha a_\beta(0)\right\}\right\rangle^{(2)}_{ab} G^b_{\alpha\beta}(0)\,, \end{aligned} \tag{2.51}$$

where, as in eq. (2.35), we use the matrix notation $A_\mu = f^{abc} A_\mu{}^b$ and account for the fact that $A_\mu(0) = 0$.

Unfortunately, it is impossible to substitute simply propagator (2.35) in eq. (2.51) since expansion (2.35) refers actually to $D_{\mu\nu}(x, 0)$. At first, we must write $D_{\mu\nu}(x, y)$, differentiate over y and only then put $y = 0$. In gluon propagator we are interested only in terms proportional to G since eq. (2.51) already contains G^2 explicitly. To this order modifications which are needed in (2.35) are trivial,

$$\begin{aligned} \int e^{iqx} dx D_{\mu\nu}(x, y) &= \int i e^{iqx} d^4x \langle T\{a_\mu{}^a(x) a_\nu{}^b(y)\}\rangle \\ &= \frac{g_{\mu\nu}}{q^2} e^{iqy} + \frac{2g}{q^4} G_{\mu\nu}(0) e^{iqy} - giy_\varphi G_{\varphi\varrho}(0) \frac{q_\varrho}{q^4} e^{iqy} g_{\mu\nu} + \cdots. \end{aligned} \tag{2.52}$$

The reader will easily derive this formula himself (herself) starting from Lagrangian (2.46).

Combining eqs. (2.52) and (2.51) we obtain

$$i \int e^{iqx} d^4x \langle T\{j_s(x)\, j_s(0)\}\rangle = -8\alpha_s{}^2 \frac{g}{q^2} \langle G^a_{\mu\nu} G^b_{\nu\beta} G^c_{\beta\mu} f^{abc}\rangle + \text{other operators}\,. \tag{2.53}$$

We have done above a series of exercises in operator expansion. The coefficients of several operators have been found. The material is organized in such a way that the main emphasis is put on routine computations. Performing them from the very beginning up to the very end we do not try to hide the theoretical "kitchen". On the other hand, we do not seek to carry the problems through, up to physical applications and concentrate on purely computational aspects, i.e. those aspects which are presented in original papers only by a few lines (if at all).

The reasons which compelled us to deviate from the settled tradition reading "the way of deriving formulae is a question of no importance" is as follows. First of all, it is impossible to find technical details elsewhere. At best, there are short comments explaining the points with peculiarities, deviations from the "standard". Here we clarify the standard itself, demonstrate the consistent procedure which proves to be extremely effective in all known cases.

Experience shows that the reader who has not got used to the language of operator expansion usually encounters with some difficulties in attempts to obtain this or that coefficient, although, in principle, no additional knowledge is needed except for the Feynman diagram technique. True, the diagrams look somewhat unusually — some lines are cut and "inserted" in vacuum — and this fact seems to discourage. Certainly, not for a long time. As a rule one quickly succeeds in adapting the Feynman diagrammar. Unfortunately, this language is not the most economic one in the case at hand. The Schwinger approach, much more convenient in this range of problems, is much less known, and, therefore, the main aim which we pursued in this review is pedagogical. In correspondence with this are the most typical examples chosen above, and they are supplied with rather lengthy explanations.

Appendix A

In this appendix the results for a few successive commutators are collected for the sake of reference:

$$[P^2, P_\mu] = -2igG_{\mu\alpha}P_\alpha;$$

$$\big[P^2[P^2, P_\mu]\big] = -4g^2G_{\mu\beta}G_{\beta\alpha}P_\alpha + 2igD^2G_{\mu\alpha}P_\alpha - 4gD_\beta G_{\alpha\mu}P_\beta P_\alpha;$$

$$\begin{aligned}\big[P^2[P^2[P^2, P_\mu]]\big] = {} & -8igD_\gamma D_\beta G_{\alpha\mu}P_\gamma P_\beta P_\alpha \\ & + \big(-8ig^2D_\beta(G_{\mu\gamma}G_{\gamma\alpha}) - 4gD_\beta D^2G_{\mu\alpha} + 4gD^2D_\beta G_{\mu\alpha} \\ & + 8ig^2D_\gamma G_{\alpha\mu}G_{\gamma\beta} + 8ig^2(D_\beta G_{\gamma\mu})\,G_{\gamma\alpha}\big)\,P_\beta P_\alpha + \big(4g^2D^2(G_{\mu\beta}G_{\beta\alpha}) \\ & + 8ig^3G_{\mu\beta}G_{\beta\gamma}G_{\gamma\alpha} - 2igD^4G_{\mu\alpha} + 4g^2(D^2G_{\mu\gamma})\,G_{\gamma\alpha} \\ & - 8g^2D_\beta G_{\gamma\mu}D_\beta G_{\gamma\alpha}\big)\,P_\alpha.\end{aligned}$$

In the following two commutators we omit terms containing the field operators of dimension higher than 6:

$$\begin{aligned}\big[P^2[P^2[P^2[P^2, P_\mu]]]\big] = {} & 16gD_\delta D_\gamma D_\beta G_{\alpha\mu}P_\delta P_\gamma P_\beta P_\alpha \\ & + 8igD^2D_\gamma D_\beta G_{\alpha\mu}P_\gamma P_\beta P_\alpha - 16g^2D_\gamma D_\beta G_{\alpha\mu}(G_{\gamma\delta}P_\delta P_\beta P_\alpha \\ & + P_\gamma G_{\beta\delta}P_\delta P_\alpha + P_\gamma P_\beta G_{\alpha\delta}P_\delta) + \big(16g^2D_\delta D_\beta(G_{\mu\gamma}G_{\gamma\alpha}) \\ & - 8igD_\delta D_\beta D^2G_{\mu\alpha} + 8igD_\delta D^2D_\beta G_{\alpha\mu} - 16g^2D_\delta(D_\gamma G_{\alpha\mu}G_{\gamma\mu}) \\ & - 16g^2D_\delta\big((D_\beta G_{\gamma\mu})\,G_{\gamma\alpha}\big)\big)\,P_\delta P_\beta P_\alpha + \cdots;\end{aligned}$$

$$\Big[P^2\big[P^2[P^2[P^2[P^2, P_\mu]]]\big]\Big] = 32igD_\varrho D_\delta D_\gamma D_\beta G_{\alpha\mu}P_\varrho P_\delta P_\beta P_\alpha + \cdots.$$

Let us give also one more set of commutators

$$[P^2, \sigma G] = -D^2\sigma G + 2iD_\alpha\sigma G P_\alpha;$$

$$[P^2, [P^2, \sigma G]] = D^4\sigma G - 2iD_\alpha D^2\sigma G P_\alpha$$
$$- 4gD_\beta\sigma G G_{\beta\alpha}P_\alpha - 2iD^2D_\alpha\sigma G P_\alpha - 4D_\beta D_\alpha\sigma G P_\beta P_\alpha;$$

$$[P^2[P^2[P^2, \sigma G]]] = -8iD_\gamma D_\beta D_\alpha\sigma G P_\gamma P_\beta P_\alpha + \cdots.$$

Appendix B

In this Appendix we present the expression for the effective Lagrangian in the instanton background field

$$G^a_{\mu\nu} = -\frac{4}{g}\,\frac{\eta_{a\mu\nu}\varrho^2}{[(x - x_0)^2 + \varrho^2]^2},$$

where x_0 stands for the instanton center, ϱ is its size and $\eta_{a\mu\nu}$ are the 'T HOOFT symbols [18]. Straightforward calculations lead to the following expressions for invariants of the field:

$$\mathrm{Tr}\; G_{\mu\nu}G_{\mu\nu} = \left(\frac{4}{g}\right)^2 \pi^2;$$

$$\mathrm{Tr}\; G_{\mu\alpha}G_{\alpha\beta}G_{\beta\mu} = \frac{3i}{2}\left(\frac{4}{g}\right)^3 \frac{\pi^2}{5}\,\frac{1}{\varrho^2};$$

$$\mathrm{Tr}\; (G_{\mu\nu}G_{\mu\nu})^2 = 3\left(\frac{4}{g}\right)^4 \frac{\pi^2}{7}\,\frac{1}{\varrho^4};$$

$$\mathrm{Tr}\; \{G_{\mu\nu}G_{\nu\alpha}\}_+^2 = 3\left(\frac{4}{g}\right)^4 \frac{\pi^2}{7}\,\frac{1}{\varrho^4};$$

$$\mathrm{Tr}\; [G_{\mu\nu}G_{\nu\alpha}]_-^2 = -4\left(\frac{4}{g}\right)^4 \frac{\pi^2}{7}\,\frac{1}{\varrho^4};$$

$$\mathrm{Tr}\; [G_{\mu\nu}G_{\alpha\beta}]_-^2 = -8\left(\frac{4}{g}\right)^4 \frac{\pi^2}{7}\,\frac{1}{\varrho^4},$$

where integration over d^4x is included in Tr. As a result, the effective lagrangian turns out to be

$$L_{\mathrm{eff}}(\varrho, x_0) = \frac{d\varrho}{\varrho^5}\, d(\varrho) \left\{-\frac{1}{3}\ln\frac{M_R^2}{m^2} - \frac{4}{75}\,\frac{1}{(m\varrho)^2} - \frac{4\cdot 17}{3\cdot 5\cdot 7^2}\,\frac{1}{(m\varrho)^4} + \cdots\right\}$$

where $d(\varrho)$ is the density of instantons with size ϱ.

Appendix C

The Borel Transformation

Physically observed quantities are connected with n-point functions in Euclidean domain by dispersion relations. After the Borel transformation the latter convert into the sum rules with exponential weight which much more effectively fix parameters of low-lying resonances. Here we give some useful definitions and relations.

Consider an arbitrary function $f(x)$. The Borel transformation of $f(x)$ looks as follows:

$$\tilde{f}(\lambda) = \frac{1}{2\pi i} \int_{c-i\infty}^{c+i\infty} e^{\lambda/x} f(x)\, x d(1/x), \tag{C.1}$$

where the integration contour runs to the right of all singularities of $f(x)$. The inverse transformation is

$$f(x) = \int_0^\infty \tilde{f}(\lambda)\, e^{-\lambda/x}\, d\lambda/x. \tag{C.2}$$

The meaning of the procedure is simple. If $f(x)$ is given in the form of a series in x (may be, asymptotic one),

$$f(x) = a_0 + a_1 x + a_2 x^2 + \cdots + a_k x^k + \cdots \tag{C.3}$$

then the corresponding expansion for $\tilde{f}(\lambda)$

$$\tilde{f}(\lambda) = a_0 + \frac{a_1\lambda}{1!} + \frac{a_2\lambda^2}{2!} + \cdots + \frac{a_k\lambda^k}{k!} + \cdots$$

contains coefficients factorially suppressed with respect to (C.3). Convergence of the series improves. In order to emphasize the fact we often refer to the procedure as to the Borel improvement.

The general definition (C.1) — the integral transformation in the complex plane — is rather inconvenient when one deals with n-point functions known in the form of (asymptotic) expansion in Q^{-2}. In ref. [1] an equivalent differential operator was introduced. Now we denote it by symbol $\hat{B}$. For analytical functions with normal dispersion representation $\hat{B}$ looks as follows

$$\hat{B}\Pi(Q^2) = \lim_{\substack{Q^2, n\to\infty \\ Q^2/n = M^2 = \text{const}}} \frac{1}{(n-1)!} (Q^2)^n \left(-\frac{d}{dQ^2}\right)^n \Pi(Q^2), \tag{C.4}$$

where we have taken a two-point function $\Pi(Q^2)$ as an example. In principle, the definition can be generalized to include n-point functions (see [28]).

If $\Pi(Q^2)$ possesses normal analytical properties $\Pi(Q^2) = 1/\pi \int \operatorname{Im} \Pi(s)\, ds/(s + Q^2)$ + possible subtractions then functions $\Pi(M^2) = \hat{B}\Pi(Q^2)$ and $\{Q^2(-d/dQ^2)\, \Pi(Q^2)\}$ are connected with each other by the Borel transformation (C.1), and variable $1/Q^2$ plays the role of x [1].

It is instructive to give here a table of examples showing the action of operator $\hat{B}$ on different functions encountered with in practice. These examples make explicit the factorial suppression of series in $1/Q^2$.

It is easy to check that

$$\hat{B}\left(\frac{1}{Q^2}\right)^k = \frac{1}{(k-1)!}\left(\frac{1}{M^2}\right)^k. \tag{C.5}$$

Moreover, for positive integer k

$$\hat{B}(Q^2)^k = 0.$$

One more transformation figuring in the sum rules refers to expressions of the type $(Q^2)^k \ln Q^2$. Applying (C.4) we get

$$\hat{B}(Q^2)^k \ln Q^2 = (-1)^{k+1}\,\Gamma(k+1)\,(M^2)^k. \tag{C.6}$$

A logarithmic Q^2 dependence sometimes appears due to anomalous dimensions of operators, and then the coefficients in the operator expansion are typically as follows

$$(1/Q^2)^k\,(\ln Q^2/\mu^2)^{-\varepsilon} \tag{C.7}$$

where ε, generally speaking, is a non-integer number. What is the Borel transform of (C.7)? To the leading order in $1/(\ln Q^2/\mu^2)$ one can easily answer this question if one uses the following auxiliary relation:

$$\left(\frac{1}{Q^2}\right)^k (\ln Q^2/\mu^2)^{-\varepsilon} = \frac{1}{\varepsilon\Gamma(\varepsilon)}\,\frac{1}{(Q^2)^k}\int\limits_0^\infty dz \exp\left(-z^{1/\varepsilon}\ln\frac{Q^2}{\mu^2}\right). \tag{C.8}$$

Applying operator $\hat{B}$ and using (C.8) we get

$$\hat{B}\left[\left(\frac{1}{Q^2}\right)^k\left(\ln\frac{Q^2}{\mu^2}\right)^{-\varepsilon}\right] = \frac{1}{\Gamma(k)}\left(\frac{1}{M^2}\right)^k\left(\ln\frac{M^2}{\mu^2}\right)^{-\varepsilon}\left[1+O\left(\left(\ln\frac{M^2}{\mu^2}\right)^{-1}\right)\right]. \tag{C.9}$$

As a rule there is no need in better accuracy in applications since the coefficients of operator expansion are usually calculated to the leading order in α_s. The exact Borel transform of (C.7) is known in the literature, though (see [29]).

The authors of several works [30, 31] deal with hadrons contaning heavy quarks. It turned out that the original sum rules [1] have a parallel in nonrelativistic limit. In particular, the non-relativistic analogue of the operator (C.4) is

$$(\hat{B})_{\text{non-rel}} = \lim_{\substack{E\to\infty,\,n\to\infty\\ E/n=\varepsilon=\text{const}}} \frac{1}{(n-1)!}\,E^n\left(-\frac{d}{dE}\right)^n, \tag{C.10}$$

where is the energy measured from the threshold. The physical meaning of the Borel transformation here is even more transparant than within the relativistic approach. Proceeding from E to ε we pass, in essence, from Green function at a given energy to the time Green function (or, which is the same, the evolution operator) and $(-i/\varepsilon)$ plays the role of time [30, 32].

Appendix D

The Fourier Transformation

Here we give a collection of formulae allowing one to proceed quickly from the coordinate to momentum space in the correlation function of currents.

Throughout the review we adopt the following standard definition of the Fourier transformation

$$\begin{aligned} F(p) &= \int d^4x e^{ipx} F(x),\\ F(x) &= \int \frac{d^4p}{(2\pi)^4}\, e^{-ipx} F(p). \end{aligned} \tag{D.1}$$

3*

The most frequently encountered transformations are well-known:

$$\frac{1}{p^2} \leftrightarrow \frac{i}{4\pi^2}\frac{1}{x^2}, \quad \frac{p_\mu}{p^2} \leftrightarrow \frac{1}{2\pi^2}\frac{x_\mu}{x^4}, \quad \frac{p_\mu}{p^4} \leftrightarrow \frac{1}{8\pi^2}\frac{x_\mu}{x^2}. \tag{D.2}$$

In analysing the sum rules we deal with more general functions

$$\int \frac{d^4x}{(x^2)^n} e^{ipx}, \tag{D.3}$$

where n are integer numbers, $n \geqq 4$. Integrals of this type are divergent in the ultraviolet domain and have the structure const $(p^2)^{n-2} \ln(-p^2) + P_{n-2}(p^2)$ where P_{n-2} is a polynomial of power $n - 2$. The coefficients of this polynomial are ill-defined (infinite). However, the constant in front of the logarithmic term is finite and fixed unambiguously. The presence of the divergent polynomial never affects analysis of the sum rules. It has no singularities in p^2 and disappears after borelization (see Appendix C). In all subsequent formulae we write out only logarithms and consistently suppress subtraction polynomials.

Integral (D.3) can be done by virtue of analytic continuation in dimensionality of space — this method is most economic.

Start from

$$A_D(p^2) = \int d^Dx e^{ipx} \frac{1}{(x^2)^n} \xrightarrow{\text{Eucl.}} (-i)(-1)^n \int d^Dx e^{-ipx} \frac{1}{(x^2)^n},$$

where we have proceeded to the Euclidean space. Integrating over angles yields

$$\begin{aligned} A_D(p^2) &= -i(-1)^n (2)^{D/2} (\pi)^{D/2} |p|^{1-D/2} \int |x|^{(D/2)-2n} d|x| \, J_{(D/2)-1}(|p|\,|x|) \\ &= (-i)(-1)^n 2^{D-2n} \pi^{D/2} |p|^{2n-D} \frac{\Gamma\left(\frac{D}{2} - n\right)}{\Gamma(n)}. \end{aligned} \tag{D.4}$$

If $D \to 4$ this expression is singular — it has a pole — the corresponding residue, however, is proportional to $(p^2)^{n-2}$. As was already noted, any polynomial in p^2 is inessential and can be omitted. The non-singular part of eq. (D.4) evidently contains $\ln p^2$. Separating the nonsingular part and returning to the Minkowski space we have

$$\int \frac{d^4x}{(x^2)^n} e^{ipx} = \frac{i(-1)^n 2^{4-2n}\pi^2}{\Gamma(n-1)\,\Gamma(n)} (p^2)^{n-2} \ln(-p^2). \tag{D.5}$$

This expression generates a series of other useful relations, for instance

$$\begin{aligned} I_{n,l} &\equiv \int d^4x e^{ipx} \frac{(x\xi)^n (x\eta)^n}{(x^2)^{2n+l}} \\ &= \frac{i\pi^2 2^{4-3n-2l} n!}{(2n+l-1)!\,(n+l-2)!} (\xi\eta)^n (-p^2)^{n+l-2} \ln(-p^2), \end{aligned} \tag{D.6}$$

where ξ, η are auxiliary vectors,

$$\xi^2 = \eta^2 = 0, \qquad \xi p = \eta p = 0, \qquad \xi\eta \neq 0.$$

We encounter with transformation (D.6) in the problem of mesons with spin n. Its derivation is given in ref. [23].

The key observation on which the derivation is based is the following

$$I_{n,l} = \frac{(n!)^2}{(2n)!} \int d^4x e^{ipx} \frac{(x(\xi + \eta))^{2n}}{(x^2)^{2n+l}},$$

while integral

$$\int d^4x e^{i(p+\xi+\eta)x} \frac{1}{(x^2)^{2n+l}}$$

is already known.

References

[1] A. Vainshtein, V. Zakharov, M. Shifman, Zh. eksper. teor. Fiz. Pis'ma **27**, 60 (1978);
M. Shifman, A. Vainshtein, M. Voloshin, V. Zakharov, Phys. Lett. **77B**, 80 (1978);
M. Shifman, A. Vainshtein, V. Zakharov, Nucl. Phys. **B 147**, 385, 448 (1979).

[2] V. Novikov, M. Shifman, A. Vainshtein, V. Zakharov, Nucl. Phys. **B 191**, 301 (1981);
M. B. Voloshin, Nucl. Phys. **B 154**, 365 (1979); Yad. Fiz. **29**, 1366 (1979);
B. L. Ioffe, Nucl. Phys. **B 188**, 317; **B 191**, 591 (1981);
L. J. Reinders, H. R. Rubinstein, S. Yazaki, Nucl. Phys. **B 186**, 109 (1981);
B. L. Ioffe, A. V. Smilga, Nucl. Phys. **B 216**, 373 (1983);
V. Nesterenko, A. Radyushkin, Phys. Lett. **B 115**, 410 (1982);
E. Shuryak, Nucl. Phys. **B 198**, 83 (1984);
B. L. Ioffe, A. V. Smilga, Zh. eksper. teor. Fiz. Pis'ma **37**, 250 (1983); Nucl. Phys. **B 232**, 109 (1984).

[3] H. R. Rubinstein, in Proc. XXI Internat. Confer. on High Energy Phys.: Paris (1982).

[4] J. Schwinger, Phys. Rev. **82**, 664 (1951);
J. Schwinger, Particles, Sources and Fields, vols. 1 and, 2 Addison-Wesley (1963).

[5] V. Novikov, M. Shifman, A. Vainshtein, V. Zakharov, Nucl. Phys. **B 174**, 378 (1980).

[6] (a) M. S. Dubovikov, A. V. Smilga, Nucl. Phys. **B 185**, 109 (1981);
(b) E. Shuryak, A. Vainshtein, Nucl. Phys. **B 199**, 451; **B 201**, 143 (1982);
(c) A. Smilga, Yad. Fiz. **35**, 473 (1982).

[7] V. A. Fock, Sowj. Phys. **12**, 404 (1937);
V. A. Fock, Works on Quantum Field Theory, Leningrad University Press, Leningrad, page 150 (1957).

[8] M. Shifman, Nucl. Phys. **B 173**, 13 (1980);
T. M. Aliev, M. Shifman, Phys. Lett. **112B**, 401 (1982);
S. N. Nikolaev, A. V. Radyushkin, Phys. Lett. **110B**, 476 (1982); **124B**, 243 (1983); Nucl. Phys. **B 213**, 305 (1982);
W. Hubsmid, A. Mallik, Nucl. Phys. **B 207**, 29 (1982).

[9] K. G. Wilson, Phys. Rev. **179**, 1499 (1969).

[10] W. Zimmermann, in Lectures on Elementary Particles and Quantum Field Theory, ed. S. Deser, M. Grisaru, K. Pendleton (MIT Press, Cambridge, Mass., 1971), vol. 1.

[11] K. G. Chetyrkin, S. G. Gorishny, F. V. Tkachov, Phys. Lett. **119B**, 407 (1983).

[12] K. G. Chetyrkin, Phys. Lett. **126B**, 371 (1983).

[13] F. David, Nucl. Phys. **B 234**, 237 (1984);
M. Soldate, Preprint SLAC-PUB-3054, Stanford, 1983; Ann. Phys. (N. Y.) in press.

[14] A. Vainshtein, V. Zakharov, V. Novikov, M. Shifman, Uspechi Fiz. Nauk **136**, 553 (1982).

[15] A. Vainshtein, V. Zakharov, V. Novikov, M. Shifman, Yad. Fiz. **39**, 9 (1983).

[16] S. N. Nikolaev, A. V. Radyshkin, Preprint JINR P2-82-914, Dubna, 1982.

[17] N. Andrei, D. J. Gross, Phys. Rev. **D 18**, 468 (1978).

[18] G. t'Hooft, Phys. Rev. **D 14**, 3432 (1976).

[19] M. T. Grisaru, W. Siegel, H. Roček, Nucl. Phys. **B 159**, 429 (1979);
V. Novikov, M. Shifman, A. Vainshtein, V. Zakharov, Nucl. Phys. **B 223**, 445 (1983).

[20] V. Fateev, A. Schwarz, Yu. Tyupkin, Preprint Lebedev Physical Institute No. 155, Moscow (1976);
C. Cronstrom, Phys. Lett. **90B**, 267 (1980);
M. S. Dubovikov, A. V. Smilga, Nucl. Phys. **B 185**, 109 (1981).
[21] S. Nikolaev, A. Radyushkin, Nucl. Phys. **B 213**, 305 (1981).
[22] M. Shifman, Nucl. Phys. **B 173**, 13 (1980).
[23] M. Shifman, Yad. Fiz. **36**, 1290 (1982).
[24] B. S. De Witt, Phys. Rev. **162**, 1195, 1239 (1967).
[25] J. Honerkamp, Nucl. Phys. **B 36**, 130 (1971); **B 48**, 269 (1972);
R. Kallosh, Nucl. Phys. **B 78**, 293 (1974).
[26] G. 't Hooft, Nucl. Phys. **B 62**, 444 (1973).
[27] M. T. Grisaru et al., Phys. Rev. **D 12**, 3203 (1975);
L. F. Abbott, Nucl. Phys. **B 185**, 189 (1981).
[28] A. V. Smilga, M. A. Shifman, Yad. Fiz. **37**, 1613 (1983).
[29] S. Narison, E. de Rafael, Phys. Lett. **103B**, 57 (1981).
[30] M. B. Voloshin, Yad. Fiz. **29**, 1368 (1979); Nucl. Phys. **B 154**, 365 (1979); Preprint ITEP-21, Moscow 1980;
J. S. Bell, R. A. Bertlmann, Nucl. Phys. m **B 177**, 218 (1981);
R. A. Bertlmann, Nucl. Phys. **B 204**, 387 (1982).
[31] E. V. Shuryak, Nucl. Phys. **B 198**, 83 (1982);
V. L. Chernyak, A. R. Zhitnitsky, I. R. Zhitnitsky, Yad. Fiz. **38**, 1277 (1983);
T. M. Aliev, V. L. Eletsky, Yad. Fiz. **38**, 1537 (1983).
[32] A. Vainshtein, V. Zakharov, V. Novikov, M. Shifman, Yad. Fiz., **32**, 1622 (1980).

3.1.2. Non-singular terms in the Green functions and the operator product expansion

Exercise: Calculate one- and two-loop β functions in spinor and scalar electrodynamics using the technique described above.

This exercise will help the reader to get some experience in computations in external fields and will acquaint him with a subtlety not mentioned thus far, namely, the possible existence of terms non-singular in x in $S(x, 0)$. This terms are equivalent to delta functions and their derivatives in momentum space and are not seen in the q^{-1} expansion (see page 611 of ref. [14], reproduced above).

The approach presented below is based on two works of Vainshtein and myself [18].

First of all, a few words about generalities and definitions. It is convenient to rescale the photon field and write the Lagrangians of spinor and scalar electrodynamics in the following form

$$\mathscr{L}_{\text{spinor QED}} = -\frac{1}{4e_0^2}F_{\mu\nu}F_{\mu\nu} + \bar{\psi}\mathrm{i}\mathscr{D}\!\!\!/\psi, \tag{3.18}$$

$$\mathscr{L}_{\text{scalar QED}} = -\frac{1}{4e_0^2}F_{\mu\nu}F_{\mu\nu} + (\mathscr{D}_\mu\phi)^*(\mathscr{D}_\mu\phi), \tag{3.19}$$

$$\mathscr{D}_\mu = \partial_\mu - \mathrm{i}A_\mu, \tag{3.20}$$

where e_0 is the bare coupling constant and $F_{\mu\nu}$ is the photon field strength tensor

$$F_{\mu\nu} = \partial_\mu A_\nu - \partial_\nu A_\mu.$$

Starting from eqs. (3.18) and (3.19) we calculate vacuum diagrams in the external electromagnetic field as a functional of the external field. This quantity is called Γ, the generator of one-particle irreducible vertices. We find Γ in the one-loop, two-loop, etc. approximation treating Γ as a functional of the external electromagnetic field and limiting the loop integrations over the virtual momenta to the domain

$$\mu \le p \le M_0.$$

Here μ is the lower cutoff (the normalization point) and M_0 is the ultraviolet cutoff in momentum space. More exactly, we are interested *only* in the coefficient in front of $F_{\mu\nu}^2$, where $F_{\mu\nu}$ is the background photon field. This coefficient, $-1/(4e^2(\mu))$, represents the effective charge, and in this way we get an explicit relation between $\alpha(\mu)$ and α_0:

$$\frac{1}{\alpha(\mu)} = \frac{1}{\alpha_0} + \beta^{(1)}\ln\frac{M_0}{\mu} + \beta^{(2)}\alpha_0\ln\frac{M_0}{\mu} + (M_0\text{-independent terms}); \tag{3.21}$$

$\beta^{(i)}$ are numerical constants. Further, the Gell-Mann–Low function is defined as follows

$$\beta(\alpha) = [\mathrm{d}\alpha(\mu)/\mathrm{d}\ln\mu]_{\alpha_0, M_0\,\text{fixed}}. \tag{3.22}$$

Comparing eqs. (3.21) and (3.22) we arrive at

$$\beta(\alpha) = \alpha^2(\beta^{(1)} + \alpha\beta^{(2)} + \cdots). \tag{3.23}$$

No special efforts are needed in order to find the one-loop coefficient $\beta^{(1)}$. Actually, one can readily extract it from eq. (2.33) in ref. [14]. Proceeding from QCD to QED we get from the first term on the right-hand side

$$\beta^{(1)}_{\text{spinor QED}} = 2/3\pi; \quad \beta^{(1)}_{\text{scalar, QED}} = 1/6\pi. \tag{3.24}$$

More interesting and instructive is the analysis of the two-loop coefficient $\beta^{(2)}$. The two-loop vacuum diagram Γ is depicted in fig. 3.1, where the solid line describes a charged particle in an external field (the neutral particle, the photon, does not interact with A_{ext} directly). As a matter of fact, in the expansion of the Green functions in the background field one should keep only the terms $O(F_{\mu\nu})$ and $O(F_{\mu\nu}^2)$ since all others will not lead to the structure $F_{\mu\nu}^2$ in Γ. In two loops, evidently,

$$\Gamma^{(2)} = \tfrac{1}{2}\int d^4x\left(-i\mathscr{D}_{\mu\nu}(x)\right)\Pi_{\mu\nu}^{(F^2)}(x), \tag{3.25}$$

where $\mathscr{D}_{\mu\nu}(x)$ is the free photon propagator,

$$\mathscr{D}_{\mu\nu}(x) = g_{\mu\nu}\frac{ie_0^2}{4\pi^2}\frac{1}{x^2}, \tag{3.26}$$

$\Pi_{\mu\nu}^{(F^2)}$ is the two-point function

$$\Pi_{\mu\nu} = i\langle T\{j_\mu(x)j_\nu(0)\}\rangle, \tag{3.27}$$

$$j_\mu = \begin{cases} \bar{\psi}\gamma_\mu\psi, & \text{spinor QED,} \\ i\left(\phi^*\mathscr{D}_\mu\phi - (\mathscr{D}_\mu\phi)^*\phi\right), & \text{scalar QED,} \end{cases} \tag{3.28}$$

and the superscript F^2 reminds us that we are interested only in the F^2 term in the expansion of $\Pi_{\mu\nu}$:

$$\Pi_{\mu\nu}(x) = \cdots + C_{\mu\nu}(x)F^2(0) + \cdots. \tag{3.29}$$

Moreover, it is obvious that the coefficient $C_{\mu\nu}$ is to be calculated only in the one-loop approximation. In this approximation the expression for $C_{\mu\nu}$ is well defined – it does require neither an ultraviolet nor an infrared cutoff. To get $C_{\mu\nu}$ we simply multiply $S(x, 0)$ by $S(0, x)$ (see eq. (2.32) in ref. [14] and the preceding discussion). In this way we find that $\Pi_{\mu\mu}^{(F^2)}$ in eq. (3.25) has the form

$$\text{const.}\; x^{-2}F^2(0)$$

and $\ln(M_0/\mu)$ in eq. (3.21) appears from the integration over d^4x. Indeed, according to eq. (3.25)

$$\Gamma^{(2)} \propto \int (d^4x\, x^{-4})\; F^2(0) \propto \left(\ln\frac{|x_{\min}|}{|x_{\max}|}\right)F^2(0), \quad |x_{\min}| \sim M_0^{-1}, \quad |x_{\max}| \sim \mu^{-1}. \tag{3.30}$$

Specific calculations must be done separately for the spinor and scalar cases. To begin with, let us discuss the simpler case of spinor QED. Actually, all elements needed for the final answer are already prepared. Take the second term in eq. (2.32) in ref. [14], replace $g^2/2$ by

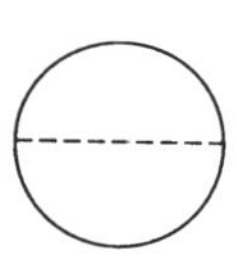

Fig. 3.1. Two-loop contribution to the generator of 1PI vertices Γ. The dashed line is the photon propagator, the solid line represents the matter field propagator in an external electromagnetic field.

e^2 and G^2 by F^2 and multiply the result by the factor i entering the definition of $\Pi_{\mu\nu}$ (see eq. (3.27)),

$$\Pi_{\mu\nu}^{(F^2)}(x) = -\mathrm{i}\frac{1}{192\pi^4}F^2(0)\frac{2x_\mu x_\nu + x^2 g_{\mu\nu}}{x^4}. \tag{3.31}$$

The last step is also trivial. Substituting eqs. (3.31) and (3.26) into (3.25) and passing to Euclidean space ($x_0 \to -\mathrm{i}x_4$) we find

$$\Gamma^{(2)} = -\int_{|x_{\min}|}^{|x_{\max}|}\frac{\mathrm{d}^4x}{x^4}\frac{e_0^2}{256\pi^6}F_{\alpha\beta}F_{\alpha\beta} = -\frac{e_0^2}{2^7\pi^4}\ln\frac{M_0}{\mu}F^2(0), \tag{3.32}$$

where eq. (3.31) has been used. In terms of $\beta^{(2)}$ this result evidently means that

$$\beta^{(2)} = 1/2\pi^2 \quad \text{(spinor QED)}. \tag{3.33}$$

For the scalar loop there emerge new elements, one of them of technical nature and the other more serious.

First of all, since the current j_μ now contains the covariant derivative (see eq. (3.28)), in order to calculate $\Pi_{\mu\nu}^{(F^2)}$ it is necessary to know the propagation function of the scalar particle from point y to point x, and only after differentiating one can put $y = 0$. The Green function of the massless scalar field is

$$G(x, y) = \langle x|\frac{1}{P^2}|y\rangle. \tag{3.34}$$

Doing the Fourier transformation we arrive at ($z \equiv x - y$)

$$G(q, y) = \int \mathrm{d}^4z\, \mathrm{e}^{\mathrm{i}qz}\,\langle x|\frac{1}{P^2}|y\rangle = \int \mathrm{d}^4z\,\langle z|\frac{1}{(P+q)^2}|y\rangle.$$

Next, we expand q^{-1}:

$$G(q, y) = \frac{1}{q^2} + \int \mathrm{d}^4z\,\langle z|\frac{1}{q^2}(-2Pq)\frac{1}{q^2} + \frac{1}{q^2}(-P^2)\frac{1}{q^2} + \cdots |y\rangle. \tag{3.35}$$

The only difference as compared to the previous consideration is that A_μ acting on $|y\rangle$ yields $A_\mu(y)|y\rangle$, not zero (cf. eq. (3.16)). Following this procedure we find

$$G(q, y) = \frac{1}{q^2} - \frac{1}{q^4}y_\rho F_{\rho\mu}(0)q_\mu + \cdots, \tag{3.36}$$

where the dots denote irrelevant terms ($\mathrm{O}(y^2)$, or terms containing F with derivatives, or $F_{\alpha\beta}F_{\alpha\gamma} - \frac{1}{4}F^2 g_{\beta\gamma}$). In coordinate space

$$G(x, y) = \frac{\mathrm{i}}{4\pi^2}\frac{1}{(x-y)^2} + \frac{x_\mu y_\rho F_{\mu\rho}(0)}{8\pi^2(x-y)^2} + \cdots. \tag{3.37}$$

Here I hastly make a remark: surprising though it is, this expression is not complete, although we, certainly, have made no arithmetic mistakes. This assertion becomes obvious after calculating $\Pi_{\mu\nu}^{(F^2)}$ starting from eqs. (3.37) and (3.11). Specifically, one gets

$$-\frac{\mathrm{i}}{192\pi^4 x^4}\left(x^2 g_{\mu\nu} - x_\mu x_\nu\right)F^2(0), \tag{3.38}$$

a result which is not even transveral! (j_μ is conserved and $\Pi_{\mu\nu}$ definitely should be transversal).

Where is the flaw in our arguments?

The construction of propagators in the background field by expanding in q^{-1}, as described above, reveals only those photonic operators that are associated with short distances. The corresponding coefficients are singular in coordinate space, e.g., $1/q^2 \leftrightarrow 1/x^2$; $1/(q^2)^n \leftrightarrow (x^2)^{n-4} \ln x^2$, $n \geq 4$, etc. Such an analysis says nothing about the possible presence of terms regular in x in the Green functions. These terms are proportional to $\delta(q)$ and derivatives of the δ function in momentum space, and thus show up only at vanishing momenta. (Delta functions appear as a result of the expansion in the external field in the approximation $p_{\text{ext}} = 0$ where p_{ext}, the momentum carried by $(A_\mu)_{\text{ext}}$, tends to zero. Physically, these delta functions are smeared in the domain $q \sim \mu$, where μ is the characteristic momentum of the background field.)

Shortly we will explain the meaning of the regular terms in the operator product expansion, but first let us check that they are indeed present in the case at hand. The most straightforward way is to analyze the equation for the Green function $G(x, 0)$,

$$-\mathscr{D}_\mu^2 G(x, 0) = \delta(x). \tag{3.39}$$

On general grounds

$$G(x, 0) = \frac{\mathrm{i}}{4\pi^2 x^2} + Cx^2F^2(0) + \mathrm{O}(F^3), \tag{3.40}$$

and our task is to fix the constant C. Equation (3.39) can be identically rewritten as

$$-\mathscr{D}^2 G(x, 0) = \left[-(\partial/\partial x)^2 + (A_\mu(x))^2 + 2\mathrm{i}A_\mu\, \partial_\mu\right]G(x, 0) = \delta(x), \tag{3.41}$$

where we have used the fact that in the Fock–Schwinger gauge $\partial A_\mu(x)/\partial x_\mu = 0$. Moreover, the third term in eq. (3.41), $A_\mu(x)\, \partial_\mu G(x, 0)$, can be omitted due to the fact that there is no term linear in F in $G(x, 0)$. As a result we are left with the relation

$$\left\{\left[-(\partial/\partial x)^2 + \tfrac{1}{16}x^2F^2(0)\right]G(x, 0)\right\}_{F^2} = 0, \tag{3.42}$$

which implies

$$C = \frac{\mathrm{i}}{512\pi^2}. \tag{3.43}$$

Summarizing the discussion we can write out the final and complete answer for $G(x, y)$ to be used instead of eq. (3.37):

$$G(x, y) = \frac{\mathrm{i}}{4\pi^2}\frac{1}{(x-y)^2} + \frac{x_\mu y_\rho F_{\mu\rho}(0)}{8\pi^2(x-y)^2} + \frac{\mathrm{i}}{512\pi^2}(x-y)^2F^2(0) + \cdots, \tag{3.44}$$

where the dots, as usually, denote the terms irrelevant for $\Pi_{\mu\nu}^{(F^2)}$. With this expression in hand we immediately get

$$\Pi_{\mu\nu}^{(F^2)} = \frac{-\mathrm{i}}{192\pi^4 x^4}\left(x^2 g_{\mu\nu} + 2x_\mu x_\nu\right)F^2(0), \tag{3.45}$$

numerically exactly the same answer we had in spinor QED (cf. eq. (3.31)). Hence, we conclude that $\beta^{(2)}$ in the scalar case coincides with that in the spinor case,

$$\beta^{(2)} = 1/2\pi^2 \quad \text{(scalar QED)}. \tag{3.46}$$

One should not think that the emergence of non-singular terms in the propagator is specific for scalars in an external electromagnetic field. True, thus far in discussing spinor

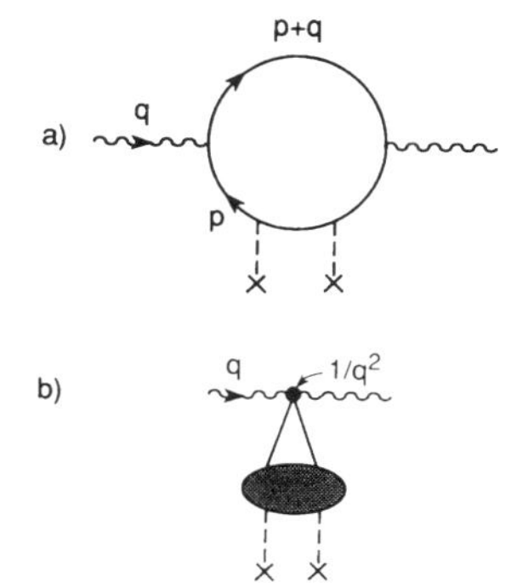

Fig. 3.2. Calculation of $\Pi_{\mu\nu}^{(F^2)}$ in scalar QED. Solid lines denote here the *bare* propagator of the scalar particles, dashed lines photons and wavy lines currents. (a) The contribution depicted here corresponds to the non-singular piece in $G(x, y)$, (b) interpretation of graph (a) within OPE.

particles – quarks or electrons – I mentioned no such terms. The point is that we were always interested in operators like G^2 (or F^2 in QED) whose vacuum expectation value is non-vanishing and this operator does not appear in $S(x, 0)$. If we considered, say, $G\tilde{G}$ we would discover just in the same way as above that this term does appear in $S(x, 0)$ with a non-singular coefficient proportional to x. More specifically, requiring that $\mathrm{i}\not{\mathscr{D}}S(x, 0) = \delta(x)$ and keeping this extra contribution we get instead of eq. (2.31) in ref. [14],

$$S(x, 0) = \frac{1}{2\pi^2}\frac{\not{x}}{(x^2)^2} - \frac{1}{8\pi^2}\frac{x_\alpha}{x^2}\tilde{G}_{\alpha\phi}\gamma_\phi\gamma_5 + \frac{\mathrm{i}}{256\pi^2}\not{x}\gamma_5 G_{\alpha\beta}\tilde{G}_{\alpha\beta} + \cdots .$$

The last aspect to be elucidated is the interpretation of these non-singular terms in the operator product expansion. Can one find a place for such terms within the consistent procedure of separation of short- and large-distance contributions?

To answer this question let us return to the consideration of $\Pi_{\mu\nu}$ in scalar QED. The puzzling last term in eq. (3.44) corresponds to the diagram presented in fig. 3.2a. Since the propagation function in the lower part of the graph is proportional to $\delta''(p)$, following the ideology of OPE we must actually cut the lower line and then the upper line shrinks to a point (fig. 3.2b). Thus, the calculation of fig. 3.2a is a two-step process. First, within the standard OPE approach, we calculate the coefficient in front of the operator $(\mathscr{D}_\mu\phi)^*(\mathscr{D}_\mu\phi)$ (see fig. 3.2b) – this coefficient is determined entirely by short distances – and, then, the conversion of $(\mathscr{D}_\mu\phi)^*(\mathscr{D}_\mu\phi)$ into $F_{\mu\nu}^2$ which can be ascribed to large distances. In other words, the second step is obviously the calculation of a photonic matrix element of $(\mathscr{D}_\mu\phi)^*(\mathscr{D}_\mu\phi)$.

The first two terms in eq. (3.44) lead to the "normal" OPE for $\Pi_{\mu\nu}^{(F^2)}$ with the coefficient function determined entirely by short distances.

Formally, the operator $\int\mathrm{d}x\,(\mathscr{D}_\mu\phi)^*(\mathscr{D}_\mu\phi)$ vanishes because of the equations of motions. One can readily convince onseself, however, that in the external gauge field there is an "anomalous" relation

$$\left\langle(\mathscr{D}_\mu\phi)^*(\mathscr{D}_\mu\phi)\right\rangle = \frac{1}{64\pi^2}F_{\alpha\beta}F_{\alpha\beta} \tag{3.47}$$

stemming from eq. (3.44).

The full Green function $G(x, y)$ certainly satisfies the equation of motion $-\mathscr{D}^2 G(x, y) = \delta(x-y)$. In calculating $\Pi_{\mu\nu}^{(F^2)}$, however, we split it in two pieces: the piece singular in x is used for determination of the coefficient function while the piece regular in x, $G^{\mathrm{reg}}(x, y)$, is

Fig. 3.3. The coefficient in front of $(\mathscr{D}_\mu\phi)^*(\mathscr{D}_\mu\phi)$ in $\Pi_{\mu\nu}$. The vertices are equal to $(2p+q)_\mu$ and $(2p-q)_\mu$, respectively.

interpreted as a matrix element. Then $\langle(\mathscr{D}_\mu\phi)^*(\mathscr{D}_\mu\phi)\rangle \neq 0$. More specifically, eq. (3.44) implies

$$\left\langle(\mathscr{D}_\mu\phi)^*(\mathscr{D}_\mu\phi)\right\rangle = \lim_{x\to 0}\left(-\mathrm{i}\mathscr{D}_\mu^2 G^{\mathrm{reg}}(x,0)\right) = \frac{1}{64\pi^2}F_{\alpha\beta}F_{\alpha\beta}.$$

Now we are finally able to explain the difference between our first result for $\Pi_{\mu\nu}$ quoted in eq. (3.38) and the complete answer presented in eq. (3.45). Equation (3.38) has been obtained by substituting in $\Pi_{\mu\nu}$ the singular part of $G(x, y)$. Thus, eq. (3.38) corresponds, in the language of OPE, to the genuine contribution of the operator F^2.

One should not forget, however, another dim = 4 operator, $(\mathscr{D}_\mu\phi)^*(\mathscr{D}_\mu\phi)$, whose matrix element in the background electromagnetic field reduces to the same structure F^2 (see eq. (3.47)). The coefficient of $(\mathscr{D}_\mu\phi)^*(\mathscr{D}_\mu\phi)$ can be trivially extracted from the diagrams of fig. 3.3. With no effort we arrive at

$$\Pi_{\mu\nu}(g) = -2\left\{\frac{g_{\mu\nu}}{q^2} - 2\frac{q_\mu q_\nu}{q^4}\right\}\left\langle(\mathscr{D}_\mu\phi)^*(\mathscr{D}_\mu\phi)\right\rangle.$$

Invoking eq. (D.2) in ref. [14] for the Fourier transformation and eq. (3.47) we then find that the contribution of fig. 3.2b is equal to

$$\frac{-\mathrm{i}}{64\pi^4}\frac{x_\mu x_\nu}{x^4}F^2(0),$$

precisely the difference between eqs. (3.45) and (3.38).

References

[1] V. Novikov et al., Nucl. Phys. B 174 (1980) 378.
[2] E. Shuryak and A. Vainshtein, Nucl. Phys. B 199 (1982) 451.
[3] J. Schwinger, Phys. Rev. 82 (1951) 664.
[4] V.A. Fock, Sov. J. Phys. 12 (1937) 404; Works on Quantum Field Theory (Leningrad Univ. Press, Leningrad, 1957) p. 150.
[5] J. Schwinger, Particles, Sources and Fields, Vols. 1, 2 (Addison-Wesley, Reading, MA, 1970).
[6] V. Fateev, A. Schwarz and Yu. Tyupkin, Lebedev Physical Inst. report No. 155, Moscow, 1976 (unpublished).
[7] C. Cronström, Phys. Lett. B 90 (1980) 267.
[8] M. Dubovikov and A. Smilga, Nucl. Phys. B 185 (1981) 109.
[9] V. Ogievetsky and E. Sokatchev, Yad. Fiz. 32 (1980) 862 [Sov. J. Nucl. Phys. 32 (1980) 433].
[10] E. Shuryak and A. Vainsthein, Nucl. Phys. B 201 (1982) 141.
[11] M. Shifman, Nucl. Phys. B 173 (1980) 13.
[12] A. Smilga, Yad. Fiz. 35 (1982) 473 [Sov. J. Nucl. Phys. 35 (1982) 271].
[13] S. Nikolaev A. Radyushkin, Phys. Lett. B 110 (1982) 476; B 116 (1982) 469 (E); Nucl. Phys. B 213 (1983) 285.
[14] V. Novikov, M.A. Shifman, A.I. Vainshtein and V.I. Zakharov, Fortschr. Phys. 32 (1984) 585.
[15] V.A. Novikov, M.A. Shifman, A.I. Vainshtein and V.I. Zakharov, Nucl. Phys. B 249 (1985) 445.
[16] W. Hubschmid and S. Mallik, Nucl. Phys. B 207 (1982) 29.
[17] A. Vainshtein et al., Yad. Fiz. 39 (1984) 124 [Sov. J. Nucl. Phys. 39 (1984) 77].
[18] A. Vainshtein and M.A. Shifman, Yad. Fiz. 44 (1986) 498 [Sov. J. Nucl. Phys. 44 (1986) 321]; Nucl. Phys. B 277 (1986) 456.

4. Heavy Quarks in the Sum Rules. Pre-Coulombic Behavior

Introduction

Historically, heavy quarkonium was the first testing ground of the sum rule method. All basic ideas concerning the possibility of parametrizing the non-perturbative vacuum fields by a few condensate parameters were born in an attempt to describe charmonium in a clean and model-independent way. The program completed in ref. [1] was a natural continuation of previous work [2] treating charmonium from related viewpoints, but without introducing the condensates. (One can say that the approach presented in ref. [2] is the parent and the zeroth approximation of the QCD sum rules in their modern form.)

The first estimates of the gluon condensates starting from the charmonium sum rules appeared in refs. [3,4]. The latter work devoted to the 0^- charmonium level, η_c, has a particularly dramatic history. The point is that the η_c particle was first found experimentally at a wrong place (2.83 GeV). Within the sum rule method we first extracted the value of $\langle G^2 \rangle$ from the analysis of J/ψ and then applied it in the 0^- channel. The result of this investigation was formulated as a categoric prediction for the η_c mass [4], $(m_c)_{\text{theor}} = 3.00 \pm 0.03$ GeV. The later discovery of the genuine η_c state with mass 2.98 GeV at Stanford [5,6] was one of the greatest successes of quantum chromodynamics and the strongest impetus for further work on the sum rules.

At the end of this chapter we reproduce two papers [7,8] explaining how one can put calculations of the charmonium parameters on "industrial basis". Ref. [7] was the first work on the QCD sum rules of this type. It gives also a good idea of possible uncertainties in the determination of the condensate parameter $\langle \alpha_s G^a_{\mu\nu} G^a_{\mu\nu} \rangle$. Ref. [8] presents important comments on the role of higher gluon condensates (of dim 8) in charmonium.

The beginning of this chapter is devoted to more theoretical aspects. Work [9], opening the chapter, treats the general questions of motion of heavy quarks in non-perturbative vacuum fields. The vacuum fields are considered as a medium with certain characteristic frequencies, and then the effect of this medium on heavy quarks is studied. A finding which might seem surprising to non-experts is that the gluon condensates, generally speaking, cannot be represented by any potential, even for *arbitrary heavy quarks*. A brief reflection shows, however, that this theorem is absolutely natural and has a transparent interpretation. Indeed, if the expansion in the condensate parameters has any meaning at all it must be an expansion in frequencies characterizing the vacuum gluon fields; thus, a built-in assumption is that these frequencies are small on the natural scale inherent to the valence quarks. In other words, the valence quarks are, in a sense, fast, and the response of the vacuum medium is slow. The existence of a genuine (instantaneous) potential would require, on the other hand, the opposite situation – very high characteristic frequencies of the gluon fields, very fast response of the medium *. Ref. [9] formalizes this heuristic argument. Instructive concrete examples

* Of course, one can always consider a "fake potential", an equivalent potential representing the effects of the gluon condensates on this or that individual level. Such fake potentials will typically depend on parameters inadmissible for normal instantaneous potentials.

explicitly demonstrating that the condensate corrections are incompatible with the genuine potential are considered in ref. [10]. For a recent discussion of the issue see ref. [11].

In the remainder the so called pre-Coulombic asymptotic behavior in heavy quarkonium is considered. If the quark mass m is large enough, the quarks are bound primarily by the Coulomb force, and the size of the bound state is of order of k_n^{-1} where the momentum k_n is defined by

$$k_n = \frac{m}{n}\frac{2}{3}\alpha_s(k_n), \tag{4.1}$$

n is the principal quantum number and α_s is the gauge coupling constant. At large values of k_n the size of the system is small in comparison with the typical wavelength of the vacuum fluctuations, so that

$$|r\mathscr{D}_\mu G_{\alpha\beta}| \ll |G_{\alpha\beta}|, \tag{4.2}$$

where r is the size of the system. Roughly speaking, $k \sim m\alpha_s$ and for sufficiently large m both the momentum k and typical energies $E \sim m\alpha_s^2$ are large compared to Λ_{QCD}. As a result, one can disregard non-homogeneity of the field, and consider quarks as a Coulomb bound state subject to a static external field. This field is (i) weak, (ii) randomly oriented, and (iii) chromoelectric. (The corrections due to the chromomagnetic field are suppressed by two powers of $\alpha_s(k_n)$.) The last condition – typical energy splittings of the system $E \sim m\alpha_s^2 \gg \Lambda_{\mathrm{QCD}}$ – is very important since in this case one can approximately neglect the time spread of the external field.

The formulation of the problem outlined above is elegant by itself [12–14], so is the solution proposed in refs. [12–14]. I think it is one of the cleanest problems in non-perturbative QCD. In ref. [13] the hyperfine splittings in the S levels of heavy quarkonium due to the gluon condensate are found. Refs. [12,14] present the general formulae for the shifts in masses and e^+e^- widths induced by $\langle G^2\rangle$, for instance

$$M_{nl} = 2m - \frac{k_n^2}{m}\left(1 - \frac{m^2}{18k_n^6}n^2 a_{nl}\langle\pi\alpha_s G^2\rangle\right). \tag{4.3}$$

Here l is the orbital angular momentum and a_{nl} is a known coefficient function of order one.

All three papers are reproduced below. There is a slight discrepancy between refs. [12] and [14] in the prediction for e^+e^- widths. The reasons causing the discrepancy are explained in detail in ref. [14].

All further details are nicely explained in the original publications. Only a few additional remarks are in order.

In ref. [15] the calculation of ref. [13] is extended to include the hyperfine splittings in the P levels of heavy quarkonium. The interrelation between the Leutwyler–Voloshin asymptotics and the potential description is thoroughly investigated in a recent paper [16]. As results from the discussion above, the coefficients a_{nl} cannot be *analytically* reproduced by any genuine instantaneous potential but are well approximated numerically by a fake potential proposed in ref. [10].

As we move from asymptotically large quark masses to lower values (or, for a given mass, to higher values of n, the principal quantum number), the characteristic energy splittings become smaller and the quark frequencies lower. As a result, the Coulomb quark system at hand starts feeling *

$$\left\langle G^a_{\alpha\beta}(t_1) G^b_{\mu\nu}(t_2) \right\rangle, \quad t_1 \neq t_2, \tag{4.4}$$

rather than $\langle G^a_{\mu\nu}(0) G^a_{\mu\nu}(0) \rangle$. The finite correlation time of the gluon medium comes into play [17,18].

Finally, I would like to mention that the Leutwyler–Voloshin result [12,14] tacitly assumes that the effect of the vacuum medium just shifts the position of the pole and renormalizes its residue but does not introduce, say, cuts due to multiparticle states (quarkonium + glueballs). This assumption – quite reasonable, of course – is equivalent to a certain constraint on higher gluon condensates.

General conditions under which the asymptotic Leutwyler–Voloshin formulae are valid are discussed in refs. [19–21]. Explicitly introducing a finite correlation time the authors explain how different regimes (the potential regime versus the Leutwyler–Voloshin one) emerge in different limiting cases. These works (see also preceeding papers [17,18]) is an attempt to extend the Leutwyler–Voloshin expressions to smaller masses. The price we have to pay for the possibility of getting more information than contained in the pre-Coulombic formulae [12,14] is rather high, though. In addition to the gluon condensate $\langle G^2 \rangle$ one now has to introduce more detailed characteristics of the vacuum fields, such as a finite correlation time for the gluon medium and, moreover, a specific function parametrizing the time evolution in eq. (4.4). Correspondingly, the predictions emerging in this way [19–21] are more model-dependent.

References

[1] M.A. Shifman, A.I. Vainshtein and V.I. Zakharov, Nucl. Phys. B 147 (1979) 385.
[2] V.A. Novikov, L.B. Okun, M.A. Shifman, A.I. Vainshtein, M.B. Voloshin and V.I. Zakharov, Phys. Rep. 41 (1978) 1.
[3] A.I. Vainshtein, V.I. Zakharov and M.A. Shifman, Pis'ma Zh. Eksp. Teor. Fiz. 27 (1978) 60 [JETP Lett. 27 (1978) 55].
[4] M.A. Shifman, A.I. Vainshtein, M.B. Voloshin and V.I. Zakharov, Phys. Lett. B 77 (1978) 80.
[5] T.M. Himel et al., (Mark II Collab.), Phys. Rev. Lett. 45 (1980) 1146.
[6] R. Partridge et al., (CB Collab.), Phys. Rev. Lett. 45 (1980) 1150.
[7] L. Reinders, H. Rubinshtein and S. Yazaki, Nucl. Phys. B 186 (1981) 109.
[8] L. Reinders, H. Rubinshtein and S. Yazaki, Phys. Lett. B 138 (1984) 425.
[9] M.B. Voloshin, Nucl. Phys. B 154 (1979) 365.
[10] J.S. Bell and R.A. Bertlmann, Nucl. Phys. B 227 (1983) 435.
[11] R.A. Bertlmann, Nucl. Phys. B (Proc. Suppl.) 23A (1991) 307.
[12] H. Leutwyler, Phys. Lett. B 98 (1981) 447.
[13] M.B. Voloshin, Yad. Fiz. 35 (1982) 1016 [Sov. J. Nucl. Phys. 35 (1982) 592].
[14] M.B. Voloshin, Yad. Fiz. 36 (1982) 247 [Sov. J. Nucl. Phys. 36 (1982) 143].
[15] A.Yu. Morozov, Yad. Fiz. 36 (1982) 1302 [Sov. J. Nucl. Phys. 36 (1982) 755].
[16] R.A. Bertlmann and P. Hofer, Phys. Lett. B 233 (1989) 210.
[17] D. Gromes, Phys. Lett. B 115 (1982) 482.
V.N. Baier and Yu.F. Pinelis, Phys. Lett. B 116 (1982) 179.
[18] I.I. Balitsky, Nucl. Phys. B 254 (1985) 166.
[19] U. Marquard and H.G. Dosch, Phys. Rev. D 35 (1987) 2238.
[20] H.G. Dosch and U. Lisenfeld, Phys. Lett. B 219 (1989) 493.
H.G. Dosch and Yu.A. Simonov, Z. Phys. 45 (1989) 147.
[21] A. Kramer, H.G. Dosch, R.A. Bertlmann, Phys. Lett. B 223 (1989) 105; Fortschr. Phys. 40 (1992) 93.

* The Wilson line factor connecting the points t_1 and t_2 and making this expression gauge invariant is implied.

Nuclear Physics B154 (1979) 365–380

ON DYNAMICS OF HEAVY QUARKS IN A NON-PERTURBATIVE QCD VACUUM

M.B. VOLOSHIN
Institute of Theoretical and Experimental Physics, B. Cheremushkinskaya Ul. 25, 117259 Moscow, USSR

Received 11 December 1978

Non-perturbative large scale fluctuations of gluonic fields, which are present in the exact vacuum of QCD, result in a modification of the interaction between quarks in quarkonium. Such effects are considered for heavy quarks in cases when the size of the quarkonium system is much smaller than that of the dominating gluonic fluctuations. The Green function of relative motion in heavy quarkonium is calculated for energies below threshold taking account of leading non-perturbative effects. It is shown that once non-perturbative gluonic degrees of freedom are taken into consideration they can never be absorbed into a sort of interaction potential between quarks.

1. Introduction

Since the discovery of instanton solutions in QCD [1] (see also [2] and references therein) and of gauge ambiguities [3], it has been realized that the spectrum and structure of eigenstates in QCD should be drastically different from the customary perturbation theory pattern. In particular, these peculiarities of the theory should result in a reconstruction of the vacuum state. At the level of present understanding the exact vacuum of QCD should contain large scale fluctuations not describable in usual perturbation theory. In ref. [2] such vacuum fluctuations caused by instanton solutions of moderate scale size were discussed, so that in a sense weak coupling theory remained justified. However, a complete investigation of the problem at all distances requires overcoming the strong coupling barrier and is inaccessible at present.

On the other hand, it is widely believed (see, e.g., refs. [2, 4]) that non-perturbative fluctuations of the gluonic field cause an essential modification of the interaction between a heavy quark and an antiquark. Therefore, such fluctuations might play an important role in the dynamics of a heavy quark-antiquark bound system called quarkonium. In fact the same is true for light quarks (u, d, s) as well, but there is an important difference between heavy and light quarks. The former can be considered as external objects placed in a gluonic vacuum, while the fields of the latter also undergo non-perturbative vacuum fluctuations themselves [2], and that is why the situation is far more complicated for light quarks.

In this paper it will be shown that for heavy enough quarks the effect of non-perturbative fluctuations of the gluonic field on certain properties of heavy quarkonium can be calculated, to some extent, without a detailed knowledge of the structure of the vacuum or excited states in QCD. In the approach to be developed below, non-perturbative effects in the dynamics of heavy quarkonium are expressed in terms of vacuum mean values of local operators constructed from gluonic field operators. In particular, the leading effects are proportional to a vacuum matrix element of the form

$$\langle 0|F^a_{\mu\nu}(0)F^a_{\mu\nu}(0)|0\rangle\,, \tag{1}$$

where $F^a_{\mu\nu}(x)$ is the gluonic field tensor.

In fact the procedure outlined below is very close to the Wilson operator expansion [5], but provides a somewhat more general view of the properties of quarkonium. In particular, in sect. 3, the non-relativistic Green function of relative motion in quarkonium will be calculated up to terms proportional to the term in eq. (1). The Green function calculated in sect. 3 generates expressions for a set of two-point vacuum amplitudes generated by currents of heavy quarks including corrections due to the matrix element (1).

Effects of the vacuum mean value (1) in the phenomenology of charmonium and light mesons were first considered in refs. [6, 7] in terms of the Wilson operator expansion. From comparison of the sum rules for charm production in e^+e^- annihilation with experimental data, the numerical estimate was found [6, 7] to be

$$\langle 0|F^a_{\mu\nu}F^a_{\mu\nu}|0\rangle \simeq 0.12\ \mathrm{GeV}^4\,, \tag{2}$$

with a possible 20–40% uncertainty*.

Alternatively one can try to estimate the numerical value of the vacuum matrix element (1) by considering the instanton contribution. Such a calculation [8] involves integration over the scale size of the instantons, ρ, and the corresponding integral is badly divergent at large ρ. In other words the term in eq. (1) receives a dominant contribution from instantons of large scale size which in turn fail to be good approximations to actual field fluctuations [2]. Thus, one is again led into the strong coupling region and the problem of theoretical calculation of the numerical value of eq. (1) remains unsolved.

* Throughout this paper a somewhat non-standard but convenient normalisation for $F^a_{\mu\nu}$ is used. It is fixed by the convention that the Yang-Mills Lagrangian is written in the form

$$\mathcal{L}_{\mathrm{YM}} = -\frac{1}{g^2}F^a_{\mu\nu}F^a_{\mu\nu}\,,$$

with $g^2 \equiv 4\pi\alpha_s$. $F^a_{\mu\nu}$ is related to the conventionally normalized field tensor $G^a_{\mu\nu}$ as follows: $F = \frac{1}{2}gG$.

To avoid possible confusion two points should be noted. Firstly, the (ultraviolet divergent) perturbation theory part of the vacuum average (1) is implied, as usual, to be subtracted by definition, so that a non-zero value of the matrix element (1) is a purely non-perturbative effect. In other words, one can understand the operator product in eq. (1) as an N-product :FF: Secondly, the operator $F^a_{\mu\nu}(x)F^a_{\mu\nu}(x)$ has zero anomalous dimension [7], so that one need not specify the normalization point for the gluonic operators. (In this sense the value (1) is a universal constant.)

The major starting point of this paper is the simple fact that for large mass m of a quark, one can consider a quark-antiquark pair localized at a relative distance r such that, on the one hand, $r \gg m^{-1}$, so that the momenta of the quarks are non-relativistic, $p \sim r^{-1} \ll m$, while, on the other hand, r is small in the scale of strong interactions, so that methods of short-distance QCD are applicable. Surely, such a situation requires the quark mass to be large; even charmed quarks are not enough heavy in this sense, while for quarks composing the Υ family ($m \simeq 4.5$ GeV) this situation seems to be conceivable. The relevance of the approach developed below improves with increasing mass of the quark and it can be helpful for a description of still heavier quarkonium systems which it is widely believed will be discovered in the future.

The expansion parameters in the approach considered are typically the "old" $\alpha_s(k \sim 1/r)$ and the "new" one $(r/R)^{(\text{some power})}$ where R is an intrinsic scale size of non-perturbative vacuum fluctuations. The former parameter describes usual perturbation theory effects while the latter one refers to the non-perturbative corrections which we are mainly interested in. There is in fact no formal definition of R. In terms of refs. [2, 8] it is of the order of the critical size of instantons ρ_c. (According to estimates of ref. [8] $\rho_c^{-1} \simeq 0.17$ GeV.) In terms of the approach of this paper, the scale of the expansion parameter is given by the ratio of vacuum matrix elements of different dimension, say

$$R^{-2} \sim \left| \frac{\langle 0|\mathrm{D}F\mathrm{D}F|0\rangle}{\langle 0|FF|0\rangle} \right| ,$$

with D being the covariant derivative.

For a colourless quark-antiquark pair separated by a distance $r \ll R$, the interaction with fluctuations of scale size $\sim R$ can be expanded into a multipole series, the dipole term being the leading one. Furthermore, for non-relativistic quarks the interaction can also be expanded in powers of $v/c \sim (mr)^{-1}$, so that only the electric-like dipole interaction survives in the leading approximation. Therefore, if a quantum mechanical object is considered in which heavy quarks are constrained to be non-relativistic, and separated by a distance short in comparison with R, a sort of perturbation theory in powers of (r/R) and $(mr)^{-1}$ emerges.

In sect. 2 we shall consider the contribution of non-perturbative vacuum fluctuations to the characteristics of the lowest levels of a superheavy quarkonium. By superheavy is meant a quarkonium for which asymptotic freedom is valid for the lowest bound states which are dominantly determined by the Coulomb-like gluonic exchange. Obviously, this requires the following condition to be satisfied:

$$\alpha_s(k \simeq m\alpha_s) \ll 1 . \tag{3}$$

This inequality sets in for a quark mass equal to a few tens of GeV. The wave function of the nth level is proportional to $\exp(-k_0 r/n)$, with $k_0 = \frac{2}{3} m\alpha_s(k_0)$ being the Bohr momentum. Hence, for $(k_0/n)^{-1}$ small in the scale of R, our approach becomes applicable since a contribution of distances $r \gg (k_0/n)^{-1}$ is exponentially small. Surely, this situation does not seem to be of paramount physical interest, since firstly there is enough time to wait for the discovery of quarkonium with tens of GeV mass, and secondly the effects we calculate are only small corrections to essentially Coulomb-like characteristics of quarkonium levels comparable with corrections of a Breit-Fermi type, and even with purely radiative corrections having relative magnitude of order α_s^3. Nevertheless, it seems reasonable to start with a discussion of this somewhat idealized example, since in this case the basic assumptions and the method are illustrated most transparently.

In sect. 3 we consider the Green function of relative motion of quarks in quarkonium at non-physical energies far enough below threshold. To this end, the quarks are constrained at short distances by the condition that the energy is negative (i.e., below the threshold) and large:

$$\varepsilon = -k^2/m, \qquad k \gg R^{-1}, \qquad \alpha_s(k) \ll 1 . \tag{4}$$

The latter condition assures validity of usual perturbation theory.

The physical interest of considering the Green function is firstly, that through spectral representation it generates sum rules for spectral densities of various operators constructed from the fields of the heavy quarks in hand, and secondly, that knowing the Green function even for unphysical energies one can judge the quark-antiquark interaction, say in terms of a potential picture.

The consideration presented below also brings an interesting byproduct. Namely, our approach inevitably requires an existence of states composed from a quark-antiquark pair in a colour octet state and an explicitly coloured gluonic excitation, which will be discussed in sect. 4. Such states should be located in a narrow energy band around the quark-antiquark threshold. However, with our method we can make no quantitative conclusions about such states since this problem again requires overcoming the strong coupling barrier. The Green function calculated in sect. 3 can already be applied for discussion of properties of the Υ family, say, if one takes $k \simeq 1.5$–2 GeV. In this case the sum rules generated by the Green function coincide with the non-relativistic limit of the sum rules derived in refs. [6, 7, 9, 10] for the case of charmonium. Thus the approach developed here can also be considered as a simple non-relativistic method of derivation of the sum rules.

2. Non-perturbative effects in superheavy quarkonium

In this section we consider a superheavy quark-antiquark pair so that condition (3) is satisfied and in the lowest bound states the quark and the antiquark are located at distances

$$r \sim [m\alpha_s(k \simeq m\alpha_s)]^{-1}, \tag{5}$$

which are small compared to the scale of the strong interaction.

At short distances a non-relativistic quark and antiquark interact with each other by the Coulomb-like potential

$$V(r) = -\frac{4}{3}\frac{\alpha_s(1/r)}{r}P_0 + \frac{2}{3}\frac{\alpha_s(1/r)}{r}P_8, \tag{6}$$

where P_0 and P_8 are the projection operators for colour singlet and colour octet states respectively. The lowest levels of the superheavy quarkonium are determined by the colour singlet attractive part of the potential (6).

We are interested here in corrections due to non-perturbative gluonic fluctuations in the exact vacuum of QCD. To account for them one must consider the total Hilbert space including states of heavy quarks as well as of gluons. The Hamiltonian acting in this space can be classified into three terms:

$$\mathcal{H} = \mathcal{H}_Q + \mathcal{H}_g + \mathcal{H}_{int} \tag{7}$$

where $\mathcal{H}_Q$ is the Hamiltonian acting on dynamical variables of quarks only, including the Coulomb-like gluonic exchange interaction between them. In the leading non-relativistic approximation it can be written in the form

$$\mathcal{H}_Q = -\frac{\nabla^2}{m} + V(r), \tag{8}$$

where $V(r)$ is given by eq. (6) and the ∇ operator is acting on the relative coordinate $\boldsymbol{r} = \boldsymbol{r}_Q - \boldsymbol{r}_{\bar{Q}}$. (We exclude the overall motion of quarkonium by placing the coordinate system origin in the centre of mass of the system.) $\mathcal{H}_g$ is the Hamiltonian for gluonic degrees of freedom, and finally, $\mathcal{H}_{int}$ is the Hamiltonian describing the interaction of quarkonium with gluonic degrees of freedom.

The most unknown term here is $\mathcal{H}_g$, and the only property of it that we assume for a while is that it enjoys the full rights of a Hamiltonian operator in the absence of heavy quarks.

In the case where the $\mathcal{H}_{int}$ term is neglected, the total Hilbert space of the problem splits into a direct product of two independent subspaces, one for quarkonium states and the other for gluonic states. Thus, in this approximation the states of the whole system are factorized in the form

$$|N\rangle = |n_Q\rangle|n_g\rangle, \tag{9}$$

$|n_Q\rangle$ being the eigenstates of the Hamiltonian (8) and $|n_g\rangle$ being the same for $\mathscr{H}_g$. In fact as has already been mentioned in sect. 1, little is known about the spectrum and properties of $|n_g\rangle$ and there is enough room for assumptions. We make here two assumptions about the spectrum of gluonic states. The first one is that they have the lowest energy state $|0\rangle$ called vacuum, and that this state is invariant under Lorentz transformations as well as under global colour rotations. The Hamiltonian $\mathscr{H}_g$ is taken to be normalized in such a way that

$$\mathscr{H}_g|0\rangle = 0. \tag{10}$$

The second assumption will be described later.

Thus, without the $\mathscr{H}_{int}$ term the total spectrum of the problem contains the pure quarkonium states $|n_Q\rangle|0\rangle$ and states with explicit gluonic excitations $|n_g\rangle$. Now we are going to switch on the interaction $\mathscr{H}_{int}$ and by considering it as a perturbation calculate the energy shifts of quarkonium states, i.e., of $|n_Q\rangle|0\rangle$. To do this we need an explicit form of $\mathscr{H}_{int}$. The leading non-relativistic term is given by the interaction with the time-like component of the gluonic 4-potential A^a_μ. As it has been described in sect. 1 we also expand the interaction in multipoles. The first two terms of this expansion are the "charge"

$$\mathscr{H}_0 = 2t^a A^a_0(0)\,, \tag{11}$$

and the dipole

$$\mathscr{H}_1 = -\xi^a \boldsymbol{r} \cdot \boldsymbol{E}^a(0)\,, \tag{12}$$

where $t^a = t^a_1 + t^a_2$ ($a = 1, 2, \ldots, 8$) are the colour SU(3) generators for the quark-antiquark system (t^a_1 and t^a_2 refer to the quark and the antiquark respectively; for a quark $t^a_1 = \frac{1}{2}\lambda^a$), ξ^a are operators orthogonal to t^a: $\xi^a = t^a_1 - t^a_2$, and finally,

$$E^a_i = -F^a_{0i}. \tag{13}$$

(Note, that the coupling constant $\frac{1}{2}g$ is included in the normalization of the gluonic field.)

We will consider the perturbation expansion in $\mathscr{H}_{int}$ up to the second order, starting from colourless states $|n_Q\rangle$. Therefore the "charge" term (11) will be not operative, since

$$t^a|\text{colourless state}\rangle = 0\,,$$

and only the dipole (12) survives.

For the same reasons, in the approximation considered we will have no problems with gauge invariance since (12) contains only the operator $\boldsymbol{E}^a$. Therefore we need not specify the gauge used.

Now we have everything to write down the energy shift due to $\mathscr{H}_{int}$ of a quarkonium level, which in the zeroth order has the form $|N^0\rangle = |n_Q\rangle|0\rangle$ with $|n_Q\rangle$ being colourless. To the first order we have

$$\delta\epsilon_n = \langle N^0|\mathscr{H}_1|N^0\rangle = -\langle n_Q|\xi^a r_i|n_Q\rangle\langle 0|E^a_i|0\rangle = 0\,.$$

There are many reasons for this expression to be equal to zero. One is that $\langle 0|E_i^a|0\rangle = 0$ because of the vacuum state invariance under space and colour rotations. The second-order expression is

$$\begin{aligned} \delta\epsilon_n &= \sum_{|M\rangle} \frac{|\langle N^0|\mathcal{H}_1|M\rangle|^2}{\epsilon_{N^0} - \epsilon_M} \\ &= \sum_{|m_Q\rangle|m_g\rangle} \frac{\langle n_Q|\xi^a r_i|m_Q\rangle\langle m_Q|r_k\xi^b|n_Q\rangle\langle 0|E_i^a|m_g\rangle\langle m_g|E_k^b|0\rangle}{\epsilon_{n_Q} - \epsilon_{m_Q} - \epsilon_{m_g}}, \end{aligned} \tag{14}$$

where ϵ_{n_Q}, ϵ_{m_Q} are energy eigenvalues of $\mathcal{H}_Q$ and ϵ_{m_g} is the same for $\mathcal{H}_g$. Note, that the operators ξ^a link colour singlets to colour octets only. Therefore the states $|m_Q\rangle$ refer to the continuum spectrum of the Hamiltonian (8), (6) in the colour octet sector, thus, in (14) $\epsilon_{m_Q} > 0$. Therefore, if $|n_Q\rangle$ is taken to be one of the lowest bound colourless states, the energy difference $\epsilon_{n_Q} - \epsilon_{m_Q}$ is negative and large for large enough mass of the quark.

Now we are in position to make the second assumption about the spectrum of gluonic states $|m_g\rangle$. Namely we assume that in the situation considered the r.h.s. of eq. (14) can be expanded in powers of $\epsilon_{m_g}/(\epsilon_{n_Q} - \epsilon_{m_Q})$. Then the first term of this expansion has the form

$$\begin{aligned} \delta\epsilon_n &= \sum_{|m_g\rangle} \langle 0|E_i^a|m_g\rangle\langle m_g|E_k^b|0\rangle \\ &\times \sum_{|m_Q\rangle} \frac{\langle n_Q|\xi^a r_i|m_Q\rangle\langle m_Q|\xi^b r_k|n_Q\rangle}{\epsilon_{n_Q} - \epsilon_{m_Q}} \\ &= \langle 0|E_i^a E_k^b|0\rangle \sum_{|m_Q\rangle} \frac{\langle n_Q|\xi^a r_i|m_Q\rangle\langle m_Q|\xi^b r_k|n_Q\rangle}{\epsilon_{n_Q} - \epsilon_{m_Q}}. \end{aligned} \tag{15}$$

Thus one sees that with the latter assumption the sum over the unknown spectrum of gluonic states, $|m_g\rangle$ is reduced to the vacuum mean value $\langle 0|E_i^a E_k^b|0\rangle$. One can employ colour and Lorentz invariance of the vacuum state to relate the latter to the manifestly Lorentz-invariant value (1). Indeed, one has

$$\langle 0|E_i^a E_k^b|0\rangle = \tfrac{1}{3}\delta_{ik}\tfrac{1}{8}\delta^{ab}\langle 0|\boldsymbol{E}^c\boldsymbol{E}^c|0\rangle .$$

Now it is convenient to introduce the 4-vector q_μ which is the total 4-momentum of the quark-antiquark system considered. In the rest frame to which all the previous relations refer q has only a time-like component q_0, so that from eq. (13) one has

$$\boldsymbol{E}^a\boldsymbol{E}^a = -(q_\alpha F_{\mu\alpha})(q_\beta F_{\mu\beta})/q^2 .$$

Due to the Lorentz invariance of the vacuum,

$$\langle 0|F_{\mu\alpha}^a F_{\mu\beta}^a|0\rangle = \tfrac{1}{4} g_{\alpha\beta}\langle 0|F_{\mu\nu}^a F_{\mu\nu}^a|0\rangle ,$$

and thus we have

$$\begin{aligned}\langle 0|E_i^a E_k^b|0\rangle &= \tfrac{1}{24}\delta_{ik}\delta^{ab}\langle 0|\boldsymbol{E}^c\boldsymbol{E}^c|0\rangle \\ &= -\tfrac{1}{96}\delta_{ik}\delta^{ab}\langle 0|F^c_{\mu\nu}F^c_{\mu\nu}|0\rangle .\end{aligned} \tag{16}$$

Note the negative sign of the last expression. It shows that with positive value of the quantity (1) (see eq. (2)) the fluctuations of the gluonic field correspond to an imaginary "electric" component, or in the other words that it is more adequate to treat the fluctuations in Euclidean space, which fits very well with the instanton philosophy.

The assumption of the validity of the expansion in powers of $\epsilon_{m_g}/(\epsilon_{n_Q}-\epsilon_{m_Q})$ refers, in fact, to the structure of gluonic fluctuations. Indeed, the assumption implies that the sum over the intermediate gluonic states in eq. (15) is saturated by states $|m_g\rangle$ with energies low as compared to $|\epsilon_{n_Q}|$. In this sense the gluonic system is assumed to be "soft" in comparison with the quark-antiquark system. This conjecture can be analyzed by considering subsequent terms of the expansion. The first of them is proportional to

$$\langle 0|E_i^a \mathscr{H}_g E_i^a|0\rangle , \tag{17}$$

which is zero because of kinematical reasons. Indeed, using eq. (10) one has

$$\begin{aligned}\langle 0|E_i^a \mathscr{H}_g E_i^a|0\rangle &= \langle 0|E_i^a[\mathscr{H}_g, E_i^a]|0\rangle \\ &= i\langle 0|E_i^a\left(\frac{\partial}{\partial t}E_i^a\right)|0\rangle = -i\frac{q_\alpha q_\gamma q_\beta}{(q^2)^{3/2}}\langle 0|F^a_{\mu\alpha}\partial_\gamma F^a_{\mu\beta}|0\rangle ,\end{aligned}$$

and the vacuum mean value in the last expression in this chain is vanishing since it should have an odd number of Lorentz indices. The next term is proportional to

$$\langle 0|E_i^a \mathscr{H}_g \mathscr{H}_g E_i^a|0\rangle = \langle 0|(\partial_0 E_i^a)(\partial_0 E_i^a)|0\rangle ,$$

which is reduced to the vacuum mean value of the $d=6$ operator $\langle 0|\mathrm{D}F^a\mathrm{D}F^a|0\rangle$. Thus our assumption is valid in the case

$$|\langle 0|\mathrm{D}F^a\mathrm{D}F^a|0\rangle/\langle 0|F^aF^a|0\rangle| \ll \epsilon_{n_Q}^2 .$$

In the vein of ref. [2] the ratio on the l.h.s. is of order ρ_c^{-2}, where ρ_c is the cutoff parameter for instanton scale size.

Thus the perturbation theory in $\mathscr{H}_{\rm int}$ and the expansion in powers of ϵ_{m_g} (which is in a sense a perturbation expansion in $\mathscr{H}_g$) both result in a series if vacuum mean values of operators of increasing dimension. This expansion seems to make sense as long as the vacuum matrix elements are determined by large distances, or in other words as long as the gluonic system can be considered as soft.

Certainly, the proposed expansion also fails when the quarkonium system becomes soft itself. This takes place for the levels near threshold when the energy gap $\epsilon_{m_Q}-\epsilon_{n_Q}$ is comparable with the characteristic values of ϵ_{m_g}. We can argue that in this situation a mixing of states of the type $|n_Q\rangle|0\rangle$ with $|m_Q\rangle|m_g\rangle$ containing explicit gluonic excitations becomes essential. In fact it is quite obvious, if one considers the first correction to the wave function, that

$$|N^0\rangle + |\delta N\rangle = |N^0\rangle + \sum_{|M\rangle} \frac{\langle M|\mathcal{H}_1|N^0\rangle}{\epsilon_{N^0} - \epsilon_M}|M\rangle$$

$$= |n_Q\rangle|0\rangle - \sum_{m_Q, m_g} \frac{\langle m_Q|\xi^a r_i|n_Q\rangle\langle m_g|E_i^a|0\rangle}{\epsilon_{n_Q} - \epsilon_{m_Q} - \epsilon_{m_g}}|m^Q\rangle|m_g\rangle .$$

This expression also shows that as far as the dipole interaction (12) is concerned, the mixing occurs with states containing a colour octet quark-antiquark pair and a colour octet gluonic excited state.

There is one serious problem concerning such states. The problem emerges if one invokes the idea of confinement. In this case there are no free states with explicit colour and one should avoid considering them. This however seems impossible at the present level of understanding as long as gluonic operators like $\boldsymbol{E}^a$ or $F^a_{\mu\nu}$ are dealt with. These operators require colour octet states since they have no matrix elements between colour singlets. In fact we do not know the language in which QCD should be formulated to include both the idea of gauge colour and confinement. In the approach considered there is no sign of confinement and therefore we are forced to consider explicitly coloured gluonic states and the same for quarks. Obviously we can and actually do restrict ourselves to consideration of states with no net colour. With the idea of confinement, the colourless states made from a coloured quark pair and a coloured gluonic excitation should be bound, thus adding a new type of state to quarkonium spectroscopy.

On phenomenological grounds such states were first considered in ref. [11] and there are also calculations considering states with valence gluons in the framework of the MIT bag model [12].

Now we will postpone a further discussion of quarkonium states with gluonic excitations till sect. 4.

Returning to eq. (15) it is worth noting that it can be applied in particular calculations of energy shifts of superheavy quarkonium levels, if this problem becomes actual some day. With eq. (16), after noting that

$$\sum_{|m_Q\rangle} \frac{|m_Q\rangle\langle m_Q|}{\epsilon_{m_Q} - \epsilon_{n_Q}}$$

is the Green function for the repulsive potential $+\frac{2}{3}\alpha_s(r^{-1})/r$ denoted here as $G^{(8)}(\epsilon_{n_Q})$, ($|m_Q\rangle$ are colour octets) and using the relation

$$\langle \text{colourless state}|\xi^a \xi^b|\text{colourless state}\rangle = \tfrac{2}{3}\delta^{ab} ,$$

one can rewrite eq. (15) in the form

$$\delta\epsilon_n = 8\kappa^4 \int d^3r\, d^3r' \psi_n^*(\boldsymbol{r}) r_i G^{(8)}(\boldsymbol{r}, \boldsymbol{r}', \epsilon_n) r_i' \psi_n(\boldsymbol{r}') , \tag{18}$$

where

$$\kappa^4 = \tfrac{1}{144}\langle 0|F^a_{\mu\nu}F^a_{\mu\nu}|0\rangle \simeq (170\ \mathrm{MeV})^4\,, \qquad (19)$$

and $G^{(8)}(\boldsymbol{r}, \boldsymbol{r}', \epsilon)$ is the above mentioned Green function in the coordinate representation obeying the equation:

$$\left(-\frac{\boldsymbol{\nabla}^2}{m}+\frac{2}{3}\frac{\alpha_s(r^{-1})}{r}-\epsilon\right)G^{(8)}(\boldsymbol{r}, \boldsymbol{r}', \epsilon) = \delta(\boldsymbol{r}-\boldsymbol{r}')\,.$$

In the same approximation one can also find the corrections to matrix elements of a colour singlet operator $\mathcal{O}$ acting only on quark variables (say, matrix elements for radiative transitions) between lowest states of a superheavy quarkonium. For this one should calculate the corrections to the wave functions up to second order and then take the matrix element. The final answer is

$$\begin{aligned}\langle N'|\mathcal{O}|N\rangle = {} & \langle n'_Q|\mathcal{O}|n_Q\rangle[1+4\kappa^4(\langle n_Q|r_i\cdot G^{(8)'}(\epsilon_{n_Q})\cdot r_i|n_Q\rangle \\ & +\langle n'_Q|r_i\cdot G^{(8)'}(\epsilon_{n'_Q})\cdot r_i|n'_Q\rangle)] \\ & -8\kappa^4[\langle n'_Q|r_i G^{(8)}(\epsilon_{n'_Q})\cdot\mathcal{O}\cdot G^{(8)}(\epsilon_{n_Q})\cdot r_i|n_Q\rangle \\ & +\langle n'_Q|\mathcal{O}\cdot G^{(0)}(\epsilon_{n_Q})\cdot r_i\cdot G^{(8)}(\epsilon_{n_Q})\cdot r_i|n_Q\rangle \\ & +\langle n'_Q|r_i\cdot G^{(8)}(\epsilon_{n'_Q})\cdot r_i\cdot G^{(0)}(\epsilon_{n'_Q})\cdot\mathcal{O}|n_Q\rangle]\,, \qquad (20)\end{aligned}$$

(the pole terms of $G^{(0)}(\epsilon_{n_Q})$ at $\epsilon=\epsilon_{n_Q}$ and of $G^{(0)}(\epsilon_{n'_Q})$ are here implied to be subtracted) where $G^{(8)\prime}(\epsilon)$ is the derivative of $G^{(8)}(\epsilon)$ over ϵ: $G^{(8)\prime}(\epsilon) \equiv (\partial/\partial\epsilon)G^{(8)}(\epsilon)$, and $G^{(0)}(\epsilon)$ is the Coulomb-like Green function for the colour singlet states:

$$\left(-\frac{\boldsymbol{\nabla}^2}{m}-\frac{4}{3}\frac{\alpha_s(r^{-1})}{r}-\epsilon\right)G^{(0)}(\boldsymbol{r}, \boldsymbol{r}', \epsilon) = \delta(\boldsymbol{r}-\boldsymbol{r}')\,.$$

All the products in eq. (20) inside the Dirac brackets should be understood as operator products.

One can readily verify that the relative magnitude of the corrections discussed is of order $\kappa^4 m^2/k_0^6$. Therefore, the formulae (15), (18) and (20) can be trusted as long as this parameter is small.

3. Quarkonium Green function

In this section we proceed to the consideration of a more realistic case of a quarkonium made from quarks of a moderate though large enough mass, say, of the Υ family or somewhat heavier. In this case the quarks inside the bound states are no longer located at sufficiently short relative distances and to assure the applicability of our approach we must consider the system at energies where there are no bound states. An adequate object in this situation is the Green function. Indeed, if one considers the propagation of quarks off a stationary local source producing quark-antiquark pairs with an energy far below threshold, $\epsilon = -k^2/m$, then the wave function of the pairs is fading as $\exp(-kr)$ and only distances $r \lesssim k^{-1}$ are relevant.

For quarks interacting with gluons we must start with the total Green function which is an operator in the total Hilbert space involving states of heavy quarks as well as of gluons,

$$\mathscr{G}(\epsilon) = \sum_N \frac{|N\rangle\langle N|}{\epsilon_N - \epsilon}, \tag{21}$$

with the sum running over the complete set of eigenstates $|N\rangle$ of the total Hamiltonian,

$$(\mathscr{H}_Q + \mathscr{H}_g + \mathscr{H}_{int})|N\rangle = \epsilon_N |N\rangle . \tag{22}$$

The Green function (21) is in fact a resolvent operator

$$\mathscr{G}(\epsilon) = \frac{1}{\mathscr{H}_Q + \mathscr{H}_g + \mathscr{H}_{int} - \epsilon}, \tag{23}$$

and we are going to calculate it by considering the term $\mathscr{H}_g + \mathscr{H}_{int}$ as a perturbation:

$$\begin{aligned} \mathscr{G}(\epsilon) = {} & \frac{1}{\mathscr{H}_Q - \epsilon} - \frac{1}{\mathscr{H}_Q - \epsilon}(\mathscr{H}_g + \mathscr{H}_{int})\frac{1}{\mathscr{H}_Q - \epsilon} \\ & + \frac{1}{\mathscr{H}_Q - \epsilon}(\mathscr{H}_g + \mathscr{H}_{int})\frac{1}{\mathscr{H}_Q - \epsilon}(\mathscr{H}_g + \mathscr{H}_{int})\frac{1}{\mathscr{H}_Q - \epsilon} + \dots . \end{aligned} \tag{24}$$

The perturbation expansion in $\mathscr{H}_g$ is justified in the case where the relevant sums over intermediate gluonic states are determined by those with energy much less than $|\epsilon|$. We suppose as previously that that is the case if one projects $\mathscr{G}(\epsilon)$ out on the gluonic vacuum state, thus leaving the Green function to be an operator acting only on quark variables. It is also convenient to project it out on the colourless sector of heavy quark-antiquark states and define

$$\begin{aligned} G(\epsilon) &= P_o \langle 0|\mathscr{G}(\epsilon)|0\rangle P_o \\ &= \sum_N P_o \frac{\langle 0|N\rangle\langle N|0\rangle}{\epsilon_N - \epsilon} P_0 . \end{aligned} \tag{25}$$

Note, that $G(\epsilon)$ is still an operator acting on colour singlet states of quarkonium.

Substituting the expansion (24) into eq. (25), one finds that in terms written out explicitly in eq. (24) $\mathscr{H}_g$ is not operative due to eq. (10) and since it commutes with $\mathscr{H}_Q$. Moreover, the vacuum mean value of $\mathscr{H}_{int}$ also vanishes, and thus the term linear in $\mathscr{H}_g+\mathscr{H}_{int}$ in (24) brings zero contribution to $G(\epsilon)$. Therefore, with the explicit form of $\mathscr{H}_{int}$ given by eq. (12) one finds

$$G(\epsilon)=P_0\frac{1}{\mathscr{H}_Q-\epsilon}P_0 \\ +\langle 0|E_i^a E_k^b|0\rangle P_0\frac{1}{\mathscr{H}_Q-\epsilon}\xi^a r_i\frac{1}{\mathscr{H}_Q-\epsilon}\xi^b r_k\frac{1}{\mathscr{H}_Q-\epsilon}P_0\,. \tag{26}$$

Now this expression can be readily adapted for actual calculations if one notes that $(\mathscr{H}_Q-\epsilon)^{-1}$ is the Green function for the Coulomb-like problem with the potential (6):

$$\frac{1}{\mathscr{H}_Q-\epsilon}=G^{(0)}(\epsilon)P_0+G^{(8)}(\epsilon)P_8\,.$$

Thus, using eq. (16) and simple properties of ξ^a operators we find in the coordinate representation

$$G(\boldsymbol{x},\boldsymbol{y},\epsilon)=G^{(0)}(\boldsymbol{x},\boldsymbol{y},\epsilon)P_0 \\ -8\kappa^4\int d^3r\, d^3r'\, G^{(0)}(\boldsymbol{x},\boldsymbol{r},\epsilon)r_i G^{(8)}(\boldsymbol{r},\boldsymbol{r}',\epsilon)r_i' G^{(0)}(\boldsymbol{r}',\boldsymbol{y},\epsilon)P_0\,. \tag{27}$$

(Henceforth the projector P_0 will be omitted, but implied.)

A good piece of physical information can be derived from eq. (27). For instance this expression gives, in the non-relativisitc limit, the correction due to the vacuum mean value (1) to the whole set of amplitudes of the form

$$\Pi(k^2)=i\int d^4x\, e^{iqx}\langle 0|T\{J^{(a)}(x),J^{(a)}(0)\}|0\rangle\,, \\ q^2\equiv s\equiv 4m^2-4k^2\,, \tag{28}$$

$J^{(a)}(x)=\bar{Q}(x)\mathcal{O}^{(a)}Q(x)$ being a set of local operators constructed from the field of the heavy quarks $Q(x)$.

One can relate the Green function to, e.g., the electromagnetic vacuum polarization function $P(k^2)$ for $k^2\ll m^2$. In the case when $P(k^2)$ is defined as follows,

$$(q_\mu q_\nu-q^2g_{\mu\nu})P(k^2) \\ =i\int dx\, e^{iqx}\langle 0|T\{\bar{Q}(x)\gamma_\mu Q(x),\,\bar{Q}(0)\gamma_\nu Q(0)\}|0\rangle\,, \tag{28a}$$

it is related to $G(\epsilon)$ by the equation

$$\frac{d}{dk^2}P(k^2)=\frac{3}{2m^2}\lim_{\substack{x\to 0\\ y\to 0}}\frac{\partial}{\partial k^2}G(\boldsymbol{x},\boldsymbol{y},-k^2/m)\,, \tag{29}$$

where the factor 3 comes from the summation over colour in the electromagnetic current, while the $2m^2$ in the denominator arise because of different normalization of the relativistic and non-relativistic amplitudes. Note also, that eq. (29) contains derivatives over k^2. On the l.h.s. the differentiation eliminates the unphysical additive constant which depends on the renormalization convention, while on the r.h.s. the kinematical singularity at $\boldsymbol{x}=\boldsymbol{y}$ also drops out after differentiation.

The integral in eq. (27) can be readily done in the energy region $|\epsilon| \gg m\alpha_s^2$, ($k \gg m\alpha_s$), where the Coulomb-like interaction (6) can be also treated perturbatively, so that neglecting it one can write

$$G^{(0)}(\boldsymbol{x}, \boldsymbol{y}, \epsilon) \simeq G^{(8)}(\boldsymbol{x}, \boldsymbol{y}, \epsilon) \simeq \frac{m}{4\pi} \frac{e^{-kr}}{r} \tag{30}$$

($k^2=-m\epsilon$), with $r=|\boldsymbol{x}-\boldsymbol{y}|$. Technically it is more convenient to perform calculations in the momentum representation in which the operator $\boldsymbol{r}$ is reduced to a derivative, $\boldsymbol{r}=i\partial/\partial\boldsymbol{p}$, and then transforming back into the coordinate representation we have

$$G(\boldsymbol{x}, \boldsymbol{y}, \epsilon) = \frac{m\, e^{-kr}}{4\pi r}\left\{1-\frac{\kappa^4 m^2 r}{4}\left[\frac{3}{k^5}+\frac{3r}{k^4}+\frac{2r^2}{k^3}+\frac{r^3}{k^2}\right]\right\}. \tag{31}$$

Now the calculation of the r.h.s. of eq. (29) is trivial and one finds

$$\frac{d}{dk^2}P(k^2) = -\frac{3}{16\pi mk}\left(1-\frac{15}{4}\frac{m^2\kappa^4}{k^6}\right). \tag{32}$$

This expression precisely coincides with the non-relativistic limit (i.e., at $k^2 \ll m^2$) of the results obtained in refs. [6, 7] by rather tedious calculations of Feynman diagrams.

One can readily find the analogs of relation (29) for amplitudes of vacuum polarization by other currents $J^{(a)}$ giving to the quark-antiquark pair different angular momenta. Say, for a P-wave current, e.g., $\bar{Q}Q$ or $\bar{Q}\gamma_5\overleftrightarrow{\partial}_\mu Q$ one finds that the corresponding amplitude of the type (28) is proportional to the expression

$$\lim_{\substack{\boldsymbol{x}\to 0\\ \boldsymbol{y}\to 0}} \frac{\partial}{\partial x_i}\frac{\partial}{\partial y_k}\left(\frac{\partial}{\partial k^2}\right)^2 G\left(\boldsymbol{x}, \boldsymbol{y}, \frac{-k^2}{m}\right),$$

which is immediately calculable with eq.(31).

The physical interest of considering amplitudes of the type (28) is that through spectral representation they provide QCD sum rules for spectral densities of quarkonium states with different quantum numbers [6, 7, 9, 10]. For instance, with the quantity $P(k^2)$ (see eq. 28(a)) one explores the vector states of quarkonium which are directly observed in e^+e^- annihilation. Eq. (32) generates the following sum rules for Q-flavour production in e^+e^- annihilation

$$\int \frac{R_Q(s)\, ds}{(s-4m^2+4k^2)^2} = \frac{9\pi Q_Q^2}{16mk}\left(1+\frac{4\alpha_s(k)m}{3k} -\frac{15}{4}\frac{m^2\kappa^4}{k^6}\right), \qquad (m^2\kappa^4)^{1/3} \ll k^2 \ll m^2, \tag{33}$$

where $R_Q(s) = \sigma(e^+e^- \to \text{Q-flavour})\, 3s/4\pi\alpha^2$ and Q_Q is the electric charge of the quarks in hand. In eq. (33) the obvious first Coulomb correction is also included.

4. Connection with the potential model approach and the spectrum of quarkonium states

We start this section with a remark that eq. (27) can be taken as the result of the first iteration of the equation

$$(\mathcal{H}_Q + U(\epsilon) - \epsilon)G(\epsilon) = 1\,, \tag{34}$$

where U is a non-local interaction potential

$$U(\epsilon)\varphi \equiv 8\kappa^4 \int d^3r'(\boldsymbol{r}, \boldsymbol{r}')G^{(8)}(\boldsymbol{r}, \boldsymbol{r}', \epsilon)\varphi(\boldsymbol{r}')\,. \tag{35}$$

Note also that the expression (18) for the energy shift is also a diagonal matrix element of U. Therefore, the potential U might be inferred as a non-perturbative part of the interaction potential between quarks in quarkonium. It is interesting that there is no sign of the local linear potential usually employed in the non-relativistic potential model (see, e.g., [13]). This implies that at least at the distances in hand the linear potential is irrelevant.

However, even with the non-local interaction (35) it is impossible to go further by suppressing the gluonic degrees of freedom. This is already seen from eq. (20), only a part of which can be derived from the first iteration of the interaction (35). In fact eq. (20) cannot be obtained in any approach dealing only with quark degrees of freedom. Therefore it seems entirely inadequate to discuss the properties of quarkonium in isolation from gluonic states.

The incompleteness of the description dealing with only quark-antiquark dynamical variables becomes evident if one tries to write down a spectral representation for the Green function (27) of the form

$$G(\boldsymbol{x}, \boldsymbol{y}, \epsilon) = \sum_n \frac{\psi_n(\boldsymbol{x})\psi_n^*(\boldsymbol{y})}{\epsilon_n - \epsilon}\,, \tag{36}$$

with a complete set of orthogonal functions ψ_n. It is rather trivial to show *ad absurdum* that such a representation is impossible for the function given by (27). Indeed, from (36) and the orthogonality ansatz one finds that a function represented by eq. (36) should obey the necessary condition

$$\frac{\partial}{\partial\epsilon}G(\boldsymbol{x}, \boldsymbol{y}, \epsilon) - \int G(\boldsymbol{x}, \boldsymbol{z}, \epsilon)G(\boldsymbol{z}, \boldsymbol{y}, \epsilon)\, d^3z = 0\,. \tag{37}$$

On the other hand, calculating the same expression for the function (27) or even for that given by eq. (31), one finds that for them the condition (37) is violated by a term proportional to $G^{(0)}(\epsilon)(\partial U/\partial\epsilon)G^{(0)}(\epsilon)$.

Thus, once the relevance of the results (27) and (31) to the approximation considered is trusted, one is led inevitably to the conclusion that the description of quarkonium in terms of quark variables is incomplete and the gluonic degrees of freedom should be explicitly accounted for. Surely, after inclusion of the latter variables, the Green function $G(\epsilon)$ has a perfect spectral representation as given by eq. (25). In other words, once the non-perturbative fluctuations of the gluonic field are taken into consideration they can never be absorbed into a sort of interaction potential between quarks.

The physical interpretation of the situation described is that in the vein of the confinement idea there should be states made of both heavy quarks and gluons and these states, as it has been argued in sect. 2, should be located in a narrow energy band near the Q-flavour threshold where the quarkonium system and gluonic spectrum are equally "soft". Note, that without assuming such states to exist it is impossible to reconcile the spectral representation (25) for $G(\epsilon)$ with the absence of representation (36).

5. Conclusions

In this paper an approach is developed to consider the effects of non-perturbative gluonic fluctuations in the exact vacuum state on the non-relativistic dynamics of heavy quarkonium. The approach is applicable in the case where the quark and the antiquark are constrained to be separated by distances small compared to the characteristic scale of vacuum fluctuations, so that the interaction of the quarkonium system with the gluonic field can be expanded in multipoles. The leading corrections considered are proportional to the vacuum matrix element $\langle 0|F^a_{\mu\nu}(0)F^a_{\mu\nu}(0)|0\rangle$.

For the case of superheavy quarkonium in which the lowest bound states are determined by the short-distance Coulomb-like exchange potential, the energy shifts dùe to non-perturbative fluctuations are calculated (see eq. (18)) as well as corrections to matrix elements of a colourless operator acting on quark variables between the lowest levels (eq. (20)).

In the case of moderately heavy quarks, the Green function of relative motion in quarkonium is found up to terms proportional to $\langle 0|F^a_{\mu\nu}F^a_{\mu\nu}|0\rangle$ at energies below threshold.

An investigation of the spectral representation for the Green function led us to the conclusion that the effects of the non-perturbative fluctuations of the gluonic field cannot be expressed in terms of an interaction potential between quarks and that one should consider states which contain both the quark-antiquark pair and a gluonic exitation including those in which each of them is a colour octet.

There is nothing observed to prevent one from extending the approach proposed to terms of order $v^2/c^2 \simeq k^2/m^2$ as well as to contributions of vacuum mean values of operators of higher dimension than F^2. For this one needs to consider: (i) higher multipoles in the quarkonium interaction with the gluonic field, as well as terms of orders v/c and v^2/c^2 in the interaction; (ii) perform further iterations of $\mathcal{H}_g + \mathcal{H}_{int}$.

It can be believed that a consideration of this sort will be helpful in studying properties of the Υ family and of analogous systems made from still heavier quarks.

I am indebted to Prof. L. B. Okun' for valuable discussions and comments.

References

[1] A. A. Belavin et al., Phys. Lett. 59B (1975) 85.
[2] C. Callan, R. Dashen and D. Gross, Phys. Rev. D17 (1978) 2717.
[3] V.N. Gribov, Proc. 12th LNPI Winter School, Leningrad (1977) p 147.
[4] F. Wilczek and A. Zee, Phys. Rev. Lett. 40 (1978) 83.
[5] K. Wilson, Phys. Rev. 179 (1969) 1499.
[6] A. Vainshtein, V. Zakharov and M. Shifman, ZhETF Pisma 27 (1978) 60; Nucl. Phys. B147 (1979) 385, 448.
[7] M. Shifman et al., Phys. Lett. 77B (1978) 80; Yad. Fiz. 28 (1978) 465.
[8] M. Shifman et al., Phys. Lett. 76B (1978) 477.
[9] V. Novikov et al., Phys. Rev. Lett. 38 (1977) 626; Phys. Lett. 67B (1977) 409.
[10] V. Novikov et al., Phys. Reports 41 (1978) 1.
[11] A. Vainshtein and L. Okun, Yad. Fiz. 23 (1976) 1347.
[12] R. Jaffe and K. Johnson, Phys. Lett. 60B (1976) 201;
I.Yu. Kobzarev et al., Yad. Fiz. 27 (1978) 506.
[13] E. Eichten et al., Cornell preprint CLNS-375 (1978).

HOW TO USE HEAVY QUARKS TO PROBE THE QCD VACUUM ☆

H. LEUTWYLER
Institute for Theoretical Physics, University of Bern, CH-3012 Bern, Switzerland

Received 7 November 1980

We calculate the leading nonperturbative contributions to the energy levels and wavefunctions of a heavy $q\bar{q}$ pair.

For sufficiently heavy quarks the $q\bar{q}$ bound state picture based on a nonrelativistic positronium-like description is consistent: one gluon exchange leads to small bound state velocities of order $\beta = \frac{4}{3}\alpha_s = g^2/3\pi$. The root mean square radius of a state with principal quantum number n is of order $n^2(m\beta)^{-1}$; if n is not too large heavy quarks move at short distances where a perturbative analysis of QCD makes sense. If the quarks are not very heavy the description fails not so much because higher order effects in α_s (relativistic corrections) become sizeable, but because it neglects nonperturbative effects which are important if the size of the system is not small in comparison to the scale of QCD. We have argued [1] that the leading nonperturbative effects can be seen in the short distance behaviour of the wavefunctions. The purpose of the present paper is to show that for a heavy $q\bar{q}$ pair the leading nonperturbative effect is governed by the gluon pair density

$$\langle B^2\rangle = \tfrac{1}{4}\langle 0|G^a_{\mu\nu}G^{\mu\nu a}|0\rangle, \tag{1}$$

which produces a shift of the Schrödinger levels by

$$M_{nl} = 2m - \frac{1}{4}\frac{m\beta^2}{n^2} + m\frac{\langle B^2\rangle}{(m\beta)^4}n^6\epsilon_{nl}. \tag{2}$$

In this formula n is the principal quantum number, l the orbital angular momentum of the state, and $\beta = \frac{4}{3}\alpha_s$. [We include the strong coupling constant in the definition of the gluon field: $\nabla_\mu = \partial_\mu + iG^a_\mu\lambda^a/2$, $G^a_{\mu\nu} = \partial_\mu G^a_\nu - \partial_\nu G^a_\mu - f^{abc}G^b_\mu G^c_\nu$.] The dimensionless quantity ϵ_{nl} is of order one (see below). The formula (2) is accurate up to terms of order $m\beta^4$, $m \times \langle\nabla_\mu G^2_{\rho\sigma}\rangle(m\beta)^{-6}$, $m\langle G^3_{\rho\sigma}\rangle(m\beta)^{-6}$ and up to effects due to the pair density of light quarks.

The gluon condensate also modifies the wavefunctions. In radiation gauge the leading nonperturbative effect on the ground state wavefunction is given by

$$\psi(x-y) = (8\pi)^{-1/2}(m\beta)^{3/2}e^{-\rho/2} \times [1 - \tfrac{1}{3825}\gamma(17\rho^4 + 156\rho^3 + 936\rho^2 - 26712)],$$
$$\gamma = m^{-4}\beta^{-6}\langle B^2\rangle, \tag{3}$$

where $\rho = m\beta|x-y|$. In particular, the wavefunction at the origin, measured in e^+e^--decay is modified as follows:

$$\psi(0) = (8\pi)^{-1/2}(m\beta)^{3/2}(1 + 6.98\gamma). \tag{4}$$

Finally, we have calculated the leading nonperturbative correction to the hyperfine splitting of the ground state with the result

$$M_{\uparrow\uparrow} - M_{\uparrow\downarrow} = \tfrac{1}{3}m\beta^4(1 + 18.3\gamma). \tag{5}$$

After this work was essentially completed I learned that Voloshin [2] has carried out a similar analysis in the framework of QCD sum rules. I am indebted to Professor S. Iwao for bringing this work to my attention.

To derive the above results we observe that the quarks move in a medium containing soft gluons. If the radius r of the orbit is small compared to the average wave length of the fluctuating chromoelectric and chromomagnetic vacuum fields we have $r|\nabla_\mu G_{\rho\sigma}| \ll |G_{\mu\nu}|$. In this situation the space–time variation of

☆ Supported in part by the Schweizerischer Nationalfonds.

the fields may be neglected: terms proportional to $\langle 0|\nabla_\mu G^2_{\rho\sigma}|0\rangle$ are suppressed by two inverse powers of $m\beta$ in comparison to those retained in (2). In leading order we are therefore faced with the problem of calculating the level shift due to constant random fields. Lorentz invariance of the vacuum requires that the electric and magnetic pair densities are of equal magnitude and opposite sign ($E_i = G_{0i}, B_i = -G_{kl}$):

$$\langle 0|E^a_i E^b_k|0\rangle = -\tfrac{1}{24}\delta_{ik}\delta^{ab}\langle B^2\rangle,$$

$$\langle 0|B^a_i B^b_k|0\rangle = \tfrac{1}{24}\delta_{ik}\delta^{ab}\langle B^2\rangle. \qquad (6)$$

Since we are dealing with a nonrelativistic system this implies that the level shift due to the magnetic field is smaller by two powers of β than the shift due to the electric field: the leading nonperturbative contribution is the quadratic Stark effect due to the fluctuating chromoelectric vacuum field (there is no linear Stark effect, because the expectation value of the field vanishes). For a colour singlet state the interaction hamiltonian is

$$H_1\psi^0(\boldsymbol{x},\boldsymbol{y}) = -\boldsymbol{E}^a\cdot(\boldsymbol{x}-\boldsymbol{y})\tfrac{1}{2}\lambda^a\psi^0(\boldsymbol{x},\boldsymbol{y}).$$

To lowest order the level shift is therefore given by

$$\Delta M_{nl} = -(\psi^0_{nl}, \boldsymbol{E}\cdot\boldsymbol{z}\,(H_0 - E_n)^{-1}\,\boldsymbol{E}\cdot\boldsymbol{z}\,\psi^0_{nl}).$$

The state $\boldsymbol{E}\cdot\boldsymbol{z}\,\psi^0_{nl}$ is a colour octet. When acting on such a state one gluon exchange is repulsive:

$$H_0 \to \hat{H}_0 = -m^{-1}\Delta + \tfrac{1}{8}\beta/|z|. \qquad (7)$$

Using (6) we get

$$\Delta M_{nl} = \tfrac{1}{18}\langle B^2\rangle(\psi^0_{nl}, z^i(\hat{H}_0 - E_n)^{-1}z^i\psi^0_{nl}).$$

To evaluate the matrix element we make use of a method described in standard textbooks [3]: the equation

$$(\hat{H}_0 - E_n)\,\chi^i_{nl} = z^i\psi^0_{nl},$$

can be solved explicitly with an ansatz of the form

$$\chi = \text{polynomial} \times \exp(-\tfrac{1}{2}|z|m\beta/n).$$

The result for the level shift is given in (2) with

$$\epsilon_{nl} = \frac{2}{9}\frac{1}{n^3(2l+1)}\{(l+1)[F(n,l) - F(-n,l)]$$
$$+ l[F(n,-l-1) - F(-n,-l-1)]\},$$

Table 1
Coefficients ϵ_{nl} relevant for the leading nonperturbative mass shift.

n	l	ϵ_{nl}
1	0	1.468
2	0	1.585
2	1	0.998
3	0	1.661
3	1	1.361
3	2	0.804

$$F(n,l) = 2n[n^2 - (l+1)^2] + (n+l+2)(n+l+1)$$
$$\times\left[\frac{(n-l)(n+l+3)}{9n+16} + \frac{4(2n-l)^2}{9n+8}\right]. \qquad (8)$$

For the lowest levels the numerical values are given in table 1. The main feature of the result is that the shift explodes with increasing principal quantum number: $\Delta M_{nl} \sim n^6$. This implies that even for very heavy quarks the nonperturbative effects are large on excited levels. The calculation described above is of course only valid if the nonperturbative effect is smaller than the Schrödinger binding energy, i.e. $4 \times \beta^{-6}m^{-4}n^8\langle B^2\rangle < 1$. What counts is the ratio m/n^2. A quark four times as heavy as the b-quark suffers the same (relative) perturbation by the gluon condensate in the first excited state the b-quark experiences in the ground state. (The reason for the rapid growth with n is easy to understand: the perturbation is proportional to $z \sim n^2(m\beta)^{-1}$ while the energy denominator is of order $m\beta^2 n^{-2}$.) Inserting the estimate $\langle B^2\rangle = 0.34\,m^4_\rho \approx (600\text{ MeV})^4$ given by Shifman et al. [4] one finds that for the ground state of the $b\bar{b}$-system the nonperturbative effect reduces the Schrödinger binding energy by 13% if $\alpha_s = 0.35$, by 34% if $\alpha_s = 0.3$ and by 100% if $\alpha_s = 0.25$. For the $c\bar{c}$-system for which the quark mass is smaller by more than a factor 3 the nonperturbative effect completely overwhelms the Coulomb force even for the ground state. This observation explains why the Balmer formula fails even for the $b\bar{b}$-system: only the ground state sits in the region of the potential for which a short distance expansion is adequate. The leading nonperturbative effect lifts the degeneracy of the states belonging to the same principal quantum number. For $n = 2$ e.g. we find

$$M_{20} - M_{21} = 37.6\,m(m\beta)^{-4}\langle B^2\rangle. \qquad (9)$$

The effect does indeed have the sign observed in the $c\bar{c}$-system ($M_{\psi'} > M_{\chi}$) provided $\langle B^2 \rangle$ is positive as it is indicated by the estimate of Shifman et al. [4] and as it is required by the stability of the vacuum [5]. For the reason given above the prediction (9) cannot be used quantitatively for the first excited levels of the $b\bar{b}$-system. Since the state $n = 2, l = 1$ is the lowest lying P-wave one may hope to nevertheless obtain a reliable prediction for its mass by using variational methods instead of the above perturbative scheme. Indeed Voloshin [2] predicts $M_{21} - M_{10} = 370 \pm 30$ MeV on the basis of his analysis of the QCD sum rules.

The gluon pair density of course also modifies the wavefunctions. Again the first order effect vanishes upon averaging over the stochastic electric field. The second order perturbation is given by

$$\psi = \psi^0_{nl} + P(H_0 - E_n)^{-1} P \boldsymbol{E}\cdot\boldsymbol{z}\, (H_0 - E_n)^{-1} \boldsymbol{E}\cdot\boldsymbol{z}\, \psi^0_{nl}, \quad (10)$$

where P projects on the subspace orthogonal to ψ^0_{nl}. The physical wavefunction is the average of this expression over the electric field. It is important to realize that the first order contribution to the wavefunction $\sim (H_0 - E_n)^{-1} \boldsymbol{E}\cdot\boldsymbol{z}\, \psi^0$ is of no physical significance. In the language of the Bethe–Salpeter equation this contribution corresponds to diagrams containing an arbitrary number of ladder exchanges plus one solft gluon insertion – the physical amplitude contains only diagrams with none or with two such insertions. In the analysis described above the fictitious diagrams containing only one soft gluon insertion play the role of a mere calculational device. It would be wrong to include the contribution from these diagrams in the normalization of the state.

The expression (10) may be evaluated explicitly by applying the technique described above once more. The resulting ground state wavefunction is given in (3). This expression shows that the nonperturbative effects increase the value of the wavefunction at the origin and reduce the radius of the orbit. Since the wavefunction at the origin is measured in the decay into an e^+e^--pair

$$\Gamma_{\Upsilon \to e^+e^-} = (16\pi/9)(\alpha^2/M^2_{\Upsilon}) |\psi(0)|^2 (1 - 16\alpha_s/3\pi). \quad (11)$$

We may use the known decay rate [‡1] $\Gamma_{\Upsilon \to e^+e^-} = 1.16 \pm 0.15$ keV to extract the value of the b-quark mass and an estimate of $\langle B^2 \rangle$. Unfortunately, the result for $\langle B^2 \rangle$ is sensitive to the value of α_s. For $\alpha_s = 0.35, 0.3, 0.25$ we get $m_b = (4.805 \pm 0.010, 4.765 \pm 0.010, 4.730 \pm 0.005)$ GeV and $\langle B^2 \rangle = (0.85 \pm 0.20, 0.52 \pm 0.10, 0.27 \pm 0.05) m^4_\rho$, respectively. Unless α_s is below 0.25 we invariably obtain a positive binding energy, $m_b > \frac{1}{2} M_{\Upsilon}$.

Note that the leading nonperturbative contributions cannot be described by a potential $V(|\boldsymbol{z}|)$. They amount to an attractive nonlocal interaction of the form

$$V = \tfrac{1}{18} \langle B^2 \rangle \int_0^\infty \mathrm{d}\lambda\, z^i\, e^{-\lambda \hat{H}_0} z^i\, e^{\lambda H_0}, \quad (12)$$

where H_0 is the attractive hamiltonian relevant in the color singlet channel whereas $\hat{H}_0$ is the repulsive hamiltonian (7) governing the color octet channel.

We now turn to the calculation of the leading nonperturbative contribution to the hyperfine splitting of the ground state. Since this splitting is of order $m\beta^4$ we have to analyze the relativistic corrections to the bound state hamiltonian. These corrections may be derived from the Bethe–Salpeter equation

$$(-i\gamma^\mu \nabla_\mu + m)\varphi(x,y)(i\gamma^\nu \overleftarrow{\nabla}_\nu + m) = -ig^2 \gamma^\mu \tfrac{1}{2}\lambda^a \varphi(x,y) \gamma^\nu \tfrac{1}{2}\lambda^b D^{ab}_{\mu\nu}(x,y), \quad (13)$$

where ∇_μ is the covariant derivative with respect to the slowly varying "external" vacuum field and $D^{ab}_{\mu\nu}(x,y)$ denotes the gluon propagator in the presence of this external field. The propagator may be worked out from the differential equation

$$\nabla^\mu(\nabla_\mu a_\nu - \nabla_\nu a_\mu) - i[G_{\mu\nu}, a^\mu] = j_\nu, \qquad a^a_\mu(x) = -\int \mathrm{d}y\, D^{ab}_{\mu\nu}(x,y) j^{\nu b}(y). \quad (14)$$

In radiation gauge $\nabla_i a_i = 0$ the instantaneous term D_{00} only involves the magnetic field, whereas the electric field appears in D_{0i}, D_{i0} (the components D_{ik} are only needed to lowest order, i.e. for $\boldsymbol{E} = \boldsymbol{B} = 0$). The calculation simplifies considerably for S-wave states for which the spin–orbit interaction is absent. It turns out that in this case the magnetic effects come exclusively from the magnetic term whereas the electric effects arise from the interference of the familiar local spin–spin interaction term $\sim \boldsymbol{s}_1 \cdot \boldsymbol{s}_2$

‡1 The value quoted represents an average value of the data reviewed at the Madison Conference (1980).

$\times\, \delta^3(\boldsymbol{x}-\boldsymbol{y})$ with the electric potential $\boldsymbol{E}\cdot\boldsymbol{z}$. The electric and magnetic terms have the same sign, the electric contribution is numerically more important. The result is given in (5), a detailed account of the calculation will be given elsewhere. Using the value of $\langle \boldsymbol{B}^2 \rangle$ given by Shifman et al. [4] we predict $M_{\Upsilon} - M_{\eta_b} = 90 \pm 15$ MeV. The result is remarkably stable with respect to uncertainties in α_s. Note that the nonperturbative contribution is larger than the perturbative term.

If there is a t quark with a mass of order 20 GeV or more the $t\bar{t}$-system will provide an ideal probe to measure perturbative as well as nonperturbative QCD effects. The lowest lying states should provide enough information to extract a reliable value of α_s as well as an accurate value for the leading nonperturbative parameter $\langle \boldsymbol{B}^2 \rangle$.

It is a pleasure to thank J. Glasser, S. Iwao, M. Lüscher, P. Minkowski and P. Schwab for stimulating discussions as well as for help with numerical problems.

References

[1] H. Leutwyler, Proc. Intern. Summer Institute on Theoretical physics (Kaiserslautern, 1979), ed. W. Rühl (Plenum, New York, 1980).
[2] M.B. Voloshin, Nonrelativistic bottomium in the physical vacuum of QCD. Mass of the 1P level, preprint ITEP 21 (Moscow, 1980); Nucl. Phys. B154 (1979) 365.
[3] M. Kotani, Quantum mechanics, Vol. 1 (Yuwanami, Tokyo, 1951) p. 127;
L.I. Schiff, Quantum mechanics (McGraw-Hill, New York, 1968) p. 263.
[4] M.A. Shifman et al., Nucl. Phys. B147 (1979) 385, 448; A.I. Vainshtein et al., Sov. J. Nucl. Phys. 27(2) (1978); V.A. Novikov et al., Phys. Rep. 41 (1978) 1.
[5] H. Leutwyler, Phys. Lett. 96B (1980) 154.

On the hyperfine splitting of the S-levels of heavy quarkonium

M. B. Voloshin

Institute of Theoretical and Experimental Physics, State Commission on Use of Atomic Energy
(Submitted 3 August 1981)
Yad. Fiz. **35**, 1016–1020 (April 1982)

We have found an expression for the hyperfine splitting of the S-levels of heavy quarkonium taking into account nonperturbative effects. The latter arise as a result of the interaction of the quarks with the large-scale fluctuations of the gluon field in the physical vacuum of QCD, which cannot be described by perturbation theory. We present an estimate of the splitting of the mass of the Υ resonance and the η_b: $M(\Upsilon) - M(\eta_b) \lesssim 30$ MeV. We have also obtained estimates for the analogous splittings in $t\bar{t}$-quarkonium.

PACS numbers: 12.35.Eq

The problem of the hyperfine splitting of the quarkonium levels is without doubt of interest for the understanding of the quark-spin-dependent forces which act in bound quark systems. Many readers will probably remember the lively discussion concerning the mass difference between the J/ψ resonance and the paracharmonium state—the η_c meson—in terms of the potential model (see, e.g., the review in Ref. 1) as well as in the framework of the basic principles of QCD using dispersion sum rules.[2] Recall that the mass of the η_c was uniquely predicted[2] by the latter method as 3.00 ± 0.03 GeV in complete agreement with the mass of the η_c (2.98) which was subsequently found at SLAC.[3]

There is at present a similar interest in the prediction of the mass splitting between the Υ resonance and the ground state of the spin singlet state of the $b\bar{b}$ system—the η_b. We know from the experience with the calculation of the mass of the η_c that the potential model contains too many free parameters and is too crude to give a reliable quantitative prediction of the hyperfine splitting of the $b\bar{b}$ quarkonium. Calculations by a method which is similar to the one used in Ref. 2 lead to the estimate

$$M_\Upsilon - M_{\eta_b} \leqslant 30 \text{ MeV}. \tag{1}$$

This result was presented (without a derivation) by the present author on several occasions (see for example the review in Ref. 4) as an illustration of the fact that the search for the η_b in the decay $\Upsilon \to \eta_b \gamma$ is practically hopeless since with such a small decay energy the branching ratio does not exceed 5×10^{-5}. Recently, however, Leutwyler[5] has (also without a detailed derivation) given the estimate $M_\Upsilon - M_{\eta_b} = 90 \pm 15$ MeV, which he found on the basis of ideas about the nonperturbative effects in heavy quarkonium which are similar to those used previously by the present author.

In view of the above discrepancy it seems appropriate to give a detailed derivation of the estimate (1). This is one of the aims of the present paper. At the same time we shall get an estimate of the hyperfine splitting of the S-states of heavy quarkonium. The validity of this estimate improves as the mass of the quark increases. This result is expressed by formula (15) (see below).

The physical picture within whose framework we have studied the characteristics of heavy quarkonium was proposed in Refs. 2 and 6 and has been developed subsequently in several papers (see for example Refs. 5–10). In this picture it is assumed that in addition to the exchange of gluons between the quarks at short range the characteristics of the heavy quarkonium are also determined by the interaction of the quarkonium with the large-scale fluctuations of the gluon field which are present in the physical QCD vacuum. (Here the term 'physical vacuum' is used in distinction to the term 'vacuum of perturbation theory.') In heavy quarkonium the leading effects which arise from the interaction with these fluctuations are proportional to the mean square of the tension of the field of the gluon fluctuations: $\langle F^a_{\mu\nu}(0) F^a_{\mu\nu}(0)\rangle$. For a numerical value of this mean square we use the estimate[9] (see also Refs. 2 and 6)

$$\eta = \frac{1}{18}\langle F^2\rangle \simeq 0.01 \text{ GeV}^4. \tag{2}$$

[Here our normalization of the field tension corresponds to the Lagrangian in the form $\mathscr{L} = -(1/g^2)F^2 + \dots$, where $g^2 = 4\pi\alpha_s$.]

In a sufficiently heavy quarkonium the Coulomb-like forces of the one-gluon exchange confine the quarks to small distances. We can therefore neglect the contribution of the vacuum expectation values of the gluon operators of dimension $d > 4$ compared with the terms which are proportional to the mean value (2).

Thus the problem of the hyperfine splitting of heavy quarkonium is reduced to taking into account the effects which depend on the spin of the quarks in the short-range gluon-exchange forces as well as in the interaction with the vacuum fluctuations. This problem can be solved systematically in the limit of very heavy quarks. In this limit the wave function of the quarkonium is mainly defined by the Coulomb-like potential of the one-gluon exchange and is localized at short distances. Under these conditions it is also justified to use the expansion in powers of the speed of the quarks.

The latter expansion, as is well known, leads in perturbation theory to a spin-dependent potential of the Breit-Fermi type which gives for the pure spin-spin interaction between the quark and antiquark in QCD the expression

$$V_{B-F}=\frac{2\pi}{3}\frac{\alpha_s(m)}{m^2}\sigma_1\sigma_2\delta(\mathbf{r})\left[\frac{4}{3}P_S-\frac{1}{6}P_A\right],\qquad(3)$$

where m is the mass of the quark, σ_1 and σ_2 are the spin operators (Pauli matrices) which act on the spin variables of the quark and antiquark, respectively, and P_S and P_A are the projection operators for the color singlet and color octet (adjoint) state of the quark-antiquark pair (the factors 4/3 and −1/6 are the color factors which describe the strength of the Coulomb interaction in these two states). Note that in the expression (3) we have the coupling constant α_s which is normalized at distances of the order of the Compton wavelength of the quark: $\alpha_s(m)$, since it is at these distances that the spin-spin interaction (3) is localized.

It is not difficult to find the hyperfine splitting of the n^3S_1 and n^1S_0 states of the quarkonium in the Coulomb limit of heavy quarks, which is caused by the interaction (3) (in this case the term P_A drops out since the system is colorless):

$$\Delta E^{(0)}_{n\,B-F}=\frac{32\pi}{9}\frac{\alpha_s(m)}{m^2}|\psi^{(0)}_n(0)|^2=\frac{32}{9}\alpha_s(m)\frac{k_n^3}{m^2},\qquad(4)$$

where $\psi_n^{(0)}(r)$ is the Coulomb wave function of the nS state and

$$|\psi_n^{(0)}(0)|^2=k_n^3/\pi,\qquad(5)$$

while k_n is found from the relation

$$k_n=\frac{1}{n}\frac{2}{3}m\alpha_s(k_n).\qquad(6)$$

Formula (4) describes the hyperfine splitting in the limit of very heavy quarks. In this limit the terms which are proportional to the vacuum expectation value (2) are negligibly small. However, these terms are very important in the case of a moderate mass of the quark and even more important for the $b\bar{b}$ system.

The interaction with the vacuum fluctuations leads first of all to an additional contribution to the splitting due to the spin-dependent terms in this interaction, and secondly gives corrections to the matrix element of the Breit-Fermi Hamiltonian.

It is not difficult to estimate the latter effect in the leading order in v/c. In this approximation the diagonal matrix element of the operator $\delta(\mathbf{r})P_S$ also determines the decay width of the decay of the $^3S_1$ state into an e^+e^- pair:

$$\Gamma(n^3S_1\to e^+e^-)=4\pi Q^2\frac{\alpha^2}{m^2}\langle n|\delta(\mathbf{r})P_S|n\rangle\left|1+\frac{A_Z}{A_\gamma}\right|^2,\qquad(7)$$

where Q is the electric charge of the quark and A_Z/A_γ is the ratio of the annihilation amplitude through a Z boson and through a photon. (Usually the quantity $\langle n|\delta(\mathbf{r})P_S|n\rangle$ is denoted by $|\psi_n(0)|^2$; however, the handling of the wave function of the quarkonium in the presence of nonperturbative fluctuations requires certain qualifications,[8] and therefore we avoid this notation.) In formula (7) we have neglected a known correction of order $\alpha_s(m)$ since we have not taken into account a similar correction in formula (3).

As far as the operator $\delta(\mathbf{r})P_A$ in formula (3) is concerned, we note that its contribution must be strongly suppressed. Indeed, the mean value of this operator vanishes when the interaction with the vacuum fluctuations is neglected, since the system which we are considering is colorless. Furthermore, the quark pair is in an octet state for some of the time as a result of the interaction with the fluctuations, but in the leading order the transition from the singlet state into the octet state is caused by the dipole operator[8-10]

$$V_{int}=-\xi^a r_i E_i^a,\qquad(8)$$

where $\mathbf{E}^a$ is the chromoelectric component of the field of fluctuations, $\mathbf{r}$ is the distance between the quarks, and ξ^a is the difference between the color generators acting on the color variables of the quark and antiquark: $\xi^a=t_1^a-t_2^a$. The operator (8) gives rise to the transition from the colorless S-state of the quarks into the octet P-state in which the matrix element of $\delta(\mathbf{r})P_A$ vanishes because of the factor $\delta(\mathbf{r})$. Finally, the operator $\delta(\mathbf{r})P_A$ enters the expression (3) with a small coefficient [of the order of N_c^{-2} in comparison with $\delta(\mathbf{r})P_S$, where N_c is the number of colors, $N_c=3$]. Thus it appears natural to ignore completely the contribution of the Breit-Fermi interaction in the octet state. In this case the nonperturbative contribution which we have discussed above reduces to the replacement in formula (4) of $|\psi_n^{(0)}(0)|^2$ by the quantity $\langle n|\delta(\mathbf{r})P_S|n\rangle$ which is measured in e^+e^--annihilation [see formula (7)]. This quantity can also be found theoretically[10] in the limit of heavy quarks:

$$\langle n|\delta(\mathbf{r})P_S|n\rangle=\frac{k_n^3}{\pi}\left(1+c_n n^2\frac{m^2\eta}{k_n^6}\right)\qquad(9)$$

where the c_n are numerical coefficients which depend weakly on n and which are in magnitude approximately equal to 5–6. Explicit formulas for c_n were found in Ref. 10. In the cases $n=1$ and 2 which are of direct physical interest the c_n are: $c_1=4.934$ and $c_2=5.484$.

Let us now consider the contribution of the spin-dependent terms proper in the interaction of the quarkonium with the field of the vacuum fluctuations. For an estimate of this contribution to the hyperfine splitting, which is proportional to the vacuum expectation value (2), it is sufficient to consider only those terms in the interaction which do not contain any derivatives of the field tensions $\mathbf{E}^a$ and $\mathbf{H}^a$. Considering the standard expansion (see for example Ref. 11) of the interaction between the quarkonium and the external field in powers of v/c, it is easy to see that in the order $(v/c)^2$ the spin-spin splitting arises only on account of the second order in the interaction of the chromomagnetic moment of the quark with $\mathbf{H}^a$:

$$V_{int}^{(1)}=\xi^a\mathbf{\Delta}\mathbf{H}^a/m,\qquad(10)$$

where $\mathbf{\Delta}=(1/2)(\sigma_1-\sigma_2)$.

The contribution of this interaction to the hyperfine splitting is obtained by standard methods[8-10] and is

$$[M(n^3S_1)-M(n^1S_0)]_{int}=\frac{2\eta}{m^2}\langle nS|(H_A-E_n)^{-1}|nS\rangle\qquad(11)$$

(in the derivation of Eq. (11) we have taken into account that

$$\langle H_iH_k\rangle={}^1/_3\delta_{ik}\langle H^2\rangle={}^1/_{12}\delta_{ik}\langle F^2\rangle).$$

The operator $(H_A - E_n)^{-1}$ in formula (11) is the nonrelativistic Green's function of the quarkonium in the color octet state with the energy $E_n = -k_n^2/m$. In the matrix element in Eq. (11) only the S-wave part of this operator makes a contribution. It is therefore particularly convenient, when calculating this matrix element in the Coulomb limit of heavy quarks, to use the representation for the partial Green's functions in a Coulomb-like potential which were given in Ref. 10. For $l = 0$ this representation is of the form

$$\langle x|(H_A-E_n)^{-1}|y\rangle_{l=0}=\frac{mk_n}{2\pi}\exp[-k_n(x+y)]\sum_{s=0}^{\infty}\frac{L_s^1(2k_nx)L_s^1(2k_ny)}{[s+1+(m\alpha_s/12k_n)](s+1)}, \quad (12)$$

where $L_s^1(z)$ are the Laguerre polynomials,

$$L_s^1(z)=\frac{1}{s!}\frac{e^z}{z}\left(\frac{d}{dz}\right)^s(e^{-z}z^{s+1}). \quad (13)$$

Using this representation and the explicit form of the functions of the nS state, it is not difficult to find the matrix element in Eq. (11) for arbitrary n [nonvanishing integrals arise only on account of terms with $s = (n-2)$, $(n-1)$, and n in the sum (12)]. The result is of the form

$$[M(n^3S_1)-M(n^1S_0)]_{\rm int}=\frac{\eta}{mk_n^2}\frac{1}{4}\left[\frac{n-1}{n-1+n/8}+\frac{n+1}{n+1+n/8}+\frac{32}{9}\right]. \quad (14)$$

[We have taken into account that by Eq. (6) $m\alpha_s(k_n)/(12k_n) = n/8$.]

Combining the formulas (4), (9), and (14), we find the final expression for the pre-Coulomb behavior of the hyperfine splitting of the S-states:

$$M(n^3S_1)-M(n^1S_0)\simeq\frac{32}{9}\alpha_s(m)\frac{k_n^3}{m^2}\left(1+c_nn^2\frac{m^2\eta}{k_n^6}+\frac{9d_n}{128}\frac{m^2\eta}{k_n^5(m\alpha_s(m))}\right)\left(1+O\left(\frac{k}{m}\right)+O(\alpha_s)\right), \quad (15)$$

where d_n denotes the expression in the square brackets in Eq. (14), $d_n \simeq 4.5-5.5$. Recall that the quantity η which appears in this formula has been defined above in Eq. (2); it characterizes the strength of the vacuum fluctuations. k_n is the Bohr momentum in the n-th state and is determined from Eq. (6).

Formula (15) is our final result for the hyperfine splitting of the S-states of heavy quarkonium taking into account the contribution which is proportional to the vacuum expectation value (2). Note that for the first levels the contributions of the first and second correction terms in Eq. (15), which are proportional to η, are parametrically of approximately the same order, but with increasing n the last term is $\sim n^{-3}$ in comparison with the first correction term.

Let us now consider the hyperfine splitting of the Υ and η_b. For $b\bar{b}$-quarkonium we have[9] $m = m_b \simeq 4.8$ GeV and $k_1 \simeq 0.96$ GeV, and therefore we can use formula (15) only for a very rough estimate, since the nonperturbative corrections exceed the pure Coulomb contribution (the Υ-resonance is essentially a non-Coulomb-like system). In this case it appears reasonable to express the matrix element of the Breit-Fermi interaction—as we have discussed above—in terms of the e^+e^--width of the Υ resonance, and to make a rough estimate of the contribution of the spin-dependent interaction with the vacuum fluctuations to the hyperfine splitting using formula (14). In this way we get the estimate

$$M(\Upsilon)-M(\eta_b)\simeq 8\Gamma_{ee}(\Upsilon)\frac{\alpha_s(m_b)}{\alpha^2}+1.12\frac{\eta}{m_bk_1^2}. \quad (16)$$

With $\Gamma_{ee}(\Upsilon) \simeq 1.1$ keV and $\alpha_s(m_b) \simeq 0.15$, the first term is 25 MeV while the second term is 2.5 MeV. This shows that even if the contribution of the second kind is doubled (as happens with the Breit-Fermi contribution on account of the significant departure of the Υ from a Coulomb-like state), nevertheless in view of the smallness of the second term the full splitting between the Υ and the η_b will not exceed ~30 MeV. It is the estimate obtained in this way which we have presented previously.

Let us note in conclusion that formulas of the type (15) and (16) are useless for the estimate of the splitting of the excited states Υ' and η_b' in view of the smallness of k_2 in $b\bar{b}$-quarkonium (in other words, the Υ' and η_b' are too large and interact strongly with the vacuum fluctuations). Formulas like (15) or (16) are, however, fully applicable for discussion of the splitting of the $1S$- and $2S$-states of the $t\bar{t}$-quarkonium ($m_t > 18$ GeV). In this connection we note that it follows from the estimate of the e^+e^--decay widths[10] that the $1^3S_1-1^1S_0$ splitting in $t\bar{t}$-quarkonium should be 20–25 MeV if $2m_t \lesssim 50$ GeV, while for a greater mass this splitting can be estimated using the pure Coulomb formula (4). One can also conclude from the results of Ref. 10 and formulas of the type of Eq. (16) that the splitting of the excited levels $2^3S_1-2^1S_0$ is for $t\bar{t}$-quarkonium is ~10 MeV for $2m_t \lesssim 180$ GeV.

The author is grateful to M.A. Shifman for questions which have stimulated the writing of this paper.

Addendum. After the submission of this paper for publication, M.A. Shifman informed the author that Refs. 12 and 13 have been submitted to the Bonn Conference on the interactions of photons and leptons. In these papers the correction of order $\alpha_s(m)$ to the Breit-Fermi interaction has been calculated. Taking into account this correction as well as corrections to Γ_{ee}, we can write the analogue of formula (16) in the form

$$M(n^3S_1)-M(n^1S_0)=\frac{8\Gamma_{ee}}{(3Q)^2}\frac{\alpha_s(m)}{\alpha^2}\left(1+6.1\frac{\alpha_s(m)}{\pi}\right)+\frac{d_n}{4}\frac{\eta}{mk_n^2}.$$

As a result of taking into account the correction factor $(1 + 6.1\alpha_s(m)/\pi)$ the estimate for the $\Upsilon-\eta_b$ splitting is increased by approximately 8 MeV.

[1]V. A. Novikov *et al.*, Phys. Rep. **41C**, 1 (1978).
[2]A. I. Vaĭnshteĭn, M. B. Voloshin, V. I. Zakharov, and M. A. Shifman, Yad. Fiz. **28**, 465 (1978) [Sov. J. Nucl. Phys. **28**, 237 (1978)].
[3]E. D. Bloom, in: Proceedings of the International Symposium on Lepton and Photon Interactions, Fermilab, eds. T. B. W. Kirk and H. Abarbanel, 1979, p. 92.
[4]M. B. Voloshin and V. I. Zakharov, DESY Report F-15-80/03, 1980.
[5]H. Leutwyler, University of Bern Preprint, November 1980; Phys. Lett. **98B**, 447 (1981).
[6]A. I. Vaĭnshteĭn, V. I. Zakharov, and M. A. Shifman, Pis'ma Zh. Eksp. Teor. Fiz. **27**, 60 (1978) [JETP Lett.

27, 55 (1978)].
[7]M. A. Shifman, A. I. Vaĭnshteĭn, and V. I. Zakharov, Nucl. Phys. B147, 385, 448 (1979).
[8]M. B. Voloshin, Nucl. Phys. B154, 365 (1979).
[9]M. B. Voloshin, Preprint ITEP-21, 1980.
[10]M. B. Voloshin, Preprint ITEP-30, 1981.
[11]V. B. Berestetskiĭ, E. M. Lifshits, and L. P. Pitaevskiĭ, Kvantovaya élektrodinamika (Quantum Electrodynamics), Moscow, Nauka, 1980, §33.
[12]W. Buchmüller, Y. Ng, and S. Tye, Fermilab Pub-81-49, 1981.
[13]R. Barbieri, R. Gatto, and E. Remidi, Contributed paper to the International Symposium on Lepton and Photon Interactions, Bonn, August 1981.

Translated by W. von Schlippe

Pre-Coulombic asymptotic behavior of heavy-quarkonium levels

M. B. Voloshin

Institute of Theoretical and Experimental Physics, State Atomic Energy Commission
(Submitted 8 October 1981)
Yad. Fiz. **36**, 247–255 (July 1982)

We calculate the leading nonperturbative corrections to the binding energies and the e^+e^- widths of the levels of heavy quarkonium, which are determined mainly by a Coulomb-like interaction at small distances. We discuss the question of the normalization of the quark wave function in a fixed color state and give arguments that this normalization is not observable because of color exchange with nonperturbative fluctuations of the gluon field in the QCD vacuum. We discuss the physical reasons for the conclusion drawn previously that it is not possible to describe the nonperturbative effects in terms of an interaction potential, and we establish the connection with other work in which attempts are made to calculate such a potential.

PACS numbers: 12.35.Cn, 14.80.Dq

1. INTRODUCTION

It is well known that in perturbative QCD the interaction between nonrelativistic heavy quarks and antiquarks at small distances r leads to a two-particle Coulomb-like potential of the type

$$V(r)=-\frac{4}{3}\frac{\alpha_s(r^{-1})}{r}P_S+\frac{1}{6}\frac{\alpha_s(r^{-1})}{r}P_A, \quad (1)$$

where P_S and P_A are the projectors on a color singlet and octet (adjoint) state of a quark-antiquark pair, and $\alpha_s(k)$ is the QCD coupling constant,

$$\alpha_s(k)=2\pi/[b\ln(k/\Lambda_{QCD})]. \quad (2)$$

If the mass m of the quark is sufficiently large, the first few levels of quarkonium can be described in terms of the potential (1). Indeed, the wave function of the levels with principal quantum number n in the singlet sector of the potential (1) is localized at distances of order k_n^{-1} where k_n is determined by the relation

$$k_n=\frac{m}{n}\frac{2}{3}\alpha_s(k_n), \quad (3)$$

and for a sufficiently large value of the ratio m/n at distances $r\sim k_n^{-1}$ QCD perturbation theory can be applied and leads to the interaction (1).

In this paper we will consider the leading nonperturbative corrections to the Coulomb values of the properties (binding energies, widths, e^+e^- widths) of the levels of heavy quarkonium. Our analysis can only be applied directly to systems which consist of very heavy quarks ($m\gtrsim 15$ GeV for $n=1$), and may turn out to be of great interest with the discovery of $t\bar{t}$ quarkonium and, possibly, of even heavier systems. At the same time our discussion establishes the limits of applicability of the potential (1) and solves the question of the leading modifications of the interaction between a quark and antiquark at small distances.

Some time ago it was shown[1-3] that nonperturbative corrections to the dynamics of heavy quarks at small distances arise because of the interaction of the quarks with fluctuations of the gluon field which are present in the physical vacuum of QCD and cannot be described by perturbation theory. In this case the leading effects at small distances in heavy quarkonium are proportional to the vacuum expectation value

$$\eta={}^1/_{18}\langle 0|\pi\alpha_s G^a_{\mu\nu}(0)G^a_{\mu\nu}(0)|0\rangle, \quad (4)$$

where $G^a_{\mu\nu}(x)$ is the gluon field tensor.

In Ref. 3 a systematic expansion was obtained for the properties of nonrelativistic quarkonium at small distances in terms of vacuum averages of gluonic operators of the type (4). This permits, in particular, the determination of nonperturbative corrections to the Coulombic asymptotic behavior of the properties of heavy quarkonium levels. An approximate quantitative analysis of the corrections to the binding energies and e^+e^- widths of the ground levels was made in Ref. 4. (An analysis of the expansion similar to the one obtained in Ref. 3 has also been made by Peskin in terms of path integrals.[5])

In Sec. 2 we briefly review the expansion[3] for nonperturbative effects in heavy quarkonium and demonstrate the method of calculating the level shifts and the corrections to the e^+e^- widths which are proportional to the vacuum average (4). In Sec. 3 we give the details of a concrete calculation of these corrections by means of a special representation for the Green function of the Coulomb system and give explicit expressions for the level shifts and the corrections to the e^+e^- widths as well as numerical estimates of these widths. In Sec. 4 we discuss the question of the normalization of the quarkonium wave function for a fixed color state of the quark pair, because of the difference between our results for the corrections to the e^+e^- widths and the results of Leutwyler.[6] The normalization conditions turn out to be highly nontrivial, and taking them into account in the method of Ref. 6 removes the discrepancy between our results and those of Ref. 6. Finally, in Sec. 5 we discuss the impossibility, shown in Ref. 3, of describing the nonperturbative effects in terms of a potential (local or nonlocal) for the quark-antiquark interaction and discuss the connection with other work in which attempts are made to calculate a nonperturbative potential.

2. DESCRIPTION OF THE METHOD

The method applied in Ref. 3 to calculate effects in nonrelativistic quarkonium which are proportional to vacuum averages of the type (4) is based on an expansion of the interaction of quarkonium with the gluon field of vacuum fluctuations in a multipole series. In this case in leading order in the ratio of the quarkonium size and the characteristic dimension of the fluctuations one can restrict oneself to the dipole term

$$H_{int}=-{}^1/_2 g\xi^a \mathbf{r}\mathbf{E}^a, \tag{5}$$

where $\mathbf{r}=\mathbf{x}_Q-\mathbf{x}_{\bar{Q}}$, $E_i^a=G_{i0}^a$, $g^2=4\pi\alpha_s$, and $\xi^a=t_1^a-t_2^a$ is the difference of the color generators acting on the quark and antiquark respectively. The expressions for the corrections to a given quantity which are proportional to the vacuum average (4) are obtained by calculating to second order in the interaction (5) and by taking the average of $\mathbf{E}^2$ with respect to the gluonic vacuum (the first order terms in H_{int} vanish when averaged). In this case, it is useful to use the following relation, which is a consequence of Lorentz invariance of the vacuum (Refs. 3 and 6):

$$\langle 0|E_i^a E_k^b|0\rangle=-{}^1/_{96}\delta_{ik}\delta^{ab}\langle 0|G_{\mu\nu}^c G_{\mu\nu}^c|0\rangle. \tag{6}$$

A convenient object for the application of this method of calculation is the nonrelativistic propagator (Green function) of the system projected on the colorless sector of a quark-antiquark pair and the gluonic vacuum

$$G_S(E)=P_S\langle 0|(H-E)^{-1}|0\rangle P_S,$$

where H is the total Hamiltonian of the system: heavy quarkonium + gluons, and E is the energy measured from the threshold $2m$. By the method described above we get for $G_S(E)$ the following expansion[3]

$$G_S(E)=(H_S-E)^{-1}-\eta(H_S-E)^{-1}r_i(H_A-E)^{-1}r_i(H_S-E)^{-1}+\ldots, \tag{7}$$

where we have not written down the contributions of vacuum averages of gluonic operators of dimension $d>4$, and H_S and H_A are the Coulomb Hamiltonians for singlet and octet states respectively:

$$H_S=-\frac{\Delta}{m}-\frac{4}{3}\frac{\alpha_s}{r},\quad H_A=-\frac{\Delta}{m}+\frac{1}{6}\frac{\alpha_s}{r}. \tag{8}$$

Below threshold, i.e., at negative energies $E=-k^2/m$ the quarks are localized at distances $r\sim k^{-1}$, which guarantees the applicability of the expansion (7) for a sufficiently large value of $|E|$.

By means of the expansion (7) for the propagator G_S one can solve the question of the leading nonperturbative effects in quarkonium at small distances. In particular, if one considers the expression (7) at energy values close to the nth Coulomb level $E_n=-k^2/m^2$ [where $G_S(E)$ has a pole] and if one uses the spectral representation

$$(H_S-E)^{-1}=\sum_n\frac{|n\rangle\langle n|}{E_n-E}, \tag{9}$$

where $|n\rangle$ is the Coulomb wave function, it is not difficult to arrive at an expression for the energy shift of the nth Coulomb level[3]

$$\delta E_n=\eta\langle n|r_i(H_A-E)^{-1}r_i|n\rangle. \tag{10}$$

[Of course, this expression can also easily be obtained directly in second order in the interaction (5).]

The e^+e^--decay widths of the $^3S_1$ levels of quarkonium are proportional to the residues of the poles of the matrix element $\langle \mathbf{x}=0|G_S(E)|\mathbf{y}=0\rangle$ (Refs. 4 and 7):

$$\Gamma(^3S_1\to e^+e^-)=4\pi Q^2\frac{\alpha^2}{m^2}\left(1-\frac{16\alpha_s(m)}{3\pi}\right)\times\left|1+\frac{A_Z}{A_\gamma}\right|^2\operatorname*{Res}_{E=E_{pole}}\langle \mathbf{x}=0|G_S(E)|\mathbf{y}=0\rangle. \tag{11}$$

Here, Q is the quark charge, A_Z/A_γ is the ratio of the amplitudes for annihilation via a Z boson and via a photon. We stress that in the perturbative correction $(1-16\alpha_s(m)/3\pi)$ the value of α_s is normalized at the quark mass. For potential systems formula (11) coincides with the well known formula which expresses Γ_{ee} in terms of $|\psi(0)|^2$. As shown in Ref. 3 (and as will be discussed below) the nonperturbative effects do not correspond to a potential interaction. In this situation formula (11) is due to the fact that close to threshold the amplitude of the electromagnetic vacuum polarization of heavy quarks is proportional[3,4,7] to the quantity $\langle \mathbf{x}=0|G_S(E)|\mathbf{y}=0\rangle$.

In the Coulomb limit when only the first term remains in the expansion (7) we have $G_S(E)=(H_S-E)^{-1}$, and in this case

$$\rho_n\equiv\operatorname*{Res}_{E=E_n}\langle \mathbf{x}=0|G_S(E)|\mathbf{y}=0\rangle=\frac{k_n^3}{\pi}=\frac{1}{\pi n^3}\left(\frac{2}{3}m\alpha_s(k_n)\right)^3, \tag{12}$$

which when substituted into Eq. (11) gives the well known expression for the e^+e^- width Γ_n of Coulomb levels. The contribution of the second term in (7) determines the nonperturbative correction we want to calculate. In order to calculate the correction to the residue explicitly we note that the second term in (7) has a pole of second and first order at $E=E_n$. If we compare this with the expansion

$$\langle \mathbf{x}=0|G_S(E)|\mathbf{y}=0\rangle=\frac{\rho_n+\delta\rho_n}{E_n+\delta E_n-E}+O[(E_n+\delta E_n-E)^0]$$
$$=\frac{\rho_n}{E_n-E}-\frac{\rho_n\delta E_n}{(E_n-E)^2}+\frac{\delta\rho_n}{E_n-E}+O[(E_n-E)^0]+O[(\delta E_n)^2], \tag{13}$$

with $\delta\rho_n$ and δE_n proportional to η we arrive at the conclusion that the correction to the residue is given by the term with the first order pole.

3. EXPLICIT FORMULAS AND NUMERICAL ESTIMATES

Technically it is convenient to calculate the corrections given by the formulas (7) and (10) by means of a special representation for the partial-wave Green functions which correspond to the Hamiltonian $H(\alpha)=-\Delta/m-\alpha/r$. Namely, at negative energy $E=-k^2/m$ one has for the partial-wave Green functions defined in the standard way:

$$\langle \mathbf{x}|(H(\alpha)-E)^{-1}|\mathbf{y}\rangle=\sum_{l=0}^{\infty}(2l+1)G_l^{(\alpha)}(x,y;E)P_l((\mathbf{xy})/xy), \tag{14}$$

the representation

$$G_l^{(\alpha)}(x,y;-k^2/m)=\frac{mk}{2\pi}(2kx)^l(2ky)^l e^{-k(x+y)}\sum_{s=0}^{\infty}\frac{L_s^{2l+1}(2kx)L_s^{2l+1}(2ky)s!}{(s+l+1-m\alpha/2k)(s+l+1)!}, \tag{15}$$

where $x=|\mathbf{x}|$, $y=|\mathbf{y}|$, the $P_l(z)$ are Legendre polynomials, and finally the L_s^p are Laguerre polynomials defined by the formula

$$L_s^p(z)=\frac{e^z z^{-p}}{s!}\left(\frac{d}{dz}\right)^s(e^{-z}z^{s+p}). \tag{16}$$

The representation (15) can be verified by direct substitution using the orthogonality relation

$$\int_0^\infty e^{-z}z^pL_n^p(z)L_m^p(z)dz=p!\binom{n+p}{n}\delta_{nm}.$$

Formula (15) may be considered as the spectral representation in the complex plane of the coupling constant α. In this case, at fixed k the Green function has a pole at $\alpha=2k_n/m$. Conversely, at fixed $\alpha G_l^{(\alpha)}$ has poles at the energy values corresponding to the Coulomb bound states, and the residues factorize into products of Coulomb wave functions. (Probably the representation (15) is well known in the theory of the Coulomb problem.)

To calculate the shift of an energy level with principal quantum number n and orbital angular momentum l from formula (10) one has to substitute the explicit form of the Coulomb wave function ψ_{nl} and use the representation (15) for $(H_A-E)^{-1}$, i.e., at $\alpha=-\alpha_s/6$. From the sum (14) there remain in this case only the terms with orbital angular momenta $l-1$ and $l+1$, and the radial integrals leave at most five terms from the sum (15) for each of the corresponding partial-wave Green functions. The corresponding integrals can be calculated directly from the definition (16) of the Laguerre polynomials. As a result the formula for the energy shift due to the vacuum average (4) has the form

$$\delta E_{nl}/|E_n|=(m^2\eta/k_n^6)\,n^2a_{nl}, \tag{17}$$

where the coefficients a_{nl} are given by the expression

$$a_{nl}=\frac{1}{4n^3(2l+1)}\{(l+1)[F(n,l)-F(-n,l)]+l[F(n,-l-1)-F(-n,-l-1)]\},$$

where

$$F(n,l)=2n[n^2-(l+1)^2]+(n+l+2)(n+l+1)\times\left[\frac{(n-l)(n+l+3)}{9n+16}+\frac{4(2n-l)^2}{9n+8}\right], \tag{18}$$

and $E_n=-k_n^2/m$ is defined by relation (3). (This form for the coefficients a_{nl} is due to Leutwyler.[6])

To obtain the corrections to the e^+e^- widths of $^3S_1$ states it is necessary to calculate the quantity $\langle\mathbf{x}=0|G_S(E)|\mathbf{y}=0\rangle$ from formula (7). Here, one can use the representation (15) in the second term for $(H_A-E)^{-1}$ (only the partial wave with $l=1$ gives a contribution), while for $(H_S-E)^{-1}$ it is more convenient to use the expression[4,7]

$$\langle\mathbf{r}|\left(H_s+\frac{k^2}{m}\right)^{-1}|\mathbf{y}=0\rangle=\langle\mathbf{x}=0|\left(H_s+\frac{k^2}{m}\right)^{-1}|\mathbf{r}\rangle=\frac{mk}{2\pi}e^{-kr}\int_0^\infty e^{-2krt}\left(\frac{1+t}{t}\right)^\lambda dt,$$

where $\lambda=(2/3)m\alpha_s(k)/k$. After these substitutions the integrals over the intermediate variables in the second term in (7) are calculated very easily, and the result has the form

$$\langle\mathbf{x}=0|G_s\left(-\frac{k^2}{m}\right)|\mathbf{y}=0\rangle=-\frac{mk}{4\pi}+\frac{mk}{2\pi}\sum_{n=1}^{\infty}\frac{k_n}{(k-k_n)}-\frac{9\eta m^3}{2\pi k^5}\lambda^2\sum_{s=0}^{\infty}\frac{(s+1)(s+2)(s+3)}{(s+2+\lambda/8)}\left[\frac{\Gamma(s-\lambda)}{\Gamma(s+5-\lambda)}\right]^2. \tag{19}$$

Comparing the expansion of this expression in the vicinity of the pole at $E=E_n$ with the expansion (13), we find for the correction to the e^+e^- widths of $^3S_1$ states:

$$\frac{\delta\Gamma_n}{\Gamma_n}=\frac{m^2\eta}{k_n^6}n^2(b_n+2a_{n0}), \tag{20}$$

where Γ_n is the Coulomb value of the width determined by formulas (11) and (12), and the coefficients b_n are given by the expression

$$b_n=\frac{(n+1)(n+2)(n+3)}{48n(9n+16)}\left[25-\frac{6}{9n+16}\right]+\frac{2(n+1)(n+2)}{3(9n+8)}\left[5-\frac{3}{9n+8}\right]-\frac{n^2-1}{18n^2}-\frac{2(n-1)(n-2)}{3(9n-8)}\left[5+\frac{3}{9n-8}\right]-\frac{(n-1)(n-2)(n-3)}{48n(9n-16)}\left[25+\frac{6}{9n-16}\right]. \tag{21}$$

Numerical values for the coefficients a_{nl} and b_n for the first few levels are given in Table I. We note that the correction to the width is relatively larger than the correction to the binding energy: with good accuracy the equality $\delta\Gamma_n/\Gamma_n\simeq 3\delta E_n/|E_n|$ is satisfied. This becomes exact in the (academic) limit $n\to\infty$: $b_{n\to\infty}=a_{n\to\infty,0}=35/18$.

As has been noted previously in Ref. 4, the $m^2k_n^{-6}$ behavior of the correction guarantees that with growing quark mass the properties of the levels sharply approach the asymptotic Coulomb behavior. At the parameter values $\eta\simeq 0.01$ GeV^4 and $\Lambda_{QCD}\simeq 0.1$ GeV (see Ref. 7) the combined effect of the Coulomb interaction and the nonperturbative correction is that in the region of quarkonium masses where the correction does not exceed about 30%, $2m\simeq 30$–50 GeV for the $1S$ level and $2m\simeq 100$–180 GeV for the $2S$ level, the coefficients $\gamma_{1,2}$ in the formula $\Gamma(n^3S_1\to e^+e^-)=\gamma_n(3Q)^2|1+A_z/A_\gamma|^2$ are practically constant and are approximately $\gamma_1\simeq 1.0$ keV and $\gamma_2\simeq 0.38$ keV. For higher quark mass they follow the Coulomb formula (the correction becomes negligibly small). Since these values practically coincide with the e^+e^- widths of the T and T' resonances (γ_2 is apparently somewhat smaller), it is natural to expect that also for intermediate values of the quark mass these coefficients change little. Therefore, for the first two levels of $t\bar{t}$ quarkonium ($Q_t=2/3$), which is still hypothetical, one can expect the following values for the e^+e^- widths:

$$\Gamma(1^3S_1(t\bar{t})\to e^+e^-)=(4.0\pm0.2)\text{ keV}\,|1+A_Z/A_\gamma|^2 \text{ for } 2m_t\leqslant 50\text{ GeV},$$
$$\Gamma(2^3S_1(t\bar{t})\to e^+e^-)=(1.5\pm0.1)\text{ keV}\,|1+A_Z/A_\gamma|^2 \text{ for } 2m_t\leqslant 180\text{ GeV}.$$

TABLE I.

level	a_{nl}	b_n	level	a_{nl}	b_n
1S	1.652	1.630	3S	1.869	1.943
2S	1.783	1.918	2P	1.123	–

4. NORMALIZATION OF THE QUARK WAVE FUNCTION

When the calculations presented above were finished the author learned that recently Leutwyler[6] has performed an analysis which is similar to the one performed in Ref. 3. He also arrived at formula (10) for the level shift, and by a method which is slightly different from the one given above he found the explicit expressions (17) and (18) for the level shifts. He also considered the correction to the wave function of the 1S level and to its e^+e^- width. The latter result of Leutwyler differs from ours. The reason for this difference is very fundamental and deserves a special discussion.

In Ref. 6 it is assumed that the wave function of a quark-natiquark pair in a color singlet state is normalized to unity also when the interaction with vacuum fluctuations is taken into account. In fact, however, this is not true since the quark pair exchanges color with the vacuum fluctuations and spends part of the time in a color octet state. [Of course, the color of the quarks is in this case compensated by the color of the soft gluon states which saturate the vacuum average (4).] The impossibility of describing quarkonium with nonperturbative effects taken into account in terms of orthonormal wave functions in the color singlet sector change is a theorem which has been proven in Ref. 3. The question of the correction to the normalization of the wave function in the singlet sector is resolved by means of an expression for the matrix element of an arbitrary operator $\mathcal{O}$ which acts only on the dynamical variables of the quark and antiquark (but not on the gluon degrees of freedom). Taking the corrections into account, the expression has the form[3]

$$\begin{aligned}\langle M|\mathcal{O}|N\rangle=\langle m|\mathcal{O}|n\rangle[1+{}^1/_2\eta(\langle n|r_i(H_A-E_n)^{-2}r_i|n\rangle\\ +\langle m|r_i(H_A-E_m)^{-2}r_i|m\rangle)]\\ +\eta[\langle m|\mathcal{O}\overline{(H_S-E_n)^{-1}}r_i(H_A-E_n)^{-1}r_i|n\rangle\\ +\langle m|r_i(H_A-E_m)^{-1}r_i\overline{(H_S-E_m)^{-1}}\mathcal{O}|n\rangle]\\ -\eta\langle m|r_i(H_A-E_m)^{-1}\mathcal{O}(H_A-E_n)^{-1}r_i|n\rangle,\end{aligned} \tag{22}$$

where $\langle M|$ and $\langle N|$ are exact eigenstates, $\langle m|$ and $\langle n|$ are the unperturbed (Coulomb) states, and the bar on $(H_S-E_n)^{-1}$ denotes that the pole at $E=E_n$ has been subtracted:

$$\overline{(H_S-E_n)^{-1}}=\sum_{m\neq n}\frac{|m\rangle\langle m|}{E_m-E_n}$$

[for the operator $(H_A-E)^{-1}$ such a subtraction is not necessary since the spectrum of H_A lies at $E>0$ and here we discuss discrete levels at $E<0$].

It is clear from formula (22) that the correction to the diagonal matrix element of the unit operator vanishes as it should. However, if we perform the decomposition $1=P_S+P_A$ and calculate the diagonal matrix elements of the projectors P_S and P_A separately, we obtain

$$\langle N|P_S|N\rangle=1+\eta\langle n|r_i(H_A-E_n)^{-2}r_i|n\rangle, \tag{23}$$

$$\langle N|P_A|N\rangle=-\eta\langle n|r_i(H_A-E_n)^{-2}r_i|n\rangle \tag{24}$$

(the matrix element of P_A is determined by the last term in Eq. (22) and that of P_S by all other terms).

Formula (23) contains the correction to the normalization of the wave function in the singlet sector. If this correction is taken into account in the approach of Leutwyler,[6] than the value of $\Gamma_{ee}(1S)$ calculated by his method coincides with the determination by formulas (20) and (21).

At first sight formulas (23) and (24) contain the paradox that the probability of finding a quark-antiquark pair in a color-singlet state is larger than one, and that this probability is negative for an octet state. The origin of this seemingly paradoxical sign of the correction is easy to trace. It is apparently due to the negative sign of $\langle 0|\mathbf{E}^a\mathbf{E}^a|0\rangle$ (see formula 6). This sign also shows up in the fact that the shifts of all levels are positive, which for the ground state is in apparent contradiction with the well known theorem of quantum mechanics. By and large the picture is that the gluon states which saturate the average (4) should formally be assigned a negative norm (the appearance of fields with negative norm in gauge theories is also known in the framework of perturbation theory).

The physical question is whether this seemingly incorrect sign does not violate unitarity in observable quantities. It is very natural to expect that unitarity is not violated in gauge-invariant quantities. We do not intend to prove this general claim here, but we note that formulas (23) and (24) are not in direct conflict with unitarity since the projectors P_S and P_A, which are multiplied by a unitarity operator (in general by any nonlocal operator) in coordinate space, are not separately observable; only their sum is observable. In fact, any gauge-invariant source which produces quark-antiquark pairs at a finite distance contains also a source of gluons, and can therefore interact directly (but not via quarks) with vacuum gluon fluctuations. In this case the color state of the quark-antiquark pair itself is not fixed, and therefore because of the color exchange with the vacuum fluctuations there is no sense in speaking of the probability of producing a quark pair in a given color state. [In connection with these comments we note that in e^+e^- annihilation in the nonrelativistic limit one measures the exchange amplitude between two local sources of the form $P_S\delta(\mathbf{r})$.]

5. NONPERTURBATIVE CORRECTIONS IN TERMS OF MODIFICATIONS IN THE INTERACTION

The notion has spread widely that nonperturbative effects of QCD in heavy quarkonium may perhaps be described by means of a potential for the quark-antiquark interaction. Much effort has been spent to find the form of this potential from phenomenological fits to the data on charmonium and Υ resonances[8] of calculations of the Wilson loop in an instanton field.[9] However, as has been shown in Ref. 3, the effects of the interaction of quarkonium with vacuum gluon fluctuations in principle cannot be described by any potential (local or nonlocal, and also nonlocal in time). Here we will once more discuss the physical basis for this conclusion without repeating the formal proof.[3] A general feature of the work in which a nonperturbative potential is derived is the assumption (explicit or implicit) that such a potential exists. After this assumption is made, one can calculate some property of quarkonium taking nonperturbative effects into account and find the form of a potential which reproduces these effects. It is precisely the assumption that a potential exists which is wrong in this procedure. It is therefore not surprising that different forms for the potential are obtained by considering different properties of quarkonium.

A characteristic example of a paper in which the programme described above is carried out is the recent paper of Ref. 10. In this paper it is concluded that the effects of the interaction with vacuum fluctuations leads to a potential for the interaction between quarks which is proportional to $\eta m r^4$:

$$U(r)=\text{const}\cdot\eta m r^4. \tag{25}$$

Let us explain how one can arrive at such a conclusion. It can be seen from formula (10) for the level shift [or from (7)] that a nonlocal operator which depends on the energy enters into the expression for the corrections:

$$\hat{U}(\mathbf{r})=\eta r_i(H_A-E)^{-1}r_i. \tag{26}$$

It is not difficult to understand why this operator may simulate a potential of the form (25). The quantities which enter in (26) have an operator meaning from which one derives that to an order of magnitude the operator (26) is

$$U(r,E)\sim\eta r^2/E.$$

For a nonrelativistic system one can estimate from the uncertainty relation that

$$E\sim p^2/m\sim 1/mr^2.$$

Therefore, the effects of the operator (26) coincide parametrically with the action of the potential (25). It is not difficult, however, to convince oneself that if a potential has been derived, for instance, from the corrections to the energies [see formulas (17) and (18)], then the numerical coefficient in front of $\eta m r^4$ depends on (n, l). A different dependence on (n, l) is obtained if the potential is derived from the corrections to the e^+e^- widths. Thus, a potential of the form (25) cannot be used as an effective nonperturbative potential.

One can go further and prove[3] that the nonlocal operator (26) which depends on the energy (the dependence on the energy in the time-dependent formalism also implies nonlocality in the time) also cannot be used as a kernel in the evolution equation for the wave function. This can be seen from formula (22), where the last term does not correspond to an iteration of the interaction (26). [This circumstance has been overlooked in the oversimplified analysis of Ref. 6, and it is therefore concluded in Ref. 6 that the operator (26) is the effective interaction operator.]

The question has been solved in principle in Ref. 3, where it has been shown that the effects of the interaction with nonperturbative fluctuations in general do not lead to a two-particle interaction. Here we want to explain once more the physical reasons which lead to this behavior. The parameters of the expansion in the vacuum averages of gluon operators[3] are, roughly speaking, the ratios r_Q/r_g and τ_Q/τ_g, where r_Q and τ_Q (r_g and τ_g) are the characteristic dimensions and the evolution time of the quark system (of the gluon fluctuations), $\tau_Q\sim mr_Q^2\sim 1/|E|$, and $\tau_g\sim r_g\sim\eta^{1/4}$. The assumption that τ_Q is small compared to τ_g implies that the retardation of the interaction through vacuum fluctuations is large on the time scale of the quark evolution. In other words, the quarks can make a few revolutions between the moments at which the interaction with the fluctuations described by the Hamiltonian (5) takes place. A potential description corresponds to the opposite situation when the time for the interaction to propagate is much less than τ_Q. Therefore, one can say that from the very start the expansion used in the approach described here is an antipotential expansion.

From well known similar phenomena one can for instance mention the (dominating) part of the Lamb shift in hydrogen which arises from the emission and absorption of photons by an electron and also cannot be described by a potential. This process can be viewed as the interaction of an electron with perturbative vacuum fluctuations of photons. In the case of the Lamb shift this effect is only of order $(v/c)^3$, since in QED the vacuum fluctuations in perturbation theory do not have a characteristic scale. In QCD the nonperturbative fluctuations, as is the basic assumption, have a characteristic scale (of order Λ_{QCD}) and show a non-potential character also in the leading nonrelativistic approximation. Here, it is appropriate to note that in QCD the non-potential character also appears in the leading nonrelativistic approximation in perturbation theory[11] because of the emission and absorption of gluons by the Coulomb field between a quark and an antiquark. However, this effect is screened by two powers of the coupling constant α_s and is therefore small. Moreover, it only depends logarithmically on the dimensions of quarkonium and here we are interested in power effects.

Finally we want to discuss the extensively used method of calculating a potential by considering the Wilson loop (see, for instance Ref. 9). By this method one calculates the energy of a quark and antiquark which are at rest at a relative distance r. If a nonperturbative potential exists, as is usually assumed, then such a calculation would reproduce it since in this case $E(r)=U(r)$. It is not difficult to understand, however, that the quantity $E(r)$ calculated in this way is strictly speaking only the average value of the Hamiltonian over a specially chosen state without kinetic energy. It is possible that such a situation takes place for states with energies $|E|\ll\Lambda_{QCD}$ which are very close to the threshold, and in this region an approximate potential description may be valid. In particular, examining the Wilson loop may turn out to be useful in solving the basic problem of confinement.

In the situation of Coulomb bound states we are interested in here, the velocities of arbitrarily heavy quarks remain finite: $v/c\sim\alpha_s$. Therefore, the kinetic energy is $T\sim m\alpha_s^2$ and cannot be neglected. Moreover, as has already been noted, our expansion assumes that the kinetic energy of the quarks is much larger than the characteristic mass scale ($T \sim |E| \gg \Lambda_{QCD}$) of the vacuum fluctuations, and it is clear that precisely this situation is realized for the ground levels of quarkonium for heavy enough quarks. As we have seen, in this situation a description in terms of some nonperturbative potential has no basis at all.

[1]A. I. Vaĭnshteĭn, V. I. Zakharov, and M. A. Shifman, Pis'ma Zh. Eksp. Teor. Fiz. **27**, 60 (1978) [JETP Lett. **27**, 55 (1978)]; M. A. Shifman *et al.*, Phys. Lett. **77B**, 80 (1978).
[2]M. A. Shifman, A. I. Vaĭnshteĭn, and V. I. Zakharov, Nucl. Phys. **B147**, 385, 448 (1979).
[3]M. B. Voloshin, Nucl. Phys. **B154**, 365 (1979).
[4]M. B. Voloshin, Preprint ITEF-54 (1979).
[5]M. Peskin, Nucl. Phys. **B156**, 365 (1979).
[6]H. Leutwyler, Univ. Bern preprint, November 1980.
[7]M. B. Voloshin, Preprint ITEF-21 (1979).
[8]M. Krammer and H. Kraseman, Acta Phys. Austr. Suppl. XXI, 259 (1979).
[9]C. Callan, R. Dashen, and D. Gross, Phys. Rev. **D17**, 2717 (1978).
[10]J. S. Bell and R. A. Bertlmann, Preprint TH-2986 CERN (1980).
[11]T. Appelquist, M. Dine, and I. Muzinich, Phys. Rev. **D17**, 2074 (1978).

Translated by L. J. Reinders

Nuclear Physics B186 (1981) 109–146

QCD SUM RULES FOR HEAVY QUARK SYSTEMS

L.J. REINDERS, H.R. RUBINSTEIN[1] and S. YAZAKI[2]

Rutherford and Appleton Laboratories, Chilton, Didcot, Oxfordshire, England

Received 26 November 1980
(Final version received 2 March 1981)

In this paper we present in a detailed and coherent fashion our work on QCD sum rules for equal mass heavy quark meson states. We discuss the technical procedures used to calculate the perturbative and non-perturbative contributions to the vacuum polarization, which have been calculated for all currents up to and including spin 2^{++}. Using dispersion relations, sum rules are derived. Extensive applications are made to the lowest lying states of the charmonium and upsilon systems. The masses of the S- and P-wave charmonium levels are reproduced to a high degree of accuracy, and the mass of the $^1P_1$ level is predicted at 3.51 GeV. For the upsilon system it only appears to be possible to predict the Υ-η_b splitting which gives 60 MeV. Very accurate values are given for the current quark masses at $p^2 = -m_q^2$: $m_c = 1.28$ GeV and $m_b = 4.25$ GeV.

1. Introduction

There is by now impressive evidence that hadrons are composite states built from quarks. More surprising is that these constituents act in their bound states with relatively large constituent masses while their interactions are relatively weak. This is reflected in the additive nature of many properties like magnetic moments [1].

Around this idea flourished the potential model epitomized by the QCD form of the Fermi-Breit hamiltonian in which residual *one*-gluon interactions are enough to give a good description of the baryon spectrum [2].

Less established but extremely attractive is that at short distances the interactions are indeed solely mediated by vector gluons coupled to colour. The theory is in a confining phase, which manifests itself by strong infrared singularities and is a natural way of producing the bound states: when quarks separate they create pairs to screen the intolerable force. Calculations in this phase are extremely difficult. Some success has been achieved by Padé approximants in pure Yang-Mills lattices [3] but the task is especially difficult as fermions are not easily included.

The idea to approach the bound-state problem from the asymptotic freedom side is due to Shifman, Vainshtein, Zakharov (SVZ) and co-workers [4, 5]. Consider the vacuum polarization operator as induced by some current with well defined quantum numbers in flavour and spacetime. This function can be analyzed in terms of

[1] Present and permanent address: the Weizmann Institute, Rehovot, Israel.
[2] On leave of absence from the University of Tokyo, Japan.

moments given by

$$M_n = \int \frac{\mathrm{Im}\,\Pi(s)\,\mathrm{d}s}{(s+Q^2)^n}\,. \tag{1.1}$$

It is important to note that large n is testing large distances and small n short distances. Therefore, starting with a value of Q^2 large enough compared to the characteristic scale of confinement, one is really in a regime in which asymptotic freedom prevails.

When n grows, new forces must come in to prevent the quarks from separating. Normal abelian multi-exchanges cannot bind and indeed can be calculated (and minimized) by an appropriate choice of the quark mass definition $m_c = m_c(p^2 = -m_c^2)$ [4]. Therefore, it is argued that new components must be seen at intermediate distances which reflect the strong vacuum fluctuations which, in turn, confine. It is then conjectured that the appearance of these components (power corrections) sets the limit of asymptotic freedom and must be at the origin of resonance formation. That the two regimes overlap and constraint equations are operating is not obvious. Nevertheless, there are several limits in hadronic physics that these transitions occur in this way:

(a) s-t channel duality. Here the averaging goes to very low energy and, indeed, it is believed that dual models are intimately connected with QCD.

(b) Lattice Monte Carlo calculations [6] show that there is no intermediate desert and that strong and weak coupling join suddenly.

Nevertheless, it is the success of a variety of applications that will give credibility to the idea as long as the theory cannot be solved.

If the idea works, one can relate lagrangian parameters of QCD to physical observable hadronic quantities. For this to be possible one needs a further assumption. Once the dispersion relation (1.1) is saturated with physical states, the convergence is fast. This is again like duality and, moreover, in some cases (like vector currents) this can be checked. Already at moderately high n the lowest lying resonance in a particular channel gives more than 90% of the moments M_n.

The general picture, which will be treated in more detail in sect. 2, is the following. The polarization operator is constructed from the time ordered product of any two currents and the short-distance expansion is assumed to be valid. For charmonium, which is our main target here, the quarks are heavy and small Q^2 (even $Q^2 = 0$!) is enough to ensure asymptotic freedom. One calculates the perturbative corrections to the polarization amplitude at short distances as well as the contribution of large-distance effects. The latter originate from higher dimensional operators in the operator product expansion and are the basic new features introduced by SVZ. Lacking a consistent scheme to calculate the vacuum expectation values of the local operators induced by non-perturbative effects, these are treated as phenomenological parameters and their values are to be determined from the data.

At very low n one is not able to isolate a single resonance, since the weight function in eq. (1.1) is not strong enough. For higher n the system gives physical information until the calculation to first order in α_s (both for perturbative and non-perturbative contributions) breaks down as the coefficients of these contributions grow with n (the non-perturbative even like n^3) and the equations become invalid at high n. By tuning the equations with Q^2 one can enlarge the validity region of the moments. This is one of the advantages of our calculations compared to SVZ who indeed had a very small stability region at $Q^2 = 0$.

Shifman et al. have applied this scheme to a large range of data starting with the vector and pseudoscalar channels of charmonium. The detailed experimental data in the vector channel are used to fix the values of the various parameters. Application to the pseudoscalar channel [7] led to the prediction for the mass of the η_c at 3.0 GeV long before the data of the Crystal Ball experiment [8] became known. Detailed applications have also been made with considerable success to light quark systems, in particular to the ρ and ϕ mesons, whose masses and electronic widths have been calculated with remarkable accuracy [5].

In two recent letters [9, 10] we have briefly reported on extensive calculations to extend the scheme to include all possible currents in the S- and P-wave section of heavy quarks [9] and the application to the charmonium spectrum [10]. The purpose of this paper is to present in a detailed and coherent fashion our work on QCD sum rules for the equal mass heavy quark meson states, including all the technical details, which were hitherto unexplained, on the derivation of the cumbersome formulae. This is necessary for completeness and also because we still disagree with some of the results for the currents obtained by SVZ. We will present more physical results and discuss the application to the upsilon-like systems.

The outline of the paper is as follows. In sect. 2 we review the general procedure and sects. 3 and 4 are devoted to the calculation of the Wilson coefficients of the perturbative and non-perturbative operators, respectively. In sect. 5 the various tests performed to check the results obtained in sects. 3 and 4 are discussed. Finally, sects. 6 and 7 treat the applications to the charmonium and upsilon systems.

2. General formalism

Consider a current of the form $\bar{q}_i \Gamma_n q_i$ where the index i on q denotes the flavour of the quark and the index n on the operator Γ_n its tensor structure and charge conjugation. The correlation function of two such currents determines the vacuum polarization operator given by

$$T_{\mu\nu\ldots} \Pi^J(q^2) = i \int \mathrm{d}x \, \mathrm{e}^{iqx} \langle 0 | \mathrm{T}(J_{\mu\ldots}(x) J_{\nu}\ldots(0)) | 0 \rangle \,. \tag{2.1}$$

Following SVZ we start at short distances and it is natural to make a Wilson expansion [11] of (2.1):

$$i\int \mathrm{d}x\, \mathrm{e}^{iqx} \mathrm{T}\left(J^A_{\mu\ldots}(x) J^B_{\nu\ldots}(0)\right) = T_{\mu\nu\ldots} C^{AB}_n(q) O_n, \tag{2.2}$$

where the C^{AB}_n are the Wilson coefficients and the O_n are local operators constructed from quark and gluon fields.

In QED and in theories where the vacuum has a simple structure, one can give good arguments for the validity of such an expansion. Its physical meaning is well known: c-number coefficients determine the short-distance propagation and the large-distance part is contained in the matrix elements of the operators O_n. The questions we have to deal with, are of course:

(a) Is this separation meaningful in QCD?

(b) Is the convergence of the series and therefore the classification in terms of naive operator dimensions meaningful?

(c) Is the series valid up to some power of the inverse mass as is, for example, learnt from instantons?

We will discuss these points as far as we think they are relevant to our situation. A general, but certainly not definitive discussion has been given by SVZ [5, 12].

Since in (2.1) we are interested in the vacuum expectation values of the operators O_n, we have only to consider spin-zero operators. A list of all operators with dimensions $d \leqslant 6$ has been given in ref. [5]. The expansion (2.2) can be considered as an expansion in inverse powers of q^2 where the higher dimensional operators give power corrections to the perturbative contribution from the unit operator ($d=0$). Perturbation theory to all orders is contained in the Wilson coefficient of the unit operator. New operators are, for example, $G^a_{\mu\nu}G^a_{\mu\nu}$, which in perturbation theory has a vanishing vacuum expectation value, but in QCD non-perturbative forces change the nature of the vacuum and induce non-vanishing vacuum expectation values. This is by no means a new phenomenon. In the framework of current algebra [13] the expectation value $\langle 0|\bar{q}q|0\rangle$ for light quark fields q is known to be unequal to zero and responsible for the breakdown of chiral symmetry. These matrix elements contain all the non-perturbative contributions. Consequently Feynman diagram techniques can be used to calculate its Wilson coefficients. The vacuum expectation values of the non-perturbative operators like $\langle 0|G^a_{\mu\nu}G^a_{\mu\nu}|0\rangle$ are treated phenomenologically. If they are independently calculated one must be sure that the way they are evaluated and our phenomenological value stem from identical definitions. One possibility would be to calculate it from a Monte Carlo simulation on a lattice. There, one must first redefine the operator to eliminate the relevant perturbative divergences.

As we will consider only heavy quark systems in this paper, we can safely neglect operators with dimension $d>4$ and apart from the identity operator will only consider the gluon condensate operator $G^a_{\mu\nu}G^a_{\mu\nu}$ and the two fermion operators $\bar{\psi}M\psi$.

For heavy quarks like the c quark the latter will involve $\langle 0|\bar{c}c|0\rangle$. In a heavy quark expansion this is reduced to the matrix elements of gluon operators [5] which are suppressed by a factor of m_c and can therefore also be neglected.

This makes the situation for heavy quarks relatively simple as we only have to calculate the Wilson coefficients for the identity operator and the operator $G^a_{\mu\nu}G^a_{\mu\nu}$. Diagramatically the operator product expansion for the product of two currents can be written as follows (for heavy quarks, to first order in α_s)

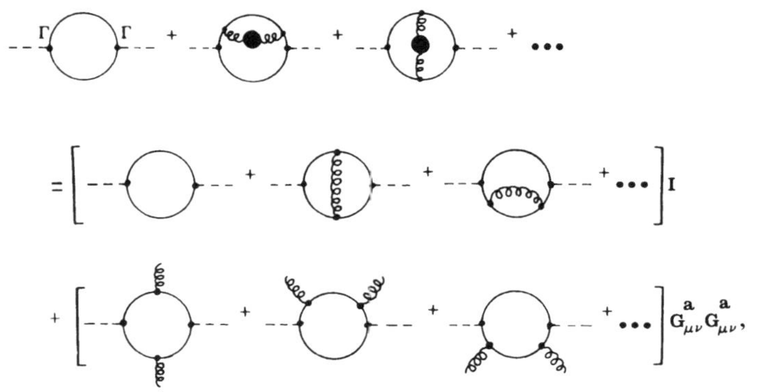

(2.3)

where the first set of diagrams on the right-hand side are the normal perturbative ones and the second set arises in the following two steps:

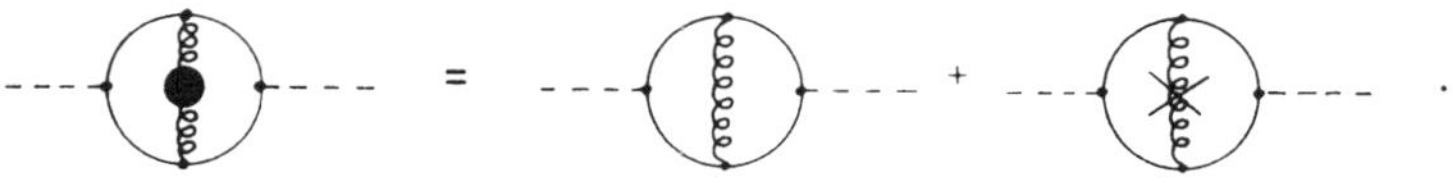

(2.4)

In the last diagram of this equation 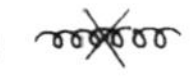 stands for the modification of the gluon propagator due to the non-perturbative vacuum and in the OPE this diagram factorizes as follows:

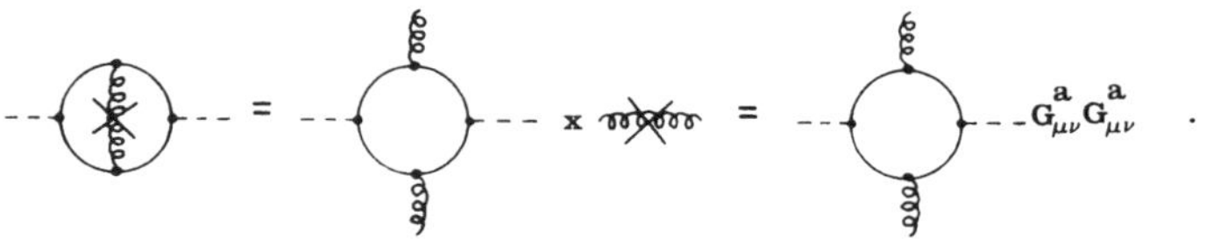

(2.5)

In sects. 3 and 4 we will describe in detail how the various diagrams have been calculated.

As has been reported earlier [9], we have calculated the Wilson coefficients for all possible currents with $J^{PC} = 0^{--}, 1^{--}, 0^{++}, 1^{++}, 1^{+-}$ and 2^{++} that couple to the observed physical states, i.e. for the following set of currents:

$$\begin{aligned}
J_{\rm S}^i &= \bar{q}_i q_i & J^{PC} &= 0^{++} \\
J_{\rm P}^i &= i\bar{q}_i \gamma_5 q_i & &= 0^{--} \\
J_{\rm V}^i &= \bar{q}_i \gamma_\mu q_i & &= 1^{--} \\
J_{\rm A}^i &= \eta_{\mu\nu} \bar{q}_i \gamma_\mu \gamma_5 q_i & &= 1^{++} \\
J_{\rm A'}^i &= \bar{q}_i \overleftrightarrow{\partial}_\mu \gamma_5 q_i & &= 1^{+-} \\
J_{\rm T}^i &= i\bar{q}_i(\gamma_\mu \overleftrightarrow{\partial}_\nu + \gamma_\nu \overleftrightarrow{\partial}_\mu + \tfrac{2}{3}\eta_{\mu\nu} \overleftrightarrow{\not\partial}) q_i & &= 2^{++}
\end{aligned} \tag{2.6}$$

where $\eta_{\mu\nu} = q_\mu q_\nu / q^2 - g_{\mu\nu}$. The various currents differ only by a different current-quark vertex Γ.

Having calculated all relevant diagrams which contribute to the Wilson coefficients $C_{\rm I}$ (identity operator) and $C_{\rm G}$ (gluon condensate) we have an expression for the vacuum polarization $\Pi^J(q^2)$ in the deep euclidean region in terms of the fundamental parameters of QCD ($\alpha_{\rm s}, m_{\rm c}$) and $\langle 0|(\alpha_{\rm s}/\pi)G_{\mu\nu}^a G_{\mu\nu}^a|0\rangle$. The latter is a renormalization group invariant quantity and will be treated phenomenologically while $m_{\rm c}$ and $\alpha_{\rm s}$ are running parameters.

On the other hand $\Pi^J(q^2)$ is related to its imaginary part by a dispersion relation

$$\Pi^J(q^2) = \frac{1}{\pi} \int_{4m_{\rm c}^2}^{\infty} \frac{\operatorname{Im}\Pi^J(s)}{s - q^2}\, {\rm d}s, \tag{2.7}$$

and in turn $\operatorname{Im}\Pi^J(s)$ is related to a cross section. In particular, for the vector current we have

$$\operatorname{Im}\Pi^{\rm V}(s) = \frac{1}{12\pi Q_{\rm c}^2} R_{\rm c}(s) = \frac{9}{64\pi^2\alpha^2} s\sigma({\rm e}^+{\rm e}^- \to \text{charm}), \tag{2.8}$$

with $Q_{\rm c} = \frac{2}{3}$ the charge of the charmed quark and $\sigma({\rm e}^+{\rm e}^- \to \text{charm})$ includes final states with open and hidden charm.

At this point one obtains an equation that relates the parameters of the QCD lagrangian via the asymptotic freedom expression which follows from calculating (2.3) with the hadronic masses and couplings. If all states plus the continuum are fed into $\operatorname{Im}\Pi^J(s)$ and the theory is correct, the experimental side (2.7) should match the calculated side exactly. However, because of duality, saturation with a few resonances should be effective and the equations then determine the parameters of the lowest lying states. We will follow Shifman et al. in using the narrow resonance approximation and write

$$\operatorname{Im}\Pi^J(s) = \sum_{\text{res}} \frac{9\pi m_R^2}{4g_R^2}\delta(s - m_R^2) + \text{continuum}, \tag{2.9}$$

where m_R is the mass of the resonance and g_R the coupling of the resonance to the appropriate current. For the vector current, g_R is related to the electronic width of the resonance:

$$\Gamma_R(e^+e^-) = \frac{4\pi\alpha^2}{3g_R^2} m_R .$$

For actual calculations we define the moments (1.1) by taking the nth derivative of the polarization operator

$$M_n^J(Q_0^2) = \frac{1}{n!}\left(-\frac{d}{dQ^2}\right)^n \Pi^J(Q^2)|_{Q^2=Q_0^2}, \qquad Q^2 = -q^2$$

$$= \frac{1}{\pi}\int_{4m_c^2}^{\infty} \frac{\operatorname{Im}\Pi(s)\,ds}{(s+Q_0^2)^{n+1}} . \tag{2.10}$$

We have already noted in sect.1 that large n tests large distances which can easily be seen by differentiating the right-hand side of eq. (2.1). An important observation in ref. [5] is that for heavy quarks there is a region of Q_0^2 starting at about $Q_0^2 = 0$ for which asymptotic freedom holds. In fact all calculations for heavy quarks by Shifman et al. [5, 7] have been performed at $Q_0^2 = 0$. The asymptotic freedom expression for the moments (2.10) can be written as

$$M_n^J(\xi) = A^J(n)\left[1 + a_n(J;\xi)\alpha_s + b_n(J,\xi)\phi\right], \tag{2.11}$$

where $\xi = Q_0^2/4m_c^2$ and

$$\phi = \frac{4\pi^2}{9}\left\langle 0\left|\frac{\alpha_s}{\pi} G_{\mu\nu}^a G_{\mu\nu}^a\right|0\right\rangle \Big/ (4m_c^2)^2 \tag{2.12}$$

is the matrix element of the gluon condensate. The dimensionless coefficients $a_n(J,\xi)$ and $b_n(J,\xi)$ are tabulated in table 1. Their calculation will be presented in detail in sects. 3 and 4. They are the moments of the perturbative and non-perturbative contributions respectively, discussed above; a_n and b_n are functions of ξ (or Q_0^2), which has to be chosen such that for a large range of n $a_n\alpha_s, b_n\phi \ll 1$ for first-order perturbation theory to make sense. This will also ensure that the convergence of the operator product expansion series (2.2) is meaningful. We will see later

TABLE 1

In the following tables $\rho = \xi/(1+\xi)$ and $F(a,b,c;x)$ is the hypergeometric function ${}_2F_1(a,b,c;x)$; m is the running quark mass $m = m_c(p^2 = -(\xi+1)m_c^2)$.

Pseudoscalar current ($^1S_0$)

$$A^P(n) = \frac{3}{8\pi^2}\frac{2^n(n-1)!}{(2n+1)!!}(4m^2)^{-n}(1+\xi)^{-n}F\left(n,\tfrac{1}{2},n+\tfrac{3}{2};\rho\right),$$

$$a_n(P) = \frac{(2n+1)!!}{3\cdot 2^{n-1}n!}\left[\pi - \frac{1}{2(n+1)}\left(\tfrac{1}{2}\pi - \frac{3}{4\pi}\right)F(n,1,n+2;\rho)\right]$$

$$\times\frac{1}{F\left(n,\frac{1}{2},n+\frac{3}{2};\rho\right)} - \left(\tfrac{1}{2}\pi - \frac{3}{4\pi}\right)$$

$$+\frac{1}{\pi}\left[\frac{8}{3} - \frac{4}{n}\frac{F\left(n,\frac{3}{2},n+\frac{3}{2};\rho\right)}{F\left(n,\frac{1}{2},n+\frac{3}{2};\rho\right)} - \frac{5}{6}\frac{1}{n+\frac{3}{2}}\frac{F\left(n,\frac{3}{2},n+\frac{5}{2};\rho\right)}{F\left(n,\frac{1}{2},n+\frac{3}{2};\rho\right)}\right]$$

$$-2n\frac{\ln(2+\xi)}{\pi}\frac{(2+\xi)}{(1+\xi)^2}\frac{F\left(n+1,\frac{1}{2},n+\frac{3}{2};\rho\right)}{F\left(n,\frac{1}{2},n+\frac{3}{2};\rho\right)},$$

$$b_n(P) = -\frac{n(n+1)(n+2)(n+3)}{2n+3}(1+\xi)^{-1}$$

$$\times\left[\frac{F\left(n+1,-\frac{3}{2},n+\frac{5}{2};\rho\right)}{F\left(n,\frac{1}{2},n+\frac{3}{2};\rho\right)} - \frac{6}{n+3}\frac{F\left(n+1,-\frac{1}{2},n+\frac{5}{2};\rho\right)}{F\left(n,\frac{1}{2},n+\frac{3}{2};\rho\right)}\right]$$

Vector current ($^3S_1$)

$$A^V(n) = \frac{3}{4\pi^2}\frac{2^n(n+1)(n-1)!}{(2n+3)!!}(4m^2)^{-n}(1+\xi)^{-n}F\left(n,\tfrac{1}{2},n+\tfrac{5}{2};\rho\right),$$

$$a_n(V) = \frac{(2n+1)!!}{3.2^{n-1}n!}\left(\frac{2n+3}{2n+2}\right)\frac{1}{F\left(n,\frac{1}{2},n+\frac{5}{2};\rho\right)}$$

$$\times\left[\pi - \left(\tfrac{1}{3}\pi + \tfrac{1}{2}\left(\tfrac{1}{2}\pi - \frac{3}{4\pi}\right)\right)F(n,1,n+2;\rho)\frac{1}{n+1}\right.$$

$$\left.+\frac{1}{3}\frac{1}{(n+1)(n+2)}\left(\tfrac{1}{2}\pi - \frac{3}{4\pi}\right)F(n,2,n+3;\rho)\right]$$

$$-\left(\tfrac{1}{2}\pi - \frac{3}{4\pi}\right) - 2n\frac{\ln(2+\xi)}{\pi}\frac{(2+\xi)}{(1+\xi)^2}\frac{F\left(n+1,\frac{1}{2},n+\frac{5}{2};\rho\right)}{F\left(n,\frac{1}{2},n+\frac{5}{2};\rho\right)},$$

$$b_n(V) = -\frac{n(n+1)(n+2)(n+3)}{2n+5}(1+\xi)^{-2}\frac{F\left(n+2,-\frac{1}{2},n+\frac{7}{2};\rho\right)}{F\left(n,\frac{1}{2},n+\frac{5}{2};\rho\right)}.$$

TABLE 1 (continued)

Scalar current ($^3P_0$)

$$A^S(n) = \frac{3}{8\pi^2}\frac{3\cdot 2^n(n-1)!}{(2n+3)!!}(4m^2)^{-n}(1+\xi)^{-n}F\left(n,\tfrac{3}{2},n+\tfrac{5}{2};\rho\right),$$

$$a_n(S) = \tilde{a}_n + \frac{1}{\pi}\left[\frac{16}{3} - \frac{20}{3}\frac{1}{n}\frac{F\left(n,\frac{5}{2},n+\frac{5}{2};\rho\right)}{F\left(n,\frac{3}{2},n+\frac{5}{2};\rho\right)}\right.$$

$$-\frac{20}{3}\frac{1}{n+\frac{5}{2}}\frac{F\left(n,\frac{5}{2},n+\frac{7}{2};\rho\right)}{F\left(n,\frac{3}{2},n+\frac{5}{2};\rho\right)} + \frac{4}{3}\frac{1}{n(n-1)}(1+\xi)\frac{F\left(n-1,\frac{5}{2},n+\frac{5}{2};\rho\right)}{F\left(n,\frac{3}{2},n+\frac{5}{2};\rho\right)}$$

$$\left.-\frac{2}{\left(n+\frac{5}{2}\right)\left(n+\frac{7}{2}\right)}\frac{F\left(n,\frac{7}{2},n+\frac{9}{2};\rho\right)}{F\left(n,\frac{3}{2},n+\frac{5}{2};\rho\right)}\right],$$

where

$$\tilde{a}_n = \frac{4}{9}\frac{(2n+3)!!}{2^{n+1}(n+1)!}\left[\pi - \frac{\pi-6/\pi}{n+2}\frac{F(n,2,n+3;\rho)}{F(n,1,n+2;\rho)}\right]$$

$$\times\frac{F(n,1,n+2;\rho)}{F\left(n,\frac{3}{2},n+\frac{5}{2};\rho\right)} - \frac{2}{3}\left(\frac{1}{2}\pi - \frac{3}{\pi}\right) - 2n\frac{\ln(2+\xi)}{\pi}\frac{(2+\xi)}{(1+\xi)^2}\frac{F\left(n+1,\frac{3}{2},n+\frac{5}{2};\rho\right)}{F\left(n,\frac{3}{2},n+\frac{5}{2};\rho\right)},$$

$$b_n(S) = -\frac{3n(n+1)(n+2)(n+3)}{2n+5}(1+\xi)^{-1}$$

$$\times\left[\frac{F\left(n+1,-\frac{1}{2},n+\frac{7}{2};\rho\right)}{F\left(n,\frac{3}{2},n+\frac{5}{2};\rho\right)} - \frac{2}{3(n+3)}\frac{F\left(n+1,\frac{1}{2},n+\frac{7}{2};\rho\right)}{F\left(n,\frac{3}{2},n+\frac{5}{2};\rho\right)}\right].$$

Axial vector current ($^3P_1$)

$$A^A(n) = \tfrac{2}{3}A^S(n),$$

$$a_n(A) = \tilde{a}_n + \frac{1}{\pi}\left[\frac{4}{3} - \frac{2}{n}\frac{F\left(n,\frac{5}{2},n+\frac{5}{2};\rho\right)}{F\left(n,\frac{3}{2},n+\frac{5}{2};\rho\right)}\right.$$

$$-\frac{4}{3}\frac{1}{n+\frac{5}{2}}\frac{F\left(n,\frac{5}{2},n+\frac{7}{2};\rho\right)}{F\left(n,\frac{3}{2},n+\frac{5}{2};\rho\right)} + \frac{2}{n(n-1)}(1+\xi)\frac{F\left(n-1,\frac{5}{2},n+\frac{5}{2};\rho\right)}{F\left(n,\frac{3}{2},n+\frac{5}{2};\rho\right)}$$

$$\left.-\frac{8}{3\left(n+\frac{5}{2}\right)\left(n+\frac{7}{2}\right)}\frac{F\left(n,\frac{7}{2},n+\frac{9}{2};\rho\right)}{F\left(n,\frac{3}{2},n+\frac{5}{2};\rho\right)}\right],$$

$$b_n(A) = -\frac{3n(n+1)(n+2)(n+3)}{(2n+5)}(1+\xi)^{-1}\frac{F\left(n+1,-\frac{1}{2},n+\frac{7}{2};\rho\right)}{F\left(n,\frac{3}{2},n+\frac{5}{2}\ \rho\right)}.$$

TABLE 1 (continued)

Axial vector current ($^1P_1$)

$$A^B(n) = \tfrac{1}{3} A^S(n)$$

$$a_n(B) = \tilde{a}_n + \frac{1}{\pi}\left[-\frac{2}{3n}\frac{F(n,\frac{5}{2},n+\frac{5}{2};\rho)}{F(n,\frac{3}{2},n+\frac{5}{2};\rho)} + \frac{2}{3(n+\frac{5}{2})}\frac{F(n,\frac{5}{2},n+\frac{7}{2};\rho)}{F(n,\frac{3}{2},n+\frac{5}{2};\rho)} + \frac{4}{3n(n-1)}(1+\xi)\frac{F(n-1,\frac{5}{2},n+\frac{5}{2};\rho)}{F(n,\frac{3}{2},n+\frac{5}{2};\rho)} - \frac{1}{3(n+\frac{5}{2})(n+\frac{7}{2})}\frac{F(n,\frac{7}{2},n+\frac{9}{2};\rho)}{F(n,\frac{3}{2},n+\frac{5}{2};\rho)}\right],$$

$$b_n(B) = b_n(S).$$

Tensor current ($^3P_2$)

$$A^T(n) = \frac{3}{8\pi^2}\frac{2^{n+2}(2n+3)(n-1)!}{(2n+5)!!}(4m^2)^{-n}(1+\xi)^{-n}\tilde{F}(n,\rho),$$

where

$$\tilde{F}(n,\rho) = \frac{n+\frac{5}{2}}{n+\frac{3}{2}}\left[F(n,\tfrac{3}{2},n+\tfrac{5}{2};\rho) - \frac{1}{n+\frac{5}{2}}F(n,\tfrac{5}{2},n+\tfrac{7}{2};\rho)\right],$$

$$a_n(T) = \frac{4}{9}\frac{(2n+3)!!}{2^{n+1}(n+1)!}\frac{2n+5}{2n+3}\frac{F(n,1,n+2;\rho)}{\tilde{F}(n,\rho)} \times\left[\pi - \frac{1}{n+2}\left(\frac{4\pi}{5} + 2\left(\tfrac{1}{2}\pi - \frac{3}{\pi}\right)\right)\frac{F(n,2,n+3;\rho)}{F(n,1,n+2;\rho)} + \frac{1}{(n+2)(n+3)}\frac{12}{5}\left(\tfrac{1}{2}\pi - \frac{3}{\pi}\right)\frac{F(n,3,n+4;\rho)}{F(n,1,n+2;\rho)}\right]$$

$$-\tfrac{2}{3}\left(\tfrac{1}{2}\pi - \frac{3}{\pi}\right) - \frac{4}{3\pi}$$

$$+\frac{1}{\pi}\left[\frac{16}{9n}\frac{n+\frac{5}{2}}{n+\frac{3}{2}}\left\{F(n,\tfrac{5}{2},n+\tfrac{5}{2};\rho) - \frac{1}{n+\frac{5}{2}}F(n,\tfrac{7}{2},n+\tfrac{7}{2};\rho)\right\} + \frac{24}{7}\frac{1}{n+\frac{7}{2}}\left\{\frac{2n+2}{2n+3}F(n,\tfrac{5}{2},n+\tfrac{9}{2};\rho) + \frac{1}{2n+3}F(n,\tfrac{7}{2},n+\tfrac{9}{2};\rho)\right\} - \frac{286}{525}\frac{1}{n+\frac{3}{2}}F(n,\tfrac{5}{2},n+\tfrac{7}{2};\rho)\right]\Big/\tilde{F}(n,\rho)$$

TABLE 1 (continued)

$$-2n\frac{\ln(2+\xi)}{\pi}\frac{n+\frac{5}{2}}{n+\frac{3}{2}}\frac{(2+\xi)}{(1+\xi)^2}$$

$$\times\left[F\left(n+1,\tfrac{3}{2},n+\tfrac{5}{2};\rho\right)-\frac{1}{n+\frac{5}{2}}F\left(n+1,\tfrac{5}{2},n+\tfrac{7}{2};\rho\right)\right]\Big/\tilde{F}(n,\rho),$$

$$b_n(\mathrm{T})=-\frac{3n(n+1)(n+2)(n+3)}{2n+3}(1+\xi)^{-1}$$

$$\times\left[F\left(n+1,-\tfrac{1}{2},n+\tfrac{7}{2};\rho\right)+\frac{2}{(2n+7)}F\left(n+1,\tfrac{1}{2},n+\tfrac{9}{2};\rho\right)\right.$$

$$-\frac{13}{3}\frac{1}{(2n+7)(n+3)}F\left(n+1,\tfrac{3}{2},n+\tfrac{9}{2};\rho\right)$$

$$\left.+\frac{16}{3}\frac{1}{(2n+7)(n+3)(n+2)}F\left(n+1,\tfrac{5}{2},n+\tfrac{9}{2};\rho\right)\right]\Big/\tilde{F}(n,\rho).$$

that for the calculations to be reliable Q_0^2 has to be chosen unequal to zero. The normalization factors $A^J(n)$ which correspond to the moments of the bare loop graph in (2.3) are also tabulated in table 1.

By inserting (2.9) into (2.10) we will find an expression for $M_n^J(\xi)$ in terms of the hadronic masses and couplings, in particular we can write

$$M_n^J(\xi)=\frac{9}{4}\frac{m_\mathrm{R}^2}{g_\mathrm{R}^2}\frac{1}{\left(m_\mathrm{R}^2+Q_0^2\right)^{n+1}}\left[1+\delta_n^J(\xi)\right], \tag{2.13}$$

where m_R and g_R are the parameters of the lowest lying resonance and $\delta_n^J(\xi)$ contains the contributions from higher resonances and the continuum. For high n $\delta_n^J(\xi)$ will go to zero due to the factors $((m_\mathrm{R}^2+Q_0^2)/(m_{\mathrm{R}'}^2+Q_0^2))^{n+1}$ it contains and because $m_{\mathrm{R}'}>m_\mathrm{R}$. So, from a certain value of n onwards $M_n^J(\xi)$ in (2.13) will be practically equal to the contribution of the first resonance. This will be even more the case if we consider ratios of the moments (2.13),

$$r_n^J(\xi)=\frac{M_n^J(\xi)}{M_{n-1}^J(\xi)}=\frac{1}{m_\mathrm{R}^2+Q_0^2}\frac{1+\delta_n^J(\xi)}{1+\delta_{n-1}^J(\xi)}, \tag{2.14}$$

which immediately give the mass m_R of the lowest resonance if we are at sufficiently high n where $\delta_n^J(\xi)\simeq\delta_{n-1}^J(\xi)$. It can easily be shown that for the charmed vector sector where a large number of resonances is known and $M_n^V(\xi)$ can be calculated to high accuracy, the J/ψ gives about 50% of $M_1^V(\xi=0)$ and already 90% of $M_4^V(\xi=0)$. For $\xi \neq 0$ the convergence of $\delta_n^J(\xi)$ to zero will be less fast and for very large ξ it will be difficult to extract the parameters of the lowest lying resonances from (2.13) or (2.14). Large (spacelike) Q_0^2 means moving away from the resonance region in the Q^2 plane up to a point from where it will be impossible to distinguish individual resonances.

The asymptotic freedom side of the moment equation, (2.11), is valid for n and ξ such that a_n and b_n are not too large; a_n and b_n grow with n but decrease with ξ as will be seen in sects. 3 and 4. For a certain value of ξ there will be a range of values for which the experimental side of the moment equation [or the ratio (2.14)] is dominated by a single resonance, while the asymptotic freedom side is still valid. For small n there occurs a breakdown as the effect of higher states should become noticeable, and at high n the expansion to first order in α_s (for the perturbative as well as the non-perturbative contributions) does not hold any longer. The stability region in n will change with ξ and one should study the stability of the moments for growing ξ. In sects. 6 and 7 we will apply this method to the charmonium and upsilon systems to extract the parameters of the lowest lying resonance in each channel.

A second method is a variation of the first in which ratios of moments like (2.14) for two different currents are considered. Since it is reasonable to believe that the parameters δ_n^J and $\delta_n^{J'}$ are very similar, in particular in the same partial wave, e.g. for the $^3S_1$ and $^1S_0$ channels or the three P-wave channels of charmonium, the corrections from higher resonances tend to cancel each other and we have

$$\frac{r_n^J}{r_n^{J'}} \simeq \frac{m_{R'}^2 + Q_0^2}{m_R^2 + Q_0^2}\,. \tag{2.15}$$

This method is, unlike the first, incapable of giving the absolute normalization of the levels but should give the ratios of the levels with high accuracy. From dimensional considerations [see (2.10)] it is clear that $A^J(n)$ must be proportional to $1/(m_c^2 + Q_0^2)^n$; therefore r_n^J is proportional to $1/(m_c^2 + Q_0^2)$ and very sensitive to the quark mass m_c, while the ratio (2.15) is independent of m_c.

3. Calculation of the perturbative part of $\Pi(q^2)$

The diagrams which contribute to the perturbative part of $\Pi(q^2)$, i.e. to the Wilson coefficient of the identity operator in the OPE (2.2), are given in fig. 1. The diagrams with four-point vertices are only present for the $J=1^{+-}$ and $J=2^{++}$ currents of (2.6), which contain derivative couplings. We have calculated the

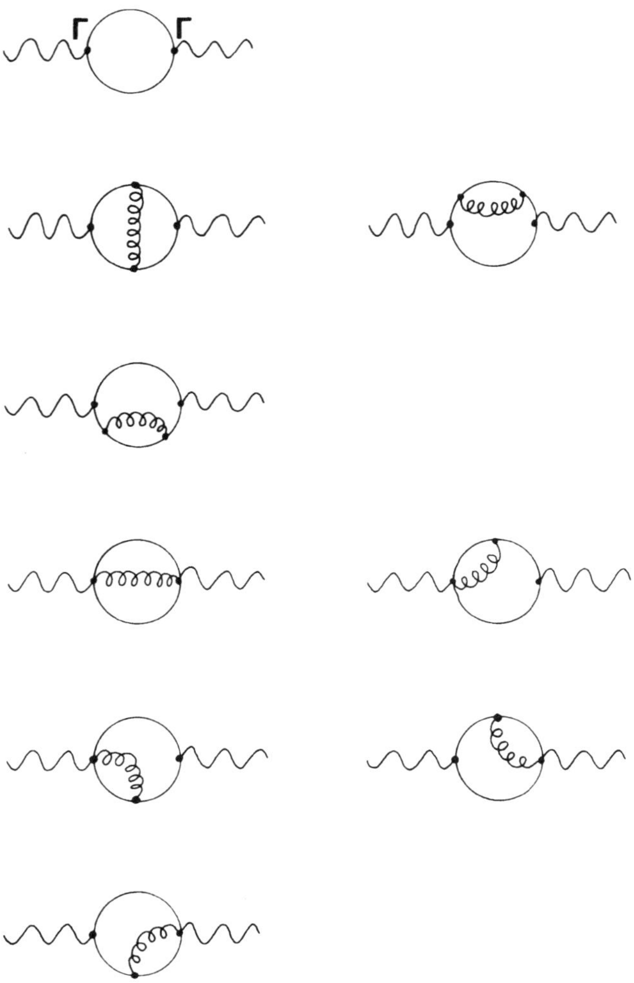

Fig. 1. Diagrams that contribute to the vacuum polarization to first order in α_s. Curly lines depict gluons, continuous lines quarks, and wavy lines currents. Γ, as in the first diagram, is the current vertex. Graphs with 4-point vertices are only present for cases with derivative couplings.

imaginary part of each diagram, following Schwingers method [14] of on-mass-shell renormalization.

For the vector case $J_V = \bar{q}\gamma_\mu q$ we can simply copy Schwingers result and use his interpolation formula:

$$\operatorname{Im}\Pi^{(V)}_{\text{pert}}(s) = \frac{1}{8\pi} v(3 - v^2)\left[1 + \tfrac{4}{3}\alpha_s\left(\frac{\pi}{2v} - \tfrac{1}{4}(v+3)\left(\tfrac{1}{2}\pi - \frac{3}{4\pi}\right)\right)\right]. \tag{3.1}$$

For the pseudoscalar current $J_P = i\bar{q}\gamma_5 q$ we get for the imaginary part

$$\mathrm{Im}\,\Pi^{(P)}_{\mathrm{pert}}(s) = \frac{3sv}{8\pi}\left\{1+\frac{4}{3}\frac{\alpha_s}{\pi v}\left[(1+v^2)\right.\right.$$
$$\times\left(\tfrac{1}{6}\pi^2 + \ln\frac{1+v}{1-v}\ln\frac{1+v}{2}\right.$$
$$\left. + 2l\left(\frac{1-v}{1+v}\right) + 2l\left(\frac{1+v}{2}\right) - 2l\left(\frac{1-v}{2}\right) - 4l(v) + l(v^2)\right)$$
$$+\left(\tfrac{19}{16} - \tfrac{7}{8}v^2 + \tfrac{3}{16}v^4 - v(3-v)\right)\ln\frac{1+v}{1-v} + 6v\ln\frac{1+v}{2}$$
$$\left.\left. - 4v\ln v + \tfrac{29}{8}v - \tfrac{3}{8}v^3\right]\right\}, \tag{3.2}$$

where $v = (1-4m_c^2/s)^{1/2}$ and $l(x) \equiv -\int_0^x \mathrm{d}t\, t^{-1}\ln(1-t)$. Our result differs slightly from the result given by Shifman et al. in ref. [7] in the polynomial part. The singularity in (3.2) for $v \to 1$ ($m_c \to 0$ or $s \to \infty$) is a consequence of the fact that the pseudoscalar current is not conserved. The coefficient $\frac{3}{2}$ multiplying this logarithmic singularity equals the anomalous dimension for this current.

To demonstrate the calculation in some detail, we take the $J = 1^{+-}$ P-wave current as an example.

In lowest order, i.e. the bare loop diagram in fig. 1, the imaginary part of the polarization function for this current takes the form

$$\mathrm{Im}\,\Pi^{(0)}_{\mathrm{pert}}(q^2) = q^2\frac{1}{8\pi}v^3. \tag{3.3}$$

The calculation of the first-order correction to $\mathrm{Im}\,\Pi^{(A')}(q^2)$ can be separated into two parts, namely the vertex correction part (virtual gluon) and the gluon emission part. The wave-function renormalization is q^2 independent and included in the vertex correction as a normalization factor.

3.1. THE VERTEX CORRECTION

The axial vector vertex is corrected as follows:

$$\bar{\psi}\overleftrightarrow{\partial}_\mu\gamma_5\psi \to \bar{\psi}\overleftrightarrow{\partial}_\mu\gamma_5\psi F_{A'}(q^2), \tag{3.4}$$

where the form factor $F_{A'}(q^2)$ is given by

$$F_{A'}(q^2) = F_{A'}(0) + \frac{4}{3}\frac{\alpha_s}{\pi}\frac{1}{2v}\left[f(v) - 2v + (1+v^2)\ln\frac{1+v}{1-v}\right]. \tag{3.5}$$

The function $f(v)$ commonly appears in any type of vertex and is defined by

$$f(v) = v\int_0^1 \mathrm{d}v' \frac{1}{v'^2 - v^2}(1 + v'^2)\ln\left(\frac{4m^2}{\lambda^2}\frac{v'^2}{1-v'^2}\right)$$

$$= \left[(1+v^2)\ln\frac{1+v}{1-v} - 2v\right]\ln\frac{\lambda}{m}$$

$$+\tfrac{1}{2}(1+v^2)\left[\pi^2 + \tfrac{1}{2}\ln^2\frac{1+v}{2} - \tfrac{1}{2}\ln^2\frac{1-v}{2}\right.$$

$$\left. + l\left(\frac{1+v}{2}\right) - l\left(\frac{1-v}{2}\right) - 4l(v) + l(v^2)\right], \quad (3.6)$$

where λ is the virtual gluon mass and the Spence function $l(x)$ is defined above.

In terms of this function, the virtual gluon correction to $\mathrm{Im}\,\Pi^{(A')}(q^2)$ can be written as

$$\mathrm{Im}\,\delta\Pi^{\mathrm{virt}}_{\mathrm{pert}}(q^2) = \mathrm{Im}\,\Pi^{(0)}_{\mathrm{pert}}(q^2) \times \frac{4}{3}\frac{\alpha_s}{\pi}\frac{1}{v}\left[f(v) - 2v + (1+v^2)\ln\frac{1+v}{1-v}\right], \quad (3.7)$$

where $\mathrm{Im}\,\Pi^{(0)}_{\mathrm{pert}}$ is the bare loop part given by (3.3).

3.2. THE REAL GLUON EMISSION

The one gluon emission contribution is obtained by calculating the diagrams shown in fig. 2, which are obtained by cutting the relevant diagrams in fig. 1 in all possible ways, and is given by the formula

$$q^2\,\mathrm{Im}\,\delta\Pi^{\mathrm{real}}_{\mathrm{pert}}(q^2) = \tfrac{2}{3}g^2\left\langle\left[(q^2)^2 v^2(-1)\left(\frac{p_1}{p_1\cdot k} - \frac{p_2}{p_2\cdot k}\right)^2\right.\right.$$

$$\left.\left. + 2q^2\frac{(p_1 + p_2\cdot k)^2}{(p_1\cdot k)(p_2\cdot k)} - 16(p_1 + p_2\cdot k)\right]\right\rangle \quad (3.8)$$

Here, we use the notation $\langle\cdots\rangle$ to indicate the three-body phase-space integral

$$\langle\cdots\rangle = \int\int\int \frac{\mathrm{d}^3k}{2\nu(2\pi)^3}\frac{\mathrm{d}^3p_1}{2E_1(2\pi)^3}\frac{\mathrm{d}^3p_2}{2E_2(2\pi)^3}(2\pi)^4\delta^{(4)}(q - k - p_1 - p_2)\ldots. \quad (3.9)$$

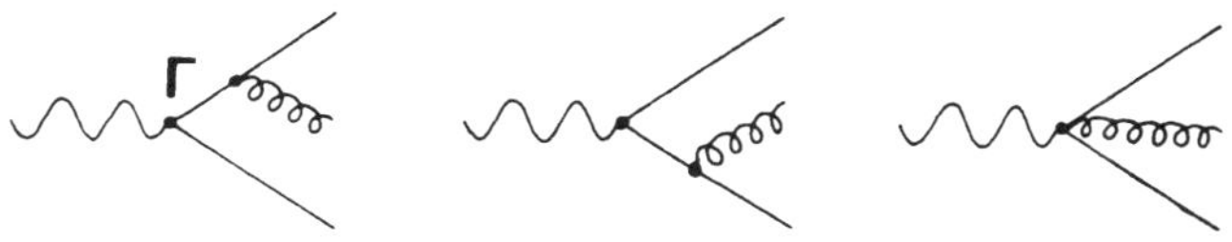

Fig. 2. Diagrams that contribute to the one-gluon emission part of $\mathrm{Im}\,\Pi_{\mathrm{pert}}$. They are obtained by cutting the relevant diagrams in fig. 1 in all possible ways.

In (3.8) the first term which is common to all currents, gives the result

$$\left\langle (-1)\left(\frac{p_1}{p_1\cdot k}-\frac{p_2}{p_2\cdot k}\right)^2\right\rangle=\frac{1}{8\pi}\frac{1}{4\pi^2}\left\{\mathfrak{A}(v)-f(v)+4v+\ln\frac{1+v}{1-v}\right\} \tag{3.10}$$

with $f(v)$ given by (3.6) and $\mathfrak{A}(v)$ defined as

$$\begin{aligned}\mathfrak{A}(v)=(1+v^2)\Bigg[&\tfrac{1}{6}\pi^2+\ln\frac{1+v}{1-v}\ln\frac{1+v}{2}+2l\left(\frac{1-v}{1+v}\right)\\&+2l\left(\frac{1+v}{2}\right)-2l\left(\frac{1-v}{2}\right)-4l(v)+l(v^2)\Bigg]\\&+2v\ln\frac{(1-v^2)^{3/2}}{8v^2}\,.\end{aligned} \tag{3.11}$$

The other two terms in (3.8) are specific to each current and result in polynomial terms and terms proportional to $\ln[(1+v)/(1-v)]$.

Summing the real gluon contribution (3.8) and the vertex correction (3.7), we get for the total first-order correction to $\mathrm{Im}\,\Pi^{(\mathrm{A}')}_{\mathrm{pert}}(q^2)$,

$$\begin{aligned}\mathrm{Im}\,\Pi^{(1)}_{\mathrm{pert}}(q^2)=\mathrm{Im}\,\Pi^{(0)}_{\mathrm{pert}}(q^2)v^{-3}\frac{4}{3}\frac{\alpha_s}{\pi}\\ \times\left[v^2\mathfrak{A}(v)+P^{(\mathrm{A}')}(v)\ln\frac{1+v}{1-v}+Q^{(\mathrm{A}')}(v)\right].\end{aligned} \tag{3.12}$$

The polynomial functions $P^{(\mathrm{A}')}(v)$ and $Q^{(\mathrm{A}')}(v)$ are given by

$$P^{(\mathrm{A}')}(v)=\tfrac{13}{16}+\tfrac{28}{16}v^2+\tfrac{17}{16}v^4-\tfrac{2}{16}v^6, \tag{3.13a}$$

$$Q^{(\mathrm{A}')}(v)=-\tfrac{13}{8}v+\tfrac{47}{24}v^3+\tfrac{1}{4}v^5, \tag{3.13b}$$

respectively.

By performing the same steps, a form similar to (3.12) can be obtained for the other P-wave currents with the spin-dependent polynomial functions $P^{(\Gamma)}(v)$ and $Q^{(\Gamma)}(v)$ listed below. For the scalar current

$$P^{(S)}(v) = \tfrac{3}{16} + \tfrac{34}{16}v^2 - \tfrac{13}{16}v^4, \tag{3.14a}$$

$$J^{PC} = 0^{++},$$

$$Q^{(S)}(v) = -\tfrac{3}{8}v + \tfrac{45}{8}v^3, \tag{3.14b}$$

for the axial current

$$P^{(A)}(v) = \tfrac{21}{32} + \tfrac{59}{32}v^2 + \tfrac{19}{32}v^4 - \tfrac{3}{32}v^6, \tag{3.15a}$$

$$J^{PC} = 1^{++},$$

$$Q^{(A)}(v) = -\tfrac{21}{16}v + \tfrac{46}{16}v^3 + \tfrac{3}{16}v^5, \tag{3.15b}$$

and for the tensor current*

$$P^{(T)}(v) = \tfrac{1}{45} + \tfrac{313}{360}v^2 + \tfrac{23}{96}v^4 - \tfrac{409}{1440}v^6 + \tfrac{31}{1440}v^8 - \frac{v^{10}}{480}, \tag{3.16a}$$

$$J^{PC} = 2^{++},$$

$$Q^{(T)}(v) = -\tfrac{2}{45}v - \tfrac{47}{540}v^3 + \tfrac{229}{10800}v^5 - \tfrac{1}{24}v^7 + \tfrac{1}{240}v^9. \tag{3.16b}$$

To take the moments for each P-wave current by inserting $\operatorname{Im}\Pi^J_{\text{pert}}(s)$ in (2.10), it is convenient to introduce the interpolation formula for the common part in each P-wave expression derived by Schwinger [14]:

$$v^2\mathfrak{A}(v) + P(v)\ln\frac{1+v}{1-v} + Q(v) \sim v^2\left[\tfrac{1}{2}\pi^2 - v\frac{1+v}{2}\left(\frac{1}{2}\pi^2 - 3\right)\right] \tag{3.17}$$

with the function $\mathfrak{A}(v)$ of (3.11) and the polynomial functions $P(v)$ and $Q(v)$ defined as

$$P(v) = 5\left(\frac{1+v^2}{2}\right)^2 - 2, \tag{3.18a}$$

$$Q(v) = +\tfrac{3}{2}v(1+v^2). \tag{3.18b}$$

* For the tensor ($J^{PC} = 2^{++}$) current, the formula corresponding to (3.12) is, $\operatorname{Im}\Pi^{(1)} = \operatorname{Im}\Pi^{(0)}[\tfrac{1}{3}v^3(1 - \tfrac{2}{5}v^2)]^{-1}\,\tfrac{4}{3}(\alpha_s/\pi)[\tfrac{1}{3}v^2(1-\tfrac{2}{5}v^2)\mathfrak{A}(v) + P^{(T)}(v)\ln[(1+v)/(1-v)] + Q^{(T)}(v)]$.

We have first calculated the moments at $Q_0^2=0$, i.e. separating off the bare loop coefficients $A^J(n)$ at $Q_0^2=0$, we get the coefficients $a_n(J,\xi=0)$ as defined by (2.11). Both these coefficients have been tabulated in ref. [9]. Some of these expressions were already obtained by Shifman et al. [5]. Above, in the calculation of the imaginary parts we have used the on-mass-shell renormalization prescription of Schwinger, i.e. the renormalization point is $p^2=+m_c^2$. It appears reasonable to choose the mass renormalization such that higher order corrections in α_s are minimal. With this in mind, we have replaced $m_c(p^2=+m_c^2)$ by the running quark mass, and (for $Q_0^2=0$) approximated this by $m_c(p^2=-m_c^2)$, in view of the fact that in quark loops virtual momenta $p^2\simeq -m_c^2$ give the dominant contribution. We have used the Landau gauge and corrected for the replacement by the running quark mass by a factor $-4n\ln 2/\pi$ in $a_n(J,\xi=0)$ (see ref. [4] for the details of this procedure). For $Q_0^2\neq 0$, the quark mass has to be renormalized at $p^2=-(\xi+1)m_c^2$, which gives the correction factor proportional to $\ln(2+\xi)$ in table 1. Similarly α_s has to be replaced by the running quark mass in $\operatorname{Im}\Pi^J_{\text{pert}}$, and has been approximated by $\alpha_s(4m_c^2)$ in performing the integral (2.10) at $Q_0^2=0$.

The coefficients $a_n(J,\xi)$ for $\xi\neq 0$ ($Q_0^2\neq 0$) have been listed in table 1. They can, of course, be obtained from $a_n(J,\xi=0)$ by

$$a_n(J,\xi)=\sum_{k=n}^{\infty}\frac{A^J(k,\xi=0)}{A^J(n,\xi)}a_k(J,0)\frac{k!}{n!(k-n)!}(-\xi)^{k-n}, \tag{3.19}$$

and $A^J(n,\xi)$ follows from $A^J(n,\xi=0)$ by a similar formula. The coefficients $A^J(n,\xi)$ have also been tabulated in table 1. We emphasize that the parameter m in $A^J(n,\xi)$ is the running quark mass $m\equiv m_c(p^2=-(\xi+1)m_c^2)$, and α_s in (2.11) is the running coupling constant $\alpha_s=\alpha_s((\xi+1)4m_c^2)$.

4. Calculation of the non-perturbative part

In this section we will calculate the coefficient function C_G of the gluon field strength operator $O_G=G^a_{\mu\nu}G^a_{\mu\nu}$. To begin with, let us go back to the starting point of the operator expansion. After averaging over the spin of a product of currents the operator expansion has, in general, the form

$$i\int d^4x\, e^{iqx}\,T(J_\Gamma(x)J_\Gamma(0))=C_I(q)I+\cdots$$
$$+C_G(q)G^a_{\mu\nu}G^a_{\mu\nu}+D_G(q)\left\{G^a_{\mu\lambda}q_\lambda G^a_{\mu\nu}q_\nu-\tfrac{1}{4}q^2G^a_{\mu\nu}G^a_{\mu\nu}\right\}. \tag{4.1}$$

In the above expansion, we are interested in the operator $G^a_{\mu\nu}G^a_{\mu\nu}$ which can survive even after sandwiching between states with vacuum quantum numbers. Since (4.1) is an operator identity we can choose any state to single out a particular coefficient function. For the present purpose, it is most convenient to take the matrix element of (4.1) between one-gluon states. The left-hand side of (4.1) is then the forward gluon scattering amplitude on a colour singlet current,

$$T_{\alpha\beta}\left((q+k)^2,q^2,k^2\right)=\left\langle k,\alpha\middle|i\int d^4x\,e^{iqx}\mathrm{T}(J_\Gamma(x)J_\Gamma(0))\middle|k,\beta\right\rangle, \quad (4.2)$$

which by gauge invariance can be decomposed into two invariant amplitudes C and D:

$$T_{\alpha\beta}\left((q+k)^2,q^2,k^2\right)=L_{1\alpha\beta}C\left((q+k)^2,q^2,k^2\right)+L_{2\alpha\beta}D\left((q+k)^2,q^2,k^2\right), \quad (4.3)$$

where the Lorentz tensors $L_{i\alpha\beta}$ have the form

$$L_{1\alpha\beta}=2\left(k^2g_{\alpha\beta}-k_\alpha k_\beta\right)=\langle k,\alpha|G^a_{\mu\nu}G^a_{\mu\nu}|k,\beta\rangle, \quad (4.4a)$$

$$\begin{aligned}L_{2\alpha\beta}&=q^2k_\alpha k_\beta-k\cdot qk_\alpha q_\beta-k\cdot qq_\alpha k_\beta+k^2q_\alpha q_\beta\\&\quad-\tfrac{1}{2}q^2\left(k^2g_{\alpha\beta}-k_\alpha k_\beta\right)\\&=\langle k,\alpha|\left\{G^a_{\mu\lambda}q_\lambda G^a_{\mu\nu}q_\nu-\tfrac{1}{4}q^2G^a_{\mu\nu}G^a_{\mu\nu}\right\}|k,\beta\rangle. \quad (4.4b)\end{aligned}$$

Comparing (4.3) with the right-hand side of (4.1) we have

$$C_\mathrm{G}(q^2)=C\left((q+k)^2,q^2,k^2\right)|_{k_\mu=0}, \quad (4.5a)$$

$$D_\mathrm{G}(q^2)=D\left((q+k)^2,q^2,k^2\right)|_{k_\mu=0}. \quad (4.5b)$$

By using the relation

$$\langle 0|G^a_{\alpha\beta}G^a_{\gamma\delta}|0\rangle=\tfrac{1}{12}(g_{\alpha\gamma}g_{\beta\delta}-g_{\alpha\delta}g_{\gamma\beta})\langle 0|G^a_{\alpha\beta}G^a_{\alpha\delta}|0\rangle,$$

one can see that the second term in (4.1) will give zero when sandwiched between states with vacuum quantum numbers. Therefore, we are only interested in the amplitudes C_G.

The amplitude $C(s,q^2,k^2)$ can be expressed in its imaginary part by a dispersion relation

$$C(s,q^2,k^2)=\frac{1}{\pi}\int ds'\frac{1}{s'-s}\,\mathrm{Im}\,C(s',q^2,k^2). \quad (4.6)$$

In lowest order in g^2, the Feynman diagrams relevant for $\operatorname{Im} C$ are

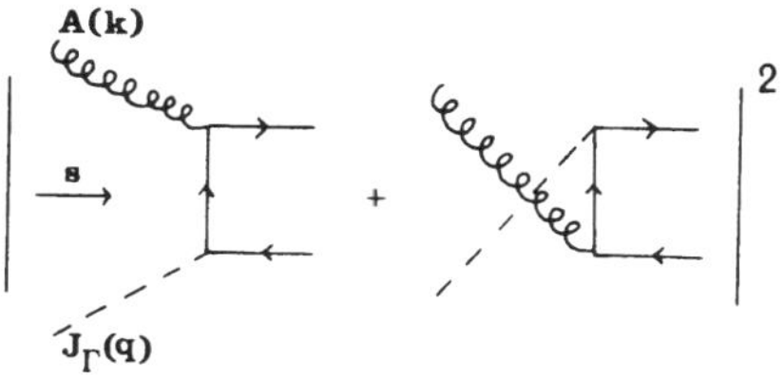

(4.7)

This is very similar to the gluon emission contribution to the perturbative part calculated in sect. 3 (before the three-body phase-space integration has been carried out). For currents with derivative couplings we get an extra graph with a four-point vertex in (4.7).

For example, $\operatorname{Im} C$ in the pseudoscalar case has the form

$$\operatorname{Im} C^{(\mathrm{P})}(s,q^2,0) = \tfrac{1}{2}\alpha_s \frac{q^2}{(s-q^2)^2}\left[-\ln\frac{1+v}{1-v} + \frac{2q^2 s}{(s-q^2)^2}\left(\frac{5v-v^3}{1-v^2} - \frac{5+v^2}{2}\ln\frac{1+v}{1-v}\right)\right]. \tag{4.8}$$

Then, using the dispersion relation (4.6), we can obtain the coefficient function $C_{\mathrm{G}}^{(\mathrm{P})}(q^2)$ as a power expansion in $q^2/4m_c^2$:

$$\begin{aligned} C_{\mathrm{G}}^{(\mathrm{P})}(q^2) &= C(q^2,q^2,0) \\ &= \frac{1}{\pi}\int \mathrm{d}s' \frac{1}{s'-q^2}\operatorname{Im} C(s',q^2,0) \\ &= \sum_n A_n^{(\mathrm{P})} b^{(\mathrm{P})}(n)\left(\frac{q^2}{4m_c^2}\right)^n, \end{aligned} \tag{4.9}$$

which is the same as calculating the moments at $Q_0^2 = 0$ by means of formula (2.10). $A_n^{(\mathrm{P})}$ is again the bare loop coefficient and $b^{(\mathrm{P})}(n)$ is the non-perturbative coefficient as defined by (2.11) for $\xi = 0$. We have carried out this programme for all currents listed in (2.6). The coefficients $b^{(J)}(n, \xi = 0)$ have been tabulated in ref. [9]. For some currents these expressions were already obtained by Shifman et al. [5]. In table 1 we have listed the coefficients at $\xi \neq 0$ for all currents. They can be obtained from the coefficients at $\xi = 0$ by a formula similar to (3.19).

The non-perturbative coefficients can also be calculated in the following way. All relevant diagrams are obtained by cutting the gluon line in the diagrams of fig. 1. The full diagrams are then calculated after performing an expansion around $k=0$, following the suggestion in ref. [5]. The calculations are quite complex, but can be handled completely by computer and provide a useful check to the first method.

5. Consistency checks

The calculations presented in the last two sections leading to the expressions of table 1 are rather complicated and cumbersome. Moreover, as mentioned, we disagree with some of the expressions given by SVZ. Therefore, we consider it useful to present in this section several checks of our calculations for both the perturbative and non-perturbative parts.

For the perturbative part we confront our results with those of the non-relativistic one-gluon exchange potential at low energy on one hand and with the solutions of the renormalization group equation for the current correlation function at high energy on the other hand.

For the non-perturbative part, as explained in sect. 4, the coefficients $C_G(q^2)$ are related to the gluon-current scattering amplitude. Therefore, we can check our results by looking at the low energy behaviour of these amplitudes.

For the tensor current ($^3P_2$) a separate check has been made by applying the projection operator $g_{\rho\sigma}$ to the vertex $\bar{\psi}\overleftrightarrow{\partial}^{\rho}\gamma^{\sigma}\psi$ which reproduces the result of the scalar current ($^3P_0$).

5.1. LOW-ENERGY LIMIT

The real gluon emission contribution to $\mathrm{Im}\,\Pi^{(\Gamma)}_{\mathrm{pert}}$ which involves a three-body phase-space integration is less singular at $v \sim 0$ than the two-body phase-space integration of the $\bar{q}q$ intermediate state. In fact the behaviour of $\mathrm{Im}\,\delta\Pi^{\mathrm{real}}_{\mathrm{pert}}$ at $v \sim 0$ is (see subsect. 3.2)

$$\mathrm{Im}\,\delta\Pi^{\mathrm{real}}_{\mathrm{pert}} \sim \mathrm{Im}\,\Pi^{(0)}_{\mathrm{pert}} \times \mathrm{O}\left(\alpha_s v^2 \ln v\right). \tag{5.1}$$

The virtual gluon contribution [see (3.7)], on the contrary, behaves like the Coulomb correction

$$\mathrm{Im}\,\delta\Pi^{\mathrm{virt}}_{\mathrm{pert}} \sim \mathrm{Im}\,\Pi^{(0)}_{\mathrm{pert}} \times \frac{4}{3}\frac{\pi\alpha_s}{2v}. \tag{5.2}$$

This can be seen by tracing back the behaviour of the form factor $F^{(\Gamma)}(q^2)$ near threshold,

$$F^{(\Gamma)}(q^2) = F^{(\Gamma)}(0) + \frac{4}{3}\frac{\alpha_s}{\pi}\frac{1}{2v}\left[f(v) + g^{(\Gamma)}(v)\right], \tag{5.3}$$

with the spin-independent function $f(v)$ of eq. (3.6) which behaves like $f(v) \sim (1+v^2)[\frac{1}{2}\pi^2 - 2v]$ and the spin-dependent function $g^{(\Gamma)}(v)$. For the various P-wave currents these spin-dependent functions $g^{(\Gamma)}$ are given by

$$g^{(\mathrm{S})}(v) = +2v + (1-v^2)\ln\frac{1+v}{1-v} \sim 4v - \tfrac{4}{3}v^3,$$

$$g^{(\mathrm{A})}(v) = -v + \frac{2+v^2}{2}\ln\frac{1+v}{1-v} \sim v + \tfrac{7}{3}v^3,$$

$$g^{(\mathrm{A}')}(v) = 2v + (1+v^2)\ln\frac{1+v}{1-v} \sim 0 + \tfrac{8}{3}v^3. \tag{5.4}$$

To compare our results [including the tensor current ($^3\mathrm{P}_2$)] with the one-gluon exchange potential, we have to go back to the imaginary part of each form factor and take the low-energy limit properly. In particular, the tensor vertex is corrected as

$$\gamma_\mu \partial_\nu \to \gamma_\mu \partial_\nu F_1^{(\mathrm{T})}(q^2) + \frac{1}{4m}[\gamma_\mu, \gamma_\rho] q^\rho \partial_\nu F_2^{(\mathrm{T})}(q^2). \tag{5.5}$$

At low energy we find the following relation between these two form factors and the form factor for the axial current $^1\mathrm{P}_1$:

$$\mathrm{Im}\left(F_{\mathrm{T}}^{(1)} - F_{\mathrm{T}}^{(2)}\right) - \tfrac{2}{5}v^2 \,\mathrm{Im}\, F_{\mathrm{T}}^{(2)} = \mathrm{Im}\, F^{(\mathrm{A}')} + \frac{g^2}{16\pi}\frac{4}{3}v \times \left(-\tfrac{7}{5}\right). \tag{5.6a}$$

In addition we have for the scalar $^3\mathrm{P}_0$ and axial $^3\mathrm{P}_1$:

$$\mathrm{Im}\, F^{(\mathrm{S})} = \mathrm{Im}\, F^{(\mathrm{A}')} + \frac{g^2}{16\pi}\frac{4}{3}v \times (4), \tag{5.6b}$$

$$\mathrm{Im}\, F^{(\mathrm{A})} = \mathrm{Im}\, F^{(\mathrm{A}')} + \frac{g^2}{16\pi}\frac{4}{3}v \times (1). \tag{5.6c}$$

These spin-dependent parts [relative to $\mathrm{Im}\, F^{(\mathrm{A}')}(q^2)$] should be compared with the respective spin-dependent parts of the non-relativistic one-gluon exchange potential. There the P-state splittings originate from spin-orbit and spin-tensor terms [2] (the spin-spin term has a δ-function behaviour and does not contribute to P-states). The relative spacings agree with the coefficients of v in (5.4) and (5.6).

The spin-independent part for P-states is reflected in the coefficient $\tilde{a}_n$ in table 1. This is equal for all P-waves (except for the tensor) and originates from Schwinger's interpolation formula [14].

We also checked the spin-dependent part of the S-wave currents in the same way, introducing the scalar quark system as a theoretical tool to be able to separate clearly the spin independent parts. In this case there are only contributions from spin-spin terms in the potential model approach, and again our numbers agreed.

5.2. HIGH-ENERGY LIMIT

To check the high-energy limit of our calculations, it is convenient to compare them with the solutions of the renormalization group equation for the current correlation functions.

The renormalization group equation for the correlation function of the current $J^{(\Gamma)}(x)=\bar{\psi}(x)\Gamma\psi(x)$ can be written in the form [15],

$$\left[\mu\frac{\partial}{\partial\mu}+\beta(g)\frac{\partial}{\partial g}+\gamma_m(g)m\frac{\partial}{\partial m}+2\gamma^{\Gamma}(g)\right]\Pi^{(\Gamma)}(q^2,g^2,m^2,\mu^2)=D^{\Gamma}(g), \quad (5.7)$$

with $\beta(g)$ the renormalization group function, the anomalous dimensions $\gamma_m(g)$ for the mass and $\gamma^{\Gamma}(g)$ for the current. The inhomogeneous term $D^{\Gamma}(g)$ is calculated by the subtractive renormalization part in a similar manner as for the vector current case [16]. By solving the above equation in the space-like q^2 region and taking the imaginary part after analytic continuation, we obtain the following form for the current polarization function to leading order and at high energy:

$$\operatorname{Im}\Pi(q^2,m^2,g^2,\mu^2)\to\operatorname{Im}\Pi^{(0)}(q^2)\times\left\{1+g^2b_0\ln(q^2/\mu^2)\right\}^{\gamma_0^{\Gamma}/b_0}. \quad (5.8)$$

Here, b_0 and γ_0^{Γ} are the leading coefficients of the β-function and the anomalous dimension $\gamma^{\Gamma}(g)$:

$$\beta(g)=-b_0g^3+\cdots,$$
$$\gamma^{\Gamma}(g)=\gamma_0^{\Gamma}g^2+\cdots. \quad (5.9)$$

To compare these results with our calculations, we rewrite the leading log summation part as

$$\left\{1+g^2b_0\ln(q^2/\mu^2)\right\}^{\gamma_0^{\Gamma}/b_0}=\left\{1+g^2b_0\ln(4m^2/\mu^2)\right\}^{\gamma_0^{\Gamma}/b_0}$$
$$\times\left[1+g^2(4m^2)\gamma_0^{\Gamma}\ln(q^2/4m^2)+\cdots\right], \quad (5.10)$$

with the running coupling constant $g^2(4m^2)$,

$$g^2(4m^2)=g^2/\left(1+g^2b_0\ln(4m^2/\mu^2)\right).$$

The anomalous dimensions $\gamma^{\Gamma}(g)$ for the various currents have been calculated [17] and are given by

$$\begin{array}{lcccccc} \Gamma: & \gamma_\mu & \gamma_\mu\gamma_5 & 1 & \gamma_5 & \overleftrightarrow{\partial}_\mu\gamma_5 & \overleftrightarrow{\partial}_\mu\gamma_\nu \\ 4\pi^2\gamma_0^{\Gamma}: & 0 & 0 & -2 & -2 & \frac{2}{3} & \frac{16}{9}. \end{array} \tag{5.11}$$

The anomalous dimensions are equal to the coefficient of the logarithmic singularity for $v \to 1$ in $\operatorname{Im}\Pi_{\text{pert}}(q^2)$. It can be verified that the expressions in sect. 3 indeed reproduce the numbers (5.11). In principle they should also be equal to the residue for $n \to 0$ of the coefficient a_n in table 1. However, to be able to give closed expressions in terms of hypergeometric functions at $Q_0^2 \neq 0$, we had to rearrange various terms and made some approximations. We have checked that in the original expressions at $Q_0^2 = 0$ the residues for $n \to 0$ agree with (5.11).

We can use the relations between the current polarization functions to check the consistency of our results. For example, up to g^2, we have the following relation between the axial vector, vector and pseudoscalar current polarization functions:

$$\operatorname{Im}\Pi^{(\mathrm{A})}_{\text{pert}}(q^2) = \operatorname{Im}\Pi^{(\mathrm{V})}_{\text{pert}}(q^2) - (1-v^2)\operatorname{Im}\Pi^{(\mathrm{P})}_{\text{pert}}(q^2)$$
$$+\frac{1}{8\pi}\frac{\alpha_s}{\pi}\frac{4}{3}\left[(1-v^2)^3\ln\frac{1+v}{1-v} + \tfrac{1}{2}v(1-v^2)^2\right]. \tag{5.12}$$

In the chiral limit ($v \to 1$) this reduces to the well-known result $\operatorname{Im}\Pi^{(\mathrm{A})}(q^2) = \operatorname{Im}\Pi^{(\mathrm{V})}(q^2)$. For the bare loop only ($\alpha_s \equiv 0$ in 5.12) we find an extra piece proportional to $m_c^2 \operatorname{Im}\Pi^{(\mathrm{P})}$ since the divergence of the axial current is proportional to the pseudoscalar current. Moreover, to order α_s all corrections to the first two terms of (5.12) should be of order m_c^4 or higher. It can easily be seen that this is indeed the case in (5.12).

5.3. THE NON-PERTURBATIVE PART

To check the calculations of the non-perturbative part, we investigate the amplitude for gluon scattering off the various currents in detail. In the low energy region, the most singular part of the coefficient $C_G^{(\mathrm{L})}(q^2)$ for an L-wave current takes the form

$$C_G^{(\mathrm{L})}(q^2) = 16\frac{g^2}{\pi}(q^2)^{L+1}\int_{4m^2}^{\infty} \mathrm{d}s \frac{1}{(s-q^2)^3}$$
$$\times\left[\frac{s-4m^2}{(s-q^2)^2} - \frac{L}{(s-q^2)} + \frac{L(2L+1)}{4}\frac{1}{(s-4m^2)}\right]$$
$$\times \operatorname{Im}\Pi^{(\mathrm{L})}(s)/s^L. \tag{5.13}$$

Here, $\operatorname{Im}\Pi^{(L)}(s)$ is the imaginary part of the bare loop polarization function of an L-wave current which has the typical threshold behaviour at $q^2 = 4m^2$:

$$\operatorname{Im}\Pi^{(L)}(s)/s^L \sim \left[1 - \frac{4m^2}{s}\right]^{L+1/2} \tag{5.14}$$

Substituting the expression (5.14) into (5.13), we find for the coefficient $C_G^{(L)}(q^2)$

$$C_G^{(L)}(q^2) \sim (4m^2)^{-2}\frac{g^2}{\pi}(2L+1)v^{-6}$$
$$\times \frac{\Gamma(L+\frac{3}{2})}{\Gamma(L-\frac{3}{2})}(q^2)^L v^{2L+1}\ln\frac{1+v}{1-v}, \tag{5.15}$$

or, at high n, the moments can be expressed as

$$C_G^{(L)}(q^2) \sim (4m^2)^{-2}\frac{g^2}{\pi}(2L+1)\sum_n\left(\frac{q^2}{4m^2}\right)^n n^3 A_n^{(L)}, \tag{5.16}$$

with $A_n^{(L)}$ the moment of the bare loop of $\Pi^{(L)}(q^2)$. This threshold behaviour [or n^3 behaviour relative to $A_n^{(L)}$ in the moment] corresponds to an effective exchange potential which grows like r^3.

Our results for the moments of both S-wave and P-wave currents have the n^3 behaviour of (5.16) and also the coefficient $2L+1$, as can be seen from table 1.

We checked that the spin-dependent part of the moment for each P-state is consistent with the level spacing obtained by an effective potential of the form

$$V = a(\bar{\psi}\gamma_\mu\psi)(\bar{\psi}\gamma_\mu\psi)r^3. \tag{5.17}$$

Finally we have the relation between the vector, axial vector and pseudoscalar coefficients

$$A_n^{(A)}b_n^{(A)} = A_n^{(V)}b_n^{(V)} - A_{n+1}^{(P)}b_{n+1}^{(P)}, \tag{5.18}$$

which is valid to leading order for the same reasons as (5.12).

6. The charmonium system

Since the discovery of the J/ψ particle, potential models have been applied extensively to the charmonium spectrum. In the early days their success was impressive [18] especially in the predictions for the radial recurrences of the J/ψ, with a very small number of parameters. With the discovery of the P-states it became necessary to modify the simple and physically appealing picture of the Cornell model by assuming that the confining potential is a mixture of a Lorentz scalar and a Lorentz vector [19]. This considerably increases the number of parameters. The flexibility of potential models is apparent from the fact that they even succeeded in accommodating the χ (2.83) as the pseudoscalar η_c. Since the recent analysis in the Crystal Ball experiment [8] this state has disappeared and the η_c is now believed to be at 2.98 GeV. A splitting of 120 MeV, however, does also not follow naturally in a potential model approach, which would give about 30 MeV. This does strongly suggest that non-perturbative effects play a role in the charmonium spectrum and the prediction of the η_c at 3.00 GeV by Shifman et al. [7, 20] [at the same time ruling out the χ (2.83) as a possible candidate] can indeed be seen as one of the most impressive successes of the dispersion approach to spectroscopy.

We will use the moment methods described in sect. 2 to calculate the full charmonium spectrum including the P-states. We have briefly reported on the results before [10]. As an example we have plotted the coefficients a_n and b_n for the axial vector current as a function of n for various values of ξ in figs. 3 and 4. It is immediately clear from these figures that only a few moments can be used at $\xi = 0$ if we adopt the value $\phi = 1.35 \times 10^{-3}$ as obtained by Shifman et al. and α_s about 0.3. Since the values of $b(n,\xi)$ for P-waves are on average a factor three larger than the $b(n,\xi)$ coefficients for S-waves, ξ has to be chosen larger for P-waves than for S-waves. In fig. 5 we show the stability region for the ratio (2.15) of the pseudoscalar to vector current for various values of ξ. It is obvious from this picture that determination of the η_c resonance parameters at $\xi = 0$ as is done by SVZ [7, 20] is quite hazardous. On the other hand, if we choose ξ very large, the contributions from higher resonances to the sum rule will be so large that we cannot extract the

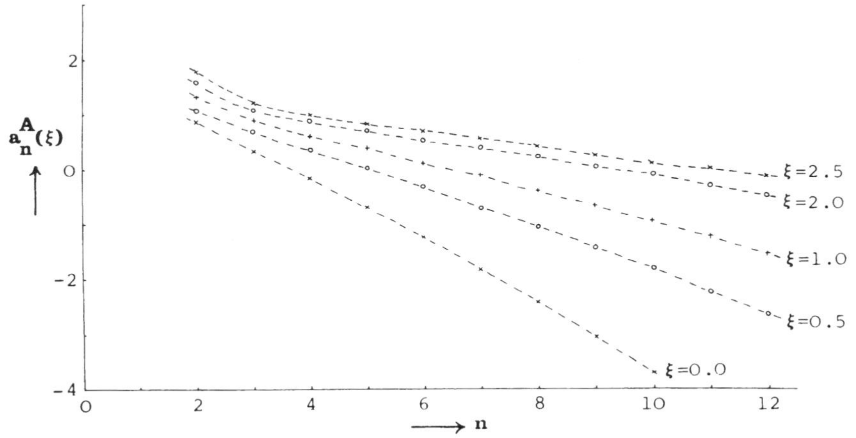

Fig. 3. The coefficient $a_n^A(\xi)$ for the axial vector current as a function of n for various values of ξ.

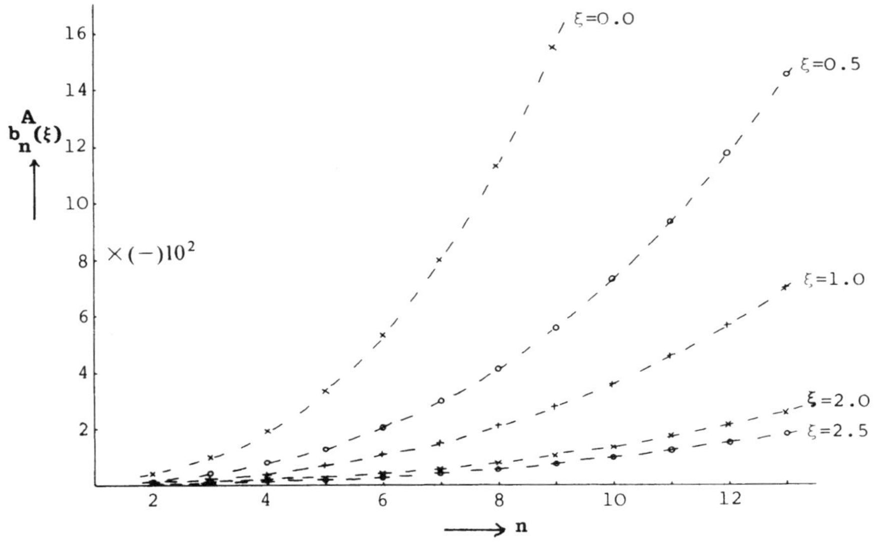

Fig. 4. The coefficient $b_n^A(\xi)$ for the axial vector current as a function of n for various values of ξ.

parameters of the lowest lying resonance in a reliable way. We have no criterion to decide which ξ to choose as for most channels there are no higher resonances known, and we cannot calculate its contribution to the moments. For each channel we have determined ξ empirically, but apart from the vector channel we have no guarantee that the contribution from higher resonances goes quickly enough to zero with n. It seems reasonable, however, to assume that the spectra in other channels are similar

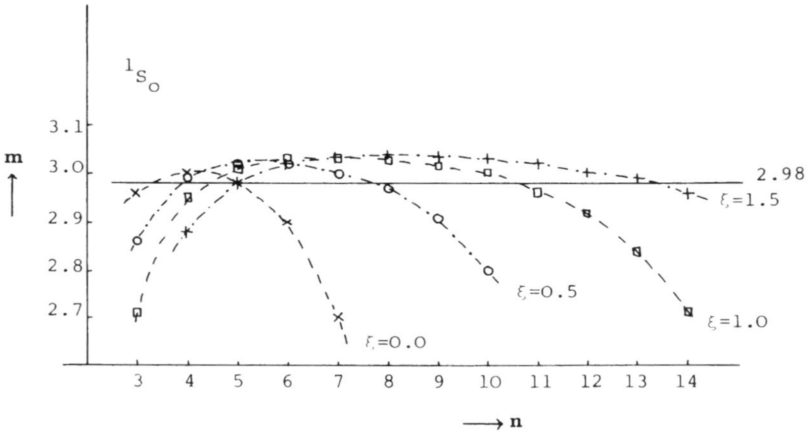

Fig. 5. Stability regions in n for the pseudoscalar S-state of charmonium for various values of ξ. For all ξ values the parameters are as in (6.5).

to the vector channel. In the next section we will see that for the upsilon system our method indeed fails. Although we can find a large stability region for the moments the contamination from higher resonances is so large that it does not appear to be possible to distinguish individual resonances.

Apart from higher resonances a continuum contribution has to be introduced in each channel. Following Shifman et al. [5] we adopt a θ-function dependence for this part of the cross section

$$\operatorname{Im}\Pi^{V}(s)_{\text{cont}}=\frac{1}{4\pi}\left(1+\frac{\alpha_s}{\pi}\right)\theta(s-s_0). \tag{6.1}$$

Due to kinematical factors which can be read off immediately from the bare loop coefficients $A^J(n)$ in table 1, the normalization is different for different J-values even in the same partial wave. Apart from this we have chosen the same behaviour as (6.1) without taking into account that various currents have anomalous dimensions and therefore, in principle, a different coefficient of α_s in (6.1). The value for s_0 is of course different for different partial waves but can to a good approximation be chosen the same for all J-values in a partial wave. Due to the θ-function behaviour in (6.1) we do not account for the kinematical "tail" in a proper way, i.e. we have to choose s_0 slightly higher than the data suggest. Since s_0 is always higher than the first few resonances, the continuum contribution goes very rapidly to zero with n and its contribution to the $\delta_n^J(\xi)$ in (2.13) will be negligible. Instead of treating the continuum in this way we have chosen to include it explicitly on the *theoretical* side of the sum rule, i.e. in (2.11) and consequently the $\delta_n^J(\xi)$ do not contain contributions from the continuum. Although we have to choose a definite continuum threshold for each partial wave, we have to make sure that the final result is not sensitive to the exact location of the threshold.

One should also note that α_s (and m_c), being running parameters, are ξ-dependent since the region which dominates the dispersion integral in (2.10) will be different for different Q_0^2. Moreover, α_s also depends on n as noted by Shifman et al. At present we are not in a position to be able to give quantitative estimates of these dependences. Instead, at a certain value of ξ we have chosen α_s constant at $\alpha_s(Q_0^2+4m_c^2)$, with its value calculated from the asymptotic freedom formula

$$\alpha_s(Q_0^2+4m_c^2)=\alpha_s(4m_c^2)\Big/\left(1+\frac{25}{12\pi}\alpha_s(4m_c^2)\ln\frac{Q_0^2+4m_c^2}{4m_c^2}\right), \tag{6.2}$$

with $\alpha_s(4m_c^2)\simeq 0.3$ as determined from the ratio R in e^+e^- annihilation [21].

Similarly, m_c depends on Q_0^2. In sect. 3 we have discussed the choice of the renormalization point. At Q_0^2 the quark mass is chosen to be renormalized at $m_c(p^2=-(\xi+1)m_c^2)$, which in the Landau gauge is expressed in $m_c(p^2=-m_c^2)$

by the following equation [4]:

$$\frac{m_c(\xi)}{m_c} = 1 - \frac{\alpha_s}{\pi}\left\{\frac{2+\xi}{1+\xi}\ln(2+\xi) - 2\ln 2\right\}, \tag{6.3}$$

where $m_c \equiv m_c(p^2 = -m_c^2)$.

The results for the masses of the S-states and the P-states are given in the accompanying figures. Our parameters are ϕ [as defined in (2.12)], α_s and m_c. The values chosen for ξ are $\xi = 1.0$ for S-waves and $\xi = 2.5$ for P-waves. For these values the stability region for the moments is sufficiently large to make reliable estimates for the masses of the states. Using eq. (6.2) the α_s values for S- and P-waves have been calculated.

Figs. 6 and 7 show the masses as determined from the moments (2.14). These moments are very sensitive to the quark mass. Even a slight change in the quark mass would spoil the beautiful agreement with the experimental mass value for the vector case (fig. 6a). We believe that this calculation provides the best evaluation of

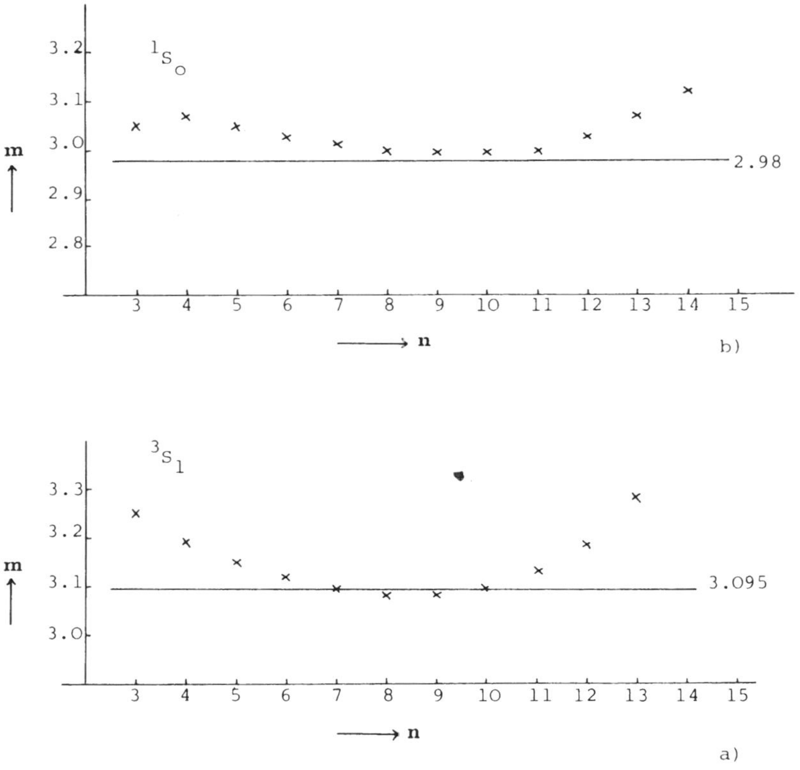

Fig. 6. Results for (a) the vector and (b) the pseudoscalar states of charmonium from the moments (2.14). All parameters as in (6.5). For comparison the experimental mass values have also been indicated.

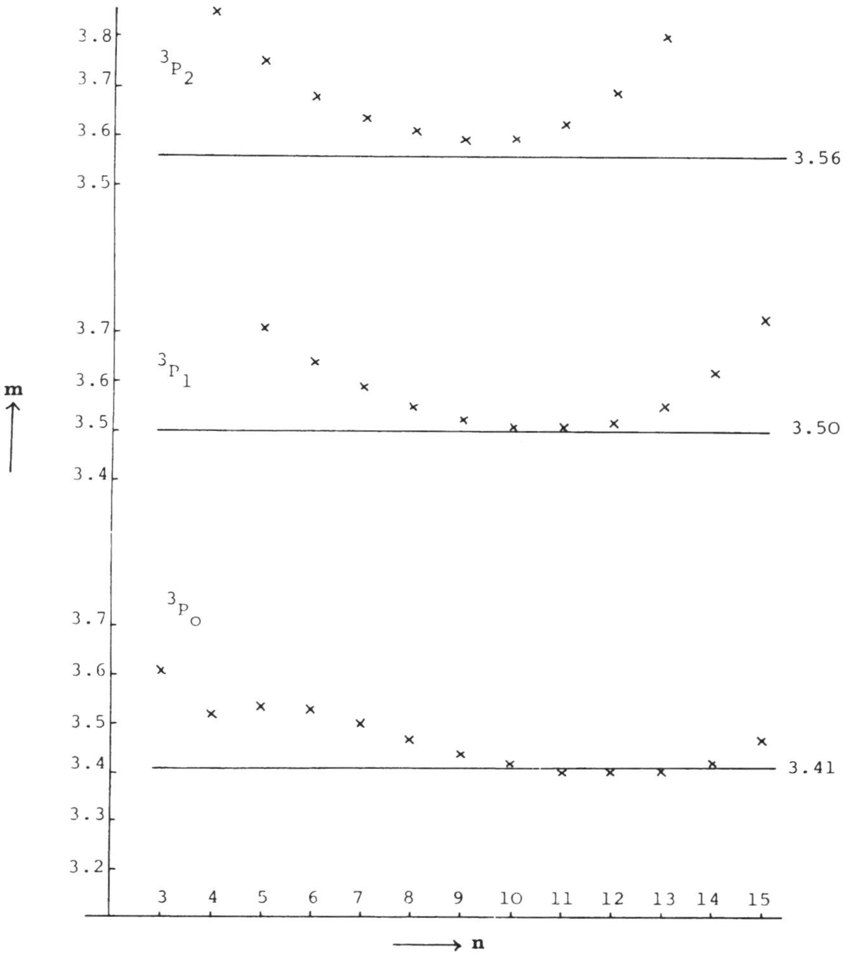

Fig. 7. Results for the P-states of charmonium from the moments (2.14). All parameters as in (6.5). For comparison the experimental mass values have also been indicated.

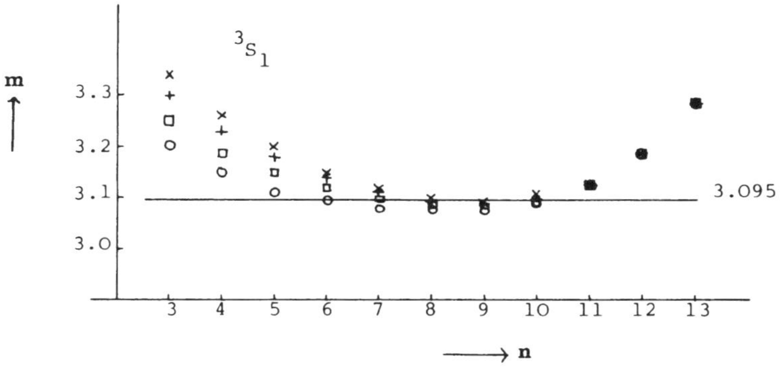

Fig. 8. Results for the vector state of charmonium from the moments (2.14) for various choices of the starting point s_0 of the continuum contribution (x: $\sqrt{s_0} = 4.4$ GeV; $+$: $\sqrt{s_0} = 4.2$ GeV; □: $\sqrt{s_0} = 4.0$ GeV and ○: $\sqrt{s_0} = 3.8$ GeV).

the charmed current quark mass at this value of Q_0^2. Values at other Q_0^2 can then be obtained from formula (6.3). In particular, the value of the quark mass to be used at $\xi = 2.5$ for the P-waves is now fixed, and we can also calculate the quark mass at the euclidean point $p^2 = -m_c^2$, i.e. $Q^2 = 0$:

$$m_c(p^2 = -m_c^2) \simeq 1.28 \text{ GeV}. \tag{6.4}$$

The following set of parameters has been obtained

S-waves ($\xi = 1.0$)	P-waves ($\xi = 2.5$)	
$m_c = 1.25$ GeV	$m_c = 1.22$ GeV	
$\phi = 0.14 \times 10^{-2}$	$\phi = 0.14 \times 10^{-2}$	
$\alpha_s = 0.27$	$\alpha_s = 0.25$	
$\sqrt{s_0} = 4.0$ GeV	$\sqrt{s_0} = 4.8$ GeV	(6.5)

We stress that only ϕ and $m_c(\xi = 1.0)$ are free parameters. The values for α_s follow from eq. (6.2) with $\alpha_s(4m_c^2) \simeq 0.3$ and eq. (6.3) has been used to calculate m_c at $\xi = 2.5$. The value for ϕ is slightly larger than found by SVZ who determined all parameters exclusively from the vector sector. But the results are not very sensitive to a slight change in ϕ; ϕ can be changed between 0.12×10^{-2} and 0.14×10^{-2} without changing the results in any significant way. SVZ use $\alpha_s = 0.2$ in ref. [5], as found from the J/ψ decay rates. We cannot allow $\alpha_s = 0.2$ without increasing ϕ significantly which would upset the P-levels. Like ϕ, α_s can be changed by ~ 10%

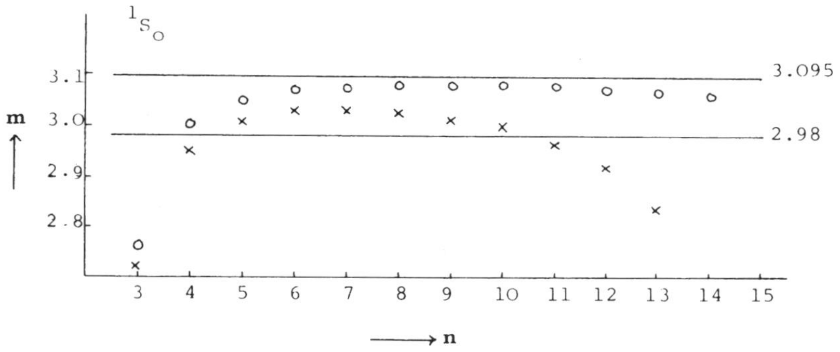

Fig. 9. Results for the pseudoscalar S-state of charmonium from the ratios (2.15) with the vector state for $\phi = 0.14 \times 10^{-2}$ (crosses) and $\phi = 0.0$ (open circles). All other parameters as in (6.5). The experimental values for the J/ψ and η_c have also been indicated.

without changing the results significantly. It should be noted that the values (6.5) for the parameters, are not fixed from a single level as done by SVZ, but chosen to give an overall satisfactory picture of the spectrum.

Fig. 8 illustrates that the results of the vector case are insensitive to the exact starting point s_0 of the continuum contribution. Similar figures can be obtained also for the other currents. It shows that the results would have been the same if we had included the continuum contribution on the experimental side of the sum rule.

As expected the stability increases significantly when we consider the ratios (2.15). The results are displayed in fig. 9 for the pseudoscalar case and fig. 10 for the axial and tensor states. We have used these figures to get the following mass values for the states:

$$^1\mathrm{S}_0:\ m = 3.01 \pm 0.02\ \mathrm{GeV},$$

$$^3\mathrm{P}_2:\ m = 3.56 \pm 0.01\ \mathrm{GeV},$$

$$^3\mathrm{P}_1:\ m = 3.50 \pm 0.01\ \mathrm{GeV}. \qquad (6.6a)$$

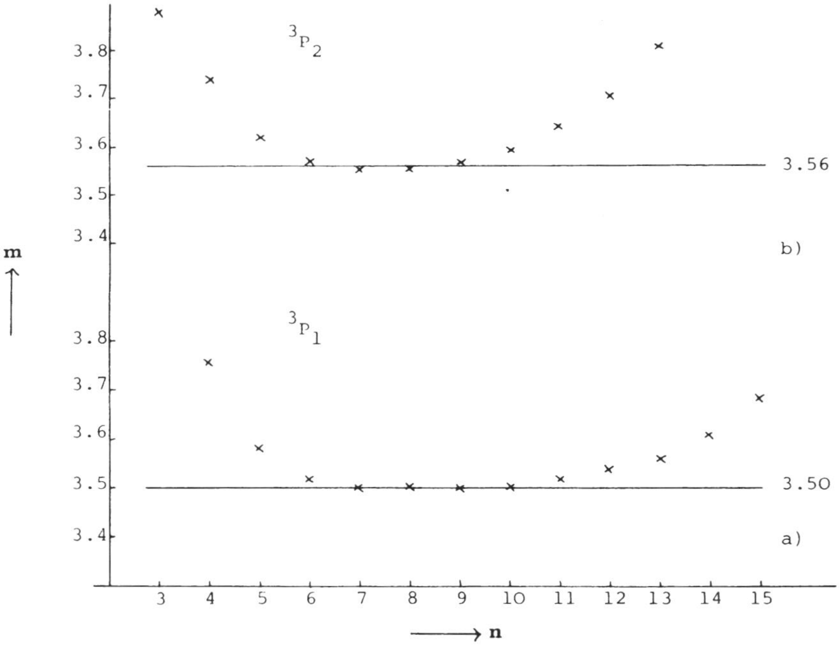

Fig. 10. Results for (a) the axial and (b) the tensor P-states of charmonium from the ratios (2.15) with the scalar state for the parameter values (6.5). The experimental mass values of the states have also been indicated.

From figs. 5a and 6a we find for the vector and scalar masses

$$^3S_1:\ m = 3.10 \pm 0.01\ \text{GeV},$$

$$^3P_0:\ m = 3.40 \pm 0.01\ \text{GeV}. \qquad (6.6b)$$

The agreement with known masses is extremely good and the number of parameters is certainly smaller than in potential models. The value obtained by Shifman et al. for the η_c mass in ref. [7] agrees with ours, but it is seen that for different values of ξ a more reliable estimate can be made and that the calculation can be extended to include the P-waves.

Finally fig. 11 shows the prediction on the mass of the axial vector state with negative charge conjugation. From fig. 11b, the ratio of its mass to the scalar mass, we find

$$^1P_1:\ m = 3.51 \pm 0.01\ \text{GeV}. \qquad (6.6c)$$

This ratio is so stable that we consider this prediction as a strong test of the theory.

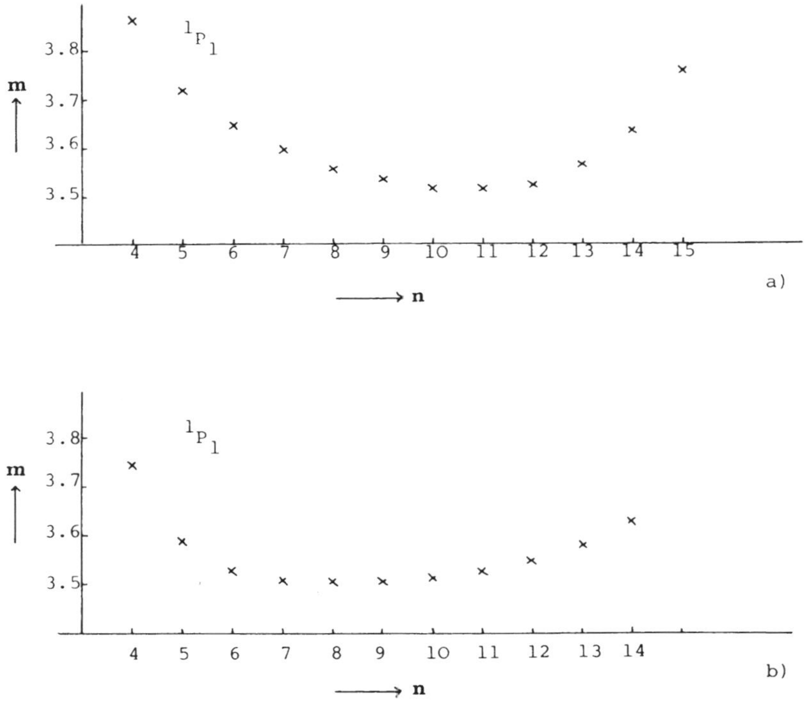

Fig. 11. Predictions for the axial vector $^1P_1$ state of charmonium from (a) the moments (2.14) and (b) the ratios (2.15) with the scalar state. Parameter values as in (6.5).

Having determined the masses, we can use the moments (2.13) directly to calculate the couplings g_R. Contamination by higher resonances will be larger than for the ratios (2.14) or (2.15) and we do not expect to obtain as accurate results as for the masses. Only for the vector channel g_R is related to the physical width $\Gamma(e^+e^-)$. Neglecting all higher resonances, we find

$$\Gamma_{J/\psi}(e^+e^-) = 5.34\ \text{keV}, \tag{6.7}$$

which compares quite favourably with the experimental number $\Gamma_{J/\psi}(e^+e^-) = 4.4 \pm 0.6$ keV. We estimate that the error in (6.7) is 10–20% due to higher resonances.

7. The upsilon system

The physics of the upsilon system as seen from these sum rules is markedly different from the physics of charmonium. As was seen in sect. 6 several remarkable conditions hold simultaneously for the charmonium system:

(a) $a_n\alpha_s \sim b_n\phi$ for a region of n,

(b) for these n the lowest lying state provides for a large part of the moment integral (> 90%),

(c) different J states in a given partial wave (in fact all states) have their plateau at about the same values of n, which makes the ratios (2.15) meaningful.

If we turn to the upsilon system we face a new situation. The system is now much more concentrated at the origin and one must force a large relative separation to feel the confinement. More precisely, the level structure is mainly determined by gluon exchanges and the influence of confinement is not effective at the relevant moments to determine the bound state position. This is reflected in the fact that the parameter ϕ defined for the charmonium system by (2.12) will now be a factor $(m_b/m_c)^4 \sim 100$ smaller. It is therefore necessary to probe very large n to feel the presence of ϕ in our sum rules. This leads to a region of n where the perturbative series is in doubt [4].

The other factor that determines the efficiency of the sum rules is the saturation ability. For charmonium, contributions from higher resonances to the moments went very rapidly to zero with growing n. For the upsilon system the level spacings are relatively smaller, and we have to go to much higher n to minimize the contamination of higher resonances. This can easily be seen by comparing the ratios $(m_\psi/m_{\psi'})^{2n}$ and $(m_\Upsilon/m_{\Upsilon'})^{2n}$, which at $n = 10$ are 0.03 and 0.32, respectively. Therefore, one has little hope to be able to determine the parameters of the lowest lying resonances in this way.

One can see that all these fears are justified by looking at the plots. As an example, fig. 12 shows the results for the axial channel as a function of n for various values of ξ. Although a stable plateau is reached, it is not independent of ξ and, in fact, a plateau independent of ξ cannot be reached before perturbation theory breaks down. We note that increasing ξ lowers the level of the plateau, i.e. the prediction for

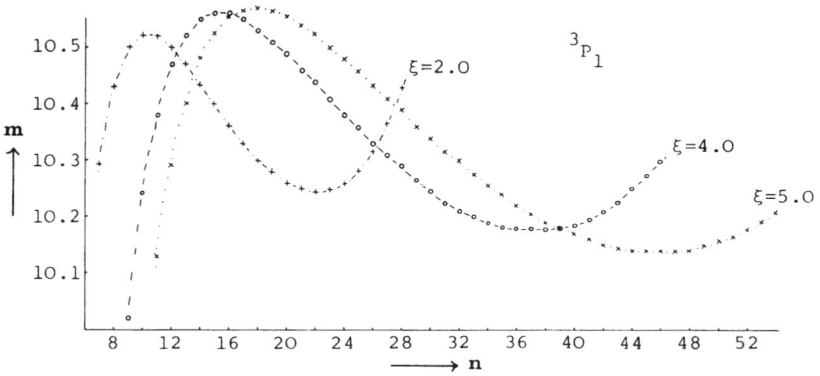

Fig. 12. Results for the axial vector channel of upsilonium from the moments (2.14) for various values of ξ. The parameter values are $\alpha_s = 0.16$, $\phi = 0.14 \times 10^{-4}$, $m_b = 4.17$ and $\sqrt{s_0} = 13.2$ GeV.

the mass. This is due to the fact that with growing ξ the contamination by higher resonances at the plateau becomes smaller.

Studying the moments (2.14) for S-waves, one can see that the plateaus for the pseudoscalar and vector channels partly overlap for $\xi = 1.0$, and although the plateau levels are still going down with growing ξ, a reliable value for the ratio of the pseudoscalar and vector mass can be obtained from the ratios (2.15). The results are plotted in fig. 13 and give

$$m_{\Upsilon} - m_{\eta_b} \simeq 60 \text{ MeV}. \tag{7.1}$$

From the vector channel we are able to give an accurate estimate of the current b-quark mass, since higher resonances are known and the experimental side of the sum rule can be calculated. Using formula (6.3) to transform to the point $p^2 = -m_b^2$,

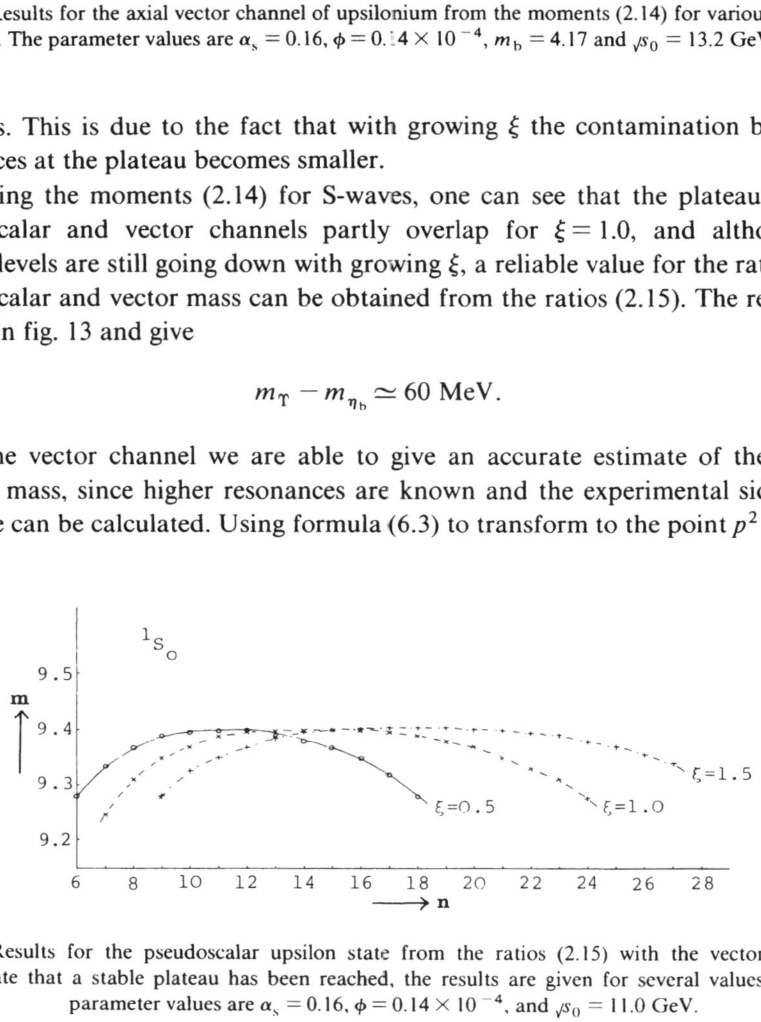

Fig. 13. Results for the pseudoscalar upsilon state from the ratios (2.15) with the vector state. To demonstrate that a stable plateau has been reached, the results are given for several values of ξ. The parameter values are $\alpha_s = 0.16$, $\phi = 0.14 \times 10^{-4}$, and $\sqrt{s_0} = 11.0$ GeV.

we find

$$m_b(p^2 = -m_b^2) \simeq 4.26 \text{ GeV}. \tag{7.2}$$

For this channel we can also determine quite accurately the influence of contamination by higher resonances on the prediction of the upsilon mass. Neglecting higher resonances we find $m_\Upsilon \simeq 9.58$ GeV, and although the contamination is not much more than 1% at the values of n for which stability is reached for $\xi = 1$, it has a significant influence on the prediction of the mass.

In the $L = 1$ case, no ξ-independent plateau can be achieved for low enough n to justify perturbation theory, but also the stability regions for different J-values occur at widely different values of n, which makes it impossible to use the ratios (2.15), and no reliable mass predictions can be made.

Other authors [22] have tried to exploit the "tail" of the wave function where the non-perturbative effects are large, by other methods. We refer to these papers for their results. This section emphasizes the unique properties of charmonium where every condition is met to make the sum rules productive.

8. Conclusions

In this paper we have presented detailed calculations for the $L = 0$ and $L = 1$ heavy quark systems and applied the results to the charmonium and upsilon states. Enlarging the original techniques of SVZ [5] we have found that QCD does indeed constrain the resonance spectrum in a significant way.

Our results can be summarised as follows:

(a) At the phenomenological level we have made contact with all levels of charmonium and predicted the elusive 1^{+-} state. We have explored the Υ spectrum and found that the limitations of one level saturation of the dispersion relations prevent us to make predictions on level splittings for P-waves. A reliable value for the η_b-Υ splitting is obtained.

We have checked that very few parameters (m_q, α_s and ϕ) control the spectrum to good accuracy. Seen from another viewpoint we have an excellent determination for the current quark masses m_c and m_b.

We have opened the possibility of studying other resonances and couplings by the same methods, especially the masses of the open charm states D, D*, F and F* and their electromagnetic mass differences. These states are now completely determined in terms of known parameters. Calculations of the moment equations for these states are in progress.

(b) At the technical level we have extended the theory to $\xi(= Q^2/4m_q^2) \gg 0$, i.e. to space-like Q^2, which not only gives a different view of the physics, but also checks the insensitivity of the results with respect to variations in Q^2.

We have calculated the polarization operator induced by other currents than done so far by SVZ and verified the known ones. We report corrections to the latter and hope that the expressions will be useful elsewhere.

The non-perturbative part has been calculated by an independent method (gluon-current Compton scattering) and we found agreement with the other method.

Apart from applications to open charm states, other extensions are possible like the study of quark mixings and the baryon spectrum. The non-relativistic potential model approach is especially successful for baryons [2], and it would be extremely interesting to see how non-perturbative effects as investigated in this paper effect these results.

The possibility is also open to calculate $\langle 0|G^a_{\mu\nu}G^a_{\mu\nu}|0\rangle$ by means of Monte Carlo lattice calculations, which should provide further tests of the theory.

We thank the Rutherford and Appleton Laboratories for their hospitality and the generous facilities to carry out this work. We also extend our thanks to our colleagues for useful discussions, and acknowledge the support of the Advanced Research Projects Agency and the Massachusetts Institute of Technology Math Lab Group for the use of the programme MACSYMA.

References

[1] O.W. Greenberg, Ann. Rev. Nucl. Sci. 28 (1978) 327;
J. Rosner, Phys. Reports 11 (1974) 190
[2] A. de Rújula, H. Georgi and S.L. Glashow, Phys. Rev. D12 (1975) 147;
N. Isgur and G. Karl, Phys. Rev. D18 (1978) 4187;
L.J. Reinders, J. Phys. G4 (1978) 1241
[3] J.M. Drouffe and C. Itzykson, Phys. Reports 38 (1978) 134
[4] V.A. Novikov, L.B. Okun, M.A. Shifman, A.I. Vainshtein, M.B. Voloshin and V.I. Zakharov, Phys. Reports 41 (1978) 1
[5] M.A. Shifman, A.I. Vainshtein and V.I. Zakharov, Nucl. Phys. B147 (1979) 385; 448; 519
[6] M. Creutz, L. Jacobs and C. Rebbi, Phys. Rev. D20 (1979) 1915;
M. Creutz, Phys. Rev. Lett. 43 (1979) 553
[7] M.A. Shifman, A.I. Vainshtein and V.I. Zakharov, Phys. Lett. 77B (1978) 80
[8] T.M. Himmel et al., Phys. Rev. Lett. 45 (1980) 1146;
R. Partridge et al., Phys. Rev. Lett. 45 (1980) 1150
[9] L.J. Reinders, H.R. Rubinstein and S. Yazaki, Phys. Lett. 94B (1980) 203
[10] L.J. Reinders, H.R. Rubinstein and S. Yazaki, Phys. Lett. 95B (1980) 103
[11] K.G. Wilson, Phys. Rev. 179 (1969) 1499
[12] V.A. Novikov, M.A. Shifman, A.I. Vainshtein, and V.I. Zakharov, Nucl. Phys. B174 (1980) 378
[13] M. Gell-Mann, R. Oakes and B. Renner, Phys. Rev. 175 (1968) 2195
[14] J. Schwinger, Particles, sources, and fields, vol. II, (Addison-Wesley, 1973)
[15] N. Sakai, Phys. Rev. D22 (1980) 2220;
T. Inami and T. Kubota, preprint UT-Komaba 80-10; Nucl. Phys. B179 (1981) 171
[16] T. Appelquist and H. Georgi, Phys. Rev. D8 (1973) 4000;
A. Zee, Phys. Rev. D8 (1973) 4038

[17] D.J. Gross, Proc. 1975 Les Houches Summer School, Methods in field theory, ed. R. Balian and J. Zinn-Justin, (North-Holland, 1976)
[18] E. Eichten, K. Gottfried, T. Kinoshita, K.D. Lane and T.-M Yan, Phys. Rev. D17 (1978) 3090; D21 (1980) 203; and references therein
[19] A. Henriques, B. Kellett, and R. Moorhouse, Phys. Lett. 64B (1976) 85
[20] M. Shifman, Z. Phys. C4 (1980) 289
[21] M. Dine and J. Sapirstein, Phys. Rev. Lett. 43 (1979) 668
[22] M. Voloshin, Preprint ITEP-21 (1980);
M. Voloshin and V. Zakharov, DESY report F-15-80/03 (May, 1980)

STABILITY OF CHARMONIUM SPECTRUM CALCULATIONS FROM QCD SUM RULES

L.J. REINDERS, H.R. RUBINSTEIN [1] and S. YAZAKI [2]
CERN, Geneva, Switzerland

Received 16 January 1984

We show that contrary to recent claims the QCD sum rules for charmonium remain unchanged when dimension-six and -eight operators are included. These results explain the remarkable accuracy of the mass determinations.

In two recent papers [1,2], Nikolaev and Radyushkin extended previous calculations [3,4] of the polarization function of heavy quark–antiquark systems. Using the fixed-point gauge technique [5], they calculated the Wilson coefficients of all operators of dimension six and eight in the operator product expansion for S and P wave currents. This is a difficult and important calculation. Their conclusions, however, are misleading as we will show in the following.

At $Q^2 = 0$ (Q^2 being the external reference scale) the nth derivative of $\Pi^V(Q^2)$, usually called the nth moment, can be written as

$$M_n^V(Q^2=0) = A_n^V(Q^2=0)[1 + Q_n^V(Q^2=0)\alpha_s + b_n^V(Q^2=0)\phi^{(2)} + c_n^V(Q^2=0)\phi^{(3)} + d_n^V(Q^2=0)\phi^{(4)}] \qquad (1)$$

where $A_n^V(Q^2=0)$ is the moment of the bare loop

$$A_n^V(Q^2=0) = \frac{3}{4\pi^2}\,\frac{2^n(n+1)(n-1)!}{(2n+3)!!}\,\frac{1}{(4m_c^2)^n} \qquad (2)$$

and

$$b_n^V(Q^2=0) = -\frac{n(n+1)(n+2)(n+3)}{2n+5} \qquad (3a)$$

is the coefficient of the gluon condensate defined as

$$\phi^{(2)} = \tfrac{4}{9}\pi^2\langle(\alpha_s/\pi)G^a_{\mu\nu}G^a_{\mu\nu}\rangle/(4m_c^2)^2 \simeq 0.14\times 10^{-2}\,. \qquad (3b)$$

For our analysis we have chosen the vector current, but the behaviour is similar for other currents.

The value quoted for $\phi^{(2)}$ is determined in ref. [4] from the sum rules for charmonium. It agrees well with the value obtained from massless quark systems [3,6], although in these systems it can vary by about 50%. In charmonium the accuracy is about 20% if higher-order corrections are small. In ref. [4], this was assumed to be the case.

The coefficients a_n^V and c_n^V do not concern us here. The a_n^V vary very slowly with n and Q^2 compared to the power corrections, and $c_n^V\,\phi^{(3)}$ gives negligible contributions for $n \leqslant 10$ already at $Q^2 = 0$ and decreases further with Q^2. The expressions for a_n^V and c_n^V can be found in refs. [3] and [2], respectively.

Here, we therefore concentrate on the contribution from dimension-eight operators. The general structure of the coefficients is

$$d_n^V(Q^2=0)\phi^{(4)} = \frac{n(n+1)(n+2)(n+3)}{(2n+5)(2n+7)(2n+9)} \times \sum_{I=1}^{7}\left(\sum_{J=0}^{6} A(I,J)n^J\right)\phi_I^{(4)}\,, \qquad (4a)$$

where the matrix $A(I,J)$ is given in table 1,

$$\phi_I^{(4)} = \tfrac{4}{81}\langle O_I^{(4)}\rangle/(4m_c^2)^4 \qquad (4b)$$

and the operators $O_I^{(4)}$ are given in table 2.

[1] On leave of absence from the Weizmann Institute, Rehovot, Israel.
[2] On leave of absence from the University of Tokyo, Japan.

0.370-2693/84/$ 03.00 © Elsevier Science Publishers B.V. (North-Holland Physics Publishing Division)

Table 1
The values of the coefficients $A(I, J)$ in eq. (4a) (taken from ref. [2]).

I	J = 0	1	2	3	4	5	6
1	−396	−1075/4	−811/24	38/3	29/8	1/4	0
2	−42	−257/56	3011/105	7891/480	2203/672	671/3360	−1/224
3	1638	15703/10	35209/60	1643/15	217/20	1/2	0
4	−510	6541/140	46569/140	12229/80	46519/1680	3443/1680	23/560
5	−2232	−161939/70	−35121/35	−27883/120	−8157/280	−1367/840	−3/280
6	0	−6344/35	−6686/35	−1171/15	−533/35	−289/210	−3/70
7	1080	14709/14	8797/21	10801/120	1937/168	247/280	9/280

Using a few models for the vacuum to estimate the vacuum expectation values of $O_I^{(4)}$, Nikolaev and Radyushkin found that at $Q^2 = 0$ the contribution from eight-dimensional operators to the moments (1) is large and rapidly growing with n. In particular, they found that for $n = 6$ this contribution is already about 50% of the gluon condensate contribution. In refs. [1, 2] and in a recent paper by the ITEP group [7], a gloomy view is taken of these results. Nikolaev and Radyushkin remark correctly that when the reference scale Q^2 is equal to zero these coefficients are too large and the original ITEP calculation [3] collapses. They then go on to cast doubt on the over-all validity of the method. In ref. [7], a solution of the problem is suggested by claiming that the vacuum saturation hypothesis (which was used by Nikolaev and Radyushkin to estimate the vacuum expectation values of the operators $O_I^{(4)}$) is not valid for gluonic operators. The argument which is based on the large N_c expansion is not very convincing since the resulting values for the operator matrix elements given in ref. [2] are of the expected order of magnitude. In fact, as was pointed out in ref. [4], the original calculation of ref. [3] was on shaky grounds to start with. $Q^2 = 0$ is not a suitable reference point. For the calculations to be reliable, Q^2 has to be chosen unequal to zero (space-like). Fig. 1 which is taken from ref. [4] shows this very clearly. For $Q^2 = 0$, i.e., $\xi = 0$ there is no stability region in which a reliable determination of the resonance mass (in this case the η_c) is possible.

Table 2
The operators $O_I^{(4)}$ expressed in terms of the gluon field tensor $G^a_{\mu\nu}$, $j^a_\mu = \Sigma_{u,d,s}\,\bar\psi\gamma_\mu(\lambda^a/2)\psi$, and the covariant derivative D_α.

I	$O_I^{(4)}$
1	$g^4(d^{abc}G^a_{\mu\nu}G^b_{\alpha\beta})^2 + \frac{2}{3}g^4(G^a_{\mu\nu}G^a_{\alpha\beta})^2$
2	$g^4(f^{abc}G^a_{\mu\nu}G^b_{\alpha\beta})^2$
3	$g^4(d^{abc}G^a_{\mu\alpha}G^b_{\alpha\nu})^2 + \frac{2}{3}g^4(G^a_{\mu\alpha}G^a_{\alpha\nu})^2$
4	$g^4(f^{abc}G^a_{\mu\alpha}G^b_{\alpha\nu})^2$
5	$g^5 f^{abc}G^a_{\mu\nu}j^b_\mu j^c_\nu$
6	$g^3 f^{abc}G^a_{\mu\nu}G^b_{\nu\lambda}D_\alpha D_\alpha G^c_{\lambda\mu}$
7	$g^4 j^a_\mu D_\alpha D_\alpha j^a_\mu$

Using the expresssions at $Q^2 = 0$ [eqs. (1)–(4)], the moments at $Q^2 \neq 0$ can easily be calculated [4]. For the bare loop we obtain

$$A_n^{\mathrm{V}}(\xi) = \frac{3}{4\pi^2}\,\frac{2^n(n+1)(n-1)!}{(2n+3)!!}\,(4m_c^2)^{-n}(1+\xi)^{-n}$$

$$\times F(n, \tfrac{1}{2}, n + \tfrac{5}{2}; \rho)\,, \qquad (5)$$

where $\xi = Q^2/4m_c^2$, $\rho = \xi/(1+\xi)$, and $F(a, b, c; x)$ is the hypergeometric function ${}_2F_1(a, b, c; x)$. The coefficients b_n^{V} become

$$b_n^{\mathrm{V}}(\xi) = -\frac{n(n+1)(n+2)(n+3)}{(2n+5)}\,(1+\xi)^{-2}$$

$$\times \frac{F(n+2, -\frac{1}{2}, n+\frac{7}{2}; \rho)}{F(n, \frac{1}{2}, n+\frac{5}{2}; \rho)}\,. \qquad (6)$$

And for $d = 8$, we get

$$d_n^{\mathrm{V}}(\xi)\phi^{(4)} = \frac{n(n+1)(n+2)(n+3)}{(2n+5)(2n+7)(2n+9)}(1+\xi)^{-4}$$

$$\times \sum_{I=1}^{7}\left(\sum_{J=0}^{6} A(I,J)\,\frac{G(J,n)}{F(n,\frac{1}{2},n+\frac{5}{2};\rho)}\right)\phi_I^{(4)}\,, \qquad (7a)$$

where

$$G(J,n) = \frac{(n+1+J)!}{(n+1)!}\,F(n+4,\tfrac{7}{2}-J,n+\tfrac{5}{2};\rho)$$

$$-\sum_{k=0}^{J-1}\alpha^J(k)G(k,n)\,, \qquad (7b)$$

with $\alpha^J(0) = (J+1)!$, $\alpha^J(J-1) = (J+1) + \alpha^{J-1}(J-2) (J \geqslant 2)$, and $\alpha^J(k) = (J+1)\alpha^{J-1}(k) + \alpha^{J-1}(k-1) (k < J-1)$.

In ref. [4], we have discussed at length the problem of choosing Q^2 for the case of charmonium and emphasized the role of the reference scale Q^2 in determining the validity region of the method for a particular problem. More precisely, this reference scale is arbitrary; the predictions for physical observables cannot depend on its value provided the following conditions are satisfied: (a) asymptotic freedom is valid, (b) the lowest-lying resonance saturates the sum rule, (c) the Wilson coefficients which are functions of Q^2 must always be small in the sense that the next order correction can be neglected.

Condition (a) is the basic starting point of the theory. Fig. 1 shows that for the case of charmed quarks, $Q^2 = 0$ is not large enough. There is essentially no plateau where on the one hand the QCD expression is reliable ($n = 6$ is already outside the fiducial region) and on the other hand the lowest-lying resonance saturates the sum rule. Condition (b) expresses the limitation on choosing Q^2 too large. The larger Q^2 the larger is the weight of higher resonances in the sum rule. The contribution to the moment $M_n^{\mathrm{V}}(Q^2)$ of a higher resonance with mass m_{R}', relative to the ground state m_{R} is $\sim[(m_{\mathrm{R}}^2 + Q^2)/(m_{\mathrm{R}'}^2 + Q^2)]^n$ and tends to one for large Q^2. Condition (c) is required since the Wilson coefficients are calculated perturbatively. In the case of open charm [8], for instance, there is open conflict between conditions (b) and (c). The power correction proportional to $m_c\langle\bar{q}q\rangle$ (m_c the charmed quark mass and $\langle\bar{q}q\rangle$ the light quark chiral condensate) is so large that single resonance saturation becomes impossible when Q^2 is chosen large enough to make the correction small.

In ref. [4], we have calculated the S- and P-wave levels of charmonium, including only the power corrections from the four-dimensional operator $G_{\mu\nu}^a G_{\mu\nu}^a$. Fig. 1 shows that at $\xi = 0$ the calculation is marginal. This is the reason why we explored the dependence on ξ. At $\xi = 1$, there is a large plateau for determining the resonance mass. A large region of overlap between

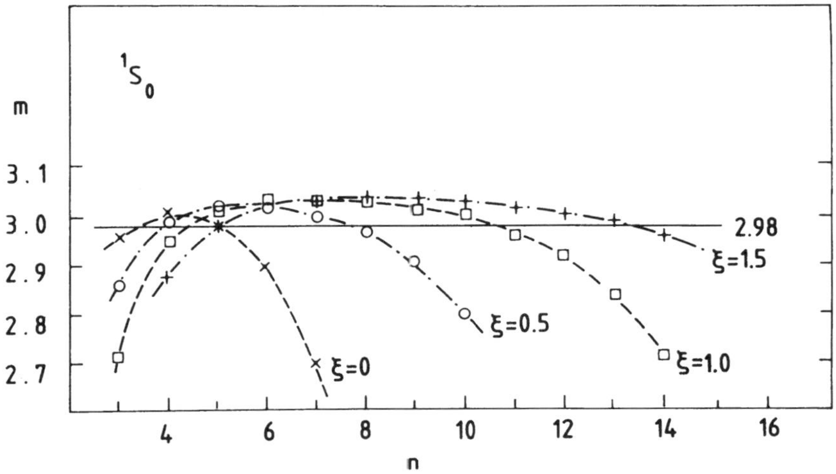

Fig. 1. Stability region in n for the pseudoscalar S-state of charmonium for various values of ξ. Figure taken from ref. [4].

the resonance and QCD representation of the amplitude is crucial. In the region of stability the two representations agree and resonance parameters can be extracted. Therefore, the mass of the state at the plateau should not change when ξ is varied. This is non-trivial since the coefficients of the operators vary substantially with ξ.

The breakdown of the plateau at large n and at very small n can also be understood. At large n ($n > 10$ for $\xi = 1$) condition (c) is not valid, and for small n ($n < 5$) condition (b) on saturation by the lowest-lying resonance is not satisfied. However, as expected, saturation starts at lower n for smaller Q^2 (or ξ).

The calculations of Nikolaev and Radyushkin give us the opportunity to check whether condition (c) is fulfilled in the plateau region by comparing the contribution (7a) from eight-dimensional operators with the gluon condensate contribution. To do this we have to know the matrix elements of the operators $O_I^{(4)}$. Unfortunately, there exist no reliable models of the vacuum for calculating these matrix elements. One of the models used in refs. [1,2] is to express these matrix elements in terms of lower-dimensional ones using the vacuum saturation hypothesis. The results have been tabulated in table 3.

In fig. 2, we have plotted the $d = 4$ and $d = 8$ contributions to the moments $M_n^V(\xi)$ at $\xi = 0$ and $\xi = 1$. As can be seen, at $\xi = 0$, the $d = 8$ contribution is not small compared to the $d = 4$ contribution. Already at $n = 6$, the $d = 8$ contribution is larger than the uncertainty in the $d = 4$ contribution which is ~20% in the determination of $\phi^{(2)}$. This is not surprising since the

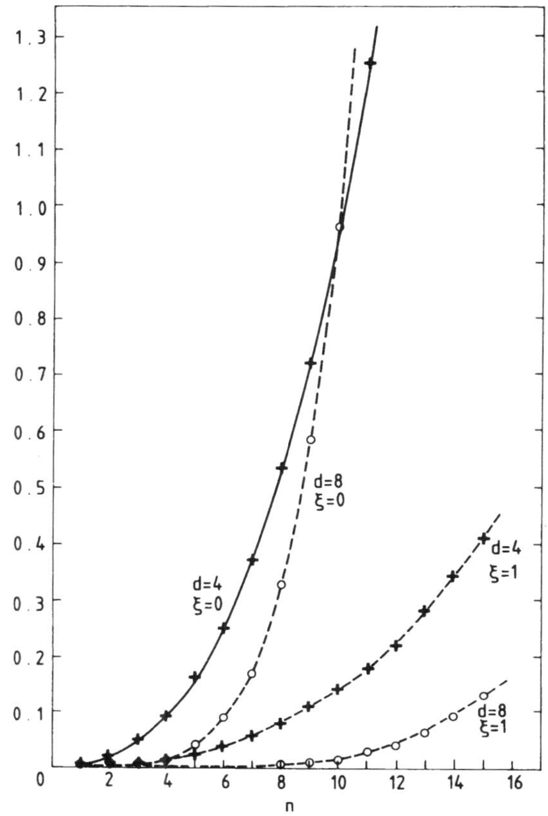

Fig. 2. The contribution of $d = 4$ and $d = 8$ operators to the moments $M_n^V(\xi)$ (relative to the bare loop) for $\xi = 0$ and $\xi = 1$. We have plotted the absolute value of the $d = 4$ contribution.

Table 3
The expressions and approximate values of the matrix elements $\langle O_I^{(4)}\rangle$ using the vacuum saturation hypothesis [2]. $\langle g^2G^2\rangle \simeq (0.83\ \mathrm{GeV})^4$, the parameter $M^2 \simeq 0.3\ \mathrm{GeV}^2$ characterizes the momentum of the vacuum gluons, and $\langle \bar{q}q\rangle$ is the light quark condensate.

I	$\langle O_I^{(4)}\rangle$	Values [GeV^8]
1	$(65/144)\langle g^2G^2\rangle^2$	0.10
2	$(45/144)\langle g^2G^2\rangle^2$	0.07
3	$(38/144)\langle g^2G^2\rangle^2$	0.06
4	$(9/144)\langle g^2G^2\rangle^2$	0.014
5	$-(3/4)g^4\langle\bar{q}q\rangle\langle g\bar{q}\sigma_{\mu\nu}G_{\mu\nu}q\rangle$	-0.7×10^{-3}
6	$M^2\langle g^3 f^{abc} G^a_{\mu\alpha} G^a_{\alpha\beta} G^c_{\beta\nu}\rangle$	0.014
7	$-(4/3)g^4M^2\langle\bar{q}q\rangle^2$	-0.63×10^{-3}

calculation with only dimension-four operators was marginal at $Q^2 = 0$. The conclusion is that $Q^2 = 0$ is too close to threshold to be considered in the asymptotic freedom region.

For $\xi = 1$ the situation is changed drastically. Both contributions are suppressed but the $d = 8$ contribution by a much larger factor than the $d = 4$ one. At $n = 6$ the $d = 8$ terms are essentially zero and up to moment $n = 11$ their contribution is less than 20% of the $d = 4$ one. This confirms the result of fig. 1 where it can be seen that an $\xi = 1$ the plateau extends up to $n = 10$. Including dimension-eight operators can only extend the stability to higher n. The conclusion of the authors of refs. [1,2] on the general failure of the sum rules is therefore unwarranted. In fact, we consider this as a strong corroboration of the results and procedure in ref. [4] for the S and P levels of charmonium as well as for the standard value of the gluon condensate.

References

[1] S.N. Nikolaev and A.V. Radyushkin, Phys. Lett. 124B (1983) 242.
[2] S.N. Nikolaev and A.V. Radyushkin, JINR Preprint R2-82-914 [in Russian].
[3] M.A. Shifman, A.I. Vainshtein and V.I. Zakharov, Nucl. Phys. B147 (1979) 385.
[4] L.J. Reinders, H.R. Rubinstein and S. Yazaki, Nucl. Phys. B186 (1981) 109.
[5] A.V. Smilga, Yad. Fiz. 35 (1982) 473 [Sov. J. Nucl. Phys. 35 (1982) 271].
[6] L.J. Reinders, H.R. Rubinstein and S. Yazaki, Nucl. Phys. B196 (1982) 125.
[7] V.A. Novikov, M.A. Shifman, A.I. Vainshtein, M.B. Voloshin and V.I. Zakharov, Preprint ITEP-71 (1983).
[8] L.J. Reinders, H.R. Rubinstein and S. Yazaki, Phys. Lett. 104B (1981) 305.

5. Baryons

Introduction

The method based on the QCD sum rules is explicitly approximate. The possibility of describing the effects of the vacuum medium on the valence quarks in terms of a few condensate parameters is not universal. Each time one intends to apply the sum rule method in a new hadronic system one has to decide the question of its applicability anew. Needless to say that expanding the range of problems solved in this way is advantageous both for hadron phenomenology and for QCD itself, since we accumulate valuable information on the QCD vacuum.

In early and mid-eighties three major breakthroughs predetermined the success of the sum rule method practically in all spheres of low-energy hadronic physics. I mean the generalization to baryons and to three-point functions (form factors) at intermediate and small momentum transfers. There are many subtle points in these problems which, at first sight, could make them intractable. The works which were the first to show that all difficulties could be overcome are discussed and reproduced in this and two subsequent chapters. Calculation of the baryon parameters and, especially, form factors is not just a mere repetition of what has been already done for the meson systems. The reader will see that new ideas are involved and, even to a larger extent, a new computational technique (chapter 3).

This chapter is devoted to the simplest two-point function baryon sum rules. They have been proposed by Ioffe [1] and Chung et al. [2] independently and approximately at the same time. (Ref. [1] is reproduced here.) In those days it was not at all clear that the method would work for the baryon family. Actually, it turned out to be very successful, producing a wealth of predictions for the baryon masses and residues. If in the pioneering papers [1,2] the attention was focussed mainly on the nucleon and Δ, later works [3–6] present a more comprehensive analysis of the intricacies of the baryon familly, including all members of the $1/2^+$ and $1/2^-$ octets and $3/2^-$ decuplet.

What is even more important, the picture of the QCD vacuum populated by condensates has been confirmed, and the values of the quark and mixed condensates have been extracted with a better accuracy.

Of particular importance is the mixed quark–gluon condensate,

$$\left\langle -\tfrac{1}{2}g\bar{\psi}\sigma_{\alpha\beta}\lambda^a G^a_{\alpha\beta}\psi\right\rangle \equiv m_0^2\langle\bar{\psi}\psi\rangle, \quad m_0^2 \approx 0.8\ \mathrm{GeV}^4, \tag{5.1}$$

which was first found in ref. [3]. Recent analysis [7] fully confirms the original estimate quoted above.

As for other condensate parameters, for instance, $\langle\bar{\psi}\Gamma\psi\bar{\psi}\Gamma\psi\rangle$, I must admit that on this issue there is no consensus in the current literature devoted to the baryon sum rules. On one hand, refs. [1–4] assume the standard value of the quark condensate $\langle\bar{\psi}\Gamma\psi\rangle$ and exact factorization for the four-fermion operators $\langle\bar{\psi}\Gamma\psi\bar{\psi}\Gamma\psi\rangle$. Extremely impressive agreement is achieved under this assumption between the predictions for the baryon spectrum and the observed regularities. On the other hand, a value of $\langle\bar{\psi}\Gamma\psi\rangle$ substantially lower than the standard one is advocated in refs. [5,6]. To compensate for the "lost" factor of ~ 2 in $\langle\bar{\psi}\Gamma\psi\rangle$ the authors of refs. [5,6] have to introduce a significant violation of factorization in $\langle\bar{\psi}\Gamma\psi\bar{\psi}\Gamma\psi\rangle$ in order to maintain the four-quark condensate at the standard level. The controversy over the issue of factorization is aggravated by a polemic on a related question – the proper choice of interpolating current for the octet baryons. The pattern of the quark condensates is sensitive to this choice provided, of course, that the baryon spectrum is properly reproduced.

Due to the approximate character of the sum rule method the right choice of interpolating current is of a paramount importance. In the meson channels there is always an obvious natural choice leaving, in essence, no ambiguity. Limiting ourselves to the current with the minimal number of derivatives in all practically interesting cases we find only one current with the given quantum numbers. The situation becomes more complicated for baryons. Here typically one can built several currents with the given quantum numbers which have no derivatives. The task is to optimize the structure of the interpolating current in such a way that the coupling to the baryon under consideration is maximally strong. In this way one can efficiently suppress contamination due to higher mass resonances.

Different interpolating currents have been chosen in refs. [1,3,4] on the one hand, and in refs. [2,5,6], on the other. The question of the best choice is still unsettled and, unfortunately, I cannot present here the final solution. The criteria ensuring the most desirable and favorable features of the baryon current are summarized in ref. [8] which, to my mind, formulates the problem in the most transparent way. (We reproduce this paper below.) Needless to say that the original choice of Ioffe is advocated in this paper. An additional argument showing that the current of ref. [1] has certain unique properties has been published recently [9]. It has been shown [9] that the effect of the so-called direct instantons [10] on one of two kinematical structures appearing in the nucleon sum rules (F_1) is absent for Ioffe's current but rapidly grows (as a matter of fact, explodes) even for small admixtures of the second current. Direct instantons are poorly controllable and contaminate the theoretical part of the sum rules to the extent that usually no reliable theoretical calculations are possible. The authors of ref. [9] conclude that due to a drastic contribution of the direct instantons the sum rules for F_1 have no confidence domain at all, and they should be excluded from consideration for all choices of the interpolating current except that made in ref. [1].

It is worth mentioning that a dedicated analysis of meson sum rules [11] revealed no considerable deviations from factorization in the four-fermion operators $\langle\bar{\psi}\gamma_\mu\psi\bar{\psi}\gamma_\mu\psi\rangle$ and $\langle\bar{\psi}\gamma_\mu\gamma_5\psi\bar{\psi}\gamma_\mu\gamma_5\psi\rangle$, a fact which cannot be disregarded. (The works quoted above treat the matrix elements over pions and kaons, which are readily reducible, however, to the vacuum matrix elements.)

Given all these indications it seems reasonable to accept, for the time being, the conservative point of view formulated in refs. [1,3,4,8] which implies Ioffe's choice for the nucleon current and exact factorization of the four-fermion operators. A critically minded reader should necessarily familiarize himself/herself, however, with the alternative position of papers [5,6]. It is clear that further discussion is needed, especially concerning the role of the direct instantons. Their significant contribution in the baryon sum rules, if confirmed, might completely change at least some aspects of our present understanding. Factorization of the four-fermion operators is an extremely important question crucial for many other problems in hadron physics. Reliable independent estimates of possible deviations from factorization are badly needed. The stakes are very high.

A few remarks concerning the immediate consequences of the baryon sum rules [1,2]:

(i) They are indispensable in the studies of the flavor symmetry breaking effects in the quark condensates (the meson sum rules are sensitive to a much lesser extent). The corresponding parameter, crucial for many other applications,

$$\gamma = \frac{\langle \bar{s}s \rangle}{\langle \bar{u}u \rangle} - 1, \tag{5.2}$$

has been roughly estimated * in ref. [1] and later analyzed in detail in refs. [3,12].

(ii) The baryon residues, a byproduct of the sum rule analysis, turn out to be important in phenomenology. Indeed, the proton residue enters as a key element in the calculation [13] of the proton lifetime in the schemes of the Grand Unification (see also ref. [14]). Moreover, these residues are crucial elements in the subsequent stages of the program: they appear when one tries to determine the baryon magnetic moments within the sum rule method (chapter 7). And finally, the neutron residue determines the constant in front of Q^{-4} in the asymptotics of the neutron magnetic form factor [15]. Surprising though it is, this constant has a different sign [15] and is much smaller in absolute value than what is measured today experimentally at Q^2 of the order of 20 GeV^2, implying that the present domain of Q^2 is extremely far from the genuine asymptotics. For further details see the review paper [16].

The values of the residues found in the first publications seem now somewhat underestimated, perhaps by a factor ~ 1.5. The point is that $O(\alpha_s)$ terms in the baryon sum rules, not included in the original analysis, have been calculated [17] and turned out to be rather large (in the case of F_1). These terms enhance the residues, but practically do not affect the predictions for the masses, a situation quite standard for perturbative corrections.

In the end of this chapter we reproduce the work of Reinders et al. [12] devoted to the calculation of the vacuum parameters from an overall fit of the baryons in the 56-plet. One should not think that this work is the ultimate truth, the more so since a wrong sign of the gluon condensate was used in the original formulae (see Erratum immediately after this paper). Rather, it is intended to give an idea of how the sum rule machinery can be launched in the opposite direction – the known spectrum as the input, the vacuum parameters as the output. This is a very concise report, and I am sure that particular numbers given there will be subject to checks and double checks and will fluctuate with time. (This is inevitable since QCD is the genuine strong coupling theory, and high precision is not on the agenda in the forseeable future.) Therefore, it is not the specific numerical results presented in ref. [12] which are most instructive, but the approach itself.

One of the early reviews devoted to the baryon sum rules was written by Ioffe [16]. Another review on this topic which might be recommended is the invited talk [18].

References

[1] B.L. Ioffe, Nucl. Phys. B 188 (1981) 317; (E) Nucl. Phys. B 191 (1981) 591.

[2] Y. Chung, H.G. Dosch, M. Kremer and D. Schall, Phys. Lett. B 102 (1981) 175; Nucl. Phys. B 197 (1982) 55.

[3] V.M. Belyaev and B.L. Ioffe, Zh. Eksp. Teor. Fiz. 83 (1982) 876 [JETP 56 (1982) 493]; Zh. Eksp. Teor. Fiz. 84 (1983) 1236 [JETP 57 (1983) 716].

[4] D. Espriu, P. Pascual and R. Tarrach, Nucl. Phys. B 214 (1983) 285.

[5] Y. Chung, H.G. Dosch, M. Kremer and D. Schall, Z. Phys. C 25 (1984) 151.

[6] H.G. Dosch, M. Jamin and S. Narison, Phys. Lett. B 220 (1989) 251.

[7] A.A. Ovchinnikov and A.A. Pivovarov, Yad. Fiz. 48 (1988) 1135 [Sov. J. Nucl. Phys. 48 (1988) 721].

[8] B.L. Ioffe, Z. Phys. C 18 (1983) 67.

[9] A.E. Dorokhov and N.I. Kochelev, Z. Phys. C 46 (1990) 281.

* The fact that γ is negative has been observed independently in ref. [2].

[10] V. Novikov, M. Shifman, A. Vainshtein and V. Zakharov, Nucl. Phys. B 191 (1981) 301.
[11] K.G. Chetyrkin et al., Phys. Lett. B 147 (1986) 104.
L.J. Reinders and S. Yazaki, Nucl. Phys. B 288 (1987) 789.
K.G. Chetyrkin and A.A. Pivovarov, Nuovo Cimento A 100 (1988) 899.
[12] L. Reinders, H. Rubinstein and S. Yazaki, Phys. Lett. B 120 (1983) 209; (E) Phys. Lett. B 122 (1983) 487.
[13] V.S. Berezinsky, B.L. Ioffe and Ya.I. Kogan, Phys. Lett. B 105 (1981) 33.
[14] V. Khatsimovsky, Yad. Fiz. 43 (1986) 465 [Sov. J. Nucl. Phys. 43 (1986) 297].
[15] V.A. Avdeenko, V.L. Chernyak and S.A. Korenblit, Asymptotic Behavior of Nucleon Form Factors in QCD, preprint 23-79, Irkutsk, 1979 [unpublished]; Yad. Fiz. 33 (1981) 481 [Sov. J. Nucl. Phys. 33 (1981) 252].
[16] B.L. Ioffe, Acta Phys. Pol. B 16 (1985) 543.
[17] A.A. Ovchinnikov, A.A. Pivovarov and L.P. Surguladze, Yad. Fiz. 48 (1988) 562 [Sov. J. Nucl. Phys. 48 (1988) 358].
M. Jamin, Z. Phys. C 37 (1988) 635.
[18] H.G. Dosch, Invited talk at Workshop on Nonperturbative Methods, Montpellier, France, 1985. Published in Montpellier Workshop 1985: 40 (QCD 161: W685: 1985).

Nuclear Physics B188 (1981) 317–341

CALCULATION OF BARYON MASSES IN QUANTUM CHROMODYNAMICS

B.L. IOFFE

Institute of Theoretical and Experimental Physics, Moscow 117259, USSR

Received 10 March 1981

The polarization operator of quark currents with baryon quantum numbers is considered in quantum chromodynamics. The non-zero mean vacuum values of the field operator products are taken into account. The sum rules are obtained assuming that in the virtuality region ~1 GeV, among the mean vacuum values violating the chiral invariance, the most important is $\langle 0|\bar{\psi}\psi|0\rangle$. Saturating these sum rules by the lowest baryonic states one is able to calculate the masses of the isobar Δ and nucleon N, $M_\Delta = 1.4$ GeV, $M_N = 1$ GeV, up to 15% through the known value $\langle 0|\bar{\psi}\psi|0\rangle$. The mass splitting in the baryonic decuplet $M_{\Sigma^*} - M_\Delta = 125$ MeV is calculated in first order in the current strange quark mass $m_s = 150$ MeV. Certain results for other baryonic resonances have also been obtained.

1. Introduction

As is well known, the masses of the current u, d, s quarks are comparatively small and if they are disregarded, the QCD lagrangian (without heavy quarks) is chiral invariant [1–4]. The physical states of hadrons are not, however, chiral invariant, which indicates a spontaneous symmetry breaking and is exhibited, in particular, in the appearance of non-zero, chiral invariance violating, mean vacuum values of the products of quark field operators. Of such mean vacuum values the most popular is $\langle 0|\bar{\psi}\psi|0\rangle$ which was calculated within the low-energy limit on the basis of the hypothesis of partial conservation of the axial current (PCAC) and turned out to be [4]

$$\langle 0|\bar{\psi}\psi|0\rangle \underset{m_\pi^2, m_u, m_d \to 0}{=} -\frac{1}{2}\frac{f_\pi^2 m_\pi^2}{m_u + m_d}, \tag{1}$$

where ψ is the field of the u- or d-quark, f_π is the pion decay constant, $f_\pi = 130$ MeV, m_π is the pion mass and m_u and m_d are the masses of u- and d-quarks. The ratios m_u/m_d and m_d/m_s can be found using the PCAC hypothesis and the strange quark mass m_s can be estimated from the mass differences in baryonic multiplets. The standard choice [4], $m_u = 4.2$ MeV, $m_d = 7.5$ MeV, leads to the numerical value

$$\langle 0|\bar{\psi}\psi|0\rangle = -(240\ \text{MeV})^3\,. \tag{2}$$

The value of $\langle 0|\bar{\psi}\psi|0\rangle$, (2), is not small, which should be expected if the chiral symmetry were violated owing only to the quark mass, but has a typical hadronic scale.

From the viewpoint of general physics, the most striking effect of spontaneous violation of chiral invariance is, of course, the fact that baryons are massive. Clearly, if there were no spontaneous violation of chiral symmetry in massless QCD, the baryon masses would be zero, but they are really by no means small. A question arises, whether it is possible to put the values of the baryon masses into quantitative agreement with the value of the quark field correlator $\langle 0|\bar{\psi}\psi|0\rangle$. At first sight, a direct relationship between these quantities seems unlikely since from dimensional considerations it must have the form

$$M^3 = c\langle 0|\bar{\psi}\psi|0\rangle\,, \tag{3}$$

where c is a numerical coefficient. At $M \geq 1$ GeV, c must be of order 10^2 – a value which seems to be improbably large. Nevertheless, the answer to this question is positive and we will show in this paper that QCD does involve relations of the type (3) and that these relations make it possible, at a given numerical value of the correlator $\langle 0|\bar{\psi}\psi|0\rangle$ (2), to calculate baryon masses with good enough accuracy and in good agreement with experiment.

The basis for the approach proposed is the study of the polarization operator of three-quark currents

$$\begin{aligned} \Pi(p) &= i\int \mathrm{d}^4x\, \mathrm{e}^{ipx}\langle 0|\mathrm{T}\{\eta(x),\, \bar{\eta}(0)\}|0\rangle\,, \\ \eta_{\alpha\beta\gamma} &= \psi_\alpha\psi_\beta\psi_\gamma\,, \end{aligned} \tag{4}$$

and obtaining sum rules analogous to those used by Shifman, Vainshtein and Zakharov [5] (hereafter SVZ) for studying the properties of mesonic resonances. Following the general method developed by SVZ, I will suppose that in the region of space-like $p^2 \sim -1\ \mathrm{GeV}^2$ taking account of non-perturbative effects in QCD is mainly reduced to the appearance in $\Pi(p)$ of power terms proportional to the vacuum mean values of the products of quark and gluonic fields. Of such mean vacuum values the most important is the correlator $\langle 0|\bar{\psi}\psi|0\rangle$ since it results in the appearance in $\Pi(p)$ of new (compared with the calculation in massless QCD) structure functions corresponding to spontaneous violation of chiral symmetry. The same correlator is most essential also from dimensional considerations. For this reason terms proportional to $\langle 0|\bar{\psi}\psi|0\rangle$ and terms with higher dimension $\langle 0|\bar{\psi}\psi|0\rangle^2$ will be taken into account in addition to the simplest quark loop when calculating $\Pi(p)$ in QCD. (The terms $\langle 0|\bar{\psi}\psi|0\rangle^2$ contain large numerical factors, which is why their contribution is important.) Another correlator $\langle 0|\bar{\psi}\sigma_{\mu\nu}\lambda^a G^a_{\mu\nu}\psi|0\rangle$ violating chiral invariance will also give some contribution in the nucleon case. In such a way we will get in QCD the expression for $\Pi(p)$ in the region of space-like $p^2 \sim -1\ \mathrm{GeV}^2$ – the left-hand part of the desired sum rules. The right-hand, phenomenological part of the sum rules will be determined by saturation of dispersion integrals for each of the structure functions contained in $\Pi(p)$ by the contribution of the low-lying baryonic states. Because of the presence of the mass term, the baryonic state contributes to at least two structure functions and this leads to the appearance of two or more equations which enable one to exclude (and thereby determine) the unknown residue and get an equation for the baryon mass almost without any parameters except $\langle 0|\bar{\psi}\psi|0\rangle$. In the isobar case there are several structure functions with definite relationships between them at the phenomenological level. For this reason we have an important possibility of checking the whole approach since the same relations must hold between the structure functions calculated in QCD.

This method will be applied to the calculation of masses of the isobar Δ and nucleon N. Besides, taking account of the strange quark mass m_s (in first order in the current quark mass m_s) makes it possible to calculate the mass splitting in the baryonic decuplet, i.e. the difference $M_{\Sigma^*} - M_\Delta$.

The material in the paper is organized as follows.

In sect. 2 we present the method. Thereupon, because in the isobar case a number of additional critical tests for this approach appear, in sect. 3 we deal with the polarization operator of currents with the isobar quantum numbers and calculate the isobar Δ mass. The mass difference $M_{\Sigma^*} - M_\Delta$ is calculated in sect. 4. The sum rule for the polarization operator of the current with nucleon quantum number is obtained and the nucleon mass is calculated in sect. 5. In sect. 6 some other baryonic resonances are considered and the values of residues related to coefficients in the nucleon form factor asymptotics and proton decay amplitudes are given. The role of effects not accounted for in the sum rules and our conclusions are presented in sects. 7 and 8.

2. Description of the method

We will start with the set of diagrams which will be accounted for in the calculation of the polarization operator $\Pi(p)$ (4) in QCD. One and the same set of diagrams will appear in both the isobar and nucleon cases. We suppose that in the region $-p^2 \equiv s \sim 1\ \mathrm{GeV}^2$ the effective interaction constant in QCD, $\alpha_s(s)$, is small enough $\alpha_s(1\ \mathrm{GeV}^2) \approx 0.3$–$0.4$ and that terms of the order of α_s can be neglected. The terms $\sim[\alpha_s(\mu^2) \ln (s/\Lambda^2)]^n$ corresponding to the leading logarithmic approximation will be accounted for, though they will not be important.

To define the set of significant diagrams, it is reasonable to proceed from the operator expansion for $\Pi(p)$, assuming $1/s$ to be small enough and neglecting for a while the leading logarithmic corrections. Then it is evident that in the first approximation in $1/s$ $\Pi(p)$ is described by the free quark loop of fig. 1a. The terms of next order in $1/s$ correspond to the operator $\bar{\psi}\psi$ in the operator expansion and are presented in fig. 1b. These terms contain the mean vacuum value $\langle 0|\bar{\psi}\psi|0\rangle$, violate the chiral invariance and will be responsible for the appearance of the baryon mass. (It should be emphasized that unlike the operator $\Pi(p)$ with mesonic quantum numbers considered by SVZ, in the discussed case of the polarization

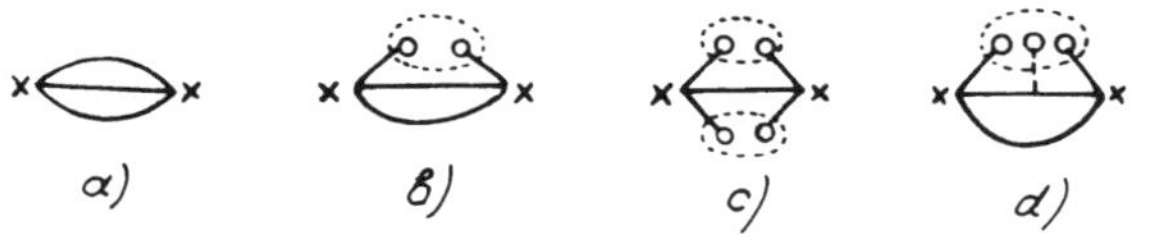

Fig. 1 Set of diagrams accounted for in the calculation of the polarization operator $\Pi(p)$: solid lines correspond to quarks, dashed lines to gluons, circles outlined by dots stand for the mean vacuum values of the field operators, crosses for interactions with the external current.

operator $\Pi(p)$ with baryonic quantum numbers, the quantity $\langle 0|\bar{\psi}\psi|0\rangle$ does not enter with a small factor proportional to the quark mass. Owing to the appearance of this small factor, the correlator $\langle 0|\bar{\psi}\psi|0\rangle$ was not as important in the description of mesonic resonances as will be the case here.) It can be seen from a comparison of figs. 1a and 1b that these diagrams contribute to different structures of the polarization operator: fig. 1a to terms with an odd number of the Dirac γ-matrices and fig. 1b to terms with an even number of γ-matrices.

Of operators of higher dimension, the most important are those containing products of four quark fields of the type $\bar{\psi}\Gamma\psi \cdot \bar{\psi}\Gamma\psi$. The diagram corresponding to the mean vacuum value of such operators is given in fig. 1c. The operators $\bar{\psi}\Gamma\psi \cdot \bar{\psi}\Gamma\psi$ with dimension $d=6$ are not next to the operator $\bar{\psi}\psi$ as far as their dimension is concerned; the operators $G^a_{\mu\nu}G^a_{\mu\nu}(d=4)$ and $g_s\bar{\psi}\sigma_{\mu\nu}\lambda^a\psi G^a_{\mu\nu}(d=5)$ have lower dimension. The special role of the operators $\bar{\psi}\Gamma\psi \cdot \bar{\psi}\Gamma\psi$ is connected with the fact that their mean vacuum values appear in the polarization operator with a large numerical factor. Indeed, compare figs. 1a and 1c. Fig. 1a has an integration over two quark loops, while fig. 1c has not. The integration in each loop inserts the factor $(2\pi)^2$ in the denominator so that fig. 1c is enhanced compared to fig. 1a by the numerical factor $(2\pi)^4$. There is no such numerical enhancement in fig. 2, corresponding to accounting for the vacuum mean value of the square of the gluonic field strength $\langle 0|G^a_{\mu\nu}G^a_{\mu\nu}|0\rangle$ and so the terms of such a type will be neglected.

Considering the mean vacuum value $\langle 0|\bar{\psi}\Gamma\psi\bar{\psi}\Gamma\psi|0\rangle$, we shall assume that the vacuum state in all channels dominates in the intermediate state sum, so that the 4-quark correlator effectively reduces to the correlator square $\langle 0|\bar{\psi}\psi|0\rangle^2$. (The same assumption was previously used by SVZ.) Physically such an assumption means that only one simplest combination of quark fields condenses. Analogous assumption, as is known, is widely used in the physics of a condensed medium (for instance, in the superconductivity theory). Formally, when calculating the mean vacuum

Fig. 2. The diagram proportional to the contribution $\langle 0|G^h_{\mu\nu}G^h_{\mu\nu}|0\rangle$ to the polarization operator.

value from four ψ-operators, one has to perform all possible non-perturbative pairings of $\bar{\psi}$ and ψ and at each pairing to assume that

$$\langle 0|: \bar{q}_\alpha^a q_\beta^b : |0\rangle = \tfrac{1}{12}\delta^{ab}\delta_{\alpha\beta}\langle 0|\bar{\psi}\psi|0\rangle\,. \tag{5}$$

Here q_α^a stands for a quark field of given flavour u, d, s; $a, b = 1, 2, 3$ are the colour, $\alpha, \beta = 1, 2, 3, 4$ are the spinor indeces and the symbol :: means subtraction of the perturbation theory contribution, i.e. the normal operator product in the sense of perturbation theory. (In eq. (5) we suppose exact flavour SU(3) symmetry in $\langle 0|\bar{q}q|0\rangle$; for a discussion of deviations from it see sect. 4.) It can easily be seen that fig. 1c contributes to the structure tensor with odd numbers of γ-matrices, i.e. to the same structures as fig. 1a and consequently can be considered as a correction to the latter.

The first correction in $1/s$ to the tensor structures of the polarization operator with an even number of γ-matrices, i.e. to fig. 1b, is described by fig. 1d and is proportional to the mean vacuum value $g_s\langle 0|\bar{\psi}\sigma_{\mu\nu}(\lambda^h/2)\psi G_{\mu\nu}^h|0\rangle$ violating the chiral invariance. Here $G_{\mu\nu}^h$ is the gluonic field strength tensor, λ^h are the Gell-Mann matrices in the colour space, $h = 1, \ldots 8$, $\sigma_{\mu\nu} = \frac{1}{2}i(\gamma_\mu\gamma_\nu - \gamma_\nu\gamma_\mu)$. This correction will be also taken into account but its role will be rather small, especially for the isobar case. When calculating fig. 1d, it is convenient to use the formula

$$\langle 0|: q_\alpha^a \bar{q}_\beta^b G_{\lambda\sigma}^h : |0\rangle = -\frac{1}{2^6\cdot 3}\langle 0|\bar{\psi}\sigma_{\mu\nu}\frac{\lambda^m}{2}\psi G_{\mu\nu}^m|0\rangle(\sigma_{\lambda\sigma})_{\alpha\beta}\frac{\lambda_{ab}^h}{2}\,. \tag{6}$$

Thus [yet without corrections $\sim(\alpha_s \ln(s/\Lambda^2))^n$], the polarization operator will be calculated on the basis of the set of diagrams of fig. 1. The corrections $\sim\alpha_s^n$ can easily be taken into account under the leading logarithmic approximation since the anomalous dimensions of the operator $\bar{\psi}\psi$ and of the current operators η_Δ and η_N are known. As usual, if the anomalous dimension of the operator $O_n(x)$ in the operator expansion of the T-product,

$$\mathrm{T}\{\eta(x), \bar{\eta}(0)\} = \sum_n C_n(x^2)O_n(0)\,, \tag{7}$$

is γ_n and the anomalous dimension of the current is γ_η, then the coefficient functions $C_n(p^2)$ in the momentum representation contain the factor

$$\left[\frac{\alpha_s(p^2)}{\alpha_s(\mu^2)}\right]^{2\gamma_\eta-\gamma_n}\,. \tag{8}$$

The anomalous dimensions will not be taken into account in terms proportional to the mean vacuum value from the operator $g_s\bar{\psi}\sigma_{\mu\nu}\lambda^h\psi G_{\mu\nu}^h$ since the contribution of these terms is comparatively small and the numerical value $g_s\langle 0|\bar{\psi}\sigma_{\mu\nu}(\lambda^h/2)\psi G_{\mu\nu}^h|0\rangle$ contains some ambiguity.

One more remark is relevant to anomalous dimensions γ of the four-fermion operators. Hereafter, it will be assumed that because of the factorization $\gamma = 2\gamma_{\bar{\psi}\psi}$, where $\gamma_{\bar\psi\psi}$ is the anomalous dimension of the operator $\bar\psi\psi$, $\gamma_{\bar\psi\psi}=\frac{4}{9}$ for three flavours. Rigorously speaking, such an assumption is invalid since, provided the factorization condition holds in the low-energy limit, it must be violated with increase of p^2 due to different anomalous dimensions of different operators $\bar\psi\Gamma\psi\cdot\bar\psi\Gamma\psi$. However (see appendix to the first paper by SVZ), the appearing deviations are small and, as calculations show, result, in our case, in variations of the polarization operator of not more than 3–4% i.e. they will not affect the baryon mass values.

Now some words on the used form of the dispersion sum rules. For each of the structure functions $f(p^2)$ of the polarization operator $\Pi(p)$ the following dispersion relation can be written:

$$f(s)=\frac{1}{\pi}\int_0^\infty \frac{\operatorname{Im} f(p'^2)}{p'^2+s}\,\mathrm{d}p'^2\,,\qquad s=-p^2\,, \tag{9}$$

with the necessary number of subtractions. (In fact, since the imaginary part of fig. 1a amplitude behaves as $\hat{p}p^4$, three subtractions are necessary.) Such a form is, however, inconvenient for our purposes because we wish the low-lying state contribution to dominate the right-hand side of the dispersion sum rules whereas it is not enhanced in (9). In order to get this, let us make use of the Borel improvement of the disperson integral convergency suggested by SVZ, i.e. consider the limit

$$\mathscr{B}f(s)\equiv \lim_{\substack{n\to\infty\\ s\to\infty\\ s/n\to M^2=\mathrm{const}}} \frac{s^{n+1}}{n!}\left(-\frac{\mathrm{d}}{\mathrm{d}s}\right)^n f(s)\,. \tag{10}$$

Applying the limiting procedure (10) to (9) we obtain

$$\mathscr{B}f(s)=\frac{1}{\pi}\int_0^\infty \mathrm{e}^{-p^2/M^2}\operatorname{Im} f(p^2)\,\mathrm{d}p^2\,. \tag{11}$$

If M in (11) is chosen to be of order of the lowest hadronic state mass in the given channel, the contribution of higher states will be suppressed. Clearly, the use of the Borel transformation (10) eliminates the problem of subtraction terms and takes off the polynomial terms in $f(s)$.

The following formulae will be useful when applying the limiting procedure to the amplitude of fig. 1:

$$\mathscr{B}\left(\ln\frac{\Lambda^2}{s}\right)=M^2\,,\qquad \mathscr{B}\left(s\ln\frac{\Lambda^2}{s}\right)=-M^4\,,\qquad \mathscr{B}\left(\frac{1}{s}\right)=1\,. \tag{12}$$

3. The isobar

Among various components of the isobar $\Delta(T=\frac{3}{2}, J^P=\frac{3}{2}^+)$, the most convenient for consideration is the component Δ^{++} (or Δ^-) which consists of quarks of one and the same flavour. First of all one should choose the form of the quark current

$\eta(x)$ corresponding to Δ^{++}. The form of the current should be chosen for the amplitude $\langle\Delta^{++}|\bar{\eta}|0\rangle$ to be large enough, i.e. at saturation of the sum rules by the isobar contribution there will be no suppression due to a mischoice of the form of the current. We know that the physical isobar consists of s-wave quarks. For this reason the introduction of ψ operator derivatives into the current is undesirable. Though the quarks in view are constituents and the current is constructed from current quarks, it is nevertheless clear that a better approximation can be obtained if one starts with the s-wave current quarks but not with introduction of the p-wave current quarks *ad initium*.

If we restrict ourselves to the consideration of currents proportional to the product of three quark field operators without derivatives, then the form of the current with isobar quantum numbers can be determined unambiguously. Denote the u-quark field operator by $u^a(x)$ (a is the colour index). Then the current is

$$\eta_\mu(x) = (u^a(x) C\gamma_\mu u^b(x)) u^c(x)\varepsilon^{abc}, \tag{13}$$

where C is the charge conjugation matrix. Evidently, the current $\eta_\mu(x)$ (13) possesses the necessary quantum numbers, its isospin is $T=\frac{3}{2}$ and it contains the state with $J^P=\frac{3}{2}^+$. Let us prove that all the rest of the non-zero currents are expressed via (13). To this end, write all the possible currents in a form where the matrix γ_μ carrying the external index μ is in front of the spinor u^c whose indices are not contracted with the other spinor indices, i.e. in the form

$$\eta_\mu = (u^a(x) CO_k u^b(x))\gamma_\mu O_k u^c(x)\varepsilon^{abc}, \tag{14}$$

where O_k is the full set of γ-matrices, $O_k = \{1, \gamma_5, \gamma_\lambda, \gamma_\lambda\gamma_5, \sigma_{\lambda\nu}\}$. Transposing the spinors u^a and u^b and accounting for the relation $C^{\mathrm{T}} = -C$(T stands for transposition), it can easily be seen that

$$u^a C u^b \varepsilon^{abc} = u^a C\gamma_5 u^b \varepsilon^{abc} = u^a C\gamma_\lambda\gamma_5 u^b \varepsilon^{abc} = 0\,. \tag{15}$$

The transposition of spinors u^b and u^c and the use of the Fierz relations and (15) gives

$$(u^a C\sigma_{\lambda\nu} u^b)\gamma_\mu\sigma_{\lambda\nu} u^c \varepsilon^{abc} = -\tfrac{1}{2}(u^a C\sigma_{\lambda\nu} u^b)\gamma_\mu\sigma_{\lambda\nu} u^c\varepsilon^{abc} = 0\,. \tag{16}$$

With the help of the Fierz transformation and (15) the equality

$$(u^a C\gamma_\lambda u^b)\sigma_{\lambda\mu} u^c\varepsilon^{abc} = i(u^a C\gamma_\mu u^b)u^c\varepsilon^{abc} \tag{17}$$

can be easily obtained which expresses the last non-zero variant in (14) via the current η_μ (13). The current η_μ (13) automatically satisfies the requirements which could be imposed on it *a priori*. Let us introduce the left-hand and right-hand components of spinors $u_{\mathrm{L}} = \frac{1}{2}(1+\gamma_5)u$ and $u_{\mathrm{R}} = \frac{1}{2}(1-\gamma_5)u$ writing (13) as

$$\eta_\mu = (u_{\mathrm{L}}^a C\gamma_\mu u_{\mathrm{R}}^b + u_{\mathrm{R}}^a C\gamma_\mu u_{\mathrm{L}}^b)(u_{\mathrm{L}}^c + u_{\mathrm{R}}^c)\varepsilon^{abc}\,. \tag{18}$$

It is seen from (18) that for massless quarks in the c.m.s. of quarks u^a and u^b, the sum of projections of their spins on the direction of their relative momentum is 1, i.e. the spin of the system $u^a u^b$ is 1 and the angular momentum according to the Pauli principle is zero (or two). Then, taking into account momentum and parity conservation, the quark u^c is also in an s- or d-state, i.e. the current η_μ corresponds to the description of the isobar in terms of constituent quarks, which, as was mentioned above, is very desirable. Finally, there is a non-relativistic limit for the current η_μ (13). This is also desirable, though not a compulsory requirement.

Using the Fierz transformation it can be shown that the current η_μ (18) is symmetric relative to transpositions of spinor indices of left-hand and right-hand spinors separately. This allows one to identify η_μ with one of the three-quark operators considered in ref. [7] and to use the value of the current anomalous dimension calculated in this paper.

On the basis of the method described in sect. 2, the contributions to the polarization operator

$$\Pi_{\mu\nu}(p) = i\int \mathrm{d}^4x\, \mathrm{e}^{ipx}\langle 0|\mathrm{T}\{\eta_\mu(x), \bar{\eta}_\nu(0)\}|0\rangle \tag{19}$$

of the diagrams of fig. 1 is calculated directly. In this case, for fig. 1a it is more convenient to calculate the imaginary part of the polarization operator at positive p^2 and afterwards to reconstruct the real part using the dispersion relation. As for figs. 1b–d, $\Pi_{\mu\nu}(p)$ can be immediately calculated at negative p^2.

I present here the results of the calculations of the contributions of figs. 1a–d to the polarization operator (Λ_U is the UV cut-off parameter):

$$\mathrm{Im}\,\Pi^{(a)}_{\mu\nu} = -\frac{p^4}{20(2\pi)^3}\Big\{\delta_{\mu\nu}\hat{p} - \tfrac{5}{16}\gamma_\mu\gamma_\nu\hat{p} + \tfrac{5}{16}(\gamma_\mu p_\nu - \gamma_\nu p_\mu) - \frac{p_\mu p_\nu \hat{p}}{p^2} + \tfrac{1}{16}(\gamma_\mu p_\nu + \gamma_\nu p_\mu)\Big\}, \tag{20}$$

$$\Pi^{(b)}_{\mu\nu} = \frac{1}{3\pi^2}\langle 0|\bar{\psi}\psi|0\rangle p^2 \ln\frac{\Lambda_\mathrm{U}^2}{-p^2}\Big\{\delta_{\mu\nu} - \tfrac{5}{16}\gamma_\mu\gamma_\nu + \tfrac{1}{4}(\gamma_\mu p_\nu - \gamma_\nu p_\mu)\frac{\hat{p}}{p^2} - \frac{1}{2}\frac{p_\mu p_\nu}{p^2}\Big\}, \tag{21}$$

$$\Pi^{(c)}_{\mu\nu} = \tfrac{4}{3}\langle 0|\bar{\psi}\psi|0\rangle^2 \frac{1}{p^2}\{\delta_{\mu\nu}\hat{p} - \tfrac{3}{8}\gamma_\mu\gamma_\nu\hat{p} + \tfrac{3}{8}(\gamma_\mu p_\nu - \gamma_\nu p_\mu) - \tfrac{1}{8}(\gamma_\mu p_\nu + \gamma_\nu p_\mu)\}, \tag{22}$$

$$\Pi^{(d)}_{\mu\nu} = -\frac{1}{3\pi^2}\frac{1}{8}\ln\frac{\Lambda_\mathrm{U}^2}{-p^2}\, g_s\Big\langle 0\Big|\bar{\psi}\sigma_{\lambda\sigma}\frac{\lambda^h}{2}\psi G^h_{\mu\nu}\Big|0\Big\rangle(\delta_{\mu\nu} - \tfrac{1}{4}\gamma_\mu\gamma_\nu). \tag{23}$$

The p^2 polynomials in (21), (23), vanishing under Borel transformation, are omitted. In addition, the terms $\sim p_\mu p_\nu/p^2$ in (23), which are inessential for the following, are also omitted. At this stage of the consideration it is appropriate to compare the relations between the different tensor structures which appeared in the polarization operator calculated in QCD with the relations which should be expected if one takes into account in the r.h.s. of the dispersion sum rules the contribution from the isobar only. In this case the r.h.s. of the dispersion sum rule is proportional to

$$\sum_r \langle 0|\eta_\mu(0)|\Delta(p,r)\rangle\langle\Delta(p,r)|\bar{\eta}_\nu(0)|0\rangle\,, \tag{24}$$

where summation is made over the spin projections r of the isobar Δ with the momentum p, $p^2=M_\Delta^2$. The matrix element contained in (24) is of the form

$$\langle 0|\eta_\mu(0)|\Delta(p,r)\rangle=\lambda v_\mu^{(r)}(p)\,, \tag{25}$$

where $v_\mu^{(r)}(p)$ is the wave function (spin vector of the isobar in Rarita–Schwinger formalism and λ is a constant. Indeed, because of the subsidiary conditions $\gamma_\mu v_\mu(p)=0$, $p_\mu v_\mu(p)=0$, which $v_\mu(p)$ obeys in the Rarita–Schwinger formalism, and the Dirac equations $(\hat{p}-M_\Delta)v_\mu=0$, (25) is a unique Lorentz-invariant expression for this matrix element. Substituting (25) into (24) and summing over r, we find (the functions $v_\mu^{(r)}$ are normalized by $\bar{v}_\mu^{(r)}v_\mu^{(r)}=-2M_\Delta$)

$$\begin{aligned}\sum_r &\langle 0|\eta_\mu(0)|\Delta(p,r)\rangle\langle\Delta(p,r)|\bar{\eta}_\nu(0)|0\rangle\\ &=-\lambda^2\left\{\left[\delta_{\mu\nu}\hat{p}-\tfrac{1}{3}\gamma_\mu\gamma_\nu\hat{p}+\tfrac{1}{3}(\gamma_\mu p_\nu-\gamma_\nu p_\mu)-\frac{2}{3}\frac{p_\mu p_\nu\hat{p}}{M_\Delta^2}\right]\right.\\ &\quad\left.+\left[\delta_{\mu\nu}-\tfrac{1}{3}\gamma_\mu\gamma_\nu+\tfrac{1}{3}(\gamma_\mu p_\nu-\gamma_\nu p_\mu)\frac{\hat{p}}{M_\Delta^2}-\frac{2}{3}\frac{p_\mu p_\nu}{M_\Delta^2}\right]M_\Delta\right\}.\end{aligned} \tag{26}$$

On the r.h.s. of (26) the structures with even and odd numbers of γ-matrices are written separately in order to compare them in a more convenient way with the expressions for $\Pi_{\mu\nu}^{(\mathrm{a,c})}$ and $\Pi_{\mu\nu}^{(\mathrm{b,d})}$, respectively. It can be seen by comparing (20) and (22) with the first square bracket in (26), and (21) with the second, that the relations between different structure functions in the polarization operator reproduce very well analogous relations in the isobar on-mass-shell. The resemblance between (26) and the main term of the operator expansion $\Pi_{\mu\nu}^{(\mathrm{a})}$ is especially strong and it deteriorates a bit when going over to the next terms, $\Pi_{\mu\nu}^{(\mathrm{b})}$, $\Pi_{\mu\nu}^{(\mathrm{c})}$. (It should be borne in mind that one has to compare with (26) not the expressions (20)–(22) directly but the results of substitutions of the latter into (11) at $M^2=M_\Delta^2$. In doing so, for instance, an extra factor $\frac{1}{2}$ as compared to the term $\delta_{\mu\nu}\hat{p}$ will appear in the term $p_\mu p_\nu\hat{p}$ in (20), so that $-\frac{1}{2}p_\mu p_\nu\hat{p}/M_\Delta^2$ in (20) should be compared with $-\frac{2}{3}p_\mu p_\nu\hat{p}/M_\Delta^2$ in (26).) It is interesting to emphasize that the structure which is absent in (26) appears in (20), (22) with small coefficients. In $\Pi_{\mu\nu}^{(\mathrm{d})}$ the relation between structures differs from that in (26) but this term is itself a very small correction (~5%) to $\Pi_{\mu\nu}^{(\mathrm{c})}$.

Let us now compare (20)–(22) with the expression which would appear on the r.h.s. of the dispersion sum rules if the resonance N* with $J^P = \frac{1}{2}^-$ dominated it. The general expression for the matrix element $\langle 0|\eta_\mu|N^*\rangle$ would then be

$$\langle 0|\eta_\mu(0)|N^*(p, r)\rangle = (\alpha p_\mu + \beta\gamma_\mu)v^{(r)}, \tag{27}$$

where $v^{(r)}$ is the spinor characterizing the state N*, and the r.h.s. of the dispersion sum rule would be proportional to (m is the resonance N* mass):

$$\begin{aligned}\sum_r \langle 0|\eta_\mu(0)|N^*(p, r)\rangle\langle N^*(p, r)|\bar{\eta}_\nu(0)|0\rangle \\ = -\beta^2[\gamma_\mu\gamma_\nu\hat{p} - (\gamma_\mu p_\nu - \gamma_\nu p_\mu)] + \beta(\beta + m\alpha)(\gamma_\mu p_\nu + \gamma_\nu p_\mu) \\ + \alpha^2 p_\mu p_\nu \hat{p} + m\beta^2\gamma_\mu\gamma_\nu + \alpha\beta(\gamma_\mu p_\nu - \gamma_\nu p_\mu) + \alpha(2\beta + m\alpha)p_\mu p_\nu .\end{aligned} \tag{28}$$

On the r.h.s. of (28) the term $\delta_{\mu\nu}$ is absent (which was evident *a priori*) and the relations between the remaining structures do not look like (20)–(22).

Note that the relations between the structures appearing in $\Pi_{\mu\nu}$ are cardinally connected with the fact that the current η_μ has isospin $T = \frac{3}{2}$ but are not a simple consequence of η_μ being a spin vector. Indeed, if instead of the current η_μ (13) we consider an analogous one, but with isospin $T = \frac{1}{2}$,

$$\eta_\mu^{T=1/2} = [(u^a C\gamma_\mu u^b)d^c - (u^a C\gamma_\mu d^b)u^c]\varepsilon^{abc} \tag{29}$$

(d is the d-quark field) and calculate the polarization operator according to fig. 1, the relations between the structure functions will be totally different from those in (20)–(22) [see eqs. (65)–(67)]. This is a reflection of the fact that the baryonic channel with $T = \frac{3}{2}$ has the resonance Δ, while the channel with $T = \frac{1}{2}$ has not such a strong resonance with $J = \frac{3}{2}$.

It follows from the above discussion that there are very good chances for the polarization operator $\Pi_{\mu\nu}$, which corresponds to the current η_μ in the dispersion sum rule, to be saturated by the isobar Δ contribution.

Though the contribution of states with $J = \frac{1}{2}$ to the dispersion sum rules for $\Pi_{\mu\nu}$ is small enough to exclude it totally, these sum rules should be written for the structure functions at $\delta_{\mu\nu}\hat{p}$ and $\delta_{\mu\nu}$. Hereafter, we will consider in this section only such sum rules. Before writing them, it is reasonable to estimate the contribution of $\Pi_{\mu\nu}^{(d)}$ as compared to $\Pi_{\mu\nu}^{(c)}$. The mean vacuum value $g_s\langle 0|\bar{\psi}\sigma_{\lambda\sigma}(\lambda^h/2)\psi G_{\lambda\sigma}^h|0\rangle$ has been estimated in ref. [6]. If we denote

$$g_s\langle 0|\bar{\psi}\sigma_{\mu\nu}\frac{\lambda^h}{2}\psi G_{\mu\nu}^h|0\rangle \equiv m_0^2\langle 0|\bar{\psi}\psi|0\rangle, \tag{30}$$

then, as was shown in [6], the sign of $g_s\langle 0|\bar{\psi}\sigma_{\mu\nu}\lambda^h\psi G_{\mu\nu}^h|0\rangle$ coincides with the sign of $\langle 0|\bar{\psi}\psi|0\rangle$, i.e. $m_0^2 > 0$ and m_0^2 was estimated as 0.5–1 GeV^2. It follows from these estimates that the ratio of the coefficients of $\delta_{\mu\nu}$ in (23) and (21) at $-p^2 \sim 1$ GeV^2 is approximately $\frac{1}{8}$–$\frac{1}{16}$, i.e. small.

We now turn to the sum rules and first restrict discussion to a rather rough approximation, disregarding the anomalous dimensions of the operators, and accounting for only the isobar Δ contribution on the r.h.s. of the sum rules. Making use of eqs. (11), (12), (20)–(23), (30), the sum rules at the structure factors $\delta_{\mu\nu}\hat{p}$ and $\delta_{\mu\nu}$ can be written as

$$M^6 + \tfrac{20}{3}a^2 = \tilde{\lambda}^2\, e^{-M_\Delta^2/M^2}$$
$$\tfrac{20}{3}M^4 a\left(1 - \frac{1}{8}\frac{m_0^2}{M^2}\right) = \tilde{\lambda}^2 M_\Delta\, e^{-M_\Delta^2/M^2} \qquad (31)$$

where the isobar Δ contribution is expressed via (26) and is denoted by

$$a = -(2\pi)^2\langle 0|\bar{\psi}\psi|0\rangle\,,$$
$$\tilde{\lambda}^2 = 5\cdot(2\pi)^4\lambda^2\,. \qquad (32)$$

The right- and left-hand sides of eqs. (31) must coincide with each other as functions of variable the M in some interval of M. It is natural to expect that this interval contains the point $M = M_\Delta$. Let us put $M = M_\Delta$ and exclude the unknown parameter $\tilde{\lambda}^2$ from the set of equations (31). If the small correction $\frac{1}{8}m_0^2/M^2$ is neglected, there appears the quadratic equation for M_Δ^3:

$$(M_\Delta^3)^2 - \tfrac{20}{3}M_\Delta^3 a + \tfrac{20}{3}a^2 = 0\,. \qquad (33)$$

The solution of this equation gives the isobar mass

$$M_\Delta = [\tfrac{10}{3}a(1+\sqrt{\tfrac{2}{5}})]^{1/3} = 1.43\ \text{GeV}\,, \qquad (34a)$$

or with the correction due to the term $m_0^2/8M^2$ (at $m_0^2 = 0.5\ \text{GeV}^2$)

$$M_\Delta = 1.41\ \text{GeV} \qquad (34b)$$

at the numerical value $\langle 0|\bar{\psi}\psi|0\rangle = -(240\ \text{MeV})$ [3]. (We took the largest root of the quadratic equation (31); the second root does not satisfy the system (33) – see table 1). The values M_Δ of (34) differ from the experimental value $M_{\Delta\text{exp}} = 1.23$ GeV by 15% only, which is a fine agreement with respect to the comparatively rough approximation. In particular, it should be borne in mind that experimentally the baryonic channel with $T = \frac{3}{2}$, $J^P = \frac{3}{2}^+$ has two resonances with masses 1.23 and 1.69 GeV and what is determined theoretically is an intermediate value between both masses (but closer to the lowest state because the highest state is suppressed by the exponential factor $\exp[-(M_{\Delta_2}^2 - M_{\Delta_1}^2)/M_{\Delta_1}^2]$).

Now it is necessary to go over to the next, more rigorous, level of consideration, to account for different corrections to the sum rules (31) and, what is more important, to test whether the r.h.s. and l.h.s. of (31) do coincide with each other as functions of M at a fixed M_Δ. The following two types of correction to the sum rules (31) will be accounted for:

(i) The corrections due to anomalous dimensions of the operators discussed in sect. 2. The anomalous dimension of the current η_μ (13) is known [7]: $\gamma_\eta = \frac{2}{27}$. Therefore, the left-hand sides of (31) should be multiplied by $[\alpha_s(M)/\alpha_s(\mu)]^{4/27}$.

Besides, taking account of anomalous dimensions, a in (31) is replaced by $a \to a[\alpha_s(M)/\alpha_s(\mu)]^{-4/9}$.

(ii) On the r.h.s. of the sum rules, in addition to the isobar contribution, we will take into account the contribution of the hadronic continuum and will assume that it is equal to the contribution of Im $\Pi^{(a)}_{\mu\nu}$ and Im $\Pi^{(b)}_{\mu\nu}$ to the structures $\delta_{\mu\nu}\hat{p}$ and $\delta_{\mu\nu}$, respectively, and starts with some value $p^2 = W^2$.

After introducing these two corrections (and transferring the continuum contribution to the l.h.s.) the sum rules will take the form

$$I_1(M) \equiv M^6 \left(\frac{\ln (M/\Lambda)}{\ln (\mu/\Lambda)}\right)^{-4/27} \left[1 - e^{-W^2/M^2}\left(\frac{W^4}{2M^4} + \frac{W^2}{M^2} + 1\right)\right] + \tfrac{20}{3} a^2 \left(\frac{\ln (M/\Lambda)}{\ln (\mu/\Lambda)}\right)^{20/27}$$
$$= \tilde{\lambda}^2 e^{-M_\Delta^2/M^2}, \qquad (35)$$
$$I_2(M) \equiv \frac{20}{3M_\Delta} aM^4 \left\{\left(\frac{\ln (M/\Lambda)}{\ln (\mu/\Lambda)}\right)^{8/27} \left[1 - e^{-W^2/M^2}\left(\frac{W^2}{M^2} + 1\right)\right] - \frac{m_0^2}{8M^2}\right\} = \tilde{\lambda}^2 e^{-M_\Delta^2/M^2},$$

where Λ is the parameter entering the definition of the effective charge, $\alpha_s(M) = 4\pi/9 \ln (M^2/\Lambda^2)$, and μ is the normalization point at which, by assumption, $\langle 0|\bar{\psi}\psi|0\rangle$ coincides with the low-energy limit (1).

The sum rules (35) permit two critical tests: first, the ratio of the l.h. sides $I_1(M)/I_2(M)$ must equal 1; second, the M dependence of the r.h.s. and l.h.s. of (35) at fixed M_Δ must be the same at large enough variation scale of each function. Fig. 3 represents the ratio $I_1(M)/I_2(M)$ for $M_\Delta = 1.4$ GeV. (The M_Δ dependence of I_1/I_2 is relatively weak; $I_1/I_2 \sim M_\Delta$.) For comparison, an analogous curve for the nucleon is presented in the same figure (see sect. 5). Fig. 4 shows the curves $I_1(M)$, $I_2(M)$ and $e^{-M_\Delta^2/M^2}$ for $M_\Delta = 1.4$ GeV, matched at the point $M = M_\Delta = 1.4$ GeV. The calculation was made with the following values of the parameters: $\langle 0|\bar{\psi}\psi|0\rangle = -(240\ \text{MeV})^3$ ($a = 0.546\ \text{GeV}^3$), $\Lambda = 150$ MeV, $\mu = 500$ MeV, $m_0^2 = 0.5\ \text{GeV}^2$, $W = 2.5$ GeV.

As can be seen from fig. 3, within $1.03 < M < 1.45$ GeV the equality $I_1/I_2 = 1$ is fulfilled to an accuracy of better than 25%. It should be emphasized that the fulfillment of this equality is extremely non-trivial: it really contains no free

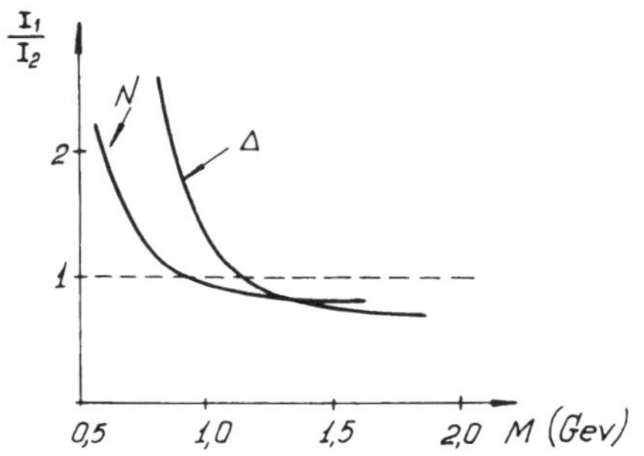

Fig. 3. The ratio $I_1(M)/I_2(M)$ for the isobar and nucleon [formulae (35), (58)].

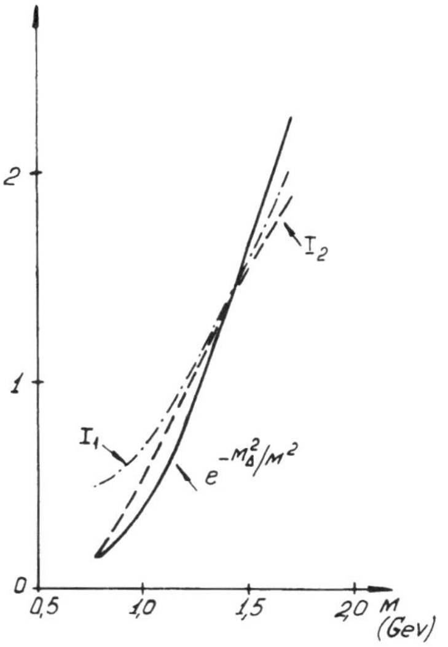

Fig. 4. Functions $I_1(M)$, $I_2(M)$, $\exp(-M_\Delta^2/M^2)$ for the isobar at $M_\Delta = 1.4$ GeV.

parameters except a, which is taken from the low-energy relation (the effect of the other parameters is rather small) and the ratio I_1/I_2 "does not know" that it is to be saturated by a single resonance. Thereby, fulfillment of the ratio $I_1/I_2 \approx 1$ proves two facts at once: (i) In QCD the role of corrections not accounted for on the left-hand sides of (35) is small, not more than several tens of percent. (ii) The role of other resonances on the right-hand sides of (33) is also comparatively small. (In this case the limitation is weaker since the ratio of the right-hand sides of (35) is not very sensitive to higher resonances.)

Fig. 4 shows that within $1 < M < 1.6$ GeV, while each of the functions $I_1(M)$, $I_2(M)$ and $\exp(-M_\Delta^2/M^2)$ is varied by a factor of 3–4, their M dependence turns out to be equal up to 20–25%, i.e. this test is also fulfilled well enough. Discussing the parameter effect it should be noticed that the parameters Λ, μ and m_0 weakly affect the curves of fig. 3, 4 within a permissible interval of their variation. The effect of the continuum threshold W is somewhat larger. This parameter is, however deliberately chosen sufficiently large because as it decreases the situation improves: the curve of fig. 3 approaches 1 while the curves of fig. 4 approach each other.

Up to now we have considered one value, $M_\Delta = 1.4$ GeV, which coincided with the value obtained from our rougher consideration [eq. (34)]. To estimate the permissible interval of M_Δ variation one must choose definite accuracy criteria. We will introduce three such criteria:

(i) $I_1/I_2 = 1 \pm 25\%$.

(ii) The ratio of the left- and right-hand sides of both eqs. (33) must be equal to the constant up to 25%.

TABLE 1

The M variation intervals at given isobar mass M_{Δ}, between which conditions (i)–(iii) are fulfilled: $M_{\min} < M < M_{\max}$

M (GeV)	$M_{\min}$ (GeV)	$M_{\max}$ (GeV)
1	0.93	1.06
1.2	0.97	1.25
1.4	1.03	1.45
1.6	1.10	1.45

(iii) The continuum contribution must not be more than 30% from the contribution of all the rest on the l.h.s. of (35).

Table 1 shows the variation intervals of $M_{\min} < M < M_{\max}$ at given M_{Δ} inside which criteria (i)–(iii) are fulfilled (the parameters are the same as before). At $M_{\Delta} = 1.4$ and 1.6 GeV $M_{\max}$ is determined by the third criterion and if W decreases from 2.5 to 2 GeV, $M_{\max}$ decreases to 1.2 GeV. As is seen from table 1, the widest "gates" for the variable M exist at $M_{\Delta} = 1.4$ GeV, at $M_{\Delta} \leqslant 1$ GeV and $M_{\Delta} \geqslant 1.6$ GeV the gates slam (especially hard at $W = 2$ GeV).

Summarizing the consideration of the sum rules for the isobar, it should be repeated that the rules obtained satisfy all the tests and lead to the isobar mass value $M_{\Delta} = 1.4$ GeV $\pm$ 15%. The error mentioned also includes an uncertainty due to QCD corrections not accounted for (it should not be forgotten that taking the cubic root improves the accuracy three times).

Besides the isobar mass, the sum rules make it possible to determine the value of the constant λ, introduced in (25), which is the transition amplitude of the isobar into three quarks which are at the same point. The knowledge of this constant may be useful in considering small distance processes involving the isobar. From (33) it follows that

$$\lambda^2 = 2.5 \cdot 10^{-3}\ \mathrm{GeV}^6 \tag{36}$$

with a precision of about 50%.

4. Mass splitting in the decuplet

Up to now quarks have been assumed to be massless, i.e. all states of the decuplet degenerate. One can try to make similar considerations for the decuplet members which contain strange quarks, for instance, Σ^*, to account for the strange quarks mass m_s in first order and to get the mass splitting in the baryonic decuplet. (According to the Gell-Mann–Okubo formula $M_{\Omega} - M_{\Xi^*} = M_{\Xi^*} - M_{\Sigma^*} = M_{\Sigma^*} - M_{\Delta}$). The calculation of the mass splitting in the decuplet is less reliable than that of the Δ isobar mass because we have to take the sum rule difference for Σ^* and Δ while with such a subtraction the corrections not accounted for can be more essential.

The current with quantum numbers Σ^* can be obtained from (13) using the SU(3) symmetry relations

$$\eta_\mu^{\Sigma^*} = \sqrt{\tfrac{1}{3}}[2(u^a C\gamma_\mu s^b)u^c + (u^a C\gamma_\mu u^b)s^c]\varepsilon^{abc}, \tag{37}$$

where $s^a(x)$ is the s-quark field.

A serious danger in considering the current with a massive quark is the fact that, because of the s-quark mass in the three-quark system, the states which were absent in case of massless quarks may appear. The presence of these states can result in appearance on the r.h.s. of the sum rules of a contribution from excited states with orbital excitations. In this case, when calculating the sum rule difference for Σ^* and Δ, the contribution of such states can give rise to an effect capable of masking the mass difference effect $M_{\Sigma^*} - M_\Delta$. Luckily, the current (35) is such that there are none of these states in it. In fact, considering the system of two massless u-quarks and making the Fierz transformation in the first term of (37), one can show that the statements made after formula (18) for this system apply as before, i.e. the spin of u-quarks is 1 and the orbital moment is zero (or two). Then the s-quark also possesses orbital moment equal to zero (or two), i.e. the orbital moments remain the same as for massless quarks.

All the calculations of $\Pi_{\mu\nu}$ are completely analogous to those performed in sect. 3. We present the results as extra terms in figs. 1a–d proportional to m_s:

$$\Delta \operatorname{Im} \Pi_{\mu\nu}^{(a)} = -m_s \frac{p^4}{8(2\pi)^3}\left\{\delta_{\mu\nu} - \tfrac{1}{3}\gamma_\mu\gamma_\nu + \tfrac{1}{3}(\gamma_\mu p_\nu - \gamma_\nu p_\mu)\frac{\hat{p}}{p^2} - \frac{2}{3}\frac{p_\mu p_\nu}{p^2}\right\}, \tag{38}$$

$$\Delta\Pi_{\mu\nu}^{(b)} = \frac{m_s}{3\pi^2}\langle 0|\bar{\psi}\psi|0\rangle \ln\frac{\Lambda_U^2}{-p^2}\{\delta_{\mu\nu}\hat{p} - \tfrac{3}{8}\gamma_\mu\gamma_\nu\hat{p} + \tfrac{3}{8}(\gamma_\mu p_\nu - \gamma_\nu p_\mu) - \tfrac{1}{8}(\gamma_\mu p_\nu + \gamma_\nu p_\mu)\}, \tag{39}$$

$$\Delta\Pi_{\mu\nu}^{(c)} = \frac{2}{3}\frac{m_s}{p^2}\langle 0|\bar{\psi}\psi|0\rangle^2(\delta_{\mu\nu} - \tfrac{1}{4}\gamma_\mu\gamma_\nu), \tag{40}$$

$$\Delta\Pi_{\mu\nu}^{(d)} = 0. \tag{41}$$

With SU(3) symmetry violation, $\langle 0|\bar{s}s|0\rangle \neq \langle 0|\bar{u}u|0\rangle$. This circumstance was not taken into account in (38)–(40) since in this case it will go beyond the accuracy of eqs. (38)–(40) but it must be taken into account in the zero-order terms in m_s, (21), (22), before using them in the sum rules for Σ^*. Necessary changes reduce to the following: in (21) for $\Pi_{\mu\nu}^{(b)}$

$$\langle 0|\bar{\psi}\psi|0\rangle \to \tfrac{2}{3}\langle 0|\bar{u}u|0\rangle + \tfrac{1}{3}\langle 0|\bar{s}s|0\rangle,$$

and in (22) for $\Pi_{\mu\nu}^{(c)}$,

$$\langle 0|\bar{\psi}\psi|0\rangle^2 \to \langle 0|\bar{u}u|0\rangle[(\tfrac{1}{3})\langle 0|\bar{u}u|0\rangle + (\tfrac{2}{3})\langle 0|\bar{s}s|0\rangle].$$

Let us write the sum rules for the structures at $\delta_{\mu\nu}\hat{p}$ and $\delta_{\mu\nu}$. We confine ourselves to the following rough approximation: disregard the continuum contribution on the r.h.s. and the anomalous dimensions and the term $m_0^2/8M^2$ on the l.h.s. Introducing the notations

$$\langle 0|\bar{s}s|0\rangle = \langle 0|\bar{u}u|0\rangle(1+\gamma)\,, \qquad a = (2\pi)^2\langle 0|\bar{u}u|0\rangle\,, \qquad \delta = m_s/M\,,$$

using (20)–(23), (38)–(40) and making the same transformations as in the derivation of formula (31), we have

$$M^6 + \tfrac{20}{3}M^3 a\delta + \tfrac{20}{3}a^2(1-\tfrac{2}{3}\gamma) = \tilde{\lambda}^2_{\Sigma^*}\,\mathrm{e}^{-M_{\Sigma^*}/M^2}$$

$$\tfrac{20}{3}aM^4(1-\tfrac{1}{3}\gamma) + \tfrac{5}{2}M^7\delta + \tfrac{10}{3}M\delta a^2 = \tilde{\lambda}^2_{\Sigma^*}M_{\Sigma^*}\,\mathrm{e}^{-M_{\Sigma^*}/M^2}\,. \tag{42}$$

Putting $M = M_{\Sigma^*}$ and excluding $\tilde{\lambda}^2_{\Sigma^*}$ from (42) we arrive at the equation for M_{Σ^*}:

$$M^6_{\Sigma^*}(1-\tfrac{5}{2}\delta) - \tfrac{20}{3}aM^3_{\Sigma^*}(1-\tfrac{1}{3}\gamma-\delta) + \tfrac{20}{3}a^2(1-\tfrac{2}{3}\gamma-\tfrac{1}{2}\delta) = 0\,. \tag{43}$$

This equation should be compared with that for M_Δ, (33), obtained under the same assumptions. The following relation between M_{Σ^*} and M_Δ comes from (43) and (33) in first order in γ and δ:

$$M_{\Sigma^*} - M_\Delta \approx \tfrac{2}{3}m_s + \tfrac{1}{9}\gamma M_\Delta\,. \tag{44}$$

The quantity γ is unknown, there are only arguments in favour of $\gamma > 0$ [8]. It is natural to suppose that γ is of the order of the usual parameters of SU(3) violation, say, $\gamma \approx \frac{1}{5}$. Then the contribution from the second term to (44) is rather small and at $m_s = 150$ MeV we have

$$M_{\Sigma^*} - M_\Delta = 125\ \text{MeV}\,, \tag{45}$$

compared with the experimental value 150 MeV. It should be repeated once more that the calculation of the mass difference is based on stronger assumptions than the calculation of M_Δ. In particular, the role of the continuum could be more significant here if the ratio of the continuum contribution to the resonance contribution in the sum rules for Σ^* and Δ differs by a value of order M_s/M_Δ.

5. The nucleon

The consideration for the nucleon case is quite similar to that for the isobar case. First one should choose a three-quark current such that the transition amplitude of a nucleon into it is not small. There is every reason to believe that the current (for the proton)

$$\eta(x) = (u^a(x)C\gamma_\mu u^b(x))\gamma_5\gamma_\mu d^c(x)\varepsilon^{abc} \tag{46}$$

possesses such a property.

Considerations analogous to those for the isobar show that with such a choice of the current and with massless u-quarks, all quarks are in an s (or d) state. Again, the current (46) has a non-relativistic limit. In the nucleon case, however, unlike the isobar, there is one more current without quark field derivatives and with the above properties. Discussion of this current will be postponed for a while and later some arguments will be given why it should be expected that the nucleon transition amplitude into it is small.

For the current (46) the contributions of the diagrams of fig. 1 to the polarization operator are calculated directly and turn out to be

$$\mathrm{Im}\, \Pi^{(\mathrm{a})} = \frac{p^4}{8(2\pi)^3}\hat{p}\,, \tag{47}$$

$$\Pi^{(\mathrm{b})} = -\frac{1}{(2\pi)^2}\langle 0|\bar{\psi}\psi|0\rangle p^2 \ln\frac{\Lambda_{\mathrm{U}}^2}{-p^2}\,, \tag{48}$$

$$\Pi^{(\mathrm{c})} = -\tfrac{2}{3}\langle 0|\bar{\psi}\psi|0\rangle^2 \frac{\hat{p}}{p^2}\,, \tag{49}$$

$$\Pi^{(\mathrm{d})} = -\frac{1}{2(2\pi)^2} g_s\langle 0|\bar{\psi}\sigma_{\mu\nu}\frac{\lambda^h}{2}\psi G^h_{\mu\nu}|0\rangle \ln\frac{\Lambda_{\mathrm{U}}^2}{-p^2}\,. \tag{50}$$

From (47)–(50) using the Borel transformation we get two sum rules (at the structures $\hat{p}$ and 1):

$$M^6 + \tfrac{4}{3}a^2 = \tilde{\lambda}_{\mathrm{N}}^2\, \mathrm{e}^{-M_{\mathrm{N}}/M^2}\,, \tag{51}$$

$$2aM^4\left(1+\frac{1}{2}\frac{m_0^2}{M^2}\right) = \tilde{\lambda}_{\mathrm{N}}^2 M_{\mathrm{N}}\, \mathrm{e}^{-M_{\mathrm{N}}/M^2}\,, \tag{52}$$

where m_0 and a are determined according to (30), (32). On the r.h.s. of (51), (52) only a one-nucleon state contribution is left. The matrix element of the nucleon transition into the current (46) has the form

$$\langle 0|\eta(0)|\mathrm{N}(p,r)\rangle = \lambda_{\mathrm{N}} v^{(r)}(p)\,, \tag{53}$$

where $v^{(r)}(p)$ is the spinor which describes the state of a nucleon with momentum p and polarization r (the renormalization is $\bar{v}^r(p)v^r(p) = 2M_{\mathrm{N}}$). The constant $\tilde{\lambda}_{\mathrm{N}}$ in (51), (52) is related to λ_{N} in (53) by

$$\tilde{\lambda}_{\mathrm{N}}^2 = 2(2\pi)^4\lambda_{\mathrm{N}}^2\,. \tag{54}$$

Putting $M = M_{\mathrm{N}}$ in (51), (52) and excluding $\tilde{\lambda}_{\mathrm{N}}^2$ from them, we arrive at the equation for M_{N}:

$$(M_{\mathrm{N}}^3)^2 - 2aM_{\mathrm{N}}^3\left(1+\frac{m_0^2}{2M_{\mathrm{N}}^2}\right) + \tfrac{4}{3}a^2 = 0\,, \tag{55}$$

the solution of which gives the nucleon mass

$$M_{\mathrm{N}} = \{(2\pi)^2|\langle 0|\bar{\psi}\psi|0\rangle|^2[1+\varepsilon+\sqrt{(1+\varepsilon)^2-\tfrac{4}{3}}]\}^{1/3}\,, \tag{56}$$

where $\varepsilon = m_0^2/2M_N^2$. (The second root does not satisfy eqs. (51), (52); see table 2). Numerically, with the same value of $\langle 0|\bar{\psi}\psi|0\rangle$ as before and $m_0^2 = 0.5\ \mathrm{GeV}^2$

$$M_N = 985\ \mathrm{MeV}\,. \tag{57}$$

The value $m_0^2 = 0.5\ \mathrm{GeV}^2$ is chosen near the lower boundary of the estimate of ref. [6] for better agreement with experiment. At $m_0^2 = 1\ \mathrm{GeV}^2$, instead of (57) we would have $M_N = 1.08$ GeV, i.e. the variation of m_0^2 within permissible limits does not affect the nucleon mass value very strongly. It should be remarked, however, that since the roots of eq. (55) do not strongly differ from each other (the correction $\sim a^2$ is approximately equal to the first term M_N^6), the effect of different small corrections to the sum rules on the value of M_N is considerably larger than in the isobar case. This circumstance becomes essential in the calculation of the mass differences in the baryon octet.

Let us now turn to the more detailed analysis of the sum rules (51), (52), taking into account the anomalous dimensions and the contribution of the hadronic continuum. The anomalous dimension of the current η (46) is $\gamma_\eta = -\frac{2}{9}$ [7] and the sum rules (51), (52) take the form

$$I_1(M) \equiv M^6\left(\frac{\ln(M/\Lambda)}{\ln(\mu/\Lambda)}\right)^{4/9}\left[1 - \mathrm{e}^{-W^2/M^2}\left(\frac{W^4}{2M^4} + \frac{W^2}{M^2} + 1\right)\right] + \tfrac{4}{3}a^2\left(\frac{\ln(M/\Lambda)}{\ln(\mu/\Lambda)}\right)^{4/3}$$
$$= \tilde{\lambda}_N^2\,\mathrm{e}^{-M_N^2/M^2}\,, \tag{58}$$
$$I_2(M) \equiv 2a\frac{M^4}{M_N}\left\{\left(\frac{\ln(M/\Lambda)}{\ln(\mu/\Lambda)}\right)^{8/9}\left[1 - \mathrm{e}^{-W^2/M^2}\left(\frac{W^2}{M^2} + 1\right)\right] + \frac{m_0^2}{2M^2}\right\}$$
$$= \tilde{\lambda}_N^2\,\mathrm{e}^{-M_N^2/M^2}\,,$$

in the same notations as in (53).

The ratio $I_1(M)/I_2(M)$ for $M_N = 1$ GeV is presented in fig. 3. (The continuum threshold W is taken to be 2 GeV, the other parameters are the same as in sect. 3). As is seen from fig. 3, the equality $I_1(M)/I_2(M) = 1$ within $0.8 < M < 1.5$ GeV is fulfilled with accuracy better than 20%. Fig. 5 shows the curves $I_1(M)$, $I_2(M)$ and $\exp(-M_N^2/M^2)$ for $M_N = 1$ GeV matched at the point $M = M_N = 1$ GeV. The curves coincide well with each other, each of them being varied by a factor of 4–5. Just the same as in the isobar case, some diminishing of W improves the coincidence of the curves. Table 2 gives the variation intervals of $M_{\min} < M < M_{\max}$ with the given M_N between which the criteria introduced in sect. 3 are fulfilled. In the calculations W was taken to be 2 GeV. When W diminishes to 1.7 GeV, $M_{\max}$ diminishes to 1 GeV. Table 2 exhibits that the best value of the nucleon mass following from the sum rule is $M_N = 1$ GeV with an error including an uncertainty in m_0 and QCD corrections not accounted for of order 15–20%.

In the nucleon case, the knowledge of the constant λ_N defined in (53) is of considerable interest, since this quantity emerges as a factor in the proton decay

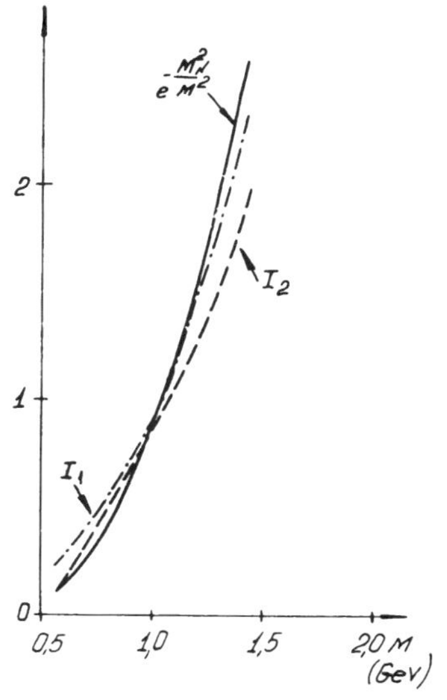

Fig. 5. Functions $I_1(M)$, $I_2(M)$, $\exp(-M_N^2/M^2)$ for nucleon at $M_N = 1$ GeV.

amplitude in unified gauge theories and is also related to the numerical coefficient of the proton form factor asymptotics. From (58) it follows that

$$\lambda_N^2 = 1.2 \cdot 10^{-3}\,\mathrm{GeV}^6, \tag{59}$$

with an error of order ±50%.

Let us now discuss the question of possible types of currents with proton quantum numbers. As follows from (15), besides the current (46), there exists only one current without quark field derivatives,

$$\eta_1(x) = (u^a(x) C\sigma_{\mu\nu} u^b(x))\sigma_{\mu\nu}\gamma_5 d^c(x)\varepsilon^{abc}. \tag{60}$$

To compare the currents $\eta(x)$ and $\eta_1(x)$ it is convenient to write them in another

TABLE 2

The M variation intervals at given nucleon mass M_N, between which conditions (i)–(iii) of sect. 3 are fulfilled $M_{min} < M < M_{max}$

M_N (GeV)	M_{min} (GeV)	M_{max} (GeV)
0.8	0.75	1.00
0.9	0.77	1.10
1.0	0.78	1.20
1.2	0.90	1.20

way, making the Fierz transformation and introducing the right- and left-hand spinors:

$$\eta(x) = 2[(u^aCd^b)\gamma_5 u^c - (u^aC\gamma_5 d^b)u^c]\varepsilon^{abc}$$
$$= 4[(u^a_{\rm R}Cd^b_{\rm R})u^c_{\rm L} - (u^a_{\rm L}Cd^b_{\rm L})u_{\rm R}]\varepsilon^{abc}, \qquad (61)$$

$$\eta_1(x) = 2[(u^aCd^b)\gamma_5 u^c + (u^aC\gamma_5 d^b)u^c]\varepsilon^{abc}$$
$$= -4[(u^a_{\rm R}Cd^b_{\rm R})u^c_{\rm R} - (u^a_{\rm L}Cd^b_{\rm L})u^c_{\rm L}]\varepsilon^{abc}. \qquad (62)$$

From (62) one can immediately see the properties of the current $\eta_1(x)$ sharply differing from the current $\eta(x)$; namely, for $\eta_1(x)$ all the diagrams containing the mean vacuum values of quark fields, figs. 1b–d, are zero. Indeed, the non-zero mean vacuum values $\langle 0|\bar\psi\psi|0\rangle$ or $\langle 0|\bar\psi\sigma_{\mu\nu}\lambda^h\psi G^h_{\mu\nu}|0\rangle$ appear only if the right-hand spinor is paired with the left-handed one. In the polarization operator with the current $\eta_1(x)$ $\Pi(\eta_1, \bar\eta_1)$ this is possible only if three spinors are paired at once, i.e. starting from the terms $\langle 0|\bar\psi\psi|0\rangle^3$. Thus, the terms in $\Pi(\eta_1, \bar\eta_1)$ violating chiral invariance are strongly suppressed. In accordance with the duality principle, this means that the contribution of the baryonic resonances to $\Pi(\eta_1, \bar\eta_1)$ must be small, i.e. the amplitudes of the baryon transition into the current $\eta_1(x)$ are small. The hadronic continuum contribution with low threshold must here dominate in the dispersion sum rule for $\Pi(\eta_1, \bar\eta_1)$. Thus, the application of the current $\eta_1(x)$ in the sum rules is useless.

We will also dwell on the possibility of calculating the mass splitting in the baryonic octet. As has been already mentioned, because of the properties of the sum rules (51), (52) and of the equation for the nucleon mass (55), taking account of small corrections is in this case much less reliable than in the decuplet case. Nevertheless, one can try to do it. For the Σ hyperon [the replacement $d \to s$ in the current (46)], the extra terms in the polarization operator which are proportional to $m_{\rm s}$ are

$$\Delta\,\mathrm{Im}\,\Pi^{(a)} = \frac{m_{\rm s}p^4}{4(2\pi)^3},$$

$$\Delta\Pi^{(c)} = -\frac{4}{3}\frac{m_{\rm s}}{p^2}\langle 0|\bar\psi\psi|0\rangle^2, \qquad (63)$$

$$\Delta\Pi^{(b)} = \Delta\Pi^{(d)} = 0.$$

The mean vacuum values should be replaced by $\langle 0|\bar ss|0\rangle$ in (48) and by $\langle 0|\bar uu|0\rangle^2$ in (49). As a result we have instead of eq. (55) ($\delta = m_{\rm s}/M_\Sigma$):

$$M^6_\Sigma(1-2\delta) - 2aM^3_\Sigma\left(1+\frac{m_0^2}{2M^2_\Sigma}\right)(1+\gamma) + \tfrac{4}{3}a^2(1-2\delta) = 0. \qquad (64)$$

The solution of this equation at $\gamma = 0.15$, $m_{\rm s} = 150$ MeV yields $M_\Sigma - M_{\rm N} = 240$ MeV, a number which is very close to the experimental value. Such a good coincidence seems to be accidental. For Ξ and Λ hyperons the calculations do not lead to reasonable results: the values of their masses strongly depend on the parameters m_0 and γ and, moreover, turn out to be lighter than the Σ and nucleon, respectively, at almost every m_0 and γ. Perhaps, the failure in the calculation of the Ξ and Λ masses can be attributed to the fact that in the case of Ξ and Λ non-perturbative effects due to instantons may be important (see sect. 7).

6. Other baryonic resonances

The approach under consideration allows one to make definite conclusions about some other baryonic resonances.

First of all, it follows from the sum rules of sect. 3 that there is no strong baryonic resonance R of comparatively small mass ($M_R < 1.6$ GeV) with isospin $T = \frac{3}{2}$ and $J^P = \frac{3}{2}^-$. Suppose for a moment, that the resonance Δ does not exist but the resonance R does. Then because of the opposite (as compared to Δ) parity of the resonance R, the terms with an even number of γ-matrices and, in particular, proportional to $\delta_{\mu\nu}$ in (26) have another sign and the sum rules (31), (35) cannot be satisfies. If there are two resonances Δ and R and $M_R - M_\Delta \leqslant 300$ MeV, then saturating the sum rules by these two resonances and doing some calculations, one can see that there will be no contradiction to the sum rule (i.e. too large $M_\Delta > 1.5-1.6$ GeV) only if the ratio of residues is $\lambda_R^2/\lambda_\Delta^2 < 0.1$. The statements made are inapplicable to comparatively heavy R resonances, $M_R > 1.6$ GeV which correspond to the region of M in the sum rules where the continuum contribution is dominant.

Consider now the sum rules for the polarization operator corresponding to the current (29) with isospin $T = \frac{1}{2}$. The imaginary part of this polarization operator is contributed by the states with $J = \frac{3}{2}$ and $J = \frac{1}{2}$, so its study will permit one to get information on these states. In particular, from these sum rules one can find the residue which determines the coefficient in the asymptotics of the electromagnetic proton form factor. The calculation of the polarization operator for the current $\eta_\mu^{T=1/2}$ results in

$$\operatorname{Im} \Pi_{\mu\nu}^{(a)} = -\frac{3}{5 \cdot 2^8 \pi^3} \left\{ \delta_{\mu\nu}\hat{p} + \tfrac{5}{8}\gamma_\mu\gamma_\nu\hat{p} - \tfrac{5}{8}(\gamma_\mu p_\nu - \gamma_\nu p_\mu) - \frac{p_\mu p_\nu \hat{p}}{p^2} - \tfrac{7}{8}(\gamma_\mu p_\nu + \gamma_\nu p_\mu) \right\}, \tag{65}$$

$$\Pi_{\mu\nu}^{(b)} = -\frac{1}{16\pi^2}\langle 0|\bar{\psi}\psi|0\rangle p^2 \ln\frac{\Lambda_U^2}{-p^2} \left\{ \delta_{\mu\nu} - \tfrac{5}{4}\gamma_\mu\gamma_\nu + (\gamma_\mu p_\nu - \gamma_\nu p_\mu)\frac{\hat{p}}{p^2} + \frac{p_\mu p_\nu}{p^2} \right\}, \tag{66}$$

$$\Pi^{(c)}_{\mu\nu} = -\frac{1}{4p^2}\langle 0|\bar{\psi}\psi|0\rangle^2\{\delta_{\mu\nu}\hat{p} - \tfrac{3}{2}\gamma_\mu\gamma_\nu\hat{p}$$

$$+\tfrac{3}{2}(\gamma_\mu p_\nu - \gamma_\nu p_\mu) - \tfrac{1}{2}(\gamma_\mu p_\nu + \gamma_\nu p_\mu)\}\,, \tag{67}$$

$$\Pi^{(d)}_{\mu\nu} = \frac{1}{2^7\pi^4} g_s\langle 0|\bar{\psi}\sigma_{\mu\nu}\frac{\lambda^h}{2}\psi G^h_{\mu\nu}|0\rangle$$

$$\times\left\{\ln\frac{\Lambda_U^2}{-p^2}(\delta_{\mu\nu} + 2\gamma_\mu\gamma_\nu) + (\gamma_\mu p_\nu - \gamma_\nu p_\mu)\frac{\hat{p}}{p^2} + 4\frac{p_\mu p_\nu}{p^2}\right\}. \tag{68}$$

The comparison of (66) with (21) shows that the term $\sim\delta_{\mu\nu}$ in (66) is different in sign from the analogous term in (21). This means that when the sum rules at the structures at $\delta_{\mu\nu}\hat{p}$ and $\delta_{\mu\nu}$ are saturated by resonances, the basic contribution is given by the state with $J^P = \frac{3}{2}^-$ but not with $J^P = \frac{3}{2}^+$. We come to an interesting conclusion that in the channel with $T = \frac{1}{2}$, unlike the channel with $T = \frac{3}{2}$, among states with spin $\frac{3}{2}$ the lightest and strongest is the resonance with negative parity $J^P = \frac{3}{2}^-$. This conclusion is consistent with experiment: experimentally, the resonance $N^*(T = \frac{1}{2}, J^P = \frac{3}{2}^+)$ lies far higher than $N^*(T = \frac{1}{2}, J^P = \frac{3}{2}^-)$.

Under an approximation disregarding the continuum contribution and leading logarithmic corrections, the sum rules for the structure functions at $\delta_{\mu\nu}\hat{p}$ and $\delta_{\mu\nu}$ saturated by resonance R with $J^P = \frac{3}{2}^-$ have the form

$$\begin{gathered} M^6 - \tfrac{10}{3}a^2 = 5\cdot 2^7\pi^4\lambda_R^2\, e^{-M_R^2/M^2} \\ \tfrac{10}{3}aM^4\left(1 - \frac{m_0^2}{8M^2}\right) = 5\cdot 2^7\pi^4\lambda_R^2 M_R\, e^{-M_R^2/M^2}\,. \end{gathered} \tag{69}$$

Putting $M = M_R$ and solving the appearing equation for M_R, we find $M_R = 1.30$ GeV. Taking account of the continuum contribution and of logarithmic corrections must, perhaps, increase M_R and push it closer to the experimental value $M_{R\,exp} = 1.52$ GeV.

The contribution of states with $J^P = \frac{1}{2}^\pm$ to the imaginary part of the polarization operator can be investigated using the sum rules for other structures. The matrix element $\langle 0|\eta_\mu^{T=1/2}|N^*, J^P = \frac{1}{2}^-\rangle$ and the contribution of these states to the r.h.s. of the sum rules is determined by formulae (27), (28). For states with $J^P = \frac{1}{2}^+$ one can write

$$\langle 0|\eta_\mu^{T=1/2}|N, J^P = \tfrac{1}{2}^+\rangle = (\alpha_N p_\mu + \beta_N\gamma_\mu)\gamma_5 v^{(r)}\,, \tag{70}$$

and the contribution of such states to the sum rules can be obtained by replacing $m \to -m$, $\alpha \to \alpha_N$, $\beta \to \beta_N$ in (28).

Consider first the sum rule at the structure proportional to $\gamma_\mu\gamma_\nu$. Subtracting from this structure the contribution of the resonance R with $J^P = \frac{3}{2}^-$, one can see that the sign of the l.h.s. of the sum rule after subtraction is such that the leading contribution to the r.h.s. is given by the resonance with $J^P = \frac{1}{2}^+$ (nucleon) but not with $J^P = \frac{1}{2}^-$, which is, of course, natural. Therefore, it is reasonable to saturate the r.h.s. of the sum rules by the contributions of the resonance R with $J^P = \frac{3}{2}^-$ and the nucleon. In the nucleon case the constants α_N and β_N introduced in (70) are related to the constant λ_N determined in (53) by

$$\lambda_N = \tfrac{2}{3}(4\beta_N - \alpha_N M_N) . \tag{71}$$

It should be borne in mind that the current $\eta_\mu^{T=1/2}$ (29) is not a renormcovariant quantity. The renormcovariant currents which transform through themselves at the transformations of the renormalization group are

$$\eta_{1\mu} = [(u^a C d^b)\gamma_\mu u^c - (u^a C\gamma_5 d^b)\gamma_\mu\gamma_5 u^c]\varepsilon^{abc} ,$$
$$\eta_{2\mu} = [(u^a C\sigma_{\rho\nu} d^b)\sigma_{\rho\nu}\gamma_\mu u^c - (u^a C\sigma_{\rho\nu} u^b)\sigma_{\rho\nu}\gamma_\mu d^c]\varepsilon^{abc} .$$

The current $\eta_\mu^{T=1/2}$ is related to the currents $\eta_{1\mu}$ and $\eta_{2\mu}$ by $\eta_\mu^{T=1/2} = \frac{1}{4}(3\eta_{1\mu} + \frac{1}{2}\eta_{2\mu})$. Therefore, the calculation of the leading logarithmic corrections to the polarization operator $\Pi(\eta_\mu^{T=1/2}, \bar\eta_\nu^{T=1/2})$ requires the calculation of the polarization operators $\Pi(\eta_{1\mu}, \eta_{1\nu})$, $\Pi(\eta_{2\mu}, \bar\eta_{2\nu})$, $\Pi(\eta_{1\mu}, \bar\eta_{2\nu})$ under the same approximation. We will present only the results of a rather rough consideration, disregarding the leading logarithmic corrections and the continuum contribution. Under this approximation all the eight sum rules for the polarization operator $\Pi(\eta_\mu^{T=1/2}, \bar\eta_\nu^{T=1/2})$ within the interval $1.1 < M < 1.3$ GeV can be satisfied up to 50% at the following values of the parameters:

$$M_R = 1.30\ \text{GeV} , \qquad M_N = 1.1\ \text{GeV} , \qquad \lambda_R^2 = 3.10^{-4}\ \text{GeV}^6 ,$$
$$\alpha_N = 1.1 \cdot 10^{-2}\ \text{GeV}^2 , \qquad \beta_N = 2.1 \cdot 10^{-2}\ \text{GeV}^3 . \tag{72}$$

The nucleon mass value obtained here agrees well with that calculated in sect. 5. Among the residues α_N, β_N and λ_R, α_N, which determines the numerical coefficient in the asymptotics of the nucleon form factor at large momentum transfers is of the most interest. Unfortunately, the accuracy in α_N and β_N seems to be rather low, ~30–40%. [With such an accuracy, for instance, the equality (71) is fulfilled.]

7. The role of corrections not accounted for

In the above consideration we have taken into account only several terms of the operator expansion: two terms of each structure of the polarization operator with odd and even numbers of γ matrices. The second term of the expansion $\Pi^{(c)}$ in the first structure was (especially for the nucleon case) not much smaller than the first, $\Pi^{(a)}$. A question arises, whether the results would be strongly changed by the next terms of the operator expansion.

It can be argued that it is not the case. As has been mentioned in the introduction, the term $\Pi^{(c)}$ is anomalously large because of the large numerical factor in front of it. There are no such factors in front of the other terms of the operator expansion. Let us take different unconsidered terms of the operator expansion in turn.

The first correction to the structures of the polarization operator with an odd number of γ-matrices contains the mean vacuum value $\langle 0|G^h_{\mu\nu}G^h_{\mu\nu}|0\rangle$ and is shown in fig. 2. The ratio of this diagram to that of fig. 1a must be of order $4\pi\alpha_s k\langle 0|G^h_{\mu\nu}G^h_{\mu\nu}|0\rangle/M^4$, where k is the numerical coefficient which can be estimated as $k=\frac{1}{12}$ from the contribution to the polarization operator of the vector current $\bar{\psi}\gamma_\mu\psi$, calculated in ref. [5]. Using this value of k and the known value $(\alpha_s/\pi)\langle 0|G^h_{\mu\nu}G^h_{\mu\nu}|0\rangle = 10^{-2}\ \mathrm{GeV}^4$ we find that the ratio of fig. 2 diagram to fig. 1a will be of order 5% at $M=1$ GeV. Even if the numerical coefficient k will be several times larger in our case because of the combinatoric factors, it is unlikely for this ratio to exceed 10–20%. The same smallness with respect to fig. 1b can be expected from the contribution of diagrams corresponding to the emission of one more, going to vacuum, gluon in fig. 1d. Another correction to the structures with an even number of γ-matrices appears from fig. 6, proportional to $\langle 0|\bar{\psi}\psi|0\rangle^3$. The ratio of the contribution of this diagram to that of fig 1b can be estimated as $(\alpha_s/\pi)[(2\pi)^2\langle 0|\bar{\psi}\psi|0\rangle]^2/M^6 \leq 0.1$. Perhaps, the corrections $\sim\alpha_s/\pi$ to figs. 1a, b will be numerically more essential, but they also hardly exceed 20–30%. Even if these estimates are too optimistic and the calculation shows that the corrections not accounted for are several times larger than the above estimates, they will only weakly affect the baryonic mass values since the sum rules determine not M but M^3.

Let us make one more remark on a possible role of non-perturbative corrections due to instantons. For massless quarks these corrections are most essential when the zero modes give a contribution. It follows from the form of quark currents (18) and (61) that only one u-quark for the case of the current (18), one u- and one d-quark in the case of the current (61) can be in the state corresponding to the zero mode. This means that the zero modes cannot compensate the smallness stemming from the quark determinant, i.e. for the isobar and nucleon the instanton contribution must be small. Note that the situation changes when instead of the nucleon, Ξ or Λ hyperons are considered (but not Σ) where the quark determinant appears to be proportional to the strange quark mass only, i.e. essentially larger.

8. Conclusion

We repeat once more the basic results obtained in this paper:

(i) The isobar Δ mass has been calculated.

Fig. 6. The diagram proportional to the contribution $\langle 0|\bar{\psi}\psi|0\rangle^3$ to the polarization operator.

(ii) The nucleon mass has been calculated.

(iii) The Σ^* and Δ mass difference has been determined.

(iv) It has been shown that in the channel with isotopic spin $T=\frac{3}{2}$ the resonance with $J^P=\frac{3}{2}^-$ lies significantly higher than that with $J^P=\frac{3}{2}^+$.

(v) It has been shown that in the channel with $T=\frac{1}{2}$ the resonance with $J^P=\frac{3}{2}^-$ lies lower than that with $J^P=\frac{3}{2}^+$ and the mass of the resonance with $J^P=\frac{3}{2}^-$ has been calculated.

(vi) The residues have been obtained which determine the coefficient at the asymptotics of the nucleon form factor and the coefficient at the proton decay amplitude in unified theories.

The calculations made show that the sum rule method using the basic concepts of QCD is completely applicable to considerations of baryons and is successful. It is seen that the nature of phenomena leading to violation of chiral invariance in the mean vacuum values of the quark field product $\langle 0|\bar{\psi}\psi|0\rangle$ and to the appearance of baryonic masses is, in fact, just the same and manifests itself in the presence of a relationship between these quantities. It can also be seen that $\langle 0|\bar{\psi}\psi|0\rangle$ is fundamental to the QCD description of low-energy processes involving baryons. This means that in our understanding of non-perturbative QCD as the strong interaction theory we are on the right path.

The possibilities of the method used are not exhausted by the problems considered in this paper; it can be extended to a number of other problems: calculation of mases of charmed baryons, determination of other statical parameters of baryons, etc.

I would like to thank V.L. Chernyak, V.A. Khoze, L.N. Lipatov, M.V. Terentyev, M.B. Voloshin for useful discussions. I am especially obliged to V.A. Novikov, M.A. Shifman and A.I. Vainshtein whose remarks and criticism promoted the success of my work.

References

[1] M. Gell-Mann, R.J. Oakes and B. Renner, Phys. Rev. 175 (1968) 2195
[2] S.L. Glashow and S. Weinberg, Phys. Rev. Lett. 20 (1968) 224
[3] M. Gell-Mann, Acta Phys. Austr. Suppl. IX (1972) 733;
H. Fritzsch, M. Gell-Mann and H. Leutwyler, Phys. Lett. 47B (1973) 365
[4] S. Weinberg, A. Festschrift for I.I. Rabi, ed. L. Motz (Academy of Sciences, N.Y., 1977)
[5] M.A. Shifman, A.I. Vainshtein and V.I. Zakharov, Nucl. Phys. B147 (1979) 385, 448
[6] V. Novikov, M. Shifman, A. Vainshtein, M. Voloshin and V. Zakharov, Proc. Int. Conf. 'Neutrino' 78, Lafayette, 1978, p. C278
[7] M.E. Peshkin, Phys. Lett. 88B (1979) 128
[8] V. Novikov, M. Shifman, A. Vainshtein and V. Zakharov, Preprint ITEP-48 (1981)

Nuclear Physics B191 (1981) 591–592
© North-Holland Publishing Company

ERRATUM

B.L Ioffe, Calculation of baryon masses in quantum chromodynamics, Nucl. Phys. B188 (1981) 317.

In calculation of the polarization operator the mean vacuum value of quark field operators was approximately set as

$$\langle 0|q_\alpha^a(x)\bar{q}_\beta^b(0)|0\rangle \approx -\tfrac{1}{12}\delta^{ab}\delta_{\alpha\beta}\langle 0|\bar{q}(0)q(0)|0\rangle\,.$$

In fact, in the framework of the used approximations, the correct expression [in the gauge $x_\mu A_\mu(x)=0$] is

$$\langle 0|q_\alpha^a(x)\bar{q}_\beta^b(0)|0\rangle = -\tfrac{1}{12}\delta^{ab}(\delta_{\alpha\beta}-\tfrac{1}{4}im_q\hat{x}_{\alpha\beta})\langle 0|\bar{q}(0)q(0)|0\rangle - \frac{x^2}{2^6\cdot 3}g_s\langle 0|\bar{q}(0)\sigma_{\mu\nu}\frac{\lambda^h}{2}G_{\mu\nu}^h(0)q(0)|0\rangle\delta^{ab}\delta_{\alpha\beta}\,. \qquad (1)$$

Taking account of terms erroneously omitted in (1) does not affect the main results of the paper, because these terms influence only the small corrections to the sum rules proportional to the mean vacuum value $g_s\langle 0|\bar{\psi}\sigma_{\mu\nu}(\lambda^h/2)G_{\mu\nu}^h\psi|0\rangle$ and to the current quark masses. In almost every case the corresponding amendments improve the agreement between theoretical expectations and experiment. The necessary changes are the following. In the case of the isobar, the coefficient $\frac{1}{8}$ in front of eq. (23) must be replaced by $\frac{1}{2}$ and the same change must be made in eq. (31) for the term $\sim m_0^2/M^2$. As a result, the best fit for the isobar mass is $M_\Delta = 1.30$ GeV. For the nucleon case, eq. (50) must be substituted by $\Pi^{(d)}=0$ and the best fit for the nucleon mass gives $M_N = 0.90$ GeV. The curves of fig. 3 are practically unchanged. The curves of figs. 4 and 5 change in a way that, if matched at $M = M_\Delta(M_N)$, to the right of this point $I_2(M)$ lies 25% higher than $I_1(M)$ at $M = 1.6$ GeV in fig. 4 and at $M = 1.4$ GeV in fig. 5. The most important differences arise in the case of the mass splitting in the baryon octet. For the Σ hyperon eq. (63) $\Delta\Pi^{(b)}=0$ is replaced by

$$\Delta\Pi^{(b)} = \frac{m_s}{4\pi^2}\langle 0|\bar{\psi}\psi|0\rangle \ln\frac{\Lambda_U^2}{-p^2}\hat{p}\,.$$

For the Λ hyperon we have now

$$\Delta\,\mathrm{Im}\,\Pi^{(a)} = -\frac{m_s p^4}{12(2\pi)^3}\,, \qquad \Delta\Pi^{(b)} = -\frac{m_s}{3(2\pi)^2}\langle 0|\bar{\psi}\psi|0\rangle \ln\frac{\Lambda_U^2}{-p^2}\hat{p}\,,$$

$$\Delta\Pi^{(c)} = -\tfrac{8}{9}m_s\langle 0|\bar{\psi}\psi|0\rangle^2\frac{1}{p^2}\,.$$

(The terms, proportional to $m_s g_s \langle 0|\bar{\psi}\sigma_{\mu\nu}\lambda^h G^h_{\mu\nu}\psi|0\rangle$ are neglected.) As a result, at $m_s = 150$ and $\gamma = \langle 0|\bar{s}s|0\rangle / \langle 0|\bar{u}u|0\rangle - 1 \approx -0.2$, it is possible to describe the mass splitting in the baryon octet satisfactorily: $M_\Lambda - M_N \approx 150$ MeV, $M_\Sigma - M_N \approx 300$ MeV, $M_\Xi - M_N \approx 380$ MeV $\pm$ (20–30)%. In the formulae for the mass splitting in the decuplet in essential terms proportional to $\delta_{\mu\nu}$, the term $\delta_{\mu\nu}\hat{p}$ in (39) must be multiplied by $\frac{3}{4}$. With the same values of m_s and γ we have now $M_{\Sigma^*} - M_\Delta = 190$ MeV $\pm$ 30%. Finally, eq. (68) is to be replaced by

$$\Pi^{(d)}_{\mu\nu} = \frac{1}{2^5 \pi^2} g_s \langle 0|\bar{\psi}\sigma_{\mu\nu}\frac{\lambda^h}{2} G^h_{\mu\nu}\psi|0\rangle \left\{ \ln \frac{\Lambda_U^2}{-p^2} (\delta_{\mu\nu} - \tfrac{1}{4}\gamma_\mu\gamma_\nu) - \tfrac{5}{4}(\gamma_\mu p_\nu - \gamma_\nu p_\mu)\hat{p}/p^2 - \tfrac{1}{2} p_\mu p_\nu / p^2 \right\} .$$

This change does not practically affect the value of the mass of the resonance $R(J^P = \frac{3}{2}^-, T = \frac{1}{2})$.

I am indebted to D.I. Dyakonov and E.V. Shuryak for discussions which helped me to find the error.

Z. Phys. C - Particles and Fields 18, 67-68 (1983)

Zeitschrift für Physik C Particles and Fields

On the Choice of Quark Currents in the QCD Sum Rules for Baryon Masses

B.L. Ioffe

Institute of Theoretical and Experimental Physics, SU-117259 Moscow, USSR

Received 2 December 1982

Abstract. The cryteria for the choice of the most suitable quark currents in the QCD sum rules for baryon masses are discussed. The currents used in [1, 2] are preferable comparing with those adopted in [3, 4] since the latters result in large power corrections.

When considering the QCD sum rules for determination of baryon masses, a question arises on the choice of the most suitable quark currents. There exist three independent, quark currents with proton quantum numbers and no derivatives

$$\eta^{(1)} = (u^a C \gamma_\mu u^b) \gamma_5 \gamma_\mu d^c \varepsilon^{abc}$$
$$\eta^{(2)} = (u^a C \sigma_{\mu\nu} u^b) \sigma_{\mu\nu} \gamma_5 d^c \varepsilon^{abc}$$
$$\eta_\mu^{(3)} = [(u^a C \gamma_\mu u^b) \gamma_5 d^c - (u^a C \gamma_\mu d^b) \gamma_5 u^c] \varepsilon^{abc} \quad (1)$$

such that the matrix elements $\langle 0| \eta^{(i)} |p\rangle$, $i=1,2,3$ are nonzero. Therefore, the sum rules for the proton mass calculation can be formulated using anyone of the currents (1) or an arbitrary linear combination.

In [1] it was argued that the most suitable current for the determination of the proton mass is $\eta^{(1)}$, the sum rules for the polarization operator $\Pi(\bar{\eta}^{(1)}, \eta^{(1)})$ are obtained and the proton mass is calculated. Furthermore these calculations have been improved in [2] where higher power corrections were taken into account and the sum rules for the current $\eta_\mu^{(3)}$ were also considered.

Chung, Dosch, Kremer and Schall in their papers [3, 4] choose for proton mass determination some linear combination $\eta^{(1)}$, $\eta^{(2)}$ and $\eta_\mu^{(3)}$ as a current which enters the sum rule. The proton mass value obtained with such a current appeared to be somewhat different from the results of [1, 2].

The purpose of this note is to explain the origin of the discrepancy and to show why the choice made in [1] is preferable. The quark current used in the baryonic sum rules should be chosen proceeding from two requirements: 1) the contribution of the lowest baryonic state in the sum rules must exceed the contributions of higher states (continuum); 2) disregarded power corrections in the operator expansion must be small in the domain of characteristic values of the Borel mass parameter M^2.

In [3, 4] the quark current was chosen proceeding only from the first criterion: the coefficients in the linear combination $\eta^{(1)}$, $\eta^{(2)}$, $\eta^{(3)}$ were found by minimizing the continuum contribution comparing with the contribution of the lowest baryonic state (proton). It is clear, however, that the fulfillment of the first cryterion only must inevitably result in the violation of the second one, i.e., in a strong increase of higher order power corrections. Indeed, suppose we have succeeded in choosing such a current that the continuum contribution is very small. Then, the r.h.s of the sum rule after Borel transformation will be proportional to $\exp(-m_B^2/M^2)$, where m_B is the mass of the lowest baryonic state. At the same time, the l.h.s. is a series of the form $AM^6 + bM^2 + B + C/M^2 + \ldots$ (for the first sum rule) and it is clear that the sum rule will be fulfilled with a good enough accuracy only when far-distant terms will be essential in this series. This general argument is confirmed by a specific calculation of coefficients in the proton sum rules for the current obtained in [3, 4] by minimizing the continuum contrubution. In approximation of [1, 3, 4] when only coefficients A and $B \sim \langle 0| \bar{q} q |0\rangle^2$ were taken into account, for characteristic values $M^2 \sim 1\,\mathrm{GeV}^2$ the ratio B/AM^6 is ≈ 4, i.e. the second term in the operator expansion in [3, 4] is much larger than the first one. (At higher $M^2 > 2\,\mathrm{GeV}^2$, when $B/AM^6 < 0.5$ the continuum contribution becomes enormously large).

There are no reasons to think that the next terms of the operator expansion will be much smaller. In any case, until they will be taken into account, the results obtained from the sum rules with the current chosen in [3, 4] cannot give rise to confidence. In particular, the statement made in [4] on the absence of the $N-\Delta$ mass splitting due to chiral symmetry violating quark condensate seems to have no justification.

For the current $\eta^{(1)}$ chosen in [1, 2] the above formulated criteria are fulfilled: for $M^2 \approx 1\,\mathrm{GeV}^2$ the ratio $B/AM^6 \approx 0.4$ and the ratio of the continuum contribution to proton contribution is not more than 50-30% (depending on the value of the continuum threshold $W = 1.5$-$2.0\,\mathrm{GeV}$). The higher power corrections calculated in [2] in the characteristic M^2 region appeared to be smaller than 0.1, i.e., inessential.

For completeness I repeat here the arguments in favour of the choice of the current $\eta^{(1)}$ for the proton sum rules. The use of the pure current $\eta^{(2)}$ in the sum rules is unreasonable for the following reason. The current $\eta^{(2)}$ has such a property that chirality violation structures of the polarization operator $\Pi(\bar{\eta}^{(2)}, \eta^{(2)})$ are non-zero starting from higher order terms of the operator expansion proportional to $\alpha_s \langle 0|\bar{q}q|0\rangle^3$. This can be easily seen writing $\eta^{(2)}$ as

$$\eta^{(2)} = -4[(u_R^a C d_R^b) u_R^c - (u_L^a C d_L^b) u_L^c]\,\varepsilon^{abc} \quad (2)$$

and noting that chirality violating terms can appear only at simultaneous pairing of three spinors. Therefore, while the structure proportional to $\not{p}$ in $\Pi(\bar{\eta}^{(2)}, \eta^{(2)})$ has the same order of magnitude as in $\Pi(\bar{\eta}^{(1)}, \eta^{(1)})$, the structure proportional to I is relatively suppressed. The contributions of all states enter the structure proportional to $\not{p}$ with the same sign while the contributions of states with opposite parities enter the structure proportional to I with different signs. Hence it follows that in the structure $\sim I$ in $\Pi(\bar{\eta}^{(2)}, \eta^{(2)})$ the proton contribution must be strongly compensated by the contributions of states with negative parity ($J^P = 1/2^-$). Thereby the pure current $\eta^{(2)}$ is not suitable for the calculation of the proton mass.

The current $\eta_\mu^{(3)}$ is inconvenient for the calculation of the proton mass for two reasons: 1) the polarization operator $\Pi(\bar{\eta}_\mu^{(3)}, \eta_\nu^{(3)})$ contains besides the proton, also contributions from states with spin 3/2. In order to exclude them one must separate certain structures, which deteriorates the accuracy; 2) the current $\eta_\mu^{(3)}$ is not renormcovariant. The latter difficulty can be obviated writing $\eta_\mu^{(3)}$ as

$$\eta_\mu^{(3)} = \tfrac{1}{4}[-\tfrac{3}{2}\gamma_\mu \eta^{(1)} + \tfrac{1}{2}\eta_\mu^{(4)}] \quad (3)$$

where currents $\eta^{(1)}$ and

$$\eta_\mu^{(4)} = [(u^a C \sigma_{\nu\lambda} d^b)\gamma_5 \sigma_{\nu\lambda}\gamma_\mu u^c - (u^a C \sigma_{\nu\lambda} u^b)\gamma_5 \sigma_{\nu\lambda}\gamma_\mu d^c]\cdot \varepsilon^{abc} \quad (4)$$

are renormcovariant, and considering in addition to $\Pi(\bar{\eta}^{(1)}, \eta^{(1)})$ the polarization operators $\Pi(\bar{\eta}_\mu^{(4)}, \eta_\nu^{(4)})$ and $\Pi(\bar{\eta}^{(1)}, \eta_\mu^{(4)}) + \Pi(\bar{\eta}_\mu^{(4)}, \eta^{(1)})$. The corresponding calculations have been made in [2]. The results for the proton mass are in an agreement with those obtained from the sum rules for the current $\eta^{(1)}$ but their accuracy is somewhat worse (especially for the case of the sum rules for $\Pi(\bar{\eta}_\mu^{(4)}, \eta_\nu^{(4)})$. Thus, the use of the pure currents $\eta^{(2)}$ and $\eta_\mu^{(3)}$ in the sum rules for the proton mass is unsuitable. It is, of course, possible to consider a mixture of currents $\eta^{(1)}$, $\eta^{(2)}$, $\eta^{(3)}$, but in this case the admixture of two latters must be comparatively small (unlike the current of [3, 4] where it is not the case).

In case of the sum rules determining the isobar mass, analogous situation takes place. Here the current

$$\eta_\mu^{\Delta} = (u^a C \gamma_\mu u^b) u^c \varepsilon^{abc} \quad (5)$$

is preferable, for which the higher power corrections are smaller than for the current used in [3, 4]. But in case of Δ the difference between the results obtained with the current (5) and the current used in [3, 4] is not so large and is within the accuracy of both approaches.

References

1. B.L. Ioffe: Nucl. Phys. **B188**, 317 (1981); E: **B191**, 591 (1981)
2. V.M. Belyaev, B.L. Ioffe: Preprint ITEP-59 (1982), JETP **83**, 876 (1982) (in Russian)
3. Y. Chung, H.G. Dosch, M. Kremer, D. Schall: Nucl. Phys. **B197**, 55 (1982)
4. Y. Chung, H.G. Dosch, M. Kremer, D. Schall: Z. Phys. C – Particles and Fields **15**, 367 (1982)

BARYONS IN QCD AND CHIRAL SYMMETRY BREAKING PARAMETERS

L.J. REINDERS
CERN, Geneva, Switzerland

H.R. RUBINSTEIN
The Weizmann Institute of Science, Rehovot, Israel

and

S. YAZAKI
University of Tokyo, Japan

Received 6 October 1982

We calculate all baryons in the 56 representation using QCD sum rules. All masses are well predicted and require stringent values of the chiral parameters $\langle 0|\bar{u}u|0\rangle = -(230 \pm 15\ \mathrm{MeV})^3$ and $\gamma = \langle\bar{s}s\rangle/\langle\bar{u}u\rangle - 1 = -0.19 \pm 0.02$. The determination of γ is the most precise to date, the strange quark mass and the quark condensate are also accurately fixed.

Recently there has been remarkable progress in calculating baryon masses in QCD. Ioffe [1] has pointed out that in the QCD sum rule scheme $M_B^3 \sim \langle 0|\bar{q}q|0\rangle$ in the limit of massless quarks. In more recent papers his original results have been corrected [2] and expanded [3] to include several other states of the 56 representation.

In this letter we present a calculation of all states in the 56 representation. We confirm that all states can be obtained with only one new parameter $\gamma = \langle 0|\bar{s}s|0\rangle/\langle 0|\bar{u}u|0\rangle - 1$. It turns out that the octet baryons provide the best case for determining this chiral symmetry breaking parameter which was hitherto undetermined. Its value $\gamma = -0.19$ is interesting and unexpected.

We start with the choice for the nucleon current made by Ioffe [1]

$$\eta_N(x) = \epsilon_{abc}[u^a(x)\, C\gamma_\mu u^b(x)]\,\gamma_5\gamma_\mu d^c(x)\,. \tag{1}$$

Considering only currents without derivatives there is one more current with the same quantum numbers and with a non-relativistic limit

$$\eta'_N(x) = \epsilon_{abc}[u^a(x)\, C\sigma_{\mu\nu} u^b(x)]\,\sigma_{\mu\nu}\gamma_5\, d^c(x)\,. \tag{2}$$

Ioffe argues that the coupling of the nucleon to eq. (2) should be small since the polarization operator induced by this current does not get any contribution from the lowest dimensional chiral symmetry breaking operators in the operator product expansion.

Starting from eq. (1) it is straightforward to derive the currents which correspond to the other $J^P = \frac{1}{2}^+$ octet baryons

$$\eta_\Sigma(x) = \epsilon_{abc}[u^a(x)\, C\gamma_\mu u^b(x)]\,\gamma_5\gamma_\mu s^c(x)\,, \tag{3a}$$

$$\eta_\Lambda(x) = \sqrt{\tfrac{2}{3}}\,\epsilon_{abc}\{[u^a(x)C\gamma_\mu s^b(x)]\,\gamma_5\gamma_\mu d^c(x) - [d^a(x)\, C\gamma_\mu s^b(x)]\,\gamma_5\gamma_\mu u^c(x)\}\,, \tag{3b}$$

$$\eta_\Xi(x) = -\epsilon_{abc}[s^a(x)\, C\gamma_\mu s^b(x)]\,\gamma_5\gamma_\mu u^c(x)\,. \tag{3c}$$

The lowest dimensional currents for the decuplet resonances are unique and can be written as follows

$$\eta^{\mu}_{\Delta^{+}} = \epsilon_{abc}[\mathrm{u}^a(x)\, C\gamma_\mu\, \mathrm{u}^b(x)]\, \mathrm{u}^c(x)\,, \tag{4a}$$

$$\eta^{\mu}_{\Sigma^*} = \sqrt{\tfrac{1}{3}}\,\epsilon_{abc}\{2[\mathrm{u}^a(x)\, C\gamma_\mu\, \mathrm{s}^b(x)]\, \mathrm{u}^c(x) + [\mathrm{u}^a(x)\, C\gamma_\mu\, \mathrm{u}^b(x)]\, \mathrm{s}^c(x)\}\,, \tag{4b}$$

$$\eta^{\mu}_{\Xi^*} = \sqrt{\tfrac{1}{3}}\,\epsilon_{abc}\{2[\mathrm{s}^a(x)\, C\gamma_\mu\, \mathrm{u}^b(x)]\, \mathrm{s}^c(x) + [\mathrm{s}^a(x)\, C\gamma_\mu\, \mathrm{s}^b(x)]\, \mathrm{u}^c(x)\}\,, \tag{4c}$$

$$\eta^{\mu}_{\Omega} = \epsilon_{abc}[\mathrm{s}^a(x)\, C\gamma_\mu\, \mathrm{s}^b(x)]\, \mathrm{s}^c(x)\,. \tag{4d}$$

To derive sum rules for these baryons we apply as usual [4] the operator product expansion to the two-point functions induced by the currents given above. It has been observed by Ioffe [1] and Chung et al. [5] that the dominant power corrections come from chiral symmetry breaking two- and four-quark operators. Let us first consider the nucleon case. The two-point function has two invariant amplitudes

$$\mathrm{i}\int \mathrm{d}^4x\, \exp(\mathrm{i}px)\langle 0|T(\eta_{\mathrm{N}}(x)\,\bar{\eta}_{\mathrm{N}}(0))|0\rangle = \not{p}F_1(p^2) + F_2(p^2)\,. \tag{5}$$

The total number of dimensions on the left-hand side is odd. This implies that $F_1(p^2)$ has an even number of dimensions and $F_2(p^2)$ an odd number. Therefore, contributions to $F_2(p^2)$ from even dimensional operators in the OPE will necessarily be multiplied by a small quark mass while the odd dimensional quark condensate operator $\bar{\mathrm{q}}\mathrm{q}$ will get no small quark mass factor and will dominate. This constitutes an important difference with the meson case (see refs. [4,6]). The invariant function $F_1(p^2)$ gets contributions from all operators. The contribution from $G^a_{\mu\nu}G^a_{\mu\nu}$ has explicitly been calculated by Smilga in ref. [2]. It is responsible for about 10% of the mass of the nucleon and should therefore be taken into account. At present we neglect first order α_s perturbative corrections, which we hope to calculate later. However, they can be absorbed in the continuum contribution and reliable mass estimates can be obtained without them.

Our procedure is identical to the one used before for light quark mesons [4,7]. Since there are two independent tensor structures, we get two independent sum rules. We have calculated the various Wilson coefficients, saturated the left-hand side of eq. (5) by the nucleon with a coupling λ_{N} to the current and taken the Borel transform, which gives [2]

$$M^6 - bM^2 + \tfrac{4}{3}a^2 = 2(2\pi)^4\lambda_{\mathrm{N}}^2\exp(-M_{\mathrm{N}}^2/M^2)\,, \quad 2aM^4 = 2(2\pi)^4 M_{\mathrm{N}}\lambda_{\mathrm{N}}^2\exp(-M_{\mathrm{N}}^2/M^2)\,, \tag{6a,b}$$

where $a = -(2\pi)^2\langle 0|\bar{\mathrm{q}}\mathrm{q}|0\rangle$ and $b = \pi^2\langle 0|(\alpha_s/\pi)\, G^a_{\mu\nu}G^a_{\mu\nu}|0\rangle$. The coupling λ_{N} can be eliminated by taking the ratio of eqs. (6b) and (6a) which results in an equation for M_{N} as a function of M^2. To analyse the sum rules (6) we introduce a continuum contribution with a threshold s_0 in the form of a θ-function to account for higher resonances in the sum rule. We only include a continuum contribution in eq. (6a). The reasons for this are that (a) in sum rule (6b) opposite parity states tend to cancel [1,5] and (b) the continuum contribution can be absorbed in a slight change in a. In the sum rule (6a) we put the explicit Borel transform of the perturbative quark cross section which (a) gives again a small contribution and (b) mimics the perturbative contribution from gluon exchange. Therefore we get

$$M_{\mathrm{N}}(M^2) = \frac{2aM^4}{M^6[1 - \exp(-s_0/M^2)(s_0^2/2M^4 + s_0/M^2 + 1)] - bM^2 + \tfrac{4}{3}a^2}\,. \tag{7}$$

In fig. 1 we have plotted M_{N} as a function of M^2 for a few values of s_0 and for $a = 0.5$ [i.e. $\langle 0|\bar{\mathrm{q}}\mathrm{q}|0\rangle = -(230\ \mathrm{MeV})^3$] and $b \simeq 0.13$ as determined from charmonium [4,6] or lattice calculations [8]. We see from fig. 1 that a plateau develops for $s_0 \simeq 5\ \mathrm{GeV}^2$ at a mass value around 0.9 GeV and that inclusion of the continuum changes the prediction for the mass by about 5%. This determines the accuracy of our calculations, and in the following we will not take any continuum contributions into account in the sum rules, and give the maximum of the curve as our prediction for the mass.

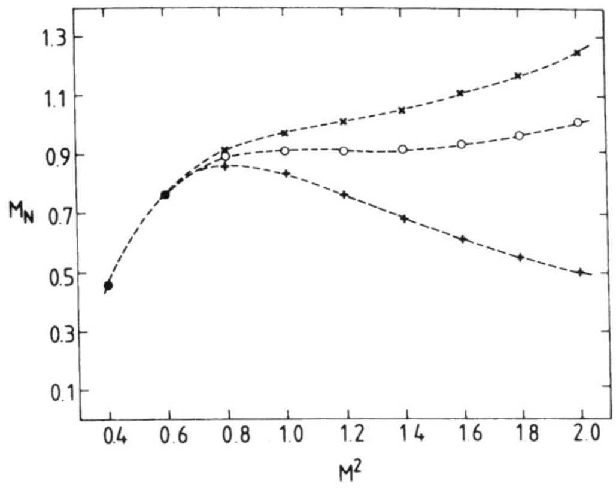

Fig. 1. Plot of the nucleon mass M_N as a function of M^2 [eq. (7)] for a few values of the continuum threshold s_0 (×: s_0 = 4.5 GeV^2, ○: s_0 = 5.25 GeV^2, +: $s_0 = \infty$); a = 0.5 GeV^3 and b = 0.13 GeV^4.

Taking also corrections to first order in the strange quark mass m into account, we can derive sum rules like (6) for the other members of the baryon octet which results in the equations:

$$M_N(M^2) = \frac{2aM^4}{M^6 - bM^2 + \frac{4}{3}a^2}, \tag{8a}$$

$$M_\Lambda(M^2) = \frac{-2mM^6 + 2a(3-\gamma)M^4 + \frac{8}{3}a^2m(2-\gamma)}{3M^6 + 2am(1-3\gamma)M^2 - 3bM^2 + \frac{4}{3}a^2(3+4\gamma)}, \tag{8b}$$

$$M_\Sigma(M^2) = \frac{2mM^6 + 2a(1+\gamma)M^4 + \frac{8}{3}a^2m}{M^6 - 2am(1+\gamma)M^2 - bM^2 + \frac{4}{3}a^2}, \tag{8c}$$

$$M_\Xi(M^2) = \frac{2aM^4 + 4a^2m(1+\gamma)}{M^6 - bM^2 + \frac{4}{3}a^2(1+\gamma)^2}, \tag{8d}$$

where $\gamma = [\langle 0|\bar{s}s|0\rangle - \langle 0|\bar{u}u|0\rangle]/\langle 0|\bar{u}u|0\rangle$. A check of these formulae is given by keeping the octet pieces of the symmetry breaking which gives the Gell-Mann–Okubo mass formula. Investigating these formulae one finds that the mass splittings are very sensitive to the parameter γ. We will come back to this later. We have kept γ-corrections in the first order m terms for completeness although they are of higher order.

Before matching the equations for the octet baryons to the experimental masses by varying m, a, and γ, we also include the decuplet resonances given by the currents (4). Since we deal with $J = \frac{3}{2}$ states the tensor structure is more complicated. We saturate the phenomenological side of the sum rule by a single resonance and use the Rarita–Schwinger formalism for the propagator. On the theoretical side there is a large number of invariant functions but the two proportional to $g_{\mu\nu}$ and $\not{p}g_{\mu\nu}$ receive contributions from $J = \frac{3}{2}$ states only (at least for massless quarks). We will therefore write sum rules by equating the $g_{\mu\nu}$ and $\not{p}g_{\mu\nu}$ structures on both sides. For more details see ref. [1]. The sum rules for the Δ read

$$M^6 - \tfrac{25}{18}bM^2 + \tfrac{20}{3}a^2 = 5(2\pi)^4\lambda_\Delta^2 \exp(-M_\Delta^2/M^2), \quad \tfrac{20}{3}aM^4(1 - \tfrac{1}{2}m_0^2/M^2) = 5(2\pi)^4 M_\Delta \lambda_\Delta^2 \exp(-M_\Delta^2/M^2), \tag{9a,b}$$

where the parameter m_0 is defined as

$$\langle 0|g_s \bar{q}\sigma_{\mu\nu}\tfrac{1}{2}\lambda^a q G^a_{\mu\nu}|0\rangle \equiv m_0^2\langle 0|\bar{q}q|0\rangle. \tag{10}$$

For the octet baryons the coefficient of this operator vanishes [2]. In ref. [9] m_0^2 has been estimated to be 0.5–1 GeV^2 from analysing open charm states, so its contribution in eq. (9b) cannot be neglected. We will treat it as a free parameter restricted between 0.5 and 1 GeV^2. In eqs. (9) we have again included the gluon condensate contribution calculated by Smilga [2].

Taking the ratio of eqs. (9b) and (9a) we find the following equation for M_Δ as a function of M^2

$$M_\Delta(M^2) = \frac{\frac{20}{3}aM^2(1-\frac{1}{2}m_0^2/M^2)}{M^6 - \frac{25}{18}bM^2 + \frac{20}{3}a^2}\,. \tag{11a}$$

Similarly we find for the other decuplet resonances

$$M_{\Sigma^*}(M^2) = \frac{\frac{20}{3}aM^4(1+\frac{1}{3}\gamma)(1-\frac{1}{2}m_0^2/M^2) + \frac{5}{2}mM^6 + \frac{10}{3}a^2m}{M^6 + 5am(1-\frac{1}{3}\gamma)M^2 - \frac{25}{18}bM^2 + \frac{20}{3}a^2(1+\frac{2}{3}\gamma)}\,, \tag{11b}$$

$$M_{\Xi^*}(M^2) = \frac{\frac{20}{3}aM^4(1+\frac{2}{3}\gamma)(1-\frac{1}{2}m_0^2/M^2) + 5mM^6 + \frac{20}{3}a^2m(1+\gamma)}{M^6 + 10am(1+\frac{1}{3}\gamma)M^2 - \frac{25}{18}bM^2 + \frac{20}{3}a^2(1+\gamma)(1+\frac{1}{3}\gamma)}\,, \tag{11c}$$

$$M_\Omega(M^2) = \frac{\frac{20}{3}aM^4(1+\gamma)(1-\frac{1}{2}m_0^2/M^2) + \frac{15}{2}mM^6 + 10a^2m(1+\gamma)^2}{M^6 + 15am(1+\gamma)M^2 - \frac{25}{18}bM^2 + \frac{20}{3}a^2(1+\gamma)^2}\,. \tag{11d}$$

The simplicity of the structure of these formulae reflects the fact that for the decuplet all quark pairs are in a spin one state.

At this stage we have expressions for eight masses as a function of M^2 and the parameters a, b, m, m_0 and γ. These parameters are not completely free; a is restricted by current algebra to lie between 0.3 and 0.7 GeV^3 [$\langle 0|\bar{u}u|0\rangle^{1/3} \simeq -(200\text{–}260)$ MeV], m_s should be about 150 MeV, b has been determined before from charmonium [4,6] and should be about 0.13 GeV^4, m_0^2 has been estimated to be 0.5–1 GeV^2, and the absolute value of γ is expected to be not larger than 0.3, since it measures SU(3) symmetry breaking. As we neglect all continuum contributions to the sum rules we cannot expect our mass values to agree better than within five per cent with the experimental values.

As pointed out above the octet mass splittings are extremely sensitive to the value of γ. In fig. 2 we have plotted the octet masses as a function of γ. It can be seen that the correct range of levels can only be obtained for $0.17 < -\gamma < 0.21$. This is essentially independent of reasonable changes in the other parameters. The decuplet masses are not sensitive to γ.

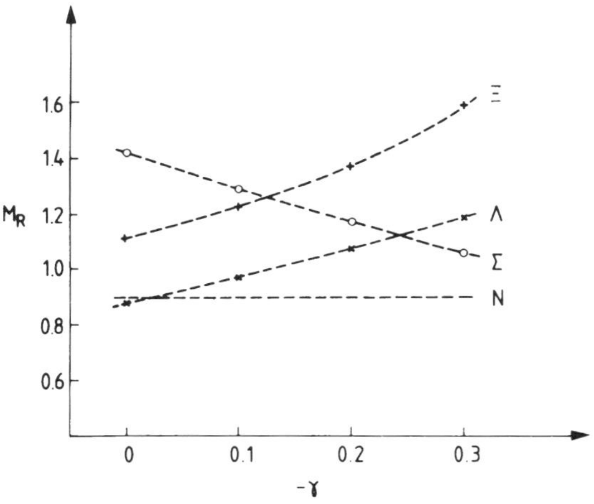

Fig. 2. Plot of the octet masses as a function of γ (a = 0.6 GeV^3, m = 140 MeV, and b = 0.15 GeV^4).

Table 1
Results for the octet and decuplet masses for a few values of a; m_0^2 has been adjusted to give the best possible decuplet masses; m is fixed at 140 MeV, b at 0.15 GeV^4 and γ at −0.19.

	(a, m_0^2)			exp. (GeV)
	(0.4, 0.5)	(0.5, 0.7)	(0.6, 0.9)	
M_N	0.85	0.88	0.90	0.94
M_Λ	1.03	1.03	1.06	1.12
M_Σ	1.24	1.19	1.18	1.19
M_Ξ	1.39	1.37	1.35	1.32
M_Δ	1.22	1.23	1.24	1.23
M_{Σ^*}	1.34	1.36	1.38	1.38
M_{Ξ^*}	1.48	1.52	1.54	1.53
M_Ω	1.67	1.70	1.74	1.67

In table 1 we have listed the results for the masses for a few values of a. For each value of a, m_0 has been adjusted to give the best possible values for the decuplet masses. The mass values in table 1 correspond to the maxima (as a function of M^2) of the sum rules (8) and (11). It can be seen from the table that it is not difficult to get predictions for the masses within 5% of the experimental values. The results are not sensitive to the mass of the strange quark and the value of b; γ has been kept fixed at -0.19. We arrive at the following set of parameter values

$$a = 0.4-0.6\ \mathrm{GeV}^3\,, \qquad \text{i.e.,}\ \langle 0|\bar{u}u|0\rangle = -(230 \pm 15\ \mathrm{MeV})^3\,,$$

$$b = 0.15\ \mathrm{GeV}^4\,, \qquad \text{i.e.,}\ \langle 0|(\alpha_s/\pi)\, G^a_{\mu\nu} G^a_{\mu\nu}|0\rangle = (350\ \mathrm{MeV})^4\,,$$

$$m = 140 \pm 20\ \mathrm{MeV}\,, \quad m_0^2 = 0.5-1.0\ \mathrm{GeV}^2\,, \quad \gamma = -0.19 \pm 0.02\,, \quad \text{i.e.,}\ \langle 0|\bar{s}s|0\rangle = 0.8\langle 0|\bar{u}u|0\rangle\,. \tag{12}$$

The parameter γ has been argued to be positive [10]. The argument relies on positivity and the first order mass effect in chiral perturbation theory. However, our definition of the condensate in which all perturbative contributions are subtracted is not the same [‡1]. It is amusing that the heavy quark mass expansion of $\langle 0|\bar{s}s|0\rangle$ also gives $\gamma < 0$ at the value of m_s. Also, using other sum rules Mallik [11] has obtained a rough value of γ in agreement with our result in sign and magnitude.

We conclude that QCD sum rules naturally explain the levels in baryon systems of light quarks to a remarkable accuracy (5%) in the lowest order approximation in the operator expansion. Our main result here, however, is the observation that the appearance of baryon masses and mass splittings is very sensitive to some parameters that characterize spontaneous chiral symmetry breaking (a) and SU(3) symmetry breaking (γ). The theory fixes $\langle 0|\bar{u}u|0\rangle = -(230 \pm 15\ \mathrm{MeV})^3$ and $\gamma = -0.19 \pm 0.02$. These quantities can be reinterpreted in the f and d couplings which are the basis of octet dominance for symmetry breaking. In fact, the negative sign of γ is due to the fact that the splitting between Ξ and N (f-type) is much larger than the one between Σ and Λ (d-type). QCD is therefore able to predict dynamically the strength of the two octet components, as well as the absolute baryon masses, the distance between the octet and decuplet, m_s and γ. Confirmation of the values of these parameters elsewhere is important.

‡1 One of us (HRR) thanks H. Leutwyler and P. Minkowski for enlightening discussions on this point.

References

[1] B.L. Ioffe, Nucl. Phys. B188 (1981) 317; B191 (1981) 591 (E).
[2] A.V. Smilga, Yad. Fiz. 35 (1982) 473.
[3] P. Pascual and R. Tarrach, Barcelona preprint UBFT-FP-5-82 (1982).
[4] M.A. Shifman, A.I. Vainshtein and V.I. Zakharov, Nucl. Phys. B147 (1979) 385.
[5] Y. Chung et al., Phys. Lett. 102B (1981) 175; Nucl. Phys. B197 (1982) 55.
[6] L.J. Reinders, H.R. Rubinstein and S. Yazaki, Nucl. Phys. B186 (1981) 109.
[7] L.J. Reinders, H.R. Rubinstein and S. Yazaki, Nucl. Phys. B196 (1982) 125.
[8] T. Banks et al., Nucl. Phys. B190 (FS3) (1981) 692.
[9] V.A. Novikov et al., Proc. 1978 Neutrino Conf. (Purdue, 1978), ed. E.C. Fowler, p. C278.
[10] V.A. Novikov et al., Nucl. Phys. B191 (1981) 301.
[11] S. Mallik, Bern preprint BUTP 82-13 (March 1982), to be published in Nucl. Phys. B.

L.J. Reinders, H.R. Rubinstein, S. Yazaki, Baryons in QCD and chiral symmetry breaking parameters, Phys. Lett. 120B (1983) 209.

In ref. [2] [A.V. Smilga, Yad. Fiz. 35 (1982) 473] the gluon condensate contributions to octet baryons have the wrong sign. This implies that in eqs. (6a), (7), on page 210, and (8a)–(8d), on page 211, the sign of the b term has to be positive. Qualitatively this does not change our results. Quantitatively it involves the following change in the values of the parameters (see page 213)

$$\langle 0|\bar{u}u|0\rangle = -(300 \pm 30\ \text{MeV})^3 ,$$

$$\gamma = -0.23 \pm 0.02, \quad m_s = 180 \pm 20\ \text{MeV} .$$

With this value for $\langle 0|\bar{u}u|0\rangle$ ($a \simeq 1.0$) the power corrections to the decouplet states are too large for the approximations to be valid.

Secondly, ref. [10] has been misquoted. It should be referred to after the statement on the heavy quark mass expansion (page 213, fourth line below eq. (12)).

* * *

6. Form Factors at Intermediate Q^2

Introduction

In the initial stages the main efforts have been invested in the analysis of the two-point functions. A large amount of results concerning the meson and baryon masses and residues has been obtained. On the theoretical side, our knowledge of the QCD vacuum became much deeper and its description in terms of different condensates much more complete. Considerable experience in both conceptual and technical issues has been accumulated.

All these factors, taken together, made possible the next step – generalization of the method to form factors at intermediate values of q^2 (q is the momentum transfer). To calculate the form factors it is necessary to consider three-point functions; two currents produce the initial and final particles and the third current induces the vertex we are interested in. For instance, the electromagnetic form factor of a pion can be extracted from

$$\Gamma_{\mu\alpha\nu} = -\int \mathrm{d}^4x\ \mathrm{d}^4y\ \mathrm{e}^{\mathrm{i}p'x-\mathrm{i}qy}\left\langle 0\right|T\left[a_\mu(x)j_\alpha^{\mathrm{em}}(y)a_\nu^+(0)\right]\left|0\right\rangle, \tag{6.1}$$

where $a_\mu = \bar{u}\gamma_\mu\gamma_5 d$, and p' and $p = p' - q$ are the momenta flowing through the axial currents a_μ and a_ν^+, respectively. It is quite clear that for purely kinematical reasons the corresponding analysis is more sophisticated and labor-consuming than in the case of two-point functions. Many technical devices call for essential development. Double Borelization (with respect to p^2 and p'^2 independently) is one of the examples. The natural temptation to adjust the single Borelization procedure with $p^2 = p'^2$ to the form factor situation (e.g. ref. [1]), like the one considered in the previous chapters, leads basically nowhere. Although the single Borelization procedure for the three-point functions (6.1) can be justified theoretically [2] it does not allow one to reliably separate the amplitudes of interest from contamination. If reliable theoretical control is desirable one cannot avoid using the double Borelization procedure in the form factor situation.

Why we discuss here only intermediate values of Q^2, say $1 < Q^2 < 3$ GeV^2, where $Q^2 \equiv -q^2$?

The theoretical prediction for the correlation function (6.1) is obtained on the basis of the operator product expansion. In the case at hand the behavior of the condensate terms is very peculiar: at large Q^2 they are constant or even grow with Q^2 while the relative weight of the perturbative term decreases, a specific feature which was absent in the problems considered in the previous chapters. Such a disastrous behavior means that we lose control over the condensate terms of higher dimension starting from a certain boundary value of Q^2 close to 3 GeV^2. Another, more technical reason blocking the analysis at $Q^2 \geq 3$ GeV^2 is as follows. The pion contribution to eq. (6.1) rapidly falls off relative to that of excited states with increasing Q^2 and above ~ 3 GeV^2 the separation of the pion term from the contamination due to higher resonances in the a_μ channels becomes extremely difficult, if possible at all.

At small Q^2 the characteristic distance between the points y and x (and y and 0) is large, the standard OPE approach cannot be used, and $\Gamma(p, p', q)$ is not calculable in this way *. Technically, this fact manifests itself in singularities in $\Gamma(p, p', q)_{\mathrm{theor}}$ at $Q^2 \to 0$.

* The latter difficulty can be circumvented, though, at a price of introducing new condensates. A modification of the sum rule method allowing one to calculate static properties at $Q^2 \to 0$ is discussed in chapter 7.

Therefore, without a serious restructuring of the method one cannot go beyond the intermediate domain of Q^2.

The first works [3,4] on form factors in the QCD sum rules have been done approximately at the same time, independently. Two of them are reproduced in this chapter. They nicely supplement each other. The authors calculate the pion electromagnetic form factor and the results are strikingly close to experimental data in the domain where the method is applicable. The theoretical curve in this domain is close to the vector dominance model but lies much higher than the extrapolation of the naive asymptotic prediction [5]. The same situation occurs for the ρ meson – the ratio of the form factors, which should be large asymptotically, turns out to be small and agrees with the vector dominance model.

Let us parenthetically note that attempts to establish asymptotic ($Q^2 \to \infty$) power-like behavior of form factors for composite hadrons can be traced back to the prehistoric pre-QCD era (e.g. ref. [6]). The so-called quark counting rules [7], obtained essentially on the basis of dimensional and scaling arguments, were the turning point in this field. The dynamical realization of the quark counting rules in QCD has been worked out in ref. [5]. Only after these works it became possible to estimate a boundary value of Q^2 marking the onset of the asymptotic regime. The corresponding formulae which are now usually referred to as "the naive asymptotic predictions" are valid in the academic limit where $\log Q^2/\Lambda^2_{\rm QCD}$ becomes much larger than 1. As a matter of fact, this situation is even worse since actually all relevant expressions contain a relatively small power of this logarithm. No surprise that the sum rule calculations of the form factors do not overlap with the naive asymptotic predictions. For a discussion of the relation between the QCD sum rule results and asymptotic formulae from modern viewpoints see ref. [8].

Following the lines sketched above the transition electromagnetic form factor $\rho\pi$ and some other vertices have been found [9,10] later. Further evolution and the desire to master the domain of low Q^2 resulted in new ideas, which will be discussed in chapter 7.

References

[1] L.J. Reinders, H. Rubinstein and S. Yazaki, Nucl. Phys. B 213 (1983) 109.
[2] A.V. Smilga and M.A. Shifman, Yad. Fiz. 37 (1983) 1613 [Sov. J. Nucl. Phys. 37 (1983) 958].
[3] V.A. Nesterenko and A.V. Radyushkin, Phys. Lett. B 115 (1982) 410.
[4] B.L. Ioffe and A.V. Smilga, Phys. Lett. B 114 (1982) 353; Nucl. Phys. B 216 (1983) 373.
[5] G.P. Lepage and S.J. Brodsky, Phys. Lett. B 87 (1979) 359.
V.L. Chernyak and A.R. Zhitnitskii, Pis'ma Zh. Eksp. Teor. Fiz. 25 (1977) 544 [JETP Lett. 25 (1977) 510]; Phys. Rep. 112 (1984) 173.
G.R. Farrar and D.R. Jackson, Phys. Rev. Lett. 43 (1979) 246.
A.V. Efremov and A.V. Radyushkin, Phys. Lett. B 94 (1980) 245.
[6] D. Amati et al., Phys. Lett. B 27 (1968) 38.
[7] V.A. Matveev, R.M. Muradyan and A.N. Tavkhelidze, Nuovo Cimento Lett. 7 (1973) 719.
S.J. Brodsky and G.R. Farrar, Phys. Rev. Lett. 31 (1973) 1153; Phys. Rev. D 11 (1975) 1309.
[8] A.V. Radyushkin, Hadronic Form Factors: Perturbative QCD Versus QCD Sum Rules, Plenary Talk at 2nd European Workshop on Hadronic Physics with Electrons Beyond 10 GeV, Dourdan, France, 1990, CEBAF-TH-91-07.
[9] V.M. Belyaev and Ya.I. Kogan, Z. Phys. C 20 (1983) 357.
[10] V.L. Eletsky, B.L. Ioffe and Ya.I. Kogan, Phys. Lett. B 122 (1983) 423.

SUM RULES AND THE PION FORM FACTOR IN QCD

V.A. NESTERENKO and A.V. RADYUSHKIN
Laboratory of Theoretical Physics, JINR, Dubna, USSR

Received 12 May 1982

We propose a new approach to the investigation of the pion electromagnetic form factor in QCD based on the systematic use of the QCD sum rule technique. The theoretical curve obtained for $F_\pi(Q^2)$ is in good agreement with existing experimental data.

1. Introduction. The discovery that perturbation theory can be applied to study exclusive high momentum transfer processes (see, e.g., refs. [1–3] and references therein) has been an important step in the development of perturbative QCD. In particular, asymptotic freedom enables one to easily reproduce in the asymptotic $Q^2 \to \infty$ region the well-known quark counting rules for electromagnetic form factors of hadrons [4]. However, for experimentally accessible momentum transfers Q the agreement between existing theory [1–3] and experimental data for pion form factor is very poor (see, e.g., ref. [5]).

This observation, nevertheless, should not be treated as an evidence against QCD itself because the perturbative QCD approach [1–3] is applicable only for asymptotically large Q^2, and the extrapolation of the asymptotic QCD formulas into the region of moderately large Q^2 is not justified. For pion, e.g., the main $1/Q^2$-contribution in the region $Q^2 \to \infty$ is due to the hard rescattering (one-gluon exchange) subprocess (fig. 1a). However, a straightforward use of the asymptotic formalism in the $Q^2 \lesssim 10$ GeV^2 region leads to the conclusion that the mean virtuality of the gluon (fig. 1a) is much smaller than 1 GeV^2 [6]. In such a situation it is, of course, misleading to rely on asymptotic freedom. According to the standard formalism [1–3], the gluon line corresponding to virtuality smaller than some $\lambda^2 \sim 1$ GeV^2 should be absorbed by the soft pion wavefunction. The resulting diagram looks like that shown in fig. 1b. In the asymptotic analysis [1–3] this diagram is simply ignored because

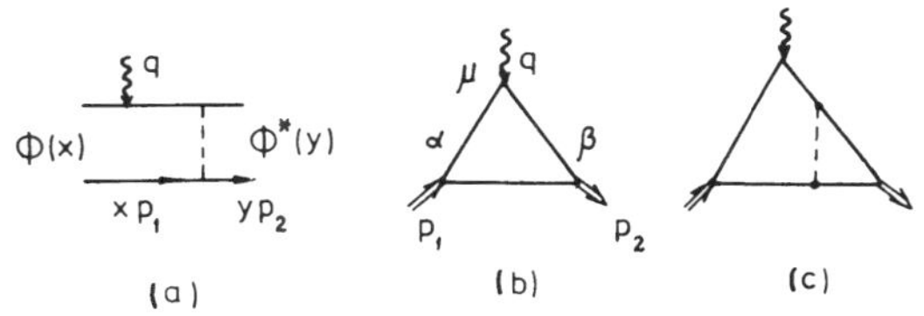

Fig. 1. Diagrams relevant to calculation of the pion form factor in QCD: (a) asymptotic perturbative QCD diagram; (b) lowest-order diagram of the QCD sum rule approach; (c) one of two-loop diagrams of the QCD sum rule approach.

the upper estimate for its large-Q^2 behaviour is only $1/Q^4$ [1]. However, a complete evaluation of its contribution within the perturbative QCD approach is impossible, since the region of small quark virtualities is dominated by nonperturbative effects. Earlier one of the authors (A.R.) attempted to take these effects into account within the framework of a QCD inspired model [7]. In the present letter we outline a new approach [‡1] to exclusive processes in QCD that has much wider applicability than the asymptotic analysis [1–3]. The basic idea of the approach is the duality between quarks and hadrons: we show, in particular, that one can obtain the pion form factor by calculating the quark diagrams (the lowest-order ones are shown in figs. 1b,c) with local quark–"pion" vertices and averaging the result over the appropriate duality interval.

‡1 In the course of our investigations we have learned that a similar approach was independently proposed by Ioffe and Smilga.

2. Derivation of sum rules. Technically, our analysis is based on the QCD sum rule approach [8] that has proved to be a very effective tool for studying the "static" properties of hadrons, such as masses, leptonic widths, etc. To analyse the pion form factor, we consider the three-point amplitude

$$T_{\mu\alpha\beta}(p_1,p_2) = \mathrm{i}^2 \int \exp(-ip_1x + ip_2y)$$
$$\times \langle 0|T\{j_\beta(y)J_\mu(0)j_\alpha^+(x)\}|0\rangle\, \mathrm{d}^4x\, \mathrm{d}^4y \qquad (1)$$

(for notation see fig. 1b), where J_μ is the electromagnetic current and $j_\alpha = \bar{\mathrm{d}}\gamma_5\gamma_\alpha \mathrm{u}$ is the axial current. The latter satisfies the necessary condition that it should have nonzero projection onto the pion state $|P\rangle$

$$\langle 0|j_\alpha(0)|P\rangle = \mathrm{i}f_\pi P_\alpha, \qquad (2)$$

where $f_\pi \simeq 133$ MeV is the pion decay constant. In principle, one may use also a pseudoscalar combination $\bar{\mathrm{d}}\gamma_5\mathrm{u}$. In this case, however, there appear serious complications due to direct interactions of quarks with instantons [9,10], such interactions being absent for the axial current.

The amplitude $T_{\mu\alpha\beta}(p_1,p_2)$ is the sum of various structures, and the corresponding invariant amplitudes T_i depend on three variables: $p_1^2, p_2^2, q^2 = (p_1 - p_2)^2$. Owing to asymptotic freedom, one may calculate $T_i(p_1^2,p_2^2,q^2)$ in the deep euclidean region $p_1^2,p_2^2,q^2 < -\lambda^2 \sim -1\ \mathrm{GeV}^2$. To extract the desired information about the form factors of physical states, we use the double dispersion relation

$$T_i(p_1^2,p_2^2,q^2)$$
$$= \frac{1}{\pi^2}\int_0^\infty \mathrm{d}s_1 \int_0^\infty \mathrm{d}s_2 \frac{\rho_i(s_1,s_2,q^2)}{(s_1-p_1^2)(s_2-p_2^2)} + \ldots, \qquad (3)$$

The terms not written explicitly in eq. (2) are polynomials in p_1^2 and/or p_2^2. They disappear after one applies to eq. (3) the Borel procedure $\hat{B}_{12}$

$$\hat{B}_{12} = \lim_{n_1\to\infty}\ \lim_{n_2\to\infty} \left\{(-1)^{n_1+n_2} \frac{(p_1^2)^{n_1}(p_2^2)^{n_2}}{(n_1-1)!(n_2-1)!}\right.$$
$$\left. \times \left(\frac{\mathrm{d}}{\mathrm{d}p_1^2}\right)^{n_1} \left(\frac{\mathrm{d}}{\mathrm{d}p_2^2}\right)^{n_2}\Bigg|_{p_1^2=n_1M_1^2,\, p_2^2=n_2M_2^2}\right\}, \qquad (4)$$

which is a straightforward generalization of that used in ref. [8]. Applying $\hat{B}_{12}$ to eq. (3) gives

$$\Phi_i(M_1^2,M_2^2,Q^2) = \frac{1}{\pi^2}\int_0^\infty \frac{\mathrm{d}s_1}{M_1^2}\int_0^\infty \frac{\mathrm{d}s_2}{M_2^2}$$
$$\times \rho(s_1,s_2,q^2)\exp(-s_1/M_1^2 - s_2/M_2^2), \qquad (5)$$

where $Q^2 = -q^2$ and $\Phi_i = \hat{B}_{12}T_i$.

The pion contribution into the spectral density is proportional to

$$\langle 0|j_\beta|p_2\rangle\langle p_2|J_\mu|p_1\rangle\langle p_1|j_\alpha^+|0\rangle$$
$$\sim f_\pi^2 F_\pi(Q^2) p_2^\beta p_1^\alpha (p_1^\mu + p_2^\mu), \qquad (6)$$

and the first idea is to extract from $T_{\mu\alpha\beta}$ the structure $p_1^\alpha p_2^\beta(p_1^\mu + p_2^\mu)$. However, there exist also other structures $(p_1^\beta p_2^\alpha, p_1^\alpha p_1^\beta, p_2^\alpha p_2^\beta)$ that coincide with $p_1^\alpha p_2^\beta$ for $q = 0$. This complication disappears if all the basic structures are expanded in $P = p_1 + p_2$ and $q = p_2 - p_1$. Then the relevant structure is $P_\alpha P_\beta P_\mu$ and the simplest way to extract the corresponding invariant amplitude (hereafter referred to as T) is to multiply $T_{\alpha\beta\mu}$ by $n^\alpha n^\beta n^\mu$, where n is a lightlike vector orthogonal to $q(n^2 = 0, (nq) = 0, (nP) \neq 0)$.

The calculation of $\Phi = \hat{B}_{12}T$ is most simply performed by the Feynman parametrization or its exponential version ("α-representation", see, e.g. ref. [11]). Then it is straightforward to obtain the contribution of the diagram 1b:

$$\Phi^{(1\mathrm{b})}(M_1^2,M_2^2,Q^2) = \frac{3}{2\pi^2(M_1^2+M_2^2)}$$
$$\times \int_0^1 x(1-x)\exp\left(-\frac{xQ^2}{(1-x)(M_1^2+M_2^2)}\right)\mathrm{d}x$$
$$\times\left\{\frac{2}{3}\exp\left[-\left(\frac{m_\mathrm{u}^2}{1-x}+\frac{m_\mathrm{d}^2}{x}\right)\left(\frac{1}{M_1^2}+\frac{1}{M_2^2}\right)\right]\right.$$
$$\left. + \frac{1}{3}\exp\left[-\left(\frac{m_\mathrm{u}^2}{x}+\frac{m_\mathrm{d}^2}{1-x}\right)\left(\frac{1}{M_1^2}+\frac{1}{M_2^2}\right)\right]\right\}. \qquad (7)$$

The x-variable may be interpreted as the fraction of the total pion momentum carried by the passive quark in the infinite momentum frame. Note, that for massless quarks $\Phi^{(1\mathrm{b})} \sim 1/Q^4$ as $Q^2 \to \infty$, in agreement with the general analysis described in refs. [1,12]. In what follows, we neglect $m_u, m_d \sim 10$ MeV $\ll \lambda$.

Notice further that eq. (5) is the (double) Laplace transformation in $1/M_i^2$. Applying the inverse transformation to eq. (7) gives the free-quark spectral function

$$\rho_0(s_1,s_2,q^2) = \tfrac{3}{2}(Q^2)^2 [(\mathrm{d}/\mathrm{d}Q^2)^2 + \tfrac{1}{3}Q^2(\mathrm{d}/\mathrm{d}Q^2)^3] \times [(s_1+s_2+Q^2)^2 - 4s_1s_2]^{-1/2}. \quad (8)$$

In the real world $\rho(s_1,s_2,q^2)$ differs, of course, from $\rho_0(s_1,s_2,q^2)$. The difference is most pronounced for small s_1,s_2. In particular, $\rho(s_1,s_2,q^2)$ contains the pion term

$$\rho_{\pi\pi}(s_1,s_2,q^2) = \pi^2 f_\pi^2 F_\pi(Q^2)\,\delta(s_1 - m_\pi^2)\,\delta(s_2 - m_\pi^2). \quad (9)$$

Furthermore, $\rho(s_1,s_2,q^2)$ vanishes below the 3π-threshold (i.e. in the region $m_\pi^2 < (s_1,s_2) < 9m_\pi^2$), and only in the region, where both s_1 and s_2 are sufficiently large, ρ is close to ρ_0. This means that $\Phi(M_1^2, M_2^2, Q^2)$ also differs from the free-quark value (7). As it was argued in ref. [8], nonperturbative corrections $(1/M^2)^N$ are much more important than the perturbative ones [‡2]. Taking into account the contributions proportional to $(\alpha_s/\pi)\langle G_{\mu\nu}^a G_{\mu\nu}^a\rangle$ and $\alpha_s\langle\bar{q}q\rangle^2$ (cf. ref. [8]) and using eqs. (5), (9), we arrive at the following representation for the pion form factor

$$f_\pi^2 F_\pi(Q^2) = \frac{1}{\pi^2}\int_0^{s_0} \mathrm{d}s_1 \int_0^{s_0} \mathrm{d}s_2\, \rho_0(s_1,s_2,q^2)\exp[-(s_1+s_2)/M^2] + \frac{\alpha_s\langle G_{\mu\nu}^a G_{\mu\nu}^a\rangle}{12\pi M^2} + \frac{208\pi\alpha_s\langle\bar{q}q\rangle^2}{81M^4}\left(1 + \frac{2}{13}\frac{Q^2}{M^2}\right) - \Delta(s_0,M^2,Q^2) + \mathrm{O}(1/M^6) + \mathrm{O}(\alpha_s(s_0)/\pi), \quad (10)$$

where $\Delta(s_0,M^2,Q^2) \equiv B(M^2,Q^2) \sim B_0(s_0,M^2,Q^2)$ is the difference between the "true" background contribution $B(M^2,Q^2)$

$$B(M^2,Q^2) = \frac{1}{\pi^2}\int^{\infty}\mathrm{d}s_1\int^{\infty}_{\max\{s_1,s_2\}>9m_\pi^2}\mathrm{d}s_2 \times \rho(s_1,s_2,q^2)\exp[-(s_1+s_2)/M^2] \quad (11)$$

and its free-quark analogue $B_0(s_0,M^2,Q^2)$

$$B_0(s_0,M^2,Q^2) = \frac{1}{\pi^2}\int^{\infty}\mathrm{d}s_1\int^{\infty}_{\max\{s_1,s_2\}>s_0}\mathrm{d}s_2 \times \rho_0(s_1,s_2,q^2)\exp[-(s_1+s_2)/M^2]. \quad (12)$$

3. Quark–hadron duality. To get rid of power corrections in eq. (10), one should take $M^2 = \infty$. Furthermore, for any fixed $Q^2 = Q_0^2$ the Δ-term can be also eliminated by an appropriate choice of the s_0-parameter. As a result, one obtained the following representation for $F_\pi(Q_0^2)$:

$$f_\pi^2 F_\pi(Q_0^2) = \frac{1}{\pi^2}\int_0^{s_0}\mathrm{d}s_1\int_0^{s_0}\mathrm{d}s_2\,\rho_0(s_1,s_2,q^2 = -Q_0^2) \quad (13)$$

that is nothing else but a (generalized) finite-energy sum rule (FESR, cf. refs. [13–15]) or a duality relation between the resonance (pion) and free-quark contributions. The functions $\rho_{\pi\pi}(s_1,s_2,q^2)$ and $\rho_0(s_1,s_2,q^2)$ are quite different but if one averages them over the relevant duality interval, the result is the same in both cases. The parameter s_0 can be interpreted then as the boundary between the pion duality interval and the next one related to the A_1-resonance. In such an interpretation s_0 is the same number (for all Q^2 values) at the midpoint of the interval between $m_\pi^2 = 0$ and $m_{A_1}^2 \approx 1.2$ GeV^2. This gives the estimate $s_0 \approx 0.6$ GeV^2 [15].

Within the QCD sum rule approach [8] s_0 is not a free parameter. Rather, it is determined by the magnitude of the quark and gluon condensates $\langle\bar{q}q\rangle$ and $\langle GG\rangle$ [‡3]. To extract s_0 from the sum rule (10), we analysed the M^2-dependence of its rhs for a chosen Q_0^2 value and various s_0-values (taking $\Delta(s_0,M^2,Q^2) = 0$, as argued above). It is easy to establish (see fig. 2) that for sufficiently large M^2 our theoretical "prediction"

‡2 To treat initial and final states on equal footing and to simplify the calculation, we take henceforth $M_1 = M_2 = M$.

‡3 In numerical estimates we use the values $(\alpha_s(\pi)\langle G_{\mu\nu}^a G_{\mu\nu}^a\rangle = 0.012$ GeV^4 and $\alpha_s\langle\bar{q}q\rangle^2 = 1.83 \times 10^{-4}$ GeV^6 taken from ref. [8].

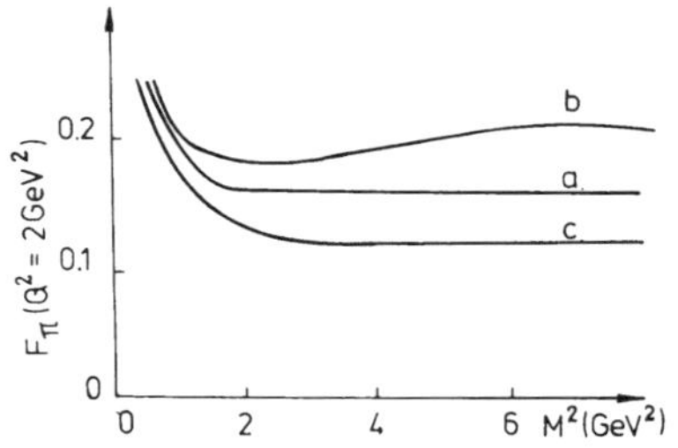

Fig. 2. Typical dependence of $F_\pi(Q^2 = 2\ \text{GeV}^2)$ on the auxiliary parameter M^2: (a) $s_0 = 0.7\ \text{GeV}^2$; (b) $s_0 = 0.8\ \text{GeV}^2$; (c) $s_0 = 0.6\ \text{GeV}^2$.

for $F_\pi(Q_0^2)$ has a very weak dependence on the auxiliary ("unphysical", cf. ref. [16]) parameter M^2, but the onset of the asymptotic regime strongly depends on s_0. The "true" value of s_0 is evidently that for which the region of weak sensitivity to variations of M^2 is the largest one. For $Q_0^2 = 1, 2$ and $3\ \text{GeV}^2$ this criterion gives values close to $s_0 = 0.7\ \text{GeV}^2$. Furthermore, to check the self-consistency of the whole approach, we estimated the s_0-parameter also from the analysis of the two-point amplitude [‡4] related to $\langle T(j_\beta j_\alpha^+)\rangle$ and obtained the same result $s_0 = 0.7\ \text{GeV}^2$. The accuracy of the method, however, should not be overestimated: it is not better than 10–20%.

4. Discussion of results. Using the explicit form of eq. (8), one can reduce eq. (13) to

$$F_\pi(Q^2) = (s_0/4\pi^2 f_\pi^2)$$
$$\times\ [1 - (1 + 6s_0/Q^2)/(1 + 4s_0/Q^2)^{3/2}]\,. \qquad (14)$$

Another formula can be obtained, if one substitutes the integration over the square $(0 \leqslant s_1 \leqslant s_0, 0 \leqslant s_2 \leqslant s_0)$ in eq. (13) by integration over the triangle $(0 \leqslant s_1 + s_2 \leqslant S_0 = \sqrt{2}s_0)$ of equivalent area. This gives

$$F_\pi^{(\mathrm{TR})}(Q^2) = S_0/8\pi^2 f_\pi^2 (1 + Q^2/2S_0)^2. \qquad (15)$$

However arbitrary this substitution may seem for $Q^2 \gtrsim 1\ \text{GeV}^2$ eq. (15) reproduces eq. (14) with an accuracy better than 10%. So, eq. (15) is very convenient for a quick estimate of predictions for eq. (14).

Before comparing eq. (14) with experimental data, we want to stress that our analysis is reliable only in the $Q^2 \gtrsim 1\ \text{GeV}^2$ region where the asymptotic freedom guarantees the absence of large corrections (of, say, $1/Q^2$ type). One should also remember that there exists some freedom in choosing s_0 : $s_0 = 0.7 \pm 0.1\ \text{GeV}^2$. Our curve corresponding to eq. (14) for $s_0 = 0.7\ \text{GeV}^2$ is shown in fig. 3 (solid line). In the region $Q^2 > 1\ \text{GeV}^2$ it is in excellent agreement with the existing data [17].

It is also interesting to analyse the predictions for $F_\pi(Q^2)$ resulting from the basic sum rule (10) for finite M^2. As argued in ref. [8], one should choose M^2 so as to reduce to 30% both the power corrections (that blow up as $M^2 \to 0$) and the background contribution $B(M^2, Q^2)$ (that grows as $M^2 \to \infty$). For two-point functions considered by SVZ [8] such a compromise is really possible for $M^2 \simeq s_0$. Alas, this strategy fails for eq. (10): to reduce the power corrections to 30% at the reference momentum $Q_0^2 = 2\ \text{GeV}^2$, one should choose M^2 in the region $M^2 \geqslant 1.8\ \text{GeV}^2$, where $B(M^2, Q^2)$ yields more than 70% compared to the total sum. Moreover, since the first term in the rhs of eq. (10) rapidly diminishes with growing Q^2, the power corrections (that are almost constant for $Q^2 \leqslant 6\ \text{GeV}^2$) exceed the 30% boundary just above the reference point. If one takes $M^2 = 1.8\ \text{GeV}^2$, the power corrections reach 100% for $Q^2 = 6\ \text{GeV}^2$. This means

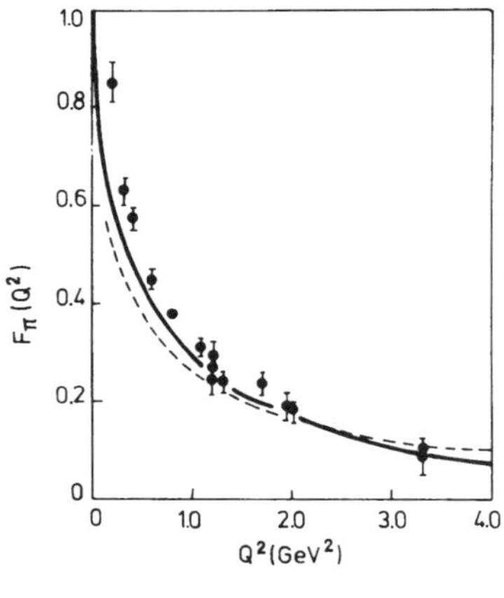

Fig. 3. Comparison between experimental data (taken from ref. [17]) and our theoretical predictions based on eq. (10): $M^2 = \infty$ (solid line); $M^2 = 1.8\ \text{GeV}^2$ (broken line).

‡4 This amplitude was analysed first by SVZ in ref. [8] using a slightly different method. In their fits SVZ used $s_0 = 0.75\ \text{GeV}^2$.

that for sufficiently large Q^2 one looses the control over the $1/M^2$-expansion. There exists a simple explanation of this phenomenon. Note, that the main contribution into $\Phi^{(1b)}$ [eq. (7)] for large Q^2 gives the region $x \sim M^2/Q^2$, where the passive quark has the virtuality $k^2 \sim M^4/Q^2$. Thus, the region $Q^2 \gtrsim M^4/m_\rho^2$ is in fact beyond the scope of the asymptotic freedom. Of course, taking M^2 sufficiently large, one can control the power corrections for arbitrarily high Q^2 values. The most radical way out is to take $M^2 = \infty$. Then the power corrections are absent altogether. However, this is achieved at the price of the (infinite) growth of the $B(M^2, Q^2)$-term. Still, it is the difference $\Delta = B - B_0$ that matters. As argued above, the choice $M^2 = \infty$ reduces the whole M^2-business to finding the appropriate "effective threshold" s_0. It is worth noting also that the finite-M^2 results for $F_\pi(Q^2)$, in the Q^2-region where power corrections are under control, do agree with the $M^2 = \infty$ result, eq. (14) (see fig. 3).

5. Concluding remarks. Thus, the one-gluon-exchange contribution for available Q^2 is of little importance. However, the situation changes drastically in the asymptotic $Q^2 \to \infty$ region, because fig. 1c gives in this region $1/Q^2$-contribution (corresponding to quark counting rules [4] and asymptotic QCD analysis [1–3]) whereas eq. (14) behaves asymptotically as $1/Q^4$. Thus, extrapolating the asymptotic QCD formulas [1–3] into the region of moderately large Q^2 one should not expect a good description of experimental data, and vice versa: good fits of existing data extrapolated into the large-Q^2 region may be irrelevant to the "true" asymptotic behaviour of $F_\pi(Q^2)$. In particular, we consider the result obtained by Dubnicka et al. [18] highly encouraging. Analysing all data on $F_\pi(Q^2)$ (both in spacelike and timelike regions) and taking properly into account the analytic properties of $F_\pi(Q^2)$, they obtained the curve that behaves just like $1/Q^4$ in the asymptotic region as a best fit.

We are grateful to A.V. Efremov and V.A. Meshcheryakov for their interest in this work and support. We acknowledge B.L. Ioffe and A.V. Smilga for stimulating discussions, especially, for that concerning the background contribution.

References

[1] A.V. Efremov and A.V. Radyushkin, Phys. Lett. 94B (1980) 245.
[2] S.J. Brodsky and G.P. Lepage, Phys. Rev. D22 (1980) 2157.
[3] A.H. Mueller, Phys. Rep. 73 (1981) 237.
[4] V.A. Matveev, R.M. Muradyan and A.N. Tavkhelidze, Lett. Nuovo Cimento 7 (1973) 719;
S.J. Brodsky and G.R. Farrar, Phys. Rev. Lett. 31 (1973) 1153.
[5] C.A. Dominguez, report DOE/ER/05223-48, Texas University, College Station (1981).
[6] F.M. Dittes and A.V. Radyushkin, Yad. Fiz. 34 (1981) 529.
[7] A.V. Radyushkin, JINR report P2-80-687 (Dubna, 1980); A.V. Efremov and A.V. Radyushkin, JINR report E2-80-521 (Dubna, 1980).
[8] M.A. Shifman, A.I. Vainshtein and V.I. Zakharov, Nucl. Phys. B147 (1979) 385, 448.
[9] B.V. Geshkenbein and B.L. Ioffe, Nucl. Phys. B166 (1980) 340.
[10] V.A. Novikov, M.A. Shifman, A.I. Vainshtein and V.I. Zakharov, Nucl. Phys. B191 (1981) 301.
[11] N.N. Bogolubov and D.V. Shirkov, Introduction to the theory of quantized fields (Interscience, New York, 1959).
[12] A.V. Efremov and A.V. Radyushkin, Teor. Mat. Fiz. 42 (1980) 147.
[13] K.G. Chetyrkin, N.V. Krasnikov and A.N. Tavkhelidze, Phys. Lett. 76B (1978) 83.
[14] R.A. Bertlmann, Acta Phys. Austr. 53 (1981) 305.
[15] N.V. Krasnikov and A.A. Pivovarov, report NBI-HE-81-38 (Kopenhagen, 1981).
[16] P.M. Stevenson, Phys. Rev. D23 (1981) 2916.
[17] C. Bebek et al., Phys. Rev. D17 (1978) 1693.
[18] S. Dubnicka, A.Z. Dubnickova and V.A. Meshcheryakov, report IC/77/155 (ICTP, Trieste, 1977).

Nuclear Physics B216 (1983) 373–407

MESON WIDTHS AND FORM FACTORS AT INTERMEDIATE MOMENTUM TRANSFER IN NON-PERTURBATIVE QCD

B.L. IOFFE and A.V. SMILGA

ITEP, Moscow, 117259, USSR

Received 14 September 1982

A general method is proposed for the QCD based calculations of form factors at intermediate Q^2 and of the partial widths of the low-lying mesonic resonances. The basic idea is to use the QCD sum rules for the vertex functions and to saturate their Borel transform by the lowest mesonic states. With this method the pion electromagnetic form factor along with electromagnetic form factors of ρ and A_1 mesons and transition form factors $\gamma\pi \to A_1$ at $0.5 \lesssim Q^2 \lesssim 3$ GeV^2 are calculated. The widths $\rho \to 2\pi$ and $A_1 \to \rho\pi$ are also determined. The results are in a good agreement with experiment.

1. Introduction

Not long ago Shifman, Vainshtein and Zakharov (SVZ) suggested a powerful method based on first QCD principles to determine the masses of low-lying resonances [1]. The idea was to consider the polarization operators of quark currents in the euclidean region, to calculate them in the framework of QCD accounting for both perturbative and non-perturbative corrections and then connect them with the lowest-lying resonances in the corresponding channels by use of the Borel improved dispersion relations.

The application of this method resulted in impressive successes. The efforts of the inventors [1] and the followers [2–6] brought about the situation where the masses of the lowest-lying meson and baryon resonances are calculated in the framework of QCD in practically all conceivable channels in terms of only two QCD parameters: quark and gluon vacuum condensate densities.

The leptonic decays widths (such as $\rho \to e^+e^-$, $\pi \to \mu\nu$, etc.) connected with the residues of the resonances with respect to the corresponding QCD currents are also determined. All the theoretical predictions are in very good agreement with experiment.

This work is devoted to the extension of the SVZ method to the problem next in complexity, namely, to the determination of dynamic characteristics of resonances such as their electromagnetic form factors and the three-hadron vertices based only on the fundamental properties of QCD.

The extension is rather straightforward for a suitably chosen quark current: one should consider three-point functions in the euclidean region in addition to the polarization operators, write down the double dispersion relations for invariant structures and saturate them by the low-lying resonances. The method proposed allows one to find any two-particle widths for meson decays $A \to B + C$ and form factors of transitions $F_{A\to B}(Q^2)$ at intermediate Q^2 for mesonic states A, B, C which are the lowest in this channel. Thereby, this method is a substitute for phenomenological approaches exploited up to now for the calculation of these quantities and in some cases can be a foundation for them. In particular, as will be shown below, with this method the vector dominance hypothesis can be justified in the circle of phenomena under consideration. As can be expected, the method opens a wide possibility for further activity. It should be noted that the amount of calculations grows considerably compared to the two-point cases investigated in refs. [1–6].

In this paper we calculate the electromagnetic form factors of π, ρ and A_1 mesons and the decay widths $\rho \to 2\pi, A_1 \to \rho\pi$. The paper presents an extended and supplemented version of a previous publication where the results for the pion form factor are presented [7] (see also [8]). The structure of the paper is the following. In the next section we describe the method. In sect. 3 we present the calculation of the three-point function of two axial currents and the electromagnetic current in the lowest order. Sects. 4 and 5 are devoted to the calculation of non-perturbative power corrections. In sect. 6 the results for the pion form factor are presented. In sect. 7 we calculate the $g_{\rho\pi\pi}$ constant and the ρ-meson width. Sects. 8 and 9 are devoted to considerations of electromagnetic form factors of ρ and A_1 mesons. In sect. 9 the $A_1 \to \rho\pi$ decay width is also calculated. The results are summarized in sect. 10.

2. The method

Now we proceed with the detailed account of our technique adopted to the main problem of interest – the pion form factor.

Consider the three-point function

$$\Gamma_{\mu\nu,\lambda}(p, p'; q) = -\int d^4x\, d^4y\, e^{i(p'x-qy)}$$
$$\times \langle 0|T\{j^+_{A\nu}(x), j^{el}_{\lambda}(y), j_{A\mu}(0)\}|0\rangle , \qquad (1)$$

where $j_{A\mu}(x)$ is the axial vector light quark current, $j_{A\mu}(x) = \bar{u}(x)\gamma_\mu\gamma_5 d(x)$; $j^{el}_\lambda(x) = e_u\bar{u}(x)\gamma_\mu u(x) + e_d\bar{d}(x)\gamma_\mu d(x)$ is the electromagnetic current, and $q = p' - p$.

In the euclidean region p^2, p'^2, $q^2 < 0$, where the virtualities are large enough, the amplitude $\Gamma_{\mu\nu,\lambda}(p, p';q)$ can be calculated in the framework of perturbative QCD, the corresponding graphs being depicted in figs. 1a, b. (Later we shall not display quite symmetrical graphs where the electromagnetic current enters the d-quark line.)

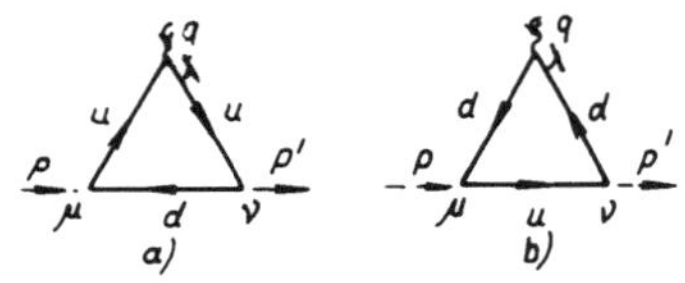

Fig. 1.

If the virtualities are not so high the corrections to the lowest order result come into play. There are two sources of corrections: the perturbative corrections (an example is given in fig. 2) and non-perturbative corrections arising from non-trivial interactions with vacuum fields. In the virtuality region $|p^2| \sim |p'^2| \sim |q^2| \sim 1\ \mathrm{GeV}^2$ of interest, the strong interaction constant $\alpha_s \approx 0.3$, perturbative corrections are small, $\alpha_s/\pi \sim 0.1$ and will be disregarded. The non-perturbative corrections decreasing as a power law with increasing virtuality are, however, very important here. It is these which determine the mass scale of the lowest hadronic states in QCD. These power corrections can be consistently accounted for based on the operator expansion. In doing this, it is clear from dimensional considerations, that just as in the mesonic mass calculations [1], the corrections due to gluon $\langle 0|G^a_{\mu\nu}G^a_{\mu\nu}|0\rangle$ and quark $\langle 0|\bar{\psi}\psi|0\rangle^2$ condensates are most important.

Intuitively, non-perturbative effects must be more relevant for the light quark systems as they are effects responsible for confinement and the mere existence of light resonances. For heavy quark systems perturbative gluon exchange is as significant or even more significant due to Coulomb effects.

The amplitude $\Gamma_{\mu\nu,\lambda}(p, p', q)$ includes a lot of different non-trivial tensor structures. Examples are $P_\mu P_\nu P_\lambda$, $P_\lambda \delta_{\mu\nu}$, $q_\lambda(P_\mu q_\nu + P_\nu q_\mu)$, etc. (The notation $P = p + p'$ is introduced here.) Consider the coefficient of some structure $f_i(p^2, p'^2, q^2)$ and write down the dispersion relation over the axial current virtualities p^2 and p'^2 holding $Q^2 = -q^2$ fixed. Generally we can write* $(r = -p^2, r' = -p'^2)$

$$f_i(r, r', Q^2) = \int_0^\infty \mathrm{d}s \int_0^\infty \mathrm{d}s' \frac{\rho_i(s, s', Q^2)}{(s+r)(s+r')}$$

$$+ \text{subtraction terms}. \qquad (2)$$

The quantity $-4\pi^2\rho_i(p^2, p'^2, Q^2)$ equals the double discontinuity of the amplitude $f_i(p^2, p'^2, Q^2)$ on the cuts $0 \leqslant p^2, p'^2 < \infty$. The subtraction terms in (2) are polynomials in one variable r or r' but arbitrary functions in two others (r', Q^2 or r, Q^2, respectively). To link the QCD-calculated structure functions $f_i(r, r', Q^2)$ with the parameters of physical states and to get rid of unknown subtraction terms, let us

* The QCD sum rules for the vertex function using double dispersion relations were originally considered by Khodjamirian [9] who investigated the radiative transitions in charmonium.

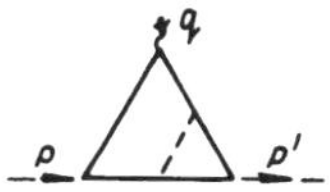

Fig. 2.

exploit the Borel transformation procedure proposed in [1] in the two variables r and r' simultaneously. In the case of the vertex function under consideration the use of the double Borel transformation is absolutely necessary, since, as may be shown, the subtraction terms are non-zero and do not reduce to a constant. Determine the Borel transform $f(M^2)$ of the function

$$f(r)=\frac{1}{\pi}\int_0^{\infty}\mathrm{d}s\,\mathrm{Im}\,f(s)/(s+r)$$

as

$$f(M^2)\equiv\mathcal{B}_{M^2}f(r)=\lim_{\substack{n\to\infty\\ r\to\infty\\ r/n\to M^2}}\frac{r^{n+1}}{n!}\left(-\frac{\mathrm{d}}{\mathrm{d}r}\right)^n f(r)$$

$$=\frac{1}{\pi}\int_0^{\infty}\mathrm{d}s\,\mathrm{e}^{-s/M^2}\,\mathrm{Im}\,f(s). \tag{3}$$

The double Borel transformation of eq. (2) yields

$$\mathcal{B}_{M^2}\mathcal{B}'_{M^2}f_i(r,r',Q^2)=\int_0^{\infty}\mathrm{d}s\int_0^{\infty}\mathrm{d}s'\,\mathrm{e}^{-(s+s')/M^2}\rho_i(s,s';Q^2). \tag{4}$$

The Borel transformation is convenient in bringing about the exponential weight in the dispersion relation and thus suppressing the contribution of higher states and continuum. It was successfully applied in the two-point correlator sum rules [1–6]. We use equal values for the Borel parameters in both channels. The value of M^2 corresponds to the effective virtualities of the amplitude. It should be optimized so that the contribution of higher states is sufficiently suppressed and simultaneously the power corrections are still under control.

The r.h.s. of (4) can be represented as a sum over the physical hadronic states. If M^2 is chosen of order 1 GeV^2, then the r.h.s. of (5) will be dominated by the lowest hadronic state contribution in the given channel while the higher state contribution will be suppressed. Thus, in our case the spectral function can be written as

$$\begin{aligned}\rho_{\mu\nu,\lambda}(p,p';q)&=\langle 0|j_{A\nu}(0)|\pi^+(p')\rangle\langle\pi^+(p')|j_\lambda^{\mathrm{el}}(0)|\pi^+(p)\rangle\\&\quad\times\langle\pi^+(p)|j_{A\mu}^+(0)|0\rangle\delta(s)\delta(s')+\cdots\\&=f_\pi^2p'_\nu p_\mu(p_\lambda+p'_\lambda)F_\pi(Q^2)\delta(s)\delta(s')\\&\quad+\text{contribution of higher states}.\end{aligned} \tag{5}$$

In eq. (5) only the one-pion contribution is written explicitly, $F_\pi(Q^2)$ is the pion electromagnetic form factor, $f_\pi = 133$ MeV and in our approximation of massless quarks we put $m_\pi^2 = 0$. The higher states include the A_1 meson, the excited pseudoscalar and axial states and a smooth continuum starting from high enough values of s and s'. If attempting to account for the A_1 contribution explicitly we should include in the r.h.s. of eq. (5) three electromagnetic A_1 form factors and also two $\langle A_1 | j_\lambda^{el} | \pi \rangle$ transition form factors. All these form factors can be determined in principle from the sum rules corresponding to 7 relevant invariant structures entering the amplitude $\Gamma_{\mu\nu,\lambda}(p, p'; q)$.

It is reasonable to divide the contribution to spectral functions into three regions in the s, s' plane (fig. 3). Region I corresponds to the lowest resonance state contributions which are taken into account explicitly. It is clear that at large s, s' the contribution to the double discontinuity of the vertex function is determined by the quark loop of fig. 1. Thus, in region II (at a chosen sufficiently large value of s_1) the spectral functions are described by the double discontinuity of fig. 1. The same approximation is adopted for the spectral functions in region III. In this region, where only one of the variables s or s' is large, an argument in favour of our description of the spectral functions is quark-hadronic duality: the contributions of quark states coincide with the contributions of real hadrons if they are both averaged over some (equal) region of phase space. At not very large $Q^2 \lesssim s_0$ the contribution of region III to ρ is small. (Indeed, as follows from current conservation, at $Q^2 \to 0$, $\rho \sim \delta(s - s')$ so that $|s - s'| \sim Q^2$; see below.) Since the contribution of this region to the sum rules is suppressed by the exponential factor as compared to the contribution of resonance states, it may be expected that at not very large Q^2 the error due to our model representation of the spectral function in region III is small. In order to verify the stability of our results relative to the chosen continuum model, we have also considered another model, where the continuum contribution described by fig. 1, corresponds to the region of fig. 3 separated from the origin by the dashed line. As it turns out, the results are practically the same in both models, if $s_0 \approx 1.5 s_1$ is chosen.

Note that in the sum rules for the vertex functions, the continuum is more important than in the sum rules for the polarization operator and the ratio of the

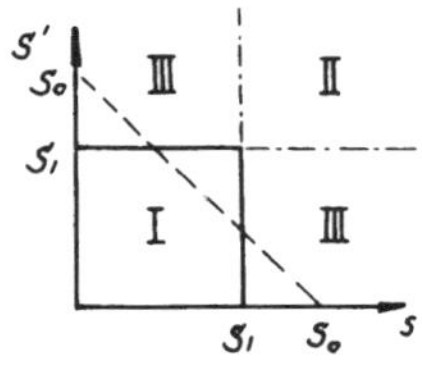

Fig. 3. s, s' integration region in formula (4). Solid and dashed lines are the boundaries of the integration region after subtraction of the continuum in "square" and "triangle" versions.

continuum contribution to the main term increases with Q^2. The reason for such a rise is physically evident: at large Q^2 the role of inelastic excitations grows. This circumstance limits the applicability region of our approach at large Q^2. Other restrictions of the applicability region in Q^2 are due to the increasing role of the power corrections: the ratio of the power corrections to the main term in the general case increases both at small and at large Q^2 (see sects. 4, 5). The fact that this method is inapplicable in the region of very small Q^2 is clear also from consideration of the zero-order diagram, fig. 1, which has a non-physical singularity at $Q^2 = 0$. The inapplicability of our approach at small Q^2 is connected with the fact that in a slowly varying electromagnetic field mean vacuum values of the type $\langle 0|\bar{\psi}\sigma_{\mu\nu}\psi|0\rangle \sim F_{\mu\nu}$ arise which we disregard.

In the case where only the pion form factor is of interest, we restrict ourselves to a study of the sum rules for one structure function and an approximation where the r.h.s of (4) is approximated by the contributions from the one-pion state and the continuum, i.e. the A_1 meson is included in the continuum. As was shown in ref. [1], the replacement of the A_1 meson contribution in the sum rules for the polarization operator by the continuum (at a corresponding choice of the continuum threshold) weakly affects the constant f_π determined from the sum rule, i.e. the quark-hadron duality principle works here well. We expect that an analogous situation will also take place in our case. Thus, our model of the spectral functions is

$$\rho_i(s, s', Q^2) = c_i f_\pi^2 F_\pi(Q^2)\delta(s)\delta(s') + [1 - \theta(s_1 - s)\theta(s_1 - s')]\rho_i^{\text{quark}}(s, s', Q^2), \qquad (6)$$

or

$$\rho_i(s, s', Q^2) = c_i f_\pi^2 F_\pi(Q^2)\delta(s)\delta(s') + [1 - \theta(s_0 - s - s')]\rho_i^{\text{quark}}(s, s', Q^2), \qquad (7)$$

where the constant c_i is determined by the choice of the spectral function.

3. The lowest order calculations

In this section we present the results of our calculation of the lowest order graphs of fig. 1 for the amplitude $\Gamma_{\mu\nu,\lambda}(p, p'; q)$:

$$\Gamma^{(0)}_{\mu\nu,\lambda}(p, p'; q) = \frac{i}{(2\pi)^4}\int \frac{\mathrm{d}^4k}{k^2(p'-k)^2(p-k)^2} \times \mathrm{Tr}\{\gamma_\nu\gamma_5\hat{k}\gamma_\mu\gamma_5(\hat{p}-\hat{k})\gamma_\lambda(\hat{p}'-\hat{k})\}. \qquad (8)$$

We are interested only in the part of the amplitude expressible in the double dispersive form eq. (2) (the pieces polynomial in one or both variables disappear after double Borel transform). The most direct way to obtain it is to calculate the double discontinuity of the amplitude, i.e. the spectral functions $\rho_i^{(0)}(s, s', Q^2)$. To perform this one has to put the quark lines in the graphs of fig. 1 on-mass-shell and substitute the denominators of the quark propagators for the δ functions according to Cutkosky's rule $k^{-2} \to -2\pi i\delta(k^2)$. (We neglect the small quark masses and constituently neglect the pion mass in the following analysis.) In such a way one is left with the expression including the integrals

$$I = \int \mathrm{d}^4k\,\delta(k^2)\,\delta\left[(p-k)^2\right]\delta\left[(p'-k)^2\right] = \pi/2\lambda^{1/2}, \tag{9a}$$

$$I_\mu = \int \mathrm{d}^4k\, k_\mu \delta(k^2)\,\delta\left[(p-k)^2\right]\delta\left[(p'-k)^2\right]$$

$$= -\frac{\pi}{2\lambda^{3/2}}\left[s'(s-s'-Q^2)p_\mu + s(s'-s-Q^2)p'_\mu\right], \tag{9b}$$

$$I_{\mu\nu} = \int \mathrm{d}^4k\, k_\mu k_\nu \delta(k^2)\,\delta\left[(p-k)^2\right]\delta\left[(p'-k)^2\right], \tag{9c}$$

$$I_{\mu\nu\lambda} = \int \mathrm{d}^4k\, k_\mu k_\nu k_\lambda \delta(k^2)\,\delta\left[(p-k)^2\right]\delta\left[(p'-k)^2\right], \tag{9d}$$

where

$$\lambda(s, s', Q^2) = (s+s'+Q^2)^2 - 4ss'. \tag{9e}$$

The expressions for $I_{\mu\nu}$ and especially for $I_{\mu\nu\lambda}$ are rather bulky and we do not present them here. The calculation of the I integral is most easily performed in a particular reference system – say, in the Breit system. The integrals I_μ, $I_{\mu\nu}$ and $I_{\mu\nu\lambda}$ can be found by a standard recursive procedure. The double spectral density $\rho_{\mu\nu,\lambda}(s, s', Q^2)$ equals to double discontinuity divided by $(2\pi i)^2 = -4\pi^2$. After a rather cumbersome calculations we obtain

$$\rho^{(0)}_{\mu\nu,\lambda}(p, p'; q) = \frac{3}{8\pi^2}\frac{Q^2}{\lambda^{5/2}}\left\{\left[\frac{10x}{\lambda}Q^2(x+Q^2)^3 - (x+Q^2)(4x^2+11xQ^2+Q^4)\right.\right.$$

$$\left. + \lambda(2x+Q^2)\right]\tfrac{1}{2}P_\lambda(p_\mu p_\nu + p'_\mu p'_\nu) - \tfrac{1}{2}P_\lambda(p_\mu p_\nu - p'_\mu p'_\nu)y$$

$$\times\left[\frac{10}{\lambda}Q^2(x+Q^2)^3-3(x+3Q^2)(x+Q^2)+\lambda\right]+P_\lambda(x+Q^2)^2p'_\nu p_\mu$$

$$\times\left[2x+5Q^2-\frac{5Q^2}{\lambda}(x+Q^2)^2\right]+P_\lambda p_\nu p'_\mu(x^2-y^2)\left[2x+3Q^2\right.$$

$$\left.-\frac{5Q^2}{\lambda}(x+Q^2)^2\right]+\tfrac{1}{2}q_\lambda(p_\mu p_\nu+p'_\mu p'_\nu)y\frac{1}{Q^2}$$

$$\times\left[-Q^4(x+Q^2)-\lambda Q^2-x\left(\frac{10}{\lambda}Q^2(x+Q^2)^3-(2x+7Q^2)(x+Q^2)+2\lambda\right)\right]$$

$$-\tfrac{1}{2}q_\lambda(p_\mu p_\nu-p'_\mu p'_\nu)\frac{1}{Q^2}\left[-Q^4x(x+Q^2)-\lambda Q^2x\right.$$

$$\left.-y^2\left(\frac{10}{\lambda}Q^2(x+Q^2)^3-2x-7Q^2+2\lambda\right)\right]-q_\lambda p'_\nu p_\mu y\frac{x+Q^2}{Q^2}$$

$$\times\left[(x+4Q^2)(x+Q^2)-\lambda-\frac{5Q^2}{\lambda}(x+Q^2)^3\right]$$

$$-p_\nu p'_\mu q_\lambda y(x^2-y^2)\frac{1}{Q^2}\left[(x+2Q^2)-\frac{5Q^2}{\lambda}(x+Q^2)^2\right]$$

$$-\tfrac{1}{2}\delta_{\mu\nu}P_\lambda(x^2-y^2)[\lambda-Q^2(x+Q^2)]+\tfrac{1}{2}\delta_{\mu\nu}q_\lambda y\frac{1}{Q^2}(x^2-y^2)$$

$$\times[\lambda-Q^2(x+Q^2)]+\tfrac{1}{2}(\delta_{\mu\lambda}p_\nu+\delta_{\nu\lambda}p'_\mu)(x^2-y^2)[\lambda+Q^2(x+Q^2)]$$

$$+\tfrac{1}{2}(\delta_{\mu\lambda}p_\nu-\delta_{\nu\lambda}p'_\mu)y(x^2-y^2)(Q^2+x)$$

$$-\tfrac{1}{2}(\delta_{\nu\lambda}p_\mu+\delta_{\mu\lambda}p'_\nu)(x+Q^2)\left[\lambda(2x+Q^2)-Q^2(x+Q^2)^2\right]$$

$$+\tfrac{1}{2}(\delta_{\nu\lambda}p_\mu-\delta_{\mu\lambda}p'_\nu)y(x+Q^2)\left[\lambda+(x+Q^2)^2\right], \tag{10}$$

where $P=p+p'$, $q=p'-p$, $x=s+s'$, $y=s-s'$. Expression (10) includes 14 different tensor structures. Not all of them are independent, of course, as the double spectral density (but not the full amplitude $\Gamma_{\mu\nu,\lambda}(p,p';q)$ due to appearance of equal-time commutator terms) satisfies the transversality conditions

$$p_\mu\rho_{\mu\nu,\lambda}(p,p';q)=p'_\nu\rho_{\mu\nu,\lambda}(p,p';q)=q_\lambda\rho_{\mu\nu,\lambda}(p,p';q)=0\,.$$

The coefficients $\rho_i(s, s', Q^2)$ of half the structures are antisymmetric under the exchange $s \to s'$ and their symmetric double Borel transform

$$\int \mathrm{d}s\,\mathrm{d}s'\,\mathrm{e}^{-(s+s')/M^2} \rho_i(s, s', Q^2)$$

vanishes. The pion contribution to the Borel transformed physical density is equal to

$$f_\pi^2 F_\pi(Q^2)\, p_\mu p'_\nu P_\lambda = \tfrac{1}{4} f_\pi^2 F_\pi(Q^2)\big[P_\mu P_\nu P_\lambda + P_\lambda(P_\mu q_\nu - P_\nu q_\mu) - q_\mu q_\nu P_\lambda\big]. \qquad (11)$$

In sect. 6 we shall determine the pion form factor from the sum rule for the most symmetric and attractive looking structure $\sim P_\mu P_\nu P_\lambda$.

4. The power corrections proportional to gluon condensate

Now we turn to the calculation of the non-perturbative corrections to the lowest order results. The vacuum in QCD has the complicated structure. Its wave function comprises the fluctuating gluon and quark fields, the nature of the fluctuations differing principally from the common perturbative fluctuations around the zero-field classical vacuum. When considering any QCD amplitude at small distances these fluctuations must be regarded as large scale classical fields serving as a background. As a result, in the vacuum matrix elements of the operator expansion, non-zero vacuum averages of the scalar colourless field contributions appear, such as $\langle 0|\bar\psi\psi|0\rangle$, $(\alpha_s/\pi)\langle 0|G^a_{\mu\nu}G^a_{\mu\nu}|0\rangle$, $\langle 0|\bar\psi\sigma_{\mu\nu}\lambda^a\psi G^a_{\mu\nu}\psi|0\rangle$, etc. These dimensional parameters enter the amplitude divided typically by the characteristic virtualities of the process and corresponding terms decrease as powers when virtualities rise. At moderate virtuality values they display themselves as corrections to perturbative results. The greater the dimension of the operator, the less significant is the correction. In our case, quite analogous to the light quark current polarization operator case [1], the most significant are the vacuum expectation values (v.e.v.) of the operators $G^a_{\mu\nu}G^a_{\mu\nu}$ and $(\bar\psi\Gamma\psi)(\bar\psi\Gamma\psi)$. (Due to chirality conservation the matrix elements $\langle 0|\bar\psi\psi|0\rangle$, $\langle 0|\bar\psi\sigma_{\mu\nu}\lambda^a\psi G^a_{\mu\nu}|0\rangle$ are multiplied by the bare light quark mass in this case and their contributions are small numerically.)

In this section we calculate the coefficient at the gluon condensate density $\langle 0|G^a_{\mu\nu}G^a_{\mu\nu}|0\rangle$ in the Borel-transformed invariant amplitude corresponding to the structure $P_\mu P_\nu P_\lambda$ in $\Gamma_{\mu\nu,\lambda}(p, p'; q)$. The most convenient way to perform this is to use the fixed-point gauge technique [10–13]. The method of calculation of non-perturbative corrections was developed in ref. [14]. Here we just outline its principal features. (For details see appendix A.)

In the fixed-point gauge $x_\mu A^a_\mu(x) = 0$ the potential $A_\mu(x)$ is expressed through the field strength and its covariant derivatives at the origin,

$$A^a_\mu(x) = -\tfrac{1}{2} x_\nu G^a_{\mu\nu}(0) - \tfrac{1}{3} x_\nu x_\alpha (D_\alpha G_{\mu\nu})^a(0) + \cdots, \qquad (12a)$$

or in momentum space,

$$A_\mu^a(k) = -\tfrac{1}{2}i(2\pi)^4 G_{\nu\mu}(0)\frac{\partial}{\partial k_\nu}\delta^4(k)$$

$$-\tfrac{1}{3}(2\pi)^4\left(D_\alpha G_{\nu\mu}(0)\right)^a \frac{\partial^2}{\partial k_\alpha \partial k_\nu}\delta^4(k) + \cdots . \qquad (12b)$$

Consider the amplitude $\Gamma_{\mu\nu,\lambda}(p, p'; q)$ in the external gluon vacuum field. The graphs comprising only one external gluon field line do not contribute as the vacuum averages of $G_{\mu\nu}^a(0)$, $D_\alpha G_{\mu\nu}^a(0), \ldots$ entering eq. (12) are equal to zero. To get the bilinear combination $\langle 0|G_{\mu\nu}^a G_{\mu\nu}^a|0\rangle$ in the answer one has to consider the graphs where the external field enters twice. They are depicted in fig. 4. The black circle designates the origin. The graphs where two gluon lines enter the line adjacent to the origin are not depicted as their contribution is zero [14]. Now we are to substitute the potential A_μ^a in the form of eq. (12) (in this case the second term and higher terms in the expansion can be neglected) in the corresponding Feynman integrals, to perform the vacuum averaging according to the rule

$$\langle 0|G_{\mu\nu}^a G_{\rho\sigma}^b|0\rangle = \tfrac{1}{96}\delta^{ab}\langle 0|G_{\alpha\beta}^h G_{\alpha\beta}^h|0\rangle(\delta_{\mu\rho}\delta_{\nu\sigma} - \delta_{\mu\sigma}\delta_{\nu\rho}),$$

$$a, b, h = 1, 2, \ldots, 8, \qquad (13)$$

and to calculate the integrals. It is convenient to work in the momentum representation (12b) where k is the vacuum gluon momentum. Integration over d^4k brings about the derivative over the internal loop momentum.

The full computation of the correction is extremely complicated (the reader must be convinced of this from looking once more at our result for zero order) and we did not manage to do it. We calculated, however, the correction to the main structure of interest $\sim P_\mu P_\nu P_\lambda$, where several simplification tricks proved to be possible. Our results for the Borel-transformed invariant amplitude for this structure are

$$\mathcal{B}_{M^2}\mathcal{B}'_{M^2}\Gamma_{\mu\nu,\lambda}^{\text{fig.4a}}(p^2, p'^2, Q^2) = \frac{1}{48M^4}\frac{\alpha_s}{\pi}\langle 0|G_{\mu\nu}^a G_{\mu\nu}^a|0\rangle$$

$$\times \int_0^\infty dx \int_0^x dy\, e^{-x/M^2}\frac{Q^2 x + y^2}{\lambda^{3/2}} P_\mu P_\nu P_\lambda + \cdots , \qquad (14a)$$

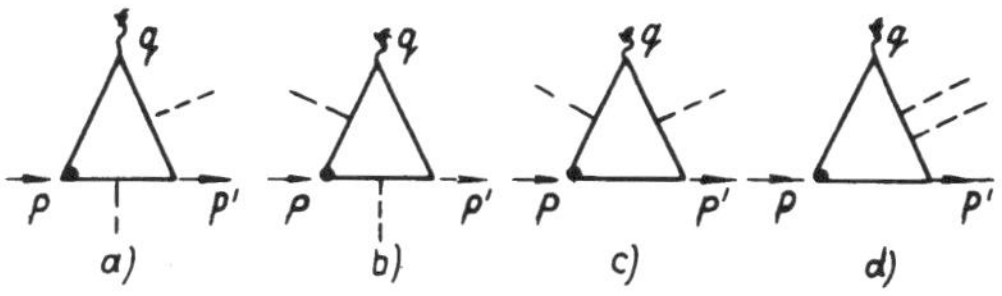

Fig. 4

$$\mathcal{B}_{M^2}\mathcal{B}'_{M^2}\Gamma^{\text{fig.4b + fig.4c}}_{\mu\nu,\lambda}(p^2, p'^2, Q^2)$$
$$= \frac{1}{12M^4}\frac{\alpha_s}{\pi}\langle 0|G^a_{\mu\nu}G^a_{\mu\nu}|0\rangle \int_0^\infty dx \int_0^x dy\, e^{-x/M^2}\frac{Q^2(x+Q^2)}{\lambda^{3/2}}P_\mu P_\nu P_\lambda + \cdots, \tag{14b}$$

$$\mathcal{B}_{M^2}\mathcal{B}'_{M^2}\Gamma^{\text{fig.4d}}_{\mu\nu,\lambda}(p^2, p'^2, Q^2) = -\frac{1}{48M^4}\frac{\alpha_s}{\pi}\langle 0|G^a_{\mu\nu}G^a_{\mu\nu}|0\rangle \int_0^\infty dx \int_0^x dy\, e^{-x/M^2}$$
$$\times\left[2\frac{\lambda + Q^2(x+Q^2)}{\lambda^{3/2}} - \frac{x+Q^2}{M^2\lambda^{1/2}}\right]P_\mu P_\nu P_\lambda + \cdots. \tag{14c}$$

Adding all the contributions and integrating we arrive at the unexpectedly simple answer

$$\mathcal{B}_{M^2}\mathcal{B}'_{M^2}\Gamma^{\langle G^2\rangle}_{\mu\nu,\lambda}(p^2, p'^2, Q^2) = \frac{\alpha_s}{\pi}\frac{\langle 0|G^a_{\mu\nu}G^a_{\mu\nu}|0\rangle}{48M^2}P_\mu P_\nu P_\lambda + \cdots. \tag{15}$$

Note that in the region of intermediate $Q^2 \sim 1\text{–}3\ \text{GeV}^2$ of interest the main contribution to (15) (more than 80%) is given by figs. 4a–c with gluon emission from different lines. Hoping that the same situation will also hold in other structures, when calculating corrections due to gluon condensate for the structures $P_\lambda\delta_{\mu\nu}$ and $p_\nu\delta_{\mu\lambda} + p'_\mu\delta_{\nu\lambda}$ we took into account only more simple graphs of figs. 4a–c. The result is

$$\mathcal{B}_{M^2}\mathcal{B}'_{M^2}\Gamma^{\text{figs.4a-c}}_{\mu\nu,\lambda} = \frac{\alpha_s}{48\pi}\frac{\langle 0|G^a_{\mu\nu}G^a_{\mu\nu}|0\rangle}{M^2}\delta_{\mu\nu}P_\lambda\int_0^\infty dx\, e^{-x/M^2}$$
$$\times\left\{-\frac{4x^2M^2}{(2x+Q^2)^3} + \frac{xQ^2}{(2x+Q^2)^2} + \frac{1}{2M^2}\right.$$
$$\left.\times\frac{x(5x+3Q^2)}{2x+Q^2} + \tfrac{1}{2}\left(1 - \frac{x}{M^2} - \frac{3}{2}\frac{Q^2}{M^2}\right)\ln\left(1+\frac{2x}{Q^2}\right)\right\}$$
$$+\frac{\alpha_s}{96\pi}\frac{\langle 0|G^a_{\mu\nu}G^a_{\mu\nu}|0\rangle}{M^2}(\delta_{\mu\lambda}p_\nu + \delta_{\nu\lambda}p'_\mu)\int_0^\infty dx\, e^{-x/M^2}$$
$$\times\left\{\frac{M^2(24x^2+10xQ^2+Q^4)}{(2x+Q^2)^3} + \frac{x(26x+11Q^2)}{(2x+Q^2)^2}\right.$$
$$\left.-\frac{x(5x+3Q^2)}{M^2(2x+Q^2)} - 2[1+(4x+Q^2)/4M^2]\ln(1+2x/Q^2)\right\}. \tag{16}$$

5. The power corrections due to quark condensate

As has been noted, in the chiral-invariant amplitudes of the type we considered, the terms proportional to the quark condensate $\langle 0|\bar{\psi}\psi|0\rangle$ may appear only multiplied by the quark mass m_q, i.e. must be small. The next in dimension are the terms of the type $\langle 0|\bar{\psi}\Gamma\psi \cdot \bar{\psi}\Gamma\psi|0\rangle$. Following [1] we shall suppose that for these terms the factorization condition is fulfilled, i.e. vacuum states dominate in the sum over intermediate states in all channels, so that there is the equality

$$\langle 0|\bar{\psi}^a_\alpha \bar{\psi}^b_\beta \psi^c_\gamma \psi^d_\delta|0\rangle = \tfrac{1}{144}\langle 0|\bar{\psi}\psi|0\rangle^2 \left(\delta^{ad}\delta^{bc}\delta_{\alpha\delta}\delta_{\beta\gamma} - \delta^{ac}\delta^{bd}\delta_{\alpha\gamma}\delta_{\beta\delta}\right),$$

$$a, b, c, d = 1,2,3, \qquad \alpha, \beta, \gamma, \delta = 1,2,3,4. \tag{17}$$

The term proportional to $\langle 0|\bar{\psi}\psi|0\rangle^2$ in the graph language correspond to simultaneous cutting of two quark lines in the diagram with one virtual gluon like fig. 5. Since the number of such terms is very large, it is convenient to divide them into several types and to calculate terms of a given type separately, using in each case the most suitable method. Graphs in which the virtual gluon is hard with virtuality p^2, p'^2 (like fig. 6) or q^2 (like fig. 7) will be attained for the first type. Terms of this

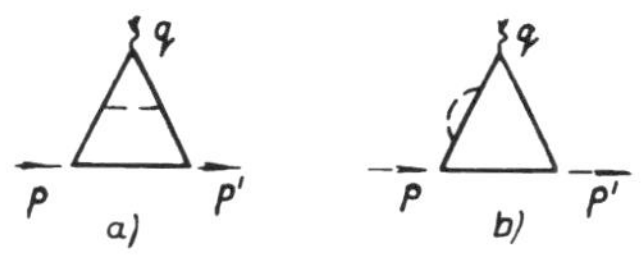

Fig. 5.

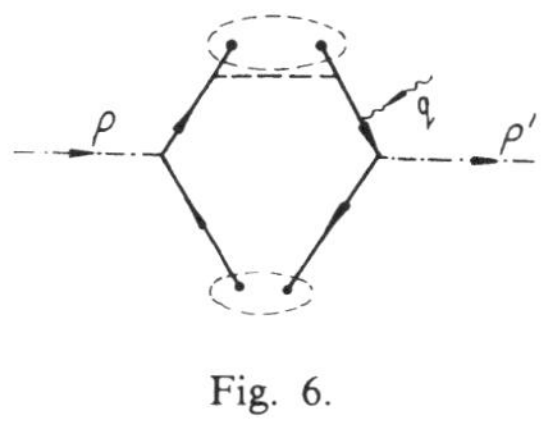

Fig. 6.

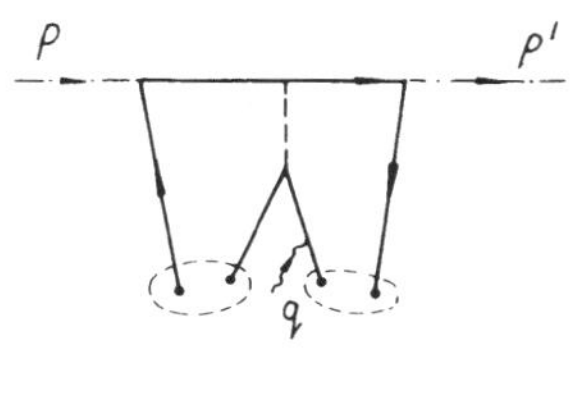

Fig. 7.

type are calculated in an elementary way by using Feynman rules accounting for (17) and lead to the following results for the Borel-transformed amplitudes:

$$\mathcal{B}_{M^2}\mathcal{B}'_{M^2}\Gamma^{\text{fig.6}}_{\mu\nu,\lambda}(p, p'; q) = \tfrac{8}{9}\pi\alpha_s\langle 0|\bar{\psi}\psi|0\rangle^2$$

$$\times \frac{1}{M^4}\left\{\left[P_\lambda(p_\mu p_\nu + p'_\mu p'_\nu) + q_\lambda(p_\mu p_\nu - p'_\mu p'_\nu)\right]\left(\tfrac{1}{2} - \frac{M^2}{Q^2}\right)\right.$$

$$+ P_\lambda p'_\nu p_\mu\left(1 + \frac{2M^2}{Q^2}\right) + (p_\mu\delta_{\lambda\nu} + p'_\nu\delta_{\lambda\mu})$$

$$\times\left(4M^2 + \frac{M^4}{Q^2} - \tfrac{1}{2}Q^2\right) + 2P_\lambda\delta_{\mu\nu}M^2$$

$$\left. - 2(p_\nu\delta_{\mu\lambda} + p'_\mu\delta_{\nu\lambda})M^2\right\}, \tag{18}$$

$$\mathcal{B}_{M^2}\mathcal{B}'_{M^2}\Gamma^{\text{fig.7}}_{\mu\nu,\lambda}(p, p'; q) = \tfrac{8}{9}\pi\alpha_s\langle 0|\bar{\psi}\psi|0\rangle^2\frac{1}{Q^4}$$

$$\times\left\{P_\lambda(p_\mu p_\nu + p'_\mu p'_\nu) + q_\lambda(p_\mu p_\nu - p'_\mu p'_\nu) - 2P_\lambda p'_\nu p_\mu\right.$$

$$\left. - Q^2(p'_\nu\delta_{\mu\lambda} + p_\mu\delta_{\nu\lambda})\right\}. \tag{19}$$

In graphs of all other types the virtual gluon is soft. Among these graphs we first consider those where only one soft quark pair is present explicitly (see fig. 8). To obtain the vacuum average of the required dimension it is necessary to account for the x-dependence of the quark vacuum field in the same way as we have done for the gluon field (see, e.g. ref. [3], errata). In the fixed-point gauge $\psi(x)$ can be expanded in a series in covariant derivatives

$$\psi(x) = \psi(0) + x_\alpha\nabla_\alpha\psi(0) + \tfrac{1}{2}x_\alpha x_\beta\nabla_\alpha\nabla_\beta\psi(0) + \tfrac{1}{6}x_\alpha x_\beta x_\gamma\nabla_\alpha\nabla_\beta\nabla_\gamma\psi(0) + \cdots,$$

$$\nabla_\alpha = \partial_\alpha + ig\frac{\lambda^a}{2}A^a_\alpha. \tag{20}$$

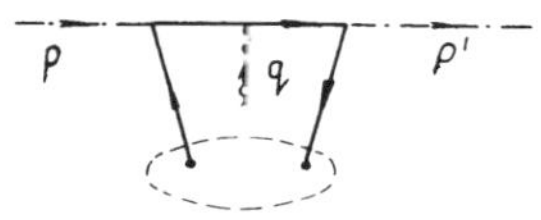

Fig. 8.

The operator of the required (6th) dimension stems from the $\sim x^3$ term in eq. (20).

Making use of equations of motion one can show (see appendix B) that the vacuum matrix element $\langle 0|\bar{\psi}^a_\alpha(0)\, \nabla_\sigma \nabla_\rho \nabla_\tau \psi_\beta(0)|0\rangle$ which appears when substituting this term into the vacuum average $\langle 0|\bar{\psi}^a_\alpha(0)\psi^b_\beta(0)|0\rangle$ has the form

$$\langle 0|\bar{\psi}^a_\alpha(0)\, \nabla_\sigma \nabla_\rho \nabla_\tau \psi^b_\beta(0)|0\rangle = -\frac{ig^2}{3^5\cdot 2^4}\delta^{ab}\langle 0|\bar{\psi}\psi|0\rangle^2(\gamma_\sigma\delta_{\rho\tau}+\gamma_\tau\delta_{\rho\sigma}-5\gamma_\rho\delta_{\sigma\tau})_{\beta\alpha}. \tag{21}$$

Formulae (20), (21) should be substituted into the expression for fig. 8 and it should be taken into account that in the chosen gauge (the origin on the l.h.s. vertex of fig. 8) $x_\sigma \equiv -i\partial/\partial p'_\sigma$ at fixed q (but not p!). Further calculations are made directly and lead to

$$\begin{aligned}\mathcal{B}_{M^2}\mathcal{B}'_{M^2}\Gamma^{\text{fig.8}}_{\mu\nu,\lambda}(p,p';q) &= \frac{4\pi\alpha_s}{81M^4}\langle 0|\bar{\psi}\psi|0\rangle^2\\ &\times\left\{P_\lambda(p_\mu p_\nu+p'_\mu p'_\nu)\left(2+\frac{Q^2}{M^2}\right)+2P_\lambda p'_\nu p_\mu\left(1+\frac{Q^2}{M^2}\right)\right.\\ &+2P_\lambda p_\nu p'_\mu - q_\lambda(p_\mu p_\nu - p'_\mu p'_\nu)\left(2-\frac{Q^2}{M^2}\right)+2\delta_{\mu\nu}P_\lambda M^2\\ &+2(\delta_{\nu\lambda}p_\mu+\delta_{\mu\lambda}p'_\nu)M^2\left(1+2\frac{Q^2}{M^2}-\frac{1}{2}\frac{Q^4}{M^4}\right)\\ &\left.+(\delta_{\mu\lambda}p_\nu+\delta_{\nu\lambda}p'_\mu)(2-Q^2/M^2)\right\}.\end{aligned} \tag{22}$$

Finally, there are graphs comprising a soft quark pair and a soft gluon (see fig. 9). If substituting the expansions for $A^a_\mu(x)$ and $\psi(x)$, eqs. (12), (20), one sees that the 6th dimensional vacuum average $\langle 0|\bar{\psi}\psi|0\rangle^2$ can be generated in two ways. First, the quark field can be taken at the origin in which case the second term of the $A^a_\mu(x)$ expansion, eq. (12), is relevant. In this case vacuum matrix elements $\sim \langle 0|\bar{\psi}^a_\alpha\psi^b_\beta D_\rho G^h_{\mu\nu}|0\rangle$ emerge which are reduced to the form (see appendix B)

$$\langle 0|\bar{\psi}^a_\alpha(0)\psi^b_\beta(0)D_\rho G^h_{\mu\nu}|0\rangle = -\frac{g}{3^3\cdot 2^5}\langle 0|\bar{\psi}\psi|0\rangle^2(\lambda^h)^{ba}(\delta_{\rho\nu}\gamma_\mu-\delta_{\rho\mu}\gamma_\nu)_{\beta\alpha}, \tag{23}$$

due to equations of motion.

The receipe for the calculation of such terms is as follows: write fig. 9 in momentum space denoting gluon momentum by k, make use of eq. (23) for the

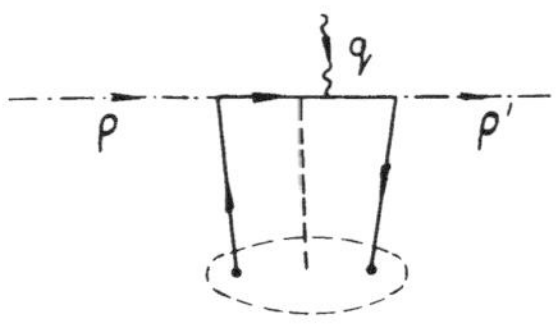

Fig. 9.

vacuum average of the product of quark and gluon operator, apply to the whole expression the differential operator $-\frac{1}{3}\partial^2/\partial k_\alpha \partial k_\beta$ and finally put k equal to zero.

The second way to obtain the required vacuum average of the operator $\langle 0|\bar{\psi}\psi|0\rangle^2$ is to take the second term in expansion (20) and the first in (12). The matrix elements $\langle 0|\bar{\psi}^a_\alpha(\nabla_\sigma \psi^b_\beta)G^h_{\rho\mu}|0\rangle$ that appear in this way are expressed through $\langle 0|\bar{\psi}\psi|0\rangle^2$ as (appendix B)

$$\langle 0|\bar{\psi}^a_\alpha(0)(\nabla_\sigma \psi^b_\beta)G^h_{\rho\mu}(0)|0\rangle = \frac{g}{3^3 \cdot 2^6}\langle 0|\bar{\psi}\psi|0\rangle^2 \times (\lambda^h)^{ba}(\delta_{\sigma\rho}\gamma_\mu - \delta_{\sigma\mu}\gamma_\rho - i\varepsilon_{\sigma\rho\mu\xi}\gamma_5\gamma_\xi)_{\beta\alpha},$$

$$\varepsilon_{0123} = 1. \tag{24}$$

Thus, to calculate the second-type terms corresponding to fig. 9 one should substitute the second term of expansion (20) into the vacuum average appearing in the graph, take into account (24), then go to the momentum representation, apply differential operators $-i\partial/\partial p'_\sigma$ and $\partial/\partial k_\rho$ to the result and put $k = 0$.

The result of the calculation of terms corresponding to fig. 9 (after Borel transformation) is

$$\mathcal{B}_{M^2}\mathcal{B}'_{M^2}\Gamma^{\text{fig.9}}_{\mu\nu,\lambda}(p, p'; q) = \frac{4\pi\alpha_s}{81}\langle 0|\bar{\psi}\psi|0\rangle^2 \frac{1}{M^4} \times \left\{ P_\lambda(p_\mu p_\nu + p'_\mu p'_\nu)\left(2 + \frac{Q^2}{M^2}\right) - 2P_\lambda p'_\nu p_\mu\left(5 - \frac{Q^2}{M^2}\right) + 14P_\lambda p_\nu p'_\mu - q_\lambda(p_\mu p_\nu - p'_\mu p'_\nu)(2 - Q^2/M^2) + 14P_\lambda \delta_{\mu\nu}M^2 + (p_\mu\delta_{\nu\lambda} + p'_\nu\delta_{\mu\lambda})M^2\left(-10 + 10\frac{Q^2}{M^2} - \frac{Q^4}{M^4}\right) + 7(p'_\mu\delta_{\nu\lambda} + p_\nu\delta_{\mu\lambda})M^2(2 - Q^2/M^2)\right. . \tag{25}$$

The full power correction due to the quark condensate is the sum of the contributions (18), (19), (22) and (25):

$$\mathcal{B}_{M^2}\mathcal{B}'_{M^2}\Gamma^{\langle\bar\psi\psi\rangle^2}_{\mu\nu,\lambda}(p,p';q) = \frac{4\pi\alpha_s}{81}\langle 0|\bar\psi\psi|0\rangle^2\frac{1}{M^4}$$

$$\times\left\{P_\lambda(p_\mu p_\nu + p'_\mu p'_\nu)\left(13 - 18\frac{M^2}{Q^2} + 18\frac{M^4}{Q^4} + 2\frac{Q^2}{M^2}\right)\right.$$

$$+P_\lambda p'_\nu p_\mu\left(10 + 36\frac{M^2}{Q^2} - 36\frac{M^4}{Q^4} + 4\frac{Q^2}{M^2}\right)$$

$$+16P_\lambda p_\nu p'_\mu + q_\lambda(p_\mu p_\nu - p'_\mu p'_\nu)$$

$$\times\left(5 - 18\frac{M^2}{Q^2} + 18\frac{M^4}{Q^4} + 2\frac{Q^2}{M^2}\right) + 52P_\lambda\delta_{\mu\nu}M^2$$

$$+(p'_\nu\delta_{\mu\lambda} + p_\mu\delta_{\nu\lambda})M^2\left(64 + 5\frac{Q^2}{M^2} - 2\frac{Q^4}{M^4}\right)$$

$$\left.+(p_\nu\delta_{\mu\lambda} + p'_\mu\delta_{\nu\lambda})M^2(-20 - 8Q^2/M^2)\right\}. \quad (26)$$

In (26) it is of importance that some structures contain a singular term in Q^2 proportional to $1/Q^2$ and $1/Q^4$ while in others they are absent. In order to expand the applicability region of the sum rules to smaller Q^2 it seems reasonable to use the structures where power corrections do not contain Q^2 singular terms. If one rewrites (26) expressing p and p' through P and q, $p = \frac{1}{2}(P - q)$, $p' = \frac{1}{2}(P + q)$, then one sees that such a property is inherent to the structure function for $P_\mu P_\nu P_\lambda$. For this reason just this structure was chosen to determine the pion form factor.

6. The pion form factor

To find the pion electromagnetic form factor, let us write the sum rule for the coefficient at the structure $P_\mu P_\nu P_\lambda$ expressing p and p' through P and q. The r.h.s. of the required sum rule is determined by eqs. (5), (6) [or (7)] and (11). As can be obtained from (10), the double spectral density of the structure $P_\mu P_\nu P_\lambda$ corresponding to the main term on the l.h.s. of the sum rule is

$$\rho^{(0)}(s,s',Q^2) = \frac{3Q^4}{16\pi^2}\frac{1}{\lambda^{7/2}}\left[3\lambda(x+Q^2)(x+2Q^2) - \lambda^2 - 5Q^2(x+Q^2)\right], \quad (27)$$

where λ is defined by eq. (9e).

Using our results for the power corrections (15), (26) and transferring the continuum contribution into l.h.s., we write the sum rule as an equality determining the pion form factor:

$$\frac{4}{f_\pi^2}\left\{\int_\Omega \mathrm{d}s\,\mathrm{d}s'\,\mathrm{e}^{-(s+s')/M^2}\rho^{(0)}(s,s',Q^2)+\frac{\alpha_s}{48\pi M^2}\langle 0|G^a_{\mu\nu}G^a_{\mu\nu}|0\rangle\right.$$
$$\left.+\frac{52\pi}{81M^4}\alpha_s\langle 0|\bar\psi\psi|0\rangle^2\left(1+\frac{2Q^2}{13M^2}\right)\right\}=F_\pi(Q^2), \tag{28}$$

where Ω is the integration region in the s, s' plane which remains after subtraction of the continuum [see eqs. (6), (7) and fig. 3].

As follows from (27), (28), the main term on the l.h.s. of (28) decreases as Q^{-4} as $Q^2\to\infty$, while the power corrections grow with Q^2. This means that sum rule (28) is inapplicable at large Q^2 (practically, at $Q^2>4$ GeV^2) where power corrections become large and uncontrollable. At small Q^2 the main term has a non-physical singularity $\sim Q^4\ln Q^2$ resulting from massless quark propagators and just like singularities in power corrections, testifying to the inapplicability of our formulae at small Q^2. The fact that this method is inapplicable at small Q^2 could be foreseen *a priori* since when approaching the physical region the operator expansion stops working and the terms exponentially decreasing at large Q^2 become essential [15].

As was mentioned, in the calculations we considered two versions of accounting for the continuum contribution, when the region Ω is a square [eq. (6)] or a triangle [eq. (7)]. Let us first consider the "triangle" version. Then the integral in (28) can be written as

$$\int_\Omega \mathrm{d}s\,\mathrm{d}s'\,\mathrm{e}^{-(s+s')/M^2}\rho_0(s,s',Q^2)=\frac{1}{8\pi^2}\int_0^{s_0}\mathrm{e}^{-x/M^2}\frac{x^2(2x+3Q^2)}{(2x+Q^2)^3}\,\mathrm{d}x. \tag{29}$$

The value s_0 can be obtained from the condition of minimum M^2 dependence of the results when accounting for restrictions due to the relative contribution of the continuum. Fig. 10 shows the M^2 dependence of the l.h.s. of (28) at the following values of the parameters: $s_0=1.2$ GeV^2, $\langle 0|(\alpha_s/\pi)G^a_{\mu\nu}G^a_{\mu\nu}|0\rangle=0.012$ GeV^4, $\alpha_s\langle 0|\bar\psi\psi|0\rangle^2=0.8\cdot 10^{-4}$ GeV^6. As is seen from fig. 10, the requirement for the l.h.s. of (28) to be M^2 independent is fulfilled within $0.7\lesssim M^2\lesssim 1.7$ GeV^2. Let us choose $M^2=1.2$ GeV^2 and find from (28) the Q^2 dependence of the pion form factor $F_\pi(Q^2)$. The result is presented in fig. 11 where experimental data are also plotted (see [16] and references therein) and, for comparison, the asymptotic dependence curve [17–21] is also given:

$$F_\pi(Q^2)_{\text{asympt}}=8\pi f_\pi^2\frac{\alpha_s(Q^2)}{Q^2}. \tag{30}$$

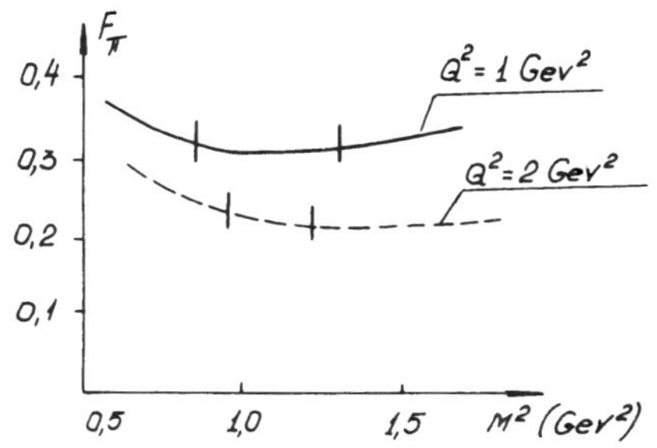

Fig. 10. The M^2 dependence of the l.h.s. of sum rule (28) at $Q^2 = 1$ GeV^2 (solid line) and $Q^2 = 2$ GeV^2 (dashed line). The left-hand vertical bars mark the M^2 values above which the ratio of the power corrections to the sum of terms does not exceed 50%. To the left of the right-hand vertical bars the ratio of the continuum contribution to all others in the l.h.s. of (28) does not exceed 50%.

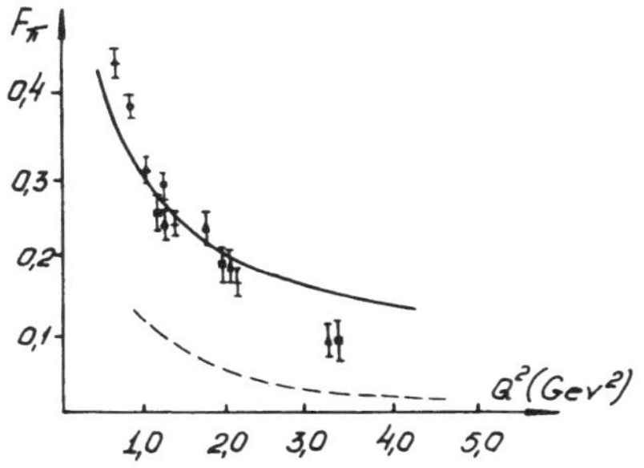

Fig. 11. The pion form factor $F_\pi(Q^2)$: solid line, the result of our calculations; dashed line, asymptotic dependence [17–21]; points, experimental data [16].

The agreement with experiment in the region $0.5 \lesssim Q^2 \lesssim 3$ GeV^2 is good enough. At $Q^2 > 3$–4 GeV^2 the accuracy of our results is comparatively low as at $Q^2 = 3$ GeV^2 the continuum contribution is already rather large ($\sim$ 50% of the rest) and the power corrections are 45% from the main term. The variation of M^2 weakly changes $F_\pi(Q^2)$ in the interval under consideration, $0.5 \lesssim Q^2 \lesssim 3$ GeV^2; for instance, at $M^2 = 1$ GeV^2 the curve for $F_\pi(Q^2)$ does not practically differ from fig. 11. The same results can be reproduced well enough in a version where the pion is dual to a square with side s_1 in the s, s' plane. The best choice of the parameters is $s_1 = 0.8$–1.0 GeV^2, $M^2 = 1.0$–1.2 GeV^2. With these values of the parameters in the interval $0.5 \lesssim Q^2 \lesssim 3$ GeV^2 the difference between form factors obtained in the "triangle" and "square" versions is not more than 20%. The value $s_1 \approx 0.8$ GeV^2 is in accordance with the continuum threshold $s_0 \approx 0.75$ GeV^2 for the axial current polarization operator [1].

Up to now when determining the pion form factor we considered the vertex function with axial currents. Instead of this one can consider the vertex function with pseudoscalar currents where matrix elements of π meson transition into quark current are also known:

$$\langle 0|i\bar{u}\gamma_5 d|\pi^-\rangle = \frac{f_\pi m_\pi^2}{m_u + m_d} = -\frac{2}{f_\pi}\langle 0|\bar{\psi}\psi|0\rangle . \tag{31}$$

The sum rule which determines the pion form factor is in this case

$$\frac{3}{16\pi^2} M^4 \left[\frac{\ln(M^2/\Lambda^2)}{\ln(\mu^2/\Lambda^2)}\right]^{-8/9} \phi(M^2, Q^2) = \frac{4}{f_\pi^2}\langle 0|\bar{\psi}\psi|0\rangle^2 F_\pi(Q^2), \tag{32}$$

where

$$\phi(M^2, Q^2) = Q^2 \int_0^{s_0} \mathrm{d}x\, \mathrm{e}^{-x/M^2} \int_0^x \mathrm{d}y \frac{x^2 - y^2}{\lambda^{3/2}}, \qquad \lambda = Q^4 + 2xQ^2 + y^2 .$$

In (32) we disregard the power corrections whose role in this case is small and the continuum contribution is transferred to the l.h.s. (The logarithmic factor appears due to the anomalous dimension of the pseudoscalar current, μ is the normalization point, $\mu \approx 0.5$ GeV.) The results of the form-factor calculations are not very different from those presented above for the axial currents but theoretically they are less confident since, as is known [22, 23], in the pseudoscalar case non-controllable, non-perturbative corrections due to instantons are significantly more important. In addition, the continuum contribution to the sum rule (32) is noticeably larger than for the axial currents and the M^2 dependence is stronger so that the accuracy of the results is worse.

7. The constant $g_{\rho\pi\pi}$ and the ρ meson width

Selecting the ρ meson contribution in the dispersion relation for $F_\pi(Q^2)$ one can, using the values of $F_\pi(Q^2)$ obtained above, find the strong coupling constant $g_{\rho\pi\pi}$ and thereby determine the ρ meson width. In the rough approximation (this is, in fact, the VDM approach), saturating the dispersion relation by the ρ meson contribution only, we have

$$F_\pi(Q^2) \simeq \frac{g_{\rho\pi\pi}}{g_\rho} \frac{m_\rho^2}{Q^2 + m_\rho^2}, \tag{33}$$

where g_ρ is the transition constant $\langle\rho|\gamma\rangle$ and the ρ meson width is neglected. From comparison of (33) with $F_\pi(Q^2)$ at $Q^2 = 1.2$ GeV^2 it follows $g_{\rho\pi\pi}/g_\rho \approx 0.9$.

This rough result can be improved if in the dispersion relation for $F_\pi(Q^2)$, besides the ρ-meson contribution, the contribution of higher states and asymptotic behaviour (30) will be taken into account.

Let us construct $F_\pi(Q^2)$ proceeding from the subtractionless dispersion relation

$$F_\pi(Q^2) = \frac{1}{\pi}\int_0^\infty \frac{\operatorname{Im} F_\pi(q^2)}{q^2+Q^2}\,dq^2, \tag{34}$$

putting in it the given $\operatorname{Im} F_\pi(q^2)$. Suppose that besides the ρ meson contribution, which is $(g_{\rho\pi\pi}/g_\rho)m_\rho^2\delta(q^2 - M_\rho^2)$, $\operatorname{Im} F_\pi(q^2)$ starts from some threshold value q_0^2. As follows from (30), in the asymptotic limit

$$\operatorname{Im} F_\pi(q^2)_{\text{asympt}} = -8\pi f_\pi \alpha_s(q^2)\frac{1}{q^2}\frac{1}{\ln(q^2/\Lambda^2)}. \tag{35}$$

Let us also take into account in $\operatorname{Im} F_\pi(q^2)$ the preasymptotic term obtained in [24] and decreasing as $1/q^4$ whose contribution to $\operatorname{Im} F_\pi(q^2)$ is approximated as

$$\operatorname{Im} F_\pi(q^2)_{\text{preasympt}} = \frac{A}{q^4}\frac{\ln\ln(q^2/m_0^2)}{\ln(q^2/m_0^2)}, \tag{36}$$

where A and m_0^2 are some parameters. (The factor $\ln\ln(q^2/m_0^2)$ is inessential for the following and is introduced for a closer coincidence of (36) with the formulae of ref. [24].) Suppose that starting from $q^2 = q_0^2$ the whole imaginary part $\operatorname{Im} F_\pi(q^2)$ reduces to asymptotic and preasymptotic terms (35), (36). The constant A can be then determined from requirement that terms $\sim 1/Q^2$ are absent asymptotically [$F_\pi(Q^2)$ contains the terms $1/Q^2 \ln Q^2$ only]. After that the ratio $g_{\rho\pi\pi}/g_\rho$ at given values of other parameters q_0^2 and m_0^2 can be obtained from the normalization condition $F_\pi(0) = 1$. The values of q_0^2 and m_0^2 are determined from the best fit of $F_\pi(Q^2)$ to that calculated above at $0.5 \lesssim Q^2 \lesssim 2$ GeV^2. As was shown by calculation, at $q_0^2 = 1.5$–1.8 GeV^2 such a fit is achieved in the wide range of m_0^2 from $\Lambda^2_{QCD} = 0.01$ to m_ρ^2 and the value of preasymptotic term at $Q^2 = 10$–100 GeV^2 agrees with that found in [24]. Note, that the value obtained, $q_0^2 \approx 1.5$ GeV^2, is the natural duality interval in the vector channel.

In the chosen interval of q_0^2 and m_0^2 the ratio $g_{\rho\pi\pi}/g_\rho$ varies from 1.14 to 1.22. Thus, our final answer for this value is

$$g_{\rho\pi\pi}/g_\rho = 1.18 \pm 0.05. \tag{37}$$

To calculate the ρ meson width

$$\Gamma_{\rho\pi\pi} = \frac{g^2_{\rho\pi\pi}}{48\pi} m_\rho \left(1 - 4m_\pi^2/m_\rho^2\right)^{3/2}, \tag{38}$$

let us make use of the theoretical values $m_\rho^2 = 0.60$ GeV^2, $4\pi/g_\rho^2 = 0.41$. found in [1].

Then from (37), (38) it follows that

$$\left(\Gamma_{\rho\pi\pi}\right)_{\text{theor}} = 175\ \text{MeV} \pm 10\%\,, \tag{39}$$

in good agreement with the experimental value $(\Gamma_{\rho\pi\pi})_{\text{exp}} = 155 \pm 5$ MeV. The thus constructed interpolation formula for $F_\pi(Q^2)$ describes the pion form factor at all Q^2 values and is in a good agreement with experiment [16]. In particular, the pion electromagnetic radius can be determined $\langle r_\pi^2 \rangle_{\text{th}} = 0.44\ \text{fm}^2$ from this formula in comparison with the experimental values $\langle r_\pi^2 \rangle_{\text{exp}} = 0.51 \pm 0.03\ \text{fm}^2$ [16] and $\langle r_\pi^2 \rangle_{\text{exp}} = 0.31 \pm 0.04\ \text{fm}^2$ [25].

8. The ρ meson form factors

The developed method makes it possible to determine the electromagnetic form factors of the vector ρ meson. To this end, one should consider the vertex function with two vector currents instead of the vertex function (1) with two axial currents considered up to now. Necessary changes are minimal: the zero-order terms and the corrections due to the gluon condensate do not change; in corrections due to the quark condensate the terms described by fig. 6 [eq. (18)] change sign. The phenomenological part of the sum rules – the r.h.s. of eq. (4) – is approximated just as before by the contribution of the lowest resonance state, ρ meson + continuum. The contribution of the ρ meson to the double spectral function is

$$\rho_{\mu\nu,\lambda}(p, p'; q) = \sum_{\text{polar}} \langle 0|j_\nu(0)|\rho'^+, \varepsilon'\rangle\langle\rho'^+, \varepsilon'|j_\lambda^{\text{el}}(0)|\rho^+, \varepsilon\rangle$$

$$\times\langle\rho^+, \varepsilon|j_\mu^+(0)|0\rangle\delta\left(s - m_\rho^2\right)\delta\left(s' - m_\rho^2\right), \tag{40}$$

$$\langle 0|j_\nu|\rho^+, \varepsilon\rangle = \sqrt{2}\,\frac{m_\rho^2}{g_\rho}\varepsilon_\nu\,, \tag{41}$$

where ε_ν is the ρ meson polarization.

The general expression for the electromagnetic vertex of a spin 1 meson can be written as

$$\langle\rho^+(p', \varepsilon')|j_\lambda^{\text{el}}|\rho^+(p, \varepsilon)\rangle$$

$$= -\varepsilon'_\sigma\varepsilon_\rho\Big\{\left[(p'+p)_\lambda\delta_{\rho\sigma} - p'_\rho\delta_{\lambda\sigma} - p_\sigma\delta_{\lambda\rho}\right]F_1(Q^2) + \left[\delta_{\lambda\rho}q_\sigma - \delta_{\lambda\sigma}q_\rho\right]$$

$$\times F_2(Q^2) + \frac{1}{m_\rho^2}p'_\rho p_\sigma(p+p')_\lambda F_3(Q^2)\Big\}\,, \tag{42}$$

where F_1, F_2 and F_3 are electric, magnetic and quadrupole form factors, $F_1(0)=1$, and $F_2(0)$ and $F_3(0)$ are connected with the static magnetic (μ) and quadrupole (D) moment of the meson:

$$\mu = 1 + F_2(0),$$

$$D = \frac{2}{m_\rho^2}\left[F_2(0) - 2F_3(0)\right]. \tag{43}$$

(The quadrupole moment is defined according to [26].)

Substituting (41) and (42) into (40) and summing over polarizations we get

$$\begin{aligned}
\rho_{\mu\nu,\lambda}(p, p'; q) = 2\frac{m_\rho^4}{g_\rho^2}\Bigg\{ & \frac{1}{2m_\rho^2}P_\lambda(p'_\mu p'_\nu + p_\mu p_\nu) \\
& \times\left[F_1 - F_2 + 2\left(1 + \frac{Q^2}{2m_\rho^2}\right)F_3\right] + p'_\nu p_\mu P_\lambda \frac{1}{m_\rho^2}\left(1 + \frac{Q^2}{2m_\rho^2}\right) \\
& \times\left[F_2 - \left(1 + \frac{Q^2}{2m_\rho^2}\right)F_3\right] - \frac{1}{m_\rho^2}p'_\mu p_\nu P_\lambda F_3 \\
& + \frac{1}{2m_\rho^2}q_\lambda(p_\mu p_\nu - p'_\mu p'_\nu)(F_1 + F_2) - \delta_{\mu\nu}P_\lambda F_1 + (\delta_{\mu\lambda}p_\nu + \delta_{\nu\lambda}p'_\mu) \\
& \times(F_1 + F_2) - (\delta_{\mu\lambda}p'_\nu + \delta_{\nu\lambda}p_\mu)\left(1 + \frac{Q^2}{2m_\rho^2}\right)(F_1 + F_2)\Bigg\} \\
& \times\delta(s - m_\rho^2)\,\delta(s' - m_\rho^2).
\end{aligned} \tag{44}$$

In order to find the ρ meson form factors it is reasonable to use 3 sum rules: for structures $\delta_{\mu\nu}P_\lambda$, $\delta_{\mu\lambda}p_\nu + \delta_{\nu\lambda}p'_\mu$, and $p'_\mu p_\nu P_\lambda$. The reason for such a choice is that the power corrections due to the quark condensate to the corresponding structure functions do not contain terms $1/Q^2$ and $1/Q^4$ which gives hope for a better approximation in the region of comparatively small Q^2.

The sum rules follow from eqs. (4), (9), (16), (18), (26) and (44) and lead to the following formulae for form factors (the continuum contribution is calculated in the "triangle" version):

$$F_1^\rho(Q^2) = \frac{g_\rho^2}{16\pi^2}\left(\frac{M}{m_\rho}\right)^4$$

$$\times e^{2m_\rho^2/M^2}\left\{\int_0^\varkappa e^{-x}\,dx\left[\frac{x(4x^2+9xk+3k^2)}{2(2x+k)^2} - \tfrac{3}{4}k\ln\left(1+\frac{2x}{k}\right)\right]\right.$$

$$+\frac{640\pi^3}{81M^6}\alpha_s\langle 0|\bar\psi\psi|0\rangle^2 - \frac{\pi^2}{24M^4}\langle 0|\frac{\alpha_s}{\pi}G_{\mu\nu}^2|0\rangle$$

$$\times\int_0^\infty e^{-x}\,dx\left[(2-2x-3k)\ln\left(1+\frac{2x}{k}\right) - \frac{16x^2}{(2x+k)^3}\right.$$

$$\left.\left.+\frac{4xk}{(2x+k)^2} + \frac{2x(5x+3k)}{2x+k}\right]\right\}, \qquad (45)$$

$$F_2^\rho(Q^2) = \frac{g_\rho^2}{16\pi^2}\left(\frac{M}{m_\rho}\right)^4$$

$$\times e^{2m_\rho^2/M^2}\left\{2\int_0^\varkappa e^{-x}\,dx\frac{x^3}{(2x+k)^2} + \frac{256\pi^3}{81M^6}\alpha_s\langle 0|\bar\psi\psi|0\rangle^2\left(4-\frac{Q^2}{M^2}\right)\right.$$

$$+\langle 0|\frac{\alpha_s}{\pi}G_{\mu\nu}^2|0\rangle\frac{\pi^2}{12M^4}\int_0^\infty e^{-x}\,dx\left[\frac{13x}{2x+k} + \frac{8x+k}{(2x+k)^2}\right.$$

$$\left.\left.-(1+2k+3x)\ln(1+2x/k)\right]\right\}, \qquad (46)$$

$$F_3^\rho(Q^2) = \frac{g_\rho^2}{4\pi}\left(\frac{M}{m_\rho}\right)^2$$

$$\times e^{2m_\rho^2/M^2}\left\{\int_0^\varkappa e^{-x}\,dx\frac{x^3}{(2x+k)^3} - \left(\frac{128\pi^3}{81M^6}\right)\alpha_s\langle 0|\bar\psi\psi|0\rangle^2\right\}, \qquad (47)$$

where

$$k = \frac{Q^2}{M^2}, \qquad \varkappa = \frac{s_0}{M^2}.$$

When calculating corrections $\sim \langle 0|G_{\mu\nu}^2|0\rangle$ to form factors F_1^ρ and F_2^ρ we took into account only diagrams with gluon emission from different lines (figs. 4a–c); the terms $\sim \langle 0|G_{\mu\nu}^2|0\rangle$ in the form factor F_3 were omitted. Therefore, the accuracy of the calculation of F_1, F_2 and especially of F_3 is not very high. Nevertheless, we believe that the calculations of the ρ-meson form factors, though of poor accuracy, are of a certain interest since there has been no information till now on these form factors at small and intermediate Q^2, either theoretical or experimental.

As was shown by the calculation, at $Q^2 = 1$–2 GeV^2 and $s_0 = 2$–2.5 GeV^2 the r.h.s. of (45)–(47) are practically (with an accuracy of better than 20%) M^2 independent within $0.8 \lesssim M^2 \lesssim 2.0$ GeV^2. At large M^2, however, the continuum contribution is very large, so that for higher confidence in the results M^2 should be chosen in the region of the lower boundary of this interval. Fig. 12 exhibits the curves $F_1^\rho(Q^2)$, $F_2^\rho(Q^2)$, $F_3^\rho(Q^2)$ for $M^2 = 1.0$ GeV^2 close to the optimal in the sense of the relative smallness of continuum contributions and power corrections (s_0 has been chosen to be 2.2 GeV^2 from physical considerations: it is ~ 1.5 times larger than the corresponding threshold in the polarization operator case).

The curves can be considered sufficiently reliable up to $Q^2 \approx 2.5$ GeV^2 where the continuum contribution reaches 60%. The power corrections in the case of form factors F_1 and F_2 are comparatively small (less than 25% of the main term for F_1 and quite small for F_2 where corrections $\sim \langle 0|G_{\mu\nu}^2|0\rangle$ and $\langle 0|\bar{\psi}\psi|0\rangle^2$ are compensated). For F_3^ρ the correction $\sim \langle 0|\bar{\psi}\psi|0\rangle^2$ is about 40% of the main term at $Q^2 = 2.5$ GeV^2. Fig. 12 also gives the ratio $F_2^\rho(Q^2)/F_1^\rho(Q^2)$. This quantity is of special interest, in particular its value at the static limit $F_2^\rho(0)/F_1^\rho(0)$. Though our approach is inapplicable at $Q^2 = 0$, as is shown in fig. 12, the value of $F_2^\rho(0)/F_1^\rho(0)$ can be well extrapolated from the region $Q^2 > 0$ and turns out to be close to unity. This fact is strong confirmation for the vector dominance hypothesis [27,28] where the electromagnetic interaction of ρ mesons is realized in the same way as for Yang-Mills bosons and $F_2^\rho(0)/F_1^\rho(0) = 1$. (Of course, such confirmation is only for the region of

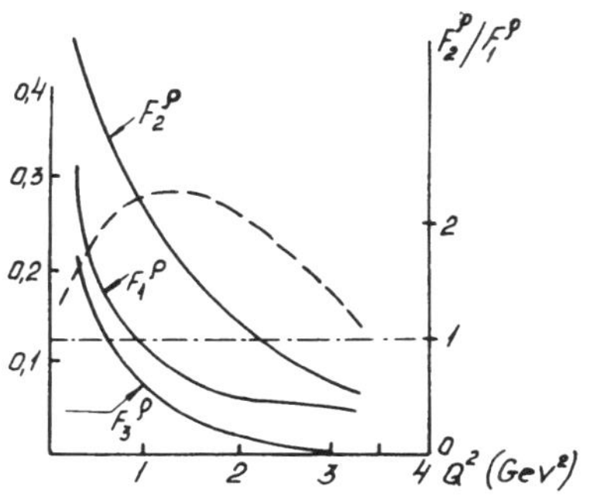

Fig. 12. Electromagnetic form factors of the ρ meson $F_1^\rho(Q^2)$, $F_2^\rho(Q^2)$, $F_3^\rho(Q^2)$ and the ratio $F_2^\rho(Q^2)/F_1^\rho(Q^2)$ (dashed curve, r.h. scale).

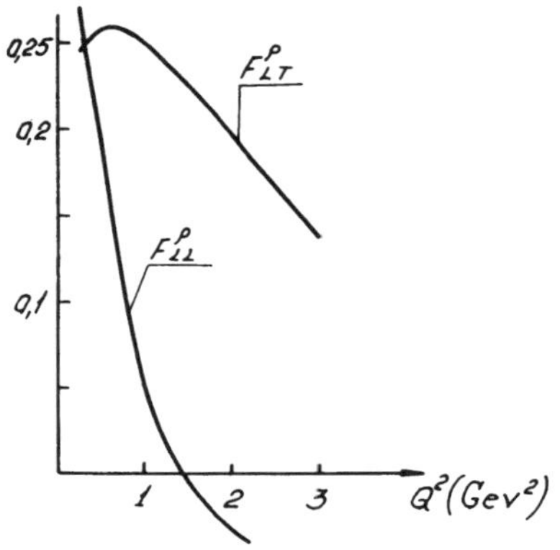

Fig. 13. Longitudinal $F^{\rho}_{LL}(Q^2)$ and transverse-longitudinal $F_{LT}(Q^2)$ form factors of the ρ meson.

small Q^2 and not for the lagrangian of refs. [27, 28] as a whole.) At a certain risk, the values of static magnetic and quadrupole moments of the ρ meson can be predicted from the curves fig. 12*. According to (43) the magnetic moment of the ρ meson (in units $e\hbar/2m_\rho c$) is close to 2, and the quadrupole moment is small, $|D| \lesssim 0.1/m_\rho^2$.

It is interesting to compare the form factors obtained above in the region of intermediate Q^2 with their asymptotic behaviour, calculated in perturbative QCD [29–31]. The asymptotics of ρ meson form factors is usually determined in terms of states with given transverse e_T and longitudinal e_L polarizations of the ρ meson but not in terms of F_1^ρ, F_2^ρ, and F_3^ρ. Let us present, therefore, the formulae relating the quantities in both approaches. In the Breit system

$$F_{TT}(Q^2) = \langle p', e_T | j_0^{el} | p, e_T \rangle / 2E = F_1(Q^2),$$

$$F_{LT}(Q^2) = \langle p', e_T | j_x^{el} | p, e_L \rangle / 2E = \frac{Q}{2m_\rho}(F_1 + F_2),$$

$$F_{LL}(Q^2) = \langle p', e_L | j_0^{el} | p, e_L \rangle / 2E = F_1 - \frac{Q^2}{2m_\rho^2} F_2 + \frac{Q^2}{m_\rho^2}\left(1 + \frac{Q^2}{4m_\rho^2}\right) F_3, \quad (48)$$

where E is the meson energy, $E = \sqrt{m_\rho^2 + \frac{1}{4}Q^2}$.

Quark counting and chirality conservation lead in the asymptotic limit to $F_{LL}(Q^2) \sim 1/Q^2$, $F_{LT}(Q^2) \sim 1/Q^3$ and $F_{TT}(Q^2) \sim 1/Q^4$, i.e. according to (48) $F_1(Q^2) \sim F_2(Q^2) \sim 1/Q^4$, $F_3(Q^2) \sim 1/Q^6$. It is seen from fig. 12 that in the region of intermediate Q^2 all three form factors decrease slower than is expected in the

* It should be taken into account that the absolute values of form factors are determined markedly worse than their relative behaviour (or ratios) since they are proportional to the constant g_ρ which is not very precisely determined from the theory.

asymptotic limit and $F_2(Q^2)$ decreases especially slowly. The difference from the asymptotic behaviour becomes even more striking if we depict the behaviour of form factors $F_{LL}(Q^2)$ and $F_{LT}(Q^2)$ (fig. 13). The form factor $F_{LL}(Q^2)$, the main one in the asymptotics, decreases very rapidly in the intermediate region, at $Q^2 \approx 1.5\ \mathrm{GeV}^2$ is close to zero, and maybe even changes sign [because of strong compensation in (48) one can make no conclusions on $F_{LL}(Q^2)$ at $Q^2 \gtrsim 1.5\ \mathrm{GeV}^2$]. On the contrary, the form factor $F_{LT}(Q^2)$, small in the asymptotics, decreases very slowly in the region of intermediate Q^2. Thus, just as for the pion form factor, the asymptotics of ρ-meson form factors seems to start very far away.

9. Form factors of the A_1 meson and the transition $\pi\gamma \to A_1$: A_1 meson width

Electromagnetic form factors of the axial A_1 meson and form factors of the transition $\pi + \gamma_{\mathrm{virt.}} \to A_1$ can be found from the same sum rules. The vertex function calculated in QCD is given by the same formulae as before [corrections due to the quark condensate should be taken according to eq. (26)]. The A_1 meson contribution to the spectral function is determined by (44) with the substitution $m_\rho \to m_A$, $g_\rho \to g_A$. In the phenomenological part of the sum rules the one-pion state should be taken into account in addition to the A_1 meson. The contribution of diagonal transitions to $\rho_{\mu\nu,\lambda}(p, p'; q)$ due to one-pion states has been calculated earlier [eq. (5)]. It is also necessary to take into account the contribution of non-diagonal transitions due to the electromagnetic current $\pi + \gamma \to A_1$. The electromagnetic vertex of such a transition is written as

$$\langle \pi^+(p')|j_\lambda^{\mathrm{el}}|A_1^+(p,\varepsilon)\rangle = -\frac{1}{m_A}\Big\{[(Pq)\delta_{\lambda\sigma} - P_\lambda q_\sigma]G_1(Q^2) + \frac{1}{m_A^2}[(Pq)q_\lambda - q^2 P_\lambda]\,p'_\sigma G_2(Q^2)\Big\}\varepsilon_\sigma, \tag{49}$$

where $G_1(Q^2), G_2(Q^2)$ are form factors of the $\pi\gamma \to A_1$ transition. The contribution of non-diagonal $\pi\gamma \to A_1$ transitions to the double spectral function turns out to be

$$\rho_{\mu\nu,\lambda}(p,p';q) = -f_\pi\sqrt{2}\,\frac{m_A}{g_A}\Big\{P_\lambda(p_\mu p_\nu + p'_\mu p'_\nu)\Big(G_1 - \frac{Q^2}{m_A^2}G_2\Big) - P_\lambda p'_\nu p_\mu\Big[\Big(2+\frac{Q^2}{m_A^2}\Big)G_1 - \frac{Q^2}{m_A^2}\Big(1+\frac{Q^2}{m_A^2}\Big)G_2\Big] + q_\lambda(p'_\mu p'_\nu - p_\mu p_\nu)G_2 + m_A^2(p'_\nu\delta_{\mu\lambda} + p_\mu\delta_{\nu\lambda})G_1\Big\}. \tag{50}$$

The tactics of determination of form factors follows from the form of (5), (44), (50): first find F_1^A, F_2^A from the sum rules for structures $\delta_{\mu\nu}P_\lambda$, $\delta_{\mu\lambda}p_\nu + \delta_{\nu\lambda}p'_\mu$ and afterwards, using the obtained F_1^A and F_2^A, define G_1 and G_2 from structures $\delta_{\mu\lambda}p'_\nu + \delta_{\nu\lambda}p_\mu$ and $q_\lambda(p'_\mu p'_\nu - p_\mu p_\nu)$.

The sum rules for F_1^A and $F_1^A + F_2^A$ differ from (45), (46) by a change of coefficients in front of the terms $\sim \langle 0|\bar{\psi}\psi|0\rangle^2$ in accord with (26), in particular, in the sum rule for F_1^A the coefficient in front of $\langle 0|\bar{\psi}\psi|0\rangle^2$ turns out to be larger by a factor of 2.6 (and of the opposite sign). Form factors G_1 and G_2 can be found from the sum rules

$$F_1^A + F_2^A + \frac{\sqrt{2}f_\pi m_A g_A}{Q^2+2m_A^2}e^{m_A^2/M^2}G_1 = \frac{g_A^2}{16\pi^2}\left(\frac{M}{m_A}\right)^4 e^{2m_A^2/M^2}\frac{m_A^2}{Q^2+2m_A^2}$$
$$\times\left\{2\int_0^\infty e^{-x}dx\,\frac{x^2(5x+3k)}{(2x+k)^2}\right.$$
$$\left.-\frac{256\pi^3}{81M^6}\alpha_s\langle 0|\bar{\psi}\psi|0\rangle^2\left(16+\tfrac{5}{4}k-\tfrac{1}{2}k^2\right)\right\}, \tag{51}$$

$$F_1^A + F_2^A + \frac{\sqrt{2}f_\pi g_A}{m_A}e^{m_A^2/M^2}G_2 = \frac{g_A^2}{16\pi^2}\left(\frac{M}{m_A}\right)^2 e^{2m_A^2/M^2}$$
$$\times\left\{3\int_0^\infty e^{-x}dx\left[\ln\left(1+\frac{2x}{k}\right)-\frac{2x}{2x+k}\right]\right.$$
$$+\frac{128\pi^3}{9Q^4M^2}\alpha_s\langle 0|\bar{\psi}\psi|0\rangle^2$$
$$\left.\times\left(1-k+\tfrac{5}{18}k^2+\tfrac{1}{9}k^3\right)\right\}. \tag{52}$$

In the sum rules for A_1 meson form factors the values of the current parameter M^2 must be considerably larger than in those for the ρ meson. This follows both from the sum rules for the polarization operators of ρ and A_1 mesons [1,2], where corresponding values of M^2 differed from each other by approximately a factor of 2, and directly from the sum rules for F_1^A: to suppress considerably larger corrections $\sim \langle 0|\bar{\psi}\psi|0\rangle^2$ larger M^2 are necessary. The values of the continuum threshold s_0 must be a factor of ~ 1.5 larger than the continuum threshold in the polarization operator (this is seen from fig. 3), i.e. $s_0 \approx 3.0$ GeV^2. The sum-rule analysis shows

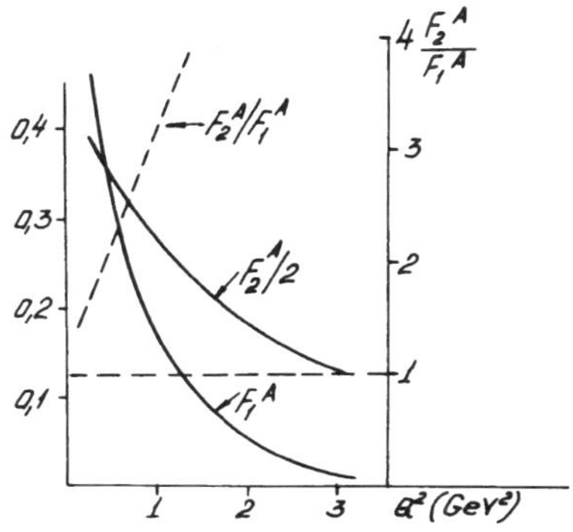

Fig. 14. A_1 meson form factors $F_1^A(Q^2)$, $F_2^A(Q^2)$ and the ratio $F_2^A(Q^2)/F_1^A(Q^2)$ (dashed curve, r.h. scale).

that at $s_0 = 3\ \mathrm{GeV}^2$ the optimal choice is $M^2 = 1.8\ \mathrm{GeV}^2$. The Q^2 dependence of $F_1^A(Q^2)$, $F_2^A(Q^2)$ and $F_2^A(Q^2)/F_1^A(Q^2)$ at $M^2 = 1.8\ \mathrm{GeV}^2$, $s_0 = 3.0\ \mathrm{GeV}^2$ is shown in fig. 14. (According to the results of ref. [2] we adopt $m_A = 1.15$ GeV, $g_A^2/4\pi = 6.0$.) As is seen from fig. 14, the behaviour of the form factors F_1^A and F_2^A weakly differs from F_1^ρ and F_2^ρ.

Fig. 15 shows the curves of transitional form factors G_1 and G_2. The form factors G_1 and G_2 were calculated disregarding corrections due to the gluon condensate. [See eqs. (51), (52).] There are two reasons why G_1 and G_2 are not so well determined from the sum rules considered as the form factors F_1 and F_2. The first is that at small Q^2 the spectral functions are proportional to $\delta(s - s')$, i.e. lie on the diagonal, fig. 3. Then the G_1 and G_2 contributions correspond to points on coordinate axes s and s'. Thereby, at small Q^2 the duality principle is not fulfilled. The other reason is that when calculating G_1 and G_2 one should subtract the previously obtained form factors $F_1^A + F_2^A$ from the sum rules (51), (52). These two circumstances lead to the fact that in the case of G_1 and G_2 we can only give semiquantitative results valid up to a factor 1.5 (or even 2). Obviously, form factors G_1 and G_2 can be determined

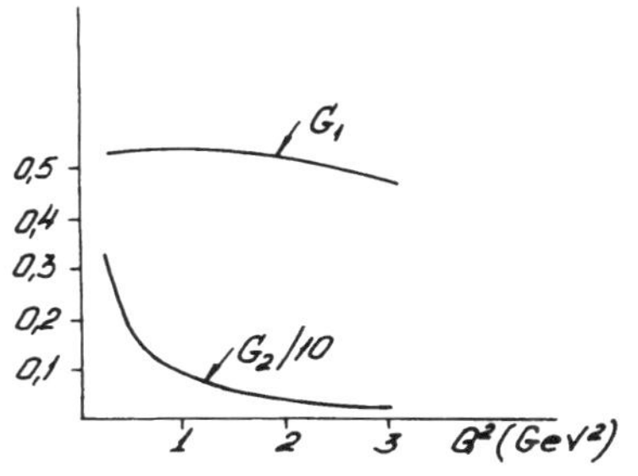

Fig. 15. Form factors of the transition $\pi + \gamma_{\mathrm{virt}} \to A_1$, $G_1(Q^2)$ and $G_2(Q^2)$.

with better accuracy if sum rules for a vertex with vector, axial and pseudoscalar currents is considered. Nevertheless, it is interesting to calculate the $A_1 \to \rho\pi$ decay width corresponding to the $G_1(Q^2)$ and $G_2(Q^2)$ found. To this end, in the spirit of VDM let us saturate $G_1(Q^2)$ and $G_2(Q^2)$ by the ρ meson contribution, i.e. put

$$G_{1,2}(Q^2) = \frac{m_\rho^2}{g_\rho} g_{1,2} \frac{1}{Q^2 + m_\rho^2} . \tag{53}$$

The $A_1 \to \rho\pi$ decay width is expressed through g_1 and g_2 as follows:

$$\Gamma(A_1 \to \rho\pi) = \frac{p}{6\pi} \{ g_1^2 (1 + \tfrac{9}{8}\gamma - \tfrac{3}{4}\gamma^2 + \tfrac{1}{8}\gamma^3)$$

$$- \tfrac{3}{4} g_1 g_2 \gamma (1-\gamma)(1 - \tfrac{4}{3}\gamma + \tfrac{1}{3}\gamma^2) + \tfrac{1}{8} g_2^2 \gamma (1-\gamma)^4 \} , \tag{54}$$

where p is the pion momentum, $\gamma = m_\rho^2/m_A^2$. Comparing (53) with the curves of fig. 15 in the region $Q^2 \sim 1$ GeV^2, we find

$$g_1 \approx 8 , \qquad g_2 \approx 12 , \tag{55}$$

which gives $\Gamma(A_1 \to \rho\pi) \approx 1$ GeV. Though this number seems essentially to exceed the experimental one, this fact cannot be considered as a discrepancy because the approximations are rough.

10. The results

In this paper we have suggested a QCD-based calculation method of two-particle meson decay widths $A \to B + C$ and of form factors of $A \to B$ transitions in the region of intermediate Q^2. The method is applicable to the consideration of any mesonic decays provided A, B, C are the lowest-lying states in a channel with the given quantum numbers. With this method, the pion electromagnetic form factor for $0.5 \lesssim Q^2 \lesssim 3$ GeV^2, the constant $g_{\rho\pi\pi}$ and the g_ρ meson decay width, ρ and A_1 meson electromagnetic form factors have been calculated and electromagnetic form factors of the $A_1 \to \pi$ transition and the $A_1 \to \rho\pi$ decay width have been semi-quantitatively estimated. It was shown that using additional information on the pion form factor asymptotics, one can calculate the constant $g_{\rho\pi\pi}$ with very high accuracy – higher than the VDM accuracy. It has been established that the ratio of the ρ (or A_1) magnetic form factor to the electric one in the region of small Q^2 is close to unity, i.e. ρ and A_1 at small Q^2 interacted with the electromagnetic field as Yang-Mills vector bosons. It has also been found that in the region of intermediate $Q^2 \lesssim 3$ GeV^2 the behaviour of electromagnetic form factors of vector mesons depending on their polarizations states is quite different from that was expected from the asymptotics: the form factor corresponding to longitudinal polarizations $F_{\mathrm{LL}}(Q^2)$ is far less and decreases faster than the transitional form factor $F_{\mathrm{LT}}(Q^2)$.

Thanks are due to B.V. Geshkenbein, M.A. Shifman, M.V. Terentyev and A.I. Vainshtein for useful discussions and remarks.

Appendix A

In this appendix some calculation details of the power corrections due to the gluon condensate are presented. As has already been mentioned, in this case the gluonic field can be considered as a given external field with potential determined by eq. (12). It is convenient to find the quark propagator $S(x, z)$ in this field. Solving the equation

$$i\gamma_\mu\left(\partial_\mu + ig\frac{\lambda^a}{2}A_\mu^a(x)\right)S(x, z) = i\delta^{(4)}(x-z), \tag{A.1}$$

and accounting for the terms to second order in A_μ, we get

$$\begin{aligned} S(x, z) = \frac{i}{(2\pi)^4}\int \mathrm{d}^4k\, \mathrm{e}^{-ik(x-z)}\Bigg\{ &\frac{\hat{k}}{k^2} + \tfrac{1}{4}g\lambda^a G_{\alpha\beta}^a \varepsilon_{\alpha\beta\sigma\rho}\gamma_5\gamma_\rho\frac{k_\sigma}{k^4} \\ &- \tfrac{1}{4}g\lambda^a G_{\alpha\beta}^a z_\beta\left(\gamma_\alpha k^2 - 2k_\alpha\hat{k}\right)\frac{1}{k^4} + \tfrac{1}{96}g^2\langle 0|G_{\mu\nu}^2|0\rangle \\ &\times\left[-z^2\frac{\hat{k}}{k^4} - 4\frac{\hat{k}}{k^6}(kz)^2 + 2\hat{z}\frac{kz}{k^4}\right]\Bigg\}. \end{aligned} \tag{A.2}$$

Here $G_{\alpha\beta}^a$ is the constant gluonic field strength and λ^a are the Gell-Mann matrices. In the last term in (A.2) corresponding to two-gluon emission from one line, the vacuum fluctuation averaging is made according to (13). The power corrections proportional to $\langle 0|G_{\mu\nu}^2|0\rangle$ can be calculated if one substitutes the propagator $S(x, z)$ (A.2) instead of each of fermionic lines in fig. 1, selects terms proportional to G^2, averages over vacuum fluctuations of the gluonic field according to (13) and goes to the momentum representation. In doing so, as can easily be verified, the contribution of the third term in the integrand in (A.2) is zero. As a result, for the vertex function with gluon emission from different lines (figs. 4a–c) we have

$$\Gamma^{\text{figs.4a-c}}_{\mu\nu,\lambda}(p, p'; q) = i\frac{g^2\langle 0|G^2_{\mu\nu}|0\rangle}{24}\frac{1}{(2\pi)^4}\int\frac{d^4k}{k^2(p-k)^2(p'-k)^2}$$

$$\times\Big\{\hat{k}\gamma_\nu[2\gamma_\lambda(p'-k, p-k) + (\hat{p}-\hat{k})\gamma_\lambda(\hat{p}'-\hat{k})]$$

$$\times\frac{1}{(p-k)^2(p'-k)^2} + [2\gamma_\nu(\hat{p}'-\hat{k})\gamma_\lambda(pk-k^2)$$

$$+(\hat{p}-\hat{k})\gamma_\nu(\hat{p}'-\hat{k})\gamma_\lambda\hat{k}]\frac{1}{k^2(p-k)^2}$$

$$+[2\gamma_\nu(p'k-k^2) + (\hat{p}'-\hat{k})\gamma_\nu\hat{k}]\gamma_\lambda(\hat{p}-\hat{k})\frac{1}{k^2(p'-k)^2}\Big\}. \tag{A.3}$$

The integrals entering (A.3) can be conveniently calculated by making use of the double spectral representation. For instance

$$\int d^4k\frac{k_\mu(p-k)_\alpha(p'-k)_\beta}{k^2(p-k)^4(p'-k)^4} = \frac{1}{4}\frac{\partial^2}{\partial p_\alpha\partial p'_\beta}\int\frac{k_\mu d^4k}{k^2(p-k)^2(p'-k)^2}$$

$$= \tfrac{1}{4}i\pi^2\frac{\partial^2}{\partial p_\alpha\partial p'_\beta}\iint\frac{ds\,ds'}{\lambda^{3/2}(s, s', q^2)}$$

$$\times\frac{s'(s-s'+q^2)p_\mu + s(s'-s+q^2)p'_\mu}{(s-p^2)(s'-p'^2)}. \tag{A.4}$$

(On the r.h.s. of (A.4) the expression for integral I_μ (8b) is substituted.) When differentiating the integrand in p_α and p'_β it should not be forgotten that $q^2 = (p'-p)^2$, i.e. it depends on p and p'. Marked simplifications emerge when selecting the structure $P_\mu P_\nu P_\lambda$. The calculation result of the contribution of figs. 4a–c to the corresponding Borel transformation function is given in the text [formulae (14a, b)].

After simple transformations the vertex function with two-gluon emission from one line (fig. 4d) can be written very compactly:

$$\Gamma^{\text{fig.4d}}_{\mu\nu,\lambda}(p,p';q)=\frac{g^2}{9\cdot 2^6}\langle 0|G^2_{\mu\nu}|0\rangle$$

$$\times\left[\frac{\partial^4}{\partial p_\beta\partial p'_\beta\partial p_\alpha\partial p'_\alpha}-\frac{\partial^2}{\partial p_\alpha\partial p'_\alpha}\frac{\partial^2}{\partial p'_\beta\partial p'_\beta}\right]\Gamma^{(0)}_{\mu\nu,\lambda}(p,p';q),\tag{A.5}$$

where $\Gamma^{(0)}_{\mu\nu,\lambda}(p,p';q)$ is the vertex function in the zero-order approximation [eq. (8)]. Representing $\Gamma^{(0)}_{\mu\nu,\lambda}$ as an expansion in all possible tensor structures, writing the spectral representation for the coefficient functions at each structure, using (10) and differentiating as shown in (A.5), one can, in principle, calculate $\Gamma^{\text{fig.4d}}_{\mu\nu,\lambda}$. But practically, it appears very tiresome and was realized only for the function with structure $P_\mu P_\nu P_\lambda$ [eq. (14c)]. In the other cases, as it seems to us, the result is not worth the labour. Note, that the second term in (A.5) is more convenient to calculate not by 4-fold differentiation of $\Gamma^{(0)}_{\mu\nu,\lambda}$ but by representing it as

$$-\frac{ig^2}{9\cdot 16}\langle 0|G^2_{\mu\nu}|0\rangle\frac{\partial^2}{\partial p_\rho\partial p'_\sigma}\int\frac{\mathrm{d}^4k}{(2\pi)^4}\frac{\mathrm{Tr}\{\gamma_\mu\hat{k}\gamma_\nu\gamma_\sigma\gamma_\lambda\gamma_\rho\}}{k^2(p-k)^2(p'-k)^2}.\tag{A.6}$$

Appendix B

CALCULATION OF THE VACUUM MATRIX ELEMENTS

In this appendix we calculate the vacuum averages from some quark and gluon operators of 6th dimension, expressing them through $\langle 0|\bar\psi\psi|0\rangle^2$.

Let us first consider the vacuum average (v.a.) $\langle 0|\bar\psi^a_\alpha\psi^b_\beta D_\rho G^n_{\mu\nu}|0\rangle$ (all operators are taken at one point, α,β are spinor indices, ρ,μ,ν are vector indices, a,b,n are colour, $a,b=1,2,3$, $n=1,2,\dots 8$). The general expression for this v.a. following from Lorentz and colour invariance is

$$\langle 0|\bar\psi^a_\alpha\psi^b_\beta D_\rho G^n_{\mu\nu}|0\rangle=A(\delta_{\rho\nu}\gamma_\mu-\delta_{\rho\mu}\gamma_\nu)_{\beta\alpha}(\lambda^n)^{ba}.\tag{B.1}$$

Multiplying (B.1) by $\delta_{\rho\mu}(\gamma_\nu)_{\alpha\beta}(\lambda^n)^{ab}$, let us sum over all repeated indices and take into account the equations of motion

$$D_\mu G^n_{\mu\nu}=\tfrac{1}{2}g\bar\psi\gamma_\nu\lambda^n\psi.\tag{B.2}$$

As a result we obtain

$$A=\frac{g}{3\cdot 2^9}\langle 0|(\bar\psi\gamma_\mu\lambda^n\psi)(\bar\psi\gamma_\mu\lambda^n\psi)|0\rangle=-\frac{g\langle 0|\bar\psi\psi|0\rangle^2}{3^3\cdot 2^5}\tag{B.3}$$

because of (17). Substitution of (B.3) into (B.1) gives formula (23). Let us now calculate the v.a. $\langle 0|\bar{\psi}^a_\alpha \nabla_\sigma \nabla_\rho \nabla_\tau \psi^b_\beta|0\rangle$. In our notations ∇_σ is a covariant derivative acting in the fundamental representation:

$$\nabla_\sigma = \partial_\sigma + ig\frac{\lambda^n}{2}A^n_\sigma, \tag{B.4a}$$

unlike D_σ the covariant derivative in the adjoint representation:

$$(D_\sigma)^{mn} = \partial_\sigma \delta^{mn} - gf^{mnq}A^q_\sigma. \tag{B.4b}$$

The general expression for v.a. is

$$\langle 0|\bar{\psi}^a_\alpha \nabla_\sigma \nabla_\rho \nabla_\tau \psi^b_\beta|0\rangle = \delta^{ab}(B\gamma_\sigma\delta_{\rho\tau} + C\gamma_\rho\delta_{\sigma\tau} + D\gamma_\tau\delta_{\rho\sigma})_{\beta\alpha}. \tag{B.5}$$

Multiplying (B.5) by $(\gamma_\tau)_{\alpha\beta}$ or by $(\gamma_\sigma)_{\alpha\beta}$ and using the equations of motion (massless quarks)

$$i\gamma_\mu \nabla_\mu \psi = 0, \tag{B.6}$$

we find the relationship between B, C and D:

$$D = B, \qquad C = -5B. \tag{B.7}$$

Multiplying of (B.5) by $(\gamma_\rho)_{\alpha\beta}\delta^{ab}\delta_{\sigma\tau}$ gives

$$B = -\frac{1}{3^3 \cdot 2^5}\langle 0|\bar{\psi}\nabla_\sigma \hat{\nabla} \nabla_\sigma \psi|0\rangle. \tag{B.8}$$

Since

$$\hat{\nabla}\nabla_\sigma = \nabla_\sigma \hat{\nabla} + [\hat{\nabla}, \nabla_\sigma], \qquad [\nabla_\rho, \nabla_\sigma] = \tfrac{1}{2}ig\lambda^n G^n_{\rho\sigma}, \tag{B.9}$$

then, accounting for the equations of motion,

$$\begin{aligned}\langle 0|\bar{\psi}\nabla_\sigma \hat{\nabla} \nabla_\sigma \psi|0\rangle &= ig\langle 0|\bar{\psi}\nabla_\sigma \gamma_\rho \frac{\lambda^n}{2} G^n_{\rho\sigma}\psi|0\rangle \\ &= ig\langle 0|\bar{\psi}\frac{\lambda^n}{2}\gamma_\rho \psi (D_\sigma G_{\rho\sigma})^n|0\rangle \\ &\quad + ig\langle 0|\bar{\psi}\frac{\lambda^n}{2}\gamma_\rho G^n_{\rho\sigma}\nabla_\sigma \psi|0\rangle.\end{aligned} \tag{B.10}$$

(Because of properties of the operator ∇_σ: $[\nabla_\sigma, \frac{1}{2}\lambda^n G^n_{\rho\sigma}] = \frac{1}{2}\lambda^n (D_\sigma G_{\rho\sigma})^n$.) On the other hand, acting in the l.h.s. of (B.10) by the operators $\nabla_\sigma \hat{\nabla}$ to the left and using (B.9) we have

$$\langle 0|\bar{\psi}\nabla_\sigma \hat{\nabla} \nabla_\sigma \psi|0\rangle = -ig\langle 0|\bar{\psi}\frac{\lambda^n}{2}\gamma_\rho G^n_{\rho\sigma}\nabla_\sigma\psi|0\rangle . \tag{B.11}$$

Comparing (B.10) and (B.11) we find

$$\langle 0|\bar{\psi}\frac{\lambda^n}{2}\gamma_\rho G^n_{\rho\sigma}\nabla_\sigma\psi|0\rangle = -\tfrac{1}{2}\langle 0|\bar{\psi}\frac{\lambda^n}{2}\gamma_\rho\psi(D_\sigma G_{\rho\sigma})^n|0\rangle , \tag{B.12}$$

and

$$\langle 0|\bar{\psi}\nabla_\sigma \hat{\nabla} \nabla_\sigma \psi|0\rangle = \tfrac{1}{2}ig\langle 0|\bar{\psi}\frac{\lambda^n}{2}\gamma_\rho\psi(D_\sigma G_{\rho\sigma})^n|0\rangle . \tag{B.13}$$

Substituting (B.2) into (B.13) and (B.13) into (B.8) we have

$$B = \frac{ig^2}{3^3\cdot 2^8}\langle 0|\bar{\psi}\gamma_\rho\lambda^n\psi\cdot\bar{\psi}\gamma_\rho\lambda^n\psi|0\rangle = -\frac{ig^2\langle 0|\bar{\psi}\psi|0\rangle^2}{3^5\cdot 2^4} , \tag{B.14}$$

because of the factorization hypothesis. Formulae (B.5), (B.7), (B.14) give eq. (21) in the text.

Let us finally calculate v.a. $\langle 0|\bar{\psi}^a_\alpha \nabla_\sigma \psi^b_\beta G^n_{\mu\nu}|0\rangle$. The general expression for it has the form

$$\langle 0|\bar{\psi}^a_\alpha \nabla_\sigma \psi^b_\beta G^n_{\mu\nu}|0\rangle = \left[E(\delta_{\sigma\mu}\gamma_\nu - \delta_{\sigma\nu}\gamma_\mu) + iF\varepsilon_{\sigma\mu\nu\lambda}\gamma^5\gamma_\lambda\right]_{\beta\alpha}(\lambda^n)^{ba} . \tag{B.15}$$

Multiplying (B.15) by $(\gamma_\sigma)_{\beta\alpha}$, using the equations of motion (B.6) and the relation

$$\varepsilon_{\sigma\mu\nu\lambda}\gamma_\sigma\gamma_\lambda\gamma^5 = i(\gamma_\mu\gamma_\nu - \gamma_\nu\gamma_\mu), \qquad \varepsilon_{0123} = 1 , \tag{B.16}$$

we find the connection between E and F: $E = -F$. In order to find E, multiply (B.15) by $\delta_{\sigma\mu}(\gamma_\nu)_{\alpha\beta}(\lambda^n)^{ab}$. Then

$$E = \frac{1}{3\cdot 2^8}\langle 0|\bar{\psi}\lambda^n\gamma_\nu\nabla_\sigma\psi G^n_{\sigma\nu}|0\rangle$$

$$= \frac{1}{3\cdot 2^8}\langle 0|\bar{\psi}\frac{\lambda^n}{2}\gamma_\rho\psi(D_\sigma G_{\rho\sigma})^n|0\rangle . \tag{B.17}$$

Substituting (B.2) into (B.17) we obtain

$$E = -\frac{g}{3\cdot 2^{10}}\langle 0|\bar{\psi}\lambda^n\gamma_\rho\psi\cdot\bar{\psi}\lambda^n\gamma_\rho\psi|0\rangle$$

$$= \frac{g}{3^3\cdot 2^6}\langle 0|\bar{\psi}\psi|0\rangle^2 , \tag{B.18}$$

which gives formula (24).

References

[1] M.A. Shifman, A.I. Vainshtein and V.I. Zakharov, Nucl. Phys. B147 (1979) 385, 448
[2] L.J. Reinders, H.R. Rubinstein and S. Yazaki, Nucl. Phys. B196 (1982) 125
[3] B.L. Ioffe, Nucl. Phys. B188 (1981) 817; (E: B191 (1981) 591)
[4] Y. Chung, H.G. Dosch, M. Kremer and D. Schall, Nucl. Phys. B197 (1982) 55
[5] E.V. Shuryak, Nucl. Phys. B198 (1982) 83
[6] V.M. Belyaev and B.L. Ioffe, JETP-83 (1982) 876
[7] B.L. Ioffe and A.V. Smilga, Phys. Lett. 114B (1982) 353
[8] A.V. Nesterenko and A.V. Radyushkin, Phys. Lett. 115B (1982) 410
[9] A.Yu. Khodjamirian, Phys. Lett. 90B (1980) 460
[10] J. Schwinger, Particles, sources and fields (Addison-Wesley, 1973)
[11] V.A. Fateev, A.S. Schwartz and Yu.V. Tyupkin, Preprint Lebedev Phys. Inst. no. 155 (1976)
[12] M.S. Dubovikov and A.V. Smilga, Nucl. Phys. B185 (1981) 109
[13] C. Cronström, Phys. Lett. 90B (1980) 267
[14] A.V. Smilga, Yad. Fiz. 35 (1982) 473
[15] M.S. Dubovikov and A.V. Smilga, preprint ITEP-42 (1982)
[16] C. Bebek et al., Phys. Rev. D17 (1978) 1793
[17] V.L. Chernyak and A.R. Zhitnitsky, ZhETF Pisma 25 (1977) 544;
V.L. Chernyak, Proc. 15th Winter School, vol. I, p. 65, Leningrad, 1980
[18] A.V. Efremov and A.V. Radyushkin, Phys. Lett. 94B (1980) 245
[19] G.P. Lepage and S.J. Brodsky, Phys. Lett. 87B (1979) 359
[20] G.R. Farrar and D.R. Jackson, Phys. Rev. Lett. 43 (1979) 246
[21] A. Duncan and A.H. Mueller, Phys. Rev. D21 (1980) 1636
[22] B.V. Geshkenbein and B.L. Ioffe, Nucl. Phys. B166 (1980) 340
[23] V.A. Novikov et al., Nucl. Phys. B191 (1981) 301
[24] B.V. Geshkenbein and M.V. Terentyev, preprint ITEP-45 (1982)
[25] E.B. Dally et al., Phys. Rev. D24 (1981) 1718
[26] L.D. Landau and E.M. Lifshitz, Field theory (Pergamon Press, 1956)
[27] T.D. Lee, S. Weinberg and B. Zumino, Phys. Rev. Lett. 18 (1967) 1029
[28] T.D. Lee, and B. Zumino, Phys. Rev. 163 (1967) 667
[29] A.I. Vainshtein and V.I. Zakharov, Phys. Lett. 72B (1978) 368
[30] V.L. Chernyak, High-energy physics, Proc. 15th Winter School of LINP, Leningrad, 1980, p. 65
[31] G. Lepage and S. Brodsky, Phys. Rev. D22 (1980) 2157

7. Hadrons in External Fields: Static Characteristics (Magnetic Moments, Charge Radii and All That)

Introduction

The response of the hadronic system to weak external fields, say, electromagnetic, is an interesting dynamic parameter reflecting the internal structure of the hadron. No theoretical scheme can pretend to be complete unless it successfully predicts such quantities as magnetic moments or charge radii. As we already know from the previous examples any additional information on the hadron structure is in the sum rule method immediately translated to certain predictions for the vacuum condensates. Thus, analyzing the electromagnetic properties of hadrons we learn a lot about the properties of the vacuum fluctuations – and this aspect is, perhaps, most important. Needless to say that all these data concerning the vacuum fields are extremely valuable for the vacuum-model builders. In a sense, the extension of the method, to be discussed below, opens a new chapter in the QCD sum rules.

To calculate, say, the magnetic moment or the charge radii one, clearly, has to deal with three-point functions – one electromagnetic current plus two currents producing the initial and final hadrons. The standard technique is applicable if all three momenta flowing in the currents lie in the deep Euclidean domain. In this way we can calculate the form factors at intermediate Q^2 (see chapter 6), but not the static properties. To find the latter an extrapolation is needed to the point where the momentum flowing in the electromagnetic current approaches zero.

The first attempts were based on a certain extrapolation procedure (of the vector dominance model type) which does not seem to be very reliable. Gradually, however, a new efficient approach has been worked out allowing one to deal directly with the static properties, without an extrapolation in Q^2. The main idea is to consider directly *two-point functions in the given external field*, which is assumed to be static or slowly varying. Certainly, the operator product expansion must be modified accordingly, because in the external fields new condensates, absent in the "pure vacuum", appear. The best known example of this type is the vacuum expectation value $\langle\bar{\psi}\sigma_{\mu\nu}\psi\rangle$ forbidden by Lorentz symmetry in a pure vacuum. In the presence of a constant electromagnetic field, however,

$$\langle\bar{\psi}\sigma_{\mu\nu}\psi\rangle = (4\pi\alpha)^{1/2}\chi F_{\mu\nu}\langle\bar{\psi}\psi\rangle, \tag{7.1}$$

where χ is a non-vanishing quantity called the vacuum magnetic susceptibility, first introduced in ref. [1]. The physical meaning of eq. (7.1) is obvious: the external magnetic field induces spin alignment of the vacuum quark pairs, and since the field is supposed to be small, the average spin is proportional to the field.

Among other crucial elements which ensure the success of the program it is worth mentioning a new version of the Borelization procedure. The point is that the old version could not be used. To explain the problem let us consider the nucleon magnetic moment. The phenomenological representation of the sum rules contains three distinct terms depicted in fig. 7.1. The momenta flowing in the electromagnetic current approach zero. As a result, the

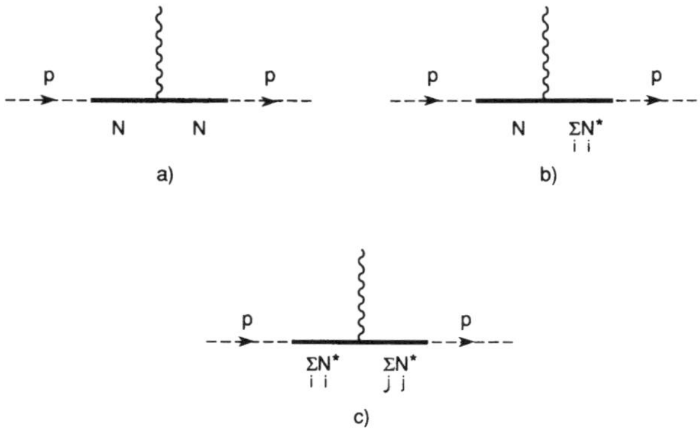

Fig. 7.1. Saturation of the sum rule for the nucleon magnetic moment. Wavy and dashed lines denote the electromagnetic and nucleon (η) currents, respectively, e.g. $\eta_n(x) = d^a(x)C\gamma_\mu d^b(x)\gamma_\mu\gamma_5 u^c(x)\epsilon^{abc}$. The momentum flowing in the electromagnetic current tends to zero while the momenta p corresponding to η are the same for both legs. The sum in (b) and (c) runs over all excited states with the appropriate quantum numbers.

momenta corresponding to the two nucleon currents necessarily coincide with each other, and, hence, the Borelization procedure one can apply is *the simultaneous Borelization with respect to the two legs*.

The first term (fig. 7.1a) is the double-pole contribution we are interested in. The third term (fig. 7.1c), with excited states produced by both nucleon currents, is exponentially suppressed after the simultaneous Borelization. The most unpleasant term is that in fig. 7.1b. It contains a single nucleon pole and is not suppressed exponentially relative to that in fig. 7.1a. (Actually, it is suppressed only with powers of the Borel parameter M^{-2}.) Thus, one has to invest special efforts to get rid of this contamination.

The fact that the operator product expansion can be modified in a calculable way for three-point functions with a small off-shellness of one of the legs has been noted in ref. [2]. For the nucleon magnetic moments the program has been first formulated and carried out, see refs. [1,3], independently by two groups. We reproduce here the detailed version of the work of Ioffe and Smilga.

The values of the magnetic susceptibility obtained originally by these two groups differed roughly speaking by a factor of two. Soon it became clear, however, that the induced vacuum condensates, and the magnetic susceptibility in particular, can be estimated with satisfactory accuracy from some auxilary sum rules, representing, in a sense, a QCD-based generalization of the vector dominance model. A careful analysis performed in refs. [4,5] leads to the result

$$\chi \approx -6\ \text{GeV}^{-2}, \tag{7.2}$$

lying approximately in the middle of the interval of the first estimates. The accuracy of this result is probably not worse than 10 to 15%. Shortly after publications [1,3] the same technique has been applied with great success to the calculation of the magnetic moments of hyperons [6]. Later, this problem – the magnetic moments of baryons – has been reanalyzed in a series of papers by Pasupathy and co-workers [7,8,9,10], who tried to take into account all previous results.

Baryons in an electromagnetic field by no means exhaust the possibilities of the method. A very promising direction of research has been outlined in ref. [11], which is reproduced in this chapter. The idea is to consider the nucleon current in an external (auxiliary) axial-vector field coupled to $\bar{u}\gamma_\mu\gamma_5 u - \bar{d}\gamma_\mu\gamma_5 d$. Just like in the case of the electromagnetic field, introduction of the axial-vector field generates new condensates, for instance $\langle\bar{\psi}\gamma_\mu\gamma_5\psi\rangle \sim A_\mu$, where A_μ is the auxiliary axial-vector field. An elegant aspect of the given problem is the fact that the soft-pion technique allows one to fix the most important condensates of this type theoretically, so that no new unknown parameters appear. In particular,

$$\langle\bar{u}\gamma_\mu\gamma_5 u\rangle = -\langle\bar{d}\gamma_\mu\gamma_5 d\rangle = f_\pi^2 A_\mu. \tag{7.3}$$

Other, less important condensates appearing in ref. [11], were previously found from independent meson sum rules (e.g. ref. [12]), so that the theoretical part of the sum rules is completely specified, and a clean prediction for the axial coupling constant emerges. The original result for the axial coupling constant g_A obtained in ref. [11] has been slightly corrected in the subsequent publication [13],

$$g_A - 1 = 0.4 \pm 0.2. \tag{7.4}$$

(The correction essentially affects only the error bars; for more details see ref. [13].) An extensive analysis of all vector and axial-vector couplings in the baryon decuplet is presented in ref. [14]. Of great interest in connection with the so-called proton spin crisis is the value of the singlet axial constant g_A^s. Naive estimates of this constant based on the assumption that there are essentially no s quarks in the proton wave function have been obtained in ref. [13]. The assumption of a small s quark admixture in nucleons is now questioned in the context of the problem of the spin crisis. Whatever the final solution of this problem might be, it is clear that a new analysis of the sum rules for g_A^s is needed. Such work is in progress now [15].

Finally, it is worth mentioning that an approach close in spirit has been suggested in ref. [16] in the problem of πNN and πNΔ coupling constants.

Returning to the electromagnetic properties of hadrons one should not forget that actually there are much more parameters than just magnetic moments, and any self-consistent dynamical approach must yield them all simultaneously. Therefore, calculation of the electric charge radii of mesons and baryons is a crucial test of both the basic ideas of the QCD vacuum and the technical devices presented in this chapter. Such calculations exist in the literature. Below, the first paper on the pion charge radius by Nesterenko and Radyushkin is reproduced [17]. Analogous investigations of the kaon and nucleon charge radii have been carried out in refs. [18,19].

The technique of calculations in external fields sketched above can be equivalently formulated in a different language – as a modification of the operator expansion which takes into account the so-called “bi-local” power corrections. The condensates in the external fields arise within this alternative approach as two-point correlation functions at zero momentum. In some applications such a language turns out to be more convenient, and it is sometimes used in the literature. As an example let me mention the calculation [20] of the power corrections to the first moments of the structure functions for deep inelastic scattering. The power corrections to the well-known parton results are proportional to reduced matrix elements, over the nucleon state, of certain complicated quark–antiquark–gluon operators.

A new development of the ideas of this chapter is presented in refs. [21,22]. The authors argue that it is possible to consider the correlation functions of hadronic currents in variable external fields – not only in constant ones – and, hence, to preserve the possibility of doing Borelization with respect to the two hadron momenta independently. The vacuum condensates in varying external fields are described by response functions. The latter can be studied using the expansion in conformal operators familiar from the QCD analysis of exclusive processes. A generalization of eq. (7.1) using this technique would read

$$\left\langle\bar{\psi}(x)\sigma_{\mu\nu}\psi(0)\right\rangle_F = (4\pi\alpha)^{1/2}\chi\langle\bar{\psi}\psi\rangle\int_0^1 \mathrm{d}u\ \phi(u)\ F_{\mu\nu}(ux), \tag{7.5}$$

where $x^2=0$. Substituting the external field in the form of a plane wave in eq. (7.5) it is easy to identify the function $\phi(u)$ with the wave function of the photon, describing its decay to the quark–antiquark pair in the infinite momentum frame. The varible u acquires the physical meaning of the fraction of the photon momentum carried by the quark. Paper [22] contains a pedagogical presentation of this approach; it treats anew the example of the nucleon magnetic moment, well studied previously using the traditional method [1,3]. In ref. [21] this technique is applied to the complicated problem of the radiative decay $\Sigma \to p\gamma$.

The evolution line which I have just sketched illustrates, probably in a most remarkable way, how the original ideas and devices of the sum rule method are transformed and expanded in the ongoing chase for new applications. The weak radiative decays like $\Sigma \to p\gamma$ is a relatively new and sophisticated branch which has an obvious "genetic" relation to the problem of the magnetic moments, although for $\Sigma \to p\gamma$ the theory has gone much further, of course. The subject has been worked out in refs. [21,23]. Another practically important problem close in spirit – but not technically – to $\Sigma \to p\gamma$ is the weak non-leptonic decays of D mesons [24]. The theoretical situation in both problems is rather messy, and the last word is definitly not said.

References

[1] B.I. Ioffe and A.V. Smilga, Pis'ma Zh. Eksp. Teor. Fiz. 37 (1983) 250; (E) 38 (1983) 42 [JETP Lett. 37 (1983) 298; (E) 38 (1983) 48]; Nucl. Phys. B 232 (1984) 109.
[2] I.I. Balitsky, Phys. Lett. B 114 (1982) 53.
[3] I.I. Balitsky and A.V. Yung, Phys. Lett. B 129 (1983) 328.
[4] V.M. Belyaev and Ya.I. Kogan, Yad. Fiz. 40 (1984) 1035 [Sov. J. Nucl. Phys. 40 (1984) 659].
[5] Ya. Balitsky, A.V. Kolesnichenko and A.V. Yung, Yad. Fiz. 41 (1985) 282 [Sov. J. Nucl. Phys. 41 (1985) 178].
[6] B.L. Ioffe and A.V. Smilga, Phys. Lett. B 133 (1983) 436.
[7] C.B. Chiu, J. Pasupathy and S.L. Wilson, Phys. Rev. D 33 (1986) 1961.
[8] S.L. Wilson, C.B. Chiu and J. Pasupathy, Phys. Rev. D 36 (1987) 1451.
[9] J. Pasupathy and J.P. Singh, Phys. Rev. D 36 (1987) 1442.
[10] C.B. Chiu, S.L. Wilson, J. Pasupathy and J.P. Singh, Phys. Rev. D 36 (1987) 1553.
[11] V.M. Belyaev and Ya.I. Kogan, Pis'ma Zh. Eksp. Teor. Fiz. 37 (1983) 611 [JETP Lett. 37 (1983) 730].
[12] V.A. Novikov, M.A. Shifman, A.I. Vainshtein, M.B. Voloshin and V.I. Zakharov, Nucl. Phys. B 237 (1984) 525.
[13] V.M. Belyaev, B.L. Ioffe and Ya.I. Kogan, Phys. Lett. B 151 (1985) 290.
[14] V.M. Belyaev, B.Yu. Blok and Ya.I. Kogan, Yad. Fiz. 41 (1985) 439 [Sov. J. Nucl. Phys. 41 (1985) 280].
[15] B.L. Ioffe and A.Y. Khodjamirian, to be published.
[16] L.J. Reinders, H.R. Rubinstein and S. Yazaki, Nucl. Phys. B 213 (1983) 109.
[17] V.A. Nesterenko and A.V. Radyushkin, Pis'ma Zh. Eksp. Teor. Fiz. 39 (1984) 576 [JETP Lett. 39 (1984) 707].
[18] Ya. Balitsky, A.V. Kolesnichenko and A.V. Yung, Phys. Lett. B 157 (1985) 309.
[19] V.A. Beilin, V.A. Nesterenko and A.V. Radyushkin, Int. J. Mod. Phys. A 3 (1988) 1183.
[20] V.M. Braun and A.V. Kolesnichenko, Nucl. Phys. B 283 (1987) 723.
[21] I.I. Balitsky, V.M. Braun and A.V. Kolesnichenko, Nucl. Phys. B 312 (1989) 509.
[22] V.M. Braun and I.E. Filyanov, Z. Phys. C 44 (1989) 157.
[23] I.I. Balitsky, V.M. Braun and A.V. Kolesnichenko, Yad. Fiz. 44 (1986) 1594 [Sov. J. Nucl. Phys. 44 (1986) 1028]. V. Khatsimovsky, Yad. Fiz. 45 (1987) 181; 46 (1987) 869 [Sov. J. Nucl. Phys. 45 (1987) 116; 46 (1987) 496].
[24] B.Y. Blok and M.A. Shifman, Yad. Fiz. 45 (1987) 211, 478, 841; 46 (1987) 1310 [Sov. J. Nucl. Phys. 45 (1987) 135, 301, 522; 46 (1987) 767].

Nuclear Physics B232 (1984) 109–142

NUCLEON MAGNETIC MOMENTS AND MAGNETIC PROPERTIES OF THE VACUUM IN QCD

B.L. IOFFE and A.V. SMILGA

Institute of Theoretical and Experimental Physics, Moscow 117259 USSR

Received 7 July 1983

Magnetic moments of a proton and a neutron are calculated in the QCD sum rule approach. The substantial role of the external electromagnetic field induced vacuum expectation values, the most important of which is connected with quark condensate magnetic susceptibility, is demonstrated. The results are $\mu_p = 3.0$, $\mu_n = -2.0(\pm 10\%)$ which is in perfect agreement with experiment. The invariant amplitudes of the $\Delta \to p\gamma$ transition are also calculated.

1. Introduction

A very powerful method for determining static and dynamic properties of the lowest-lying resonant states has been recently developed. This method, the method of QCD sum rules, was suggested in the pioneering work by Shifman, Vainshtein and Zakharov [1] where the masses of a number of meson resonances was successfully calculated. In refs. [2, 3] QCD sum rules were extended to the baryon case. Recent investigations made it possible to determine some dynamic properties of hadrons, namely, the meson electromagnetic form-factors at intermediate Q^2 values [4–7] and meson partial widths [5, 8]. In this work we report the solution of the other long-standing problem of the strong interaction theory and calculate nucleon magnetic moments*.

The problem is by no means trivial as in this case one has to consider scattering of nucleons on an external electromagnetic field *at zero momentum transfer*. In the case when even one of the external momenta is small the standard QCD sum rules method can not be applied since the crucial point of this approach is the calculation of the polarization operators $\Pi(p^2)$ or of the vertex functions $\Gamma(p^2, p'^2, q^2)$ in the euclidean region $p^2, p'^2, q^2 < 0$ at rather large virtualities in all channels where effects of the non-trivial vacuum structure can be described as controllable corrections to perturbative results. Thus, the technique of refs. [4–8] cannot be applied here. If trying to analytically continue the formulae for electromagnetic form factors, valid for the intermediate-Q^2 region to the point $Q^2 = 0$, one would encounter with singular terms $1/Q^2$, $1/Q^4$, ln Q^2 etc., or, if the power singularities are absent in

* The short communication on the results of the proton and neutron momenta calculation has been published earlier [9]. In our work we present the detailed version and add also the consideration of the $\Delta \to p\gamma$ transition.

a particular case, with unknown contributions non-described by a standard operator expansion. The appearance of such terms makes the procedure unjustified.

In order to solve the problem we consider the quarks as moving in a constant electromagnetic field $F_{\mu\nu}$.

We state that even in the case when one of the external momenta is zero, a variant of the operator expansion including new phenomenological parameters can be written. The physical meaning of these parameters is clear. They describe a response of the vacuum mean values to the presence of an external colourless field. E.g. the vacuum expectation value (v.e.v.) $\langle 0|\bar{q}\sigma_{\mu\nu}q|0\rangle$ where q is the quark field; $q = u, d$, is zero when external fields are absent according to Lorentz invariance. If quarks move in a constant electromagnetic field, then an external tensor $F_{\mu\nu}$ is present and one can write in general $\langle 0|\bar{q}\sigma_{\mu\nu}q|0\rangle_{\mathrm{F}} = \sqrt{4\pi\alpha}\,\chi_q F_{\mu\nu}\langle 0|\bar{q}q|0\rangle$, χ being a new parameter which may be thought of as the magnetic susceptibility of the quark condensate. In our problem the role of the v.e.v.'s $\langle 0|\bar{q}\sigma_{\mu\nu}q|0\rangle$ induced by the external field is very significant and their contribution to the sum rules can by no means be neglected.

There is another specific feature which distinguishes the sum rules for the polarization operator in external field from the common sum rules for determining hadron masses. When presenting $\Pi(p)$ in terms of the physical state matrix elements we are interested in the contribution $\langle 0|\eta|\mathrm{N}\rangle\,\langle \mathrm{N}|j^{\mathrm{el}}|\mathrm{N}\rangle\,\langle \mathrm{N}|\bar{\eta}|0\rangle(p^2-m^2)^{-2}$ where η is the quark current with nucleon quantum numbers. This contribution is proportional to the nucleon magnetic moment and has a double pole at $p^2 = m^2$. Besides this term, there are, however, terms of other types with just a single pole at $p^2 = m^2$. The terms of both types acquire the same exponential factor $\exp(-m^2/M^2)$ after Borel transformation. Therefore, the single-pole terms are not exponentially damped with respect to the double-pole terms and taking them into account is absolutely necessary. These terms give rise to additional phenomenological parameters in the r.h.s. of the sum rules.

Nevertheless, we succeeded in excluding all of these unknown parameters using the sum rules for invariant functions at different tensor structures and in determining nucleon magnetic moments with a good enough accuracy. After some simplifications to be specified later which do not practically affect the answer, our result for proton and neutron magnetic moments can be presented in a very simple form

$$\mu_{\mathrm{p}} = \frac{8}{3}\left(1 + \frac{1}{6}\frac{a}{m^3}\right),$$

$$\mu_{\mathrm{n}} = -\frac{4}{3}\left(1 + \frac{2}{3}\frac{a}{m^3}\right), \qquad (1)$$

where μ_{p}, μ_{n} are the magnetic moments in nuclear magneton units, $a = -(2\pi)^2\langle 0|\bar{q}q|0\rangle \approx 0.55\ \mathrm{GeV}^3$ and m is the nucleon mass. Substituting the numbers into eq. (1) we get $\mu_{\mathrm{p}} = 2.96$, $\mu_{\mathrm{n}} = -1.93$ with the estimated uncertainty $\sim 10\%$.

(The experimental values of the magnetic moments are $\mu_p = 2.79$, $\mu_n = -1.91$.) We estimate also the quark condensate magnetic susceptibility

$$\chi_q = -e_q(350 \pm 50 \text{ MeV})^{-2} . \tag{2}$$

In sect. 2 we describe our method in detail. Sect. 3 is devoted to the derivation of the sum rules of interest. In sect. 4 we determine the nucleon magnetic moments. In sect. 5 the transition $\Delta \to p + \gamma$ is considered. The general discussion and other possible applications of the method are the subjects of the last section.

2. The method

We start with the consideration of the polarization operator of currents bearing nucleon quantum numbers in the weak external constant electromagnetic field $F_{\mu\nu}$. To find nucleon magnetic moments we are interested in the terms linear in $F_{\mu\nu}$. Thus we define

$$i \int d^4x \langle 0|T\{\eta(x), \bar{\eta}(0)\}|0\rangle \, e^{ipx} = \Pi^{(0)}(p) + \sqrt{4\pi\alpha}\, \Pi_{\mu\nu}(p) F_{\mu\nu} , \tag{3}$$

where $\Pi^{(0)}(p)$ is the polarization operator when the field is absent, $\eta = \eta_p, \eta_n$ are the currents with proton and neutron quantum numbers

$$\eta_p(x) = u^a(x) C\gamma_\mu u^b(x) \gamma_5 \gamma_\mu d^c(x) \varepsilon^{abc} , \tag{4a}$$

$$\eta_n(x) = d^a(x) C\gamma_\mu d^b(x) \gamma_5 \gamma_\mu u^c(x) \varepsilon^{abc} . \tag{4b}$$

$u^a(x)$ and $d^a(x)$ are quark fields. The choice of currents with nucleon quantum numbers is not unique. The choice of eq. (4), most suitable in our opinion, was suggested in ref. [2]. Another choice was advocated in ref. [10] but we think that the current used in ref. [10] is inconvenient since it results in uncontrollably large power corrections. A detailed argument in favour of our viewpoint can be found in ref. [11].

Following the idea of the QCD sum rule method we calculate $\Pi(p)$ in the euclidean region $p^2 < 0$, $|p^2| \sim 1 \text{ GeV}^2$ as the operator expansion series in $1/p^2$ whose coefficients are expressed through the v.e.v.'s of various operators. On the other hand, we write dispersion relations for $\Pi(p)$ and saturate them by contributions of the lowest physical states. As usual, in the QCD sum rule approach, in order to suppress the contribution of excited states to sum rules, we apply the Borel transformation to the polarization operator structure functions

$$\mathcal{B}_{M^2} f_i(s) = \lim_{\substack{s\to\infty \\ n\to\infty \\ s/n = M^2}} \frac{s^{n+1}}{n!} \left(-\frac{d}{ds}\right)^n f_i(s)$$

$$= \frac{1}{\pi} \int_0^\infty dp^2 \, e^{-p^2/M^2} \, \text{Im}\, f_i(p^2) , \qquad s = -p^2 , \tag{5}$$

if the $f_i(p^2)$ are represented by the dispersion relation

$$f_i(p^2)=\frac{1}{\pi}\int_0^{\infty}\mathrm{d}p'^2\,\frac{\mathrm{Im}\,f_i(p'^2)}{p'^2-p^2}+\text{subtraction terms}\,.$$

Let us dwell on the operator expansion and classify the operator v.e.v.'s according to their dimensions d. Since we are interested in the term linear in $F_{\mu\nu}$ in eq. (3) the operator of the lowest dimension is $F_{\mu\nu}$ itself. The field-induced v.e.v. with $d=3$ $\langle 0|\bar{q}\sigma_{\mu\nu}q|0\rangle_{\mathrm{F}}$, $\sigma_{\mu\nu}=\frac{1}{2}i(\gamma_\mu\gamma_\nu-\gamma_\nu\gamma_\mu)$ is of the next dimension and this is the reason for its importance in the problem under consideration. Four-dimensional v.e.v.'s are absent. Indeed, a principally possible operator $\bar{q}(\nabla_\mu\gamma_\nu-\nabla_\nu\gamma_\mu)q$ where ∇_μ is the covariant derivative has positive C-parity and its v.e.v. cannot be induced by a C-odd electromagnetic field. Another possible operator $\varepsilon_{\mu\nu\lambda\sigma}\bar{q}\gamma_5\gamma_\lambda\nabla_\sigma q$ is C-odd, but for massless quarks it vanishes according to the equations of motion since $\varepsilon_{\mu\nu\lambda\sigma}\bar{q}\gamma_5\gamma_\lambda\nabla_\sigma q=\frac{1}{2}\bar{q}(\sigma_{\mu\nu}\hat{\nabla}+\hat{\nabla}\sigma_{\mu\nu})q=0$. Three v.e.v.'s: the quark condensate density $\langle 0|\bar{q}q|0\rangle$ multiplied by $F_{\mu\nu}$; field-induced v.e.v.'s $g\langle 0|\bar{q}(\frac{1}{2}\lambda^n)G^n_{\mu\nu}q|0\rangle_{\mathrm{F}}=\sqrt{4\pi\alpha}\,\kappa_{\mathrm{q}}F_{\mu\nu}\langle 0|\bar{q}q|0\rangle$ and $g\varepsilon_{\mu\nu\lambda\sigma}\langle 0|\bar{q}\gamma_5(\frac{1}{2}\lambda^n)G^n_{\lambda\sigma}q|0\rangle_{\mathrm{F}}=i\sqrt{4\pi\alpha}\,\xi_{\mathrm{q}}F_{\mu\nu}\langle 0|\bar{q}q|0\rangle$ have dimension $d=5$*. (Here $G^n_{\mu\nu}$ is the gluon field strength and λ^n are the colour Gell-Mann matrices). The v.e.v. $\langle 0|\bar{q}q|0\rangle=-(240\ \mathrm{MeV})^3$ (at the low-energy normalization point) has been known for a long time (see ref. [1] for a review and discussion), the rest of the v.e.v.'s have originally appeared in this problem. There are five 6-dimensional operator v.e.v.'s

$$\langle 0|\bar{q}q|0\rangle\langle 0|\bar{q}\sigma_{\mu\nu}q|0\rangle_{\mathrm{F}}\,,\qquad \langle 0|G^n_{\alpha\beta}G^n_{\alpha\beta}|0\rangle F_{\mu\nu}$$

$$g\langle 0|\bar{q}(\tfrac{1}{2}\lambda^n)(G^n_{\mu\lambda}\vec{\nabla}_\nu-\overleftarrow{\nabla}_\nu G^n_{\mu\lambda})\gamma_\lambda q-(\mu\to\nu)|0\rangle_{\mathrm{F}}$$

$$\varepsilon_{\mu\nu\rho\sigma}g\langle 0|\bar{q}(\tfrac{1}{2}\lambda^n)(G^n_{\rho\lambda}\vec{\nabla}_\sigma+\overleftarrow{\nabla}_\sigma G^n_{\rho\lambda})\gamma_\lambda\gamma_5 q|0\rangle_{\mathrm{F}}$$

$$d^{ikl}\langle 0|G^i_{\mu\lambda}G^k_{\lambda\rho}G^l_{\rho\nu}-(\mu\to\nu)|0\rangle_{\mathrm{F}}\,,$$

where d^{ikl} are SU(3) structure constants. (We adopt the factorization hypothesis saturating the product of the four-quark field operators by the contribution of an intermediate vacuum state and disregarding non-trivial four-fermion v.e.v.'s). The contribution of the last four v.e.v.'s is small (see below) and will be neglected in the following. A great number of operator v.e.v.'s correspond to dimension $d=7$, namely

$$\langle 0|\bar{q}\sigma_{\alpha\beta}(\tfrac{1}{2}\lambda^n)G^n_{\alpha\beta}q|0\rangle F_{\mu\nu}\,,\qquad \langle 0|\bar{q}\sigma_{\mu\nu}qG^n_{\alpha\beta}G^n_{\alpha\beta}|0\rangle_{\mathrm{F}}\,,\qquad \langle 0|\bar{q}\sigma_{\alpha\beta}G^n_{\alpha\beta}G^n_{\mu\nu}q|0\rangle_{\mathrm{F}}$$

$$\langle 0|\bar{q}\sigma_{\mu\alpha}G^n_{\alpha\beta}G^n_{\beta\nu}q-(\mu\to\nu)|0\rangle_{\mathrm{F}}\,,\qquad \langle 0|\bar{q}\lambda^n q\cdot\bar{q}\lambda^n\gamma_\mu\nabla_\nu q-(\mu\to\nu)|0\rangle_{\mathrm{F}}\,,$$

etc. The first of them is known

$$-g\langle 0|\bar{q}\sigma_{\alpha\beta}(\tfrac{1}{2}\lambda^n)G^n_{\alpha\beta}q|0\rangle\equiv m_0^2\langle 0|\bar{q}q|0\rangle\,, \tag{6}$$

* The authors are indebted to Ya.Ya. Balitsky who has drawn their attention to the appearance of the last v.e.v.

where $m_0^2 \approx 0.8\ \mathrm{GeV}^2$ was determined in ref. [3]. To estimate the second v.e.v. we can apply the factorization hypothesis (though it is not here as firmly based as in the case of four-fermion operators). Then

$$\langle 0|\bar{q}\sigma_{\mu\nu}qG^n_{\alpha\beta}G^n_{\alpha\beta}|0\rangle_{\mathrm{F}} = \langle 0|\bar{q}\sigma_{\mu\nu}q|0\rangle_{\mathrm{F}}\langle 0|G^n_{\alpha\beta}G^n_{\alpha\beta}|0\rangle\,. \tag{7}$$

All other 7-dimensional v.e.v.'s are non-factorizable and unknown. It can be shown particularly (see sect. 3) that the v.e.v. $\langle 0|\bar{q}\lambda^n\sigma_{\mu\rho}q\bar{q}\lambda^n\gamma_\rho\nabla_\nu q|0\rangle_{\mathrm{F}} - (\mu\leftrightarrow\nu)$ contribution to some sum rules is not small. Therefore, in order not to introduce a great number of unknown quantities into the theory we avoid using sum rules where the 7-dimensional v.e.v.'s may be important. Taking only the operator v.e.v.'s with $d \leq 6$ (and in some cases specified below the factorizable v.e.v.'s of higher dimensions) we are left with three electromagnetic field-induced v.e.v.'s:

$$\langle 0|\bar{q}\sigma_{\mu\nu}q|0\rangle_{\mathrm{F}} = \sqrt{4\pi\alpha}\,\chi_{\mathrm{q}}F_{\mu\nu}\langle 0|\bar{q}q|0\rangle\,, \tag{8}$$

$$g\langle 0|\bar{q}(\tfrac{1}{2}\lambda^n)G^n_{\mu\nu}q|0\rangle_{\mathrm{F}} = \sqrt{4\pi\alpha}\,\kappa_{\mathrm{q}}F_{\mu\nu}\langle 0|\bar{q}q|0\rangle\,,$$

$$g\varepsilon_{\mu\nu\lambda\sigma}\langle 0|\bar{q}\gamma_5(\tfrac{1}{2}\lambda^n)G^n_{\lambda\sigma}q|0\rangle_{\mathrm{F}} = i\sqrt{4\pi\alpha}\,\xi_{\mathrm{q}}F_{\mu\nu}\langle 0|\bar{q}q|0\rangle\,. \tag{9}$$

We call χ_{q} the quark condensate magnetic susceptibility. Indeed, if the quarks were non-relativistic χ_{q} (up to a constant factor) would evidently have this meaning. For relativistic quarks attending (8) the situation is rather obscure but we will maintain the denotion of χ_{q} in this case also.

We assume χ_{q}, κ_{q} and ξ_{q} to be proportional to the quark charge

$$\chi_{\mathrm{q}} = e_{\mathrm{q}}\chi\,,\qquad \kappa_{\mathrm{q}} = e_{\mathrm{q}}\kappa\,,\qquad \xi_{\mathrm{q}} = e_{\mathrm{q}}\xi\,. \tag{10}$$

This assumption corresponds to neglection of the closed-loop contribution shown in fig. 1. Evidently, for massless quarks the diagrams fig. 1 with gluon exchange are zero in any order of perturbation theory according to chirality conservation. Chirality violation might appear from instantons but, as was shown in ref. [12], instantons (in the diluted-gas approximation) do not contribute to the diagrams of fig. 1. The amplitude of fig. 1 has a certain resemblance to that of $\varphi-\omega$ mixing. The only difference between these two amplitudes is that the chirality is violated in the first and is conserved in the second. Therefore the experimental smallness of $\varphi-\omega$ mixing is also an argument for our assumption.

Now we turn to the description of the terms in the operator expansion which will be taken into account in the calculation of $\Pi_{\mu\nu}(p)$ in QCD (the l.h.s. of the desired sum rules). Three different tensor structures contribute to $\Pi_{\mu\nu}(p)$: $\hat{p}\sigma_{\mu\nu} + \sigma_{\mu\nu}\hat{p}$,

Fig. 1.

$i(p_\mu\gamma_\nu - p_\nu\gamma_\mu)\hat{p}$ and $\sigma_{\mu\nu}$. The first structure includes an odd number of γ-matrices and conserves chirality. The second and the third include even number of γ-matrices and violate chirality. (We adopt u- and d-quarks to be massless.)

Let us consider first the l.h.s. of the sum rule at the odd structure $\hat{p}\sigma_{\mu\nu} + \sigma_{\mu\nu}\hat{p}$. The lowest-dimension operator contributing to this structure is $F_{\mu\nu}$ $(d=2)$ and the corresponding Feynman diagram is depicted in fig. 2. Because of chirality conservation the $d=3$ and $d=5$ operators do not contribute to the odd structure, so the next in dimension are 6-dimensional operators. The taking into account of the field-induced v.e.v. $\langle 0|\bar{q}q|0\rangle\langle 0|\bar{q}\sigma_{\mu\nu}q|0\rangle_F$ $(d=6)$ corresponds to the diagram of fig. 3. The contribution of the operator $F_{\mu\nu}G^n_{\alpha\beta}G^n_{\alpha\beta}$ $(d=6)$ is determined by two-loop graphs (like fig. 4) which includes a small factor $(2\pi)^{-4}$. The direct calculation of the contribution of the operator $G^n_{\alpha\beta}G^n_{\alpha\beta}$ into $\Pi^{(0)}(p)$ showed that it does not practically affect the nucleon mass [13, 3]. Thereby the term proportional to the v.e.v. $F_{\mu\nu}\langle 0|G^n_{\alpha\beta}G^n_{\alpha\beta}|0\rangle$ in the operator expansion can safely be neglected*. Taking into account the field induced v.e.v. $d^{ikl}\langle G^i_{\mu\lambda}G^k_{\lambda\sigma}G^l_{\sigma\nu}|0\rangle_F$ leads to the same numerical factor. The contributions of the v.e.v.'s $g\langle 0|\bar{q}(\frac{1}{2}\lambda^n)(G^n_{\mu\lambda}\vec{\nabla}_\nu - \overleftarrow{\nabla}_\nu G^n_{\mu\lambda})\gamma_\lambda q|0\rangle_F$ and $\varepsilon_{\mu\nu\rho\sigma}g\langle 0|\bar{q}(\frac{1}{2}\lambda^n)$ $(G^n_{\rho\lambda}\vec{\nabla}_\sigma + \overleftarrow{\nabla}_\sigma G^n_{\rho\lambda})\gamma_\lambda\gamma_5 q|0\rangle_F$ are determined by the one-loop diagram, and have a numerically small factor $(2\pi)^{-2}$ and for this reason will also be disregarded.

In the sum rule at odd structure $\hat{p}\sigma_{\mu\nu} + \sigma_{\mu\nu}\hat{p}$ we take into account the contributions of v.e.v.'s of 8-dimensional operators assuming the factorization hypothesis. There are four such v.e.v.'s, $F_{\mu\nu}\langle 0|\bar{q}q|0\rangle^2$, $-g\langle 0|\bar{q}\sigma_{\alpha\beta}(\frac{1}{2}\lambda^n)G^n_{\alpha\beta}q|0\rangle\langle 0|\bar{q}\sigma_{\mu\nu}q|0\rangle_F = \sqrt{4\pi\alpha}\,e_q\chi m_0^2F_{\mu\nu}\langle 0|\bar{q}q|0\rangle^2$, $\sqrt{4\pi\alpha}\,e_q\kappa F_{\mu\nu}\langle 0|\bar{q}q|0\rangle^2$ and $i\sqrt{4\pi\alpha}\,e_q\xi F_{\mu\nu}\langle 0|\bar{q}q|0\rangle^2$. The corresponding diagrams are depicted in figs. 5–7. The estimate of 8-dimensional terms is important. Indeed, as will be seen, the contribution of a 6-dimensional term to the sum rules is rather large, even larger than the fig. 2 contribution since the magnitude of the quark condensate magnetic susceptibility χ is large. 8-dimensional terms (especially the term $\sim\chi m_0^2\langle 0|\bar{q}q|0\rangle^2$) can be considered as the first correction to the 6-dimensional term. Therefore their account is necessary since it permits checking the convergence of the operator expansion series. Of course, the results will be convincing (and really they are) if contributions of 8-dimensional terms are much smaller than those of $d=6$.

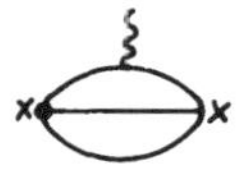

Fig. 2. The simplest quark loop contributing to the polarization operator. Here (and in the following figures) solid lines correspond to quarks, wave lines to the external electromagnetic field, crosses to quark currents η, $\bar{\eta}$.

* One of the authors (A.V.S.) has to warn the reader that the contribution of the v.e.v. $\langle 0|G^n_{\alpha\beta}G^n_{\alpha\beta}|0\rangle$ to $\Pi^{(0)}(p)$ found in ref. [13] differs from the correct value given in ref. [3] by the sign. The physics conclusions of the work in [13] do not change.

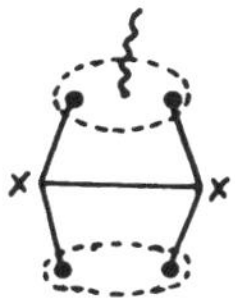

Fig. 3. The graph corresponding to the quark condensate magnetic susceptibility contribution to the odd structure of the polarization operator. Dotted lines encircle he places of non-perturbative interaction with the vacuum fields.

The lowest-dimension operator which contributes to chirality violating structures in $\Pi_{\mu\nu}(p)$ is $\bar{q}\sigma_{\mu\nu}q$ ($d=3$), see fig. 8. The next in dimension with $d=5$ are operator v.e.v.'s $\langle 0|\bar{q}q|0\rangle F_{\mu\nu}$ and those of eq. (9) corresponding to figs. 9 and 10. Assuming factorization again we are left in even structures with two 7-dimensional operator v.e.v.'s $\langle 0|\bar{q}\sigma_{\mu\nu}q|0\rangle_{\mathrm{F}}\langle 0|G^n_{\alpha\beta}G^n_{\alpha\beta}|0\rangle$ and $g\langle 0|\bar{q}\sigma_{\alpha\beta}(\frac{1}{2}\lambda^n)G^n_{\alpha\beta}q|0\rangle F_{\mu\nu}$. The related diagrams are shown in figs. 11 and 12. (Note that the higher the operator dimension the less reliable is the factorization hypothesis. We show in the following that for one of the tensor structures $-\sigma_{\mu\nu}$ the factorization hypothesis does not work well.)

As is seen, the interval of dimensions for the odd structure $d=2$–8 is larger than the interval of dimensions $d=3$–7 for even structures. Therefore one may expect a better accuracy of the results obtained from the sum rule at odd structure. As for even structure we shall use only the sum rule at the structure $i(p_\mu\gamma_\nu-p_\nu\gamma_\mu)\hat{p}$. The reasons are the following.

(i) The structure $i(p_\mu\gamma_\nu-p_\nu\gamma_\mu)\hat{p}$ contains two extra powers of momentum in the numerator in comparison with $\sigma_{\mu\nu}$. This results in the appearance of an extra factor $1/p^2$ in non-perturbative corrections. As a consequence, the Borel transformation

$$\mathscr{B}\frac{1}{(p^2)^n}=\frac{(-1)^n}{(M^2)^{n-1}(n-1)!},$$

brings about a factor $1/n$ in the higher-dimensional contribution to the structure $i(p_\mu\gamma_\nu-p_\nu\gamma_\mu)\hat{p}$ as compared to the structure $\sigma_{\mu\nu}$ which improves the convergence of the power series. At the same time, the role of the excited states (continuum) in the r.h.s. of the sum rule at the structure $i(p_\mu\gamma_\nu-p_\nu\gamma_\mu)\hat{p}$ reduces also as compared to that of $\sigma_{\mu\nu}$.

(ii) The sum rule for the structure $\sigma_{\mu\nu}$ includes infrared divergence which indicates the appearance of unknown infrared non-factorizable v.e.v.'s. Under the procedure used in the following the unknown higher-dimensional contributions affect rather significantly the result obtained from the sum rules for this structure.

Fig. 4. The gluon condensate contribution to the polarization operator; dashed lines correspond to gluons.

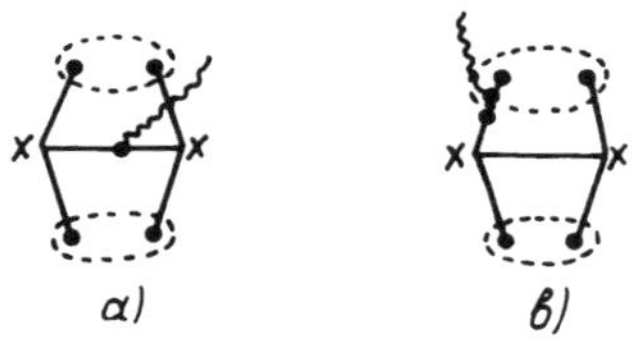

Fig. 5.

The calculation of the coefficients for the above operators is the subject of sect. 3. The remainder of this section we devote to the very important issue of saturating the sum rules by contributions of the physical states.

The part in the polarization operator $\Pi(p)$, linear in the field $F_{\mu\nu}$, can be presented phenomenologically as a sum of the contributions shown in figs. 13–15. Our interest is concentrated on the graph of fig. 13: the current $\bar{\eta}$ creates a nucleon which interacts with the electromagnetic field and is annihilated by the current η. The contribution of this graph to the polarization operator includes the double pole and is proportional to $\lambda_N^2 \mu/(p^2-m^2)^2$ where λ_N is the transition amplitude of the current η into the nucleon state, $\langle 0|\eta|N\rangle = \lambda_N v$, v is the nucleon spinor, μ is the nucleon magnetic moment (the total magnetic moment for structures $\hat{p}\sigma_{\mu\nu}+\sigma_{\mu\nu}\hat{p}$ and $\sigma_{\mu\nu}$, and the anomalous one for the structure $i(p_\mu\gamma_\nu - p_\nu\gamma_\mu)\hat{p}$). After the Borel transformation it gives the term $\sim\lambda_N^2\mu \exp(-m^2/M^2)M^{-2}$. Fig. 15 corresponds to transitions between excited states which will be approximated by a continuum in the usual way and are damped exponentially after the Borel transformation. The most dangerous for us are the graphs of fig. 14 which include a transition of the nucleon to excited states. Their contribution to $\Pi_{\mu\nu}$ is $\sim 1/(p^2-m^2)(p^2-m^{*2})$ where m^* is the mass of an excited state. The Borel transform of this expression includes the term $\sim(m^{*2}-m^2)^{-1}\exp(-m^2/M^2)$ which is not damped exponentially compared to the fig. 13 contribution. For realistic values of M^2 it is of the same order of magnitude and must be accounted for. The N→ excited state transitions can be taken into account by introducing two new phenomenological constants A_N and B_N

$$\langle 0|\eta|N\rangle_F = \int d^4x \, \langle 0|T\{j_\mu^{el}(x)A_\mu(x), \eta(0)\}|N\rangle - \text{pole term}$$

$$= \sqrt{4\pi\alpha}\,\tfrac{1}{4}\lambda_N[A_N\sigma_{\mu\nu} + iB_N(p_\mu\gamma_\nu - p_\nu\gamma_\mu)/m]F_{\mu\nu}v \,. \qquad (11)$$

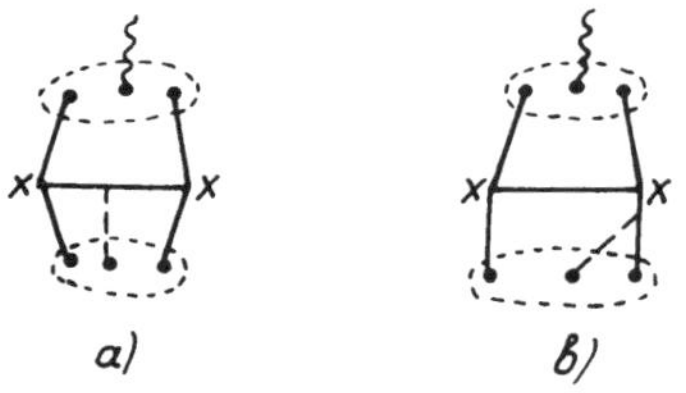

Fig. 6.

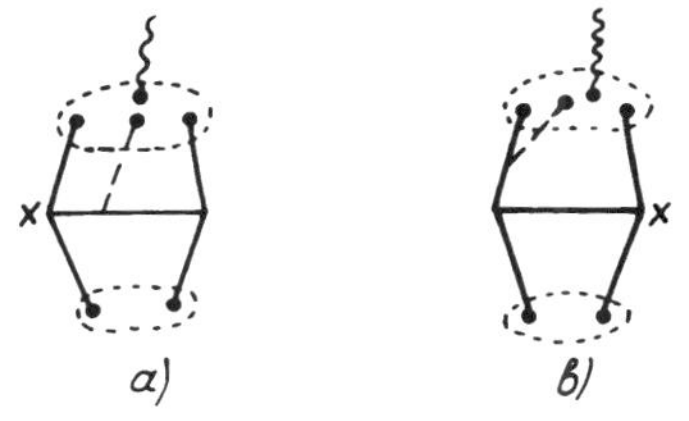

Fig. 7.

The necessity of these new phenomenological parameters corresponding to the transition nucleon $\rightarrow$ excited states in the presence of an external field can also be shown in some other way. Let us consider a three-point function $\Gamma(p^2, p'^2, q^2)$ in the limit $p^2 \rightarrow p'^2$, $q^2 \rightarrow 0$. At $q^2 < 0$ we can write for $\Gamma(p^2, p'^2, q^2)$ a double dispersion relation in p^2, p'^2

$$\Gamma(p^2, p'^2, q^2) = \int \frac{\rho(s, s', q^2)\, \mathrm{d}s\, \mathrm{d}s'}{(s-p^2)(s'-p'^2)} + P(p^2) f(p'^2, q^2) + P(p'^2) f(p^2, q^2)\,, \tag{12}$$

where $P(p^2)$ is a polynomial*. The function $f(p^2, q^2)(f(p'^2, q^2))$ can be represented by one-variable dispersion relations in $p^2(p'^2)$ with a pole contribution from the one-nucleon state. Thus, if $P \neq 0$, the last two terms in the r.h.s. of (12) in the limit $p^2 = p'^2$, $q^2 = 0$ bring about the discussed above new phenomenological constants into $\Gamma(p^2, p^2, 0) = \Pi(p^2)$. Suppose now that $P = 0$. Even in this case such constants arise. The function $\rho(s, s', q^2)$ can generally be written as a sum

$$\rho(s, s', q^2) = a(q^2)\delta(s-m^2)\delta(s'-m^2) + b(s', q^2)\delta(s-m^2) + b(s, q^2)\delta(s'-m^2) + C(s, s', q^2)\,, \tag{13}$$

where $C(s, s', q^2)$ does not have δ-singularities in s, s' at the nucleon mass. Evidently, the first term in the r.h.s. of (13) corresponds to fig. 13, the second and the third terms correspond to fig. 14. After substituting eq. (13) into eq. (12), taking the limit $p^2 = p'^2$, $q^2 = 0$ and applying the Borel transformation in p^2 we come to the

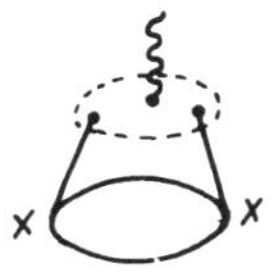

Fig. 8.

* It can be easily shown, considering the simplest triangle-loop diagram with arbitrary internal masses that at $q^2 < 0$ there are no anomalous thresholds and the double dispersion relation (12) holds.

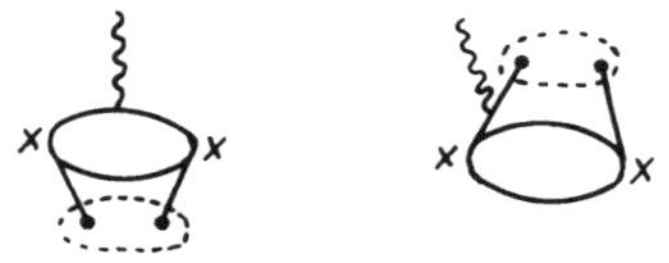

a) б)

Fig. 9.

result

$$\mathscr{B}\Pi(p^2) = a\,\mathrm{e}^{-m^2/M^2}\frac{1}{M^2} + 2\int_{m^{*2}}^{\infty} \mathrm{d}s\, b(s)\,\frac{1}{s-m^2}\,(\mathrm{e}^{-m^2/M^2} - \mathrm{e}^{-s/M^2})$$
$$- \int_{m^{*2}}^{\infty} \mathrm{d}s\,\mathrm{d}s'\,\frac{1}{s-s'}\,C(s,s')(\mathrm{e}^{-s/M^2} - \mathrm{e}^{-s'/M^2})\,. \qquad (14)$$

The term

$$2\,\mathrm{e}^{-m^2/M^2}\int \mathrm{d}s\,\frac{b(s)}{s-m^2}\,,$$

is not damped exponentially as compared with the first term in the r.h.s. of (14) and its contribution to $\mathscr{B}\Pi(p^2)$ is just of the form given by eq. (11).

In a recent work [14] an attempt was made to determine the coupling constants $g_{\pi NN}$, $g_{\pi N\Delta}$ etc. The authors employed the QCD sum rule technique applying the Borel transformation over the variable p^2 to the three-point functions $\Gamma(p^2, p^2, q^2)$. They used the standard continuum model and did not take into account the lowest state→excited state transitions. In our opinion this procedure is not theoretically founded. The effect of the lowest state→excited state transitions in an external field may fortunately prove to be small in some cases (a good agrement with experiment is reported in ref. [14] may reflect their actual smallness in the cases considered there) but one cannot expect it in advance. In our case the constants A and B contribute to sum rules with a vengeance. To ignore them would lead to a wrong result*.

a) б)

Fig. 10.

* Note that in the case where $|q^2|$ is reasonably large and $p^2 \neq p'^2$ the unwanted terms can be eliminated by a double Borel transformation [4–6] (see eqs. (12), (13)). In our case $q=0$, only one kinematic invariant is present and the double Borel transformation cannot be applied.

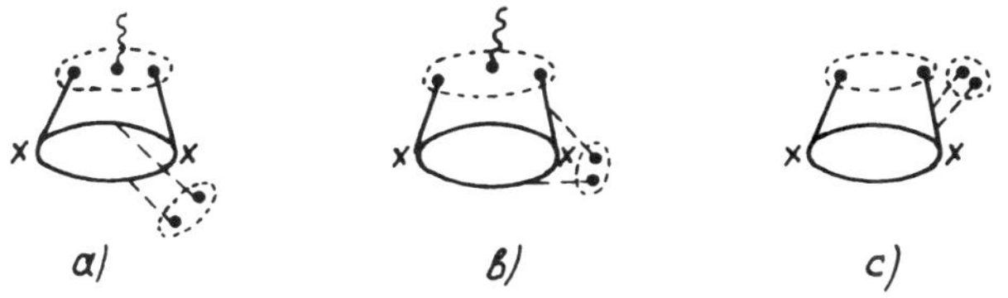

Fig. 11.

The same remark is relevant also for the recent attempts at calculating the nucleon magnetic moments in the lattice approach [15, 16] (see, however, [17]). We postpone the discussion of this problem for the last section.

3. Derivation of the sum rules

We are ready now to proceed with calculating the coefficients in the operator expansion of $\Pi_{\mu\nu}(p)$.

We use the fixed-point gauge for the electromagnetic field $x_\mu A_\mu(x)=0$. This gauge is very convenient for calculating operator expansion coefficients. (The gauge was originally suggested by Fock [18] and then reinvented independently by a number of people. The detailed description of the operator expansion coefficient calculation method in this gauge can be found in ref. [13].) In the fixed-point gauge the vector potential $A_\mu(x)$ is expressed via the field strength

$$A_\mu(x)=-\tfrac{1}{2}F_{\mu\nu}x_\nu\,. \tag{15}$$

(Generally, $A_\mu(x)$ is expressed as an infinite series over covariant derivatives of the field but in our case the field is constant and only the first term of the series survives.) The quark field $q(x)$ at small x can be expanded as

$$q(x)=q(0)+x_\alpha\nabla_\alpha q(0)+\tfrac{1}{2}x_\alpha x_\beta\nabla_\alpha\nabla_\beta q(0)+\cdots\,, \tag{16}$$

where $\nabla_\alpha=\partial_\alpha+i\sqrt{4\pi\alpha}\,e_q A_\alpha(x)$. In the following we use repeatedly the quark Green function in the external electromagnetic field. The answer written in the fixed-point gauge is

$$\begin{aligned}&\langle 0|\mathrm{T}\{q_i^a(x),\bar{q}_k^b(0)\}|0\rangle\\&=i\delta^{ab}\frac{\hat{x}_{ik}}{2\pi^2x^4}-\tfrac{1}{12}\delta^{ab}\delta_{ik}\langle 0|\bar{q}q|0\rangle+i\delta^{ab}\frac{\sqrt{4\pi\alpha}}{32\pi^2x^2}e_qF_{\alpha\beta}(\hat{x}\sigma_{\alpha\beta}+\sigma_{\alpha\beta}\hat{x})_{ik}\\&\quad-\tfrac{1}{24}\delta^{ab}(\sigma_{\alpha\beta})_{ik}\langle 0|\bar{q}\sigma_{\alpha\beta}q|0\rangle_F\\&\quad+\tfrac{1}{288}\delta^{ab}(\sigma_{\alpha\beta})_{ik}\sqrt{4\pi\alpha}\,e_q\langle 0|\bar{q}q|0\rangle(F_{\alpha\beta}x^2+2F_{\alpha\rho}x_\rho x_\beta)\,.\end{aligned} \tag{17}$$

Fig. 12.

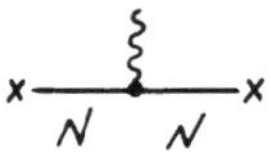

Fig. 13.

Here the first term is just the usual Green function of a free massless spinor particle, the second term corresponds to the presence of a non-polarized quark condensate. The third term reflects the action of the electromagnetic field on the particle; it can be obtained by solving the Dirac equation in an external potential $A_\mu(x)$ (15). The fourth term is due to the polarization of the quark condensate by an external field and the last term arises after expanding $q(x)$ up to the second order according to eq. (16).

For definiteness we consider the polarization operator of the proton currents $\Pi(\eta_\mathrm{p}, \bar{\eta}_\mathrm{p})$. We start with examining the odd-structure $\sigma_{\mu\nu}\hat{p}+\hat{p}\sigma_{\mu\nu}$. The lowest-order contribution is determined by the graph of fig. 2. Two of the quark correlators are given by the first term in eq. (17) while the third correlator by the third term. The result for $\Pi_{\mu\nu}(p)$ defined by (3) is

$$\Pi_{\mu\nu}(p)_{\mathrm{fig.}\,2} = -\frac{e_\mathrm{u}}{32\pi^4}(\sigma_{\mu\nu}\hat{p}+\hat{p}\sigma_{\mu\nu})p^2 \ln\frac{\Lambda^2}{-p^2}\,, \tag{18}$$

where Λ is the ultraviolet cut-off.

The contribution of graph fig. 3 corresponding to the (factorized) v.e.v. $\langle 0|\bar{q}q|0\rangle$ $\langle 0|\bar{q}\sigma_{\mu\nu}q|0\rangle_\mathrm{F}$ can be calculated in a similar way. The answer is

$$\Pi_{\mu\nu}(p)_{\mathrm{fig.}\,3} = -\frac{e_\mathrm{u}}{3p^2}\chi\langle 0|\bar{q}q|0\rangle^2(\sigma_{\mu\nu}\hat{p}+\hat{p}\sigma_{\mu\nu})\,, \tag{19}$$

where the notation of eqs. (8), (10) has been used. The fact that the r.h.s. of (19) is proportional to the u-quark charge can be easily understood. From the form of the η_p current it follows that the polarization operator contains a trace over u-quark indices. Therefore, according to chirality conservation if two $\bar{q}q$ operators are replaced by their v.e.v.'s, they must be both u-quark operators and not the one a u- and the other a d-quark operator. Since by our basic assumption $\langle 0|\bar{u}\sigma_{\mu\nu}u|0\rangle \sim e_\mathrm{u}$, the factor e_u arises.

The contribution of the 8-dimensional operator $\bar{q}q \cdot \bar{q}qF_{\mu\nu}$ is determined by the graphs of fig. 5. In fig. 5a the electromagnetic field acts on the hard (d) quark so that two quark correlators are given by the second term in eq. (17) and one (d-quark) correlator by the third term. In the graph of fig. 5b one of the u-quark fields is

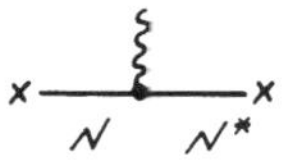

Fig. 14.

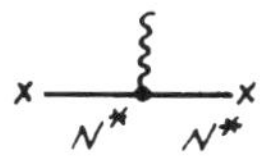

Fig. 15.

expanded in x up to the second order according to eq. (16) and the quark correlator is given by the last term in eq. (17). This corresponds to interaction of a soft u-quark with the electromagnetic field. The other u-quark correlator is given by the second term in (17) and the d-quark correlator by the first. The result of the calculation is

$$\Pi_{\mu\nu}(p)_{\text{fig.5}} = (e_{\text{d}} + \tfrac{2}{3}e_{\text{u}}) \frac{1}{6p^4} \langle 0|\bar{q}q|0\rangle^2 (\sigma_{\mu\nu}\hat{p} + \hat{p}\sigma_{\mu\nu}) . \tag{20}$$

The operator v.e.v.'s $g\langle 0|\bar{q}\sigma_{\alpha\beta}(\frac{1}{2}\lambda^n)G^n_{\alpha\beta}q|0\rangle\langle 0|\bar{q}\bar{\sigma}_{\mu\nu}q|0\rangle_{\text{F}}$, $g\langle 0|\bar{q}(\frac{1}{2}\lambda^n)G^n_{\mu\nu}q|0\rangle_{\text{F}} \times \langle 0|\bar{q}q|0\rangle$ and $\varepsilon_{\mu\nu\lambda\sigma}g\langle 0|\bar{q}\gamma_5(\frac{1}{2}\lambda^n)G^n_{\lambda\sigma}q|0\rangle_{\text{F}}\langle 0|\bar{q}q|0\rangle$ include vacuum gluon fields. The contribution of the first of them to the polarization operator corresponds to fig. 6, the second and the third to fig. 7. Technical complications arising in the calculation of these graphs are not serious and are described in appendix A. We present here only the results:

$$\Pi_{\mu\nu}(p)_{\text{fig.6}} = -\tfrac{1}{24}e_{\text{u}}\chi m_0^2 \langle 0|\bar{q}q|0\rangle^2 \frac{1}{p^4}(\sigma_{\mu\nu}\hat{p} + \hat{p}\sigma_{\mu\nu}) , \tag{21}$$

$$\Pi_{\mu\nu}(p)_{\text{fig.7}} = -\tfrac{1}{18}e_{\text{u}}(\kappa - 2\xi)\langle 0|\bar{q}q|0\rangle^2 \frac{1}{p^4}(\sigma_{\mu\nu}\hat{p} + \hat{p}\sigma_{\mu\nu}) , \tag{22}$$

in the notation of eqs. (6), (8), (9), (10).

Adding together all contributions given by eqs. (18)–(22) we obtain the final answer for the terms at odd structure in $\Pi_{\mu\nu}(p)$:

$$\Pi_{\mu\nu}(p)_{\text{odd}} = -\frac{1}{16\pi^4}(\sigma_{\mu\nu}\hat{p} + \hat{p}\sigma_{\mu\nu})\left\{\tfrac{1}{2}e_{\text{u}}p^2 \ln\frac{\Delta^2}{-p^2} \right.$$
$$\left. + \tfrac{1}{3}e_{\text{u}}\chi\frac{a^2}{p^2}\left(1 + \frac{m_0^2}{8p^2}\right) - \frac{a^2}{6p^4}[e_{\text{d}} + \tfrac{2}{3}e_{\text{u}} - \tfrac{1}{3}e_{\text{u}}(\kappa - 2\xi)]\right\} , \tag{23}$$

where

$$a = (2\pi)^2|\langle 0|\bar{q}q|0\rangle| . \tag{24}$$

As is seen from (23) the correction $\sim\chi m_0^2$ to the quark condensate magnetic susceptibility contribution is small at $|p^2| \sim 1\ \text{GeV}^2$ as compared to the term proportional to χ.

Coefficients in front of the odd-dimensional operators contributing to even structures are calculated in quite an analogous way. The lowest-order $d = 3$ contribution is determined by the graph of fig. 8. This term is proportional to the quark condensate magnetic susceptibility χ and the d-quark charge e_{d} because according to the form of the η_{p} current only d-quark v.e.v.'s can appear in the diagram of fig. 8. The v.e.v. $\langle 0|\bar{q}q|0\rangle F_{\mu\nu}$ contribution with $d=5$ is described by two diagrams of fig. 9. Fig. 9a corresponds to the case when a hard u-quark interacts with the electromagnetic field (the third term in eq. (17)) while fig. 9b corresponds to electromagnetic field interaction with a soft d-quark. The v.e.v.'s (9), $d=5$ are represented by the diagrams of fig. 10. In the first of these graphs the soft gluon is emitted by a hard quark, in the second by a soft quark. In the latter case the expansion of quark field operators in an external gluonic field must be used. The details of the calculation of fig. 10 contributions as well as 7-dimensional operator v.e.v.'s $\langle 0|\bar{q}\sigma_{\mu\nu}q|0\rangle_F \langle 0|G^n_{\alpha\beta}G^n_{\alpha\beta}|0\rangle$ are given in appendix A.

When attempting to calculate the coefficient in front of the operator $g\bar{q}\sigma_{\alpha\beta}(\frac{1}{2}\lambda^n)G^n_{\alpha\beta}qF_{\mu\nu}$ we encounter a new phenomenon. The graph contributing to the coefficient depicted in fig. 12 includes infrared divergence. The matter is that the Green function of a quark under the combined action of the electromagnetic and vacuum gluon field

$$S(k) = i\frac{\sqrt{4\pi\alpha}}{4k^6} e_q g G^n_{\alpha\beta}(\tfrac{1}{2}\lambda^n)F_{\rho\sigma}[\gamma_\beta\gamma_\sigma\gamma_\alpha\gamma_\rho + \gamma_\sigma\gamma_\beta\gamma_\rho\gamma_\alpha$$

$$-4k_\sigma\gamma_\beta\hat{k}\gamma_\alpha\gamma_\rho k^{-2} - 4k_\beta\gamma_\sigma\hat{k}\gamma_\rho\gamma_\alpha k^{-2}]\hat{k},$$

is $\sim k^{-5}$ when $k\to 0$ and the loop integral diverges logarithmically in $k^2\sim 0$ region. The influence of the small momenta region on the value of a coefficient indicates that the classification of the operators is not complete and new operators must come into play. In our case it is the four-fermion $d=7$ operators (e.g. $\langle 0|\bar{q}\lambda^n\sigma_{\mu\rho}q\bar{q}\lambda^n\gamma_\rho\nabla_\nu q|0\rangle_F - (\mu\leftrightarrow\nu)$). Such a v.e.v. cannot be estimated by any factorization hypothesis but its appearance shows that our estimate of $d=7$ contributions to $\Pi_{\mu\nu}(p)$ is not very reliable. We note however that the logarithmic divergence connected with the non-factorizable operator contribution enters the structure $\sigma_{\mu\nu}$ only and it can be hoped that for other structures our estimate of the $d=7$ operator contributions is more or less sound. It is important to note that the contributions of $d=7$ operators for which we have no reliable estimates, appear in the structure $i(p_\mu\gamma_\nu - p_\nu\gamma_\mu)\hat{p}$ multiplied by e_d (for the proton, see appendix A) and drop out by the procedure used in sect. 4 for determining magnetic moments. The same statement applies also for the (factorized) contribution of the 9-dimensional operator $\alpha_s\bar{q}\sigma_{\mu\nu}q\cdot\bar{q}q\cdot\bar{q}q$. Therefore these uncertainties affect only our estimate of χ and that is why the accuracy in determining χ is not as good as that in determining μ_p and μ_n. Omitting the details of the calculations we present here the final expression for the even structure contributions in $\Pi_{\mu\nu}(p)$:

$$\Pi_{\mu\nu}(p)_{\text{even}} = i\frac{a}{16\pi^4}(p_\mu\gamma_\nu - p_\nu\gamma_\mu)\hat{p}\left\{(e_u+\tfrac{1}{2}e_d)\frac{1}{p^2}\right.$$

$$\left.-\tfrac{1}{3}e_d\chi\left[\ln\frac{\Lambda^2}{-p^2}+\frac{\pi^2}{6p^4}\left\langle 0\left|\frac{\alpha_s}{\pi}G^n_{\alpha\beta}G^n_{\alpha\beta}\right|0\right\rangle\right]\right\}$$

$$-\frac{a}{16\pi^4}\sigma_{\mu\nu}\left\{\ln\frac{\Lambda^2}{-p^2}[e_u-\tfrac{1}{6}e_d(1+4\kappa-2\xi)]+\tfrac{1}{6}e_d\chi p^2\ln\frac{\Lambda^2}{-p^2}\right.$$

$$\left.-\frac{1}{6p^2}e_u m_0^2\ln\frac{-p^2}{\mu_0^2}\right\}, \quad (25)$$

where μ_0 is the infrared cut-off. Only the term with a logarithmic divergence in the coefficient $\sim m_0^2$ is shown. The coefficient of the operator v.e.v. $\langle 0|\bar{q}\sigma_{\mu\nu}q|0\rangle_F\langle 0|G^n_{\alpha\beta}G^n_{\alpha\beta}|0\rangle$ was calculated only for the structure $i(p_\mu\gamma_\nu-p_\nu\gamma_\mu)\hat{p}$ which we used in the following analysis.

The perturbative corrections $\sim\alpha_s^n$ must be added to expressions (23), (25). We take them into account in the leading logarithmic approximation. Anomalous dimensions of the current η and of the operators $\bar{q}q$, $\bar{q}\sigma_{\mu\nu}q$, $(\alpha_s/\pi)G^n_{\alpha\beta}G^n_{\alpha\beta}$ are known; they are $\frac{2}{9}$, $\frac{4}{9}$, $-\frac{4}{27}$ (see, e.g. [19]) and 0, respectively*. The anomalous dimension of the operator $g\bar{q}\sigma_{\alpha\beta}\lambda^n G^n_{\alpha\beta}q$ is small and we disregard it. Since the v.e.v.'s of eq. (9) are unknown and affect the results only weakly we put for simplicity their anomalous dimension equal to that of $\bar{q}q$.

After the Borel transformation (5) eqs. (23), (25) taking into account leading logarithmic corrections, give the l.h.s. the desired sum rules. In the r.h.s. of the sum rules we represent the polarization operator by the sum of the contributions of the one-proton state (double-pole terms), the proton-excited state transition amplitudes (single-pole terms) and the continuum. The one-proton contribution to $\Pi_{\mu\nu}(p)$ is given by

$$\Pi_{\mu\nu}(p)=-\frac{1}{4}\frac{\lambda_N^2}{(p^2-m^2)^2}\{\mu_p(\sigma_{\mu\nu}\hat{p}+\hat{p}\sigma_{\mu\nu})+2\sigma_{\mu\nu}m\mu_p$$

$$+\sigma_{\mu\nu}\mu_p^a(p^2-m^2)/m+2i\mu_p^a(p_\mu\gamma_\nu-p_\nu\gamma_\mu)\hat{p}/m\}, \quad (26)$$

where μ_p and μ_p^a are the proton total and anomalous magnetic moments in nuclear magneton units, λ_N is the proton transition amplitude in the current η_p as determined in refs. [2, 3]. The term resulting from the proton → excited state transition can easily be found using eq. (11). It is equal to

$$\Pi_{\mu\nu}(p)_{p\to N^*}=\tfrac{1}{4}\lambda_N^2\frac{1}{p^2-m^2}\{A_p(\sigma_{\mu\nu}\hat{p}+\hat{p}\sigma_{\mu\nu})$$

$$+2A_p\sigma_{\mu\nu}m+2iB_p(p_\mu\gamma_\nu-p_\nu\gamma_\mu)\hat{p}/m\}. \quad (27)$$

As usual, in the QCD sum rule approach the continuum contribution to the imaginary part of the polarization operator is represented by the imaginary part of the quark loops starting from some threshold value $p^2=W^2$. For convenience we transfer it into the l.h.s. of the sum rules.

* In refs. [2, 3] the sign of the anomalous dimensions of baryonic currents was wrong. (One of the authors (B.I) is indebted to Prof. H.G. Dosch for pointing out this error); taking into account this fact diminishes the residue λ_N^2 by ≈20% and affects only slightly the values of the baryonic masses.

Collecting all the pieces together we can now write the sum rules for the invariant functions for the structures $\sigma_{\mu\nu}\hat{p}+\hat{p}\sigma_{\mu\nu}$ and $i(p_\mu\gamma_\nu-p_\nu\gamma_\mu)\hat{p}$:

$$e_u M^4 E_1(M)L^{-4/9}+\frac{a^2}{3M^2}L^{4/9}[-(e_d+\tfrac{2}{3}e_u)+\tfrac{1}{3}e_u(\kappa-2\xi)$$

$$-2e_u\chi(M^2L^{-16/27}-\tfrac{1}{8}m_0^2L^{-28/27})]=\tfrac{1}{4}\tilde{\lambda}_N^2\,e^{-m^2/M^2}\left(\frac{\mu_p}{M^2}+A_p\right) \tag{28}$$

$$ma\left\{e_u+\tfrac{1}{2}e_d+\tfrac{1}{3}e_d\chi M^2\left[E_0(M)+\frac{b}{24M^4}\right]L^{-16/27}\right\}$$

$$=\tfrac{1}{4}\tilde{\lambda}_N^2\,e^{-m^2/M^2}\left(\frac{\mu_p^a}{M^2}+B_p\right), \tag{29}$$

where

$$E_0(M)=1-\exp(-W^2/M^2)\,,\qquad E_1(M)=1-\left(1+\frac{W^2}{M^2}\right)\exp\left(\frac{-W^2}{M^2}\right),$$

$$L=\ln(M/\Lambda_{QCD})/\ln(\mu/\Lambda_{QCD})\,,\qquad \tilde{\lambda}_N^2=32\pi^4\lambda_N^2\,,$$

$$b=(2\pi)^2\langle 0|(\alpha_s/\pi)G^n_{\alpha\beta}G^n_{\alpha\beta}|0\rangle\,.$$

Λ_{QCD} is the QCD parameter, $\Lambda_{QCD}\approx 100$ MeV, $\mu\approx 0.5$ GeV is the normalization point to which the used values of v.e.v.'s are referred and W is the continuum threshold, $W=1.4$–1.5 GeV for the nucleon channel [3]. For completeness we present also the sum rule for the invariant function for the structure $\sigma_{\mu\nu}$

$$aM^2\left\{[e_u-\tfrac{1}{6}e_d(1+4\kappa-2\xi)]E_0(M)+\tfrac{1}{6}e_u\frac{m_0^2}{M^2}\ln\frac{M^2}{\mu_0^2}L^{-4/9}\right.$$

$$\left.+\tfrac{1}{6}e_d M^2\chi E_1(M)L^{-16/27}\right\}=\tfrac{1}{4}\tilde{\lambda}_N^2 m\,e^{-m^2/M^2}\left[\frac{\mu_p}{M^2}-\frac{\mu_p^a}{2M^2}+A_p\right]. \tag{30}$$

The sum rules for the neutron are obtained from eqs. (28)–(30) by substituting $e_u\leftrightarrow e_d$; μ_p, $\mu_p^a\to\mu_n$; A_p, $B_p\to A_n$, B_n.

4. Determination of proton and neutron magnetic moments and the quark condensate magnetic susceptibility

In our analysis we use only the sum rules (28), (29). The sum rule (30) is disregarded for the reason explained in sect. 2. The sum rules (28), (29) include a lot of unknown parameters besides μ_p: χ, κ and ξ in the l.h.s. and A_p, B_p in the r.h.s. (For the neutron additional parameters A_n, B_n appear.) At first sight they present a serious nuisance and it seems impossible to find magnetic moments from these sum rules with a good accuracy. But really the situation is not so bad.

Note that the parameters χ, κ and ξ enter sum rule (28) multiplied by the u-quark charge $e_{\rm u}$ while in sum rule (29) they are multiplied by $e_{\rm d}$. This is a direct consequence of our neglecting closed quark loops and assuming that the magnetic susceptibility of quark condensates with a given flavour is proportional to the quark charge (see sect. 2). This circumstance makes it possible to get rid of the parameters χ, κ, ξ and also of the chirality violating v.e.v.'s of higher-dimension biquark operators induced by the action of the electromagnetic field.

Let us multiply the sum rule (28) for the proton by $e_{\rm d}$ and for the neutron by $e_{\rm u}$ and subtract the one from the other. Similarly, we multiply the sum rule (29) for the proton by $e_{\rm u}$ and for the neutron by $e_{\rm d}$ and subtract the one from the other. The obtained expressions can be represented in the form

$$\mu_{\rm p} e_{\rm d} - \mu_{\rm n} e_{\rm u} + M^2 (A_{\rm p} e_{\rm d} - A_{\rm n} e_{\rm u}) = \frac{4a^2}{3\tilde{\lambda}_{\rm N}^2} {\rm e}^{m^2/M^2} (e_{\rm u}^2 - e_{\rm d}^2) L^{4/9} , \qquad (31a)$$

$$\mu_{\rm p}^{\rm a} e_{\rm u} - \mu_{\rm n} e_{\rm d} + M^2 (B_{\rm p} e_{\rm u} - B_{\rm n} e_{\rm d}) = \frac{4amM^2}{\tilde{\lambda}_{\rm N}^2} {\rm e}^{m^2/M^2} (e_{\rm u}^2 - e_{\rm d}^2) . \qquad (31b)$$

In order to eliminate unknown single-pole contributions still remaining in the l.h.s. of eqs. (31) we apply the differential operator $1 - M^2\, \partial/\partial M^2$ to eqs. (31) and obtain

$$\mu_{\rm p} e_{\rm d} - \mu_{\rm n} e_{\rm u} = \frac{4a^2}{3\tilde{\lambda}_{\rm N}^2} (e_{\rm u}^2 - e_{\rm d}^2) \left(1 - M^2 \frac{\partial}{\partial M^2}\right) {\rm e}^{m^2/M^2} L^{4/9} , \qquad (32a)$$

$$\mu_{\rm p} e_{\rm u} - \mu_{\rm n} e_{\rm d} = e_{\rm u} + \frac{4am}{\tilde{\lambda}_{\rm N}^2} (e_{\rm u}^2 - e_{\rm d}^2) \left(1 - M^2 \frac{\partial}{\partial M^2}\right) M^2 {\rm e}^{m^2/M^2} . \qquad (32b)$$

The magnetic moments $\mu_{\rm p}$ and $\mu_{\rm n}$ can be approximately determined by setting $M = m$, disregarding anomalous dimensions and substituting for the residue $\tilde{\lambda}_{\rm N}^2$ the value

$$\tilde{\lambda}_{\rm N}^2 = \frac{2aM^4}{m} {\rm e}^{m^2/M^2} |_{M^2 = m^2} , \qquad (33)$$

which follows from the mass sum rules neglecting both anomalous dimensions and continuum contributions [2]. Solving in this approximation eqs. (32) we arrive at the quoted result of eq. (1).

We turn now to a more rigorous treatment. To do this we need the precise value of the residue $\tilde{\lambda}_{\rm N}^2$. If the nucleon mass is not a free parameter, as it was supposed to be in [2, 3] but is fixed at the experimental value $m_{\rm N} = 0.94$ GeV, the best fit is achieved by the choice $\tilde{\lambda}_{\rm N}^2 = 2.1\ {\rm GeV}^6$, $W^2 = 2.3\ {\rm GeV}^2$ (see appendix B). The solutions of eqs. (32a, b) are plotted versus M^2 in fig. 16. The variation of $\mu_{\rm p}$ and $\mu_{\rm n}$ with the change of M^2 reflects the uncertainty of our predictions. A reliable interval for M^2 is $0.8\ {\rm GeV}^2 \leqslant M^2 \leqslant 1.4\ {\rm GeV}^2$ where both the continuum contribution and power corrections are still controlable. Though the continuum does not contribute directly to the r.h.s. of eqs. (32) its cancellation is a consequence of our model of

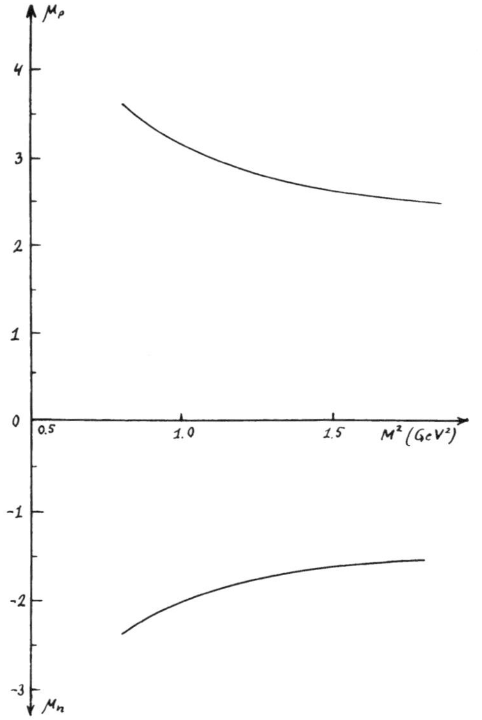

Fig. 16. The Borel parameter M^2 dependence of proton and neutron magnetic moments determined from eqs. (32).

the continuum which may not work if this contribution would be large. Besides, at large M^2 in the r.h.s. of the sum rules, single-pole terms play the main role and the procedure of extracting μ_p, μ_n in the presence of a large background is inadequate. Note also that changes in the magnitude of $\tilde{\lambda}_N^2$ do not strongly affect the values of the magnetic moments due to the term e_u in the r.h.s. of eqs. (32b) arising from the difference between the total and anomalous magnetic moments. Our final results for the nucleon magnetic moments are $\mu_p = 3.0$, $\mu_n = -2.0(\pm 10\%)$.

To estimate the value of the magnetic susceptibility we multiply the sum rules (28), (29) by $M^2 \exp(m^2/M^2)$ and apply the differential operator $1 - M^2\, \partial/\partial M^2$. The values μ_p, μ_p^a in the r.h.s. of the obtained equations depend strongly on χ. If setting $\kappa = \xi = 0$, the sum rule (28) is best fitted when $\chi = -8\ \text{GeV}^{-2}$, see fig. 17. The influence of κ and ξ on the estimate of χ is not large: $d\chi/d\kappa = -\frac{1}{2} d\chi/d\xi \approx \frac{1}{3}\ \text{GeV}^{-2}$. The sum rule (29) is best fitted when $\chi \approx -14\ \text{GeV}^{-2}$ but this estimate is less reliable because of a substantial $\chi\langle 0|G^n_{\alpha\beta} G^n_{\alpha\beta}|0\rangle$ contribution effecting the result. The matter is that the proportional to χ term in the sum rule (29) appears to be small after applying the operator $1 - M^2\, \partial/\partial M^2$. (This can be readily understood if one notes that neglecting anomalous dimensions and the continuum, its contribution to μ_p^a at the point $M^2 = m^2$ is $[(1 - M^2\, \partial/\partial M^2) M^4 \exp(m^2/M^2)]_{M^2 = m^2} = 0$.) The esti-

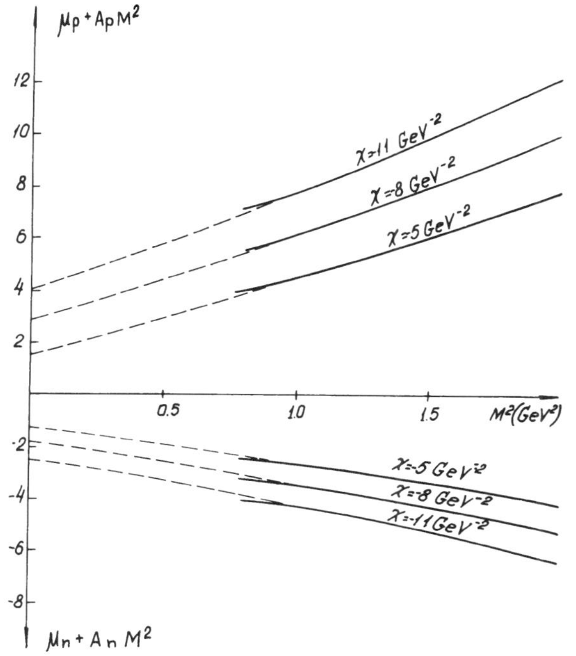

Fig. 17. The M^2 and χ dependence of $\mu_p + A_p M^2$, $\mu_n + A_n M^2$ (at $\kappa = \xi = 0$) from eq. (28) (solid lines). Crosses on the ordinate axis show the extrapolated lines of μ_p and μ_n.

mate for 7-dimensional contributions in the sum rule (29) is not very reliable because of both a great number of unaccounted non-factorizable operators (see appendix A) and uncertainty in the gluon condensate value (the estimated uncertainty in the value $\langle 0|(\alpha_s/\Pi) G^n_{\alpha\beta} G^n_{\alpha\beta}|0\rangle = 0.012\ \mathrm{GeV}^4$ obtained in ref. [1] is $\sim 50\%$). If we assume the contribution of non-factorizable four-fermion operators in the structure $i(p_\mu \gamma_\nu - p_\nu \gamma_\mu)\hat{p}$ to be small the only 7-dimensional correction to the sum rule (32b) for magnetic moments is due to the operator $g\bar{q}\sigma_{\alpha\beta}\lambda^n G^n_{\alpha\beta} q F_{\mu\nu}$. It brings about the factor $1 - (m_0^2/4M^2)L^{-4/9}$ in the second term in the r.h.s. of eq. (32b) (infrared logarithmic divergence present in the sum rule for the structure $\sigma_{\mu\nu}$ is absent here but the finite piece remains). This correction reduces the predicted absolute values for μ_p and μ_n by $\sim$10–15% and neither improves nor spoils accordance with experiment. We believe the account for this correction exceeds our accuracy as non-factorizable operators and also the uncertainty in our knowledge of the effective continuum thresholds (we assumed them to be the same for both the proton and neutron) and the residue value may bring about comparable corrections.

Significantly smaller values $|\chi| < 5\ \mathrm{GeV}^2$ are in sharp disagreement with the sum rule (29) and in order for the sum rule (28) to be fulfilled improbably large $|\kappa|$ and (or) $|\xi|$ are required. (Thus, e.g. at $\chi \approx -3\ \mathrm{GeV}^{-2}$, $\kappa - 2\xi \approx 8$.) In addition, the sum rule (28) is least satisfied as a function of M^2.

Our final result for the quark condensate magnetic susceptibility is $\chi = -(350 \pm 50\ \text{MeV})^{-2}$. As shown in sect. 5 this value is in agreement with the independent determination of χ from the sum rules for the $\Delta \to p\gamma$ transition.

The sign of the quark condensate magnetic susceptibility is "natural" in the following sense. Let us consider a simple model where the current quark in the vacuum fluctuation field acquires a mass and the resulting constituent quarks interacts with the electromagnetic field. In such a model v.e.v.

$$\langle 0|\bar{q}\sigma_{\mu\nu}q|0\rangle = \sqrt{4\pi\alpha}\, e_q \chi F_{\mu\nu} \langle 0|\bar{q}q|0\rangle$$
$$= -\int \mathrm{d}^4 p\, \mathrm{Tr}\{S(p)\sigma_{\mu\nu}\}\,, \tag{34}$$

where $S(p)$ is the constituent quark Green function in the electromagnetic field:

$$S(p) = \frac{i}{(2\pi)^4}\left\{\frac{1}{\hat{p}-m_q} - \sqrt{4\pi\alpha}\,(\tfrac{1}{2}e_q)F_{\mu\nu}\left[\frac{i}{\hat{p}-m_q}\gamma_\mu \frac{1}{\hat{p}-m_q}\gamma_\nu \frac{1}{\hat{p}-m_q}\right.\right.$$
$$\left.\left. - \frac{\mu_q^a}{2m_q}\frac{1}{\hat{p}-m_q}\sigma_{\mu\nu}\frac{1}{\hat{p}-m_q}\right]\right\}, \tag{35}$$

m_q is the quark mass, μ_q^a is its anomalous magnetic moment (in units $e_q/2m_q$). Substitution of eq. (35) into eq. (34) gives

$$\chi\langle 0|\bar{q}q|0\rangle = \frac{3}{4\pi^2} m_q \ln\left(\frac{\Lambda^2}{m_q^2}\right)(1+\tfrac{1}{2}\mu_q^a)\,, \tag{36}$$

where Λ is the ultraviolet cut-off. Since the r.h.s. of eq. (36) is positive and $\langle 0|\bar{q}q|0\rangle < 0$, χ must be negative. The numerical value of χ following from eq. (36) appears also to be rather reasonable; the value (2) for χ is obtained, for instance, at $m_q = 350$ MeV, $\ln(\Lambda^2/m_q^2) \approx 2$ and $\mu_q^a \approx 2$. Note that the same model gives also the correct sign of the quark condensate density.

The phenomenological constants A_N and B_N can also be found. They are equal to $A_p \approx 3.3\ \text{GeV}^{-2}$, $A_n \approx -1.5\ \text{GeV}^{-2}$, $B_p \approx 2.3\ \text{GeV}^{-2}$, $B_n \approx -3.7\ \text{GeV}^{-2}$. In the region $M^2 \sim 1\ \text{GeV}^2$ the single-pole term contribution to the sum rules (28), (29) is approximately as large as that of the terms proportional to the magnetic moments.

Concluding this section, a few words on the sum rules for structure $\sigma_{\mu\nu}$. As was mentioned, there is no reason to expect these sum rules to be satisfied properly. Nevertheless, it is of interest to study these sum rules and substituting into them the above obtained value of χ. In doing so it is reasonable to use a sum rule independent of the infrared cut-off parameter μ_0^2 which is obtained by applying the differential operator $\partial/\partial(1/M^2)$ to eq. (30). Unlike (28), (29), the sum rules for $\sigma_{\mu\nu}$ depend strongly on κ and ξ and suitably choosing κ and ξ one can get a more or less satisfactory agreement between the r.h. and l.h. sides in both sum rules for $\sigma_{\mu\nu}$ for the proton and neutron. Such a fit is, for instance, possible for the values of the parameters (which also fit eqs. (28), (29)) $\kappa \approx -3.2$, $\xi \approx -1.5$, $\chi \approx -8\ \text{GeV}^2$. This set of parameters leads, however, to a considerably worse agreement of the sum rules for the transition $\Delta \to p\gamma$.

5. $\Delta \to p\gamma$ transition

The method developed in this paper can be applied to many other problems besides the determination of nucleon magnetic moments. As an example we present here an estimate of invariant amplitudes of the $\Delta \to p + \gamma$ transition.

This consideration pursues two purposes: first to convince ourselves that the method can be expanded to solutions of other problems and second, to determine independently the value of the quark condensate magnetic susceptibility.

Consider the polarization operator

$$\Pi_\alpha(p) = i \int d^4x \, e^{ipx} \langle 0|T\{\eta_p(x), \bar{\eta}_\alpha(0)\}|0\rangle , \tag{37}$$

where the current η_p is determined by eq. (4a) and

$$\eta_\alpha = \sqrt{\tfrac{1}{3}}[2(u^a C\gamma_\mu d^b)u^c + (u^a C\gamma_\mu u^b)d^c]\varepsilon^{abc} , \tag{38}$$

is a current with quantum numbers of the isobar Δ^+. Write $\Pi_\alpha(p)$ in a form analogous to eq. (3):

$$\Pi_\alpha(p) = \Pi_\alpha^{(0)}(p) + \sqrt{4\pi\alpha}\,\Pi_{\alpha\mu\nu}(p)F_{\mu\nu} . \tag{39}$$

When calculating $\Pi_{\alpha\mu\nu}(p)$ we take into account the graphs of figs. 2, 3, 5 and 7 for chirality conserving and figs. 8, 9 and 10 for chirality violating structures. (In this case the chirality conserving structures contain even, and chirality violating ones – odd, number of γ-matrices.) We shall not dwell on the calculational details but give the results.

The chirality conserving structures

$$\begin{aligned}\Pi_{\alpha\mu\nu}(p)_{\text{fig.2}} = -\frac{1}{4\sqrt{3}}\frac{1}{(2\pi)^4}\{[&2\sigma_{\mu\nu}\gamma_\alpha\hat{p} - 3\sigma_{\mu\nu}p_\alpha \\ &+ \tfrac{7}{2}i(\delta_{\mu\alpha}\gamma_\nu - \delta_{\nu\alpha}\gamma_\mu)\hat{p} + 4i(\delta_{\nu\alpha}p_\mu - \delta_{\mu\alpha}p_\nu) \\ &- \tfrac{5}{2}i(p_\mu\gamma_\nu - p_\nu\gamma_\mu)\gamma_\alpha]p^2 - i(p_\mu\gamma_\nu - p_\nu\gamma_\mu)p_\alpha\hat{p}\}\gamma_5 \ln\frac{\Lambda^2}{-p^2} ,\end{aligned} \tag{40}$$

$$\begin{aligned}\Pi_{\alpha\mu\nu}(p)_{\text{fig.3}} = -\frac{1}{6\sqrt{3}}\chi\langle 0|\bar{q}q|0\rangle^2 \frac{1}{p^2}\{&3\sigma_{\mu\nu}\gamma_\alpha\hat{p} - 2\sigma_{\mu\nu}p_\alpha - 2i(p_\mu\gamma_\nu - p_\nu\gamma_\mu)\gamma_\alpha \\ &+ 4i(\delta_{\mu\alpha}\gamma_\nu - \delta_{\nu\alpha}\gamma_\mu)\hat{p}\}\gamma_5 ,\end{aligned} \tag{41}$$

$$\begin{aligned}\Pi_{\alpha\mu\nu}(p)_{\text{fig.5}} = -\frac{1}{6\sqrt{3}}\langle 0|\bar{q}q|0\rangle^2 \frac{1}{p^4}\{&2\sigma_{\mu\nu}\gamma_\alpha\hat{p} - \tfrac{4}{3}\sigma_{\mu\nu}p_\alpha + \tfrac{8}{3}i(\delta_{\mu\alpha}\gamma_\nu - \delta_{\nu\alpha}\gamma_\mu)\hat{p} \\ &+ \tfrac{16}{3}i(\delta_{\nu\alpha}p_\mu - \delta_{\mu\alpha}p_\nu) - \tfrac{8}{3}i(p_\mu\gamma_\nu - p_\nu\gamma_\mu)\gamma_\alpha\}\gamma_5 ,\end{aligned} \tag{42}$$

$$\Pi_{\alpha\mu\nu}(p)_{\text{fig.7}} = -\frac{1}{12\sqrt{3}}\langle 0|\bar{q}q|0\rangle^2 \frac{1}{p^4}\{[\sigma_{\mu\nu}\gamma_\alpha\hat{p} - \tfrac{2}{3}\sigma_{\mu\nu}p_\alpha + \tfrac{4}{3}i(\delta_{\mu\alpha}\gamma_\nu - \delta_{\nu\alpha}\gamma_\mu)\hat{p}](\kappa - 2\xi)$$
$$+\tfrac{20}{3}i(\delta_{\nu\alpha}p_\mu - \delta_{\mu\alpha}p_\nu)(\kappa - \tfrac{1}{2}\xi) - \tfrac{7}{3}i(p_\mu\gamma_\nu - p_\nu\gamma_\mu)(\kappa - \tfrac{13}{14}\xi)\}\gamma_5 . \tag{43}$$

The chirality violating structures are:

$$\Pi_{\alpha\mu\nu}(p)_{\text{fig.8}} = -\sqrt{\tfrac{1}{3}}\frac{1}{16\pi^2}\langle 0|\bar{q}q|0\rangle\Big\{-[\sigma_{\mu\nu}\gamma_\alpha + 2i(\delta_{\mu\alpha}\gamma_\nu - \delta_{\nu\alpha}\gamma_\mu)]\ln\frac{\Lambda^2}{-p^2}$$
$$+\frac{2}{p^2}[2\sigma_{\mu\nu}p_\alpha\hat{p} + \tfrac{3}{2}i(p_\mu\gamma_\nu - p_\nu\gamma_\mu)\gamma_\alpha\hat{p}$$
$$-2i(\delta_{\nu\alpha}p_\mu - \delta_{\mu\alpha}p_\nu)\hat{p} + i(p_\mu\gamma_\nu - p_\nu\gamma_\mu)p_\alpha\Big\}\gamma_5 , \tag{44}$$

$$\Pi_{\alpha\mu\nu}(p)_{\text{fig.9}} = -\sqrt{\tfrac{1}{3}}\frac{1}{16\pi^2}\langle 0|\bar{q}q|0\rangle\Big\{-2[\sigma_{\mu\nu}\gamma_\alpha + 2i(\delta_{\mu\alpha}\gamma_\nu - \delta_{\nu\alpha}\gamma_\mu)]\ln\frac{\Lambda^2}{-p^2}$$
$$+[\tfrac{8}{3}\sigma_{\mu\nu}p_\alpha\hat{p} + \tfrac{8}{3}i(p_\mu\gamma_\nu - p_\nu\gamma_\mu)\gamma_\alpha\hat{p} - \tfrac{16}{3}i(\delta_{\nu\alpha}p_\mu - \delta_{\mu\alpha}p_\nu)\hat{p}]\frac{1}{p^2}\Big\}\gamma_5 , \tag{45}$$

$$\Pi_{\alpha\mu\nu}(p)_{\text{fig.10}} = \frac{1}{3\sqrt{3}}\frac{1}{16\pi^2}\langle 0|\bar{q}q|0\rangle\frac{2}{p^2}\{2\sigma_{\mu\nu}p_\alpha\hat{p}(\kappa + \xi) - i(p_\mu\gamma_\nu - p_\nu\gamma_\mu)\gamma_\alpha\hat{p}(\kappa - 2\xi)$$
$$+3i(p_\mu\gamma_\nu - p_\nu\gamma_\mu)p_\alpha(\kappa + \tfrac{1}{2}\xi) + 5i(\delta_{\nu\alpha}p_\mu - \delta_{\mu\alpha}p_\nu)\hat{p}(\kappa - \tfrac{1}{2}\xi)\}\gamma_5 . \tag{46}$$

Note that because of the relation $\gamma_\alpha\eta_\alpha = 0$ $\Pi_{\alpha\mu\nu}$ satisfies the conditions

$$\Pi_{\alpha\mu\nu}(p)\gamma_\alpha = 0 . \tag{47}$$

The r.h.s of the desired sum rules is represented as a sum of the term of interest which has two poles at $p^2 = m^2$ and at $p^2 = m_\Delta^2$, and of terms with single poles at $p^2 = m^2$ or $p^2 = m_\Delta^2$ and of the continuum contribution. The electromagnetic transition $\Delta \to \mathrm{p} + \gamma$ amplitude for the case of the real photon is described by two constants G_1 and G_2

$$\int d^4x\,\langle p|j_\mu(x)A_\mu(x)|\Delta\rangle = i\bar{v}(\gamma_\mu G_1 - p_\mu G_2)\gamma_5 v_\alpha F_{\mu\alpha} , \tag{48}$$

which are linear combinations of magnetic dipole G_{M} and electric quadrupole G_{E} moments of transition

$$G_{\mathrm{M}} = \frac{m}{3m_\Delta}[(3m_\Delta + m)G_1 + m_\Delta(m_\Delta - m)G_2] ,$$

$$G_{\mathrm{E}} = \frac{m}{3m_\Delta}(m_\Delta - m)(G_1 + m_\Delta G_2) , \tag{49a}$$

(see [20]. In eq. (48) v and v_α are spinors of the proton and the Δ isobar, respectively.)

Experimental works present, as a rule, the data in terms of helicity amplitudes which are related to G_E and G_M by

$$A_{3/2} = -\sqrt{\frac{3\alpha\pi m_\Delta q}{4m^3}}(G_E + G_M)\,,$$

$$A_{1/2} = \sqrt{\frac{\alpha\pi m_\Delta q}{4m^3}}(3G_E - G_M)\,, \tag{49b}$$

where $q = (m_\Delta^2 - m^2)/2m_\Delta$. Using (48) one can easily write the general expression for the polarization operator contribution due to the $\Delta \to p$ transition as

$$\begin{aligned}
\Pi_{\alpha\mu\nu}(p)_{\Delta\to p} = {} & -\frac{\lambda_N\lambda_\Delta}{(p^2-m^2)(p^2-m_\Delta^2)}\Bigg\{\Bigg[\tfrac{1}{3}\sigma_{\mu\nu}\gamma_\alpha\hat{p}(m+m_\Delta) \\
& -\tfrac{1}{3}\sigma_{\mu\nu}p_\alpha\left(2m_\Delta + m - \frac{p^2}{m_\Delta}\right) - \tfrac{1}{2}i(\delta_{\nu\alpha}\gamma_\mu - \delta_{\mu\alpha}\gamma_\nu)\hat{p}(m+m_\Delta)\Bigg]G_1 \\
& + \tfrac{1}{2}i(\delta_{\nu\alpha}p_\mu - \delta_{\mu\alpha}p_\nu)[2m_\Delta G_1 - (mm_\Delta - p^2)G_2] \\
& + \tfrac{1}{6}i(p_\mu\gamma_\nu - p_\nu\gamma_\mu)\gamma_\alpha\left[-\left(4m_\Delta + m - \frac{p^2}{m_\Delta}\right)G_1 + (mm_\Delta - p^2)G_2\right] \\
& + \tfrac{1}{6}i(p_\mu\gamma_\nu - p_\nu\gamma_\mu)p_\alpha\hat{p}\frac{1}{m_\Delta^2}[-2mG_1 + (m_\Delta - m)G_2] \\
& + \Bigg[-\tfrac{1}{3}\sigma_{\mu\nu}\gamma_\alpha(p^2 + mm_\Delta) + \tfrac{1}{3}\sigma_{\mu\nu}p_\alpha\hat{p}\left(1 + \frac{m}{m_\Delta}\right) \\
& + \tfrac{1}{2}i(\delta_{\nu\alpha}\gamma_\mu - \delta_{\mu\alpha}\gamma_\nu)(p^2 + mm_\Delta)\Bigg]G_1 \\
& - \tfrac{1}{2}i(\delta_{\nu\alpha}p_\mu - \delta_{\mu\alpha}p_\nu)\hat{p}[2G_1 + (m_\Delta - m)G_2] \\
& + \tfrac{1}{2}i(p_\mu\gamma_\nu - p_\nu\gamma_\mu)\gamma_\alpha\hat{p}\left[\left(1 + \frac{1}{3}\frac{m}{m_\Delta}\right)G_1 + \tfrac{1}{3}(m_\Delta - m)G_2\right] \\
& + \tfrac{1}{6}i(p_\mu\gamma_\nu - p_\nu\gamma_\mu)p_\alpha\Bigg[-2\left(1 - \frac{p^2}{m_\Delta^2} - \frac{m}{m_\Delta}\right)G_1 \\
& - \left(2m_\Delta - m - \frac{p^2}{m_\Delta}\right)G_2\Bigg]\Bigg\}\gamma_5\,,
\end{aligned} \tag{50}$$

where λ_Δ is the Δ-transition amplitude in the current η_α $\langle 0|\eta_\alpha|\Delta\rangle = \lambda_\Delta v_\alpha$ found in ref. [3].

The single-pole terms at $p^2 = m^2$ contribute to an independent coefficient in each structure and we do not write them down. (Eqs. (47) establish certain relationships between these coefficients and G_1, G_2, but no benefit can be obtained from them.)

Because of conditions to be satisfied by the isobar propagator $\gamma_\alpha S^{-1}_{\alpha\beta}(p) = p_\alpha S^{-1}_{\alpha\beta}(p) = 0$ the single-pole terms at $p^2 = m_\Delta^2$ are described by two independent constants only and their contribution to $\Pi_{\alpha\mu\nu}(p)$ is

$$\Pi_{\alpha\mu\nu}(p)_{\Delta\to N^*} = \frac{\lambda_N\lambda_\Delta}{p^2 - m_\Delta^2}\Big\{-\tfrac{2}{3}\sigma_{\mu\nu}(\gamma_\alpha\hat{p} - p_\alpha)g + i(\delta_{\nu\alpha}\gamma_\mu - \delta_{\mu\alpha}\gamma_\nu)\hat{p}g$$

$$+\tfrac{1}{3}i(p_\mu\gamma_\nu - p_\nu\gamma_\mu)\frac{p_\alpha\hat{p}}{m_\Delta^2}(2g + m_\Delta h) + \tfrac{1}{3}i(p_\mu\gamma_\nu - p_\nu\gamma_\mu)\gamma_\alpha(g - m_\Delta h)$$

$$+ i\left(\delta_{\nu\alpha}p_\mu - \delta_{\mu\alpha}p_\nu)m_\Delta h + \tfrac{2}{3}\sigma_{\mu\nu}m_\Delta\left(\gamma_\alpha - p_\alpha\frac{\hat{p}}{m_\Delta^2}\right)g\right.$$

$$+ i(\delta_{\nu\alpha}\gamma_\mu - \delta_{\mu\alpha}\gamma_\nu)m_\Delta g - \tfrac{1}{3}i(p_\mu\gamma_\nu - p_\nu\gamma_\mu)\gamma_\alpha\frac{\hat{p}}{m_\Delta}(g - m_\Delta h)$$

$$-\tfrac{1}{3}i(p_\mu\gamma_\nu - p_\nu\gamma_\mu)\frac{p_\alpha}{m_\Delta}(2g + m_\Delta h) - i(\delta_{\nu\alpha}p_\mu - \delta_{\mu\alpha}p_\nu)\hat{p}h\Big\}\gamma_5\,, \qquad (51)$$

where g and h are constants.

Let us now consider the sum rules. Make the Borel transformation. To obviate unknown constants in the r.h.s of the sum rules induced by the single-pole terms at $p^2 = m^2$ let us apply the differential operator $-[\partial/\partial(1/M^2) + m^2]$. As a result we get the following sum rules for invariant functions of the chirality conserving structures

$$i(\delta_{\nu\alpha}\gamma_\mu - \delta_{\mu\alpha}\gamma_\nu)\hat{p}\gamma_5\,, \quad \sigma_{\mu\nu}\gamma_\alpha\hat{p}\gamma_5\,, \quad \sigma_{\mu\nu}p_\alpha\gamma_5, \quad i(p_\mu\gamma_\nu - p_\nu\gamma_\mu)p_\alpha\hat{p}\gamma_5,$$

$$i(\delta_{\nu\alpha}p_\mu - \delta_{\mu\alpha}p_\nu)\gamma_5\,, \quad i(p_\mu\gamma_\nu - p_\nu\gamma_\mu)\gamma_\alpha\gamma_5\,,$$

$$\left[\frac{\partial}{\partial(1/M^2)} + m^2\right]\left[\tfrac{7}{8}M^4E_1(M)L^{-4/27} - \tfrac{2}{3}a^2\chi L^{4/27} + \frac{4}{9}\frac{a^2}{M^2}L^{20/27}(1 + \tfrac{1}{4}\kappa - \tfrac{1}{2}\xi)\right]$$

$$= \tfrac{1}{2}\sqrt{\tfrac{3}{2}}\tilde{\lambda}_N\tilde{\lambda}_\Delta(m + m_\Delta)[G_1 + 2g(m_\Delta - m)]\,e^{-m_\Delta^2/M^2}\,, \qquad (52)$$

$$\left[\frac{\partial}{\partial(1/M^2)} + m^2\right]\left[\tfrac{3}{4}M^4E_1(M)L^{-4/27} - \tfrac{3}{4}a^2\chi L^{4/27} + \frac{1}{2}\frac{a^2}{M^2}L^{20/27}(1 + \tfrac{1}{4}\kappa - \tfrac{1}{2}\xi)\right]$$

$$= \tfrac{1}{2}\sqrt{\tfrac{3}{2}}\tilde{\lambda}_N\tilde{\lambda}_\Delta(m + m_\Delta)[G_1 + 2g(m_\Delta - m)]\,e^{-m_\Delta^2/M^2}\,, \qquad (53)$$

$$\left[\frac{\partial}{\partial(1/M^2)} + m^2\right]\left[\tfrac{9}{8}M^4E_1(M)L^{-4/27} - \tfrac{1}{2}a^2\chi L^{4/27} + \frac{1}{3}\frac{a^2}{M^2}L^{20/27}(1 + \tfrac{1}{4}\kappa - \tfrac{1}{2}\xi)\right]$$

$$= \tfrac{1}{2}\sqrt{\tfrac{3}{2}}\tilde{\lambda}_N\tilde{\lambda}_\Delta(m + m_\Delta)[G_1 + 2g(m_\Delta - m)]\,e^{-m_\Delta^2/M^2}\,, \qquad (54)$$

$$\left[\frac{\partial}{\partial(1/M^2)}+m^2\right][\tfrac{1}{4}M^2E_0(M)L^{-4/27}]$$
$$=\tfrac{1}{3}\sqrt{\tfrac{3}{2}}\tilde{\lambda}_N\tilde{\lambda}_\Delta\frac{1}{m_\Delta^2}\left\{m\left[G_1-\frac{1}{2}\frac{m_\Delta^2}{m}\left(1-\frac{m}{m_\Delta}\right)G_2\right]+2(m_\Delta^2-m^2)\right.$$
$$\left.\times(g+\tfrac{1}{2}m_\Delta h)\right\}e^{-m_\Delta^2/M^2}, \tag{55}$$

$$\left[\frac{\partial}{\partial(1/M^2)}+m^2\right]\left[M^4E_1(M)L^{-4/27}+\frac{8}{9}\frac{a^2}{M^2}L^{20/27}(1+\tfrac{5}{8}\kappa-\tfrac{5}{16}\xi)\right]$$
$$=\sqrt{\tfrac{3}{2}}\tilde{\lambda}_N\tilde{\lambda}_\Delta m_\Delta[G_1+(m_\Delta-\tfrac{1}{2}m)G_2-(m_\Delta^2-m^2)h]\,e^{-m_\Delta^2/M^2}, \tag{56}$$

$$\left[\frac{\partial}{\partial(1/M^2)}+m^2\right]\left[\tfrac{5}{8}M^4E_1(M)L^{-4/27}-\tfrac{1}{3}a^2\chi L^{4/27}+\frac{4}{9}\frac{a^2}{M^2}L^{20/27}(1+\tfrac{7}{16}\kappa-\tfrac{13}{32}\xi)\right]$$
$$=\tfrac{1}{6}\sqrt{\tfrac{3}{2}}\tilde{\lambda}_N\tilde{\lambda}_\Delta[(3m_\Delta+m)G_1+m_\Delta(m_\Delta-m)G_2+2(m_\Delta^2-m^2)(g-m_\Delta h)]\,e^{-m_\Delta^2/M^2}. \tag{57}$$

In (52)–(57) we denoted $\tilde{\lambda}_\Delta=(2\pi)^2\lambda_\Delta$ and took into account the current η_α anomalous dimension (which is $-\frac{2}{27}$), the other notation is the same as in (28), (29). Excluding the constant h from (55)–(57) one may get the following two equalities

$$\left[\frac{\partial}{\partial(1/M^2)}+m^2\right]\left\{M^2[\tfrac{1}{4}m_\Delta^2E_0(M)+\tfrac{1}{3}M^2E_1(M)]L^{-4/27}\right.$$
$$\left.+\frac{8}{27}\frac{a^2}{M^2}L^{20/27}(1+\tfrac{5}{8}\kappa-\tfrac{5}{16}\xi)\right\}$$
$$=\tfrac{1}{3}\sqrt{\tfrac{3}{2}}\tilde{\lambda}_N\tilde{\lambda}_\Delta\{(m_\Delta+m)G_1+\tfrac{1}{2}m_\Delta^2G_2+2(m_\Delta^2-m^2)g\}\,e^{-m_\Delta^2/M^2}, \tag{58}$$

$$\left[\frac{\partial}{\partial(1/M^2)}+m^2\right]\left[\tfrac{7}{24}M^4E_1(M)L^{-4/27}-\tfrac{1}{3}\chi a^2L^{4/27}+\frac{4}{27}\frac{a^2}{M^2}(1+\tfrac{1}{16}\kappa-\tfrac{19}{32}\xi)\right]$$
$$=\tfrac{1}{3}\sqrt{\tfrac{3}{2}}\tilde{\lambda}_N\tilde{\lambda}_\Delta[\tfrac{1}{2}(m+m_\Delta)G_1-\tfrac{1}{2}m_\Delta^2G_2+(m_\Delta^2-m^2)g]\,e^{-m_\Delta^2/M^2}. \tag{59}$$

Excluding the constant g from (58), (59), we arrive at the sum rule

$$\left[\frac{\partial}{\partial(1/M^2)}+m^2\right]\left\{\tfrac{1}{4}M^4[m_\Delta^2E_0(M)-M^2E_1(M)]L^{-4/27}+\tfrac{2}{3}\chi a^2L^{4/27}\right.$$
$$\left.+\frac{1}{6}\frac{a^2}{M^2}L^{20/27}(\kappa+\tfrac{1}{2}\xi)\right\}$$
$$=\tfrac{1}{2}\sqrt{\tfrac{3}{2}}\tilde{\lambda}_N\tilde{\lambda}_\Delta m_\Delta^2G_2\,e^{-m_\Delta^2/M^2}, \tag{60}$$

from which the $\Delta\to p\gamma$ transition amplitude G_2 can be determined. And vice versa, if we exclude G_2 from (58), (59) we get

$$\left[\frac{\partial}{\partial(1/M^2)}+m^2\right]\left\{\tfrac{1}{4}M^2[m_\Delta^2 E_0(M)+\tfrac{5}{2}M^2E_1(M)]L^{-4/27}\right.$$
$$\left.-\tfrac{1}{3}\chi a^2L^{4/27}+\frac{4}{9}\frac{a^2}{M^2}L^{20/27}(1+\tfrac{7}{16}\kappa-\tfrac{13}{32}\xi)\right\}$$
$$=\tfrac{1}{2}\sqrt{\tfrac{3}{2}}\tilde{\lambda}_N\tilde{\lambda}_\Delta(m+m_\Delta)[G_1+2(m_\Delta-m)g]\,e^{-m_\Delta^2/M^2}\,. \qquad (61)$$

Of the chirality-violating structures we will consider only those containing two momenta in the numerator. This will be done for the same reasons as for the nucleon magnetic moments. After excluding all the unknown constant except G_1, G_2 and g, from the r.h.s of the sum rules for these structures we are left with two independent sum rules

$$\frac{3}{2}\left[\frac{\partial}{\partial(1/M^2)}+m^2\right]m_\Delta a[\tfrac{2}{3}L^{8/27}-\tfrac{1}{3}\chi M^2E_0(M)-\tfrac{1}{3}(\kappa+\xi)L^{-4/27}]$$
$$=\tfrac{1}{2}\sqrt{\tfrac{3}{2}}\tilde{\lambda}_N\tilde{\lambda}_\Delta(m_\Delta+m)[G_1+2(m_\Delta-m)g]\,e^{-m_\Delta^2/M^2}\,, \qquad (62)$$

$$\frac{3}{2}\left[\frac{\partial}{\partial(1/M^2)}+m^2\right]m_\Delta a[\tfrac{4}{9}L^{8/27}-\tfrac{1}{3}\chi M^2E_0(M)-\tfrac{2}{9}(\kappa+\tfrac{7}{4}\xi)L^{-4/27}]$$
$$=\tfrac{1}{2}\sqrt{\tfrac{3}{2}}\tilde{\lambda}_N\tilde{\lambda}_\Delta(m_\Delta+m)[G_1+2(m_\Delta-m)g]\,e^{-m_\Delta^2/M^2}\,. \qquad (63)$$

All the obtained sum rules except for (60) include the combination $G_1+2(m_\Delta-m)g$. This circumstance does not permit us finding each of the constants G_1 and g separately. At the same time, comparison of the l.h.s of the sum rules (51)–(53), (60), (61), (62) whose r.h.s coincide, allows one to determine the accuracy of the whole approach and to understand whether the χ value previously found when studying proton and neutron magnetic moments agree with these equalities. For such a comparison it should not be expected that these equalities will coincide with good accuracy because: (i) the action of the operator $\partial/\partial(1/M^2)+m^2$ eliminated the ground term $\exp(-m^2/M^2)$ in the r.h.s of the sum rule leaving a comparatively smaller one $\exp(-m_\Delta^2/M^2)$; (ii) continuum thresholds for different sum rules may differ which we did not take into account; (iii) it is not excluded that the contribution of disregarded power corrections may achieve ~20–30% of the ground term. The latter circumstance is significant in (52)–(54), (61) since a strong cancellation between the first term in the l.h.s and the term proportional to χ arises in these sum rules. In addition, the contribution of the last term proportional to the unknown values κ and ξ may be essential in the sum rules (62), (63). Numerical comparison of the l.h.s of (52)–(54), (61), (62), (63) shows that at the value χ given by eq. (2) they agree with each other up to a factor 2. (The continuum threshold is taken to be $W^2=2.5\,\mathrm{GeV}^2$.) The values of χ significantly differing from eq. 2 lead to a strong disagreement of the sum rules. In particular, for the set of parameters

$\chi = -8\,\text{GeV}^2$, $\kappa = -3.2$, $\xi = -1.5$ discussed at the end of sect. 4, the l.h.s of the considered sum rules differ even by a factor of 3.5. At the same time, the M^2 dependence of the r.h. and l.h. sides of the sum rules is not the same: $G_1 + 2(m_\Delta - m)g$ which must be constant, changes by a factor of 1.5 within the interval $M = 1.0$–$1.5\,\text{GeV}^2$. All this shows that the accuracy of the results obtained in this section is rather low and is featured by a factor of 1.5–2. With this accuracy

$$G_1 + 2(m_\Delta - m)g \approx 1\,\text{GeV}^{-1}\,. \tag{64}$$

Such a value seems to be reasonable since experimentally $G_1 = 2.82 \pm 0.05\,\text{GeV}^{-1}$* and there is no reason to expect that the quantity $(m_\Delta - m)g$ is small.

Return now to the sum rule (60) and determine the $\Delta \to p\gamma$ transition amplitude G_2 from it. The result is

$$G_2 \approx -2\,\text{GeV}^{-2}\,, \tag{65}$$

with the same accuracy. The value (65) does not disagree with the experimental number $G_{2\,\text{exp}} = -1.3 \pm 0.2\,\text{GeV}^{-2}$.

6. Conclusion

The method suggested in this paper on the basis of which we succeeded in calculating proton and neutron magnetic moments is quite a general one. It is applicable not only to determining the static electromagnetic properties of low-lying hadronic states but also to calculating hadronic emission and absorption amplitudes of any fields with large wavelengths much larger than hadronic dimensions. Examples are semileptonic decay amplitudes and soft pion-hadron interaction constants. If this method is applied to processes involving soft pions, the pion field $\partial\varphi^a/\partial x_\mu$ (a is the isospin index) should be considered as a constant in space and time external field interacting with an axial current with hamiltonian $(1/f_\pi)j_\mu^{\text{A}a}\,\partial\varphi^a/\partial x_\mu$. The method can also be extended to the case of particles containing heavy quarks. For instance, magnetic moments of charmed baryons, $D^* \to D\gamma$, $D^* \to D\pi$, etc., decay widths can be calculated with it. Thus, the method encourages hope that a large number of problems which were not solved up to now either in QCD of in quark models, can be solved with it. Of course, the solution of each of the problems is a serious task which demands hard labour but the way seems to be open.

When using the suggested method, two drawbacks arise: first, is the appearance of new external field-induced v.e.v.'s in the sum rules. These v.e.v.'s should be determined from the same sum rules or in some other way (e.g. for the external

* The experimental values of G_1 and G_2 can be inferred from the data on chiral amplitudes [21]: $A_{1/2} = -0.139 \pm 0.005\,\text{GeV}^{-1/2}$; $A_{3/2} = -0.256 \pm 0.005\,\text{GeV}^{-1/2}$; which are related to G_1 and G_2 by relations (49).

pion field case, using current algebra commutation relations). An approach is also possible when some v.e.v. is determined from one process and is substituted into another to get physical results. We emphasize that the existence and magnitude of such v.e.v.'s are by themselves an important physical information about the structure of the vacuum in QCD which can be employed for constructing various models (for instance, the bag model).

The second drawback, of a technical character, is due to the appearance of non-diagonal transition amplitudes from the lowest state into excited ones corresponding to the 1st-order pole in p^2-m^2 for the transition amplitude in the phenomenological part of the sum rules. These terms are not exponentially damped as compared to terms with second-order poles we are interested in and, as was shown for nucleon magnetic moments, can be very essential numerically. Therefore, to obtain reliable qualitative results they must be isolated in any particular case.

Terms of this type appear also in other calculations of vertex functions with small momentum transfer (e.g. of magnetic moments, πNN vertices etc.), in particular, in lattice calculations. In refs. [15–16], the Fourier-transform of the polarization operator of the nucleon currents in an external electromagnetic field was considered

$$\Pi(t)=\int_{-\infty}^{\infty} \Pi(E,0)\, \mathrm{e}^{iEt}\, \mathrm{d}E \tag{66}$$

and its exponential asymptotics at large euclidean times t was isolated. The part of $\Pi(t)$ linear in the field is contributed by both double and single poles. Refs. [15–16] used a different choice of current and therefore it is impossible to compare the results directly. If, nevertheless, we substitute the polarization operator (26), (27) into eq. (66), then for the coefficient of the structure $\sigma_{\mu\nu}\hat{p}+\hat{p}\sigma_{\mu\nu}$ we obtain (for the proton)

$$\Pi(t)\approx-\frac{\lambda_{\mathrm{N}}^2}{4}\frac{\pi}{2m^3}\,\mathrm{e}^{-mt}[\mu_{\mathrm{p}}(1+mt)+2A_{\mathrm{p}}m^2]\,. \tag{67}$$

At asymptotically large times $mt \gg 1$ the term with μ_{p} dominates but at the really used times in refs. [15–16] the term with A_{p} is damped insufficiently. If we take $A_{\mathrm{p}}=3.3\ \mathrm{GeV}^{-2}$, as follows from our estimates, the double- and single-pole contributions appear to be equal at $t\approx 1.3\ \mathrm{GeV}^{-1}$. The largest value of t in ref. [15] was $t=3.6\ \mathrm{GeV}^{-1}$ and the contribution from single poles is 30%. In ref. [16] the times were somewhat larger, $t\approx 5.6\ \mathrm{GeV}^{-1}$ and the single-pole contribution is estimated to be 25%. To separate the double- and single-pole contributions one must study the t-dependence of the pre-exponent which is rather difficult in lattice calculations. (Such an investigation was, perhaps, made in ref. [17].)

The authors are indebted to Ya.Ya. Balitsky, V.M. Belayev, M.A. Shifman and A. V. Yung for useful discussions and valuable remarks.

Appendix A

We describe here the calculation of the coefficients of the 8-dimensional operators $g\bar{q}\sigma_{\alpha\beta}(\frac{1}{2}\lambda^n)G^n_{\alpha\beta}q\cdot\bar{q}\sigma_{\mu\nu}q$, $g\bar{q}(\frac{1}{2}\lambda^n)G^n_{\mu\nu}q\cdot\bar{q}q$ and $\varepsilon_{\mu\nu\lambda\sigma}g\bar{q}\gamma_5(\frac{1}{2}\lambda^n)G^n_{\lambda\sigma}q\cdot\bar{q}q$ which contribute to the sum rule (28) and of the coefficients of the 7-dimensional operator $\bar{q}\sigma_{\mu\nu}qG^n_{\alpha\beta}G^n_{\alpha\beta}$ contributing to the sum rule (29). In order to make such a calculation one must take into account the vacuum gluon field effect on the quark correlator $\langle 0|\mathrm{T}\{q^a_i(x), \bar{q}^b_k(0)\}|0\rangle$ (the external electromagnetic field effect was shown earlier in eq. (17)). This may be taken into account using the expansion $q(x)$ in powers of x (16) where the covariant derivative includes now the gluonic field $\nabla_\alpha = \partial_\alpha + ig(\frac{1}{2}\lambda^n)B^n_\alpha$. (Note the sign difference between our definition of the coupling constant g and that of refs. [1–3].) It is convenient to use a fixed-point gauge $x_\mu B^n_\mu(x)=0$. Considering the vacuum gluon field as external (next it will enter other v.e.v.'s) one may obtain for the quark correlator (the terms in the x expansion required in the following are retained only and the notation of (9), (10) is used):

$$
\begin{aligned}
&\langle 0|\mathrm{T}\{q^a_i(x), \bar{q}^b_k(0)\}|0\rangle \\
&\quad = i\frac{g}{64\pi^2x^2}(\lambda^n)^{ab}G^n_{\alpha\beta}(\hat{x}\sigma_{\alpha\beta}+\sigma_{\alpha\beta}\hat{x})_{ik} \\
&\quad + \frac{\sqrt{4\pi\alpha}}{9\cdot 2^6}\delta^{ab}\langle 0|\bar{q}q|0\rangle e_q F_{\alpha\beta}[\sigma_{\alpha\beta}x^2(\kappa+\xi)-2x_\rho x_\beta\sigma_{\rho\alpha}(\kappa-\tfrac{1}{2}\xi)]_{ik} \\
&\quad + \frac{1}{3\cdot 2^7}\delta^{ab}\delta_{ik}g\langle 0|\bar{q}\sigma_{\alpha\beta}\lambda^n G^n_{\alpha\beta}q|0\rangle x^2 \\
&\quad + \frac{i}{3\cdot 2^4}(\lambda^n)^{ab}(\sigma_{\alpha\beta})_{ik}x_\rho x_\sigma\langle 0|\bar{q}\sigma_{\alpha\rho}q|0\rangle G^n_{\beta\sigma} \\
&\quad + \frac{1}{9\cdot 2^5}(\lambda^n)^{ab}\delta_{ik}g[x^2\langle 0|\bar{q}\sigma_{\alpha\beta}q|0\rangle G^n_{\alpha\beta} - x_\rho x_\beta\langle 0|\bar{q}\sigma_{\alpha\rho}q|0\rangle G^n_{\alpha\beta}] \\
&\quad - \frac{\pi^2}{3^4\cdot 2^8}\delta^{ab}(\sigma_{\mu\nu})_{ik}\left\langle 0\left|\frac{\alpha_s}{\pi}G^n_{\alpha\beta}G^n_{\alpha\beta}\right|0\right\rangle\{(1+C)x^4\langle 0|\bar{q}\sigma_{\mu\nu}q|0\rangle \\
&\quad - 2Cx^2x_\rho[\langle 0|\bar{q}\sigma_{\nu\rho}q|0\rangle x_\mu - \langle 0|\bar{q}\sigma_{\mu\rho}q|0\rangle x_\nu]\}\,. \qquad (\mathrm{A.1})
\end{aligned}
$$

The first term in (A.1) corresponds to the action of the vacuum gluon field on the hard quark propagator (cf. the 3rd term in eq. (17)). The terms from the 2nd to the 5th arise after expanding $q(x)$ in x up to the second order according to eq. (16). The last term arises after expanding $q(x)$ in x in the 4th order. Its form in (A.1) is found by using repeatedly the identities

$$
\nabla^2 = \hat{\nabla}\hat{\nabla} - \tfrac{1}{4}g\sigma_{\alpha\beta}G^n_{\alpha\beta}\lambda^n, \qquad [\nabla_\alpha, \nabla_\beta] = \tfrac{1}{2}igG^n_{\alpha\beta}\lambda^n,
$$
$$
[\nabla_\alpha, G^n_{\alpha\beta}\lambda^n] = 0,
$$

(in the lowest order in g) and neglecting non-factorizable terms. The constant C in the last term of (A.1) is not necessarily to be determined since it drops out in calculating the polarization operator. We shall need also expressions for the v.e.v.'s of the quark and gluon field product when the electromagnetic field is absent

$$\langle 0|q_i^{\mathrm{a}}\bar{q}_k^{\mathrm{b}}G_{\alpha\beta}^n|0\rangle = -\frac{1}{3\cdot 2^6}\langle 0|\bar{q}\sigma_{\mu\nu}(\tfrac{1}{2}\lambda^n)G_{\mu\nu}^n q|0\rangle(\sigma_{\alpha\beta})_{ik}(\tfrac{1}{2}\lambda^n)^{\mathrm{ab}}, \tag{A.2}$$

and in terms linear in the electromagnetic-field

$$g\langle 0|q_i^{\mathrm{a}}\bar{q}_k^{\mathrm{b}}G_{\alpha\beta}^n|0\rangle_{\mathrm{F}} = -\frac{\sqrt{4\pi\alpha}}{2^4}\langle 0|\bar{q}q|0\rangle(\tfrac{1}{2}\lambda^n)^{\mathrm{ab}}[\delta_{ik}\kappa_{\mathrm{q}}F_{\alpha\beta}-\tfrac{1}{4}i\xi_{\mathrm{q}}\varepsilon_{\alpha\beta\lambda\sigma}F_{\lambda\sigma}(\gamma_5)_{ik}], \tag{A.3}$$

where the notation of (9) is used.

We are ready now to calculate the operator coefficients in question.

The coefficient of the operator $g\bar{q}\sigma_{\alpha\beta}(\frac{1}{2}\lambda^n)G_{\alpha\beta}^n q\cdot\bar{q}\sigma_{\mu\nu}q$ is described by the graphs of fig. 6a, b. In fig. 6a the vacuum gluon field is emitted by the hard d-quark so that its propagator is given by the first term in (A.1). This gluon field is averaged with the fields of one of the u-quarks according to (A.2) while the propagator of the other u-quark is determined by the 4th term in (17). In fig. 6b the vacuum gluon field is emitted by a soft u-quark whose correlator is given by the 3rd term in (A.1). The other soft u-quark interacts with the electromagnetic field and its correlator is described by the 4th term in (17); the hard d-quark propagates freely (the 1st term in (17)). The fig. 6 contribution to the polarization operator is presented in the text (eq. (21)).

Note that the procedure of the calculation of the coefficient in front of the v.e.v. $g\langle 0|\bar{q}\sigma_{\alpha\beta}(\frac{1}{2}\lambda^n)G_{\alpha\beta}^n q|0\rangle\langle 0|\bar{q}\sigma_{\mu\nu}q|0\rangle$ is not unique since the result depends on the choice of non-factorizable v.e.v.'s which are disregarded. Our recipe of treating the four-quark v.e.v. $\langle 0|q_i^{\mathrm{a}}(x)q_k^{\mathrm{b}}(x)\bar{q}_l^{\mathrm{c}}(0)\bar{q}_m^{\mathrm{d}}(0)|0\rangle$ is: first factorization and afterwards expansion in powers of x. Other ways of calculation give slightly different values for the coefficient of interest. Because the differences are small (of the order of 20–30%) and the discussed term itself is a small correction to the ground term $\langle 0|\bar{q}\sigma_{\mu\nu}q|0\rangle\langle 0|\bar{q}q|0\rangle$ these ambiguities do not affect the results.

The coefficient of the operator $g\bar{q}(\frac{1}{2}\lambda^n)G_{\alpha\beta}^n q\cdot\bar{q}q$ is given by the graphs of fig. 7a, b. In fig. 7a the vacuum gluon field is emitted by the hard d-quark whose propagator is given by the 1st term in (A.1). The vacuum gluon field is averaged with the fields of one of the u-quarks moving in the electromagnetic field, according to (A.3). The propagator of the other u-quark is determined by the 2nd term in (17). In fig. 7b the gluon field is emitted by the soft u-quark moving in an electromagnetic field and its correlator is described by the 2nd term in (A.1). The propagator of the other soft u-quark corresponds to the 2nd term in (17), the propagator of the free hard d-quark corresponds to the 1st term in (17). The sum of the graphs 7a, b is given by eq. (22).

The contributions of the five-dimensional operators $g\bar{q}(\frac{1}{2}\lambda^n)G^n_{\alpha\beta}q$ and $g\bar{q}(\frac{1}{2}\lambda^n)\gamma_5\varepsilon_{\alpha\beta\lambda\sigma}G^n_{\lambda\sigma}q$ to the even structures of the polarization operator are calculated analogously. The corresponding diagrams are shown in fig. 10a, b. In both diagrams the u-quarks are hard, and the d-quark is soft. Not dwelling on details of the calculation, we note only that their contribution to the structure $i(p_\mu\gamma_\nu - p_\nu\gamma_\mu)\hat{p}$ cancels.

There are three different diagrams contributing to the coefficients of the operator $\bar{q}\sigma_{\mu\nu}qG^n_{\alpha\beta}G^n_{\alpha\beta}$. In fig. 11a the vacuum gluon field interacts with two hard u-quarks so that u-quark propagators in the polarization operator are given by the 1st term in (A.1) and the d-quark, by the 4th term in (17); In fig. 11b gluons interact with hard u- and soft d-quarks. The d-quark propagator is obtained by expanding the $d(x)$ field in x up to the second order and is given by the 4th and 5th terms in (A.1). (Really, the 4th term does not contribute to the polarization operator.) Note that the separation of the operator contribution $\bar{q}\sigma_{\mu\nu}qG^n_{\alpha\beta}G^n_{\alpha\beta}$ in fig. 11b is not unique since a number of non-factorizable 7-dimensional operators appear here which can be expressed in terms of one another and through the operator $\bar{q}\sigma_{\mu\nu}qG^n_{\alpha\beta}G^n_{\alpha\beta}$. (The situation here resembles the one which arises when attempting to calculate 8-dimensional contributions to the vector current operator [22].) Such an ambiguity may affect the 5th term in (A.1) which in fact was obtained neglecting the v.e.v. $\langle 0|\bar{q}G^n_{\mu\rho}\lambda^n\nabla_\rho\nabla_\nu q|0\rangle_F$. For this reason one cannot expect high accuracy for our estimate of the 7-dimensional operator contributions. It is, however, important to emphasize that all of these poorly determined terms are proportional to the d-quark charge (for the proton) and thus drop out in the procedure used in the text to determine magnetic moments. In fig. 11c the d-quark field is expanded in x up to the 4th order and its correlator is given by the last term in (A.1) while the u-quarks propagate freely. As is shown by the calculation, in our approximation the 11b, c graphs contribution to the structure $i(p_\mu\gamma_\nu - p_\nu\gamma_\mu)\hat{p}$ cancel completely and only the diagram of fig. 11a is responsible for the answer. Its contribution to the structure $i(p_\mu\gamma_\nu - p_\nu\gamma_\mu)\hat{p}$ is presented in eq. (25).

Appendix B

We dwell here on the procedure of finding the transition amplitude of the proton in the current η_p (the residue) λ_N from the sum rules for the polarization operator $\Pi^{(0)}(p)$ defined in eq. (3). The QCD sum rules for the invariant functions at the structures $\hat{p}$ and 1 have the form (see ref. [3], anomalous dimensions are corrected according to the footnote on p. 140)

$$M^6E_2(M)L^{-4/9}+\tfrac{4}{3}a^2L^{4/9}+\tfrac{1}{4}bM^2E_0(M)L^{-4/9}-\tfrac{1}{3}a^2\frac{m_0^2}{M^2}=\tilde{\lambda}_N^2\exp\left(-\frac{m^2}{M^2}\right), \quad \text{(B.1)}$$

$$2aM^4E_1(M)+\frac{272}{81}\frac{\alpha_s}{\pi}\frac{a^3}{M^2}-\tfrac{1}{12}ab=m\tilde{\lambda}_N^2\exp\left(-\frac{m^2}{M^2}\right), \quad \text{(B.2)}$$

where

$$E_2(M)=1-\left(1+\frac{W^2}{M^2}+\frac{W^4}{2M^4}\right)\exp\left(-\frac{W^2}{M^2}\right),$$

and all other notation is from (28), (29). (The value $\tilde{\beta}_N^2$ defined in ref. [3] is by a factor of 8 smaller than $\tilde{\lambda}_N^2$.) The calculations of ref. [3] contain errors in the coefficients in the last two terms in the l.h.s. of (B.2): the coefficient of $\alpha_s a^3$ was by a factor of two smaller* and the coefficient in front of ab was $\frac{1}{9}$. Since both terms are small the correction of these errors only slightly affect the results.

To obtain the value of $\tilde{\lambda}_N^2$ as accurately as possible we set $m=0.94$ GeV as observed in experiment, multiply the sum rule (B.1) by $\exp(m^2/M^2)$ and the sum rule (B.2) by $m^{-1}\exp(m^2/M^2)$ and plot the resultant values of $\tilde{\lambda}_N^2$ versus M^2. The continuum threshold W must be chosen so that the variation of $\tilde{\lambda}_N^2$ with the change of M^2 would be the weakest. The best fit is achieved under the choice $W^2=2.3\ \mathrm{GeV}^2$**. This value of W is close to the mass of the first resonance state with nucleon quantum numbers. The M^2 dependence of $\tilde{\lambda}_N^2$ obtained from sum rules (B.1), (B.2) is shown in fig. 18. We see that the values $\tilde{\lambda}_N^2$ determined from (B.1), (B.2) coincide with each other and are practically constant within $0.7<M^2<1.2\ \mathrm{GeV}^2$. Our final result for $\tilde{\lambda}_N^2$ is $\tilde{\lambda}_N^2=2.1\ \mathrm{GeV}^6$ with an uncertainty of $\approx 10\%$.

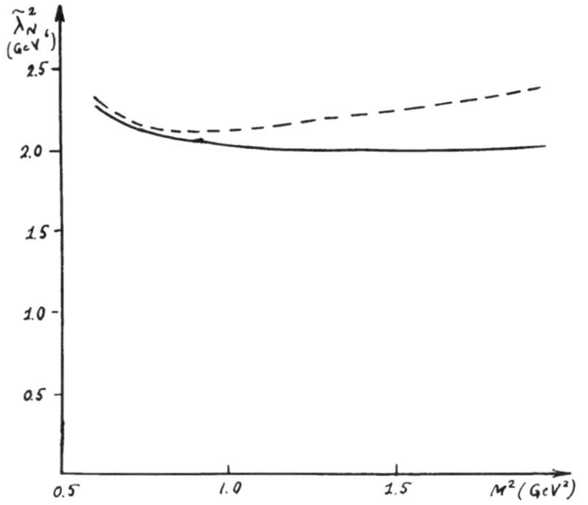

Fig. 18. The Borel parameter M^2 dependence of $\tilde{\lambda}_N^2$ determined from eq. (B.1) – solid line, and from eq. (B.2) – dashed line.

* The authors are indebted to A. Kolesnichenko who has drawn their attention to this error.

** Note the difference in this number with the first works of one of the authors (B.I.) [2] where the higher values of W were chosen. The estimate of the nucleon mass is rather stable with respect to variation of W but the value of $\tilde{\lambda}_N^2$ needed for accurate determination of magnetic moments is more sensitive. The difference with the fit of ref. [3], but within the errors of it ($m_N=1.02\pm0.12$ GeV, $\tilde{\lambda}_N^2=3.6\pm1.2\ \mathrm{GeV}^6$, $W=1.5$ GeV), stems from the fact that we took m as a given experimental number and have corrected the values of the anomalous dimensions.

Note that a significantly higher accuracy of $\tilde{\lambda}_N^2$ determination achieved here as compared with ref. [3] is due to the fact that the nucleon mass m was given but was not a free parameter as it was in the case of ref. [3]. Varying simultaneously m and $\tilde{\lambda}_N^2$ in ref. [3] one may get good agreement of the sum rule within a rather wide variation interval of both parameters.

Note added in proof

(i) When estimating the continuum contribution to sum rules, we employed the model where this contribution is described by the dispersive integral $\int_{W^2}^{\infty} \rho(s')\, ds'/(s'-s)$; the parameter W^2 is the same as in mass sum rules. Another model inspired by the double dispersive representation (see eqs. (12), (13)) is possible where the continuum is given by $\int_{W^2}^{\infty} \rho(s')\, ds'/(s'-s)^2$. It is equivalent to the first one with a reduced value of W^2. The values of nucleon magnetic moments determined from eqs. (32) (or eqs. (1)) are independent of the continuum model, while the values of χ and of $A_{p,n}$, $B_{p,n}$ are constants. For example, if the second continuum model were adopted we would obtain $\chi \approx -6\ \text{GeV}^{-2}$ and the constants $A_{p,n}$, $B_{p,n}$ would be reduced by a factor of 2.5.

(ii) Recently, a paper by Balitsky and Yung [23] has appeared, devoted, as ours is, to a calculation of nucleon magnetic moments in the sum rule approach. In this work, the Borel sum rules were also obtained for three invariant structures in the nucleon current polarization operator in an external electromagnetic field. The basic points of both approaches are similar but there are also significant differences. Whereas we used the combinations of proton and neutron sum rules for two structures $\hat{p}\sigma_{\mu\nu} + \sigma_{\mu\nu}\hat{p}$ and $i(p_\mu \gamma_\nu - p_\nu \gamma_\mu)\hat{p}$ where new field-induced vevs do not contribute, magnetic moments in ref. [23] were determined on from sum rules for the first structure, the new vevs being estimated by using the vector dominance model and subtractionless dispersive relations (note that the latter do not work doe the case of correlators when determining κ and ξ).

In our opinion, the procedure employed in ref. [23] gives a poorer accuracy than ours since in that case the obtained values of μ_p and μ_n are very sensitive to the value of χ and also to exact values of $\langle 0|\bar{q}q|0\rangle$ and W^2. Our estimate for χ obtained via the same sum rule when substituting the previously found magnetic moments in it, is about twice as large as the estimate of ref. [23] based on the vector dominance model. A good agreement of μ_p and μ_n with experiment obtained in ref. [23] is due to the fact that their value for $\langle 0|\bar{q}q|0\rangle$ is effectively 1.5 times as large as ours. Such a value of $\langle 0|\bar{q}q|0\rangle$ results in a deterioration of the baryon mass sum rule fit. If the authors of ref. [23] were to substitute our value for $\langle 0|\bar{q}q|0\rangle$, the nucleon magnetic moments would be reduced by a factor of two.

We wish to answer here the criticism of our work [9] by Balitsky and Yung. Their statement is that the power correction $\sim g\langle 0|\bar{q}\sigma_{\alpha\beta}\lambda^a G^a_{\alpha\beta} q|0\rangle F_{\mu\nu}$ to the sum rule for the structure $i(p_\mu \gamma_\nu - p_\nu \gamma_\mu)\hat{p}$ is very large and spoils the agreement with experiment of our predictions for μ_p and μ_n. In our calculation, this operator contributes to the sum rule with a coefficient which is four times smaller than that given in ref. [23] and taking account of it would reduce the predicted values of μ_p and μ_n only by 10–15% (see p. 127).

References

[1] M.A. Shifman, A.I. Vainshtein, V.I. Zakharov, Nucl. Phys. B147 (1979) 385, 448
[2] B.L. Ioffe, Nucl. Phys. B188 (1981) 317; (E: B191 (1981) 591)
[3] V.M. Belyaev, B.L. Ioffe, Sov. Phys. JETP 56 (1982) 493
[4] B.L. Ioffe, A.V. Smilga, Phys. Lett. B114 (1982) 353
[5] B.L. Ioffe, A.V. Smilga, Nucl. Phys. B216 (1983) 373
[6] V.A. Nesterenko, A.V. Radyshkin, Phys. Lett. B115 (1982) 410
[7] V.L. Eletsky, Ya.I. Kogan, preprint ITEP-170 (1982)
[8] V.L. Eletsky, B.L. Ioffe, Ya.I. Kogan, Phys. Lett. B122 (1983) 423
[9] B.L. Ioffe, A.V. Smilga, preprint ITEP-24 (1983); ZhETF Pisma 37 (1983) 250
[10] Y. Chung, H.G. Dosch, M. Kremer, D. Schall, Nucl. Phys. B197 (1982) 57; Z. Phys. C. 15 (1982) 367
[11] B.L. Ioffe, Z. Phys. C.18 (1983) 67
[12] B.V. Geshkenbein, B.L. Ioffe, Nucl. Phys. B166 (1980) 340
[13] A.V. Smilga, Yad. Fiz. 35 (1982) 473
[14] L.J. Reinders, H.R. Rubinstein, S. Yazaki, Nucl. Phys. B213 (1983) 109
[15] G. Martinelli, G. Parisi, R. Petronzio, F. Rapuano, preprint CERN TH 3334 (1982)
[16] C. Bernard, T. Draper, K. Olynyk, M. Rushton, Phys. Rev. Lett. 49 (1982) 1076
[17] F. Fucito, G. Parisi, S. Petrarca, Phys. Lett. 115B (1982) 148
[18] V.A. Fock, Sov. Phys. 12 (1937) 404
[19] M.A. Shifman, M.I. Vysotsky, Nucl. Phys. B186 (1981) 475
[20] H.F. Jones, M.D. Scadron, Ann. of Phys. 81 (1973) 1
[21] Particle Data Group, Review of particle properties. Phys. Lett. 111B (1982)
[22] M.S. Dubovikov, A.V. Smilga, Yad. Fiz. 37 (1983) 984
[23] I.I. Balitsky, A.V. Yung, Phys. Lett. 129B (1983) 328

Quantum-chromodynamics calculation of g_A

V. M. Belyaev and Ya. I. Kogan
Institute of Theoretical and Experimental Physics

(Submitted 5 May 1983)
Pis'ma Zh. Eksp. Teor. Fiz. **37**, No. 12, 611–613 (20 June 1983)

The nucleon axial coupling constant g_A is calculated in nonperturbative quantum chromodynamics. The $g_A - 1$ renormalization stems from an interaction of the pion field with a quark condensate. The calculated results agree well with experiment.

PACS numbers: 12.35.Cn

The method of quantum-chromodynamics sum rules proposed by Vaĭnshteĭn, Zakharov, and Shifman[1] is now used widely to describe the properties of hadrons. In particular, the masses of mesons[1] and baryons,[2,3] form factors, and meson coupling constants[4] have been calculated. Ioffe and Smilga[5] recently offered some sum rules for the polarization nucleon-current operator in an external electromagnetic field for use in calculating the magnetic moments of the nucleons. In the present letter we examine some corresponding sum rules in an axial external field, for use in calculating the axial coupling constant g_A. The possibility of using the external-field method for calculations was first raised by Ioffe and Smilga.[5]

1. We consider the polarization operator in an external field A_μ:

$$\Pi(q) = i \int d^4 x\, e^{iqx} < T\{\eta(x), \bar{\eta}(0)\} >_A , \tag{1}$$

where

$$\eta(x) = \epsilon^{abc}\left[(u^a\; C d^b) u^c - (u^a C\gamma_5 d^b)\gamma_5 u^c\right] \tag{2}$$

is the quark current with the quantum numbers of the proton, C is the charge-conjugation matrix, and a, b, and c are the color indices. Polarization operator (1) in the absence of an external field was used in Ref. 2 to determine the mass of the proton. Analyzing the correlation functions in the axial external field A_μ is equivalent to analyzing the correlation functions in the ordinary vacuum of quantum chromodynamics, but with an additional term in the Lagrangian:

$$\Delta \mathcal{L} = \frac{1}{\sqrt{2}} (\bar{u} \gamma_\mu \gamma_5 u - \bar{d} \gamma_\mu \gamma_5 d) A_\mu, \tag{3}$$

where A_μ is the axial external field, which is the third (neutral) component of the isotopic triplet. To evaluate g_A we must consider the approximation linear in A_μ.

Working from (3) and partial conservation of axial current, we find explicit expressions for the quark propagators and the vacuum expectation values, $\langle \psi_\alpha^a(x) \bar{\psi}_\beta^b(0) \rangle_A$ and $\langle \psi_\alpha^a(x) G_{\mu\upsilon}^i(y) \bar{\psi}_\beta^i(0) \rangle_A$, in the approximation linear in A_μ. The results are expressed in terms of the known vacuum expectation values and the pion constant $f_\pi = 133$ MeV.

By using an operator expansion we can calculate the polarization operator $\Pi(g)$ in an axial external field at $-q^2 \sim 1\ \mathrm{GeV}^2$. We can find g_A by taking into account the terms in the operator expansion up to $\langle \psi \psi \rangle^2$.

2. We now write the contribution of the physical states to polarization operator (1). The terms of interest here, which correspond to the diagram in Fig. 1a, are

$$<0 | \eta | p><p | J_\mu^A | p><p | \bar{\eta} | 0> A_\mu (q^2 - m^2)^{-2},$$

where

$$J_\mu^A = \frac{1}{\sqrt{2}} \bar{u}_p(q_1) [g_A(Q^2) \gamma_\mu \gamma_5 + f_1(Q^2) Q_\mu \gamma_5] u_p(q_2) |_{Q_\mu = q_{1\mu} - q_{2\mu} \to 0} \tag{4}$$

is the nucleon axial current. If we adopt the condition $Q_\mu A_\mu \equiv 0$, the term proportional to $Q_\mu \gamma_5$ vanishes, and the pole behavior of $f_1(Q^2)$ in the limit of $Q^2 \to 0$ presents no difficulties. In this case, in the limit of a static external field A_μ, in which we are interested, three tensor structures appear: $\hat{A}\gamma_5$, $(\hat{q}\hat{A} - \hat{A}\hat{q})\gamma_5$, $2(Aq)\hat{q}\gamma_5$. The last of these structures is the most perferable, for the following reasons: First, this structure contains the largest number of momenta, so that the contributions of the higher-lying baryon states and of the vacuum expectation values of the operators of higher dimensionalities to the sum rules are suppressed; second, it turns out that for this structure we can single out terms in the operator expansion corresponding to a sum rule with $g_A = 1$ and write sum rules for the quantity $g_A - 1$. In this paper we will consider only the structure $2(Aq)\hat{q}\gamma_5$.

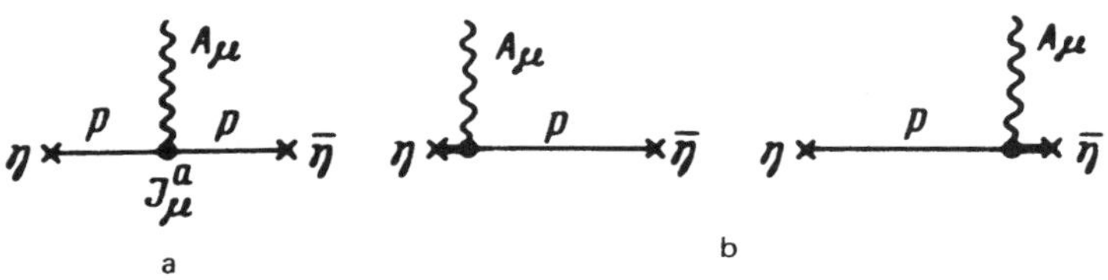

FIG. 1.

In addition to the double pole corresponding to the diagram in Fig. 1a, in which we are interested, there are also single-pole terms, corresponding to the diagrams in Fig. 1b. Ioffe and Smilga[5] have shown that the single-pole terms are not exponentially damped with respect to the two-pole terms in the Borel sum rules. In other words, in addition to the terms $(1/M^2)\exp(-m^2/M^2)$, which correspond to the double pole, there are also terms $\exp(-m^2/M^2)$, which correspond to single poles and which must be taken into account in order to find the correct value of g_A.

3. A calculation of the polarization operator leads to following Borel sum rules for the structure $2(Aq)\hat{q}\gamma_5$:

$$\frac{1}{8}M^4 + \frac{g^2<G^2>}{32} + \frac{(2\pi)^4<\bar{\psi}\psi>^2}{M^2}\left(\frac{1}{6}+\frac{1}{9}\right) + \frac{(2\pi)^2 f_\pi^2\, m_1^2}{4\cdot 18}$$

$$= \left(\frac{\tilde{\beta}^2 g_A}{M^2} + C\right)\exp(-m^2/M^2) + \sum_i \left(A_i/M^2 + D_i\right)\exp(-m_i^2/M^2), \qquad (5)$$

where

$$< 0\,|\eta|\,p> = \tilde{\beta}/(2\pi)^2\gamma_5 v(q), \qquad (\hat{q}-m)\,v(q)=0,$$

and C, A_i, and D_i are unknown constants. We might note that the value of $m_1^2 = 0.8$ GeV^2 found by Novikov[6] appears in the expression for the matrix element, $g\langle 0|\bar{u}\tilde{G}_{\mu\nu}\gamma_\nu d\,|\pi\rangle = im_1^2/4f_\pi q_\mu$. Comparing (5) with the sum rules found in Refs. 2 and 3, we see that some of the terms of the operator expansion correspond to a sum rule with $g_A = 1$.

The sum rule in Ref. 2 in which we are interest here is

$$\frac{1}{8}M^4 + \frac{g^2<G^2>}{32} + \frac{(2\pi)^4<\bar{\psi}\psi>^2}{6M^2} = \frac{\tilde{\beta}^2}{M^2}\exp(-m^2/M^2)$$

$$+ \text{(contribution of higher-lying states).} \qquad (6)$$

It can be shown that the terms in the operator expansion which are common to the two sum rules correspond to diagrams in which the external field (or pion) interacts with hard quarks. We can thus formulate some sum rules directly for the quantity $g_A - 1$. Working from (5) and (6), making use of the anomalous dimensionalities of the baryon currents, and of the terms of the operator expansion, and ignoring higher-lying states, we find some rules for $g_A - 1$:

$$(g_A - 1) + C'M^2 = \frac{1}{\tilde{\beta}^2}\exp(m^2/M^2)\left[\frac{(2\pi)^4<\bar{\psi}\psi>^2}{9}L^{4/9} + \frac{f_\pi^2(2\pi)^2 m_1^2}{4\cdot 18}L^{-8/9}\right], \qquad (7)$$

where[2,3]

$$L = \left(\frac{\alpha_s(\mu)}{\alpha_s(M)}\right), \quad \mu = 0.5\ \mathrm{GeV} \quad \text{and} \quad \tilde{\beta}^2 = 0.35\ \mathrm{GeV}^6.$$

Analysis of sum rules (7) yields the value $g_A - 1 = 0.3 \pm 0.05$, where the error is determined basically by the uncertainty in the value of the residue $\tilde{\beta}^2$. The quantity C' turns out to be about $-0.03 \pm 0.02\ \mathrm{GeV}^{-2}$. The apparent reason for the numerically small value of this quantity is the fact that the widths for the decay of the higher-lying nucleon resonances into $p\pi$ are small.

The value found for $g_A - 1$ is in excellent agreement with experiment. It should be noted that by virtue of the Goldberger-Treiman relation a calculation of g_A is also a calculation of the pion-nucleon coupling constant $g_{\pi NN}$, and by using the Adler-Weisberger sum rule we can calculate the constant of the transition $\Delta \to p\pi$.

We wish to thank B. L. Ioffe for calling our attention to this problem and for some interesting and useful discussions of many questions. We are also sincerely indebted to A. V. Smilga and M. A. Shifman for some important advice and other comments offered in the course of this study.

[1]M. A. Shifman, A. I. Vainshtein, and V. I. Zakharov, Nucl. Phys. **B147**, 385, 448 (1979).
[2]B. L. Ioffe, Nucl. Phys. **B188**, 317 (1981); (*E*) **B191**, 591 (1981); V. M. Belyaev and B. L. Ioffe, Zh. Eksp. Teor. Fiz. **83**, 876 (1982) [Sov. Phys. JETP **56**, 000 (1982)].
[3]Y. Chung *et al.*, Nucl. Phys. **B197**, 55 (1981).
[4]B. L. Ioffe and A. V. Smilga, Phys. Lett. **114B**, 353 (1982); V. A. Nesterenko and A. V. Radyushkin, Phys. Lett. **115B**, 410 (1982); V. L. Eletsky, B. L. Ioffe, and Ya. I. Kogan, Preprint ITEP-98, Institute of Theoretical and Experimental Physics, 1982.
[5]B. L. Ioffe and A. V. Smilga, Pis'ma Zh. Eksp. Teor. Fiz. **37**, 250 (1983) [JETP Lett. **37**, 298 (1983)].
[6]V. A. Novikov *et al.*, Preprint ITEP-71, Institute of Theoretical and Experimental Physics, 1983.

Translated by Dave Parsons
Edited by S. J. Amoretty

733 0021-3640/83/120733-05$01.00

Analysis of the behavior of the pion form factor at small Q^2 by the method of QCD sum rules

V. A. Nesterenko and A. V. Radyushkin
Joint Institute for Nuclear Research

(Submitted 18 April 1984)
Pis'ma Zh. Eksp. Teor. Fiz. **39**, No. 12, 576–579 (25 June 1984)

A QCD sum rule has been derived for the electromagnetic form factor of the pion in the region of a small spacelike momentum transfer, $Q^2 \lesssim m_\rho^2$. The value found for the electromagnetic radius of the pion, $\langle r_\pi^2 \rangle 1/2 = 0.66 \pm 0.03$ fm, agrees well with experimental data.

The quantum-chromodynamics sum rules[1] are actually the only calculation method available in QCD for problems in which nonperturbative effects play a governing role. Analysis of the various two-point functions $\pi(p^2)$ (correlation functions) by this method has yielded values for the masses and lepton widths of mesons and the corresponding characteristics of baryons.[2] The method was extended to three-point functions in Refs. 3 and 4. In particular, the behavior of the electromagnetic form factor of the pion was calculated for intermediate values of the momentum transfer, $m_\rho^2 \lesssim Q^2 \lesssim 3$–$4$ GeV^2, through an analysis of the amplitude $T(Q^2, p_1^2, p_2^2)$:

$$T(Q^2, p_1^2, p_2^2) = \frac{n_\mu n_\alpha n_\beta}{2(nP)^3} i^2 \int d^4x\, d^4y <0\,|\,T\,\{j_\alpha^+\left(-\frac{y}{2}\right) j_\beta\left(\frac{y}{2}\right) J_\mu(x)\,\}\,|\,0> \cdot \exp\left(i\,\frac{p_1 + p_2}{2}\, y - iqx\right) \quad (1)$$

where $j_\alpha = \bar{u}\gamma_5\gamma_\alpha d$, $J_\mu = \frac{2}{3}\bar{u}\gamma_\mu u - \frac{1}{3}\bar{d}\gamma_\mu d$ and n is a lightlike vector orthogonal to $q = p_2 - P_1 : n^2 = 0\ (nq) = 0, (np_1 + (np_2 \equiv (nP))]$. This analysis was carried out in a symmetric kinematics, $|p_1^2| \sim |p_2^2 \sim Q^2 \gtrsim m_\rho^2$ (where $Q^2 = -q^2$). The approach of Refs. 3 and 4 cannot be used at small values of Q^2 because the amplitude $T(Q^2, p_1^2, p_2^2)$ contains contributions which are strongly dependent on the dynamics over distances $\sim 1/Q$. At $Q^2 \lesssim m_\rho^2 = 0.6$ GeV^2 we cannot use asymptotic freedom (i.e., perturbation theory) to evaluate such contributions.

To refine the estimates (i.e., to incorporate nonperturbative effects) we need to carry out an additional factorization of the contributions from small and large distances. Problems of this type have been taken up previously in calculations by the method of QCD sum rules of the magnetic moments of nucleons[5,6] and of the constant[7] g_A. Our own approach to the factorization of the contributions is based on the technique of Ref. 8 for analyzing the asymptotic behavior of Feynman diagrams. Using this technique, we easily find that in the kinematics $|p_1^2| \sim |p_2^2| \gg Q^2$ the contributions to $T(Q^2, p_1^2, P_2^2)$, which are proportional to powers of $1/p_1^2$ and $1/p_2^2$, arise not only from the region in which all the intervals $[y^2, (x + y/2)^2$, and $(x - y/2)^2]$ are small but also from the region where y^2 is small but $(x \pm y/2)^2$ is not. Taking this second possibility into account gives us some additional contributions with the structure

$C(p^2)[\pi(q^2) - \pi^{\rm pert}(q^2)]$, where $\pi(q^2)$ is the "exact" correlation function of a local operator with a current J, and $\pi^{\rm pert}$ is its perturbative analog (cf. Ref. 6). The incorporation of a contribution of this type evidently corresponds to replacing the perturbative estimate of the contribution $C(p^2)\pi(q^2)$, which depends strongly on the dynamics at distances $\sim 1/Q$, by an expression incorporating nonperturbative effects. Technically, this replacement is achieved by using the dispersion relation

$$\pi(q^2) - \pi^{\rm pert}(q^2) = \frac{1}{\pi}\int_0^\infty \frac{ds}{s-q^2}(\rho(s) - \rho^{\rm pert}(s)), \tag{2}$$

where $\rho(s)$ is the "exact" spectral density and $\rho^{\rm pert}(s)$ the perturbative spectral density. The reason why there are no subtractions in (2) is that we have $\pi(q^2) - \pi^{\rm pert}(q^2) \to 0$ in the limit $|q^2| \to \infty$ by virtue of asymptotic freedom (see Appendix B in Ref. 9). Specific calculations of $\rho(s)$ have used the model $\rho(s) = f\delta(s - m_\rho^2) + \rho^{\rm pert}(s)\theta(s > s_0)$ whose parameters (f and s_0) have been extracted in the standard way from the QCD sum rules[1] for the corresponding correlation function $\pi(q^2)$. For correlation functions with $\rho^{\rm pert}(s) = 0$ this approximation reduces to vector dominance (cf. Ref. 6). For the $\langle\bar\psi\psi\rangle^2$ corrections, however, which make a numerically large contribution, we need to refine this approximation by parametrizing the contribution of higher-lying states by means of an effective resonance, (i.e., by taking[10] $\rho(s) = f_1\delta(s - m_\rho^2) + f_2\delta(s - m_{\rho'}^2)$. The sum rule found by us in this manner for $\hat{B}(p^2, M^2)T(Q^2, p_1^2, p_2^2)$ [$\hat{B}(p^2, M^2)$ is a Borel transformation[1]] is

$$\frac{f_\pi^2 F_\pi(Q^2)}{M^4} + \frac{c_\pi(Q^2)}{M^2} + e^{-m_{A_1}^2/M^2}\left\{\frac{f_{A_1}^2 F_{A_1}(Q^2)}{M^4} + \frac{c_{A_1}(Q^2)}{M^2}\right\} + \dots$$

$$= \frac{3}{2\pi^2 M^2}\int_0^1 dx \int_0^1 d\xi\, x(1-x)\exp\left(-\frac{x}{1-x}\xi(1-\xi)\frac{Q^2}{M^2}\right)$$

$$+ \frac{1}{40\pi^2 M^6}\left[Q^4 \ln\left(\frac{Q^2}{Q^2+s_0}\right) + Q^2 s_0 - \frac{s_0^2}{2}\,\frac{Q^2}{Q^2+m_\rho^2}\right]$$

$$+ \frac{\alpha_s\langle G_{\mu\nu}^a G_{\mu\nu}^a\rangle}{24\pi M^6}\left(1 + \left[\frac{m_\rho^2}{Q^2+m_\rho^2}\right]\right)$$

$$+ \frac{16}{81}\,\frac{\pi\alpha_s\langle\bar\psi\psi\rangle^2}{M^8}\left(5 + 6\left[\frac{1.6\,m_\rho^2}{Q^2+m_\rho^2} - \frac{0.6 m_{\rho'}^2}{Q^2+m_{\rho'}^2}\right]\right), \tag{3}$$

where $s_0 = 1.5$ GeV^2, $m_{\rho'}^2 = 2.0$ GeV^2, and $m_{A_1}^2 = 1.6$ GeV^2.

We would like to point out two properties of sum rule (3).

a) At $Q^2 = 0$, the right side of (3) (multiplied by M^2) agrees with the expansion for the correlation function of axial currents, $\pi \sim n^\alpha n^\beta \langle j_\alpha^+ j_\beta\rangle/(nP)^2$, so that we have

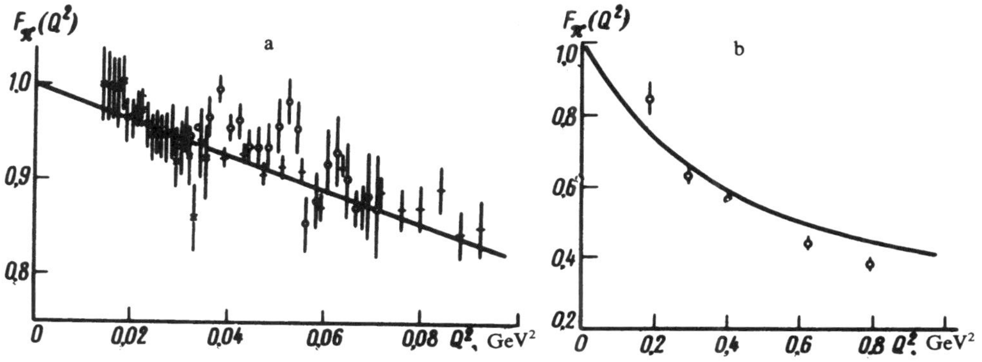

FIG. 1. Comparison of the experimental data on $F_\pi(Q^2)$ with the results of an analysis of sum rule (3). a—The region $Q^2<0.1$ GeV^2, with data from Ref. 11; b—the region $Q^2<1$ GeV^2, with data from Ref. 12.

$F_\pi(0)=1$, $F_{A_1}(0)=1$, and $C_i(0)=0$. This fact is a consequence of the Ward identity for T (cf. Ref. 7). Subtracting from (3) the sum rule for $Q^2=0$, we find a sum rule which can be used to find $[F_\pi(Q^2)-1]/Q^2$. Following the standard procedure for analyzing a sum rule of this type, 5–7 we find the value $\langle r_\pi^2\rangle 1/2 = 0.66 \pm 0.33$ fm for the electromagnetic radius of the pion, in good agreement with the experimental value $\langle r_\pi^2\rangle_{\rm expt}^{1/2} = 0.636 \pm 0.037$ fm (Fig. 1a).

b) The additional terms which we mentioned earlier [enclosed in square brackets in (3)] must "die out" with increasing Q^2. It is easy to see that this requirement is met in (3). We note, however, that an extrapolation of the $1/Q^2$ decay law of these terms beyond the formal range of applicability of sum rule (3), $Q^2 \lesssim m_\rho^2$, would be incorrect, since there are no contributions $\mathcal{O}(1/Q^2)$ (or of higher powers of $1/Q^2$) in the $\langle GG\rangle$ and $\langle\bar\psi\psi\rangle^2$ corrections in the operator expansion for $T(Q^2,p_1^2,p_2^2)$ in the kinematics $Q^2\sim|p^2|$. In other words, at $Q^2\gtrsim m_\rho^2$ the additional contributions should actually fall off more rapidly than any power of $1/Q^2$. Since these terms in (3) do not fall off rapidly enough, the value found for $F_\pi(Q^2=0.6\ \text{GeV}^2)$ turns out to be 10% higher than the experimental value (Fig. 1b). On the other hand, if we set the additional contributions equal to zero (this approach corresponds to switching to a sum rule for the kinematics $Q^2\sim|p^2|\gtrsim m_\rho^2$), we find a value $F_\pi^{\rm teor}(Q^2=0.6\ \text{GeV}^2)$ which is 10% smaller than the experimental value. This result means that the sum rules for the two kinematics ($Q^2 < m_\rho^2$ and $Q^2\gtrsim m_\rho^2$) at the point $Q^2=m_\rho^2$ agree with each other and with the experimental curve to the necessary accuracy of 10–20%.

In summary, the method of QCD sum rules has been used to generate a good description of the data on the electromagnetic form factor of the pion over essentially the entire region of space-like momentum transfer which has been studied experimentally, $0\leqslant Q^2\leqslant 3-4$ GeV^2.

In the course of our calculations we learned of a result derived by Chetyrkin *et al.*,[13] who attempted to calculate $\langle r_\pi^2\rangle$ by an analogous method but without considering the $\langle GG\rangle$ and $\langle\bar\psi\psi\rangle^2$ contributions, whose calculations were the most laborious part of our own study. Incorporating these contributions increases the value found in Ref. 13 for $\langle r_\pi^2 \rangle$ by a factor of 1.5, but the point we wish to make is not this numerical discrepancy but the circumstance that it is incorrect in principle to discard contributions of this sort when using the method of QCD sum rules, since these contributions accumulate information on the dynamics of the system in question.

We wish to thank S. G. Gorishniĭ, S. A. Larin, and K. G. Chetyrkin for useful discussions. We also thank A. V. Efremov for interest in this study.

[1]M. A. Shifman, A. I. Vaĭnshteĭn, and V. I. Zakharov, Nucl. Phys. **B147**, 385, 448 (1979).
[2]B. L. Ioffe, Nucl. Phys. **B188**, 317 (1981).
[3]V. A. Nesterenko and A. V. Radyushkin, Pis'ma Zh. Eksp. Teor. Fiz. **35**, 395 (1982) [JETP Lett. **35**, 488 (1982)]; V. A. Nesterenko and A. V. Radyushkin, Phys. Lett. **115**, 410 (1982).
[4]B. L. Ioffe and A. V. Smilga, Phys. Lett. **114B**, 353 (1982).
[5]B. L. Ioffe and A. V. Smilga, Pis'ma Zh. Eksp. Teor. Fiz. **37**, 250 (1983) [JETP Lett. **37** 298 (1983)].
[6]I. I. Balitsky and A. V. Yung, Phys. Lett. **129B**, 388 (1983).
[7]V. M. Belyaev and Ya I. Kogan, Pis'ma Zh. Eksp. Teor, Fiz. **37**, 611 (1983) [JETP Lett. **37**, 730 (1983)].
[8]A. V. Efremov and A. V. Radyushkin, Teor. Mat. Fiz. **44**, 16, 157, 327 (1980); **42**, 147 (1980); A. V. Radyushkin, Fiz. Elem. Chastits At.. Yadra. **14**, 58 (1983) [Sov. J. Part. Nucl. **14**, 23 (1983)].
[9]A. I. Vaĭnshteĭn *et al.*, Fiz. Elem. Chastits At. Yadra. **13**, 542 (1982) [Sov. J. Part. Nucl. **13**, 224 (1982)]; V. A. Novikov *et al.*, Nucl. Phys. **B191**, 301 (1981).
[10]V. M. Belyaev and Ya. I. Kogan, ITEP-12, Moscow, 1984.
[11]A. S. Vodop'yanov and É. N. Tsyganov, Fiz. Elem. Chastits At. Yadra. **15**, 5 (1984) [Sov. J. Part. Nucl. **15** (to be published)].
[12]C. Bebek *et al.*, Phys. Rev. **D17**, 1693 (1978).
[13]K. G. Chetyrkin, S. G. Gorishny, A. B. Krasulin, S. A. Larin, and V. A. Mateveev, Report *P*-340, INR, Moscow, 1984.

Translated by Dave Parsons
Edited by S. J. Amoretty

 0021-3640/84/120710-04$01.00

8. Multicolor Chromodynamics and the Sum Rules

Introduction

To make life easier theorists often "simplify" quantum chromodynamics by modifying certain aspects of the theory. Some features of the real world are either lost or distorted, of course. To make up for that these modifications ensure more freedom to manoeuvre and reveal new hidden parameters and symmetries.

In this chapter we will consider one of the most popular theoretical laboratories – multicolored QCD – with special emphasis on the topics which have a direct relation with the sum rule method. As any simplified model it cannot exactly reproduce the total physical content of the real world, and it is important to know beforehand where this or that model prediction is close to the exact answer and where one can expect deviations. In each particular case this question will be discussed separately.

Let us begin with a mini-review of the general rules of multicolored QCD suggested by 't Hooft in 1974 [1]. Combining these rules with the postulate of color confinement one can explain [2], at least at the qualitative level, known regularities of low-energy hadronic physics: the Zweig rule, the absence of the exotic mesons, etc. The description of mesons in the limit $N_c \to \infty$ is surprisingly successful in its gross features. There are, however, a few practically important exceptions: in the quark and gluon channels with total spin 0 the naive counting of powers of N_c fails – quantities small in the limit $N_c \to \infty$ turn out to be large and vice versa. This is a manifestation of the phenomenon discovered in ref. [3]: not all hadrons in the real world are equal to each other.

Multicolored QCD allows one to look from a new angle at the MIT bag model, or, more exactly, at the interpretation of the bag model which gradually emerged in QCD [4]. According to the generally accepted picture the cavity inside which the quarks are confined – the basic element of the bag model – is explained by a specific vacuum structure. The violent fluctuations of the gluon field populating the vacuum and decreasing its energy density, are frozen inside the cavity by the valence quarks. This freezing automatically leads to a positive volume energy inside the bag. The key question is the degree of freezing. If it were close to one, the bag properties as well as the bag constant B would be universal for all hadrons, irrespective of their quantum numbers, quark or gluon content, etc. The QCD sum rules, however, demonstrate that the situation is opposite, see ref. [3]. The dependence on the number of colors, to be discussed below, is an additional argument in favor of a weak freezing. The vacuum energy density, as a function of N_c grows faster than the bag constant B.

The structure of the baryon sector is another interesting aspect of multicolored QCD. The problem has been first formulated by Witten [5], who built the baryons from heavy non-relativistic quarks with Coulomb interaction. We will be more interested in the opposite limit of massless non-relativistic quarks. The baryon mass in this case can be estimated with the aid of the sum rules [6]. The result, $m_B \propto N_c^1$, is not a surprise and is not very interesting by itself. What is more important, this prediction, in conjunction with the Goldberger–Treiman and Adler–Weisberger relations at large N_c, allows one to justify the naive constituent quark model used in hadronic phenomenology for almost three decades. Let us also parenthetically note that the baryon studies in multicolored QCD resulted in a revival of the skyrmion model (see e.g. the review paper [7]), so popular at present.

8.1. The limit of a large number of colors

Following 't Hooft [1], instead of three colors, let us consider a more general theory in which the number of colors is equal to N_c, and the corresponding gauge group is SU(N_c). Thus, we get a free parameter, N_c, and in the limit $N_c \to \infty$ the structure of the theory becomes considerably simpler.

More exactly, it is assumed that $N_c \to \infty$ and simultaneously

$$g_0^2 N_c \to \text{const}. \tag{8.1}$$

Here g_0 is the bare coupling constant, i.e. the constant normalized at the ultraviolet cutoff M_0.

Why is just this limiting procedure reasonable? According to the famous formula of asymptotic freedom

$$g_0^2 = \frac{16\pi^2}{\left(\frac{11}{3}N_c - \frac{2}{3}N_f\right) \ln M_0^2/\Lambda^2}, \tag{8.2}$$

where N_f is the number of the quark flavors and Λ is the QCD scale parameter. It is obvious that the ratio M_0^2/Λ^2 is to be held fixed and N_c independent – only in this case the masses of mesons will be finite at $N_c \to \infty$, and the multicolored world will be like ours. Equation (8.2) implies then that $g_0^2 \propto N_c^{-1}$. Clearly, the effective (renormalized) coupling constant has the same N_c dependence.

Decreasing g_0^2 according to eq. (8.1) we, nevertheless, do not get a trivial free theory due to the fact that in some Feynman graphs large factors $\propto N_c$ emerge. These factors are due to a large number of the color degrees of freedom in the intermediate state which compensate for the falloff of the coupling constant.

Consider, for instance, the diagrams of fig. 8.1. The first graph is the free quark propagator. The second graph, a one-loop correction, contains an additional factor $g_0^2 N_c$ compared to the first. If we insist on the existence of a smooth limit at $N_c \to \infty$ the product $g_0^2 N_c$ should be constant.

Let us examine in more detail how the standard expansion parameter g^2 is substituted by $g_0^2 N_c$ in the case at hand. In the matrix representation the gluon field is an $N_c \times N_c$ matrix which has N_c^2 components. (Strictly speaking, this is a traceless matrix so that the number of components is $N_c^2 - 1$. If the number of colors is large, however, this difference is inessential.) Let us denote the gluon field by A_j^i where the Lorentz indices are omitted since we will trace only the color degrees of freedom (it would be more accurate to write $(A_\mu)_j^i$ or $(G_{\mu\nu})_j^i$). The indices i and j run over N_c values, 1, 2, ..., N_c. The quark and antiquark fields, q^i and $\bar{q}_j$, have N_c components each.

Fig. 8.1. One-loop correction to the free quark propagator. The dashed line denotes gluons, the solid line quarks.

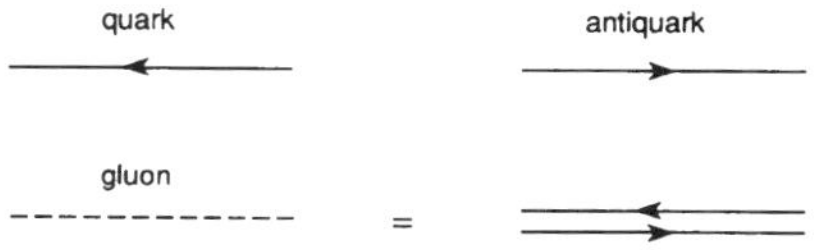

Fig. 8.2. The 't Hooft multicolored diagrams.

Thus, the gluon field is described by one color subscript, like the antiquark, and one superscript, like the quark. To count the color factors, and only for this purpose, let us, following 't Hooft, represent the gluon as a quark–antiquark pair, $A^i_j \sim q^i \bar{q}_j$ (see fig. 8.2). The interaction vertices in the 't Hooft form are depicted in fig. 8.3. Figure 8.4 gives a few examples of some simple graphs in the standard and 't Hooft representations. The 't Hooft language makes the color conservation absolutely transparent. Indeed, each color line entering the graph either forms a closed loop, or, sooner or later, leaves it.

Let us focus on the closed loops. Each closed loop in the 't Hooft graphs corresponds to summation over color and brings one power of N_c. The origin of this color factor is obvious: cutting the loop across we get N_c different terms corresponding to the color states $1\bar{1}$, $2\bar{2}, \ldots, N_c\bar{N}_c$.

To understand the advantages of the large N_c limit it is instructive to consider a simple example, say the photon polarization operator. The simplest quark loop, fig. 8.4a, is evidently proportional to N_c^1. Let us call this result "normal". One-gluon corrections are given in figs. 8.4b,c. The corresponding 't Hooft graphs contain two color loops, so that the result is proportional to $g^2N_c^2 \propto N_c$, the "normal" order of magnitude.

In the next order, among others, there appear the graphs depicted in figs. 8.4d,e. They have three color loops, and a corresponding contribution $\propto g^4N_c^3 \propto N_c$.

One can get the impression that every Feynman graph for the photon polarization operator is proportional to N_c. This impression is false. E.g. diagram 8.4f contains only one color loop and, hence, is $\propto g^4N_c^1 \propto N_c^{-1}$. Moreover, the graph depicted in fig. 8.4g, which includes two color loops in the 't Hooft representation, turns out to be proportional to $g^4N_c^2 \propto N_c^0$.

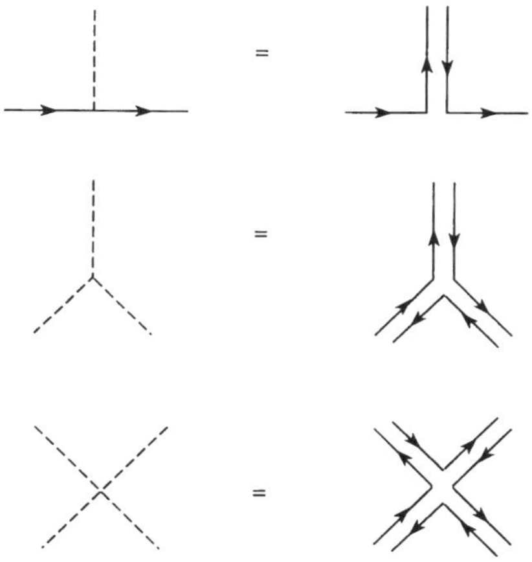

Fig. 8.3. The color structure of the interaction vertices in multicolored QCD.

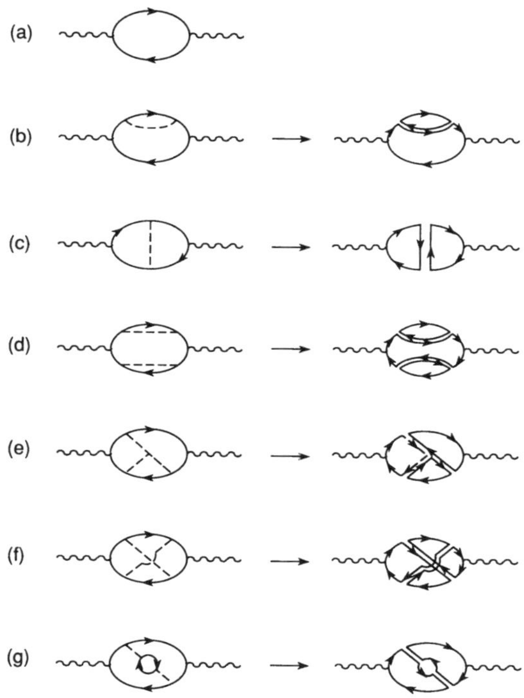

Fig. 8.4. The two-point function of the electromagnetic currents. The wavy lines denote a photon.

What is the difference between the "normal" graphs 8.4a–e from those in figs. 8.4f,g whose relative weight decreases at large N_c? One can easily convince oneself that all "normal" graphs are planar, i.e. they can be drawn on a plane without self-intersections. Diagram 8.4f, on the contrary, is not representable on a plane, and it is suppressed by an extra factor N_c^{-2}. Diagram 8.4g, although planar, contains an extra quark loop.

By inspecting several examples of this type it is not difficult to understand that the following selection rules are quite general: (i) the non-planar graphs are suppressed at least as N_c^{-2} relative to the planar ones; (ii) each additional fermion loop brings in an additional suppression $\propto N_c^{-1}$. The latter fact is absolutely trivial. The number of quark species per flavor is N_c while the number of gluons is N_c^2. Therefore, substituting a gluon loop by a quark one we lose a factor N_c. For the fermions in the adjoint representation (as in supersymmetric theories) the suppression of the fermion loops would be absent.

In certain cases there are additional, more specific selection rules [1,2], but there is no need to dwell on them now.

8.2. *Meson properties*

Thus, at $N_c \to \infty$ the whole variety of diagrams degenerates to a relatively small subset of planar graphs. The total number of the Feynman graphs is factorially large in high orders while the number of the planar graphs grows only in a power-like manner [8]. Unfortunately, even this degenerate planar theory is too complicated to allow for a quantitative solution.

However, by combining the simple rules formulated above with the color confinement postulate we are able to extract many qualitative consequences [2,9]. In this subsection we will concentrate on mesons, assuming that the number of quark flavors is fixed (does not grow with N_c). We also tacitly assume that the confinement mechanism, whatever it might be, is not ruined at large N_c.

Under these conditions at $N_c \to \infty$:

(i) the meson mass has a smooth limit;

(ii) the mesons become stable non-interacting particles, and their spectrum stretches up to infinity;

(iii) all decay amplitudes are suppressed by inverse powers of N_c, for instance, the quark meson transition into two quark mesons is $\propto N_c^{-1/2}$ while the transition into one quark meson and one glueball is $\propto N_c^{-1}$;

(iv) all meson states are pure; a $q\bar{q}$ pair of one flavor does not pass into a pair of another flavor or into gluons; two-quark states do not mix with the four-quark states, etc.;

(v) the meson scattering amplitudes are described by an infinite set of tree graphs with local vertices and free meson propagators.

The derivation of these and similar rules is extremely simple. Let us introduce external currents – colorless local operators built from the quark and/or gluon fields. Furthermore, let us consider, say, the two-point function

$$A(q) = \mathrm{i}\int \mathrm{d}^4x\ \mathrm{e}^{\mathrm{i}qx}\ \langle 0 | T[\bar{q}(x)q(x), \bar{q}(0)q(0)] | 0\rangle. \tag{8.3}$$

The Feynman graphs determining $A(q)$ in the limit $N_c \to \infty$ are similar to those displayed in figs. 8.4a–e. In accordance with the general rules, they are all planar, with a minimal number of fermion loops. Every cut of any diagram from this set contains two and only two quark lines. Moreover, no combination of the cut gluon lines can form a colorless subsystem (more exactly, one could separate such a subsystem but only at a price of losing two powers of N_c). It is easy to check the latter assertion by considering specific examples. For instance, in diagram 8.4d the structure of the color indices is as follows: $\bar{q}_i A^i_j A^j_k q^k$. Hence, the pair of gluons is in a non-singlet color representation (adjoint).

Invoking the color confinement hypothesis such a situation can be naturally interpreted as the presence of exclusively one-meson intermediate states in eq. (8.3). In other words,

$$A(q) = \sum_n \frac{|a_n|^2}{m_n^2 - q^2}, \tag{8.4}$$

where m_n and a_n are the mass and the residue of the nth meson, respectively, $a_n = \langle 0 | \bar{q}q | n\rangle$. On the other hand, asymptotic freedom requires that at large Euclidean momenta $A(q) \propto q^2 \ln q^2$. This requirement can be met only if the number of mesons in eq. (8.4) is infinite; otherwise at large q^2 the two-point function $A(q)$ would fall off. Each planar graph determining $A(q)$ is proportional to N_c. This formula can be realized in the most straightforward way under the condition

$$a_n \propto N_c^{1/2}, \quad m_n \propto N_c^0. \tag{8.5}$$

The growth of the residue with N_c is obvious by itself, though.

Equation (8.5) simultaneously shows that in the limit $N_c \to \infty$ the meson widths vanish, and they are all stable particles. It is clear, of course, that with switching on $1/N_c$ corrections we generate the decay amplitudes.

The simplest way to estimate the decay amplitudes is as follows. Let us consider the matrix element $\langle \pi^+(p_2)|J_\mu^{\text{em}}|\pi^+(p_1)\rangle$, where J_μ^{em} is the electromagnetic current. At $p_1 \to p_2$ this matrix element reduces to $(p_1+p_2)_\mu$, and since N_c counting does not depend on external momenta, $\langle \pi^+(p_2)|J_\mu^{\text{em}}|\pi^+(p_1)\rangle \propto N_c^0$ for any p_1 and p_2. On the other hand,

$$\left\langle \pi \right| J_\mu^{\text{em}} \left| \pi \right\rangle = \sum_n \frac{\langle \pi\pi | V_n \rangle \langle V_n | J_\mu^{\text{em}} | 0 \rangle}{m_n^2 - (p_1+p_2)^2}, \tag{8.6}$$

where the summation runs over all vector mesons with the appropriate quantum numbers. Since $\langle V_n | J_\mu^{\text{em}} | 0 \rangle \propto N_c^{1/2}$, see eq. (8.5), we immediately get that $\langle \pi\pi | V_n \rangle \propto N_c^{-1/2}$, as was announced above. All other results formulated above can be obtained in the same way by considering more complicated Green functions.

8.3. Experimental indications

The naive counting of N_c factors yields a lot of qualitative predictions which are in good agreement with experiment almost in all cases. This shows that as a rule $1/N_c$ is a good expansion parameter.

Let us list basic points where $1/N_c$ counting works perfectly.

(i) All normal meson states like ρ or ω (we will call them classic) are built from one pair of (constituent) quarks to a good approximation; extra pairs as well as valence gluons do not show up; the widths of the classic mesons are much smaller than their masses.

(ii) Transitions between quarks of different flavors and between quarks and gluons are suppressed. Experimentally the suppression corresponds to the so-called Zweig rule forbidding the decays like $\phi \to 3\pi$. It also explains the smallness of the glueball production cross section.

(iii) No meson was unambiguously identified thus far as a four-quark state; if such an exotic state exists it is definitely a rarity in the hadronic world, with a strongly suppressed production cross section.

All these regularities are explained by the suppression of quark loops in the large N_c limit. Take, for instance, the ρ–ω degeneracy. The simplest diagram distinguishing these channels (fig. 8.5) is suppressed by $1/N_c$ as compared to the "normal" graphs containing only one quark loop and coinciding for ρ and ω. As for exotic states, at $N_c = \infty$ the two-quark mesons become non-interacting and they cannot form bound states. To elucidate the point let us consider a local four-quark source,

$$B(x) = (\bar{\psi}(x)\Gamma\psi(x))\,(\bar{\psi}(x)\Gamma\psi(x)), \tag{8.7}$$

where Γ is some matrix in the Lorentz and flavor (but not color!) spaces. $B(x)$ produces

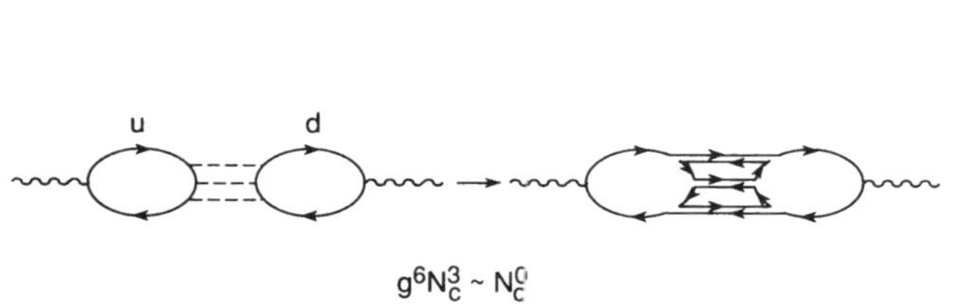

Fig. 8.5. The transition of u quarks into d in the correlation function of vector currents.

four-quark states from the vacuum. It is easy to see, however, that in the leading term $O(N_c^2)$ factorization takes place,

$$\langle 0 | T[B(x), B(0)] | 0 \rangle = \langle 0 | T[\bar{\psi}(x)\Gamma\psi(x), \bar{\psi}(0)\Gamma\psi(0)] | 0 \rangle^2. \tag{8.8}$$

In other words, the spectral density is saturated by pairs of freely propagating two-quark mesons.

We will not dwell in more detail on these and other well-known empiric regularities and the corresponding multicolor interpretation. As usual, of most interest now are exceptional cases where the theoretical predictions and experimental data deviate. In the channels with vacuum quantum numbers the valence quarks and gluons interact with the vacuum medium anomalously strongly, and the $1/N_c$ expansion is distorted. A new big numerical parameter appears, making the monotonous picture of $1/N_c$ more versatile [3].

8.4. *Where the naive N_c counting fails?*

The most spectacular example refers to the pseudoscalar mesons. In the vector and tensor nonets the neutral mesons have the quark content

$$\bar{u}u - \bar{d}d, \quad \bar{u}u + \bar{d}d, \quad \bar{s}s, \tag{8.9}$$

and the mixing between the strange quarks and u, d is small, at the level of a few percent, however the situation in the pseudoscalar nonet is quite different. Indeed, the η meson is almost pure octet while η' is almost a pure singlet,

$$\eta \sim \bar{u}u + \bar{d}d - 2\bar{s}s, \quad \eta' \sim \bar{u}u + \bar{d}d + \bar{s}s, \tag{8.10}$$

so that the transitions $\bar{u}u + \bar{d}d \leftrightarrow \bar{s}s$ are not suppressed. On the other hand, the counting of N_c factors does not depend on the quantum numbers. Then, what is the difference between 1^- and 0^-?

As a matter of fact, the zero-spin channels are specific because only there a direct mixing of the valence quarks and/or gluons with the vacuum fluctuations is possible. Interaction with the vacuum fields is so strong in these channels that the corresponding numerical parameter outweighs $1/N_c$. One can make a simple estimate of the effect taking as a toy model the one-instanton contribution. If the correlation functions like $\langle \bar{u}\gamma_5 u, \bar{d}\gamma_5 d \rangle$ at q^2 of order of 1 GeV^2 are saturated by one instanton then the structure of the matrix of the correlators is immediately fixed by the 't Hooft effective Lagrangian [10],

$$\Pi_{ab}(q^2) = \begin{bmatrix} 0 & 1 & 1 \\ 1 & 0 & 1 \\ 1 & 1 & 0 \end{bmatrix} \times \text{a common function of } q^2. \tag{8.11}$$

Here $\Pi_{ab} = \langle j_a(q), j_b(-q) \rangle$, the indices a and b mark the flavor of the quark pair (u, d or s) and j_a is the pseudoscalar quark current. The diagonal states of the matrix (8.11) are

$$\bar{u}u - \bar{d}d, \quad \bar{u}u + \bar{d}d - 2\bar{s}s, \quad \bar{u}u + \bar{d}d + \bar{s}s, \tag{8.12}$$

which exactly corresponds to the quark content of π, η, and η' mesons.

Much more advanced and sophisticated calculations along these lines are presented in ref. [11], and we refer the reader to these papers for further details.

Fig. 8.6. The correlation function (8.13): (a) the leading two-gluon contribution; (b) one of the quark–gluon diagrams corresponding to the two-pion contribution.

It is instructive to discuss here another simple and less known example, demonstrating a very considerable deviation from the large N_c limit. Consider the correlation function

$$\langle \theta_{\mu\mu}(x), \theta_{\nu\nu}(0) \rangle, \tag{8.13}$$

where $\theta_{\mu\mu}$ is the trace of the energy–momentum tensor, which reduces in the chiral limit ($m_u = m_d = m_s = 0$) to

$$\theta_{\mu\mu} = -\frac{b\alpha_s}{8\pi} G^a_{\alpha\beta} G^a_{\alpha\beta}, \quad b = \tfrac{11}{3} N_c - \tfrac{2}{3} N_f. \tag{8.14}$$

Using 't Hooft diagrams it is not difficult to find that the correlation function (8.13) is proportional to N_c^2. On the other hand, a pair of quark mesons, say 2π, gives a contribution N_c^0, a trivial assertion stemming from the fact that there are two quark loops in the diagram of fig. 8.6b. Thus, we expect that the ratio of the spectral densities

$$r = \frac{\left(\langle 0|(b\alpha_s/8\pi) G^2 |2 \text{ gluons}\rangle\right)^2}{\left(\langle 0|(b\alpha_s/8\pi) G^2 |\pi\pi + \bar{K}K + \eta\eta\rangle\right)^2} \propto N_c^2 \tag{8.15}$$

is large in the large N_c limit. (This ratio is very convenient since all normalization factors due to phase space cancel out.) Let us see whether our expectations are true.

The numerator is determined by the simplest graph of fig. 8.6a, and the denominator is known from the low-energy theorem [3]. If the invariant mass of the pair of mesons, M, is such that the low-energy expansion [3] is valid and, simultaneously $\alpha_s(M) < 1$, then

$$r = \frac{N_c^2 - 1}{16 \ln^2 M/\Lambda}. \tag{8.16}$$

In full accordance with the general rules $r \sim N_c^2 \to \infty$ in the limit $N_c \to \infty$. For $N_c = 3$, however, the ratio r is small rather than large, $r \sim 0.1 \ll 1$.

Thus we have a clear indication of the anomalously strong coupling of the gluon and quark degrees of freedom in the spin-zero channel. The same phenomenon is responsible for a relatively high sensitivity of the vacuum energy density ϵ_{vac} on the quark mass. Asymptotically, the vacuum energy density is m_q independent since m_q can enter only via the quark loops suppressed by $1/N_c$. In the three-colored world, however,

$$\left[\frac{\partial}{\partial m_s} \epsilon_{\text{vac}}\right]_{m_s = 0} = \langle \bar{s}s \rangle. \tag{8.17}$$

Substituting numbers and taking into account the fact that both ϵ_{vac} and $\langle \bar{s}s \rangle$ are negative we see that when m_s changes from zero to 150 MeV $|\epsilon_{\text{vac}}|$ increases roughly by a factor of two.

Finally, the last remark referring to the η' meson. As was noted by Witten [12], unlike all normal mesons whose masses are N_c independent, the η' mass vanishes in the limit of large

N_c. Experimentally $m_{\eta'}^2/m_\rho^2 \approx 1.8$, where the ρ meson is chosen as a typical representative of the "normal" mesons. The reason is the same as before – a very strong coupling of η′ to the gluon vacuum fields.

Summarizing, I can say that the limit of large N_c nicely explains a lot of empirical regularities. There are some specific situations, however, where $1/N_c = 1/3$ is not a good expansion parameter. These situations are precisely the same exceptional cases which are intractable or poorly tractable within the sum rule method.

8.5. Bags

The bag model, at least in its simplest form, assumes a sharp boundary between two phases – outside and inside the hadron – and a universal volume energy density B.

Alas – the actual picture seems to be more complicated since the degree of freezing of the vacuum fluctuations inside the hadrons is not high. Let us demonstrate the latter assertion in the limit of large N_c.

The first task is to establish the N_c dependence of the different matrix elements, in particular the quark and gluon condensates. To this end it is convenient to use the φ meson sum rule in multicolored QCD,

$$\frac{1}{\pi M^2}\int e^{(-s/M^2)}\,\mathrm{Im}\,\Phi(s)\,\mathrm{d}s = \frac{1}{4\pi^2}\left[\frac{N_c}{3} + \frac{8\pi^2\langle m_s\bar{s}s\rangle}{M^4} + \frac{\pi}{3}\frac{\langle\alpha_s G^a_{\mu\nu}G^a_{\mu\nu}\rangle}{M^4} + \cdots\right], \tag{8.18}$$

where

$$(q_\mu q_\nu - q^2 g_{\mu\nu})\Phi(q^2) = \mathrm{i}\int e^{\mathrm{i}qx}\,\mathrm{d}x\,\langle T[\bar{s}(x)\gamma_\mu s(x),\,\bar{s}(0)\gamma_\mu s(0)]\rangle. \tag{8.19}$$

The explicit N_c dependence enters only in the first term, corresponding to asymptotic freedom. As well known in the sum rule method the boundary value of M^2 where the power corrections become of the order of the leading term is close to the mass of the lowest-lying state. In other words,

$$\frac{\langle m_s\bar{s}s\rangle}{m_\phi^4} = \text{const.}\,N_c, \qquad \frac{\langle\alpha_s G^a_{\mu\nu}G^a_{\mu\nu}\rangle}{m_\phi^4} = \text{const.}\,N_c. \tag{8.20}$$

Since the meson mass is N_c independent in the large N_c limit we get [13]

$$\langle\bar{s}s\rangle \sim N_c \quad\text{and}\quad \langle\alpha_s G^a_{\mu\nu}G^a_{\mu\nu}\rangle \sim N_c, \tag{8.21}$$

in full agreement with what one would get from the naive counting of the number of terms in $\bar{s}s$ and G^2. Using the fact that the vacuum energy density ϵ_{vac} is proportional to the gluon condensate, one arrives at the following conclusion:

$$\epsilon_{\text{vac}} \sim \tfrac{11}{3}N_c\langle\alpha_s G^2\rangle \sim N_c^2. \tag{8.22}$$

On the other hand, the bag constant B in the standard bag should be N_c independent since otherwise the meson masses $\propto B_{\text{mes}}^{1/4}$ would have no smooth limit. (The subscript "mes" refers to mesons.) This means that

$$\frac{B_{\text{mes}}}{|\epsilon_{\text{vac}}|} \sim \frac{\epsilon_{\text{mes}} - \epsilon_{\text{vac}}}{|\epsilon_{\text{vac}}|} \sim N_c^{-2}, \tag{8.23}$$

and the suppression of the vacuum fluctuations inside the mesons is of the order N_c^{-2}.

Thus, we see that in the classic mesons, like ρ, the valence quarks affect the vacuum medium relatively weakly. As has been noted above, it is quite likely that there exist other mesons, with a stronger degree of suppression of the vacuum fluctuations inside the cavity. In the limit of large N_c the universality is restored, and all hadrons become indeed alike.

As for baryons, the number of quarks constituting the baryon in multicolored QCD is N_c, and its mass is also proportional to N_c, while the size is essentially N_c independent. If the baryon bag survives in the multicolored world, it can be only under the condition

$$B_{\text{bar}} \sim N_c, \tag{8.24}$$

since only under this condition

$$m_{\text{bar}} = \min_r\left(B_{\text{bar}} r^3 + \frac{N_c}{r}\right) \sim N_c. \tag{8.25}$$

Equation (8.24) implies, in turn, that

$$\frac{B_{\text{bar}}}{|\epsilon_{\text{vac}}|} \sim \frac{\epsilon_{\text{bar}} - \epsilon_{\text{vac}}}{|\epsilon_{\text{vac}}|} \sim N_c^{-1}. \tag{8.26}$$

In other words, the degree of suppression of the vacuum fluctuations in the baryons is larger by a factor N_c than that in the mesons, and still, is much less than one in the limit $N_c \to \infty$. The former fact is not surprising since the quark density in the baryons grows with N_c.

The above estimates can be considered as a qualitative argument against the phase transition at the hadron boundary and against the universal bag.

8.6. Factorization

Migdal was the first to notice that at large N_c any colorless operator $\mathcal{O}$ reduces to a classical field whose quantum fluctuations are negligibly small,

$$\langle \mathcal{O}^n \rangle_{\text{vac}} = \langle \mathcal{O} \rangle^n_{\text{vac}}. \tag{8.27}$$

Using the 't Hooft rules it is not difficult to check that the corrections to this relation are proportional to inverse powers of N_c. Let us consider, for instance, the operator $(\bar\psi\psi)^2$. Figure 8.7 shows that the non-factorizable contribution to $\langle(\bar\psi\psi)^2\rangle$ is of order N_c^0 while the leading factorizable term is of order N_c^2.

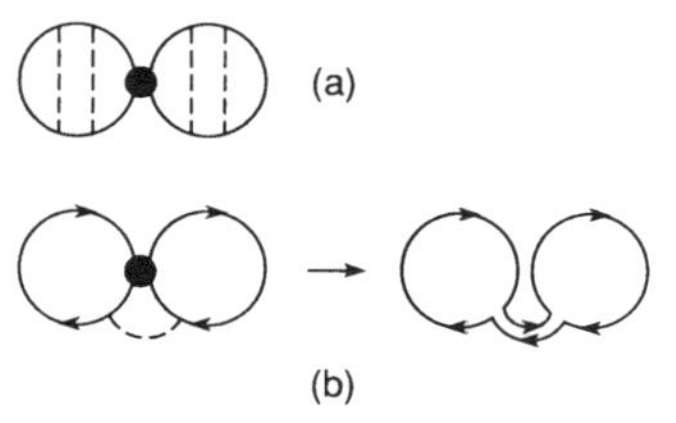

Fig. 8.7. Vacuum expectation value of the operator $(\bar\psi\psi)^2$: (a) an example of the factorizable graph $O(N_c^2)$; (b) non-factorizable graph $O(N_c^0)$.

The factorization property is crucial for our estimates of the condensates appearing in the sum rules for light quarks. The operators we encounter most often are of the type

$$(\bar{u}\gamma_\mu\gamma_5 t^a u)(\bar{d}\gamma_\mu\gamma_5 t^a d) \quad \text{and} \quad (\bar{u}\gamma_\mu\gamma_5 t^a u)(\bar{u}\gamma_\mu\gamma_5 t^a u). \tag{8.28}$$

Using the Fierz transformation for γ and t matrices * both structures can be reduced to a product of colorless operators,

$$(\bar{u}\gamma_\mu\gamma_5 t^a u)(\bar{u}\gamma_\mu\gamma_5 t^a u) = -\frac{2}{N_c}(\bar{u}\gamma_\mu\gamma_5 u)^2 + 2\Big[(\bar{u}u)^2 - (\bar{u}\gamma_5 u)^2 + \tfrac{1}{2}(\bar{u}\gamma_\mu u)^2 + \tfrac{1}{2}(\bar{u}\gamma_\mu\gamma_5 u)^2\Big], \tag{8.29}$$

$$(\bar{u}\gamma_\mu\gamma_5 t^a u)(\bar{d}\gamma_\mu\gamma_5 t^a d) = -\frac{2}{N_c}(\bar{u}\gamma_\mu\gamma_5 u)(\bar{d}\gamma_\mu\gamma_5 d) + 2\Big[(\bar{u}d)(\bar{d}u) - (\bar{u}\gamma_5 d)(\bar{d}\gamma_5 u) + \tfrac{1}{2}(\bar{u}\gamma_\mu d)(\bar{d}\gamma_\mu u) + \tfrac{1}{2}(\bar{u}\gamma_\mu\gamma_5 d)(\bar{d}\gamma_\mu\gamma_5 u)\Big]. \tag{8.30}$$

Now it is obvious that

$$\big\langle(\bar{u}\gamma_\mu\gamma_5 t^a u)(\bar{u}\gamma_\mu\gamma_5 t^a u)\big\rangle = 2\langle\bar{u}u\rangle^2 + O(N_c^0), \tag{8.31}$$

$$\big\langle(\bar{u}\gamma_\mu\gamma_5 t^a u)(\bar{d}\gamma_\mu\gamma_5 t^a d)\big\rangle = O(N_c^0), \tag{8.32}$$

and the first term in eq. (8.31) is of order of N_c^2. Thus, the $1/N_c$ expansion justifies the factorization (vacuum saturation) hypothesis.

It is worth emphasizing that eq. (8.27) does not mean that factorization is valid in all cases where the vacuum intermediate state can be inserted. Let us consider, for example, the gluon operator

$$\mathcal{O} = G^a_{\mu\nu}G^b_{\mu\nu}G^c_{\alpha\beta}G^d_{\alpha\beta}f^{ace}f^{bde}. \tag{8.33}$$

The vacuum intermediate state gives here a non-vanishing contribution,

$$\mathcal{O} \sim f^{ace}f^{bde}\Big[\big\langle G^a_{\mu\nu}G^b_{\mu\nu}\big\rangle\big\langle G^c_{\alpha\beta}G^d_{\alpha\beta}\big\rangle + \big\langle G^a_{\mu\nu}G^d_{\alpha\beta}\big\rangle\big\langle G^b_{\mu\nu}G^c_{\alpha\beta}\big\rangle\Big] \sim \frac{1}{N_c}\langle G^2\rangle^2. \tag{8.34}$$

However, there are absolutely no reasons to believe that the latter relation becomes exact in the large N_c limit. The quantum corrections can be neglected only when the singlet substructures can be singled out without $(1/N_c)$ factors (cf. eq. (8.31)). The vacuum expectation of the four-gluon operator (8.33) is proportional to N_c^3, i.e. an odd power of N_c. This fact alone signals that $\mathcal{O}$ does not reduce to a product of colorless gluon operators.

8.7. Baryons in multicolored QCD

Baryons in the large N_c limit can and actually were considered [6] using the sum rule method; all standard relations like $m_{\text{bar}} \propto N_c$ are nicely reproduced. We will not discuss this topic here in detail since at present there exists a much more elegant and convenient approach, treating the baryons in the limit $N_c \to \infty$ as solitons in the pion chiral Lagrangian,

* The latter has the form $t^a_{ij}t^a_{kl} = 2\delta_{il}\delta_{jk} - (2/N_c)\delta_{ij}\delta_{kl}$.

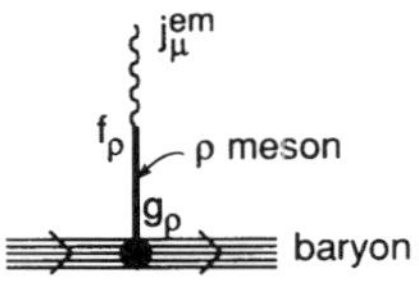

Fig. 8.8. The ρ-meson contribution in the matrix element $\langle \text{baryon} | j_\mu^{em} | \text{baryon} \rangle$.

the so-called skyrmion model (for a review and exhaustive list of references see ref. [7]). We will limit ourselves to a few remarks concerning the pion–nucleon interaction and the baryon form factors.

Let us treat both the number of colors N_c and the number of the quark flavors N_f as free parameters. Then in this theory we have $N_f^2 - 1$ Goldstone particles which will be called generically π mesons. All π mesons are massless in the chiral limit. The question we address is the relation between the vertices πNN and ρNN. We would like to find out whether the approximate SU(6) symmetry of the three-colored world survives at large N_c and arbitrary N_f.

Assume at first that $N_f > N_c$ so that one can built the baryon from all different flavors (we will lift this constraint shortly). Let the quark structure of the initial and final baryons be $q_1 \cdots q_{N-1} q_N$ and $q_1 \cdots q_{N-1} q_{N+1}$, respectively, where the subscript refers to the quark flavor. The π and ρ mesons to be considered have quark content $q_N \bar{q}_{N+1}$. The constants $g_\pi \equiv g_{\pi NN}$ and $g_\rho \equiv g_{\rho NN}$ are introduced in the standard way, $g_\pi \bar{N}' \gamma_5 N \pi$ and $g_\rho \bar{N}' \gamma_\mu N \rho_\mu$, respectively. Since the current $\bar{q}_{N-1} \gamma_\mu q_N$ is conserved, and the corresponding charge is equal to one,

$$g_\rho \sim 1/f_\rho \sim N_c^{-1/2}, \tag{8.35}$$

where f_ρ is the meson–current coupling (see fig. 8.8). This result has a natural interpretation in the constituent quark model where we have a core-spectator $(q_1 \cdots q_{N-1})$ and the "valence" quark q_N participating in the scattering act, $q_N \to q_{N+1}$. The factor $N_c^{-1/2}$ appears due to normalization of the ρ meson wave function. It seems at first sight that precisely the same answer should be valid for the πNN vertex.

One can extract information on the pion–nucleon interaction from the Goldberger–Treiman,

$$2 m_{\text{bar}} g_A = f_\pi g_\pi, \tag{8.36}$$

and the Adler–Weisberger relation,

$$g_A^2 + \frac{f_\pi^2}{\pi} \int_{m_{\text{bar}} m_\pi}^{\infty} \frac{d\nu}{\nu} [\sigma_{\pi}(\nu) - \sigma_{\bar{\pi}}(\nu)] = 1, \tag{8.37}$$

which, evidently, remain valid in multicolored QCD [6,14]. (Here g_A is the axial baryon coupling constant corresponding to the current $\bar{q}_N \gamma_\mu \gamma_5 q_{N+1}$.) Taking into account the fact that $m_{\text{bar}} \propto N_c$ and $f_\pi \propto N_c^{1/2}$, eq. (8.36) implies

$$g_\pi / g_A \sim N_c^{1/2}. \tag{8.38}$$

On the other hand, combining eqs. (8.36) and (8.37) we find

$$\frac{g_\pi^2}{4 m_{\text{bar}}^2} + \frac{1}{\pi} \int_{m_{\text{bar}} m_\pi}^{\infty} \frac{d\nu}{\nu} [\sigma_{\pi}(\nu) - \sigma_{\bar{\pi}}(\nu)] = \frac{1}{f_\pi^2} \sim \frac{1}{N_c}. \tag{8.39}$$

Here the first term on the left-hand side represents the pole contribution of the one-nucleon state. Assume now that, as in the real world, in multicolored QCD the integral on the right-hand side is saturated by a limited number of low-lying resonance states. Then, naturally, the pole and integral terms are of the same order of magnitude, and we arrive at the conclusion that [6]

$$g_\pi \sim N_c^{1/2}. \tag{8.40}$$

Comparison with eq. (8.38) yields

$$g_A = \text{const.}, \tag{8.41}$$

which agrees with the expectation of the naive quark model.

It is clear that the result $g_\pi \sim N_c^{1/2}$, $g_A \sim N_c^0$ remains valid for the baryons built from two quark flavors, say u and d, provided that the isotopic spin of the baryon is 1/2, the same as that of the nucleon. (Explicit consideration of the baryon sum rules in the limit $N_c \to \infty$ and with two quark flavors is presented in ref. [15].)

At first sight, eqs. (8.35) and (8.40) contradict the hypothesis of SU($2N_f$) symmetry, the analog of SU(6). One should take into account, however, that the πNN vertex due to γ_5 coupling contains a kinematical factor $\boldsymbol{q}/m_{\text{bar}}$ where $\boldsymbol{q}$ is the momentum transfer (of order N_c^0). The ρNN vertex has no such factor and, hence, both are proportional to $N_c^{-1/2}$, so that N_c counting and the constituent quark model do not contradict each other.

Finally, a last remark concerning the baryon form factors. Well known in modern theory are reliable methods to calculate the asymptotic behavior of form factors of any composite hadron in the limit $q^2 \to \infty$, where q is the momentum transfer. In the leading approximation a power law is predicted. (For a pedagogical presentation of the modern methods and an exhaustive list of references see e.g. the review paper [16]. Early attempts are described, for instance, in ref. [17].) In multicolored QCD the quark counting rules [18] yield for the baryon form factor

$$F_{\text{bar}}(q^2) \sim \frac{1}{(q^2)^{N_c-1}}, \quad q^2 \to \infty. \tag{8.42}$$

This means that at large N_c there exist many ($N_c - 2$) superconvergence relations of the type

$$\int \text{Im}\ F_{\text{bar}}(s) s^k \, \mathrm{d}s = 0, \quad k = 0, 1, \ldots, N_c - 3. \tag{8.43}$$

In other words, Im $F_{\text{bar}}(s)$ is an oscillating function including at least $\sim N_c$ oscillations.

On the other hand, the general analytic properties imply that the number of oscillations has a one-to-one correspondence with the number of resonances coupled to the $\overline{\text{B}}$B channel. The residues of these resonances should be comparable and have alternating signs. More exactly, the number of resonances should be $\geq N_c - 1$.

It is clear that the number of normal vector resonances strongly coupled to $\overline{\text{B}}$B, e.g. ρ, is limited. The natural conclusion is as follows: the lacking resonances are concentrated in the near-threshold domain $s \sim 4m_{\text{bar}}^2 \sim N_c^2$ and present the quasi-nuclear states, the so-called baryonium (for a review see ref. [19]). In multicolored QCD they are separated from the normal resonance domain by a huge gap.

Concluding the subsection devoted to baryons in multicolor QCD let me mention that the relation between the predictions of the baryon sum rules in the limit $N_c \to \infty$ and the Skyrme model is discussed in ref. [15].

References

[1] G. 't Hooft, Nucl. Phys. B 72 (1974) 461.
[2] E. Witten, Nucl. Phys. B 149 (1979) 285.
[3] V.A. Novikov, M.A. Shifman, A.I. Vainshtein and V.I. Zakharov, Nucl. Phys. B 191 (1981) 301.
[4] C. Callan, R. Dashen and D. Gross, Phys. Rev. D 19 (1979) 1826.
E.V. Shuryak, Phys. Lett. B 79 (1978) 135.
[5] E. Witten, Nucl. Phys. B 160 (1979) 57.
[6] B.I. Ioffe and M.A. Shifman, Nucl. Phys. B 202 (1982) 221.
[7] U.G. Meissner, Phys. Rep. 161 (1988) 213.
[8] J. Koplik, A. Neveu and S. Nussinov, Nucl. Phys. B 123 (1977) 109.
[9] G. Veneziano, Nucl. Phys. B 159 (1979) 213.
[10] G. 't Hooft, Phys. Rev. D 14 (1976) 3432.
[11] B.V. Geshkenbein and B.L. Ioffe, Nucl. Phys. B 166 (1980) 340.
E.V. Shuryak, Nucl. Phys. B 214 (1983) 237; Phys. Rep. 115 (1984) 151; Nucl. Phys. B 319 (1989) 541.
D.I. Diakonov and V.Yu. Petrov, Zh. Eksp. Teor. Fiz. 89 (1985) 751 [JETP 62 (1985) 431]; Nucl. Phys. B 272 (1986) 457.
[12] E. Witten, Nucl. Phys. B 156 (1979) 269.
[13] W. Bardeen and V.I. Zakharov, Phys. Lett. B 91 (1980) 111.
M.A. Shifman, Nucl. Phys. B 173 (1980) 13.
[14] A. Kakuto and F. Toyoda, Prog. Theor. Phys. 66 (1981) 2307.
[15] H.G. Dosch and S. Narison, Phys. Lett. B 184 (1987) 78.
[16] V.L. Chernyak and A.R. Zhitnitsky, Phys. Rep. 112 (1984) 173.
[17] D. Amati et al., Phys. Lett. B 27 (1968) 38.
[18] V.A. Matveev, R.M. Muradyan and A.N. Tavkhelidze, Nuovo Cimento Lett. 7 (1973) 719.
S.J. Brodsky and G.R. Farrar, Phys. Rev. Lett. 31 (1973) 1153; Phys. Rev. D 11 (1975) 1309.
[19] I.S. Shapiro, Phys. Rep. 35 (1978) 129.

9. Survey of New Trends

Introduction

This chapter gives an idea of the broad range of applications of the sum rule approach. Two concise reviews are kindly written especially for this volume by A. Smilga and V. Chernyak. I hasten to add that they certainly do not cover all problems solved or considered using the sum rule method. Suffice it to mention such an interesting although extremely complicated topic as exotic (or hybrid) mesons. This direction has been opened by the work [1], devoted to mesons with the quantum numbers 1^{-+}. Later this channel has been studied many times, along with other hybrid and four-quark states [2]. This subject remains controversial, and the theoretical situation is so volatile that I did not dare to discuss it in more detail here, although the results mentioned above are quite stimulating, beyond any doubt. Of specific interest is the low-energy theorem obtained in ref. [1].

The ongoing controversy in the current literature over several aspects and predictions of the method is not necessary a negative fact. On the contrary, it shows that the method is still alive, the development goes on, and new applications continue to appear – naturally – along with new question marks. One such issue, the value of the gluon condensate, has been mentioned in chapter 2. The pion (light-cone) wave function considered in detail in Chernyak's paper below is also the subject of a polemic. There exist basically two competing schools, whose origin can be traced back to works of Chernyak et al. and Radyushkin et al., respectively, in the early eighties. The essence of the contradiction between these schools is summarized in Chernyak's review. It is always risky to give the floor in scientific debates to the author representing only one side. Unfortunately, this volume cannot include a full-scale account of the arguments of both schools, and I hesitated a lot before I finally decided that one survey is better than nothing and that the issue is too important to neglect it at all. Anyhow, the interested reader can turn to the original publications mentioned in the list of references. A good supplementary reference to Radyushkin's approach is ref. [3], and recent results supporting the conclusions of ref. [3] are summarized in ref. [4]. According to these works the two-hump wave function advocated by Chernyak and Zhitnitsky is completely washed out by contributions of higher Gegenbauer polynomials. (Ref. [4] analyzes directly the value of the wave function at the center.) Additional arguments are presented in ref. [5].

The remarks above are addressed to the well-oriented reader, who presumably knows why the issue of the light-cone wave functions is so important and interesting. This question is not explicitly mentioned in Chernyak's review, and the relation between the wave functions under consideration and observable quantities is not elucidated. Perhaps, it is worth adding a few words explaining why the light-cone wave functions are needed at all.

The amplitudes of the exclusive transitions – the pion electromagnetic form factor is the clearest example – are in the focus of theoretical studies for many years; the peak of interest was probably in the seventies. It has been realized long ago that the asymptotic behavior of the form factors, in the limit $q^2 \to \infty$ where q is the momentum transfer, reflects the composite nature of hadrons. The peculiarities of the asymptotic expressions, in particular specific power laws, are in one-to-one correspondence with the number of consituents and the structure of their interactions (e.g. ref. [6]). If interactions switch off at short distances, as is the case in QCD, one can establish the so-called quark counting rules [7] stemming essentially from dimensional and scaling arguments. These rules predict the power asymptotics

$$F \sim Q^{-\nu}, \tag{9.1}$$

where $Q^2 = -q^2$ and the index ν is a calculable number related to the number of quarks constituting the hadron at hand and its quantum numbers.

As a matter of fact, in QCD one can go much further and work out a dynamical realization of the quark counting rules based on factorization of large- and short-distance contributions and explicit evaluation of relevant quark–gluon graphs [8]. In the limit $Q \to \infty$ these Feynman graphs completely control the asymptotic regime and reproduce the power law (9.1) modulo calculable logarithmic corrections,

$$F = AQ^{-\nu}(\ln Q)^{-\alpha}, \tag{9.2}$$

where A is a constant. What is remarkable, one can find not only the indices ν and α, but also the constant A, which turns out to be reducible to *an integral over the light-cone wave functions*. The latter, in turn, acquire a universal form and are trivially predictable provided that $\ln Q$ can be treated as a large parameter, and inverse powers of $\ln Q$ are neglected. The corresponding formulae are now usually referred to as the "naive asymptotic expressions". (For a pedagogical discussion of this topic see ref. [9].) Given the simplicity of the predictions one might consider this problem as one of the most elegant ever solved in QCD. Unfortunately, the solution *per se* is applicable only in the domain of very large Q^2, hardly reachable in any practical situation. Indeed to ensure the validity of the naive asymptotic formulae not only $Q^2/\Lambda^2_{\text{QCD}}$ must be large, actually it is necessary to require that a formal logarithmic parameter

$$\lambda = (\ln Q/\Lambda_{QCD})^{+\beta} \gg 1, \tag{9.3}$$

where β is typically a relatively small number. It is clear that all problems of practical importance refer to the domain – let us call it pre-asymptotic – where $\lambda \sim 1$ (this domain is still asymptotic in the sense that $Q^2/\Lambda^2_{\text{QCD}} \gg 1$). The form factors and other exclusive amplitudes in this domain are expressible in terms of the pre-asymptotic wave functions, which, in turn, can be found from the QCD sum rules. The latter task is discussed in detail in Chernyak's review.

Smilga's survey gives a broad view of the frontiers of the method. The only point which deserves a special remark is the situation with the pseudoscalar meson constants, which suddenly became a hot topic both due to the importance of the issue for numerous applications and due to recent lattice calculations contradicting previous sum rule results. The story takes a dramatic turn, as it actually happened more than once with the QCD sum rules. It might well happen that the discussion given in the review will become obsolete by the moment the volume is published due to new calculations which almost certainly are due to appear.

Concluding I would like to remind the reader that the responsibility for all assertions made in the reviews in this chapter stays with the authors of the reviews.

References

[1] I. Balitsky, D. Diakonov and A. Yung, Phys. Lett. B 112 (1982) 71; Yad. Fiz. 35 (1982) 1300 [Sov. J. Nucl. Phys. 35 (1982) 761].

[2] J. Govaerts, F. de Viron, D. Gusbin and J. Weyers, Phys. Lett. B 128 (1983) 262; Nucl. Phys. B 248 (1984) 1.
J. Govaerts, L.J. Reinders, H.R. Rubinstein and J. Weyers, Nucl. Phys. B 258 (1985) 215.
J. Govaerts, L.J. Reinders and J. Weyers, Nucl. Phys. B 262 (1985) 575.
J. Govaerts, L.J. Reinders, P. Francken, X. Ganze and J. Weyers, Nucl. Phys. B 284 (1987) 674.
J. Weyers, in: Proc. 1985 Cargese School of Physics, pp. 171–195.
V. Braun and A. Kolesnichenko, Phys. Lett. B 175 (1986) 485.
V. Braun and Yu. Shabelsky, Yad. Fiz. 50 (1989) 493 [Sov. J. Nucl. Phys. 50 (1989) 306].

[3] A.V. Radyushkin, Hadronic Form Factors: Perturbative QCD Versus QCD Sum Rules, Plenary Talk at the Second European Workshop on Hadronic Physics with Electrons beyond 10 GeV, Dourdan, France 1990.

[4] V. Braun and I. Filyanov, Z. Phys. C 44 (1989) 157.

[5] S.V. Mikhailov and A. Radyushkin, Phys. Rev. D 45 (1992) 1754.
A.P. Bakulev and A. Radyushkin, Phys. Lett. B 271 (1991) 223.

[6] D. Amati et al., Phys. Lett. B 27 (1968) 38.

[7] V.A. Matveev, R.M. Muradyan and A.N. Tavkhelidze, Nuovo Cimento Lett. 7 (1973) 719.
S.J. Brodsky and G.R. Farrar, Phys. Rev. Lett. 31 (1973) 1153; Phys. Rev. D 11 (1975) 1309.

[8] V.L. Chernyak and A.R. Zhitnitsky, Pis'ma Zh. Eksp. Teor. Fiz. 25 (1977) 544 [JETP Lett. 25 (1977) 510].
G.R. Farrar and D.R. Jackson, Phys. Rev. Lett. 43 (1979) 246.
G.P. Lepage and S.J. Brodsky, Phys. Lett. B 87 (1979) 359.
A.V. Efremov and A.V. Radyushkin, Phys. Lett. B 94 (1980) 245.

[9] V.L. Chernyak and A.R. Zhitnitsky, Phys. Rep. 112 (1984) 173.

QCD Sum Rules: New Frontiers

Andrei V. Smilga

Institute for Theoretical Physics, Bern University, CH-3012 Bern, Switzerland

The spectrum of applications of QCD sum rules is very broad. Actually, there are not so many problems in low-energy hadron physics not studied in the framework of the QCD sum rules. In some difficult cases the analysis was not so successful, but in many others new interesting and reliable results have been obtained.

The fact that some problems turned out to be too complicated for the QCD sum rules should be no surprise, of course. Even the most strong and powerful method is not omnipotent. We dare to guess, e.g., that the nucleon–nucleon scattering amplitudes will never be calculated in the QCD sum rules approach with a reasonable accuracy, and that no human being, not even the most gifted and trained one, will ever be able to cover 100 m in 5 s.

Still, the realm of problems for which the QCD sum rule method works well enough is wonderfully broad and large. Unfortunately, it is impossible for us to provide here a full and adequate discussion of everything which has been achieved. In this mini-review based mainly (but not exclusively) on recent developments, we have rather chosen several selected topics which well illustrate both the abilities and limitations of the sum rule method, and can only hope that the interested reader will be able to recover the missing information himself/herself.

1. $J/\psi \to \eta_c\gamma$ decay width

The process $J/\psi \to \eta_c\gamma$ plays a distinguished role among many other processes studied in the framework of the QCD sum rules. This is the *only* place which we know of where the sum rule predictions contradict sharply the existing experiment.

The calculations are done in a rather standard way. One has to consider the three-point function

$$T_{\mu\nu}(p, q) = \int d^4x\, d^4y\, e^{i(-px+p'y)} \langle T\{j_5(x)\, j_\mu(0)\, j_\nu(y)\}\rangle_0, \tag{1}$$

where $j_5 = \bar{c}\gamma^5 c$, $j_\mu = \bar{c}\gamma_\mu c$, in the Euclidean region p^2, $p'^2 < 0$, and calculate it in the framework of QCD taking account of all possible perturbative and non-perturbative corrections – in this case the most essential are the gluon condensate $\propto \langle \alpha_s G^2 \rangle$ and the perturbative one-loop corrections (see fig. 1) – and, on the other hand, to saturate the correlator (1) by the physical states η_c and J/ψ. In order to suppress higher-state contribu-

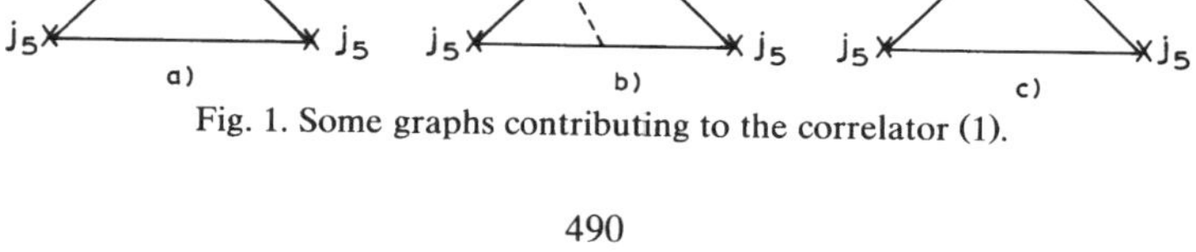

Fig. 1. Some graphs contributing to the correlator (1).

tions, one has, as usual, either to perform a double Borel transform over the parameters p^2 and p'^2 or to study the double moments M_{nk} of the corresponding spectral functions. Thus, the necessary technique is a kind of linear superposition of the technique employed earlier to study the charmonium spectrum and the double dispersion relation technique used to study the light-meson formfactors (see chapter 4 and chapter 6 of this book).

This investigation has been done in refs. [1–3]. The result for the partial decay width is [3]

$$\Gamma^{\text{theor}}_{J/\psi \to \eta_c \gamma} = (2.6 \pm 0.5)\ \text{keV}. \tag{2}$$

This value is a little bit larger than that quoted in refs. [1,2] only because, in this later work, the technically difficult calculation of the perturbative $\propto O(\alpha_s/\pi)$ corrections has also been performed. The account of perturbative corrections increases the prediction for the decay amplitude by about 10% and, correspondingly, for the decay width by ~20%. Prediction (2) agrees with the result obtained earlier within a related approach [4] and with the potential-model estimates [5]; it strongly disagrees with the experimental number [6]

$$\Gamma^{\text{exp}}_{J/\psi \to \eta_c \gamma} = (0.85 \pm 0.30)\ \text{keV}. \tag{3}$$

Since both perturbative and non-perturbative corrections as well as the excited state contribution in the confidential region of the sum rules of refs. [1–3] are reasonably small and well under control, we do not see any way to push down the theoretical prediction (2) by a factor ~3. Thus, a new independent measurement of the partial width $J/\psi \to \eta_c \gamma$ is highly desirable. Nowadays, we can only hope that the new experimental number would come into agreement with the QCD prediction (2). It *is* one of the crucial tests of QCD which still has to be done.

2. *The properties of charmed and beautiful mesons*

This subject is also interesting due to some controversy. This time, however, it is not the contradiction with experiment (relevant experimental information is still not available) but rather between different QCD sum rule calculations and different lattice calculations.

The most simple problem which must be investigated is the axial current residues of the pseudoscalar mesons, f_D and f_B,

$$\langle P|\bar{Q}\gamma_\mu\gamma^5 q|0\rangle = -\mathrm{i} f_P p_\mu, \tag{4}$$

which determine the partial leptonic widths of the B and D mesons. An enough accurate calculation of f_D and f_B in QCD sum rules has been performed by Aliev and Eletsky some years ago [7]. To this end, the correlator of two pseudoscalar currents $T\{\bar{Q}\gamma^5 q(x), \bar{q}\gamma^5 Q(0)\}$ has been considered. The perturbative $\propto \alpha_s$ as well as the non-perturbative corrections brought about by the v.e.v.'s $\langle \bar{q}q\rangle_0$, $\langle \alpha_s G^2\rangle_0$, $\langle \bar{q}\sigma_{\mu\nu} t^a G^a_{\mu\nu} q\rangle_0$ and $\alpha_s\langle \bar{q}q\rangle_0^2$ were taken into account. The following results were obtained:

$$f_D \simeq 170\ \text{MeV}, \quad f_B \simeq 130\ \text{MeV}, \tag{5}$$

with an estimated accuracy of ~20%. In a recent publication [8], this result has been confirmed.

There is, however, another work by Narison [9] where the same QCD sum rule was used but the results obtained for the B-meson are somewhat different: $f_B = (180 \pm 20)$ MeV (for f_D the results of refs. [7] and [9] coincide). In our opinion, however, the analysis of Narison is not reliable. There are two main questionable points:

(i) The continuum threshold in ref. [9] was chosen too high: $s_0 \simeq 50\ \text{GeV}^2$. As a result, stability with respect to the "technical" Borel parameter M^2 was obtained at too small values of M^2, outside the confidential region, so that the continuum contribution was practically unimportant whereas the power corrections in the "theoretical" part of sum rules were too high. As was explained in detail in the original papers (see chapter 2), theoretical estimates for the Borel-transformed correlators cease to be reliable in the small-M^2 region, since the unknown higher-order non-perturbative corrections become important. To check out whether the power and continuum corrections are still under control *is* a necessary part of any consistent sum rule analysis. The estimate of the accuracy of the predictions should be done on the basis of that.

In ref. [7] the continuum threshold was chosen in such a way that the stability in M^2 was achieved *inside* the confidential region. The continuum and power corrections were at the level of $\sim 40\%$ in the middle of the plateau, which resulted in an estimate of the accuracy $\sim 20\%$. Besides, the continuum threshold $s_0 = 33\ \text{GeV}^2$ [7] fits well the "mnemonic rule"

$$s_0 = (m_R + 0.5\ \text{GeV})^2, \tag{6}$$

which works amusingly well in any known channel.

(ii) The value of the on-shell charmed quark mass used in ref. [9] was a little bit overestimated whereas the value for m_b was a little bit underestimated. This brings about $\sim 15\%$ change in the predictions for f_P, so that f_B is increased and f_D diminished. For the B meson, these two effects (not quite correct value for m_b and overestimated continuum threshold) worked in the same direction and resulted in a $\sim 40\%$ overestimate in f_B. For f_D, on the other hand, these effects counterbalanced each other. That is why the results obtained in both papers for f_D are practically identical *.

Still, recent studies show that estimate (5) for f_b was somewhat underestimated [12]. The inaccuracy in the analysis of Aliev and Eletsky is, however, literally speaking, "of higher order". The matter is that the coefficient of α_s/π in the perturbative correction to the leading term in the sum rules [7] is unexpectedly large here, ~ 7–8, and the question of higher-order corrections and, especially, the proper normalization point for α_s becomes important.

Aliev and Eletsky substituted the coupling constant α_s normalized on the heavy-quark mass scale. An accurate re-summation of the leading logarithms in higher-order perturbative corrections [12] shows, however, that the coupling constant should be normalized rather on the normal hadronic scale –1 GeV (otherwise, these logarithmic higher-order corrections are essential). Physically, this is understandable: what counts is the characteristic size of the system, which is not small even if the mass of the heavy quark is very large. (Note the crucial difference with the heavy $\bar{Q}Q$ system whose size *is* determined by m_Q and α_s.)

As a result, estimate (5) is pushed up by 30–40 MeV. The authors of ref. [12] quote a value $f_B = 195$ MeV (with the choice $m_b = 4.8$ GeV). An extra increase is due to a somewhat higher value of the continuum threshold: $s_0 = 36\ \text{GeV}^2$. Despite the fact that this value of s_0 is (in

* In the recent paper [10] where, again, the sum rules [7] have been used, Narison obtains an even larger value for f_B: $f_B = (225 \pm 40)$ MeV. The main reason for the additional increase is the even smaller input value of m_b: $m_b = 4.6$ GeV compared to $m_b = 4.7$ GeV in ref. [9] and $m_b = 4.8$ GeV in ref. [7]. We should stress here that the difference of 200 MeV in m_b is, in fact, tremendous. What counts here is not the value of m_b as such, but rather the difference $m_B - m_b$ – the "binding energy" of the system. The value $m_b = (4.8 \pm 0.025)$ GeV (for the physical "pole" mass) was obtained long ago by Voloshin from the analysis of the sum rules for the Υ family [11]. When pushing down the value for m_b by 200 MeV, the Υ-resonance would find itself much above the $\bar{Q}Q$ threshold, which is hard to reconcile with the known fact that Coulomb effects in this system are rather important.

contrast to Narison's value) within reasonable uncertainties, the lower value $s_0 = 33\ \mathrm{GeV}^2$, adopted in ref. [7], seems to us preferable.

When choosing this lower continuum threshold, the QCD sum rule prediction for f_B would be at the level of

$$f_B \sim 170\ \mathrm{MeV}. \tag{7}$$

We like the lower value for s_0 more for two reasons. First, it conforms nicely to the empiric estimate (6). The second reason for which we prefer the lower value of 170 MeV for f_B is the following:

At large values of the quark masses, the pseudoscalar meson constant f_P should decrease as $\sim m_Q^{-1/2}$. This follows both from simple dimensional considerations [13] and from the *nonrelativistic sum rules* [14], which are valid in the limit of infinite heavy quarks. As was already discussed in the paper by Aliev and Eletsky the physics of the B meson is already rather close to this infinite heavy quark limit – the relativistic corrections in the sum rules constitute only about $\sim 25\%$. Thus, although the relativistic corrections for the D meson are 100% essential and one cannot expect the relation $(f_B/f_D) = (m_c/m_b)^{1/2}$ to hold, and it does not, an increase in the constant f_P in passing from the D to the B meson is a little bit unnatural.

However, the difference between 170 and 195 MeV is certainly within the accuracy of the whole sum rule method. We do not see the way to resolve this issue scientifically.

Still, there is another scientific but still unresolved question. As was already mentioned, the coefficient of α_s turned out to be unexpectedly large in this problem, so that perturbative corrections are at the level of $\sim 80\%$. In this situation, one feels a little bit uncomfortable until three-loop graphs are accurately calculated (not only to leading logarithmic terms). Hopefully, the coefficient of $(\alpha_s/\pi)^2$ is of "normal" order and the corresponding corrections turn out to be small.

The main general lesson to be extracted from this, perhaps too technical, discussion is that considerable care is required in handling the sum rules. Working with the right formulae in an improper way, it is very easy to obtain wrong results. We can also make the remark in this respect that to compare the predictions of, say, ref. [7] and ref. [10] and to conclude afterwards that the overall accuracy of the QCD sum rules is about 100% is, perhaps, the worst possible way to proceed.

The value (7) is rather close to the most recent lattice estimates for this value: $188 < f_B < 246$ MeV [15] and $f_B = (205 \pm 35)$ MeV [16] (the earlier lattice measurements suggested a much higher value for f_B in strong disagreement with the results of Aliev and Eletsky). If the value $f_B = 195$ MeV is accepted, there is no disagreement at all.

Note finally that the wish to know the exact value of f_B is not quite idle. It strongly affects the predictions for the top quark mass resulting from $B \leftrightarrow \bar{B}$ mixing analysis (see e.g. ref. [17]). Direct experimental measurements of this quantity are highly desirable.

The question next in complexity (after that of pure leptonic partial widths) are the semi-leptonic formfactors of the processes $D \to Ke\nu$, $B \to De\nu$, etc. These formfactors have been extensively studied using the QCD sum rules [18,19]. The technique is rather standard – one has to consider a three-point correlator and apply double Borel transformation in the meson channel (cf. chapter 6). The only novel feature is that now one can safely go to the kinematical region where q^2 (the invariant mass of the lepton pair) is positive. (Recall that it was not so in the pion formfactor problem.) The reason is, of course, simple – the cut in q^2 in the theoretical part of the sum rules starts now not from zero (as was the case for the light

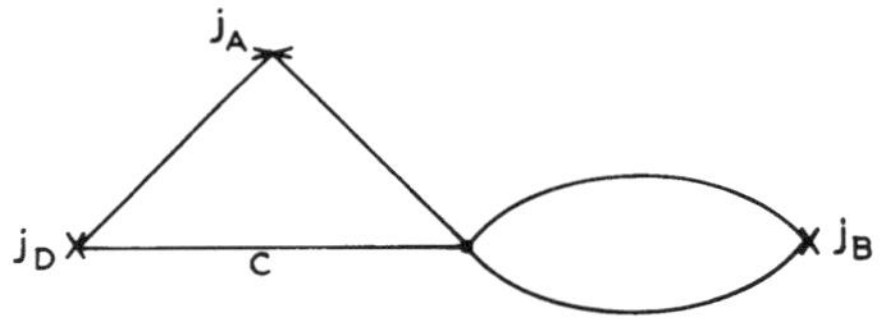

Fig. 2. A contribution to the correlator (8).

quarks) but from some threshold, which is m_c^2 for the processes $D \to \pi(K)e\nu$ and even higher in other cases. Thus, $q^2 = 0$ is no longer a distinguished dangerous point.

The results are, generally, in good agreement with the existing experimental data and with the lattice estimates, but there are also some controversies. For example, the value of the formfactor $A_2(0)$ of the decay $D \to \bar{K}^* e\nu$ estimated with the QCD sum rules [19] is much larger than the experimental value [20] and also larger than the lattice estimate [21]. We do not have at present an opinion on the reason for this disagreement and can only say that further studies (both theoretical and experimental) are required.

The heaviness of c and b quarks allows one to give also rather reliable sum rule predictions for more complicated processes, such as nonleptonic two-particle decay amplitudes of D-mesons [22]. To this end, one should consider *the four-point correlators*

$$\Pi_{D,AB}(q_A, q_B) = \int d^4x\, d^4y\, d^4z\, \exp\{iq_A x + iq_B y\} \langle T\{j_D(0)\, j_A(x)\, j_B(y)\, H_W(z)\}\rangle_0, \tag{8}$$

where $j_{D,A,B}$ are the proper currents coupled to the D meson and the mesons A and B in the final state and H_W is the weak Hamiltonian. A typical graph is depicted in fig. 2.

The four-fermion current H_W carries no external momentum, and the problem reminds a little bit of that of the nucleon magnetic moments (see chapter 7). There is, however, a considerable simplification – no new external-field-induced vacuum averages should be introduced here. The reason is, again, the large mass of the c quark. The virtual c quark line in fig. 2 turns out to be always well off mass shell and can be treated perturbatively [22].

The correlator (8) has an extra leg as compared to those treated in chapter 7, so the *double* Borel transformation should now be applied. The parasitic "single-pole" contributions of the kind we had in the magnetic moment problem are also present here. An estimate shows, however, that they are not very large in this case. The main uncertainty in the predictions [22] (about 50 decay channels of the type D, $D_s \to$ PP and D, $D_s \to$ VP have been analyzed) comes from unstudied chiral corrections (the terms proportional to the strange quark mass). In the worst cases, such as $D^0 \to K^-K^+$, $D^0 \to \bar{K}^0\eta$, the accuracy in the determination of the amplitudes is about 50% and, correspondingly, in the partial widths about 100%. In the channels not involving strange quarks, the accuracy is considerably higher – not worse than 50% for the partial widths. The results are in good accordance with the available experimental data.

Note that it is much more difficult to study similar processes for light mesons (like semi-leptonic or nonleptonic weak K-meson decays) with the QCD sum rules (new vacuum averages should be introduced). It is, after all, a general trend that the heavy mesons are usually more accessible to theoretical QCD analysis than the light ones.

3. Structure functions

Very soon after the technique of the QCD sum rules in external fields has been developed and the nucleon magnetic moments have been calculated (see chapter 7), it was used to determine the second moments of the nucleon structure functions, i.e. the fraction of nucleon energy carried by u and d quarks in the infinite momentum frame [23]. To this end, one should consider the nucleon current correlator in a weak static *gravitational field* which couples to the energy–momentum tensor $\theta_{\mu\nu}$. The results are in reasonable (though not as good as one might hope) agreement with experiment.

Much more interesting is an ambitious attempt of Belyaev and Ioffe to determine the functional dependence of the structure functions on the Bjorken scaling variable x in the framework of the QCD sum rules [24]. They considered the *four-point* correlator

$$\mathrm{i}T^{\pm}_{\mu\nu}(p, q) = \int \mathrm{d}^4x\ \mathrm{d}^4y\ \mathrm{d}^4z\ \mathrm{e}^{\mathrm{i}qx}\ \mathrm{e}^{\mathrm{i}p(y-z)} \left\langle T\left\{\eta(y)\ j^{\mp}_{\mu}(x)\ j^{\pm}_{\nu}(0)\ \bar{\eta}(z)\right\}\right\rangle_0, \tag{9}$$

where $j^{-}_{\mu} = \bar{d}\gamma_{\mu}(1+\gamma^5)u$, $j^{+}_{\mu} = \bar{u}\gamma_{\mu}(1+\gamma^5)d$ and η is the standard three-quark current with proton quantum numbers (see chapter 5). To avoid the necessity of summing up the perturbative logarithmic corrections, $Q^2 = -q^2$ was assumed to be moderately large, $Q^2 \sim 10$ GeV^2, $x = q^2/2pq$ is fixed and not too close to the kinematical boundaries $x = 0$ and 1 (see the subsequent discussion), and p^2 is Euclidean with characteristic values $|p^2| \sim 1$ GeV^2.

The structure function is connected with the imaginary part of the virtual photon–nucleon scattering amplitude at zero angle. Thus, if we want to determine it from the sum rules, we are bound to consider the imaginary part of the amplitude (9):

$$\mathrm{Im}\ T_{\mu\nu} = \frac{1}{2\mathrm{i}}\left[T_{\mu\nu}(p^2, q^2, s+\mathrm{i}\epsilon) - T_{\mu\nu}(p^2, q^2, s-\mathrm{i}\epsilon)\right]. \tag{10}$$

Now a serious theoretical problem arises: is it legitimate to use the method of operator product expansion for the imaginary part (10)? This is by no means a trivial question. We know e.g. that the operator expansion does not work for the imaginary part of the two-point correlators. Indeed, the imaginary part of higher-order terms $\sim 1/(q^2)^n$ in the operator expansion for, say, the vector current correlator $\Pi_{\mu\nu}(q)$ produces, being analytically continued in the physical region $q^2 > 0$, meaningless singularities $\propto \delta^{(n)}(q^2)$. For sure, the physical $\mathrm{Im}\ \Pi_{\mu\nu}(q) \propto \sigma(\mathrm{e}^+\mathrm{e}^- \to \text{hadrons})$ does not have singularities of this kind. The physical imaginary part grows starting from the threshold $q^2 = 4m_{\pi}^2$, shows some resonance structure and then levels off at the perturbation theory predictions.

We know what is going on in this case. The terms exponentially small and completely irrelevant at large Euclidean q^2 and not present in the operator expansion become very important in the physical region. They cancel out the singularities at $q^2 = 0$ and produce some oscillations which die out at large positive q^2 as a power of q^2. The mere requirement that the singularities should be canceled out combined with the vector-current conservation requirements produces a surprisingly good simulation of the observed resonance structures in $\mathrm{Im}\ \Pi_{\mu\nu}(q^2)$ [25].

Here, however, the situation is essentially different. There is already a large parameter $|q^2| \sim s = (p+q)^2$, and it is very reasonable to expect that all extra terms not accounted for in the operator expansion are suppressed as $|p^2|/|q^2|$. In other words, they are irrelevant

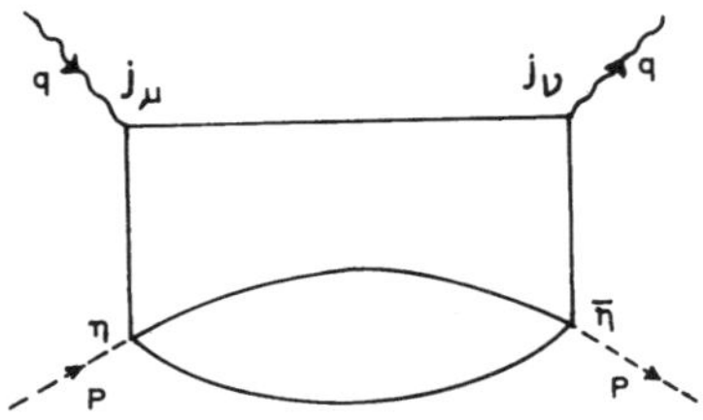

Fig. 3. The lowest-order contribution to the correlator (9).

for the leading twist structure functions and may contribute only in higher twist corrections [24].

After this crucial assumption is made, the procedure is rather standard. Belyaev and Ioffe calculated Im $T_{\mu\nu}(p^2, q^2, s)$ taking perturbative and non-perturbative corrections into account (the lowest-order contribution is depicted in fig. 3), performed the Borel transformation over the parameter p^2 (all terms $\propto |p^2|/|q^2|$ were consistently neglected) and compared the result with the phenomenological part of the sum rules. The single pole nucleon $\leftrightarrow$ continuum contributions were filtered out in the same way as in the problem of the magnetic moments.

The sum rules thus obtained have, however, a limited applicability region. One should not go too close to the kinematical boundaries $x = 0$ and 1. The reasons for this are as follows:

(i) The power corrections grow large and get out of control both at $x \sim 0$ and at $x \sim 1$.

(ii) At $x \sim 1$, the main contribution to the correlator (10) comes from the continuum, and the nucleon $\to$ nucleon contribution of interest is suppressed.

The region of applicability of the results for the valence quark distributions is estimated as $0.2 \le x \le 0.7$ for the u-quark distribution and $0.15 \le x \le 0.25$ [26] for the d quarks. The results are presented in fig. 4.

One can see that the agreement with experiment is, unfortunately, not astonishingly good. At present, the reasons for this disagreement are not completely understood and further study of this problem is certainly desirable. One can note, however, that one *could* expect in advance that the accuracy of determination of the d-quark distribution is lower than for the u quark. The reason is very simple – there are two valence u quarks in a proton and only on d quark. Accordingly, the leading term in the corresponding sum rules for the d quark is

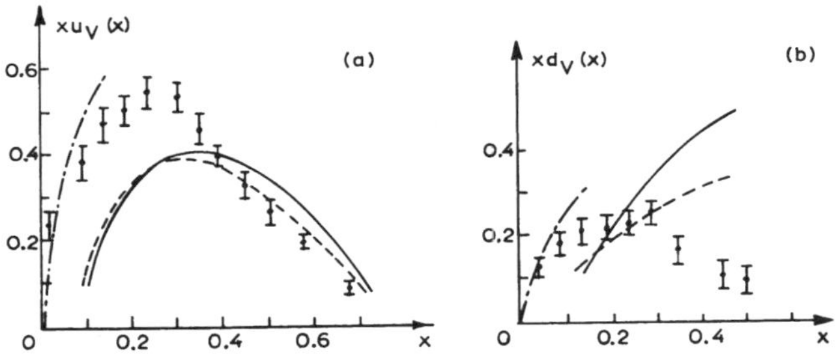

Fig. 4. Belyaev and Ioffe results for the u- and d-quark distributions as compared to experiment. Solid and dashed lines correspond to the different choices of continuum thresholds. The dotted–dashed line is a theoretical prediction for the valence quark distributions at small x obtained earlier using another method.

suppressed and the relative importance of the power corrections and also of the continuum contribution is higher [24].

The same technique has been applied later to calculate the polarized nucleon structure functions [27] and photon structure functions [28].

4. Finite temperature and density

There is quite a number of works devoted to attempts to adapt the QCD sum rule technique to the study of the hadron properties at finite temperature and density. However, no definite results have been obtained. Still, it is instructive to discuss this problem here as an example of one which clearly lies *beyond* the region of applicability of sum rules. We limit ourselves here to a discussion of the nucleon channel, where the potential hazards and relevant physics are most clearly seen.

The main idea of the early attempts to determine the nucleon properties at finite temperature [29] was as follows. At zero temperature, the most essential parameter determining the nucleon mass is the quark condensate $\langle \bar{q}q \rangle$. Thus, roughly, $m_N \propto \langle \bar{q}q \rangle^{1/3}$ (see chapter 5 for details). The condensate $\langle \bar{q}q \rangle$ depends, however, on the temperature. At the time when the work in ref. [29] was done, the character of this dependence was not yet clear, but now we know that

$$\langle \bar{q}q \rangle_T = \langle \bar{q}q \rangle_0 \left[1 - \frac{T^2}{8F_\pi^2} \right] \tag{11}$$

for $T \ll F_\pi$ [30]. Thus, we may conclude that the nucleon mass decreases with increasing temperature and, presumably, goes to zero at the point of the phase transition where the chiral symmetry is restored. One may even try to justify this conclusion: assuming that the main reason for the nucleon to be massive is the phenomenon of chirality breaking, it is reasonable to expect that the nucleon mass disappears at the point of the phase transition.

However, this is wrong. The matter is that *the very notion of mass is ambiguous at finite temperature*. Really, the mass is usually defined as the pole position of the corresponding Green function. At zero temperature, the structure of the propagator is rigidly fixed by the requirements of Lorentz invariance, and one and the same mass determines the pole position and the large-distance asymptotics of the Euclidean Green function $G(x) \propto \exp(-mx)$, which is actually probed in the QCD sum rules.

At finite temperature, however, Lorentz invariance is lost (there is a distinguished reference frame where the medium is globally at rest), and there are no reasons whatsoever to expect these Euclidean mass and pole mass to coincide. Indeed, they do not *.

If so, the Euclidean sum rules are virtually useless in the determination of the position of the nucleon pole (which is of most immediate physical meaning, although the question of physical observables is rather obscure here) in the medium at finite temperature. The technical reasons for which the naive sum rules do not work were studied in ref. [32]. The point is that the standard phenomenological model for the correlator spectral density – "resonance + continuum" – does not hold anymore at $T \neq 0$. One should take into account

* A good illustration of this general statement is provided by the properties of the quark–gluon plasma. The Debye screening mass determining the asymptotics of the correlator $\langle A_\mu(x) A_\nu(0) \rangle$ at large Euclidean distances does not coincide at all with the low-momenta "pole mass" – the plasmon frequency [31].

the effects due to scattering of the external currents by the thermal pions. A careful analysis shows that the nucleon mass is, in fact, not shifted by an amount of the order T^2/F_π^2.

A shift $\propto T^4$ is, generally speaking, present, but a much more important effect is the appearance in the same order of the imaginary part of the pole, which grows rapidly with temperature. The physical reason for the imaginary part to appear is very simple. It is just the *damping* of collective excitations with nucleon quantum numbers caused by collisions with the particles in the thermal medium (at low temperatures, they are, mostly, nearly massless pions). At temperatures $T \geq T_c$ (T_c being the phase transition point), the damping is so high that there is no reason to speak of nucleons anymore. The nucleons dissociate effectively in quarks and do not hang together.

Note that the adequate tool to study the properties of the nucleon (and other hadrons) in the region $T \ll T_c \sim F_\pi$ where the notion of hadron resonances is still meaningful, is not the sum rules but rather chiral perturbation theory. The phenomenological information on the interactions of low-energy pions with nucleons is also of a considerable help (see ref. [33] for more details).

In conclusion, let us say a few words about a recent interesting attempt to extract the properties (the equilibrium density and binding energy) of nucleons in nuclear matter from QCD sum rules [34]. (A similar attempt to exploit the QCD sum rule method to study isospin breaking effects in nuclei has been performed in ref. [35] but with less definite results.) The problem is very close in spirit to the situation at finite temperature; one should consider now the average of the nucleon current correlator over the density matrix with nonzero baryon chemical potential rather than with nonzero temperature. Like at finite temperature, the main effect comes from the change of $\langle \bar{q}q \rangle$ in the cold dense medium. In the phenomenological part of the sum rules, however, the usual "resonance + continuum" model has been used. We have seen that this model was not adequate in the finite temperature case – some extra contributions describing the scattering of the currents by the particles in the medium were present [32]. Until a similar analysis is done for the nuclear medium, no definite conclusions on the validity of the Drukarev and Levin procedure can be made *. Anyway, we think that the standard OBEP potential approach is much more adequate for the problem.

The results of ref. [34] depend strongly on the value of the σ-term. The authors use the experimental value $\sigma = 60$ MeV, which refers, however, to the so-called Cheng–Dashen kinematic point $t = 2m_\pi^2$, $\nu = 0$. For the matrix element $\langle N | \bar{u}u + \bar{d}d | N \rangle$ which enters the sum rules, the relevant point is $t = \nu = 0$, where the σ-term is rather 45 MeV [36]. If this latter value is accepted, the agreement of the predictions [34] with experiment is rather poor.

I am indebted to V.M. Braun, B.L. Ioffe and U.-G. Meissner for useful discussions. Special thanks are due to M.A. Shifman for careful reading of the preliminary draft of this paper and many important comments.

References

[1] A.Yu. Khodjamirian, Yad. Fiz. 39 (1984) 970 [Sov. J. Nucl. Phys. 39 (1984) 614].
[2] V.A. Beilin and A.V. Radyushkin, Yad. Fiz. 39 (1984) 1270 [Sov. J. Nucl. Phys. 39 (1984) 800].
[3] V.A. Beilin and A.V. Radyushkin, Nucl. Phys. B 260 (1985) 61.
[4] M. Shifman, Z. Phys. C 4 (1980) 345; (E) C 6 (1980) 282.

* In ref. [35], the continuum contribution has been altogether neglected, which is certainly not justified in our opinion.

[5] E. Eighten et al., Phys. Rev. D 21 (1980) 203.
[6] J.E. Gaiser et al., (Crystal Ball Collab.) Phys. Rev. D 34 (1986) 711.
[7] T.M. Aliev and V.L. Eletsky, Yad. Fiz. 38 (1983) 1537 [Sov. J. Nucl. Phys. 38 (1983) 936].
[8] V. Eletsky and E. Shuryak, Phys. Lett. B 276 (1992) 191.
[9] S. Narison, Phys. Lett. B 198 (1987) 104.
[10] S. Narison, Phys. Lett. B 285 (1992) 141.
[11] M.B. Voloshin, Nucl. Phys. B 154 (1979) 365.
M.B. Voloshin and Yu.M. Zaitsev, Usp. Fiz. Nauk 152 (1987) 361 [Sov. Phys. Usp. 30 (1987) 553].
[12] E. Bagan, P. Ball, V.M. Braun and H.G. Dosch, Phys. Lett. B 278 (1992) 457.
[13] Ya.I. Azimov, L.L. Frankfurt and V.A. Khoze, Pis'ma Zh. Eksp. Teor. Fiz. 24 (1976) 373 [JETP Lett. 24 (1976) 338].
[14] E.V. Shuryak, Nucl. Phys. B 198 (1982) 83.
[15] G. Alexandrou et al., Nucl. Phys. B 374 (1992) 263.
[16] A. Abada et al., Nucl. Phys. B 376 (1992) 172.
[17] N.G. Ural'tsev and V.A. Khoze, Yad. Fiz. 47 (1988) 1687 [Sov. J. Nucl. Phys. 47 (1988) 1069].
C. Dib, I. Dunietz, F. Gilman and Y. Nir, Phys. Rev. D 41 (1990) 1522.
[18] A.A. Ovchinnikov, Yad. Fiz. 50 (1989) 831 [Sov. J. Nucl. Phys. 50 (1989) 519].
A.A. Ovchinnikov and V.A. Slobodenyuk, Z. Phys. C 44 (1989) 433.
[19] P. Ball, V.M. Braun, H.G. Dosch and M. Neubert, Phys. Lett. B 259 (1991) 481.
P. Ball, V.M. Braun and H.G. Dosch, Phys. Rev. D 44 (1991) 3567.
[20] G. Crawford et al., (CLEO Collaboration) Phys. Rev. D 44 (1991) 3394.
[21] V. Lubitz, G. Martinelli and C.T. Sachrajda, Nucl. Phys. B 356 (1991) 301.
[22] B.Yu. Blok and M.A. Shifman, Yad. Fiz. 45 (1987) 211, 478, 841 [Sov. J. Nucl. Phys. 45 (1987) 135, 301, 522].
[23] A.V. Kolesnichenko, Yad. Fiz. 39 (1984) 1527 [Sov. J. Nucl. Phys. 39 (1984) 968].
[24] V.M. Belyaev and B.L. Ioffe, Nucl. Phys. B 310 (1988) 548.
[25] M.S. Dubovikov and A.V. Smilga, Yad. Fiz. 37 (1983) 984 [Sov. J. Nucl. Phys. 37 (1983) 585].
[26] B.L. Ioffe, private communication.
[27] V.M. Belyaev and B.L. Ioffe, Int. J. Mod. Phys. A 6 (1992) 1533.
[28] A.S. Gorsky, B.K. Ioffe, A.Yu. Khodjamirian and A. Oganesian, Z. Phys. C 44 (1989) 523.
[29] J. Dey, M. Dey and P. Ghose, Phys. Lett. B 165 (1985) 181.
[30] J. Gasser and Leutwyler, Phys. Lett. B 184 (1989) 83.
[31] O.K. Kalashnikov, Fortschr. Phys. 32 (1984) 525, and references therein.
[32] V.L. Eletsky, Phys. Lett. B 245 (1990) 229.
[33] H. Leutwyler and A.V. Smilga, Nucl. Phys. B 342 (1990) 302.
[34] E. Drukarev and E.M. Levin, Nucl. Phys. A 511 (1990) 679.
[35] T. Hatsuda, H. Hogaasen and M. Prakash, Phys. Rev. Lett. 66 (1991) 2851; Phys. Rev. C 42 (1990) 2212.
C. Adami and G.E. Brown, Z. Phys. A 340 (1991) 93.
[36] J. Gasser, H. Leutwyler and M.E. Sanyo, Phys. Lett. B 253 (1991) 252, 260.

Hadron Wave Functions and Hard Exclusive Processes in QCD

Viktor L. Chernyak

Institute for Nuclear Physics, Novosibirsk 630090, Russia

1. Introduction

The problem of bound states in any relativistic quantum field theory is very difficult. The more so, it is extremely difficult in QCD where the coupling constant is large in the low-energy region, and highly nontrivial non-perturbative effects (confinement, chiral symmetry breaking, etc.) are of great importance.

The conventional light-hadron states cannot, in general, be described as consisting of a few "current quarks" (i.e. those entering the fundamental QCD Lagrangian). The fact that the constituent quark model was successful in qualitatively describing the spectroscopy of light hadrons as bound states of a few "constituent quarks" is a matter of luck. The precise notion of a constituent quark, i.e. its connection with the quark field operators entering the QCD Lagrangian, had never been defined. Moreover, there are many reasons to consider the constituent quark as a complicated coherent mixture of the current quarks and gluons.

As usually, in some limiting cases the problem can be considerably simplified. In particular, it follows from simple dimensional arguments that the component of the hadron wave functions with a minimal number of constituents gives parametrically leading contributions to hard exclusive processes (i.e. those with large momentum transfers). For the conventional quark mesons we mean the wave functions of the form

$$\langle 0 | \bar{q}_\alpha(z) I q_\beta(-z) | M(p) \rangle = \phi_{\alpha\beta}(z, p), \quad I = \exp\left\{ \mathrm{i} g \int_{-z}^{z} \mathrm{d}\sigma_\mu \, A_\mu(\sigma) \right\},$$

where $q(x)$ and $A_\mu(x)$ are the quark and gluon field operators, α and β are spinor indices, $| M(p) \rangle$ is the meson state with momentum p. Moreover, the quarks q and $\bar{q}$ are produced in hard processes at small distances from each other, $z \sim 1/q \to 0$ (q is a typical large momentum transfer). Therefore, we can neglect the z^2 dependence of the wave function in the leading order (it will give power corrections) and can concentrate on the dependence on the variable $zp \sim O(1)$, $z \sim 1/q$, $p \sim q$. In other words, only the wave functions on the light cone, $z^2 = 0$, enter the leading terms of the hard amplitudes (and moreover, only the lowest-twist component of $\phi_{\alpha\beta}(z, p)$ gives a parametrically dominant contribution). Therefore, the problem is drastically simplified in the asymptotic limit *.

* In principle, the subleading power corrections can also be calculated but a general formalism which would allow one to treat this aspect accurately is not developed at present, and only the simplest examples are considered.

In the next subsections we will describe how the QCD sum rule approach allows one to approximately calculate the hadronic wave functions. It should be emphasized that, unlike the hard kernels, differing from one process to another, the wave function of a given hadron is a fundamental universal object which enters any process with the given hadron as participant.

2. *The leading twist pion wave function*

Let us consider the "two-particle" pion wave function and write it in its most general form:

$$\langle 0 | \bar{d}_\alpha(z) I u_\beta(-z) | \pi(p) \rangle = -\tfrac{1}{4}(\gamma_\mu\gamma_5)_{\alpha\beta} A_\mu + \cdots , \tag{1}$$

$$A_\mu = \langle 0 | \bar{d}(z)\gamma_\mu\gamma_5 I u(-z) | \pi(p) \rangle = \mathrm{i} f_\pi p_\mu \phi_\pi^{\mathrm{A}}(zp, z^2) + z_\mu \phi_2(zp, z^2). \tag{2}$$

Since $z \sim 1/q$, $p \sim q$, $zp \sim 1$, it is easy to see that the leading contribution to the hard amplitudes at $q \to \infty$ is provided by the wave function $\phi_\pi^{\mathrm{A}}(zp, z^2 \simeq 0)$ depending on zp only, and the omitted terms in eq. (1) will give power corrections. So, we will focus below on the properties of this leading twist wave function. Representing it in the form

$$\phi_\pi^{\mathrm{A}}(zp) = \int_0^1 \mathrm{d}x_1\, \mathrm{d}x_2\, \delta(1 - x_1 - x_2)\, \mathrm{e}^{\mathrm{i}(z\cdot x_1 p) - \mathrm{i}(z \cdot x_2 p)}\, \phi_\pi^{\mathrm{A}}(x_1, x_2) \tag{3}$$

it is not difficult to see that the wave function $\phi_\pi^{\mathrm{A}}(x_1, x_2)$ in eq. (3) describes the momentum distribution of the quarks inside the bound state (the momenta p_{u} and p_{d} carried by the u and $\bar{\mathrm{d}}$ quarks in the pion are $p_{\mathrm{u}} \approx x_1 p$ and $p_{\mathrm{d}} \approx x_2 p$, respectively; $0 \le x_1, x_2 \le 1$ are the momentum fractions). Our purpose is to investigate the properties of $\phi_\pi^{\mathrm{A}}(x_1, x_2)$.

Expanding A_μ in eq. (2) in z one finds:

$$\langle 0 | \bar{d}(0) \hat{z}\gamma_5 (z\overleftrightarrow{\mathrm{D}})^n u(0) | \pi(p) \rangle = \mathrm{i} f_\pi (zp)^{n+1} \int_{-1}^{1} \mathrm{d}\xi\, \xi^n \phi_\pi^{\mathrm{A}}(\xi), \tag{4}$$

$$\xi = x_1 - x_2, \quad x_1 + x_2 = 1, \quad z^2 = 0, \quad \mathrm{D}_\mu = \mathrm{i}\,\partial_\mu + g A_\mu, \quad \hat{z} = z_\mu \gamma_\mu . \tag{5}$$

Therefore, if we know the matrix elements on the left-hand side then we know the moments of the wave function. The QCD sum rule approach [1] allows one to find the values of these matrix elements.

Let us sketch the corresponding procedure. Consider the correlator [2]:

$$I_{n0}(z, q) = \mathrm{i} \int \mathrm{d}x\, \mathrm{e}^{\mathrm{i}qx} \langle 0 | T J_n(x) J_0^{+}(0) | 0 \rangle = (zq)^{n+1} I_{n0}(q^2), \tag{6}$$

$$J_n(x) = \bar{d}(x) \hat{z}\gamma_5 (z\overleftrightarrow{\mathrm{D}})^n u(x), \quad \overleftrightarrow{\mathrm{D}} = \overrightarrow{\mathrm{D}} - \overleftarrow{\mathrm{D}}, \quad z^2 = 0, \tag{7}$$

and calculate its asymptotic behavior at large q^2.

The leading contribution is found, of course, using perturbation theory. In the perturbative contributions the large momentum q is divided more or less equally between all lines. Besides that, there are contributions where the large momentum q flows through some selected lines only, while others are soft. For such contributions we calculate explicitly the hard (small distance) part of the diagrams using perturbation theory. The soft lines represent the non-perturbative quark and gluon fields which are averaged over fluctuations in the physical vacuum (see fig. 1). We will present here only the results of our calculations since all conceptual and technical details of the method are exhaustively discussed elsewhere.

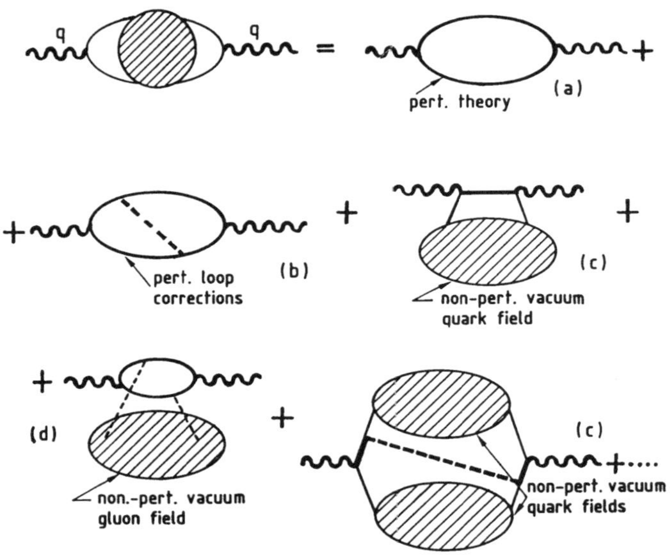

Fig. 1. Various perturbative and non-perturbative contributions to the two-point correlator.

The sum rules for the pion wave function have the form [2]

$$\frac{1}{\pi}\int_0^\infty \mathrm{d}s\ \mathrm{e}^{-s/M^2}\ \mathrm{Im}\ I_{n0}(s) = \frac{3M^2}{4\pi^2(n+1)(n+3)} + C_{\mathrm{g}}\frac{\langle 0|(\alpha_s/\pi)G^2_{\mu\nu}|0\rangle}{12M^2}$$

$$+ C_{\mathrm{q}}\frac{16\pi}{81}\frac{\langle 0|\sqrt{\alpha_s}\,\bar{u}u|0\rangle^2}{M^4} + \cdots,$$

$$\frac{1}{\pi}\mathrm{Im}\ I_{n0}(s) = f_\pi^2\langle\xi^n\rangle_\pi\delta(s) + f_{\mathrm{a}}^2\langle\xi^n\rangle_{\mathrm{a}}\delta(s-m_{\mathrm{a}}^2) + \theta(s-s_n)\frac{3}{4\pi^2(n+1)(n+3)},$$

$$C_{\mathrm{g}} = 1,\quad C_{\mathrm{q}} = (11+4n),\quad \langle\xi^n\rangle \equiv \int_{-1}^{1}\mathrm{d}\xi\ \xi^n\phi_\pi^{\mathrm{A}}(\xi). \tag{8}$$

Here the pion and the a_1 meson contributions are explicitly separated in the spectral density and s_n is the effective continuum thereshold in the given channel. It is worth noticing that we use the standard model for the continuum. The first perturbative correction to the bare-loop contribution in eq. (8) has been calculated in ref. [3] and turns out to be small.

The standard values of the vacuum condensates [1] I always use are

$$\langle 0|(\alpha_s/\pi)G^2_{\mu\nu}|0\rangle \simeq 1.2\times 10^{-2}\ \mathrm{GeV}^4,\quad \langle 0|\sqrt{\alpha_s}\,\bar{q}q|0\rangle \simeq -1.35\times 10^{-2}\ \mathrm{GeV}^3. \tag{9}$$

Applying a fitting procedure with respect to the Borel parameter M^2 (the values of s_n are chosen to obtain best fits at each n), one gets from eqs. (8), (9)

$$\pi:\quad f_\pi \approx 131\ \mathrm{MeV},\quad \langle\xi^2\rangle_{\overline{M}^2=1.5\,\mathrm{GeV}^2} \approx 0.38,\quad \langle\xi^4\rangle_{\overline{M}^2=2.2\,\mathrm{GeV}^2} \approx 0.21; \tag{10}$$

$$a_1:\quad f_{\mathrm{a}} \approx 170\ \mathrm{MeV},\quad \langle\xi^2\rangle_{\overline{M}^2=1.5\,\mathrm{GeV}^2} \approx 0.10,\quad \langle\xi^4\rangle_{\overline{M}^2=2.2\,\mathrm{GeV}^2} \approx 0.05. \tag{11}$$

Unfortunately, the accuracy of the results for higher moments, $\langle\xi^n\rangle$, decreases with increasing n since the non-perturbative corrections in the sum rules become too large, and the sum rules lose their dependence on the lowest-resonance contribution. Therefore, we are forced to limit our consideration to $n \leq 4$.

Renormalized to the normalization point $\mu_0 = 500$ MeV, the values of the moments of the pion wave function are

$$\langle \xi^0 \rangle_{\mu_0} = 1, \quad \langle \xi^2 \rangle_{\mu_0} \simeq 0.44, \quad \langle \xi^4 \rangle_{\mu_0} \simeq 0.27. \tag{12}$$

Now, to interpret these results let us compare them with the corresponding values of the perturbative (asymptotic) wave function $\phi^{\text{pert}}(\xi) = 0.75(1-\xi^2)$:

$$\langle \xi^0 \rangle = 1, \quad \langle \xi^2 \rangle = 0.20, \quad \langle \xi^4 \rangle = 0.086. \tag{13}$$

It is seen that the real pion wave function is much wider. Turning back to the sum rules and examining them in more detail one sees that the loop contribution corresponds precisely to the perturbative wave function, while the non-perturbative corrections describe the difference between the perturbative wave function and the real one. Moreover, duality teaches us that the total (integrated) hadronic contribution is the same as the perturbative one. We observe, however, a strong restructuring of the relative contributions in the spectral density: the pion moments are much larger than the perturbative ones, while those of a_1 are much smaller.

On the other hand, duality can help us to find the parametric behavior of the wave function at $|\xi| \to 1$ [4,5]. Duality reads that there exists a value of s (we will call it S_n^π) such that

$$\frac{1}{\pi}\int_0^{S_n^\pi} \mathrm{d}s \ \mathrm{Im}\ I_{n0}^{\pi}(s) = \frac{1}{\pi}\int_0^{S_n^\pi} \mathrm{d}s \ \mathrm{Im}\ I_{n0}^{\text{pert}}(s). \tag{14}$$

It seems clear that (barring cancelations) the pion fills a finite duality interval at any n, and, in particular, at $n \to \infty$. If so,

$$f_\pi^2 \langle \xi^n \rangle_\pi \simeq \frac{3}{4\pi^2} \frac{S_\infty^\pi}{n^2},$$

$$\langle \xi^n \rangle_\pi \sim \mathrm{O}(n^{-2}) \to \phi_\pi^{\mathrm{A}}(\xi) \sim (1-\xi^2) \sim \phi^{\text{pert}}(\xi) \text{ at } |\xi| \to 1. \tag{15}$$

Since this question is of importance, let us give one more independent argument in favor of such a behavior [6]. Let us consider the form factor f_+ which appears in the B-meson weak decay, $\overline{\mathrm{B}}(p) \to \pi^+(p_1)\pi^-(q)$, under the assumption of factorization:

$$\langle \pi^+(p_1) | \bar{u}(0)(1+\gamma_5)b(0) | \bar{B}(p) \rangle = \frac{M_{\mathrm{B}}^2}{M_{\mathrm{b}}} f_+(q^2 \simeq 0), \quad M_{\mathrm{b}} \to \infty. \tag{16}$$

On the one hand, the leading parametric behavior of f_+ at $M_{\mathrm{b}} \to \infty$ can be calculated by standard methods (see e.g. the review paper [5]), considering the diagram with one gluon exchange [6]:

$$f_+(q^2) \sim \mathrm{O}\left(\frac{f_\pi f_{\mathrm{B}}}{M_{\mathrm{b}}\mu_0}\right) \sim \mathrm{O}(M_{\mathrm{b}}^{-3/2}), \quad f_{\mathrm{B}}^2 \sim \mathrm{O}(\mu_0^3/M_{\mathrm{b}}), \quad q^2 \simeq \mu^2, \quad M_{\mathrm{b}} \to \infty, \tag{17}$$

where f_{B} is the B-meson decay constant analogous to f_π.

On the other hand, let us replace the B meson by the current I_R and consider the correlator:

$$K = i\int dx\, e^{iqx} \langle \pi(p_1) | TJ_L(x) I_R(0) | 0 \rangle; \tag{18}$$

$$\delta K = \frac{f_B M_B^2}{M_b} \frac{f_+(0) M_B^2}{M_b} \delta(s - M_B^2) + \cdots,$$

$$J_L(x) = \bar{u}(x)(1+\gamma_5) b(x), \quad I_R(0) = \bar{b}(0)(1-\gamma_5) d(0). \tag{19}$$

Proceeding in a standard way one obtains the sum rule [6]:

$$f_B \left(\frac{M_B}{M_b} \right)^4 f_+(0) = f_\pi \int_0^\Delta dx \frac{\phi_\pi^A(x)}{(1-x)^2} \exp\left\{ \frac{M_B^2 - M_b^2 (1-x)^{-1}}{M^2} \right\} + \text{(power corrections)}, \tag{20}$$

where S_b is the duality interval and $\Delta = (1 - M_b^2/S_b)$. Since parametrically $S_b = (M_b + \mu_0)^2$, we get $\Delta \sim O(\mu_0/M_b)$; now from eq. (20) at $\overline{M}^2 \sim O(M_b^2)$ we have

$$f_B f_+(0) \sim f_\pi \int_0^\Delta dx\, \phi_\pi^A(x) \sim f_\pi \Delta \phi_\pi^A(\bar{x} \sim \Delta). \tag{21}$$

If $\phi_\pi^A(x \to 0) \sim x \sim \phi^{\text{pert}}(x)$, then eq. (21) implies

$$f_+ \sim O\left(\frac{f_\pi}{f_b} \Delta^2 \right) \sim O(M_b^{-3/2}), \tag{22}$$

in full agreement with eq. (17). In any case, $\phi_\pi^A(x)$ cannot tend to zero more slowly than $\sim x(1-x)$ at $x \to 0$ or $x \to 1$, because the parametric behavior of f_+ obtained from eq. (21) will then exceed the estimates (17), which is impossible. This excludes a model wave function like $\phi_\pi^A(x) \sim$ const. advocated by Radyushkin or $\phi_\pi^A(x) \sim [x(1-x)]^{1/4}$ suggested by Brodsky and Lepage, because they decrease too slowly at $x(1-x) \to 0$. Let us finally point out that all qualitative considerations above concerning the end-point behavior of the pion wave function are clearly of general character and are applicable to other wave functions as well.

Taking all information together, we can say that $\phi_\pi^A(\xi)$, in comparison with $\phi^{\text{pert}}(\xi)$, has: (i) the same overall area under it (i.e. $\langle \xi^0 \rangle = 1$); (ii) the same behavior $\sim (1-\xi^2)$ at $|\xi| \to 1$; (iii) much larger values of the few lowest moments, see eqs. (12), (13). A simple model wave function which meets all these requirements has been proposed in ref. [2]; it is the so-called "two-hump" wave function,

$$\phi_\pi^A(\xi, \mu_0) = \tfrac{15}{4}(1-\xi^2)\xi^2; \tag{23}$$

$$\langle \xi^0 \rangle = 1, \quad \langle \xi^2 \rangle = 0.43, \quad \langle \xi^4 \rangle = 0.24, \quad \langle \xi^6 \rangle = 0.15.$$

Its characteristic property is that the distribution of the quark momentum is wide and the pion momentum is rarely divided equally between the two quarks; the largest part of the pion momentum is carried by one quark (see curve a in fig. 2).

The value of the second moment of the pion wave function $\langle \xi^2 \rangle_\pi^A$, has also been found from independent QCD sum rules in refs. [7,8], and the results agree with those described above.

Before proceeding further, let us point out that a method of improving the standard QCD sum rules has been proposed recently by Mikhailov and Radyushkin [9]. The idea is to

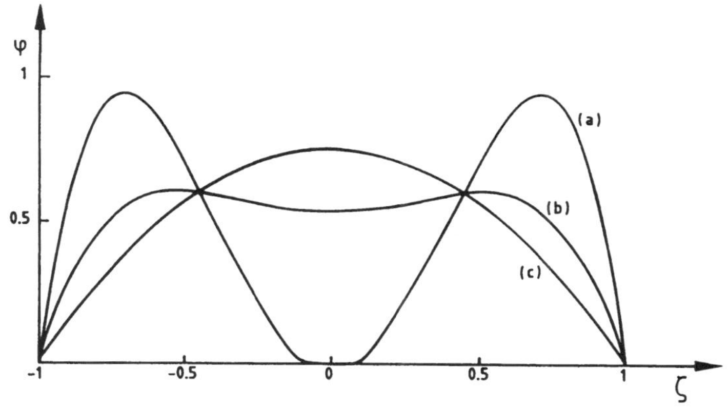

Fig. 2. Shapes of the (a) π, (b) ρ_L and (c) ϕ_L meson wave functions. Recall that the ϕ_L meson wave function coincides with the perturbative one.

introduce "non-local" vacuum condensates (for instance,

$$\langle 0 \,|\, G_{\alpha\beta}(z_2) \exp\left\{ ig \int_{z_1}^{z_2} d\sigma_\mu \, A_\mu(\sigma) \right\} G_{\alpha\beta}(z_1) \,|\, 0 \rangle$$

and some others), appearing in the operator product expansion, without expanding them in powers of $(z_2 - z_1) \sim 1/q$, as is usually done. The non-local condensates are then parametrized by functions of $z_1 - z_2$. From the viewpoint of the operator product expansion, this corresponds to summing up a subset of higher-dimension power corrections. The idea, as it is, looks reasonable in the sense that such corrections are definitely present in the sum rules; its concrete realization, however, meets with a number of major difficulties.

First of all, only a certain subset of power corrections is summed up in this approach, while at the same time all other power corrections are ignored, in spite of the fact that there are no reasons to neglect them. It is well known that, if there is no clear hierarchy, it is usually misleading to account for a part of the corrections only; it is better to drop them altogether.

Second, the multi-local (bi-local, three-local and especially four-local) vacuum condensates which appear in this approach in the "theoretical part" of the sum rules are not constants now, but functions. Even leaving aside strong additional assumptions on the factorization properties of multi-local condensates, nothing is known about even the simplest of these functions, except for the values of their first derivatives in some cases. Therefore, to obtain information from such sum rules the authors [9] are forced to introduce arbitrary models for these nonlocal vacuum condensates, and the results are sensitive to the models used.

On the whole, it remains unclear whether the approach advocated by Radyushkin and collaborators is an improvement of the standard QCD sum rules. The standard approach has given a whole lot of good solid results, in agreement with data. It seems clear that the above, improved approach to the QCD sum rules should first be tested in all those cases where the answers are known from experiment, to check wether the previous predictions of the standard QCD sum rules are not spoiled and whether the accuracy of the predictions is actually improved.

Even ignoring the above comments and examining the results only

$$\langle \xi^2 \rangle = 0.32 \pm 0.03, \quad \langle \xi^4 \rangle = 0.20 \pm 0.03, \quad \langle \xi^6 \rangle = 0.15 \pm 0.02 \tag{24}$$

(the average normalization point in eq. (24) is: $\bar{\mu}^2 \simeq 1.7$ GeV^2), one sees that there are no essential differences with eq. (10) ($\langle\xi^6\rangle \simeq 0.17$ at $\mu^2 \simeq 2.5\ \mathrm{GeV}^2$, from the standard sum rules). What is somewhat surprising is that one would expect larger differences for higher moments, but this is not the case *.

Recalling the uncertainties inherent to this approach to the sum rule method and the strong model dependence of the results obtained, it seems there are no reasons at present to give preference to the values in eq. (24) over those of eq. (10) obtained from the standard QCD sum rules.

To my mind, more stimulating results have been obtained by the same authors on the behavior of the form factor "$\gamma(q_1)+\gamma(q_2)\to\pi(p)$", [10] (see also ref. [11]). Within their approach the quantity

$$\frac{F(\omega)}{F(0)} = \int_{-1}^{1} \mathrm{d}\xi \frac{\phi_\pi^A(\xi)}{1-\omega^2\xi^2}, \quad \omega = \frac{q_1^2 - q_2^2}{q_1^2 + q_2^2} \tag{25}$$

is calculated in the interval $0 \le \omega \le 0.95$. At $|\omega| \ge 0.5$ it strongly depends on the form of $\phi_\pi^A(\xi)$.

The curve obtained in ref. [10] for $F(\omega)/F(0)$ disagrees strongly with that corresponding to $\phi^{\mathrm{pert}}(\xi)$ and to any narrow pion wave function in general. In spite of the uncertainties of this approach, the qualitative effect is beyond any doubt. It is easy to check that the curve for $F(\omega)/F(0)$ obtained in ref. [10] agrees (at $\lambda_q^2 \simeq 0.35\ \mathrm{GeV}^2$) with the wave function (23). Taking into account the uncertainties of this approach to the sum rules, however, it is unclear to what extent this can be considered as an independent quantitative check of the properties of the pion wave function.

Of course, the most direct way to find the values of the moments of the pion wave function is to do lattice calculations. Unfortunately, so far the results are inconclusive

$$\langle\xi^2\rangle_\pi = 0.34 \pm 0.17, \quad \text{ref. [12]}, \tag{26}$$

$$\langle\xi^2\rangle_\pi = 0.39 \pm 0.17, \quad \text{ref. [13]}. \tag{27}$$

$$\langle\xi^2\rangle_\pi = 0.13 \pm 0.02, \quad \text{ref. [14]}. \tag{28}$$

The last result is highly surprising because it signals a very narrow pion wave function, $\phi_\pi^A(\xi)$, and this contradicts all what is known from the sum rules and from calculations of many exclusive processes with pions. Clearly, the situation with lattice calculations should be elucidated.

3. The meson wave functions

We describe in this section the results for some wave functions of other mesons obtained using the standard QCD sum rules.

(1) For the helicity-zero ρ meson, $\rho_L = \rho_{\lambda=0}$, the leading twist wave function is defined as

$$\langle 0|\bar{d}(z)\gamma_\mu I u(-z)|\rho_L(p)\rangle = f_\rho p_\mu \int_{-1}^{1} \mathrm{d}\xi\ \mathrm{e}^{\mathrm{i}\xi(zp)}\ \phi_\rho^L(\xi) \tag{29}$$

* Using the factorization approximation we estimated the effect of the first corrections to the sum rule (8) at $n=2$. We included the terms $\langle 0|G^3|0\rangle$, $(\langle 0|G^2|0\rangle)^2$ and $\langle 0|\bar{q}q|0\rangle\langle 0|\bar{q}gGq|0\rangle$, (V.L. Chernyak, A.R. Zhitnitsky and I.R. Zhitnitsky, unpublished). Their effect is much smaller than the difference between the values of $\langle\xi^2\rangle$ in eqs. (10) and (24). This shows that this difference is due to distant terms in the power series, and, therefore, is essentially model dependent.

($\epsilon_\mu^{\lambda=0} \simeq p_\mu/M_\rho$ at $p_z \to \infty$, where $\epsilon_\mu(p)$ is the polarization vector). Moreover, the corresponding formula for the unit-helicity state, $\rho_\perp = \rho_{|\lambda|=1}$, has the form:

$$\langle 0 | \bar{d}(z)\sigma_{\mu\nu} I u(-z) | \rho_\perp(p)\rangle = f_\rho^{\mathrm{T}}\left(\epsilon_\mu^\perp p_\nu - \epsilon_\nu^\perp p_\mu\right)\int_{-1}^{1} \mathrm{d}\xi \; \mathrm{e}^{\mathrm{i}\xi(zp)} \; \phi_\rho^{\mathrm{T}}(\xi). \tag{30}$$

The leading twist wave function of any meson with $\lambda = 0$ or $|\lambda| = 1$, respectively, will have the same form. Proceeding in the same way as for the pion, one obtains the sum rules of the form (8) with the coefficients [15,16]:

$$\lambda = 0: \quad C_\mathrm{g} = 1, \quad C_\mathrm{q} = (4n - 7), \tag{31}$$

$$|\lambda| = 1: \quad C_\mathrm{g} = (n-1)/(n+1), \quad C_\mathrm{q} = -4(n-1). \tag{32}$$

Comparison of the sum rules (31), (32) and (8) shows that the contribution of the perturbative loop is universal and corresponds to the universal leading twist perturbative wave function, $\phi^{\mathrm{pert}}(\xi) = 0.75(1 - \xi^2)$. The non-perturbative contributions to the sum rules account for the difference between the real wave function and ϕ^{pert}; they depend strongly on the channel considered. Qualitatively, if the perturbative and non-perturbative contributions have the same sign, then the actual wave function will be broader than ϕ^{pert}, and vice versa. Besides, in the sum rules considered the effect of the quark condensate is usually much more important than that of the gluon. *

The contribution of the quark condensate is positive at $n \geq 2$ for the ρ_L meson, see eq. (31), but smaller than that for the pion. As a result, the ρ_L meson wave function is broader than ϕ^{pert}, but narrower than the pion one (see below). The contribution of the quark condensate is negative at $n \geq 2$ for the $\rho_\perp$ meson, see eq. (32), and its wave function appears to be even narrower than ϕ^{pert}. The results of the best fits are as follows (for the normalization point $\mu_0 \simeq 500$ MeV):

$$\rho_\mathrm{L}: \quad f_\rho \simeq 200 \text{ MeV}, \quad \langle \xi^2 \rangle \simeq 0.29, \quad \langle \xi^4 \rangle \simeq 0.17; \tag{33}$$

$$\rho_\perp: \quad f_\rho^\mathrm{T} \simeq 200 \text{ MeV}, \quad \langle \xi^2 \rangle \simeq 0.14, \quad \langle \xi^4 \rangle \simeq 0.05. \tag{34}$$

Corresponding model wave functions were proposed in refs. [15,16] (see figs. 2, 3):

$$\phi_\rho^\mathrm{L}(\xi) = \phi^{\mathrm{pert}}(\xi)\left[1 + 1.5(\xi^2 - 0.2)\right]; \quad \phi_\rho^\mathrm{T}(\xi) = \phi^{\mathrm{pert}}(\xi)\left[1 - 1.25(\xi^2 - 0.2)\right]. \tag{35}$$

(2) The wave functions of mesons containing a strange quark or – in other words – the role of SU(3) symmetry breaking effects has also been investigated [15–17]. Unlike π and ρ, the wave function of K and K* mesons are asymmetric, $\phi_{\mathrm{K,K^*}}(\xi = x_\mathrm{s} - x_\mathrm{u}) \neq \phi_{\mathrm{K,K^*}}(-\xi)$. In the theoretical part of the sum rules the main SU(3) symmetry breaking effect is usually due to the term $m_\mathrm{s}\langle 0 | \bar{s}s | 0\rangle$.

If both quarks in a bound state are heavy and the binding is weak, then the momentum fractions carried by each of them are proportional to their masses: $\langle x_1 \rangle : \langle x_2 \rangle \simeq m_1 : m_2$, i.e. heavier quarks carry larger momentum fractions. It is interesting that this property seemingly persists even for the light u and d quarks and for the "middle weight" s quark:

$$\langle x_\mathrm{s} - x_\mathrm{u} \rangle_\mathrm{K} \simeq 0.10, \quad \langle x_\mathrm{s} - x_\mathrm{u} \rangle_{\mathrm{K}^*_{\mathrm{L},\perp}} \simeq 0.15, \quad \langle x_\mathrm{d} - x_\mathrm{u} \rangle_\pi \simeq 0.5 \times 10^{-2}. \tag{36}$$

* For instance, let us differentiate the sum rule (8) with respect to M^{-2}. On the one hand, the pion contribution will disappear from the spectral density. On the other hand, the contribution of the quark condensate will become negative. This signals that the moments of the a_1-meson wave function are smaller than those of $\phi^{\mathrm{pert}}(\xi)$, and this is indeed the case, see eq. (11).

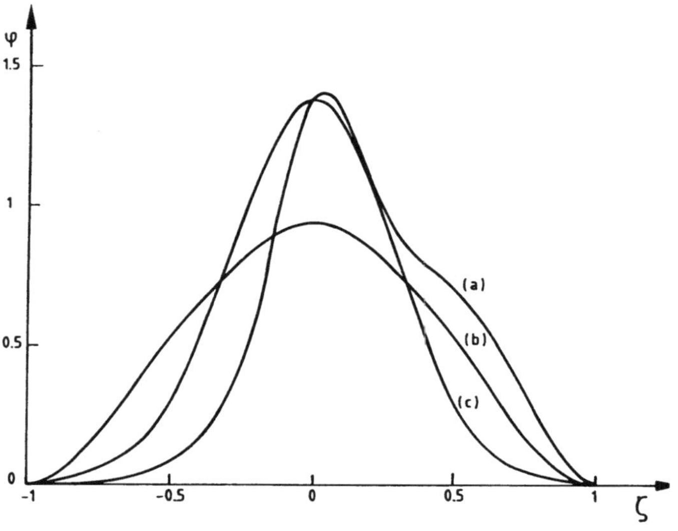

Fig. 3. Shapes of the (a) $K^*_\perp$, (b) $\rho_\perp$ and (c) $\phi_\perp$ meson wave functions.

Some regularity has also been observed for the symmetric parts of the wave functions: the larger the value of the wave function at the origin (i.e. the coupling constant f_π, f_K, etc.) the narrower is the symmetric part:

$$
\begin{aligned}
&f_\phi > f_{K^*} > f_\rho > f_K > f_\pi ;\\
&(\langle\xi^2\rangle)^L_\phi < (\langle\xi^2\rangle)^L_{K^*} < (\langle\xi^2\rangle)_K < (\langle\xi^2\rangle)_\pi ;\\
&(\langle\xi^2\rangle)^T_\phi < (\langle\xi^2\rangle)^T_{K^*} < (\langle\xi^2\rangle)^T_\rho .
\end{aligned} \tag{37}
$$

For example, a model K-meson wave function satisfying the QCD sum rules has been proposed in refs. [15,17] and has the form (see fig. 4):

$$\phi^A_K(\xi = x_s - x_u) = \tfrac{15}{4}(1-\xi^2)(0.6\xi^2 + 0.25\xi^3 + 0.08), \tag{38}$$

(cf. eq. (23)).

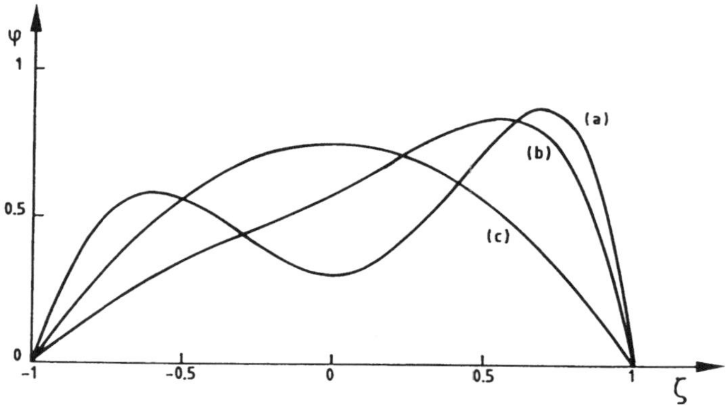

Fig. 4. Shapes of the (a) K, (b) K^*_L meson wave functions and (c) the perturbative wave function.

As a rule, the above two factors (a larger coupling constant f_i and a narrower wave function) work in opposite directions: the first one increases the amplitude while the second tends to decrease it. And which of these two factors dominates the effects of SU(3) symmetry breaking depends on the process considered. Further details can be found in refs. [15–17].

(3) While the "two-particle" leading twist wave functions, $\phi_i(\xi)$, determine the distribution of the fraction of the longitudinal momentum, x_1 and x_2, inside the mesons (at $p_z \to \infty$), the higher-twist wave functions are connected with the transverse momentum distribution of the quarks and with the many-particle components of the meson wave functions. As for the characteristic value of the intrinsic quark transverse momenta, it can be found as follows [16]. Let us consider, for instance, the matrix element:

$$\langle 0 | \bar{d}(0) \mathrm{i}\gamma_5 \vec{\mathrm{D}}_\mu \vec{\mathrm{D}}_\nu u(0) | \pi(q) \rangle = f_\pi \frac{m_\pi^2}{m_\mathrm{u} + m_\mathrm{d}} (q_\mu q_\nu A + g_{\mu\nu} B). \tag{39}$$

It is clear (take μ, $\nu = z$, $q_z \to \infty$) that $A = \langle x_\mathrm{u}^2 \rangle$, i.e. the coefficient A determines the second moment of the fraction of the u-quark momentum in the pion. It is natural to determine the second moment of the quark transverse momentum in the pion as

$$\langle 0 | \bar{d}(0) \mathrm{i}\gamma_5 \left(\vec{\mathrm{D}}_\perp\right)^2 u(0) | \pi(q) \rangle = f_\pi \frac{m_\pi^2}{m_\mathrm{u} + m_\mathrm{d}} \langle \boldsymbol{k}_\perp^2 \rangle = f_\pi \frac{m_\pi^2}{m_\mathrm{u} + m_\mathrm{d}} (-2B). \tag{40}$$

To find B (in the chiral limit $m_\mathrm{u} = m_\mathrm{d} = m_\pi = 0$) let us use the equations of motion and PCAC. Then we get

$$\begin{aligned} \langle 0 | \bar{d}(0) \mathrm{i}\gamma_5 \left(\vec{\mathrm{D}}_\mu\right)^2 u(0) | \pi(q) \rangle &= f_\pi \frac{m_\pi^2}{m_\mathrm{u} + m_\mathrm{d}} 4B = \tfrac{1}{2} \langle 0 | \bar{d} \gamma_5 \sigma_{\mu\nu} g G_{\mu\nu}^a \frac{\lambda^a}{2} u | \pi(q) \rangle \\ &\simeq \frac{1}{f_\pi} \langle 0 | \bar{u} \sigma_{\mu\nu} \mathrm{i} g G_{\mu\nu}^a \frac{\lambda^a}{2} u | 0 \rangle. \end{aligned} \tag{41}$$

Therefore, using the well-known fact that $f_\pi m_\pi^2/(m_\mathrm{u} + m_\mathrm{d}) \simeq -2\langle 0 | \bar{u}u | 0 \rangle / f_\pi$ we arive at

$$\left\langle \boldsymbol{k}_\perp^2 \right\rangle_\pi^\mathrm{P} \simeq \frac{1}{8} \frac{\langle 0 | \bar{u} \sigma_{\mu\nu} \mathrm{i} g G_{\mu\nu}^a \lambda^a u | 0 \rangle}{\langle 0 | \bar{u}u | 0 \rangle}. \tag{42}$$

It can be seen that $\langle \boldsymbol{k}_\perp^2 \rangle \neq 0$ in the chiral limit, exclusively due to interaction. This is natural because $\boldsymbol{k}_\perp \to 0$ at $m_\pi \to 0$ in the case of free quarks. The value of the ratio in eq. (42) has been found in ref. [18] from the QCD sum rules for baryons. Using this result, one has:

$$\left\langle \boldsymbol{k}_\perp^2 \right\rangle_\pi^\mathrm{P} \simeq \tfrac{1}{8}(1.5\ \mathrm{GeV}^2) \simeq (430\ \mathrm{MeV})^2. \tag{43}$$

The second moment of $\boldsymbol{k}_\perp$, $\langle \boldsymbol{k}_\perp^2 \rangle$, has been found also for the leading twist wave function [16]. The following matrix element has been considered:

$$\begin{aligned} &\langle 0 | \bar{d}(0) \gamma_\mu \gamma_5 \tfrac{1}{2} \left[\vec{\mathrm{D}}_\nu \vec{\mathrm{D}}_\lambda + \vec{\mathrm{D}}_\lambda \vec{\mathrm{D}}_\nu\right] u(0) | \pi(q) \rangle \\ &= \mathrm{i} f_\pi \left\{ q_\mu q_\nu q_\lambda \left\langle x_\mathrm{u}^2 \right\rangle + (g_{\mu\nu} q_\lambda + g_{\mu\lambda} q_\nu) A + g_{\nu\lambda} q_\mu B \right\}. \end{aligned} \tag{44}$$

Multiplying eq. (44) by $g_{\mu\nu}$ and using the equations of motion one has

$$5A + B = 0.$$

Hence,

$$\langle 0 | \bar{d}\gamma_\mu\gamma_5\big(\vec{\mathrm{D}}_\nu\big)^2 u | \pi\rangle = \mathrm{i}f_\pi q_\mu \tfrac{18}{5}B = -\tfrac{1}{2}\langle 0 | \bar{d}\gamma_\mu\gamma_5\sigma_{\alpha\beta}\mathrm{i}gG^a_{\alpha\beta}\frac{\lambda^a}{2}u | \pi\rangle . \tag{45}$$

The quantity $\langle \boldsymbol{k}_\perp^2 \rangle_\pi^{\mathrm{A}}$ is defined as follows:

$$\langle 0 | \bar{d}\gamma_\mu\gamma_5\big(\vec{\mathrm{D}}_\perp\big)^2 u | \pi\rangle \equiv \mathrm{i}f_\pi q_\mu\langle \boldsymbol{k}_\perp^2\rangle_\pi^{\mathrm{A}} = -2\mathrm{i}f_\pi q_\mu B. \tag{46}$$

Therefore, combining eqs. (45) and (46) one obtains

$$\langle 0 | \bar{d}\gamma_\mu\gamma_5\sigma_{\alpha\beta}\mathrm{i}gG^a_{\alpha\beta}\frac{\lambda^a}{2}u | \pi\rangle = \langle 0 | \bar{d}g\tilde{G}^a_{\mu\rho}\lambda^a\gamma_\rho u | \pi(q)\rangle = \mathrm{i}f_\pi q_\mu \tfrac{18}{5}\langle \boldsymbol{k}_\perp^2\rangle_\pi^{\mathrm{A}}; \tag{47}$$

$$\tilde{G}_{\mu\rho} = \tfrac{1}{2}e_{\mu\rho\alpha\beta}G_{\alpha\beta}. \tag{48}$$

The value of the matrix element in eq. (47) has been calculated independently in refs. [16,19] from different QCD sum rules, and the results agree well. The answer can be represented in the form

$$\langle 0 | \bar{d}\gamma_\mu\gamma_5\sigma_{\alpha\beta}\mathrm{i}gG^a_{\alpha\beta}\frac{\lambda^a}{2}u | \pi(q)\rangle \simeq \mathrm{i}f_\pi q_\mu \frac{1}{4}\frac{\langle 0 | \bar{u}\sigma_{\mu\nu}\mathrm{i}gG^a_{\mu\nu}\lambda^a u | 0\rangle}{\langle 0 | \bar{u}u | 0\rangle}, \tag{49}$$

$$\langle \boldsymbol{k}_\perp^2\rangle_\pi^{\mathrm{A}} \simeq \tfrac{5}{9}\langle \boldsymbol{k}_\perp^2\rangle_\pi^{\mathrm{P}} \simeq (323\ \mathrm{MeV})^2. \tag{50}$$

From eqs. (43), (50) we see that the values of the intrinsic transverse quark momenta differ somewhat for various wave functions, but the difference is not large. It appears natural that $(\langle \boldsymbol{k}_\perp^2\rangle_\pi^{\mathrm{P}}) > (\langle \boldsymbol{k}_\perp^2\rangle_\pi^{\mathrm{A}})$, because the scale of the matrix elements of spin-zero operators ($\bar{u}u$, $\bar{u}\gamma_5 u, \ldots$) is enhanced [20]. It is natural to expect that the mean quark transverse momentum inside other mesons, K, ρ, K*, etc. is close to that in the pion: $\langle \boldsymbol{k}_\perp^2\rangle_i \simeq (300\text{–}400\ \mathrm{MeV})^2$.

(4) A simple method for finding the perturbative (i.e. asymptotic) form of various wave functions of leading and non-leading twists by calculating the graphs with the lowest loops has been proposed in ref. [17] (see also refs. [21,22]).

Although as a rule the properties of the higher-twist meson wave functions are poorly known at present (see however refs. [21–23]), some general features have been revealed. Analogous to the leading twist wave function, the actual form of higher-twist wave functions differs strongly from that of perturbative wave functions. As for the values of the various wave functions at the origin, the general partern can be illustrated by the following examples [5]:

$$\langle 0 | \bar{d}(0)\gamma_\mu^\perp u(0) | \rho_\perp(q)\rangle = \epsilon_\mu^\perp f_\rho m_\rho, \quad f_\rho m_\rho \simeq 0.15\ \mathrm{GeV}^2, \tag{51}$$

$$\langle 0 | \bar{d}(0)\overleftrightarrow{\mathrm{D}}_\mu u(0) | \rho_{\mathrm{L}}(q)\rangle = f_\rho^{\mathrm{S}}\big(\epsilon_\mu^{\lambda=0}m_\rho\big) \simeq f_\rho^{\mathrm{S}}q_\mu, \quad f_\rho^{\mathrm{S}} = -f_\rho^{\mathrm{T}}m_\rho \simeq -0.15\ \mathrm{GeV}^2, \tag{52}$$

$$\langle 0 | \bar{d}(0)\mathrm{i}\gamma_5 u(0) | \pi(q)\rangle = f_\pi\frac{m_\pi^2}{m_{\mathrm{u}}+m_{\mathrm{d}}} \simeq 0.22\ \mathrm{GeV}^2, \tag{53}$$

$$\langle 0 | \bar{d}(0)\gamma_\mu^\perp\gamma_5\overleftrightarrow{\mathrm{D}}_\nu u(0) | \rho_\perp(q)\rangle = \frac{\mathrm{i}}{2}f_\rho m_\rho e_{\mu\nu\alpha\beta}\epsilon_\alpha^\perp q_\beta, \tag{54}$$

$$\langle 0 | \bar{d}(0)\gamma_\mu\gamma_5\frac{\lambda^a}{2}gG^a_{\lambda\rho}(0)u(0) | \rho_\perp(q)\rangle = f_\rho^{3\mathrm{A}}q_\mu e_{\lambda\rho\sigma\tau}q_\sigma\epsilon_\tau^\perp, \tag{55}$$

$$\langle 0 \,|\, \bar{d}(0)\gamma_\mu \overleftrightarrow{\mathrm{D}}_\nu \frac{\lambda^a}{2} g G^a_{\lambda\rho}(0) u(0) \,|\, \rho_\perp(q)\rangle = \mathrm{i} f_\rho^{3\mathrm{V}} q_\mu q_\nu \big(q_\lambda \epsilon_\rho^\perp - q_\rho \epsilon_\lambda^\perp\big), \tag{56}$$

$$\langle 0 \,|\, \bar{d}(0)\sigma_{\mu\nu}\gamma_5 g G^a_{\rho\lambda} \frac{\lambda^a}{2} u(0) \,|\, \pi(q)\rangle = f_{3\pi}\big[(q_\mu g_{\nu\lambda} - q_\nu g_{\mu\lambda}) q_\rho - (\rho \leftrightarrow \lambda)\big], \tag{57}$$

$$f_\rho^{3\mathrm{A}} \simeq 0.6 \times 10^{-2}\ \mathrm{GeV}^2; \quad f_\rho^{3\mathrm{V}} \simeq 0.3 \times 10^{-2}\ \mathrm{GeV}^2; \quad f_{3\pi} \simeq 0.4 \times 10^{-2}\ \mathrm{GeV}^2. \tag{58}$$

It can be seen that the values of the three-particle wave functions at the origin are much smaller than those for two-particle wave functions. The origin of this difference can be clearly seen from the corresponding QCD sum rules. Let us consider the two-point correlator with suitable currents, calculate only the contribution of the perturbative loop, make the Borel transformation and put finally the Borel parameter $M^2 \simeq m_\rho^2$. This is a reasonable zeroth approximation for estimating the contribution of the lowest resonance. Since the two-particle wave functions are determined by one loop while the three-particle ones by two loops, the latter are much smaller. The above values (51–58) determine the typical size of the power corrections to the leading twist contributions. We can expect, therefore, these corrections to be $\simeq (1-2) m_\rho / Q$.

Now, a few words concerning the applications of the higher-twist meson wave functions to calculations of exclusive amplitudes. Experience shows the following [5,24–26].

(i) As for the first sub-leading twist (i.e. twist-3 wave function), there are only a few of them for each meson. Those processes in which the leading twist contribution is strictly zero or highly suppressed numerically ($J/\psi \to \rho\pi$ [22], $\eta_c \to \phi\omega$ [23], etc.) can be reliably calculated in terms of these twist-3 wave functions.

(ii) There are many twist-4 wave functions for each meson and, moreover, they are related in a complicated way to each other and to the leading twist wave functions by the QCD equations of motion. These wave functions usually enter the amplitudes as power corrections to the leading terms. Rough estimates show that these corrections are large term by term. But different terms and various wave functions contribute with different signs, and, therefore, it seems extremely difficult to calculate the total power correction reliably using this approach [5,26].

4. *The baryon wave functions*

The leading twist wave functions of the baryons entering the lowest octet and decuplet have been investigated in the sum rule approach.

For obvious reasons the nucleon wave function has received most attention [4,27,28]. It is defined by the matrix element [29]:

$$\begin{aligned} \langle 0 \,|\, e^{ijk} u^i_\alpha(z_1) u^j_\beta(z_2) d^k_\gamma(z_3) \,|\, p(q)\rangle = f_{\mathrm{N}} \tfrac{1}{4} \big\{ & (\hat{q}C)_{\alpha\beta} (\gamma_5 N)_\gamma V(z_i q) \\ & + (\hat{q}\gamma_5 C)_{\alpha\beta} N_\gamma A(z_i q) \\ & - (\sigma_{\mu\nu} q_\nu C)_{\alpha\beta} (\gamma_\mu \gamma_5 N)_\gamma T(z_i q) \big\}, \end{aligned} \tag{59}$$

$$V(z_i q = 0) = T(z_i q = 0) = 1, \quad A(z_i q = 0) = 0, \qquad q \to \infty. \tag{60}$$

Here N_γ is the proton spinor, q is the proton momentum, C is the charge conjugation matrix, i, j, k are color indices and α, β, γ are spinor indices. As usual, the Fourier transform of the dimensionless wave functions,

$$V(z_i q) = \int_0^1 \mathrm{d}^3x \, \exp\left\{-\mathrm{i}\sum_{i=1}^{3} x_i(z_i q)\right\} V(x_i), \quad \mathrm{d}^3x = \mathrm{d}x_1\,\mathrm{d}x_2\,\mathrm{d}x_3\,\delta(1-x_1-x_2-x_3), \tag{61}$$

(and the same for $A(z_i q)$ and $T(z_i q)$), determines the longitudinal momentum distribution of the proton quarks, $0 \leq x_i \leq 1$, $\Sigma x_i = 1$. The identity of the two u quarks leads to the symmetry properties:

$$V(1,2,3) = V(2,1,3), \quad T(1,2,3) = T(2,1,3), \quad A(1,2,3) = -A(2,1,3). \tag{62}$$

Requiring the total isospin of the three quarks to be $1/2$, one obtains the relation

$$2T(1,2,3) = [V(1,3,2) - A(1,3,2)] + [V(2,3,1) - A(2,3,1)]. \tag{63}$$

Therefore, there is only one independent nucleon wave function of the leading twist 3, say, $\phi_{\mathrm{N}}(x) = V(x) - A(x)$, $\int_0^1 \mathrm{d}^3x \, \phi_{\mathrm{N}}(x) = 1$.

The wave function introduced above corresponds to a proton state of the form [29]:

$$|P^{\uparrow}\rangle = f_{\mathrm{N}} \int_0^1 \frac{\mathrm{d}^3x \, \phi_{\mathrm{N}}(x)}{4\sqrt{6x_1x_2x_3}} |u^{\uparrow}(x_1)[u^{\downarrow}(x_2)d^{\uparrow}(x_3) - d^{\downarrow}(x_2)u^{\uparrow}(x_3)]\rangle \tag{64}$$

($q_z \to \infty$, the arrows show spin projections onto the z-axis, for a neutron $u \to d$).

The constant f_{N} and the moments of the nucleon wave function have been found using the standard QCD sum rules for the corresponding two-point correlators analogously to the calculations described above for the pion wave function. Below the results of ref. [28] are summarized.

The value of the nucleon wave function at the origin (the constant f_{N}) is found to be

$$|f_{\mathrm{N}}| = (5.0 \pm 0.3) \times 10^{-3}\ \mathrm{GeV}^2, \tag{65}$$

which is a typical value for the three-particle wave function, see eqs. (55)–(58).

The moments of the wave function $\phi_{\mathrm{N}}(x_i)$ are given in table 1; the moments of the perturbative wave function, $\phi_{\mathrm{N}}^{\mathrm{pert}}(x_i) = 120x_1x_2x_3$, are also given there for comparison.

Table 1

Mom.	Sum rules *	$\phi_{\mathrm{N}}(x)$	$\phi^{\mathrm{pert}}(x)$	Mom.	Sum rules *	$\phi_{\mathrm{N}}(x)$	$\phi^{\mathrm{pert}}(x)$
(000)	1	1	1	(300)	$0.236^{+0.014}_{-0.026}$	0.244	0.071
(100)	$0.560^{+0.06}_{-0.02}$	0.579	0.333	(030)	$0.032^{+0.008}_{-0.004}$	0.038	0.071
(010)	$0.192^{+0.008}_{-0.012}$	0.192	0.333	(003)	$0.052^{+0.004}_{-0.004}$	0.049	0.071
(001)	$0.229^{+0.021}_{-0.029}$	0.229	0.333	(210)	$0.045^{+0.004}_{-0.004}$	0.059	0.036
(200)	$0.350^{+0.07}_{-0.03}$	0.369	0.143	(201)	$0.050^{+0.005}_{-0.006}$	0.066	0.036
(020)	$0.084^{+0.004}_{-0.019}$	0.068	0.143	(120)	$0.035^{+0.002}_{-0.008}$	0.024	0.036
(002)	$0.109^{+0.011}_{-0.019}$	0.089	0.143	(102)	$0.041^{+0.002}_{-0.004}$	0.033	0.036
(110)	$0.090^{+0.01}_{-0.01}$	0.097	0.095	(021)	$0.0^{+0.007}_{-0.004}$	0.006	0.036
(101)	$0.102^{+0.008}_{-0.012}$	0.113	0.095	(012)	$0.0^{+0.008}_{-0.005}$	0.007	0.036
(011)	$0.0^{+0.03}_{-0.03}$	0.027	0.095				

* The error bars shown include the uncertainties of the fitting procedure only.

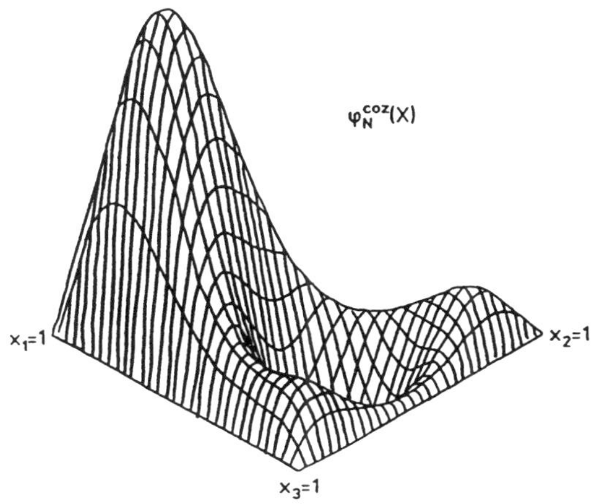

Fig. 5. The profile of the wave function $\phi_N(x)$.

It is seen that the distribution of the proton momentum over the three quarks is highly asymmetric. The largest part, $\simeq 60\%$, of the total momentum is carried by the u quark with its spin parallel to the proton spin, while each of the remaining two quarks carries $\simeq 20\%$ of the proton momentum. A simple model wave function imitating the desired moments has been proposed in ref. [28] (see fig. 5):

$$\phi_N(x) = \phi^{\text{pert}}(x)\left[23.814x_1^2 + 12.978x_2^2 + 6.174x_3^2 + 5.88x_3 - 7.098\right]. \tag{66}$$

Its moments are also given in table 1. (For other model wave functions see refs. [4,27,30].)

The lowest moments of the nucleon wave function have also been obtained in ref. [31] from independent QCD sum rules, and are in agreement with the estimates presented above.

Unexpectedly, the lattice calculation of the lowest moments of the nucleon wave function [32] resulted in a totally symmetric answer: $\langle 100\rangle = \langle 010\rangle = \langle 001\rangle = \frac{1}{3}$. It was also found that $f_N \simeq (2.9 \pm 0.6) \times 10^{-3}$ GeV^2 (compare with eq. (65)).

In my opinion, something is wrong with the lattice calculations. The method of the QCD sum rules is, of course, approximate but, to the best of my knowledge, it has never given qualitatively wrong results, if properly applied. Since it follows from the sum rules that, for instance, $\langle 010\rangle \simeq \langle 001\rangle \ll \langle 100\rangle$, it will be highly surprising if this result will be proven to be qualitatively wrong. I appeal to the experts in lattice calculations to elucidate this important point.

The properties of the leading twist wave functions for $\Delta(1232)$ have been investigated in ref. [33] using the QCD sum rules. The leading twist component of the $\Delta(\lambda = 1/2)$ state can be written in the form (compare with eq. (64)):

$$|\Delta^{++}(\lambda = 1/2)\rangle = f_\Delta \int_0^1 \frac{d^3x\ \phi_\Delta(x)}{8\sqrt{6x_1x_2x_3}} |u^\uparrow(x_1)u^\downarrow(x_2)u^\uparrow(x_3)\rangle; \quad \int_0^1 d^3x\ \phi_\Delta(x) = 1,$$

$$|\Delta^{+}(\lambda = 1/2)\rangle = \frac{f_\Delta}{\sqrt{3}} \int_0^1 \frac{d^3x\ \phi_\Delta(x)}{8\sqrt{6x_1x_2x_3}} \{2|u^\uparrow(x_1)u^\downarrow(x_2)d^\uparrow(x_3)\rangle + |u^\uparrow(x_1)u^\uparrow(x_2)d^\downarrow(x_3)\rangle\}. \tag{67}$$

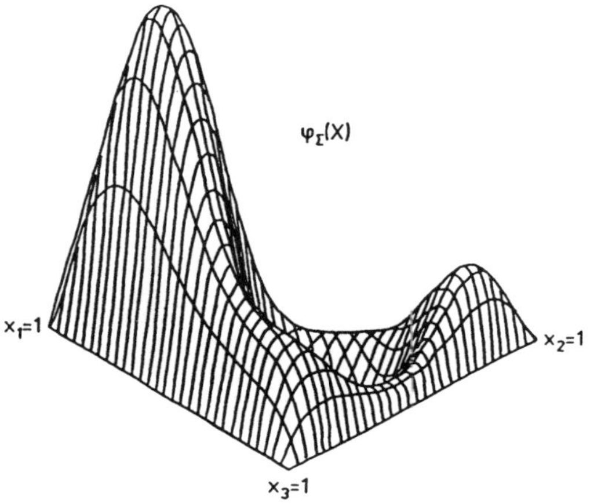

Fig. 6. The profile of the wave function $\phi_\Sigma(x)$.

It appeared that the actual form of $\phi_\Delta(x)$ differs only slightly from $\phi_\Delta^{\mathrm{pert}}(x) = 120 x_1 x_2 x_3$ and, therefore, unlike the nucleon it is nearly symmetric in all its arguments. The model proposed in ref. [33] is as follows:

$$\phi_\Delta(x) = \phi^{\mathrm{pert}}(x)\left[4.2(x_1^2 + x_3^2) + 2.52 x_2^2 - 6.72 x_1 x_3 + 0.42 x_2(x_1 + x_3)\right],$$
$$|f_\Delta| \simeq 1.2 \times 10^{-2}\ \mathrm{GeV}^2. \tag{68}$$

Furthermore, the $\Delta(\lambda = 3/2)$ wave function which is totally symmetric in its arguments and is determined by

$$|\Delta^{++}(\lambda = 3/2)\rangle = \frac{f_\Delta^{3/2}}{\sqrt{2}} \int_0^1 \frac{\mathrm{d}^3x\ \phi_\Delta^{3/2}(x)}{12\sqrt{6 x_1 x_2 x_3}}\, |u^\uparrow(x_1) u^\uparrow(x_2) u^\uparrow(x_3)\rangle;$$

$$\int_0^1 \mathrm{d}^3x\ \phi_\Delta^{3/2}(x) = 1,$$

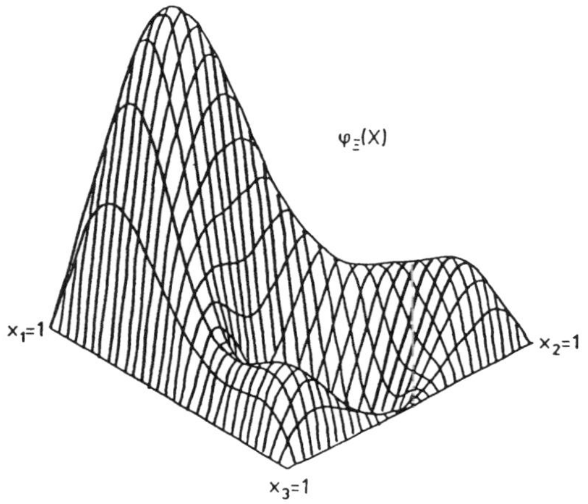

Fig. 7. The profile of the wave function $\phi_\Xi(x)$.

also practically coincides with $\phi^{\text{pert}}(x)$ [33]:

$$\phi_{\Delta}^{3/2}(x) \simeq \phi^{\text{pert}}(x) = 120 x_1 x_2 x_3, \quad |f_{\Delta}^{3/2}| \simeq 1.4 \times 10^{-2}\ \text{GeV}^2. \tag{69}$$

The wave functions of other octet and decuplet baryons (i.e. the properties of SU(3) symmetry breaking) have also been investigated [28,33,34]. These effects usually amount to 20 to 30%, and the qualitative properties of the wave functions remain the same: all octet baryons show strong asymmetry, while the asymmetry is much smaller for those from the decuplet (see figs. 6, 7).

The properties of the non-leading twist baryon wave functions have not been studied so far, although the values of some of them at the origin have been found [6 (see Appendix)].

5. *Conclusions*

The QCD sum rule approach allowed us to reveal the properties of the light-cone hadron wave functions which enter the hard amplitudes. And these properties appeared to be highly nontrivial and versatile. As I tried to illustrate above on a set of examples, the form of the light-cone wave function depends essentially on the hadron quantum numbers (and, in particular, on its helicity) and as a rule essentially differs from the corresponding perturbative (asymptotic) expression.

The properties of the hadronic wave functions described above appeared to be unexpected and even suspicious to many physicists. The old-fashioned SU(6) constituent quark model with its successful qualitative description of the hadronic spectroscopy gave rise to expectations that the light-cone wave functions of hadrons entering one and the same SU(6) multiplet are all alike and, moreover, do not differ essentially from $\phi^{\text{pert}}(x)$. Some simple prescriptions have been proposed for obtaining $\phi_i(x)$ from the SU(6) rest-frame wave functions, and the results seemed to confirm the above expectations.

What actually happens with such simplified prescriptions can be summarized as follows. Say, the pion "wave function" $\tilde{\phi}_{\pi}^{A}(x)$ which one obtains from this prescription is not the actual pion wave function $\phi_{\pi}^{A}(x)$; rather it is a mixture of the pion wave function $\phi_{\pi}^{A}(x)$, the $a_1^{\lambda=0}(1270)$ wave function $\phi_{a_1}^{L}(x)$, etc. Since $\phi_{\pi}^{A}(x)$ and $\phi_{a_1}^{L}(x)$ are, respectively, much broader and much narrower than $\phi^{\text{pert}}(x)$ (see above), it is not surprising that the "pion wave function" $\tilde{\phi}_{\pi}^{A}(x)$ obtained in this way looks like $\phi^{\text{pert}}(x)$.

The momentum distribution of the light quarks in hadrons is rather wide, generally speaking. This means that the quarks are highly relativistic, and models with non-relativistic constituents cannot be used at the quantitative level.

I cannot help mentioning a drawback of the sum rule approach (or, perhaps, its present formulation). It remains unclear what physics lies behind the results obtained. Why are the pion and $a_1^{\lambda=0}$ wave functions, respectively, much wider and much narrower than $\phi^{\text{pert}}(x)$? Why is the proton wave function so asymmetric while that of Δ^+ is nearly totally symmetric? Why are the wave functions of ρ_L and $\rho_\perp$ so different? A transparent physical explanation of these results is still lacking. Clearly, this is due to the semi-phenomenological nature of the sum rule method. Since the non-perturbative QCD dynamics is not fully understood, the basic questions remain unsolved. The main task is to build such model wave functions that we can reliably calculate any hadronic matrix element, not only those referring to very high energies. Of particular interest are low-energy hadronic matrix elements. The exact solution might be the issue in a remote future, if at all. There is a possibility, however, that a simplified model exists which still catches basic properties of the bound states. We may hope that further development will supply us with the desired models.

Finally, a few words about applications. A large number of various hard exclusive processes have been calculated in the leading-twist approximation (i.e. assuming that the power corrections are negligible at available energies). On the whole, the amplitudes calculated with our wave functions are in reasonable agreement with experiment while naive models like $\phi^{\text{pert}}(x)$ or $\phi^{\text{non-rel}}(x)$ fail to describe the data.

References

[1] M.A. Shifman, A.I. Vainshtein and V.I. Zakharov, Nucl. Phys. B 147 (1979) 385, 448.
[2] V.L. Chernyak and A.R. Zhitnitsky, Nucl. Phys. B 201 (1982) 492; (E) Nucl. Phys. B 214 (1983) 547.
[3] A. Gorsky, Yad. Fiz. 41 (1985) 430 [Sov. J. Nucl. Phys. 41 (1985) 275].
[4] V.K. Chernyak and I.R. Zhitnitsky, Nucl. Phys. B 246 (1984) 52.
[5] V.L. Chernyak and A.R. Zhitnitsky, Phys. Rep. 112 (1984) 173.
[6] V.L. Chernyak and I.R. Zhitntisky, Nucl. Phys. B 345 (1990) 137.
[7] M.S. Cragie and J. Stern, Nucl. Phys. B 216 (1983) 209.
[8] M. Lavelle, Z. Phys. C 29 (1985) 203.
[9] S.V. Mikhailov and A.V. Radyushkin, Yad. Fiz. 49 (1989) 794 [Sov. J. Nucl. Phys. 49 (1989) 494].
[10] S.V. Mikhailov and A.V. Radyushkin, Yad. Fiz. 52 (1990) 1095 [Sov. J. Nucl. Phys. 52 (1990) 697].
[11] A. Gorsky, Yad. Fiz. 50 (1989) 796 [Sov. J. Nucl. Phys. 50 (1989) 498].
[12] G. Martinelli and C.T. Sachrajda, Phys. Lett. B 190 (1987) 151; Nucl. Phys. B 306 (1988) 865.
[13] T.A. DeGrand and R.D. Loft, Phys. Rev. D 38 (1988) 954.
[14] D. Daniel, R. Gupta and D.G. Richards, Phys. Rev. D 43 (1991) 3715.
[15] V.L. Chernyak, A.R. Zhitnitsky and I.R. Zhitnitsky, Nucl. Phys. B 204 (1982) 477.
[16] A. Zhitnitsky, I. Zhitnitsky and V. Chernyak, Yad. Fiz. 38 (1983) 1074 [Sov. J. Nucl. Phys. 38 (1983) 645].
[17] A. Zhitnitsky, I. Zhitnitsky and V. Chernyak, Yad. Fiz. 38 (1983) 1277 [Sov. J. Nucl. Phys. 38 (1983) 775].
[18] V.M. Belyaev and B.L. Ioffe, Zh. Exp. Teor. Fiz. 83 (1982) 876 [JETP 56 (1982) 493].
[19] V. Novikov, M. Shifman, A. Vainshtein, M. Voloshin and V. Zakharov, Nucl. Phys. B 237 (1984) 525.
[20] V. Novikov, M. Shifman, A. Vainshtein and V. Zakharov, Nucl. Phys. B 191 (1981) 301.
[21] B.V. Geshkenbein and M.V. Terent'ev, Yad. Phys. 40 (1984) 758 [Sov. J. Nucl. Phys. 40 (1984) 487]. A. Gorsky, Yad. Fiz. 41 (1985) 430 [Sov. J. Nucl. Phys. 41 (1985) 275].
[22] V.M. Braun and I.E. Filyanov, Yad. Fiz. 52 (1990) 199 [Sov. J. Nucl. Phys. 52 (1990) 126].
[23] A. Zhitnitsky, I. Zhitnitsky and V. Chernyak, Yad. Fiz. 41 (1985) 199 [Sov. J. Nucl. Phys. 41 (1985) 127].
[24] A. Zhitnitsky, I. Zhitnitsky and V. Chernyak, Yad. Fiz. 41 (1985) 445 [Sov. J. Nucl. Phys. 41 (1985) 284].
[25] M. Benayoun, V. Chernyak and I. Zhitnitsky, Nucl. Phys. B 348 (1991) 327.
[26] A.S. Gorsky, Yad. Fiz. 41 (1985) 1592; 50 (1989) 1135 [Sov. J. Nucl. Phys. 41 (1985) 1008; 50 (1989) 708].
[27] I.D. King and C.T. Sachrajda, Nucl. Phys. B 279 (1987) 785.
[28] I.R. Zhitnitsky, A.A. Ogloblin and V. Chernyak, Yad. Fiz. 48 (1988) 841, 1398, 1410 [Sov. J. Nucl. Phys. 48 (1988) 536, 889, 896].
[29] V.A. Avdeenko, V.L. Chernyak and S.A. Korenblit, Preprint 23–70, Irkutsk, 1979, unpublished; Yad. Fiz. 33 (1981) 481 [Sov. J. Nucl. Phys. 33 (1981) 252].
[30] M. Gari and N.G. Stefanis, Phys. Rev. D 35 (1987) 1074; Phys. Lett. B 175 (1986) 462.
[31] M.J. Lavelle, Nucl. Phys. B 260 (1985) 323.
[32] G. Martinelli and C.T. Sachrajda, Phys. Lett. B 217 (1989) 319.
[33] G.R. Farrar, H. Zhang, A.A. Ogloblin and I.R. Zhitnitsky, Nucl. Phys. B 311 (1988/89) 585.
[34] A.A. Ogloblin, Yad. Fiz. 50 (1989) 189 [Sov. J. Nucl. Phys. 50 (1989) 119].

Made in United States
Orlando, FL
04 November 2026

90207791R00291